Lecture Notes in Computer Science   690
Edited by G. Goos and J. Hartmanis

Advisory Board: W. Brauer   D. Gries   J. Stoer

Claude Kirchner (Ed.)

Rewriting Techniques and Applications

5th International Conference, RTA-93
Montreal, Canada, June 16-18, 1993
Proceedings

Springer-Verlag
Berlin Heidelberg New York
London Paris Tokyo
Hong Kong Barcelona
Budapest

Series Editors

Gerhard Goos
Universität Karlsruhe
Postfach 69 80
Vincenz-Priessnitz-Straße 1
W-7500 Karlsruhe, FRG

Juris Hartmanis
Cornell University
Department of Computer Science
4130 Upson Hall
Ithaca, NY 14853, USA

Volume Editor

Claude Kirchner
INRIA Lorraine and CRIN
615 Rue du Jardin Botanique, F-54602 Villers les Nancy Cedex, France

CR Subject Classification (1991): D.3, F.3.2, F.4, I.1, I.2.2-3

ISBN 3-540-56868-9 Springer-Verlag Berlin Heidelberg New York
ISBN 0-387-56868-9 Springer-Verlag New York Berlin Heidelberg

This work is subject to copyright. All rights are reserved, whether the whole or part of the material is concerned, specifically the rights of translation, reprinting, re-use of illustrations, recitation, broadcasting, reproduction on microfilms or in any other way, and storage in data banks. Duplication of this publication or parts thereof is permitted only under the provisions of the German Copyright Law of September 9, 1965, in its current version, and permission for use must always be obtained from Springer-Verlag. Violations are liable for prosecution under the German Copyright Law.

© Springer-Verlag Berlin Heidelberg 1993
Printed in Germany

Typesetting: Camera ready by author
Printing and binding: Druckhaus Beltz, Hemsbach/Bergstr.
45/3140-543210 - Printed on acid-free paper

Preface

This volume contains the proceedings of RTA-93, the Fifth International Conference on Rewriting Techniques and Applications held June 16–18, 1993, in Montreal, Canada.

There were 91 submissions to RTA-93 authored by researchers from countries including Canada, France, Germany, Italy, India, Japan, the Netherlands, the People's Republic of China, Russia, Spain, United Kingdom, and the United States of America. Papers covered many topics: term rewriting; termination; graph rewriting; constraint solving; semantic unification, disunification and combination; higher-order logics and theorem proving, with several papers on distributed theorem proving, theorem proving with constraints, and completion.

Each submission was reviewed by at least three program committee members or their outside referees. All the members of the program committee met on February 1993 in Nancy and selected 29 papers and 6 system descriptions demonstrated during the conference and documented in this volume.

As for the proceedings of the previous conference, I welcomed the idea of presenting in the proceedings a list of open problems in the field and an update of the previous list of such open problems, showing altogether the strong activity of the term rewriting community in the large.

Three invited speakers gave a talk on their recent works related to the topics of RTA. Sergei Adian presented his work on algorithmic problems for groups and semigroups, Leo Bachmair the impact of rewriting techniques on theorem proving and Jean Gallier a general method for proving properties of typed lambda terms.

I am very grateful to the program committee for their efforts and cooperation in deciding the program and other related matters to RTA-93; to Mitsuhiro Okada for taking great care of the local arrangements for the conference; to the invited speakers Sergei Adian, Leo Bachmair and Jean Gallier, and lastly to Marian Vittek for doing everything that needed to be done to facilitate my task in organizing the program committee.

RTA-93 was sponsored by INRIA (France), the Centre de Recherche en Informatique de Nancy (France), Concordia University (Canada), the Center for Pattern Recognition and Machine Intelligence, Montreal (Canada), the Natural Science and Engineering Research Council (Canada), le Fonds pour la Formation de Chercheurs et l'Aide à la Recherche (Quebec) and the National Science Foundation (USA), and was held under the auspices of the European Association for Theoretical Computer Science.

Nancy, April 1993

Claude Kirchner
Chair, RTA-93

Program committee

Hubert Comon (Orsay)
Bruno Courcelle (Bordeaux)
Harald Ganzinger (Saarbrücken)
Jieh Hsiang (Stony Brook)
Claude Kirchner (Nancy)
Jan Willem Klop (Amsterdam)
Klaus Madlener (Kaiserslautern)
Paliath Narendran (Albany)
Mike O'Donnell (Chicago)
Mitsuhiro Okada (Montreal)
Leszek Pacholski (Wrocław)
Michael Rusinowitch (Nancy)
Mark Stickel (Menlo Park)

Organizing committee

Ronald Book (Santa Barbara)
Nachum Dershowitz (Urbana)
Jean Gallier (Philadelphia)
Deepak Kapur (Albany)
Claude Kirchner (Nancy)
Klaus Madlener (Kaiserslautern)
Pierre Lescanne (Nancy)
David Plaisted (Chapel Hill)

Local arrangements

Mitsuhiro Okada (Montreal)

Referees

A. Arnold	P. Audebaud	J. Avenhaus
H.-J. Bürckert	L. Bachmair	S. Bailey
D. Basin	H. Baumeister	A. Bockmayr
F. de Boer	M.P. Bonacina	A.M. Borzyszkowski
A. Boudet	W. Bousdira	P. Casteran
W. Charatonik	T. Chen	C. Choffrut
E.A. Cichon	E. Contejean	T. Deiss
B. Delsart	J. Denzinger	E. Domenjoud
D. Dougherty	F. Fages	M. Falaschi
D. Fehrer	M. Fernández	M.C.F. Ferreira
R. Fettig	A. Geser	R. Gilleron
G. Gonthier	B. Gramlich	S. Hölldobler
M. Hanus	D. Hofbauer	J. Hong
M. Huber	U. Hustadt	P. Jacquet
P. Johann	J.-P. Jouannaud	T. Jurdziński
X. Kühler	R. Kennaway	D. Kesner
A. Kisielewicz	H.-J. Kreowski	D. Krob
M. Kutyłowski	S. Lange	D. Lugiez
C. Lynch	F. Müller	C. Marché
R. McCloskey	W. McCune	R. McNaughton
A. Middeldorp	B. Mu	R. Nieuwenhuis
T. Nipkow	D. Niwiński	V. van Oostrom
E. Orlowska	F. Otto	M. Parigot
M. Piotrów	D. Plaisted	E. Poll
L. Puel	D. Rémy	C.R. Ramakrishnan
P. Rao	S.A. Rebelsky	B. Reinert
P. Rety	M.M. Richter	C. Ringeissen
W. Sadfi	G. Salzer	A. Sattler-Klein
M. Schmidt-Schauss	K. Schulz	H. Seidl
D.J. Sherman	A. Skowron	G. Smolka
W. Snyder	R. Socher-Ambrosius	Z. Spławski
J. Steinbach	J.-M. Steyaert	J. Stuber
S. Tison	J. Tiuryn	X. Toenne
Y. Toyama	P. Urzyczyn	S. Vorobyov
U. Waldmann	I. Walukiewicz	R. Wiehagen
C.P. Wirth	H. Zantema	H. Zhang

Table of Contents

INVITED TALK:
Rewrite Techniques in Theorem Proving
L. Bachmair (University of New York at Stony Brook) 1

Redundancy Criteria for Constrained Completion
C. Lynch and W. Snyder (Boston University) 2

Bi-rewriting, a Term Rewriting Technique for Monotonic Order Relations
J. Levy and J. Agusti (CSIC, Blanes) 17

A Case Study of Completion Modulo Distributivity and Abelian Groups
H. Zhang (The University of Iowa City) 32

A Semantic Approach to Order-Sorted Rewriting
A. Werner (University of Karlsruhe) 47

Distributing Equational Theorem Proving
J. Avenhaus and J. Denzinger (University of Kaiserslautern) 62

On the Correctness of a Distributed Memory Gröbner Basis Algorithm
S. Chakrabarti and K. Yelick (University of California at Berkeley) . 77

Improving Transformation Systems for General E-Unification
M. Moser (Technical University of Munich) 92

Equational and Membership Constraints for Infinite Trees
J. Niehren (DFKI, Saarbrücken), A. Podelski (DEC, Paris) and R. Treinen (DFKI, Saarbrücken) 106

Regular Path Expressions in Feature Logic
R. Backofen (DFKI, Saarbrücken) 121

INVITED TALK:
Proving Properties of Typed Lambda Terms: Realizability, Covers, and Sheaves
J. Gallier (University of Pennsylvania at Philadelphia) 136

Some Lambda Calculi with Categorical Sums and Products
D.J. Dougherty (Wesleyan University) 137

Paths, Computations and Labels in the λ-Calculus
A. Asperti (University of Bologna) and C. Laneve (INRIA Sophia-Antipolis) 152

Confluence and Superdevelopments
F. van Raamsdonk (CWI, Amsterdam) 168

Relating Graph and Term Rewriting via Böhm Models
  Z.M. Ariola (University of Oregon at Eugene) 183

Topics in Termination
  N. Dershowitz and C. Hoot (University of Illinois at Urbana) 198

Total Termination of Term Rewriting
  M.C.F. Ferreira and H. Zantema (University of Utrecht) 213

Simple Termination is Difficult
  A. Middeldorp (University of Tsukuba) and B. Gramlich (University
  of Kaiserslautern) 228

Optimal Normalization in Orthogonal Term Rewriting Systems
  Z. Khasidashvili (INRIA Rocquencourt) 243

A Graph Reduction Approach to Incremental Term Rewriting
  J. Field (IBM T.J. Watson Research Center) 259

Generating Tables for Bottom-up Matching
  E. Lippe (Software Engineering Research Centre, Utrecht) 274

INVITED TALK:
  On Some Algorithmic Problems for Groups and Monoids
  S. I. Adian (Steklov Mathematical Institute, Moscow) 289

Combination Techniques and Decision Problems for Disunification
  F. Baader (DFKI, Saarbrücken) and K. Schulz (University of Munich) 301

The Negation Elimination from Syntactic Equational Formula is Decidable
  M. Tajine (University Louis Pasteur, Strasbourg) 316

Encompassment Properties and Automata with Constraints
  A.-C. Caron and J.-L. Coquide and M. Dauchet (University of Lille) 328

Recursively Defined Tree Transductions
  J.-C. Raoult (IRISA, Rennes) 343

AC-Complement Problems: Satisfiability and Negation Elimination
  M. Fernàndez (LRI, Orsay) 358

A Precedence-Based Total AC-Compatible Ordering
  A. Rubio and R. Nieuwenhuis (University of Barcelona) 374

Extension of the Associative Path Ordering to a Chain of Associative Commutative Symbols
  C. Delor and L. Puel (LRI, Orsay) 389

Polynomial Time Termination and Constraint Satisfaction Tests
  D.A. Plaisted (University of North Carolina, Chapel Hill) 405

Linear Interpretations by Counting Patterns
  U. Martin *(University of St Andrews, Fife)* 421

Some Undecidable Termination Problems for Semi-Thue Systems
  G. Sénizergues *(LABRI, Bordeaux)* 434

SYSTEM DESCRIPTIONS

Saturation of First-Order (Constrained) Clauses with the *Saturate* System
  P. Nivela and R. Nieuwenhuis *(University of Barcelona)* 436

MERILL: An Equational Reasoning System in Standard ML
  B. Matthews *(University of Glasgow)* 441

Reduce the Redex → ReDuX
  R. Bündgen *(University of Tübingen)* 446

AGG — An Implementation of Algebraic Graph Rewriting
  M. Löwe and M. Beyer *(Technical University of Berlin)* 451

Smaran: A Congruence-Closure Based System for Equational Computations
  R. M. Verma *(University of Houston)* 457

LAMBDALG: Higher Order Algebraic Specification Language
  Y. Gui and M. Okada *(Concordia University, Montreal)* 462

OPEN PROBLEMS

More Problems in Rewriting
  N. Dershowitz *(University of Illinois at Urbana)*, J.-P. Jouannaud *(LRI, Orsay)* and J.W. Klop *(CWI, Amsterdam)* 468

Authors Index 488

Rewrite Techniques in Theorem Proving

Leo Bachmair
Department of Computer Science
University at Stony Brook
Stony Brook, New York, U.S.A.

The replacement of equals by equals is a common form of equational reasoning, of which rewriting is a refinement. Rewrite systems are sets of directed equations, called rewrite rules, that are used for replacements in the indicated direction only. A given expression is rewritten until a simplest possible form, a normal form, is obtained. Thus, the theory of rewriting is in essence a theory of normal forms. If a rewrite system is convergent, then all possible sequences of rewrites of equal terms result in the same normal form. In theories represented as convergent rewrite systems equality can therefore be decided rather efficiently.

Many aspects of the theory of rewriting can also be applied to resolution-style theorem provers. For instance, convergence requires that all sequences of rewrites terminate, a property that can be characterized by certain well-founded orderings called simplification orderings. If a total simplification ordering is imposed on a Herbrand base, then ground instances of a clause can be interpreted as conditional rewrite rules and refutational theorem proving may be viewed as a rewrite process: the negation of a theorem (the "goal") is rewritten until a contradiction is obtained. This method is only (refutationally) complete, though, if the set of rewrite rules extracted from the given clauses is convergent, and in general additional clauses may have to be deduced.

In this talk, I will discuss the fundamental techniques on which this rewrite approach to theorem proving is based. Two concepts are of particular interest: constraints and redundancy. Constraints provide a convenient way of describing the connection between the ground level (which embodies the interpretation of theorem proving as a rewrite process) and the general inferences that are actually applied by a prover to given clauses. In this context they have mainly been used to describe unification problems, ordering restrictions, and also certain normal-form properties of terms. Redundancy, on the other hand, allows one to optimize the proof search, as redundant formulas can be deleted and redundant inferences be ignored by a theorem prover.

Redundancy Criteria for Constrained Completion

Christopher Lynch Wayne Snyder[*]

April 2, 1993

Abstract

We study the problem of completion in the case of equations with constraints consisting of first-order formulae over equations, disequations, and an irreducibility predicate. We present several inference systems which show in a very precise way how to take advantage of redundancy notions in the context of constrained equational reasoning. A notable feature of these systems is the variety of tradeoffs they present for removing redundant instances of the equations involved in an inference. This combines in one consistent framework almost all practical critical pair criteria, including the notion of Basic Completion. In addition strict improvements of currently known criteria are developed.

1 Introduction

This paper presents a framework for exploiting redundancy notions in the context of a completion procedure for constrained equations. The constraint language consists of first-order formulae over atomic constraints consisting of equations, disequations, and an irreducibility predicate. An inference system is presented which shows precisely the tradeoffs involved in modifying constraints in order to delete unnecessary instances of the equations involved. The notion of redundancy we use is due to Bachmair and Ganzinger [1], and amounts to a semantic version of the well-known subconnectedness criterion (see [3]). Building on recent work on Basic Completion [4, 10], on constrained completion [8], and on various critical pair criteria [16, 13, 9] (see [3] for a survey), we show how a wide variety of techniques for removing redundant equations can be combined and refined in a consistent framework.

---
[*]Computer Science Department, Boston University, 111 Cummington St., Boston, MA 02215, U.S.A., lynch,snyder@cs.bu.edu.

Special cases of this inference system show how to implement a strict improvement of the technique of Basic Completion, and a stronger, hereditary version of the criterion based on subsumed critical pairs. In addition, we analyze the effect of initial constraints on the computation of critical pairs. It is hoped that this research contributes to the further development of the theory of constrained equational reasoning and to the practical improvement of existing completion procedures.

2 Preliminaries

We assume the reader is familiar with the standard definitions of terms constructed from a given set of symbols (augmented with an infinite set of Skolem constants). A *multiset* is an unordered collection with possible duplicate elements. An *equation* is a binary multiset $\{s,t\}$, conventionally represented $s \approx t$, where s and t are first-order terms over the given signature. A *substitution* is a mapping from variables to terms, e.g., $\{x_1 \mapsto t_1, x_2 \mapsto t_2, \ldots\}$, the *domain* of a substitution σ as the set $Dom(\sigma) = \{x \mid x \neq x\sigma\}$. The application of a substitution σ to a term t is denoted $t\sigma$; if τ and ρ are substitutions, then $x\tau\rho = (x\tau)\rho$, for all variables x.

We assume that a reduction ordering $\succ$ (i.e., a well-founded ordering closed under substitution and context application) total on ground terms is given. Such an ordering can be extended to a well-founded ordering $\succ_{mul}$ on finite multisets of terms in the usual way. The ordering $\succ$ on equations is simply $\succ_{mul}$ restricted to binary multisets. The *maximum* of a set S of equations, denoted $max(S)$, is defined as the smallest $S' \subseteq S$ such that $\forall B \in S, \exists B' \in S', B \preceq B'$. We denote an equation $s \approx t$ where $s \succ t$ by an expression $s \to t$ and call it a *rewrite rule*; note in this case that we must have $Var(t) \subseteq Var(s)$.

The constraint language we shall use is a modification of the one presented in [8] to account for irreducibility constraints. For additional information on constraints, see [8] and references presented there.

Definition 1 *The set of constraints C is defined inductively as the smallest set of expressions containing the* atomic constraints $\top$, $\bot$, $s = t$, and $Irr(s)$ *(for every pair of terms s, t), and such that whenever φ_1 and φ_2 are in C, then so are $(\varphi_1 \vee \varphi_2)$, $(\varphi_1 \wedge \varphi_2)$, $\neg(\varphi_1)$, $(\exists x.\,\varphi_1)$, and $(\forall x.\,\varphi_1)$.*

A constraint $\neg(s = t)$ is called a disequation.

The set of free variables in a constraint φ, denoted $Var(\varphi)$, is defined in the usual way. These are the variables that the constraint in fact constrains, and solutions are substitutions over these variables. We typically use φ and ψ to denote constraints.

Definition 2 *Let R be a ground rewrite system. We define the solutions $Sol_R(\varphi)$ of a constraint φ relative to R inductively as follows. First, $Sol_R(\bot) = \emptyset$. Then, for any ground substitution σ,*

(i) $\sigma \in Sol_R(\top)$;

(ii) $\sigma \in Sol_R(s = t)$ iff $s\sigma = t\sigma$;

(iii) $\sigma \in Sol_R(Irr(s))$ iff $s\sigma$ is R–irreducible;

(iv) $\sigma \in Sol_R(\varphi_1 \wedge \varphi_2)$ iff $\sigma \in Sol_R(\varphi_1) \cap Sol_R(\varphi_2)$;

(v) $\sigma \in Sol_R(\varphi_1 \vee \varphi_2)$ iff $\sigma \in Sol_R(\varphi_1) \cup Sol_R(\varphi_2)$;

(vi) $\sigma \in Sol_R(\neg\varphi)$ iff $\sigma \notin Sol_R(\varphi)$;

(vii) $\sigma \in Sol_R(\exists x.\varphi)$ iff there exists some ground term t such that $\{x \mapsto t\}\sigma \in Sol(\varphi)$; and

(viii) $\sigma \in Sol_R(\forall x.\varphi)$ iff for every ground term t, $\{x \mapsto t\}\sigma \in Sol(\varphi)$.

Thus, each constraint and each ground rewriting system define a set of ground substitutions; a non-ground substitution σ is said to be a solution if every ground substitution $\sigma\tau$ is a solution. A constraint is *satisfiable* relative to R if there exists some solution; if no solution exists, it is *unsatisfiable*, and equivalent to $\bot$. We say that φ is *stronger than* or a *a strengthening of* ψ if for any R, $Sol_R(\varphi) \subseteq Sol_R(\psi)$; alternately, ψ is *weaker than* or a *weakening of* φ.

Note that this is *not* a set of solutions wrt a theory R, as in [8]; the rewrite system R is only used for the irreducibility constraints. An irreducibility constraint $Irr(s)$ can be used to forbid inferences into particular subterms of an equation which are known to be irreducible, for example if they are produced by application of a substitution; this is a particular kind of redundancy check, called the Basic Strategy in [4], which here is developed further in the context of equational and disequational constraints. In addition, we shall propagate irreducibility constraints through inferences. Irreducibility constraints in completion are used in the context of an evolving rewrite system which successively approximates the limit canonical system (this limit system is represented by R in the preceeding definition); thus in practice we can only state that a constraint $Irr(s)$ in the context of a current rewrite system R' is false when s is reducible by R'; in general we could never say that such a constraint is true until the limit system is reached. However, this will be sufficient to develop an extension to the Basic Strategy in our setting.

In the sequel an idempotent substitution could be considered to be a conjunction of equations; we shall make free use of this below, for example forming a new constraint by adding a substitution, e.g., $\varphi \wedge \sigma$.

A *constrained equation* is simply an equation between two terms plus a constraint, e.g., $s \approx t\,[\varphi]$. (Later we shall extend this notation to append other constraints to the equation.) The constraint determines which ground instances of the equation are available. Since an equation A without a constraint can be considered to be a constrained equation $A[\top]$, in the sequel we use the word *equation* in general to denote a constrained equation. The symbols A, B, etc. will be used to denote either an equation with its constraint or simply the equation part, depending on the context. The *erasure* of an equation $A[\varphi]$ is defined as $A[\top]$ and similarly for sets of equations. By $\varphi\sigma$ we denote the replacement of each free occurrence of $x \in Dom(\sigma)$ in φ by $x\sigma$. We assume the normal conventions for avoiding free variable capture. Any free variable in φ which does not occur in A is assumed in $A\,[\varphi]$ to be existentially quantified at the innermost possible level.

For any ground rewriting system R, the set of *ground instances* of an equation $A[\varphi]$ *relative to* R is defined as

$$Gr_R(A[\varphi]) = \{\, A\sigma \mid \sigma \text{ ground},\ Var(A) \subseteq Dom(\sigma),\ \text{and}\ \sigma \in Sol_R(\varphi)\,\}.$$

The set of ground instances of a set E is then defined

$$Gr_R(E) = \bigcup_{A \in E} Gr_R(A).$$

Remark In order to preserve completeness, we only allow a constraint of the form $s \approx t[\ldots Irr(u)\ldots]$ if either $u \prec s$ or $u \prec t$. If this restriction does not hold, then $[\ldots Irr(u)\ldots]$ is weakened to the form $[\ldots \bot \ldots]$ if $Irr(u)$ occurs *negatively* (i.e., in the scope of an odd number of negations). If the restriction does not hold and $Irr(u)$ occurs *positively*, for u is a constant or a variable, then $[\ldots Irr(u)\ldots]$ is weakened to the form $[\ldots \top \ldots]$; but if $u = f(u_1, \ldots, u_n)$, we can weaken the constraint into the form $[\ldots Irr(u_1) \wedge \ldots \wedge Irr(u_n) \ldots]$; this decomposition of the term must be iterated just until the restricted form is attained. We shall assume in the sequel that all equations have this restricted form.

3 Redundancy and Constraints

In this paper we present a strong inference system for constrained completion. We show the various tradeoffs which can be employed when applying redundancy notions [1] to eliminate certain instances of constrained equations involved in the inferences. Intuitively, a redundant equation is an equation which is implied by smaller equations. Such equations are unnecessary in completing a set of equations. Our current formulation owes much to the paper [4].

Definition 3 *Let R be a ground rewriting system and E a set of equations. A ground instance $A \in Gr_R(E)$ is R-redundant in E if there exist equations $\{A_1, \ldots, A_n\} \subseteq Gr_R(E)$ such that $A_i \prec A$ for $1 \leq i \leq n$, and such that if each A_i is true in R, then A is true in R. If A is a non-ground equation, then A is R-redundant in E if every $A' \in Gr_R(A)$ is. If it is R-redundant, for any R, then it is simply called redundant.*

Now let $M = \{B_1, \ldots, B_k\}$ be a set of equations. We say that A is R-redundant in E upto M if for each ground instance $A\sigma \in Gr_R(A)$, there exist equations $\{A_1, \ldots, A_n\} \subseteq Gr_R(E)$ and for each $B_j \in M$ there exists a ground instance $B'_j \in Gr_R(B_j\sigma)$ such that $A_i \prec max(B'_1, \ldots, B'_k)$ for $1 \leq i \leq n$, and such that if each A_i is true in R, then A is true in R.[1]

For instance, equations with only identity instances are trivially redundant. In this paper we present a framework for representing redundancy information explicitly in an equation, by adding constraints to the equation which give more information about which instances are redundant; this information can then be propagated during inferences under certain conditions. Our notation uses an equation and a triple represented as $A[\varphi_1, \varphi_2, M]$, where A is an equation, M is a set of equations, and φ_1 and φ_2 are constraints. We can think of this as an extension of the original notation $A[\varphi]$, so that the first constraint φ_1 still represents the available instances of the equation, i.e., $Gr_R(A[\varphi_1, \varphi_2, M]) = Gr_R(A[\varphi_1])$. The other constraint and the set M record redundancy information in the following way.

Definition 4 *A constrained equation $A[\varphi_1, \varphi_2, M] \in E$ is correct for E (or simply correct if E is obvious) if for all rewrite systems R (1)$Gr_R(A[\varphi_1]) \subseteq Gr_R(A[\varphi_2])$, (2)If $B \in Gr_R(A[\varphi_2]) \setminus Gr_R(A[\varphi_1])$ then B is R-redundant in E, and (3) If $B \in Gr_R(A) \setminus Gr_R(A[\varphi_2])$ then B is R-redundant in E up to M.*

For example, an unconstrained equation has the form $A[\top, \top, \{A\}]$. We will hereafter assume that all equations are in correct form, but may eliminate a suffix of the parameters if desired. If M is missing we assume it is $\{A\}$ and a missing φ_2 is assumed equal to φ_1, and a missing φ_1 is assumed to be $\top$. The last two components are used to store information about the history of an equation. Essentially, redundancy is used in the

---

[1] The point of this rather complex definition will be made clear in a moment.

completeness proof to show when equations become true. In passing such information around the inference system, it becomes useful to separate the ordering requirements in the definition of redundancy (e.g., "$A_i \prec A$") from the logical requirements (e.g., "if each A_i is true in R ..."). We thus wish to know when A is implied by equations smaller than B, and the set M preserves information about the smallest such B. It records which axioms S (original equations) were used to construct a given equation A. Clearly, S implies A, and thus (roughly) we let $M = max(S)$ in A. This parameter does not change for any particular equation.

We will use this redundancy information to delete instances of equations. For example, it is well known that overlaps at variable positions are not necessary. This is because instances with reducible substitutions are redundanct. In our framework we make this explicit, representing the irreducibility condition in the constraint. An unconstrained equation $fx \approx gx$ would be represented in correct form here as $fx \approx gx[Irr(x), \top]$. It is sufficient to to consider cases where the constraint is false to simulate the "no overlaps at variable positions" condition and also the Basic strategy.

4 Constrained Critical Pair Generation

In this section we give a generalization of the critical pair rule from [8] and show how a variety of tradeoffs may be obtained in deleting various instances of the equations involved in an inference.

The general form of our constrained critical pair rule is

C-Deduce
$$\frac{s \to t[\varphi_1, \varphi_2, M\,] \qquad u[s'] \to v[\psi_1, \psi_2, N\,]}{u[t]\sigma \approx v\sigma[\Delta_1, \Delta_2, max(M\sigma \cup N\sigma)\,]}$$

where (1) $\sigma = mgu(s, s')$, (2) Δ_1 is a weakening of $\varphi_1\sigma \wedge \psi_1\sigma \wedge Irr(s\sigma)$, (3) Δ_1 is a strengthening of $Irr(x_1) \wedge \ldots \wedge Irr(x_n)$, where $\{x_1, \ldots, x_n\} = Var(u[t]\sigma \approx v\sigma)$, (4) the conclusion is a correct equation, and (5) after constructing the conclusion we may potentially modify some of the premise constraints as long as these are still correct equations.

In general in the inference rules we present, equations will have the form $A[Irr(s_1) \wedge \ldots \wedge Irr(s_n) \wedge \varphi_1', \varphi_2, M\,]$, where any variable in A occurs in some s_i. Note that we have not explicitly stated the condition "where s' is not a variable," but in fact this will be a consequence of the irreducibility constraints built up during the inference process. Inferences involving variable overlaps can be shown to be redundant and hence unnecessary.

The correctness criteria here basically assert that if instances of these equations are deleted by the inference, then these instances are redundant.

The general idea of the various instances of this schema we present is that certain instances of the right premise are redundant by virtue of certain instances of the left premise and the conclusion; the tradeoffs occur in considering whether we want to strengthen the right premise by deleting as many instances of the right premise as possible, in which case we need perhaps to weaken the other equations by making more instances available, or whether we wish to strengthen the conclusion as much as possible, in which case we can not delete as many instances of the right premise. Essentially these rules can be thought of as combinations of simplification and overlap rules. In addition, it is possible to define situations under which the inference itself is redundant and hence need not be performed.

Definition 5 *For any R and E, a C-Deduce inference as given above is R-redundant in E if (i) the σ-instance of either premise is R-redundant in E, or (ii) $u[t]\sigma \approx v\sigma [\varphi_1\sigma \wedge \psi_1\sigma \wedge Irr(s\sigma), \Delta_2, max(M\sigma \cup N\sigma)]$ is R-redundant in E. If it is R-redundant, for any R, then it is simply called redundant.*

To present these inference rules we need to say what the values of the constraints in the conclusion are, and how the constraints in the premises are (potentially) modified. For each case, we would need to show that the conditions of C-Deduce are satisfied; we omit these proofs from this abstract. First we present two general constraint modification rules that may be applied to strengthen the right premise after an inference has been performed.

Let **CM1** be the right premise constraint modification rule: $\psi_1 \Rightarrow \psi_1 \wedge \neg(\sigma \wedge \Delta_2 \wedge \varphi_2)$, and let **CM2** be the rule: $\psi_1 \Rightarrow \psi_1 \wedge \neg(\sigma \wedge \Delta_2)$.

It can be shown that if $s\sigma \to t\sigma \prec u[s']\sigma \to v\sigma$ then CM1 applied to the right premise yields a new correct equation.[2] If in addition we have $M\sigma \prec_{mul} \{u[s']\sigma \to v\sigma\}$ then CM2 applied to the right premise yields a correct equation.

The first inference system presented is called CCP (Constrained Critical Pairs). In this case the conclusion is as strong as possible, the left premise is not weakened, and some instances of the right premise are deleted.

Definition 6 *Let CCP be the instance of C-Deduce where $\Delta_1 = \Delta_2 = \varphi_1\sigma \wedge \psi_1\sigma \wedge Irr(s\sigma)$, and where CM1 is performed if $s\sigma \to t\sigma \prec u[s']\sigma \to v\sigma$.*

In the CCP inference Δ_1 is as strong as it can be in an inference. Given the value of Δ_1 we could try to make Δ_2 as weak as possible so we can delete more of the instances of the right premise. For example, if the conclusion

---

[2] Note that if this condition is violated then the conclusion is either unorientable or an identity.

is $fx \approx gx[Irr(x), Irr(x)]$, then we could change Δ_2 to $\top$, because all reducible instances are redundant. In general, if $\Delta_1 = Irr(x) \wedge \varphi'$ and φ' does not further constrain x, then Δ_2 can be set equal to φ' (this process can be iterated). Call the result of this iteration $NoIrrVar(\Delta_1)$. Although we shall have occasion to refer to this notion in a later section, for simplicity in this abstract, we have presented a simpler version where $\Delta_1 = \Delta_2$.

Our second instance of C-Deduce emphasizes strengthening the right premise as much as possible, essentially by simplifying as many instances of the right premise as possible by instances of the left premise. In this case we may have to weaken the left premise and construct a weaker conclusion than in the previous rule.

Definition 7 *C-Simplify is the instance of C-Deduce such that $\Delta_1 = \Delta_2 = \psi_1 \sigma$, and where in addition if $s\sigma \to t\sigma \prec u[s']\sigma \to v\sigma$ we change ψ_1 in the right premise to $\psi_1 \wedge \neg \sigma$; finally, unless $M\sigma \prec_{mul} \{u[s']\sigma \to v\sigma\}$ holds we must further modify the premise constraints so that $\varphi_1 \Rightarrow \varphi_1 \vee (\sigma \wedge \psi_1 \wedge \neg \varphi_2)$ and $\varphi_2 \Rightarrow \varphi_2 \vee (\sigma \wedge \psi_1)$.*

These two rules illustrate the range of tradeoffs available. In CCP we do not weaken the conclusion or the left premise, so that we can only eliminate some instances of the right premise. In C-Simplify we must weaken the constraints on the conclusion and the left premise in general but we can then delete all possible instances of the right premise. It is possible to define inference rules between these two extremes. In the next definition we present two rules which weaken the conclusion but not the left premise of the inference.

Definition 8 *Suppose $s\sigma \to t\sigma \prec u[s']\sigma \to v\sigma$. Then we define the rule CCP1 as the instance of C-Deduce where $\Delta_1 = \Delta_2 = \varphi_2 \sigma \wedge \psi_1 \sigma \wedge Irr(s\sigma)$ and with the strengthening $\psi_1 \Rightarrow \psi_1 \wedge \neg (\sigma \wedge \varphi_2 \wedge Irr(s\sigma))$. If in addition, we have $M\sigma \prec_{mul} \{u[s']\sigma \to v\sigma\}$, then we may define the instance CCP2 of C-Deduce where $\Delta_1 = \Delta_2 = \psi_1 \sigma \wedge Irr(s\sigma)$ and such that $\psi_1 \Rightarrow \psi_1 \wedge \neg(\sigma \wedge Irr(s\sigma))$.*

In a similar manner it is possible to define other inference rules that partially weaken the conclusion and the left premise so some instances of the right premise are deleted. For instance we can weaken the constraints on the conclusion so that just the irreducibility constraints remain, or we can weaken the constraints so that just the equational and disequational constraints remain.[3] Thus it is possible to define a spectrum of possible critical pair rules in our framework.

---

[3]To be precise we would also need to keep the irreducibility constraints on the variables of the conclusion to avoid superposing into variables.

Now we consider some examples of the above inference rules. Consider the inference

$$\frac{fa \to b \qquad fx \to gx[Irr(x), \top]}{b \approx ga[Irr(a), Irr(a), fa \to ga]}$$

on axioms. If we use the CCP rule then we may apply CM1 to the constraint of the right premise: $Irr(x) \Rightarrow Irr(x) \land (x \neq a \lor \neg Irr(a))$. If we use C-Simplify then the conclusion becomes $b \approx ga\,[\,\top\,]$ and the right premise is modified by $Irr(x) \Rightarrow Irr(x) \land x \neq a$. We can now show how these two inferences would provide additional information usable in later inferences. Assume we followed the C-Simplify inference just given with

$$\frac{fx \to gx[Irr(x) \land x \neq a, \top] \qquad gfa \to c}{gga \approx c[\bot, \bot, gfa \to c]}$$

The first thing to note is that this inference is redundant because the constraint on the conclusion is unsatisfiable. Therefore the inference does not need to be performed. However, we may be interested in simplifying the right premise, so we still perform the inference. Using C-Simplify we get $gga \approx c\,[\,\top\,]$ for the conclusion. The first constraint on the right premise becomes $\bot$ which means that none of the instances of the equation are necessary. However, the second constraint is still $\top$ which means that all the instances are redundant. Therefore we may use it to simplify an equation if we like, without weakening the constraint, but we are never required to use it in an inference. This illustrates the benefit of the second constraint. If we had not saved the second constraint we would have had to weaken the first constraint on the left premise.

To illustrate the benefit of the third component of the constraint triple we consider following the CCP inference in the first example with

$$\frac{ga \to b[Irr(a), Irr(a), fa \to ga] \qquad fga \to ga}{fb \approx ga[Irr(a), Irr(a), fga \to ga]}$$

If we want this to be a C-Simplify inference the conclusion can be weakened to $fb \approx ga[\top, \top, fga \to ga]$. Then we can use CM2 to set the first constraint of the right premise to $\bot$ as in the previous example, since all instances of left premise are true by equations smaller than the right premise.

We give one more example to illustrate a use of the irreducibility constraints. Consider the inference

$$\frac{fa \to b \qquad fa \to ga}{b \approx ga[Irr(a), Irr(a), fa \to ga]}$$

We could consider this to be a C-Simplify inference, weaken the constraint in the conclusion and change the contraint of the right premise to $\bot$. If we used

CCP instead the first constraint on the right premise becomes $\neg Irr(a)$ using CM1. Any inference using this equation as left premise is now redundant because a must be irreducible in an inference. That is, when $fa \to ga$ is used as a left premise, fa can be restricted to be in normal form (cf. the prime superposition criterion discussed below), which violates the constraint.

Naturally, other rules for simplifying rhs's of rules, orienting, etc. are necessary for a practical system, but for brevity we have presented only our critical pair rule. These are relatively straightforward adaptations of the ideas above, except for the blocking rules, and are presented in full in the long version. Irreducibility constraints give us blocking rules based on the reducibility of terms in constraints $Irr(s)$. For example, suppose we have equations $A[...Irr(u[s'])...]$ and $s \to t[\varphi]$, where $s\rho = s'$. Then the first equation can be changed to $A[...(Irr(u[s']) \wedge \neg(\varphi\rho))...]$. Clearly if all instances of $s \to t$ are available, i.e., $\varphi = \top$, then this corresponds to solving the constraint $Irr(u)$ by replacing it with $\bot$.

In the remainder of this section we show how we can set the parameters of the C-Deduce rule to give other critical pair criteria as special cases of ours. To start with we consider standard completion.

The *standard critical pair rule* can be represented in our system by letting $\Delta_1 = Irr(x_1) \wedge ... \wedge Irr(x_n)$, where $\{x_1,...,x_n\} = Var(u[t]\sigma \approx v\sigma)$, $\Delta_2 = \top$, and P be anything that yields a correct equation (since it will never be used). This is only necessary to disallow superposition into variable positions. The simplification rule can be represented by the same conclusion, with the right premise modified using CM1. Since simplification is only performed when σ is a matcher, the first constraint on the right premise becomes $\bot$ so the equation may be deleted.

Prime superposition [7] is a critical pair criterion which states that an inference is unnecessary if the lhs of the left premise is reducible. This follows directly from our redundancy criteria. An inference is redundant if $Irr(s\sigma)$ is unsatisfiable. In fact our results provide for a hereditary version of this criterion.

General superposition [16] and the critical pair criteria discussed in [9, 13, 14] are all examples of a more general principle of *subsumed critical pairs* [3]. Once an overlap on an equation A is produced, involving an *mgu* σ, then it is no longer necessary to consider overlaps on A involving *mgu*s less general or equal to σ. We simulate these critical pair criteria with disequational constraints. The constraints on the conclusion would be the same as the constraints in the standard critical pair rule. The difference is that CM1 is then performed. The first constraint of the right premise then becomes $\psi_1 \wedge \neg \sigma$. This disallows further superpositions into the right premise where the *mgu* is less general than or equal to σ, since these instances are no longer present. Again, our results provide for a hereditary version of this criterion. In other words, if a right premise has been overlapped with mgu σ, then the conclusion also never needs to be overlapped with an mgu less general or equal to σ.

In addition to naturally simulating subsumed critical pair criteria with our inference system we also naturally simulate *basic completion* [4, 10]. In this strategy, overlaps are disallowed on terms introduced by substitution. This is simulated with irreducibility constraints. In the conclusion of an inference we let $\Delta_1 = \varphi_1\sigma \wedge \psi_1\sigma$ and $\Delta_2 = NoIrrVar(\Delta_1)$. Essentially, all constraints would be conjunctions of irreducibility constraints; constraints on the variables of the premises are instantiated by the mgu which restricts us from superposing into those positions. In fact, we can obtain a stronger version, because constraints can be kept on terms not occurring in the equation. The special form of simplification required in basic completion can be simulated by our techniques for weakening the left premise.

The completion system in [8] is designed for a set of equations with initial constraints. The authors are not concerned with efficiency constraints and redundancy. As we have shown in the beginning of this section, completion is not complete with initially constrained equations unless we allow superposing into variables. In order to insure completeness [8] considered some additional inference rules which basically had the purpose of turning constrained equations into unconstrained equations. In our full paper in preparation we show how completeness can be preserved with initial constraints by allowing a limited form of variable ovelap. Our completeness proof is the first one we are aware of for equation and disequation constraints without any additional rules. We studied the combination of irreducibility constraints (to embed Basic Completion) with a subset of the constraints considered in [8]. For example we do not consider ordering constraints (see also [12] and [10]), although it seems they could be added to our system without major alterations of the framework.

We now consider the completeness of the rules presented in the previous section. For lack of space we can present no formal proofs, referring the reader to the full paper. We emphasize that we are considering only the critical pair rules here, and not the full complement of completion inference rules. It is sufficient for completeness however to consider only the critical pair rules.

Following the paradigm developed at length in the book [3], we define a *derivation* to model the process of completion.

Definition 9 *A sequence* $< S_0, S_1, \ldots >$ *of sets of equations is a* derivation *from S if $S_0 = S$ and for each $i \geq 0$, for any R, $Gr_R(S_{i+1}) = (Gr_R(S_i) \cup E_1) \setminus E_2$ where E_1 and E_2 are sets of equations such that $Gr_R(S_i) \models E_1$ and* each equation in E_2 is R-redundant in $Gr_R(S_i) \cup E_1$. Let $S_\infty = \bigcup_j \bigcap_{k \geq j} S_k$. We call S_∞ the limit of the derivation. Any equation $A \in S_\infty$ is called persisting.

Definition 10 *Let I be some instance of C-Deduce. A derivation is an I-derivation if each S_{i+1} is obtained from S_i by application of the rule I. A set S is I-saturated if every I-inference from S is redundant. An I-derivation is fair if the limit is I-saturated.*

An inference rule can be viewed as a method for adding consequences to the set and deleting redundant instances. The next result shows that this is correct in the limit.

Lemma 1 *Let R be a ground rewriting system and suppose for two sets of equations E and E', $Gr_R(E) \subseteq Gr_R(E')$. (1) Any closure (or inference) which is R-redundant in E is also R-redundant in E'. (2) If all ground instances in $Gr_R(E') \setminus Gr_R(E)$ are R-redundant in E', then any equation (or inference) which is R-redundant in E' is also R-redundant in E.*

This shows that the inference systems presented are sufficient to saturate a set of equations. We now show that saturated sets are ground canonical. In our framework, this will allow us to argue that our constrained completion systems (which are not defined as unfailing) will produce canonical sets in the limit. Our proof follows very much in the lines of the proof in the journal version of [4], with the addition of the constraint formalism. In addition, there are some delicate features of the proof which relate to the use of the irreducibility constraints defined relative to a rewrite system which is constructed from the set of constrained equations itself. First we give a method for constructing a canonical set of ground rewrite rules from a given set of equations.

Definition 11 *Let E be a set of equations and $\mathcal{EQ}$ denote the set of all ground equations. We define the ground rewriting system R_E using induction on $(\mathcal{EQ}, \succ)$ by associating with each $A \in \mathcal{EQ}$ a rewrite system R_A. Assume for a ground equation A that R_B has been defined for each ground equation B with $B \prec A$, and let $R_{\prec A}$ be defined as $\bigcup_{B \prec A} R_B$. Then $R_A = \{A\}$ if A is a member of $Gr_{R_{\prec A}}(E)$ in the form $s \to t$ and where s irreducible by $R_{\prec A}$; otherwise $R_A = \emptyset$. Finally define R_E as $\bigcup_{A \in \mathcal{EQ}} R_A$.*

Notice that the rewrite system R_E is constructed out of instances from substitutions reduced relative to smaller rewrite rules already in R_E.

Let us say that a ground instance $A\sigma$ of an equation from E is *reduced relative to R*, or an *R-reduced instance of E*, if $x\sigma$ is irreducible by R for every $x \in Dom(\sigma)$. The properties of the preceding definition we shall need are as follows.

Lemma 2 *For R_E as just defined, (i) Every equation in R_E is an R_E-reduced instance of E; (ii) R_E and $R_{\prec A}$ for any $A \in \mathcal{EQ}$ are canonical; (iii) No equation $A \in R_E$ is true in $R_{\prec A}$; and (iv) An equation $A \in Gr_{R_E}(E)$ is true in R_E iff it is true in $R_{\prec A} \cup R_A$.*

Theorem 1 *Let I be some instance of the C-Deduce rule, R_E be as above, and E be an I-saturated set of equations such that for each $A[\varphi_1, \varphi_2, M] \in E$, φ_1 is stronger than $Irr(x_1) \wedge \ldots \wedge Irr(x_n)$ for $\{x_1, \ldots, x_n\} = Var(A)$. Then R_E makes true every member of $Gr_{R_E}(E)$.*

We now state the main completeness result of the paper.

Theorem 2 *Let E be a set of unconstrained equations and S be the set of equations $A[Irr(x_1) \wedge \ldots \wedge Irr(x_n), \top]$, for $A \in E$ and $\{x_1, \ldots, x_n\} = Var(A)$. Let $< S, \ldots >$ be an I-fair derivation from S for some instance I of C-Deduce. If S_∞ contains no unorientable equations then it is ground canonical and equivalent to E. In addition the erasure of S_∞ is a canonical rewriting system equivalent to E.*

The proof that the erasure is a canonical (and not just a ground canonical) rewrite system involves a Skolemization step, and is from [4]. This shows that our inference system (which was not presented as an unfailing completion procedure) produces a canonical rewriting system in the limit. If a derivation is finite, then of course the final system is canonical. In this case it could be considered to be a constrained rewriting system, or its erasure could be produced. The adaptation of these results to the case of unfailing completion is straightforward and left to the full paper.

5 Conclusion

We have presented several inference systems which show in a very precise way how to take advantage of redundancy notions in the context of constrained equational reasoning. These systems illustrate the tradeoffs involved in this framework in a very precise way. We hope that this research contributes to the further development of the theory of constrained equational reasoning and to the practical improvement of existing completion procedures.

The method of proof used in this paper was adapted from our previous paper with Bachmair and Ganzinger on Basic Paramodulation [4] (see also [10]), which in turn adapted the results of [1] (cf. [11] and [17]). However, the inference systems are developments of the rules from the seminal paper [8] to show how irreducibility constraints can be used to express the idea of Basic Completion in combination with other kinds of equational constraints.

To apply the procedures given in this paper, one needs to have a constraint solving algorithm. Comon and Lescanne [5] have analyzed the problem of solving constraints of equations and disequations. In our framework we also consider irreducibility constraints, however, which complicates the situation. For an arbitrary canonical system R, reducibility and irreducibility tests can be made using inductive reducibility and narrowing tests, however in our setting these tests must be made with respect to an evolving rewrite system and in the presence of constrained rewrite rules. Thus only certain tests can be made. Some of these have been explained in our blocking rules. In general, for an incompletely specified rewrite system, we can only know that if a term t is reducible at some stage, it will be reducible in the limit as well; we can never state in the positive that t is irreducible before the completion process terminates.

We do not expect that this framework in its entirety would be necessarily be an efficient and useable form of completion procedure. We instead view it as a theoretical model for constrained completion, some of whose special cases may turn out to be practically useful. Our current research focusses on simple and efficient subcases of the general framework which promise to eliminate as many redundant inferences and equations as possible without excess amounts of overhead. A particular focus is on subclasses for which efficient constraint solving techniques exist. The implementation of this system, and the Basic Completion system discussed in [4], is currently being investigated at BU as part of the Masters Thesis [6].

References

[1] L. BACHMAIR AND H. GANZINGER. Rewrite-based equational theorem proving with selection and simplification. To appear in *Journal of Logic and Computation* (1992).

[2] L. BACHMAIR AND N. DERSHOWITZ. Critical Pair Criteria for Completion. *J. Symbolic Computation 6* (1988) pp.1-18.

[3] L. BACHMAIR. *Canonical Equational Proofs*. Birkhauser Boston, Inc., Boston MA (1991).

[4] L. BACHMAIR, H. GANZINGER, C. LYNCH, AND W. SNYDER. Basic Paramodulation and Superposition. In *Proc. 11th Conference on Automated Deduction*, Saratoga Springs, NY (1992) pp. 263–476. Journal version in preparation.

[5] H. COMON AND P. LESCANNE. Equational Problems and Disunification. *Journal of Symbolic Computation 7* (1989) pp. 371-426.

[6] D. Durand, *Experiments in Basic Paramodulation and Constrained Completion*, Masters Thesis, Boston University Computer Science Department (1992).

[7] D. KAPUR, D. MUSSER AND P. NARENDRAN. Only Prime Superpositions need to be considered in the Knuth-Bendix Completion Procedure. *J. Symbolic Computation 6* (1988) pp.19-36.

[8] C. KIRCHNER, H. KIRCHNER AND M. RUSINOWICH. Deduction with Symbolic Constraints. *Revue Francaise d'Intelligence Artificielle Vol 4, no. 3* (1990) pp. 9-52.

[9] W. KÜCHLIN. A Confluence Criterion based on the generalized Newman lemma. In *Proc. Eurocal '85 Lecture Notes in Computer Science 204*, pp.390-399. Berlin, Springer Verlag.

[10] R. NIEUWENHUIS AND A. RUBIO. Theorem Proving with Ordering Constrained Clauses. In *Proc. 11th Conference on Automated Deduction*, Saratoga Springs, NY (1992) pp. 477–491.

[11] J. PAIS AND G. PETERSON. Using forcing to prove completeness of resolution and paramodulation. *J. Symbolic Computation 11* (1991) pp.3-19.

[12] G. PETERSON. Complete Sets of Reductions with Constraints. In *Proc. 10th Conference on Automated Deduction*, Kaiserslautern, LNCS 449 (1990) pp. 381–395.

[13] F. WINKLER. Reducing the Complexity of the Knuth-Bendix Completion Algorithm: A Unification of Different Approaches. In *Proc. Eurocal '85 LNCS 204*, pp.378-389. Berlin, Springer Verlag.

[14] F. WINKLER AND B. BÜCHBERGER. A Criterion for eliminating unnecessary reductions in the Knuth-Bendix algorithm. In *Proc. Coll. on Algebra, Combinatorics and Logic in Computer Science*, Gyor Hungary, pp. 849–869.

[15] H. ZHANG AND D. KAPUR. Consider only General Superposition in Completion Procedures In *Proc. 3rd International Conference on Rewriting Techniques and Applications*, LNCS 355, pp.513-529. Berlin, Springer-Verlag. Saratoga Springs, NY (1992) pp. 263–476.

[16] H. ZHANG AND D. KAPUR. *Unnecessary Inferences in Associative-Commutative Completion Procedures Mathematical Systems Theory 23* (1990) pp. 175-206.

[17] H. ZHANG. *Reduction, Superposition, and Induction: Automated Reasoning in an Equational Logic.* Ph.D. Thesis, Rensselaer Polytechnic Institute (1988).

Bi-rewriting, a Term Rewriting Technique for Monotonic Order Relations*

Jordi Levy and Jaume Agustí

Institut d'Investigació en Intel·ligència Artificial (CSIC)
Camí Sta. Bàrbara s/n, 17300 Blanes, Girona, Spain.
E-mail: levy@ceab.es and agusti@ceab.es

Abstract. We propose an extension of rewriting techniques to derive inclusion relations $a \subseteq b$ between terms built from monotonic operators. Instead of using only a rewriting relation $\xrightarrow{\subseteq}$ and rewriting a to b, we use another rewriting relation $\xrightarrow{\supseteq}$ as well and seek a common expression c such that $a \xrightarrow{\subseteq}{}^* c$ and $b \xrightarrow{\supseteq}{}^* c$. Each component of the bi-rewriting system $\langle \xrightarrow{\subseteq}, \xrightarrow{\supseteq} \rangle$ is allowed to be a subset of the corresponding inclusion $\subseteq$ or $\supseteq$. In order to assure the decidability and completeness of the proof procedure we study the commutativity of $\xrightarrow{\subseteq}$ and $\xrightarrow{\supseteq}$. We also extend the existing techniques of rewriting modulo equalities to bi-rewriting modulo a set of inclusions. We present the canonical bi-rewriting system corresponding to the theory of non-distributive lattices.

1 Introduction

Rewriting systems are usually associated with rewriting on equivalence classes of terms, defined by a set of equations. However term rewriting techniques may be used to compute other relations than congruences. Particularly interesting are non-symmetric relations like pre-orders. For instance, logics of inequalities [7], rewriting logic [21], ordered algebras [8], subset logic [12, 24], unified algebras [2, 22], taxonomies [1, 23, 26], subtypes [5], refinement calculus [20], all them use some kind of pre-order on expressions. In this paper we will show the applicability of rewriting techniques to monotonic pre-order relations on first order terms (inequality logics), that is the deduction of inequalities —here we call them inclusions— from a given set of them, the axioms.

The idea of applying rewriting techniques to the deduction of inclusions between terms, like $a \subseteq b$, is very simple. We compute by repeatedly replacing both 1) subterms of a by "bigger" terms using the axioms and 2) subterms of b by "smaller" terms using the same axioms until a connection is found between a and b. Evidently there are many paths starting from a in the direction $\xrightarrow{\subseteq}$ and from b in the direction $\xrightarrow{\supseteq}$ (see figure 2). Many of them are blind alleys and others are not terminating. Thus, it is essential that the search avoids blind alleys for efficiency reasons and, specially, avoids infinite sequences of rewritings with infinite

---

* This work has been partially supported by the project TESEU (TIC 91-430) funded by the CICYT

2 Inclusions and Bi-rewriting Systems

If nothing is said, we follow the notation used in [6, 10, 16]. We shall be concerned with first-order terms over a nonempty signature. We will denote the p occurrence or position in t by $t|_p$, and the substitution of the occurrence p by s in t by $t[s]_p$. We use the relational logic notation to present the abstract bi-rewriting properties. The inverse of the relation $\longrightarrow_R$ will be denoted by $\longleftarrow_R$, its reflexive-transitive closure by $\longrightarrow_R^*$, the transitive composition by $\longrightarrow_{R_1} \circ \longrightarrow_{R_2}$, and the union by $\longrightarrow_{R_1} \cup \longrightarrow_{R_2}$.

An inclusion is an ordered pair of terms $\langle s,t \rangle$ written $s \subseteq t$. Given a finite set of inclusions I, $\subseteq_I$ will denote the monotonic (stable and compatible) closure of I. That is, $u \subseteq_I v$ iff u is $w[\sigma(s)]_p$ and v is $w[\sigma(t)]_p$ for some term w, occurrence p of w, substitution σ and inclusion $s \subseteq t$ in I. The reflexive-transitive closure $\subseteq_I^*$ defines the inclusion theory presented by I.

The orientation of a finite set of inclusions I, for rewriting purposes, may result in two sets of rewriting rules, R_1 with rules like $s \xrightarrow{\subseteq} t$ and R_2 with rules like $s \xrightarrow{2} t$. The pair $\langle R_1, R_2 \rangle$ is called a bi-rewriting system. For example, inclusions defining the union may be oriented as it is shown in figure 1.

$$I = \begin{cases} X \cup X \subseteq X \\ X \subseteq X \cup Y \\ Y \subseteq X \cup Y \end{cases} \qquad \begin{aligned} R_1 &= \left\{ r_1 : X \cup X \xrightarrow{\subseteq} X \right. \\ R_2 &= \begin{cases} r_2 : X \cup Y \xrightarrow{2} X \\ r_3 : X \cup Y \xrightarrow{2} Y \end{cases} \end{aligned}$$

Fig. 1. Orientation of the inclusion theory of the union.

In this section we suppose that each inclusion may be oriented putting it in R_1 or in R_2, or may be in both sets. In the next section we will consider the case of inclusions which can not be oriented.

Given a bi-rewriting system $\langle R_1, R_2 \rangle$ its monotonic closure results in a pair of rewriting relations $\langle \longrightarrow_{R_1}, \longrightarrow_{R_2} \rangle$. Then the relation $(\longrightarrow_{R_1} \cup \longleftarrow_{R_2})^*$ is equal to $\subseteq_I^*$.

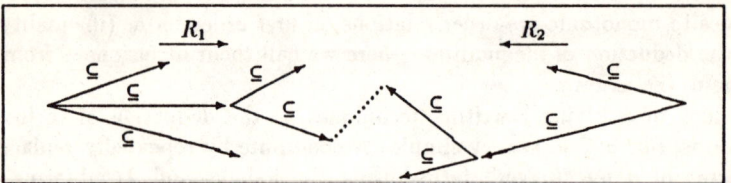

Fig. 2. An image of the bi-rewriting algorithm

Based on the pair of rewriting relations $\langle \longrightarrow_{R_1}, \longrightarrow_{R_2} \rangle$ a sound breadth-first search proof procedure for the inclusion theory $\subseteq_I^*$ can be easily defined (see figure 2). The procedure is complete and semi-decidable iff the bi-rewriting system is Church-Rosser –the branches being enumerable– and it is decidable iff the bi-rewriting system is also quasi-terminating. These two notions are studied in the different terms (infinite paths due to cycles are avoided easily). Evidently infinite different rewritings would prevent the decidability of the procedure. The solution to non-termination is to orient the axioms using a well founded ordering(s) on terms. Because the relation is non-symmetric, the orientation results in a pair of rewriting systems $\langle \xrightarrow{\subseteq}_{R_1}, \xrightarrow{\supseteq}_{R_2} \rangle$, that is, we get a bi-rewriting system. We introduce the definitions of a Church-Rosser and quasi-terminating bi-rewriting system in order to assure the decidability and the completeness of the search procedure. That is, given a set of axioms, if we can orient and complete them obtaining a confluent bi-rewriting system, then we will have a semi-decidable procedure to test $a \subseteq b$. The procedure is decidable if the bi-rewriting system is quasi-terminating.

Most of the notions of rewriting can be extended to bi-rewriting and the development of the subject follows the same pattern as rewriting: from Church-Rosser property to critical pairs lemma and then the completion process. However there are also some differences. Equational rewriting is in essence a theory of normal forms, while bi-rewriting disregards this notion since is based on quasi-termination and Church-Rosser properties. Bi-rewriting can also be seen as a generalization of equational rewriting: equations can be translated to pairs of inclusions and then we can reproduce the equational case. The price of this generalization is that bi-rewriting is based on a search procedure —which is avoided in canonical rewriting systems— and as we will see in section 2, the set of critical pairs of a non-left-linear bi-rewriting system may be infinite and then the study of confluence is case dependent.[2]

This paper proceeds as follows. In section 2 we present a version of the critical pairs theorem [10, 16, 17] for bi-rewriting systems using an extended definition of critical pairs. We also give a counter-example that invalidates this theorem stated in terms of standard critical pairs and a counter-example for the Toyama theorem [27].

In section 3 we generalize the results of section 2 to bi-rewriting systems modulo a set of (non-orientable) inclusions. We will see that the characterization of Huet for left-linear rules (in terms of α and γ properties [10, lemma 2.8]), the generalization of Peterson and Stickel [25] for non-left-linear rules (in terms of E-compatibility), and the result of Jouannaud & Kirchner [13, 14, 15] (in terms of E-coherence or confluence of cliffs), all of them are not valid for inclusions. We present a new characterization of bi-rewriting modulo a set of inclusions where stronger properties are required. We have divided section 3 in two subsections, the first devoted to abstract bi-rewriting properties and the second to term dependent properties.

In section 4 we present two examples of canonical bi-rewriting systems. We sketch a method able to handle schemes of critical pairs, which are needed in non-left-linear bi-rewriting systems. We also show some of the disadvantages of modeling inclusions with equations containing unions or intersections.

---

[2] The possibility of a infinite set of critical pairs does not apply to the translation of a set of equations into a bi-rewriting system. In fact, the set of critical pairs obtained in the translation is a subset of those obtained in the equational case, and the only disadvantage is the loss of efficiency due to the use of a search algorithm.

following paragraph, and defined as extensions of the standard definitions for term rewriting systems.

A bi-rewriting system $\langle R_1, R_2 \rangle$ is said to terminate iff $\longrightarrow^*_{R_1}$ and $\longrightarrow^*_{R_2}$ are well founded orderings. It is said to quasi-terminate (globally finite) iff the sets $\{x \mid a \longrightarrow^*_{R_1} x\}$ and $\{x \mid a \longrightarrow^*_{R_2} x\}$ are finite for any term a. It is Church-Rosser iff $(\longrightarrow_{R_1} \cup \longleftarrow_{R_2})^* \subseteq \longrightarrow^*_{R_1} \circ \longleftarrow^*_{R_2}$.

In order to test automatically the Church-Rosser property we extend the standard procedure of rewriting to bi-rewriting. So we reduce the Church-Rosser property to three simpler properties, namely bi-confluence (or commutativity), local bi-confluence and critical pairs bi-confluence.

A bi-rewriting system $\langle R_1, R_2 \rangle$ is bi-confluent or commutative iff $\longleftarrow^*_{R_2} \circ \longrightarrow^*_{R_1} \subseteq \longrightarrow^*_{R_1} \circ \longleftarrow^*_{R_2}$. It is locally bi-confluent iff $\longleftarrow_{R_2} \circ \longrightarrow_{R_1} \subseteq \longrightarrow^*_{R_1} \circ \longleftarrow^*_{R_2}$. A pair of terms s, t is bi-confluent $s \downarrow t$ iff there exists u such that $s \longrightarrow^*_{R_1} u$ and $t \longrightarrow^*_{R_2} u$. The Newman's lemma is also true in bi-rewriting systems: a terminating bi-rewriting system is Church-Rosser iff it is locally bi-confluent.

A simple extension of the standard critical pairs definition can be given for bi-rewriting systems. However, as we will see, it is not sufficient to prove the critical pairs lemma [17]. The simple definition of critical pair arises from the most general non-variable overlap between the left hand side of a rule in R_1 and the left hand side of a rule in R_2. Given $l \longrightarrow_{R_1} r$ and $s \longrightarrow_{R_2} t$, a position p of a non-variable subterm of s, and the most general unifier σ of l and $s|_p$, the pair $\sigma(t) \subseteq \sigma(s[r]_p)$ is a critical pair; and the same for critical pairs between R_2 and R_1.

Unfortunately, in the presence of non-left-linear rules, the critical pair lemma can not be proved because the confluence of variable overlaps is no longer possible. Here is a simple counter-example to the validity of this lemma. The bi-rewriting system $\langle \{f(X,X) \stackrel{\subseteq}{\longrightarrow} X\}, \{a \stackrel{\supseteq}{\longrightarrow} b\} \rangle$ has no critical pairs, and $f(a,b) \stackrel{\subseteq}{\longleftarrow} f(a,a) \stackrel{\subseteq}{\longrightarrow} a$ does not satisfy the Church-Rosser property. This problem would be avoided if $\langle a \stackrel{\subseteq}{\longrightarrow} b \rangle \in R_1$.

Non-left-linear rules also unvalidate the Toyama theorem [27] for bi-rewriting systems as the following counter-example shows. The following two bi-rewriting systems

$$R_1 = \begin{cases} X \cup X \stackrel{\subseteq}{\longrightarrow} X \\ X \cup Y \stackrel{\subseteq}{\longrightarrow} Y \cup X \\ X \cup (Y \cup Z) \stackrel{\subseteq}{\longrightarrow} (X \cup Y) \cup Z \end{cases} \qquad R_2 = \begin{cases} X \cup Y \stackrel{\supseteq}{\longrightarrow} X \\ X \cup Y \stackrel{\supseteq}{\longrightarrow} Y \end{cases}$$

and

$$R'_1 = \begin{cases} X \cap Y \stackrel{\subseteq}{\longrightarrow} X \\ X \cap Y \stackrel{\subseteq}{\longrightarrow} Y \end{cases} \qquad R'_2 = \begin{cases} X \cap X \stackrel{\supseteq}{\longrightarrow} X \\ X \cap Y \stackrel{\supseteq}{\longrightarrow} Y \cap X \\ X \cap (Y \cap Z) \stackrel{\supseteq}{\longrightarrow} (X \cap Y) \cap Z \end{cases}$$

are both Church-Rosser and have disjoint alphabets, but their union $\langle R_1 \cup R'_1, R_2 \cup R'_2 \rangle$ is not Church-Rosser as the following rewriting sequence shows.[3]

$$(A \cap B) \cup (A \cap C) \stackrel{\subseteq}{\longleftarrow}_{R_2} (A \cap (B \cup C)) \cup (A \cap C) \stackrel{\subseteq}{\longleftarrow}_{R_2} (A \cap (B \cup C)) \cup (A \cap (B \cup C)) \stackrel{\subseteq}{\longrightarrow}_{R_1} A \cap (B \cup C)$$

Using the previous definition of critical pairs, the critical pairs lemma is only true for left-linear systems: a terminating and left-linear bi-rewriting system is Church-Rosser iff all critical pairs are bi-confluent. In order to keep this lemma for non-left-linear bi-rewriting systems, we have to enlarge the set of critical pairs as follows.

---

[3] The non-confluence of this inclusion sequence is due to the addition of new symbols in the signature, not to the addition of new rules.

Definition 1. If $\langle \alpha_1 \xrightarrow{\subseteq} \beta_1 \rangle \in R_1$ and $\langle \alpha_2 \xrightarrow{\supseteq} \beta_2 \rangle \in R_2$ are two rewriting rules (with variables distinct) and p a position in α_1, then

1. if $\alpha_1|_p$ is non-variable subterm and σ is the most general unifier of $\alpha_1|_p$ and α_2 then $\langle \sigma(\alpha_1[\beta_2]_p), \sigma(\beta_1) \rangle$ is a (standard) critical pair,
2. if $\alpha_1|_p = x$ is a repeated variable in α_1, F a term $x \notin \mathcal{V}(F)$, q an occurrence in F, and $\alpha_2 \xrightarrow{*}_{R_1} \beta_2$ is not satisfied,[4] then $\langle \sigma(\alpha_1[F[\beta_2]_q]_p), \sigma(\beta_1) \rangle$ is an (extended) critical pair where σ only substitutes x by $F[\alpha_2]_q$.

The same for critical pairs between R_2 and R_1.

The set of (extended) critical pairs of the previous definition is in general infinite —$\langle \sigma(\alpha_1[F[\beta_2]_q]_p), \sigma(\beta_1) \rangle$ is a critical pair scheme— (in section 4 we will see two examples using these schemes). So the critical pairs lemma even if true with this definition of critical pairs, will be of little practical help to test bi-confluence. Then the conditions of confluence have to be studied in each case taking into account the particular shape of the non-left-linear rules.

Nevertheless, if all rules come from the translation of an equational theory then we can always have $\langle \alpha \xrightarrow{\subseteq} \beta \rangle \in R_1$ iff $\langle \alpha \xrightarrow{\supseteq} \beta \rangle \in R_2$ and the extended critical pairs schemes will not appear.[5] Notice also that an inclusion $a \subseteq b$ could be used by both R_1 and R_2 systems —as rules $a \xrightarrow{\subseteq} b$ and $b \xrightarrow{\supseteq} a$— without losing necessarily the termination property of the bi-rewriting system $\langle R_1, R_2 \rangle$.[6]

3 Bi-rewriting Modulo a Set of Inclusions

Like in equational rewriting, in bi-rewriting it is not always possible to orient all inclusions of a theory presentation in two terminating rewrite relations, as shown in the previous section. Frequently enough, we must handle three rewrite relations, the terminating relations $\longrightarrow_{R_1}$ and $\longrightarrow_{R_2}$ resulting from the inclusions oriented to the right and to the left respectively, and the non-terminating relation $\longrightarrow_I$ resulting from the non-oriented inclusions. We name these three relations a $\langle R_1, R_2 \rangle$ bi-rewriting system modulo I.[7] Figure 3 shows an example of them.

3.1 From Church-Rosser to Local Confluence

The simplest way to have a complete and decidable proof procedure for $\langle R_1, R_2 \rangle$ modulo I is reducing it to the bi-rewriting system $\langle R_1 \cup I, R_2 \cup I \rangle$ and, like in the previous section, to require of it the following properties

$$\xrightarrow{}_{R_1} \cup \xrightarrow{}_I \text{ and } \xrightarrow{}_{R_2} \cup \xleftarrow{}_I \text{ are quasi-terminating, and} \tag{1}$$

---

[4] If this condition is satisfied then we can make the pair confluent like in the equational case.
[5] Any equation $a = b$ is translated into $a \subseteq b$ and $b \subseteq a$ and these are oriented as $a \xrightarrow{\subseteq} b$ and $a \xrightarrow{\supseteq} b$, in the case we have the same orientation ordering for R_1 and R_2.
[6] Both rewriting systems can have different orientation orderings.
[7] Although we use the word "modulo", it does not mean that $\longrightarrow_I^*$ is a congruence, be aware it is a non-symmetric relation (monotonic pre-order).

$$(\xrightarrow[R_1]{} \cup \xrightarrow[I]{} \cup \xleftarrow[R_2]{})^* \subseteq (\xrightarrow[R_1]{} \cup \xrightarrow[I]{})^* \circ (\xleftarrow[R_2]{} \cup \xrightarrow[I]{})^* \qquad (2)$$

However, the quasi-termination of $\xrightarrow[R_1]{} \cup \xrightarrow[I]{}$ and $\xrightarrow[R_2]{} \cup \xleftarrow[I]{}$ is not enough to reduce the property (2) —called $\langle R_1, R_2 \rangle$ weak Church-Rosser modulo I— to the corresponding local bi-confluence (4). To do this we would need the (strong) termination of $\xrightarrow[R_1]{} \cup \xrightarrow[I]{}$ and of $\xrightarrow[R_2]{} \cup \xleftarrow[I]{}$, which are not true. The solution to this problem comes from requiring the following property stronger than (1)

$$\xrightarrow[I]{*} \circ \xrightarrow[R_1]{} \text{ and } \xleftarrow[I]{*} \circ \xrightarrow[R_2]{} \text{ are terminating, and } \xrightarrow[I]{} \text{ is quasi-terminating} \qquad (3)$$

Notice that from the fact $(\xrightarrow[I]{} \cup \xrightarrow[R]{})^* = (\xrightarrow[I]{*} \circ \xrightarrow[R]{})^* \circ \xrightarrow[I]{*}$ one can see that (3) implies (1). Using the stronger termination property (3), the weak Church-Rosser property (2) can be reduced to the following local confluence property:

$$\xleftarrow[R_2]{} \circ \xrightarrow[I]{*} \circ \xrightarrow[R_1]{} \subseteq (\xrightarrow[I]{*} \circ \xrightarrow[R_1]{})^* \circ \xrightarrow[I]{*} \circ (\xleftarrow[R_2]{} \circ \xrightarrow[I]{*})^* \qquad (4)$$

The equivalence of (2) and (4) can be proved using noetherian induction on $\xrightarrow[I]{*} \circ \xrightarrow[R_1]{}$ and $\xleftarrow[I]{*} \circ \xrightarrow[R_2]{}$. In fact, to prove this equivalence, it is not necessary for $\xrightarrow[I]{}$ to be quasi-terminating. If $\xrightarrow[I]{}$ is symmetric the above termination property (3) becomes similar to the termination property required in rewriting modulo a set of equations [3]. That is, I symmetric means we can define equivalence classes ($[s]_I \xrightarrow[R]{} [t]_I$ iff $s \xrightarrow[I]{*} \circ \xrightarrow[R]{} \circ \xrightarrow[I]{*} t$) and, the termination of $\xrightarrow[I]{*} \circ \xrightarrow[R_1]{}$ and $\xleftarrow[I]{*} \circ \xrightarrow[R_2]{}$ is equivalent to the existence of two well founded I-compatible order relations $\succ_1$ and $\succ_2$ satisfying $\xrightarrow[R_1]{} \subseteq \succ_1$ and $\xrightarrow[R_2]{} \subseteq \succ_2$; and the quasi-termination of $\xrightarrow[I]{}$ is equivalent to the finiteness of the equivalence classes.

However, we know by analogy with rewriting modulo a set of equations, that the proof procedure based on these properties is not a practical one. Like in the equational case, rewriting by $\xrightarrow[I]{*} \circ \xrightarrow[R]{}$ is inefficient, if decidable at all. Therefore we will approximate it by a weaker, but more practical notion of bi-rewriting named $\langle I \backslash R_1, I^{-1} \backslash R_2 \rangle$ by similarity to the corresponding equational definitions. As we will see later, this new rewriting relation will have to satisfy what is called a $\langle I \backslash R_1, I^{-1} \backslash R_2 \rangle$ strong Church-Rosser modulo I property, defined as follows:

$$(\xrightarrow[I \backslash R_1]{} \cup \xrightarrow[I]{} \cup \xleftarrow[I^{-1} \backslash R_2]{})^* \subseteq \xrightarrow[I \backslash R_1]{*} \circ \xrightarrow[I]{*} \circ \xleftarrow[I^{-1} \backslash R_2]{*} \qquad (5)$$

This property plus the quasi-termination of $I \backslash R_1$ and $I^{-1} \backslash R_2$ and the decidability of the I-unification are sufficient to have a more efficient complete and decidable proof procedure. The solution we will propose comes mainly from the two solutions known for the equational case [13, 25]. In the following we consider how they can be adapted to bi-rewriting.

Huet [10] and Jouannaud & Kirchner [14, 13] have proved that given a set of rules R and equations E such that $\xleftrightarrow[E]{*} \circ \xrightarrow[R]{}$ is terminating, R is strong Church-Rosser modulo E iff all peaks and cliffs are confluent: $\xleftarrow[R]{} \circ \xrightarrow[R]{} \subseteq \xrightarrow[R]{*} \circ \xleftrightarrow[E]{*} \circ \xleftarrow[R]{*}$ and $\xleftrightarrow[E]{} \circ \xrightarrow[R]{} \subseteq \xrightarrow[R]{*} \circ \xleftrightarrow[E]{*} \circ \xleftarrow[R]{*}$. Notice these are sufficient and, what is also important, necessary conditions. Besides, the finiteness of the E-equivalence classes is not required. These confluence properties, stated by Huet, are too strong and can not be reduced to the confluence of critical pairs unless the rules are left-linear. To overcome this limitation of non-left-linear systems Jouannaud & Kirchner [3, 6, 13] propose a new rewriting relation $E \backslash R$ satisfying $\xrightarrow[R]{} \subseteq \xrightarrow[E \backslash R]{} \subseteq \xleftrightarrow[E]{*} \circ \xrightarrow[R]{}$.

This relation is proved to be strong Church-Rosser modulo E iff all critical peaks $\longleftarrow_R \circ \longrightarrow_{E\backslash R}$ and critical cliffs $\longleftrightarrow_E \circ \longrightarrow_R$ are confluent. Then this confluence can be reduced to critical pairs confluence and to extended rules. We are interested in extending the same kind of result to bi-rewriting systems because on it are based the proof and completion procedures.

Then the direct translation of the previous result to the bi-rewriting case may be stated as follows. $\langle R_1, R_2 \rangle$ is strong Church-Rosser modulo I iff $\longleftarrow_{R_2} \circ \longrightarrow_{R_1} \subseteq \longrightarrow^*_{R_1} \circ \longrightarrow^*_I \circ \longleftarrow^*_{R_2}$ and $\longrightarrow_I \circ \longrightarrow_{R_1} \subseteq \longrightarrow^*_{R_1} \circ \longrightarrow^*_I \circ \longleftarrow^*_{R_2}$ and $\longleftarrow_{R_2} \circ \longrightarrow_I \subseteq \longrightarrow^*_{R_1} \circ \longrightarrow^*_I \circ \longleftarrow^*_{R_2}$ where $\longrightarrow^*_I \circ \longrightarrow_{R_1}$ and $\longleftarrow^*_I \circ \longrightarrow_{R_2}$ are terminating. Unfortunately this result is not true unless $\longrightarrow_{R_1}$ and $\longrightarrow_{R_2}$ have the same set of normal forms, which is semantically meaningless. Here is a counter-example of its validity. Let $I = \{a \xrightarrow{\subseteq} b, c \xrightarrow{\subseteq} d\}$, $R_1 = \{b \xrightarrow{\subseteq} c\}$ and $R_2 = \{c \xrightarrow{\supseteq} b\}$, then $\longrightarrow^*_I \circ \longrightarrow_{R_1}$ and $\longleftarrow^*_I \circ \longrightarrow_{R_2}$ are terminating and all peaks and cliffs are confluent, nevertheless $a \longrightarrow_I b \longrightarrow_{R_1} c \longrightarrow_I d$ is not confluent.[8]

Another way of having the strong Church-Rosser property is by means of the stronger requirement on R rewriting modulo E given by Peterson & Stickel in [25]. They define a rewriting relation between E-equivalence classes which can be modeled by $(\longleftrightarrow^*_E \circ \longrightarrow_R)^* \circ \longleftrightarrow^*_E$. They also formulate what they call an E-completeness property, equivalent to what we have called weak Church-Rosser property. They were the first to propose the mentioned relation $E\backslash R$. When this relation is E-compatible, that is, when $\longleftrightarrow^*_E \circ \longrightarrow_R \subseteq \longrightarrow_{E\backslash R} \circ \longleftrightarrow^*_E (\longleftarrow_R \circ \longleftrightarrow^*_E)^*$, then the corresponding weak and strong Church-Rosser properties both become equivalent to the peaks confluence property $\longleftarrow_{E\backslash R} \circ \longrightarrow_{E\backslash R} \subseteq \longrightarrow^*_{E\backslash R} \circ \longleftrightarrow^*_E \circ \longleftarrow^*_{E\backslash R}$. The E-compatibility is not a necessary condition although it is a sufficient one. To adapt this same result to the bi-rewriting case we will need a requirement even stronger than E-compatibility, as shown below.

Given a $\langle R_1, R_2 \rangle$ bi-rewriting system modulo I, the problem is to find which requirements two new relations $I\backslash R_1$ and $I^{-1}\backslash R_2$ have to satisfy in order to prove (5), the $\langle I\backslash R_1, I^{-1}\backslash R_2 \rangle$ strong Church-Rosser modulo I property. Since $I\backslash R_1$ and $I^{-1}\backslash R_2$ are required to satisfy at least $\longrightarrow_{R_1} \subseteq \longrightarrow_{I\backslash R_1} \subseteq \longrightarrow^*_I \circ \longrightarrow_{R_1}$ and $\longrightarrow_{R_2} \subseteq \longrightarrow_{I^{-1}\backslash R_2} \subseteq \longleftarrow^*_I \circ \longrightarrow_{R_2}$, the termination of $\longrightarrow^*_I \circ \longrightarrow_{R_1}$ and $\longleftarrow^*_I \circ \longrightarrow_{R_2}$ ensures the termination of $\longrightarrow^*_I \circ \longrightarrow_{I\backslash R_1}$ and $\longleftarrow^*_I \circ \longrightarrow_{I^{-1}\backslash R_2}$. From a computational point of view, this relations are to be based on the suppression of those applications of $\longrightarrow_I$ in $\longrightarrow^*_I \circ \longrightarrow_R$ not conducting to a new way of applying $\longrightarrow_R$ later. That is, with the new relations, all this unnecessary I-rewritings before R-rewritings could be suppressed or moved to the final I-unification. This requirement is captured by the following local commutativity property of I and $I\backslash R_1$, and of I^{-1} and $I^{-1}\backslash R_2$:

$$\longrightarrow_I \circ \longrightarrow_{I\backslash R_1} \subseteq \longrightarrow_{I\backslash R_1} \circ \longrightarrow^*_I$$
$$\longleftarrow_I \circ \longrightarrow_{I^{-1}\backslash R_2} \subseteq \longrightarrow_{I^{-1}\backslash R_2} \circ \longleftarrow^*_I \tag{6}$$

These requirements are stronger than the E-compatibility in [25] and the confluence of cliffs in [13]. Furthermore, if $\longrightarrow^*_I \circ \longrightarrow_{R_1}$ and $\longleftarrow^*_I \circ \longrightarrow_{R_2}$ are termi-

---

[8] Note that if we translate the counter-example to the classical case defining $\longrightarrow_R \stackrel{def}{=} \longrightarrow_{R_1} \cup \longrightarrow_{R_2}$, then R becomes non-terminating, and the hypothesis of the Jouannaud theorem is not satisfied.

nating then $\xrightarrow{\cdot}_I \circ \xrightarrow{}_{I\backslash R_1}$ and $\xleftarrow{\cdot}_I \circ \xrightarrow{}_{I^{-1}\backslash R_2}$ are also terminating and the local commutativities are equivalent to the global commutativities: $\xrightarrow{\cdot}_I \circ \xrightarrow{}_{I\backslash R_1} \subseteq$ $\xrightarrow{\cdot}_{I\backslash R_1} \circ \xrightarrow{\cdot}_I$ and $\xleftarrow{\cdot}_I \circ \xrightarrow{}_{I^{-1}\backslash R_2} \subseteq \xrightarrow{\cdot}_{I^{-1}\backslash R_2} \circ \xleftarrow{\cdot}_I$. These global commutativity properties lead to the equivalence of the $\langle I\backslash R_1, I^{-1}\backslash R_2\rangle$ weak Church-Rosser and the $\langle I\backslash R_1, I^{-1}\backslash R_2\rangle$ strong Church-Rosser modulo I properties. On the other hand, using the previously proved equivalence between weak Church-Rosser (2) and local bi-confluence (4) modulo I, the $\langle I\backslash R_1, I^{-1}\backslash R_2\rangle$ weak Church-Rosser property becomes equivalent to the following local bi-confluence property:

$$\xleftarrow{\cdot}_{I^{-1}\backslash R_2} \circ \xrightarrow{\cdot}_I \circ \xrightarrow{}_{I\backslash R_1} \subseteq (\xrightarrow{\cdot}_I \circ \xrightarrow{}_{I\backslash R_1})^* \circ \xrightarrow{\cdot}_I \circ (\xleftarrow{}_{I^{-1}\backslash R_2} \circ \xleftarrow{\cdot}_I)^*$$

And again, the commutativity properties and the inclusions $\xrightarrow{}_{R_1} \subseteq \xrightarrow{}_{I\backslash R_1}$ and $\xrightarrow{}_{R_2} \subseteq \xrightarrow{}_{I^{-1}\backslash R_2}$ allows us to reduce this condition to the following one

$$\xleftarrow{}_{R_2} \circ \xrightarrow{\cdot}_I \circ \xrightarrow{}_{R_1} \subseteq \xrightarrow{\cdot}_{I\backslash R_1} \circ \xrightarrow{\cdot}_I \circ \xleftarrow{\cdot}_{I^{-1}\backslash R_2} \tag{7}$$

and from this to whatever of the following ones

$$\xleftarrow{}_{I^{-1}\backslash R_2} \circ \xrightarrow{}_{R_1} \subseteq \xrightarrow{\cdot}_{I\backslash R_1} \circ \xrightarrow{\cdot}_I \circ \xleftarrow{\cdot}_{I^{-1}\backslash R_2}$$

or $\xleftarrow{}_{R_2} \circ \xrightarrow{}_{I\backslash R_1} \subseteq \xrightarrow{\cdot}_{I\backslash R_1} \circ \xrightarrow{\cdot}_I \circ \xleftarrow{\cdot}_{I^{-1}\backslash R_2}$

This results can be summarized in the following lemma:

Lemma 2. *If $\xrightarrow{\cdot}_I \circ \xrightarrow{}_{R_1}$ and $\xleftarrow{\cdot}_I \circ \xrightarrow{}_{R_2}$ are terminating, and*

$$\xrightarrow{}_{R_1} \subseteq \xrightarrow{}_{I\backslash R_1} \subseteq \xrightarrow{\cdot}_I \circ \xrightarrow{}_{R_1}$$
$$\xrightarrow{}_{R_2} \subseteq \xrightarrow{}_{I^{-1}\backslash R_2} \subseteq \xleftarrow{\cdot}_I \circ \xrightarrow{}_{R_2}$$
$$\xrightarrow{\cdot}_I \circ \xrightarrow{}_{I\backslash R_1} \subseteq \xrightarrow{}_{I\backslash R_1} \circ \xrightarrow{\cdot}_I$$
$$\xleftarrow{}_{I^{-1}\backslash R_2} \circ \xrightarrow{\cdot}_I \subseteq \xrightarrow{\cdot}_I \circ \xleftarrow{}_{I^{-1}\backslash R_2}$$
$$\xleftarrow{}_{R_2} \circ \xrightarrow{\cdot}_I \circ \xrightarrow{}_{R_1} \subseteq \xrightarrow{\cdot}_{I\backslash R_1} \circ \xrightarrow{\cdot}_I \circ \xleftarrow{}_{I^{-1}\backslash R_2}$$

then $\langle I\backslash R_1, I^{-1}\backslash R_2\rangle$ is strongly Church-Rosser modulo I.

This lemma reproduces adapted to bi-rewriting the results of Huet, Peterson, Stickel, Jouannaud and Kirchner but is based on stronger properties.

A generalization of lemma 2 was given in [18], and we summarize it bellow. The set I of non-oriented inclusions of a theory presentation is divided into two subsets I_1 and I_2 ($I = I_1 \cup I_2$). From them two non-terminating rewrite relations $\xrightarrow{}_{I_1}$ and $\xrightarrow{}_{I_2}$ can be defined such that $(\xrightarrow{}_{R_1} \cup \xrightarrow{}_{I_1} \cup \xleftarrow{}_{I_2} \cup \xleftarrow{}_{R_2})^*$ corresponds to the inclusion theory. These four relations constitute a $\langle R_1, R_2\rangle$ bi-rewriting system modulo $\langle I_1, I_2\rangle$. We say that such a system is strong Church-Rosser iff

$$(\xrightarrow{}_{R_1} \cup \xrightarrow{}_{I_1} \cup \xleftarrow{}_{I_2} \cup \xleftarrow{}_{R_2})^* \subseteq \xrightarrow{\cdot}_{R_1} \circ \xrightarrow{\cdot}_{I_1} \circ \xleftarrow{\cdot}_{I_2} \circ \xleftarrow{\cdot}_{R_2}$$

Then the generalization of lemma 2 can be stated as follows:

Lemma 3. *If $\xrightarrow[I_1]{\cdot}\circ\xrightarrow[R_1]{\cdot}$ and $\xrightarrow[I_2]{\cdot}\circ\xrightarrow[R_2]{\cdot}$ are terminating, and*

$$\xrightarrow[I_1]{\cdot}\circ\xrightarrow[I_1\backslash R_1]{} \subseteq \xrightarrow[I_1\backslash R_1]{\cdot}\circ\xrightarrow[I_1]{\cdot}$$

$$\xleftarrow[I_2\backslash R_2]{}\circ\xleftarrow[I_2]{\cdot} \subseteq \xleftarrow[I_2]{\cdot}\circ\xleftarrow[I_2\backslash R_2]{}$$

$$\xleftarrow[I_2\backslash R_2]{\cdot}\circ\xrightarrow[I_1]{\cdot} \subseteq \xrightarrow[I_1\backslash R_1]{\cdot}\circ\xrightarrow[I_1]{\cdot}\circ\xleftarrow[I_2]{\cdot}\circ\xleftarrow[I_2\backslash R_2]{}$$

$$\xleftarrow[I_2\backslash R_2]{}\circ\xrightarrow[I_1]{\cdot} \subseteq \xrightarrow[I_1\backslash R_1]{\cdot}\circ\xrightarrow[I_1]{\cdot}\circ\xleftarrow[I_2]{\cdot}\circ\xleftarrow[I_2\backslash R_2]{}$$

$$\xleftarrow[I_2\backslash R_2]{}\circ\xrightarrow[R_1]{\cdot} \subseteq \xrightarrow[I_1\backslash R_1]{\cdot}\circ\xrightarrow[I_1]{\cdot}\circ\xleftarrow[I_2]{\cdot}\circ\xleftarrow[I_2\backslash R_2]{}$$

$$\xleftarrow[I_2]{\cdot}\circ\xrightarrow[I_1\backslash R_1]{} \subseteq \xrightarrow[I_1\backslash R_1]{\cdot}\circ\xrightarrow[I_1]{\cdot}\circ\xleftarrow[I_2]{\cdot}\circ\xleftarrow[I_2\backslash R_2]{}$$

$$\xleftarrow[R_2]{\cdot}\circ\xrightarrow[I_1\backslash R_1]{} \subseteq \xrightarrow[I_1\backslash R_1]{\cdot}\circ\xrightarrow[I_1]{\cdot}\circ\xleftarrow[I_2]{\cdot}\circ\xleftarrow[I_2\backslash R_2]{}$$

then $\langle R_1, R_2\rangle$ is (strongly) Church-Rosser modulo $\langle I_1, I_2\rangle$.

The generalization comes from the fact that now the commutativity property is only required between $I_1\backslash R_1$ and I_1 and between $I_2\backslash R_2$ and I_2, and is not needed between $I_1\backslash R_1$ and I_2 or $I_2\backslash R_2$ and I_1.

Till now, we have studied Church-Rosser, termination, confluence and local confluence properties in the framework of relational algebra [4]. All proofs can be done without references to the structure of terms. In the following subsection we will consider the term structure in order to reduce the local confluence properties to the confluence of (extended) critical pairs.

3.2 From Local Confluence to (Extended) Critical Pairs

We begin defining the rewrite relations $I\backslash R_1$ and $I^{-1}\backslash R_2$ that were only axiomatically characterized by the commutativity and local confluence properties in the previous subsection.

Definition 4. We say that s rewrites to t modulo I at $[p, \sigma, \alpha\longrightarrow\beta]$, written $s\longrightarrow_{I\backslash R}t$, iff there exists a rule $\langle\alpha\longrightarrow\beta\rangle \in R$, an occurrence p in s, and a substitution σ such that $s|_p\longrightarrow_I^*\sigma(\alpha)$ and $t = s[\sigma(\beta)]_p$.

With this definition $I\backslash R$ verifies $\longrightarrow_R \subseteq \longrightarrow_{I\backslash R} \subseteq \longrightarrow_I^*\circ\longrightarrow_R$ (although in general $\longrightarrow_I^*\circ\longrightarrow_R \not\subseteq \longrightarrow_{I\backslash R}$). The relations $\longrightarrow_{I\backslash R_1}$ and $\longrightarrow_{I^{-1}\backslash R_2}$ are defined in this way.

We are using the notions of E-matching and E-unification from [25] but adapted to bi-rewriting. Given two terms s and t, we say that s I-matches t iff there exists a substitution σ such that $s\longrightarrow_I^*\sigma(t)$, and we say that s I-unify with t iff there exists a substitution σ such that $\sigma(s)\longrightarrow_I^*\sigma(t)$. Notice that, since $\longrightarrow_I$ is not necessarily symmetric s I-unify t is equivalent to t I^{-1}-unify s, but not to t I-unify s. We will suppose in the following that I-unification and I and I^{-1}-matching are decidable.

As in the equational case (to prove confluence of cliffs or E-compatibility), we will prove the commutativity properties by means of the extensionally closed property defined as follows.

Definition 5. Given a set of rules R and inclusions I, R is said to be right (left) I-extensionally closed iff whenever $\langle\alpha_1 \subseteq \beta_1\rangle \in I$, $\langle\alpha_2\longrightarrow\beta_2\rangle \in R$, $\beta_1|_p$ ($\alpha_1|_p$) and α_2 I-unify (I^{-1}-unify) with minimum unifier σ and $\beta_1|_p$ ($\alpha_1|_p$) is not a variable, then $\sigma(\alpha_1)\longrightarrow_{I\backslash R}\sigma(\beta_1[\beta_2]_p)$ (then $\sigma(\beta_1)\longrightarrow_{I^{-1}\backslash R}\sigma(\alpha_1[\beta_2]_p)$).

Since $\longrightarrow_I$ is non-symmetric, we have had to distinguish between right and left extensionally closed in the previous definition. We will suppose in the following that R_1 is right I-extensionally closed, and that R_2 is right I^{-1}-extensionally closed, or what is the same left I-extensionally closed.

Let's study now the conditions for the satisfiability of lemma 2. This conditions will be the premises of theorem 7. The rest of the section is an sketch of the proof of this theorem.

We start with the commutativity properties (6). Both properties may be generalized to $\longrightarrow_I \circ \longrightarrow_{I\backslash R} \subseteq \longrightarrow^*_{I\backslash R} \circ \longrightarrow^=_I$ where I and $I\backslash R$ stands for I and $I\backslash R_1$ in one case, and for I^{-1} and $I^{-1}\backslash R_2$ in the other. Suppose $a \longrightarrow_I b$ at $[p_1, \sigma_1, \alpha_1 \longrightarrow_I \beta_1]$ and $b \longrightarrow_{I\backslash R} c$ at $[p_2, \sigma_2, \alpha_2 \longrightarrow_R \beta_2]$, where p_i are positions, σ_i are substitutions, $\alpha_1 \longrightarrow_I \beta_1$ is an inclusion and $\alpha_2 \longrightarrow_R \beta_2$ is a rule. We have to consider the following three cases in its commutativity.

case $p_1 | p_2$ It can be easily proved that $a \longrightarrow_{I\backslash R} d \longrightarrow_I c$ where $d = a[\sigma_2(\beta_2)]_{p_2} = b[\sigma_1(\alpha_1)]_{p_1}[\sigma_2(\beta_2)]_{p_2} = c[\sigma_1(\alpha_1)]_{p_1}$.

case $p_1 \prec p_2$ Let v satisfy $p_2 = p_1 \cdot v$. We have $\beta_1|_v$ I-unify α_2. If $\beta_1|_v$ is not a variable, we are in the conditions of definition 5, and if R is right I-extensionally closed, then $a \longrightarrow_{I\backslash R} c$ at $[p_1, \sigma, \alpha_2 \longrightarrow_R \beta_2]$ for some σ.

Otherwise, there exist two occurrences v_1 and v_2 satisfying $p_1 \cdot v_1 \cdot v_2 = p_2$ and $\beta_1|_{v_1} = x$, x being a variable. If all inclusions in I are left linear (and non-erasing) then x occurs once in α_1. Let v'_1 be this occurrence. It can be proved that $a \longrightarrow_{I\backslash R} d$ at $[p'_2, \sigma_2, \alpha_2 \longrightarrow_R \beta_2]$ and $d \longrightarrow_I c$ at $[p_1, \sigma'_1, \alpha_1 \longrightarrow_I \beta_1]$ where $p'_2 = p_1 \cdot v'_1 \cdot v_2$, $\sigma'_1(y) = \sigma_1(y)$ for $y \neq x$ and $\sigma'_1(x) = \sigma_1(x)[\sigma_2(\beta_2)]_{v_2}$ and $d = c[\sigma'_1(\alpha_1)]_{p_1} = a[\sigma_2(\beta_2)]_{p'_2}$.

case $p_1 \succeq p_2$ Let v be the occurrence such that $p_2 \cdot v = p_1$. We have $a|_{p_2} \longrightarrow_I b|_{p_2}$ at $[v, \sigma_1, \alpha_1 \longrightarrow_I \beta_1]$ and therefore $a \longrightarrow_{I\backslash R} c$ at $[p_2, \sigma_2, \alpha_2 \longrightarrow_R \beta_2]$.

It must be noticed that like in [25], and differently from [13], the inclusions in I are required to be right-linear in order to prove commutativity of $\longrightarrow_I$ and $\longrightarrow_{I\backslash R_1}$, and left-linear in order to prove commutativity of $\longleftarrow_I$ and $\longrightarrow_{I^{-1}\backslash R_2}$; so, all inclusions in I have to be linear. If all inclusions are left- or right-linear, but they are not all linear, then we can oversee this problem using lemma 3 by putting right-linear inclusions in I_1 and left-linear inclusions in I_2.

Let's study now the condition (7) for the confluence of peaks. Suppose we have $a \longleftarrow_{R_2} b \longrightarrow^*_I c \longrightarrow_{R_1} d$ where reduction $c \longrightarrow_{R_1} d$ takes place at p_1 and $b \longrightarrow_{R_2} a$ at p_2. Three cases must be considered:

case $p_1 | p_2$ We can reduce the problem to the confluence of $a \longleftarrow_{R_2} b \longrightarrow_{I\backslash R_1} d'$ where reductions also take place at $[p_1, \sigma_1, \alpha_1 \longrightarrow_{R_1} \beta_1]$ and $[p_2, \sigma_2, \alpha_2 \longrightarrow_{R_2} \beta_2]$, and as in the commutativity case, both reductions can be permuted.

case $p_1 \prec p_2$ The middle I rules commute with R_2 in (7) and the problem is reduced to the confluence of $a' \longleftarrow_{I^{-1}\backslash R_2} c \longrightarrow_{R_1} d$. This case is equal to the next one if we exchange the indexes 1 by 2 and and we reverse the order of the relations in both sides of the inclusion.

case $p_1 \succeq p_2$ We commute I with R_1 in (7) and we test the confluence of $\longleftarrow_{R_2} \circ \longrightarrow_{I\backslash R_1}$. The previous case, as well as this one are generalized by the confluence of $a \longleftarrow_{R_2} b \longrightarrow_{I\backslash R_1} c$ where reductions take place at p_1 and p_2 respectively, and $p_1 \succeq p_2$. It corresponds to the equational case in the study of the confluence of $\leftarrow_{R}\circ\longrightarrow_{E\backslash R}$ where we can always suppose that the $E\backslash R$ reduction takes place below the R reduction. As we have seen in the previous section, if there is a variable overlap, and the rule used in $b\longrightarrow_{R_2}a$ is left-linear or $\alpha_1\longrightarrow^*_{R_2}\beta_1$ is satisfied, the pair is always confluent. Otherwise we have to include this kind of overlap in the critical pairs definition given below.

Definition 6. If $\langle\alpha_1\xrightarrow{\subseteq}\beta_1\rangle \in R_1$ and $\langle\alpha_2\xrightarrow{\supseteq}\beta_2\rangle \in R_2$ are two rewriting rules normalized apart, and p is a position in α_1, then

1. if $\alpha_2|_p$ is not a variable and σ is a minimum I-unifier of $\alpha_2|_p$ and α_1, then $\langle\sigma(\beta_2), \sigma(\alpha_2[\beta_1]_p)\rangle$ is a (standard) critical pair,
2. if $\alpha_2|_p = x$ is a repeated variable in α_2, F is a term $x \notin \mathcal{V}(F)$, q a position in F, and $\alpha_1 \longrightarrow^*_{R_2}\beta_1$ is not satisfied, then $\langle\sigma(\beta_2), \sigma(\alpha_2[F[\beta_1]_q]_p)\rangle$ is an (extended) critical pair where the domain of σ is $\{x\}$ and $\sigma(x) = F[\alpha_1]_q$.

$ECP(I\backslash R_1, R_2)$ denotes this set of standard and extended critical pairs. The set $ECP(R_1, I^{-1}\backslash R_2)$ can be defined similarly.

Again we have had to introduce critical pair schemes which may generate infinite critical pairs. Using this extended definition of critical pairs we can prove the following theorem which characterizes the strong Church-Rosser property of a $\langle R_1, R_2\rangle$ bi-rewriting system modulo I.

Theorem 7. *Given two sets of rules R_1 and R_2 and a set of inclusions I, if I^*R_1 and $I^{-1*}R_2$ are terminating, R_1 is right I-extensionally closed, R_2 left I-extensionally closed, all inclusions in I are linear, and all standard and extended critical pairs $ECP(I\backslash R_1, R_2)$ and $ECP(R_1, I^{-1}\backslash R_2)$ are confluent, then $\langle I\backslash R_1, I^{-1}\backslash R_2\rangle$ is (strongly) Church-Rosser modulo I.*

4 Two Examples: Towards a Completion Procedure

As we said in the previous sections, bi-rewriting compared with equational rewriting, faces the extra difficulty of a possible infinite set of critical pairs. Non-left-linear rules may generate what we called critical pair schemes (see definitions 1 and 6). The process of completion with these schemes is an open problem. In this section instead of giving the completion procedure we sketch out the possibilities of completion of two examples of bi-rewriting by means of rule schemes.

4.1 Inclusion Theory of the Union Operator

Figure 1 shows the first bi-rewriting system that we want to complete corresponding to the union operator. Its termination can be proved using the interpretation $|X \cup Y| = |X| + |Y|$. Although the standard critical pairs (scp) of this system are confluent, the presence of the non-left-linear rule $X \cup X \xrightarrow{\subseteq} X$ also makes necessary the consideration of the extended critical pairs (ecp). We will do this in two steps dividing the set of ecp in two subsets. First, we consider scp and the finite subset of ecp of the particular form $\langle \sigma(\alpha_1[\beta_2]_p), \sigma(\beta_1)\rangle$ where $\alpha_1|_p = x$ is a repeated variable in the non-left-linear rule $\langle \alpha_1 \xrightarrow{1} \beta_1\rangle \in R_1$, $\langle \alpha_2 \xrightarrow{2} \beta_2\rangle \in R_2$ being the other rule, and σ substitutes x by α_2. Between all these critical pairs we may focus into the following two sequences of oriented rules and non-oriented inclusions:

r_4 $Y \cup (X \cup Y) \xrightarrow{\subseteq} X \cup Y$ \hfill ecp from r_1 and r_3
r_5 $Y \cup X \xleftarrow{\subseteq} X \cup Y$ \hfill scp from r_2 and r_4

r_6 $(X \cup Y) \cup Y \xrightarrow{\subseteq} X \cup Y$ \hfill ecp from r_1 and r_3
r_7 $(X \cup Y) \cup (Y \cup Z) \xrightarrow{\subseteq} X \cup (Y \cup Z)$ ecp from r_2 and r_5
r_8 $(X \cup Y) \cup Z \xleftarrow{\subseteq} X \cup (Y \cup Z)$ \hfill scp from r_3 and r_7

Using the commutativity r_5 and the associativity r_8 all the other rules generated by the subset of ecp become redundant. The fact that these inclusions can not be oriented makes necessary the use of $\langle \{r_1\}, \{r_3\}\rangle$ bi-rewriting modulo $I = \{r_5, r_8\}$. Notice that in this case $\rightarrow_I^* = \leftarrow_I^*$, and so we can use the standard equational I-matching and I-unification, and also the flattened notation for $\cup$.

Let's consider now the scp and the rest of ecp $\langle \sigma(\alpha_1[F[\beta_2]_q]_p), \sigma(\beta_1)\rangle$ where F is an expression, q is an occurrence in F, and σ substitutes $\alpha_1|_p = x$ by $F[\beta_2]_q$. Using them we can obtain the sequence:

r_9 $F[X] \cup F[X \cup Y] \xrightarrow{\subseteq} F[X \cup Y]$ \hfill ecp from r_1 and r_2
r_{10} $F[X \cup Y] \xrightarrow{2} F[X] \cup F[Y]$ \hfill scp from r_2 and r_9
r_{11} $F[X \cup Y \cup Z] \xrightarrow{2} F[X \cup Y] \cup F[Y \cup Z]$ \hfill ecp from r_2 and r_9

Where the orientation in the last two rules depends on the orientation ordering used for the other symbols in the signature. Another possible orientation of r_{10} could be:

r'_{10} $F[X] \cup F[Y] \xrightarrow{\subseteq} F[X \cup Y]$ from r_2 and r_9

and, then r_9 would be subsumed by r_1 and r'_{10}, and r_{11} would become confluent.

Notice that we are dealing with rule schemes instead of ordinary rules, and that the use of rule schemes in completion is an open problem. However, in this case, the rule scheme r'_{10} may be subsumed by the following (finite) set of rules:

For any $f \in Sig^n$
$r_{12}^{(f)}$ $f(X_1, \ldots, X_n) \cup f(X'_1, \ldots, X'_n) \xrightarrow{\subseteq} f(X_1 \cup X'_1, \ldots, X_n \cup X'_n)$

where f is any n-ary symbol in the signature with $n > 0$. This results from the following compositional property:

$$F[G[X]] \cup F[G[Y]] \xrightarrow[10]{\subseteq} F[G[X] \cup G[Y]] \xrightarrow[10]{\subseteq} F[G[X \cup Y]]$$

Similarly, rules r_{10} and r_{11} are subsumed by

For any $f \in Sig^n$
$r_{13}^{(f)}$ $f(\ldots X_i \cup X'_i \ldots) \xrightarrow{2} f(\ldots X_i \ldots) \cup f(\ldots X'_i \ldots)$
$r_{14}^{(f)}$ $f(\ldots X_i \cup X'_i \cup X''_i \ldots) \xrightarrow{2} f(\ldots X_i \cup X'_i \ldots) \cup f(\ldots X'_i \cup X''_i \ldots)$

but the same does not apply to r_9. Because of this we choose r'_{10} instead of r_{10}. Finally, using this transformation we obtain the confluent $\langle R_1, R_2 \rangle$ bi-rewriting modulo I system shown in figure 3 where r_1^{ext} and r_{12}^{ext} are the I-extensions of r_1 and r_{12}, and $r_{12}^{(\cup)}$ is not necessary because is subsumed by r_1^{ext}.

$$R_1 = \begin{cases} r_1 & X \cup X \xrightarrow{\subseteq} X \\ r_1^{ext} & X \cup X \cup Y \xrightarrow{\subseteq} X \cup Y \\ r_{12}^{(f)} & f(\ldots X \ldots) \cup f(\ldots Y \ldots) \xrightarrow{\subseteq} f(\ldots X \cup Y \ldots) \\ r_{12}^{(f)\,ext} & f(\ldots X \ldots) \cup f(\ldots Y \ldots) \cup Z \xrightarrow{\subseteq} f(\ldots X \cup Y \ldots) \cup Z \end{cases}$$

$$R_2 = \left\{ r_2 \ X \cup Y \xrightarrow{\supseteq} X \right\}$$

$$I = \begin{cases} r_5 & Y \cup X \xleftrightarrow{\subseteq} X \cup Y \\ r_6 & (X \cup Y) \cup Z \xleftrightarrow{\subseteq} X \cup (Y \cup Z) \end{cases}$$

Fig. 3. A canonical bi-rewriting system for the inclusion theory of the union.

4.2 The Inclusion Theory of Non-Distributive Lattices

The presentation of non-distributive lattices may be given by the following set of inclusions:

$$X \cup X \subseteq X \qquad X \subseteq X \cap X$$
$$X \subseteq X \cup Y \qquad X \cap Y \subseteq X$$
$$Y \subseteq X \cup Y \qquad X \cap Y \subseteq Y$$

Applying to them the completion process of the previous subsection we get the confluent $\langle R_1, R_2 \rangle$ bi-rewriting modulo I system of figure 4. Notice that rule $r_4^{(\cap)}$ is subsumed by r_7, and $r_8^{(\cup)}$ is subsumed by r_3.

We don't know of any canonical rewriting system for non-distributive lattices, although they are known for distributive lattices [11] and for boolean rings [9]. So its modelization by a bi-rewriting system represents a contribution to rewriting techniques (see also [19]). The lack of disjunctive and conjunctive normal forms is the cause of non-existence of a canonical rewriting system. On the contrary, the proposed bi-rewriting system has two normalizing rules. Rules r_3 and r_7 acting in opposite directions allow to get a disjunctive normal form the first, and the other a conjunctive normal form. In a non-distributive lattice these rules are strict inclusions and they can not be used as equational rewrite rules. Furthermore, if they are put together in a unique rewriting system then we lose termination.

4.3 Why Inclusions and not Equations

In the previous subsection we discussed briefly the advantage of modeling the deduction in a non-distributive lattice by a bi-rewriting system: there is no canonical rewrite system for it. In general inclusions express weaker constraints between terms

$$R_1 = \begin{cases} r_1 & X \cup X \xrightarrow{\subseteq} X \\ r_1^{ext} & X \cup X \cup Y \xrightarrow{\subseteq} X \cup Y \\ r_2 & X \cap Y \xrightarrow{\subseteq} X \\ r_3 & X \cup (Y \cap Z) \xrightarrow{\subseteq} (X \cup Y) \cap (X \cup Z) \\ r_3^{ext} & X \cup (Y \cap Z) \cup T \xrightarrow{\subseteq} \big((X \cup Y) \cap (X \cup Z)\big) \cup T \\ r_4^{(f)} & f(\ldots X \ldots) \cup f(\ldots Y \ldots) \xrightarrow{\subseteq} f(\ldots X \cup Y \ldots) \\ r_4^{(f)ext} & f(\ldots X \ldots) \cup f(\ldots Y \ldots) \cup Z \xrightarrow{\subseteq} f(\ldots X \cup Y \ldots) \cup Z \end{cases}$$

$R_2 = r_5, r_5^{ext}, r_6, r_7, r_7^{ext}, r_8^{(f)}, r_8^{(f)ext}$ (Dual of R_1)

$$I = \begin{cases} r_9 & Y \cup X \xleftarrow{\subseteq} X \cup Y \qquad\qquad r_{11}\ Y \cap X \xleftarrow{\subseteq} X \cap Y \\ r_{10} & (X \cup Y) \cup Z \xleftarrow{\subseteq} X \cup (Y \cup Z) \quad r_{12}\ (X \cap Y) \cap Z \xleftarrow{\subseteq} X \cap (Y \cap Z) \end{cases}$$

Fig. 4. A canonical bi-rewriting system for the inclusion theory of non-distributive lattices.

than equations, allowing to use rules like r_3 and r_7 in the previous example. Even in the case of lattices where inclusions may be modeled by equations —like $a \subseteq b$ by $a \cup b = b$ or $a \cap b = a$— inclusions are more natural and have some advantages. The transitivity and monotonicity of inclusions which are captured implicitly by bi-rewriting systems, must be "implemented" explicitly by equational rewrite rules. Let's consider a little further the case of transitivity. The inclusions $a \subseteq b$ and $b \subseteq c$ can be oriented like $a \xrightarrow{\subseteq} b$ and $b \xrightarrow{\subseteq} c$ and we can prove $a \subseteq c$ rewriting a into b and b into c. However, their translation to equations results in two rules $a \cup b \longrightarrow b$ and $b \cup c \longrightarrow c$. These rules generate non-confluent critical pairs with the other rules defining the union and intersection, and the completion process leads to add the following rules $a \cap b \longrightarrow a$, $b \cap c \longrightarrow b$, $a \cup c \longrightarrow c$ and $a \cap c \longrightarrow a$. In general, the completion of a sequence $a_1 \subseteq \ldots \subseteq a_n$ lead to add rules $a_i \cup a_j \longrightarrow a_j$ and $a_i \cap a_j \longrightarrow a_i$ for any $i < j$. This means that the transitivity of inclusions is not captured by the transitivity of the equality relation or by the transitivity of the rewriting relation $\longrightarrow^*$, loosing so one of the main powers of rewriting systems.

References

1. D. M. Allester, B. Givan, and T. Fatima. Taxonomic syntax for first order inference. In *Proc. of the First Int. Conf. on Princ. of Knowledge Representation and Reasoning*, pages 289–300, 1989.
2. V. Antimirov. Term rewriting in unified algebras: an order-sorted approach. In *9th WADT - 4th Compass Workshop*, Caldes de Malavella, Spain, 1992.
3. L. Bachmair and N. Dershowitz. Completion for rewriting module a congruence. *J. of Theoretical Computer Science*, 67:173–201, 1989.
4. H. Bäumer. On the use of relation algebra in the theory of reduction systems. Technical report, Dept. Informatica, Univ. of Twente, Enschede, The Netherlands, 1992.
5. L. Cardelli. A semantics of multiple inheritance. *Information and Computation*, 76:138–164, 1988.

6. N. Dershowitz and J.-P. Jouannaud. Rewrite systems. In J. V. Leeuwen, editor, *Handbook of Theoretical Computer Science*. Elsevier Science Publishers, 1990.
7. J. Gallier. The semantics of recursive programs with function parameters of finite types: n-rational algebras and logic of inequalities. In N. Nivat and J. Reynolds, editors, *Algebraic Methods in Semantics*. Cambridge University Press, 1985.
8. I. Guesarian. *Algebraic Semantics*, volume 99 of *Lecture Notes in Computer Science*. Springer-Verlag, 1981.
9. J. Hsiang and N. Dershowitz. Rewrite methods for clausal and non-clausal theorem proving. In *10th Int. Colloquium on Automata, Languages and Programming*, Barcelona, Spain, 1983. Springer-Verlag.
10. G. Huet. Confluent reductions: Abstract properties and applications to term rewriting systems. *Journal of the ACM*, 27(4):797–821, 1980.
11. J.-M. Hullot. A catalogue of canonical term rewriting systems. Technical Report CSL-113, Computer Science Laboratory, Menlo Park, California, 1980.
12. B. Jayaraman. Impplementation of subset-equational programs. *J. of Logic Programming*, 12:229–324, 1992.
13. J.-P. Jouannaud and H. Kirchner. Completion on a set of rules modulo a set of equations. *SIAM J. computing*, 15(1):1155–1194, 1986.
14. C. Kirchner. *Methodes et Outils de Conception Systematique d'Algorithmes d'Unification dans les Theories Equationnelles*. PhD thesis, Universite de Nancy I, 1985.
15. H. Kirchner. *Preuves par Completion dans les Varietes d'Algebres*. PhD thesis, Universite de Nancy I, 1985.
16. J. W. Klop. Term rewriting systems: A tutorial. *Bulletin of the EATCS*, 32:143–183, 1987.
17. D. E. Knuth and P. B. Bendix. Simple word problems in universal algebras. In J. Leech, editor, *Computational Problems in Abstract Algebra*, pages 263–297. Pergamon Press, Elmsford, N. Y., 1970.
18. J. Levy and J. Agustí. Bi-rewriting, a rewriting technique for monotonic order relation. Technical Report IIIA 92/26, Institut d'Investigació en Intel·ligència Artificial, Blanes, Spain, 1992.
19. J. Levy and J. Agustí. Implementing inequality specifications with bi-rewriting systems. In *4th Compass Workshop*, Lecture Notes in Computer Science, Caldes de Malavella, Spain, 1992. Springer-Verlag.
20. J. Levy, J. Agustí, F. Esteva, and P. García. An ideal model for an extended λ-calculus with refinements. Technical Report ECS-LFCS-91-188, Laboratory for Foundations of Computer Science, Edinburgh, Great Britain, 1991.
21. J. Meseguer. Conditional rewriting logic as a unified model of concurrency. *J. of Theoretical Computer Science*, 96:73–155, 1992.
22. P. D. Mosses. Unified algebras and institutions. In *Principles of Programming Languages Conference*, pages 304–312. ACM Press, 1989.
23. B. Nebel. *Reasoning and Revision in Hybrid Representation Systems*. Lecture Notes in Artificial Intelligence. Springer-Verlag, 1990.
24. M. J. O'Donnell. Term-rewriting implementation of equational logic programming. In P. Lescanne, editor, *Proc. of Rewriting Techniques and Applications*, pages 1–12, Bordeaux, France, 1987. Springer-Verlag.
25. G. E. Peterson and M. E. Stickel. Complete sets of reductions for some equational theories. *Journal of the ACM*, 28(2):233–264, 1981.
26. G. Smolka and H. Aït-Kaci. Inheritance hierarchies: Semantics and unification. *Journal of Symbolic Computation*, 7:343–370, 1989.
27. Y. Toyama. On the Church-Roser property for the direct sum of term rewriting systems. *J. of the ACM*, 34(1):128–143, 1987.

A Case Study of Completion Modulo Distributivity and Abelian Groups[*]

Hantao Zhang
University of Iowa, Iowa 52242, USA
hzhang@cs.uiowa.edu

Abstract

We propose an approach for building equational theories with the objective of improving the performance of the completion procedure, even though there exist canonical rewrite systems for these theories. As a test case of our approach, we show how to build the free Abelian groups and distributivity laws in the completion procedure. The empirical results of our experiment on proving many identities in alternative rings show clearly that the gain of this approach is substantial. More than 30 identities which are valid in any alternative ring are taken from the book "Rings that are nearly associative" by K.A. Zhevlakov *et al.*, and include the Moufang identities and the skew-symmetry of the Kleinfeld function. The proofs of these identities are obtained by Herky, a descendent of RRL and a high-performance rewriting-based theorem prover.

1 Introduction

Building an equational theory in a completion procedure has been studied in the framework of (*congruence-*) *class-rewrite systems*. A class-rewrite system consists of two parts: rewrite rules R and equations S — R rewrites one class of terms modulo S to another class. The initial motivation of studying class-rewrite systems was to handle some of the important equational theories in which any orientation of the axioms yields a nonterminating system. Of great practical importance are associative and commutative (AC) rewrite systems, where S is a set of AC axioms of some binary functions.

To implement a completion procedure modulo S, usually, the special matching and unification algorithms modulo S should be available. Unification modulo various theories has been a hot topic in unification theory and has found wide applications in logic programming. The results on unification theory reveal that if S is also a canonical rewrite system, then the complete unification algorithm modulo S can be effectively constructed (using the concept of *narrowing*). Here, we saw an interesting and tightened relationship between rewrite systems and unifications: class-rewrite systems need special unifications and special unifications can be obtained from canonical rewrite systems. At this point, it is evident that if S is a canonical rewrite system, it is possible to implement a completion procedure modulo S without a special unification algorithm modulo S.

---

[*]Partially supported by the National Science Foundation Grants no. CCR-9202838 and INT-9016100.

The original motivation of studying class-rewrite systems is to avoid nonterminating equations. If S is already a canonical rewrite system, what is the purpose of a class-rewrite system modulo S? The issue is *efficiency*. The most significant advantage of building S into a completion procedure is not only that we can have faster rewriting and superposition operations, but also that we can very effectively avoid redundant computations in a completion procedure. An acute problem in every automatic reasoning system is the inability to recognize the unnecessity of a large subset of inferences explosively generated from the input. The techniques for identifying redundant computations in a completion procedure have been developed over the years (see [13] for a survey). Building S into a completion procedure allows us to integrate these techniques effectively and inexpensively.

In this paper, we report a case study of the completion procedure modulo distributivity laws and Abelian groups using the theorem prover Herky.[1] The problems of this case study are a list of identities taken from a monograph by Zhevlakov *et al.* on nonassociative rings [15]. Intuitively, a nonassociative ring is an ordinary ring in which the associativity of the multiplication is missing. A nonassociative ring is said to be *alternative* if the multiplication $*$ of the ring satisfies:

$$(x * x) * y = x * (x * y) \quad : \text{Left alternative law}$$
$$x * (y * y) = (x * y) * y \quad : \text{Right alternative law}$$

Computer proofs of these identities were considered very difficult in the past. In [8] and [9], Rick Stevens first introduced some of these identities to the community of automated reasoning. Many of the problems suggested by Stevens could be solved by T.C. Wang's Z-module method, a specialized method targeted for ring theory [10], [11]. In particular, using the Z-module method, Wang and Stevens also solved an open problem in nonassociative ring theory [12]. In [1], Anantharaman and Hsiang reported the proofs of the Moufang identities found by their general purpose theorem prover SBR2 [2]. To the best of our knowledge, the above cited references consist of an exhaustive list of the papers on automated proofs in alternative rings.

Like SBR2, Herky is a general purpose, rewriting-based theorem prover. The major inference rules of Herky and SBR2 are the same: rewriting and superposition (see [3] for a survey of the rewriting approach to equational reasoning). After we built the distributivity laws and Abelian groups into Herky, Herky is able to prove many alternative ring identities with ease. Table 1 lists the six identities reported in [1]. Equations 3-5 are the famous Moufang Identities. From the table, we can see that Herky can prove these theorems in a couple of seconds.[2]

When the free Abelian group axioms and the distributivity laws are input to Herky, Herky automatically turns on some special procedures which perform

---

[1] Herky, standing for "<u>H</u>igh-p<u>er</u>formance <u>k</u>e<u>y</u> operations" [14], is a descendent of *RRL* (*Rewrite Rule Laboratory*), a theorem proving environment for experimenting and developing reasoning methods based on rewriting techniques and equational logics [5].

[2] The time of SBR2 was reported in [1] and was measured on a Sun 3/60, and that of Herky was on a Sun Sparc station 2 (in seconds); Sparc 2 is about 6 times faster than Sun 3/60. The number of generated rules for SBR2 does not count the number of non-orientable equations and unequalities. We are told by Jieh Hsiang that the new version of SBR2 is much faster.

	SBR2			Herky		
Problem	time	eqns	rul.	time	eqns	rul.
$(x*y)*x = x*(y*x)$	32	67	15	0.58	19	9
$a(x,y,z) + a(y,x,z) = 0$	48	127	19	0.48	12	8
$x*(y*(x*z)) = ((x*y)*x)*z$	2:30:49	452	41	7.81	175	50
$((z*x)*y)*x = z*((x*y)*x)$	1:56:36	427	39	7.68	175	50
$(x*y)*(z*x) = (x*(y*z))*x$	1:56:09	638	47	7.63	175	50
$f(x,x,y,z) = 0$	2:07:58	463	39	10.55	190	59

Table 1: Statistics of SBR2 and Herky on alternative ring problems

the key operations of a completion procedure, i.e., rewriting and superposition, modulo the theories of Abelian groups and the distributivity laws. Because of its ability to automatically recognize these axioms, Herky does not lose its status as a general purpose theorem prover. Because of the special built-in procedures for these theories, Herky can gain substantial speed-ups in solving alternative ring problems.

2 Theoretical Background

We will use the standard notations on rewrite systems as given in [3]. We use x, y, z to denote variables, l, r, s, t to denote terms, and $l \to r$ to denote a rewrite rule (one-way equation). A substitution is represented by $\{x_1 \mapsto t_1, ..., x_n \mapsto t_n\}$. We also use $\{s \mapsto t\}$ to denote the operation of replacing the term s by the term t.

For any rewrite relation $\to$, let $\leftarrow$ be the inverse of $\to$; $\leftrightarrow$, $\to^+$, $\to^*$ and $\leftrightarrow^*$ be the symmetry, the transitive, the reflexive and transitive, and the reflexive, symmetric and transitive closures of $\to$, respectively. Given a set R of rewrite rules, let $\xrightarrow{R}$ denote the rewrite relation induced by R. R is said to be *terminating* if $\xrightarrow{R}^+$ is a well-founded ordering on terms. Two terms t_1 and t_2 are *joinable* if there exists t such that $t_1 \to^* t \xleftarrow{*} t_2$, or simply $t_1 \to^* \circ \xleftarrow{*} t_2$, where $\circ$ denotes the composition of relations. R is said to be *Church-Rosser* if $\leftrightarrow^*$ is equal to $\to^* \circ \xleftarrow{*}$; R is said to be *canonical* if R is both terminating and Church-Rosser.

By R/S we denote the class rewrite system composed of a rewrite system R and a set of equations S. We say s *rewrites* to t modulo S, denoted $s \xrightarrow{R/S} t$, if $s \xleftrightarrow{S}^* u[l\sigma]_p$ and $u[r\sigma]_p \xleftrightarrow{S}^* t$ for some context u, position p in u, $l \to r$ in R and substitution σ. If $\xrightarrow{R/S}$ is canonical, then $\xleftrightarrow{R \cup S}^*$ is equal to $\xrightarrow{R/S}^* \circ \xleftarrow{R/S}^*$ [3].

Jouannaud and Kirchner proved the following general result [4]: T is said to be *locally coherent modulo S with R* if $\xleftarrow{T} \circ \xrightarrow{R}$ is contained in $\xrightarrow{T}_* \xleftrightarrow{S}_* \circ \xleftarrow{T}_*$; T is said to be *locally coherent modulo S with S* if $\xleftarrow{T} \circ \xleftrightarrow{S}$ is contained in $\xrightarrow{T}_* \xleftrightarrow{S}_* \circ \xleftarrow{T}_*$.

Lemma 2.1 (Coherence Lemma) *If R/S is terminating (i.e., $\xrightarrow{R/S}^+$ is terminating), then R/S is Church-Rosser modulo S iff R/S is locally coherent modulo S with both R and S.*

When S is known to be canonical, $\overset{S}{\leftrightarrow}{}^*$ can be replaced by $\overset{S}{\rightarrow}{}^*\circ{}^*\overset{S}{\leftarrow}$, and the above lemma is reduced to the following:

Lemma 2.2 (Special Case of Coherence Lemma) *If R/S is terminating and S is canonical, then R/S is Church-Rosser modulo S iff R is locally coherent modulo S with both R and S.*

Note that we require only the local coherence of R, not R/S, in this lemma, and it is much easier to check the coherence of R than that of R/S.

A *critical pair* of two rewrite rules $l_1 \rightarrow r_1$ and $l_2 \rightarrow r_2$ of a rewrite system R is the equation $r_1\sigma = l_1[r_2]_p\sigma$, where p is a non-variable position in l_1, and σ is a most general unifier (mgu) of $l_1|_p$ and l_2. The process of computing critical pairs is called *superposition*. Let $cp(R_1, R_2)$ denote the set of all critical pairs between rules of R_1 and rules of R_2, and let $\overset{cp(R_1,R_2)}{\longleftrightarrow}$ denote its symmetric rewrite closure.

Lemma 2.3 (Extended Critical Pair Lemma) *Suppose S is canonical. If both $\overset{cp(R,R)}{\longleftrightarrow}$ and $\overset{cp(R,S)}{\longleftrightarrow}$ are contained in $\overset{R}{\rightarrow}{}^*_\diamond \overset{S}{\rightarrow}{}^*\circ{}^*\overset{S}{\leftarrow}\circ{}^*\overset{R}{\leftarrow}$ then R is locally coherent modulo S with both R and S.*

It has been known that for testing the Church-Rosser property, not every critical pair should be computed. Some criteria have been developed over the years to detect those unnecessary critical pairs (a survey on this topic can be found in [13]). We say a subset C of $cp(R, S)$ is *essential* if when $cp(R, S)$ is replaced by C in lemma 2.3, the lemma still holds. Every critical pair in $cp(R, S) - C$ is said to be *redundant*. For the sake of efficiency, C is expected to be as small as possible. Our goal is to compute a minimal (under the subset relation) subset $essential\_cp(R, S) \subset cp(R, S)$ instead of $cp(R, S)$ when building S into the completion procedure.

If some critical pairs in $cp(R, R)$ or $cp(R, S)$ are not joinable, then the Knuth-Bendix completion procedure can be used. A *completion procedure* can be described as a sequence of transformations [3]:

$$\langle E_0, R_0 \rangle \vdash \langle E_1, R_1 \rangle \vdash \cdots \vdash \langle E_n, R_n \rangle$$

such that $\langle E_i, R_i \rangle \vdash \langle E_{i+1}, R_{i+1} \rangle$ iff one of inference rules (superposition, reduction, orientation, etc. See [3] for details) applies to E_i or R_i to produce E_{i+1} and R_{i+1}. For the completion procedure modulo S, the superposition chooses a critical pair from one of the two sets, $cp(R_i, R_i)$ and $cp(R_i, S)$.

A critical pair criteria used for testing the Church-Rosser property can be also used to detect unnecessary critical pairs in the completion if these criteria are shown to be *monotonic with completion* [13]. That is, only critical pairs from $essential\_cp(R_i, R_i)$ or $essential\_cp(R_i, S)$ are added into E_{i+1}.

To facilitate the presentation, in this paper, we only try to avoid computing redundant critical pairs of $cp(R_i, S)$ which are joinable in $R_i \cup S \cup essential\_cp(R_i, S)$. It is easy to show that discarding such redundant critical pairs does not lose the completeness of any completion procedure.

3 Building Theories into Completion Procedure

In this section, we describe how the Abelian group theory and the distributivity laws are built into a completion procedure. Like any completion-based theorem prover, the completion procedure in Herky consists of two major inference rules: *rewriting* and *superposition*. The basis idea of building a theory into the completion procedure is quite simple: Because there exist canonical rewrite systems for the free Abelian groups and the distributivity laws (either respectively or collectively), we may just provide some procedures which do what a rewrite rule will do. That is, because a rewrite rule is used for rewriting and superposition, our procedures will simulate the operations of rewriting and superposition of that rule.

There are many ways to simulate rewritings by a program. For instance, we may compile the rewrite rules to obtain an equivalent program for rewriting; this approach is good when the set of rules is not fixed. When the set of rules is fixed, we may code carefully to obtain a very efficient program. Since simulating rewritings is easy, in the following, we will focus on how to simulate superpositions. We present our method in the context of proving alternative ring identities, even though the method can be used in a much wider context. In the remaining of this paper, by "experiment" we mean the experiment of proving identities of alternative rings.

3.1 Axioms of Free Alternative Rings

The theory of free alternative rings include both free Abelian groups and distributivity laws. Informally speaking, a nonassociative ring is just a ring without the associativity law for the multiplication, and a nonassociative ring is alternative if it satisfies the left and right alternative laws. Using the equational logic, the following axioms define a free alternative ring.

$$
\begin{aligned}
x + 0 &= x & x + -(x) &= 0 \\
x + y &= y + x & (x + y) + z &= x + (y + z) \\
(x + y) * z &= (x * z) + (y * z) & z * (x + y) &= (z * x) + (z * y) \\
(x * x) * y &= x * (x * y) & x * (y * y) &= (x * y) * y
\end{aligned}
$$

In [15], many properties of alternative rings are studied. While the proofs of these properties rely heavily on equality reasoning, because of lacking the associativity law for multiplication, the huge search space involved in the proofs cannot be reduced significantly by using a canonical form approach. To facilitate the proofs, the following functions are found useful:

$$
\begin{aligned}
a(x, y, z) &= ((x * y) * z) + -(x * (y * z)) : \text{The associator} \\
c(x, y) &= (x * y) + -(y * x) : \text{The commutator} \\
j(x, y) &= (x * y) + (y * x) : \text{The Jordan product} \\
f(w, x, y, z) &= a(w * x, y, z) + -(x * a(w, y, z)) + -(a(x, y, z) * w) \\
&\quad : \text{The Kleinfeld function}
\end{aligned}
$$

The associator is a measure of non-associativity and the commutator is a measure of non-commutativity in an alternative ring. Many identities of alternative rings can be expressed in terms of the above functions. For instance, the Moufang identities can be expressed in terms of the associator.

In the remaining of this paper, S_1 denotes the canonical rewrite system (modulo AC) for free Abelian groups (generated from the frist four equations without $*$); S_2 denotes the canonical rewrite system for the distributivity laws; S denotes the canonical system for free Abelian groups and distributivity laws (generated from the frist six equations) and R denotes the set of any other rewrite rules.

To prove the termination of $R \cup S$, we use an extension of Dershowitz' recursive path ordering ($\succ_{lrpo}$), with the precedence relation $f > a > j > c > * > - > + > 0$. When R (or S) contains the associative and commutative (AC) operators, then R (or S) also contains the AC extension rules (the AC extension rule is a rule of form $f(l, z) \to f(r, z)$, where $l \to r$ is a non-extension rewrite rule and the root symbol of l, i.e., f, is AC.)

3.2 Building in the Free Abelian Groups

Recall that S_1 denotes the canonical rewrite system for the free Abelian groups and contains the following rewrite rules generated by the AC completion procedure [7]:

```
[1] (0 + x) ---> x
[2] (-(x) + x) ---> 0
[3] -(0) ---> 0
[4] -(-(x)) ---> x
[5] -((x + y)) ---> (-(x) + -(y))
```

The basic idea of our approach to build the Abelian group into a completion procedure is that, besides the rewriting role of the above rules, if an essential critical pair can be generated from a rule of R and a rule of S_1, then our built-in procedure will generate the same equation.

As we pointed out in the introduction, the most significant advantage of building equations into a completion procedure is not just that we can have faster rewriting and superposition operations, but that we can very effectively avoid redundant computations in a completion procedure. Formally, let $built\_cp(R, S_1)$ denote the set of critical pairs generated by our built-in procedure, we intend that $built\_cp(R, S_1)$ be as small as possible, as long as $essential\_cp(R, S_1) \subseteq built\_cp(R, S_1)$.

Let us assume that $S \cup R$ is inter-reduced (T is inter-reduced if for any non-AC-extension rule $l \to r$ in T, l and r are irreducible by $T - \{l \to r\}$). The following rules specify what $built\_cp(R, S_1)$ contains. For every rule $l \to r$ of R which shares no variables with S_1, we have:

Rule 1: (related to $0 + x \to x$) For any variable y in l which is not introduced by the AC extension, if $y + t$ is a subterm of l, where t is a term, then $(l = r)\{(y + t) \mapsto t\}\{y \mapsto 0\}$ is in $built\_cp(R, S_1)$.

Rule 2: (related to $-(0) \to 0$) For any variable y in l, if $-(y)$ is a subterm of l, then $(l = r)\{-(y) \mapsto 0\}\{y \mapsto 0\}$ is in $built\_cp(R, S_1)$.

Rule 3: (related to $(-(-(x))) \to x$) (a) For every variable y in l such that $-(y)$ is a subterm of l, $(l = r)\{-(y) \mapsto x\}\{y \mapsto -(x)\}$ is in $built\_cp(R, S_1)$.

(b) If l is identical to $-(t)$ for some term t, then $t = -(r)$ is in $built\_cp(R, S_1)$.

Rule 4: (related to $-(x+y) \to (-(x) + -(y))$) (a) For any variable z in l, if $-(z)$ is a subterm of l then $(l = r)\{-(z) \mapsto (-(x) + -(y))\}\{z \mapsto (x+y)\}$ is in $built\_cp(R, S_1)$.

(b) If $l = (t_1 + t_2 + \cdots + t_n)$ for terms $t_1, t_2, ..., t_n$, then $-(t_1) + -(t_2) + \cdots + -(t_n) = -(r)$ is in $built\_cp(R, S_1)$.

Rule 5: (related to $(-(x) + x) \to 0$) This is the most interesting rule because it has an nontrivial AC extension.

(a) If $l = -(t)$ for some term t, then $r + t = 0$ is in $built\_cp(R, S_1)$.

(b) If $l = (t_1 + \cdots + t_i + \cdots + t_n)$ for terms $t_1, ..., t_i, ..., t_n$, then $(t_1 + \cdots + t_{i-1} + t_{i+1} + \cdots + t_n) = r + -(t_i)$ is in $built\_cp(R, S_1)$. The effect of this rule superposition is to move t_i from left to right; we will discuss this rule later.

(c) If $(t_1 + \cdots + t_n)$ is a proper subterm of l for terms $t_1, ..., t_n$ and there exist t_i, $1 \le i \le n$, and a subset $\{j_1, ..., j_k\}$ of $\{1, ..., i-1, i+1, ..., n\}$ such that t_i and $-(t_{j_1} + \cdots + t_{j_k})$ are unifiable with mgu σ, then the instance under σ of the equation resulted from deleting $t_i + t_{j_1} + \cdots + t_{j_k}$ from l is in $built\_cp(R, S_1)$.

The following lemmas justify the completeness of the above rules for generating essential critical pairs.

Lemma 3.1 *Every critical pair in $cp(R, S_1) - built\_cp(R, S_1)$ is joinable in $R \cup S_1 \cup built\_cp(R, S_1)$.*

The proof of this lemma consists of an exhaustive enumeration of every critical pair in $cp(R, S_1)$ and is omitted here because of the space limitation.

Note that it is much easier to compute $built\_cp(R, S_1)$ than to compute $cp(R, S_1)$. For instance, the effect of Rule 4(b) is to change the sign of $t_1 + \cdots + t_n$ and r at the same time. In Peterson and Stickel's AC completion procedure [7], $O(2^n)$ critical pairs will be generated from $l \to r$ and $-(x+y) \to (-(x) + -(y))$ because there are $O(2^n)$ AC-unifiers of $x + y$ and $t_1 + \cdots + t_n$. In our case, a single equation is sufficient. Moreover, our experiment shows that no term like $-(y)$ appears in R, hence, Rules 2, 3(a), 4(a) are inapplicable in our experiment. Rules 1 and 5(c) are inapplicable, too.

Rules 3(a), 5(a) and 5(b) can be further improved if the procedure for simulating rewriting (i.e., the normalization procedure) can do more work for us. Let us take a close look on Rule 5(b): If we superpose $(t_1 + \cdots + t_n) \to r$ into the AC extension of $-(x) + x \to 0$, i.e., $-(x) + x + y \to y$, the effect of this superposition is to move some arguments of $+$ from the left-hand side of $l \to r$ to its right-hand side. For instance, if we want to move t_1 to the right-hand side, we may superpose $(t_1 + \cdots + t_n) \to r$ into $x + y$ of $-(x) + x + y \to y$, with the mgu being $\sigma = \{x \mapsto t_1, y \mapsto (t_2 + \cdots + t_n)\}$. The resulting critical pair is $(t_2 + \cdots + t_n) = r + -(t_1)$. Because there are $O(2^n)$ ways to move one or more arguments of $+$ from the left-hand side of a rule to its right side, that is why ordinary superposition is very expensive when AC operators are presented. However, we can simulate this effect by the normalization procedure systematically and inexpensively.

Suppose we are allowed to move arguments of $+$ around the equality sign of an equation, now the question is: What arguments of $+$ should we move around? Certainly, to avoid keeping every variant of an equation in the system, we do not want to move every argument of $+$. For instance, suppose $t_1 + t_2 + \cdots + t_n \rightarrow r$ is in R and $t_2 \succ_{lrpo} t_1$, where $\succ_{lrpo}$ is the lexicographic recursive path ordering. If we move t_1 to the right-hand side, under the given condition, we can prove that $t_2 + \cdots + t_n \succ_{lrpo} r + -(t_1)$. So, we make the rule $t_2 + \cdots + t_n \rightarrow r + -(t_1)$, which can reduce the original rule to a trivial one. Thus, there is no need to keep an argument t_i of $+$ at the left-hand side if t_i is smaller (by $\succ_{lrpo}$) than another argument t_j.

From the above discussion, the normalization procedure can systematically do the following: for an equation $l = r$, if $+$ is the root symbol of l or r, we move maximal arguments of $+$ to the left-hand side of the equation and move non-maximal arguments of $+$ to the right-hand side. Let us call the resulting equation *standard*.

Substantial improvement of the completion procedure has been obtained when the above technique is implemented in Herky. Note that every standard equation is always made into a terminating rule. Moreover, there is no need to keep other variants of a standard equation in the system, because every critical pair, which can be generated from an variant of the equation, can be also generated from the standard equation (because of space limitation, we omit the details here).

The above technique of handling equations in an Abelian group is much more powerful than the cancellation rule. The cancellation rule allows us to derive the equation $t_1 = t_2$ from $t_1 + t = t_2 + t$ (t_2 can be an added 0) and has played an indispensable role in the proofs reported in [1]. Given $t_1 + t = t_2 + t$, our technique allows us to obtain the equation $t_1 = t_2 + t + -(t)$, which can be simplified by $x + -(x) \rightarrow 0$ to $t_1 = t_2$.

3.3 Building in the Distributivity Laws

The distributivity laws in an alternative ring are equations (5) and (6) in section 3.1. These two equations consist of a canonical rewrite system S_2 by orienting them from left to right. When these two equations are combined with (1)–(4), they can still generate a canonical rewrite system S consisting of rules [1]–[5] plus the following:

```
[6]  (x * (y + z)) ---> ((x * y) + (x * z))
[7]  ((x + y) * z) ---> ((x * z) + (y * z))
[8]  (x * 0) ---> 0
[9]  (0 * x) ---> 0
[10] (x * -(y)) ---> -((x * y))
[11] (-(y) * x) ---> -((y * x))
```

Like the previous subsection, we define the rules which generate $built\_cp(R, S_2)$ and $built\_cp(R, S)$ (assuming that $built\_cp(R, S)$ contains both $built\_cp(R, S_1)$ and $built\_cp(R, S_2)$).

Let us assume again that R and S share no variables and $R \cup S$ is inter-reduced. We say a term t is a *polynomial term* if the functions like $-$, $+$ and $*$ appear only under themselves in t; we say t is a *product term* if t is a polynomial term and does not contain $+$. For instance, $g(x)$ and $g(x) * (g(x) + g(y))$ are polynomial terms while $g(x+y)$ is not. In our experiment, both sides of each rule in R are polynomial terms.

For every rule $l \to r$ of R, we give below the rules, each of which corresponds to two of the above six rewrite rules, for generating $built\_cp(R, S_2)$ and $built\_cp(R, S)$.

Rule 6: (related to $(x * (y+z)) \to ((x*y) + (x*z))$ and $((x+y)*z) \to ((x*z) + (y*z))$)

(a) If $l = (t_1 + \cdots + t_n)$, then $(x*t_1) + \cdots + (x*t_n) = (x*r)$ and $(t_1*x) + \cdots + (t_n*x) = (r*x)$ are in $built\_cp(R, S_2)$

(b) If $t*v$ (or $v*t$) is a subterm of l, where v is variable and t is an arbitrary term, then there are two subcases to consider (the case for $v*t$ is symmetric):

- l is a polynomial term: If v appears more than once in a product term which is subterm of l, then $(l = r)\{(t*v) \mapsto (t*y)+(t*z)\}\{v \mapsto (y+z)\}$ is in $built\_cp(R, S_2)$.
- l is not a polynomial term: $(l = r)\{(t*v) \mapsto (t*y) + (t*z)\}\{v \mapsto (y+z)\}$ is in $built\_cp(R, S_2)$.

Rule 7: (related to $x*0 \to 0$ and $0*x \to 0$) For each variable y of l, if $t*y$ (or $y*t$) is a subterm of l, then $(l = r)\{(t*y) \mapsto 0, y \mapsto 0\}$ (or $(l = r)\{(y*t) \mapsto 0\}\{y \mapsto 0\}$) is in $built\_cp(R, S)$.

Rule 8: (related to $x*-(y) \to -(x*y)$ and $-(y)*x \to -(y*x)$) (a) If $l = -(t)$ for some term t, then $x*r = -(x*t)$ and $r*x = -(t*x)$ are in $built\_cp(R, S)$.

(b) For each variable z of l, if $t*z$ (or $z*t$) is a subterm of l, then $(l = r)\{(t*z) \mapsto -(t*y)\}\{z \mapsto -(y)\}$ (or $(l = r)\{(z*t) \mapsto -(y*t)\}\{z \mapsto -(y)\}$) is in $built\_cp(R, S)$.

Of the above three rules, Rule 6(b) is the most interesting. In the case when l is a polynomial term (that is the only case in our experiment), the effect of Rule 6(b) is the same as the operation *linearization* in [15] (i.e., replacing v by $(y+z)$). Substantial improvement of the completion procedure can be achieved by linearization because we linearize only non-linear variables and only once for each non-linear variable. For instance, if l is a polynomial term and every variable of l is linear, no critical pairs will be generated by Rule 6(b), because every critical pair of $l \to r$ and [6] (or [7]) is redundant.

In the previous subsection, we said that if $-(t) \to r$ is a rule of R, we then replace it by $t \to -(r)$. Hence, Rule 8(a) is inapplicable in our experiment because there are no rules in R which are of form $-(t) \to r$.

The following lemma justifies the completeness of rules 1–8 for generating essential critical pairs.

Lemma 3.2 *Every critical pair in $cp(R, S_2) - built\_cp(R, S_2)$ is joinable in $R \cup S_2 \cup built\_cp(R, S_2)$.*

Lemma 3.3 *Every critical pair in $cp(R, S) - built\_cp(R, S)$ is joinable in $R \cup S \cup built\_cp(R, S)$.*

From lemmas 2.2, 2.3, and 3.3, we have the following theorem which justifies the completeness of our approach:

Theorem 3.4 *Suppose S is a canonical system for the free Abelian groups and the distributivity laws and R is any rewrite system. If R/S is terminating and both $\overset{cp(R,R)}{\longleftrightarrow}$ and $\overset{built\_cp(R,S)}{\longleftrightarrow}$ are contained in $\overset{R}{\to}\circ\overset{S}{\to}{}^*\circ{}^*\!\overset{S}{\leftarrow}\circ{}^*\!\overset{R}{\leftarrow}$ then R is Church-Rosser modulo S.*

The above result can be easily extended to establish the correctness of the completion procedure modulo S in which $built\_cp(R,S)$ instead of $cp(R,S)$ is computed, because every critical pair of $cp(R,S) - built\_cp(R,S)$ is joinable in $R \cup S \cup built\_cp(R,S)$.

4 Automatic Proofs of Alternative Ring Identities

The rules given in the previous section for generating critical pair set $built\_cp(R,S)$ have all been implemented in Herky [14] (written in Common Lisp). In the following, we only give the proof statistics of the identities presented in Table 2. Limited space prevents us from presenting in details the proofs obtained by Herky.

Each proof is obtained as follows: We input into Herky the definitions of a, c, j and f, equations (1)–(8) (in section 3.1), plus the conjecture to be proved. The definitions of a, c, j and f are used only to simplify the conjecture so that an equivalent equation without a, c, j and f can be obtained. That is, the definitions of a, c, j and f are used only as shorthands of some inout terms. We then run the completion procedure on equations (1)–(8). The initiation procedure of Herky recognizes automatically equations (1)–(6) as the theories of free Abelian groups and the distributivity laws, and then turn on the special procedures for handling these theories. Herky then generates new equations from (1)–(8) (i.e., $built\_cp(R,S)$). During the generation process of new equations, rewriting is used whenever possible to simplify (old or new) equations by (new or old) equations. If there are sufficient rewrite rules which can reduce the conjecture to a trivial one, then a proof of the conjecture is found. Otherwise, Herky will continue to generate new equations. The termination of the completion procedure is not guaranteed when the conjecture is invalid.

In Table 2, we list some identities which are true in every alternative ring, and in Table 3, we give Herky's proof statistics of these identities (except those of (a13), (c4)–(c8) and (j4), which Herky cannot finish before running out of the space). The time was measured in Sun Common Lisp on a Sun Sparcstation 2 with 16 megabytes of main memory. For each identity listed there, the table gives the computing time, the number of new equations generated by Herky, the number of rewrite rules made from both the input equations and new equations, and the number of equations (including the input equations) in the proof of that identity. The proof of (j3) is the most complicated one, where a total number of 63 rewrite rules are generated; the other proofs need less than 63 rewrite rules.[3]

---

[3] Since all new equations (or rewrite rules) are generated from the same input equations

Name	Identity
m1	$x * ((y * z) * x) = (x * (y * z)) * x$
m2	$((x * y) * x) * z = x * (y * (x * z))$
m3	$z * ((x * y) * x) = ((z * x) * y) * x$
m4	$(x * y) * (z * x) = (x * (y * z)) * x$
a1	$a(x, x, y) = 0$
a2	$a(x, y, y) = 0$
a3	$a(x, y, x) = 0$
a4	$a(x, y, z) + a(y, x, z) = 0$
a5	$a(x, y, z) + a(z, y, x) = 0$
a6	$a(x, y, z) + a(x, z, y) = 0$
a7	$a(x, (y * x), z) = x * a(x, y, z)$
a8	$a(x, (x * y), z) = a(x, y, z) * x$
a9	$a(x * x, y, x) = 0$
a10	$a(x * y, x, y) = 0$
a11	$(x * y) * a(x, y, z) = y * (x * a(x, y, z))$
a12	$a((w * x), y, z) + -a(w, (x * y), z) + a(w, x, (y * z)) =$ $(w * a(x, y, z)) + (a(w, x, y) * z)$
a13	$a(z, x, (y * x) * y) = a(z, (x * y) * x, y)$
c1	$c(x * y, z) + -(x * c(y, z)) + -(c(x, z) * y) =$ $a(x, y, z) + -a(x, z, y) + a(z, x, y)$
c2	$c(a(x, y, z), w) = a(x * y, z, w) + a(y * z, x, w) + a(z * x, y, w)$
c3	$a((x * y), z, w) + a(x, y, c(z, w)) = (x * a(y, z, w)) + (a(x, z, w) * y)$
c4	$a(x, y, a(x, y, z)) = c(x, y) * a(x, y, z)$
c5	$a(a(x, y, z), x, y) = -(a(x, y, z) * c(x, y))$
c6	$c(x, y) * a(c(x, y) * c(x, y), z, w) = 0$
c7	$a(c(x, y) * c(x, y), z, w) * c(x, y) = 0$
c8	$a(c(x, y) * c(x, y) * c(x, y) * c(x, y), z, w) = 0$
j1	$a(x * x, y, z) = a(x, j(x, y), z)$
j2	$a(x * x, y, z) = j(x, a(x, y, z))$
j3	$j(a(x, y, z), c(x, y)) = 0$
j4	$a(j(z * x, x * z), y, z) = a(x * z * x, y, z * z)$
f1	$f(x, y, z, w) + f(y, x, z, w) = 0$
f2	$f(x, y, z, w) + f(z, y, x, w) = 0$
f3	$f(x, y, z, w) + f(w, y, z, x) = 0$
f4	$f(x, y, z, w) + f(x, z, y, w) = 0$
f5	$f(x, y, z, w) + f(x, w, z, y) = 0$
f6	$f(x, y, z, w) + f(x, y, w, z) = 0$
f7	$f(x, x, y, z) = 0$
f8	$f(x, y, x, z) = 0$
f9	$f(x, y, z, x) = 0$
f10	$f(y, x, x, z) = 0$
f11	$f(y, x, z, x) = 0$
f12	$f(y, z, x, x) = 0$

Table 2: Some identities of alternative rings.

Problem	CPU time	Equations	Rules	|proof|
m1	0.18	22	11	5
m2	7.81	175	50	32
m3	7.68	175	50	32
m4	7.63	175	50	32
a1	0.09	6	6	3
a2	0.07	5	5	3
a3	0.27	22	11	6
a4	0.20	15	10	4
a5	0.42	29	13	7
a6	0.14	13	9	4
a7	13.45	193	61	43
a8	13.31	193	61	41
a9	0.44	26	12	7
a10	1.31	69	23	11
a11	9.14	372	43	17
a12	29.07	296	62	47
c1	0.19	2	2	3
c2	31.98	296	63	47
c3	10.66	190	59	35
j1	10.49	193	61	44
j2	10.62	193	61	41
j3	31.86	365	63	47
f1	11.13	198	62	42
f2	12.04	198	62	45
f3	10.89	190	59	42
f4	11.33	189	59	39
f5	9.55	178	53	29
f6	3.60	129	36	18
f7	10.55	190	59	42
f8	10.93	193	61	42
f9	7.70	175	50	38
f10	7.87	175	50	34
f11	6.14	172	47	23
f12	0.16	5	5	4

Table 3: Statistics of Herky on some identities of alternative rings.

4.1 Comparison with Other Alternative Identities Proofs

To manifest the power of our approach, we put the following restrictions in our experiment.

- Instead of inputing the negated identities (with Skolem constants) to the prover, we use the pure forward search strategy and stop the theorem prover only when the identity to be proved is generated by the prover. That is, we did not take any advantage of the backward search strategy.

- Instead of cumulating proof efforts of the prover by retaining all the previously proved lemmas in the prover, we prove everything directly from the original (minimal) axiom set. This excludes the possibility of using any auxiliary functions or auxiliary lemmas in a proof.

To the best of our knowledge, none of the previous proofs (either by human or by computer) of the identities in Table 3 met the above two restrictions.

Because of the above restrictions, we could not prove all the identities listed in table 2 (i.e., (a13), (c4)–(c8) and (j4)). It has been shown in the literature of theorem proving that the backward search strategy is very powerful in refutational theorem proving and that using auxiliary lemmas can significantly cut the search space down. We have combined the techniques reported in this paper with the backward search strategy; some of experimental results have been reported in [6] on the commutativity problem in ring theory. We plan to integrate other useful techniques into Herky in order to attack more difficult or open problems in alternative rings.

4.2 Comparison with the Gröbner-Base Algorithm

It is well-known that the Gröbner-base algorithm is targeted for polynomials. The major difference of a polynomial in an alternative ring and a polynomial in an ordinary algebra system is that the variables of a polynomial in the latter are not instantiatable. For instance, in the Gröbner-base algorithm, we cannot derive $a + a + a = a$ from $x + x + x = x$. Because of this, the Gröbner-base algorithm can be viewed as a special case of the completion procedure for ground (i.e., variable-free) equations.

No proofs of alternative identities are ever reported by the Gröbner-base algorithm, partially because the multiplication (i.e., $*$) is assumed to be commutative and associative in the Gröbner-base algorithm. However, because of the similarity of the Gröbner-base algorithm and the completion procedure modulo the Abelian group and the distributivity laws, some ideas developed for the Gröbner-base algorithm can be also used in our case. The next subsection provides such an example.

4.3 A Very Useful Heuristic

It is interesting to observe that every axiom of a free alternative ring is *homogeneous* in the sense that the degree of each product in any given axiom is the

---

and the default search strategy is used in every proof, the rewrite rules used in every proof is a subset of the 63 rules generated in the proof of (j3).

same.[4] We can prove easily the following properties:

1. The inference rules of rewriting and superposition preserve the homogeneity, that is, homogeneous identities generate only homogeneous identities.

2. The degrees of the new identities are not smaller than that of their parents.

3. If the axiom set is homogeneous and the degree of the conjecture to be proved is d, then no identities with degree greater than d can be used in the proof of this conjecture.

The above properties allows us to design a useful heuristic for theorem provers: If the axiom set is homogeneous and the degree of the conjecture to be proved is d, then discard any identities whose degree is greater than d. For instance, the degree of the Moufang identities is 4, so we keep only identities whose degree is less than or equal to 4 when searching a proof of the Moufang identities. Using this heuristic, the search space involving these identities is significantly reduced.

This heuristic has been worked extremely well in proving identities reported in this paper. Without this heuristic, we could not find the proofs of many identities. This heuristic should work well for any set of homogeneous axioms, not just alternative rings.

5 Conclusion

The idea of building equations into a completion procedure is not new, see the references given in [3]. Most of these works aim at extending the scope of the problems which can be attacked by the completion procedure by building non-terminating or non-convergent equations into the procedure. Our objective of building equations into the completion procedure is to improve the performance of the completion procedure, even though there exist canonical rewrite systems for these equations. Our case-study shows clearly that the gain of this approach is substantial.

As a future work, we will continue our experiments on building theories into the completion procedure. For example, currently the user can declare whether the multiplication operator in the distributivity law is commutative or associative or both. We plan to study the cases when the inverse operator is not presented in the Abelian group theory. That is, we will deal with commutative monoids instead of Abelian groups. There are many interesting problems in this theory. We are considering to build in Herky theories other than the Abelian groups and the distributivity laws, as long as they play an important role in a class of interesting problems.

References

[1] Anantharaman, S., Hsiang, J.: (1990) Automated proofs of the Moufang identities in alternative rings. J. of Automated Reasoning **6** 79-109.

---

[4]The *degree* of a product is 0 if it contains 0, otherwise, is the number of the elements (other than 1) in it.

[2] Anantharaman, S., Hsiang, J., Mzali, J.: (1989) SbReve2: A term rewriting laboratory with (AC)-unfailing completion. *Proc. of the third International Conference on Rewriting Techniques and its Applications* (RTA-89). Lecture Notes in Computer Science, Vol. 355, Springer-Verlag, Berlin.

[3] Dershowitz, N., Jouannaud J.P.: (1990) Rewriting systems. In Leuven, V. (ed.): Handbook of Theoretical Computer Science. North Holland.

[4] Jouannaud, J.P., Kirchner, H.: (1986) Completion of a set of rules modulo a set of equations. *SIAM J. on Computing*, 15:1155–1194.

[5] Kapur, D., Zhang, H.: (1989) An overview of RRL: Rewrite Rule Laboratory. *Proc. of the third International Conference on Rewriting Techniques and its Applications* (RTA-89). Lecture Notes in Computer Science, Vol. 355, Springer-Verlag, Berlin, 513-529.

[6] Kapur, D., Zhang, H.: (1991) A case study of the completion procedure: Proving ring commutativity problems. In J.-L. Lassez and G. Plotkin (eds.): *Computational Logic: Essays in Honor of Alan Robinson*, MIT Press, Cambridge, MA.

[7] Peterson, G.L., Stickel, M.E.: (1981) Complete sets of reductions for some equational theories. J. ACM 28:2 233-264.

[8] Stevens, R.L.: (1987) Some Experiments in nonassociative ring theory with an automated theorem prover. J. of Automated Reasoning **3** 211-221.

[9] Stevens, R.L.: (1988) Challenge problems from nonassociative rings for theorem provers. Proc. of 9th Conference on Automated Deduction, Lecture Notes in Computer Science, Springer-Verlag, Berlin, Vol. 230. pp. 730-734.

[10] Wang, T.C.: (1987) Case studies of Z-module reasoning: proving benchmark theorems from ring theory. J. of Automated Reasoning **3** 437-451.

[11] Wang, T.C.: (1988) Elements of Z-module reasoning. Proc. of 9th Conference on Automated deduction, Lecture Notes in Computer Science, Springer-Verlag, Berlin, Vol. 230. pp. 21-40.

[12] Wang, T.C., Stevens, R.L.: (1989) Solving open problems in right alternative rings with Z-module reasoning. J. of Automated Reasoning **5** 141-165.

[13] Zhang, H.: (1991) Criteria of critical pair criteria: a practical approach and a comparative study. To appear in J. of Automated Reasoning.

[14] Zhang, H.: (1991) Herky: High performance rewriting in RRL. In Kapur, D.: (ed.): Proc. of 1992 International Conference of Automated Deduction. Saratoga, NY. Lecture Notes in Artificial Intelligence, 607, Springer-Verlag. pp. 696–700.

[15] Zhevlakov, K.A., *at al.*: (1982) Rings that are nearly associative (translated by H.F. Smith from Russia). Academic Press, New York.

A Semantic Approach to Order-Sorted Rewriting*

Andreas Werner

SFB 314, University of Karlsruhe, P.O. Box 6980,
D-76128 Karlsruhe, Germany, E-mail: werner@ira.uka.de

Abstract. Order-sorted rewriting builds a nice framework to handle partially defined functions and subtypes (see [Smolka & al 87]). In the previous works about order-sorted rewriting the term rewriting system needs to be sort decreasing in order to be able to prove a critical pair lemma and Birkhoff's completeness theorem. However, this approach is too restrictive.
Therefore, we generalize well-sorted terms to semantically well-sorted terms and well-sorted substitutions to some kind of semantically well-sorted substitutions. Semantically well-sorted terms with respect to a set of equations E are terms that denote well-defined elements in every algebra satisfying E.
We prove a critical pair lemma and Birkhoff's completeness theorem for so-called range unique signatures and arbitrary order-sorted rewriting systems. A transformation is given which allows to obtain an equivalent range unique signature from each non-range-unique one. We also show some decidability results.

1 Introduction

Order-sorted rewriting builds a nice framework to handle partially defined functions and subtypes (see [Smolka & al 87]). For instance, one can use Pascal-similar function declarations such as $+(Integer, Integer) : Integer$ where $Integer$ is a subtype of $Real$.

Differing from many-sorted rewriting, Birkhoff's completeness theorem does not hold for order-sorted rewriting which is illustrated by the following example:

Example 1. $A < B.$ $a, a' : A,$ $b : B,$ $f(A) : A.$ $\mathcal{R} :$ $a \rightarrow b$ $b \rightarrow a'$

Hence, we have $a \stackrel{*}{\leftrightarrow} a'$, but not $f(a) \stackrel{*}{\leftrightarrow} f(a')$, since $f(b)$ is ill-formed. On the other hand, $f(a) = f(a')$ holds in every model of $\mathcal{R}$. To overcome this problem we consider $f(b)$ as a *semantically well-sorted term* since a and b are equal in each model of $\mathcal{R}$. For example no mathematician would reject $(4-3)!$ because it is ill-formed, since he knows that $4-3$ and 1 are equal. Using this kind of terms allows us to prove Birkhoff's completeness theorem without the additional

---

* This work was supported by the Deutsche Forschungsgemeinschaft as part of the SFB 314.

assumption in [Smolka & al 87] that the given rewriting system is confluent (cf. [With 92]). This view also allows us to handle meaningful but ill-form terms without using retracts [Goguen & al 85].

If we want to prove a critical pair lemma, we are concerned with two problems: First, rewriting is not monotone with the term structure:

Example 2. $a, a_1, a_2 : A,\ \ f(A) : A,\ \ b : B,\ \ \ \ A < B.$

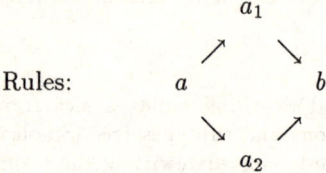

Although the critical pair $\langle a_1, a_2 \rangle$ converges to b, the terms $f(a_1)$ and $f(a_2)$ do not converge to $f(b)$ if we do not consider ill-formed terms such as $f(b)$. The usage of semantically well-sorted terms also solves this problem.

Second, there are problems with variable overlaps (confer Example 5 in [Smolka & al 87]):

Example 3.   Sorts: $Nat < Int.$

Function symbols: $0 : Nat$ $\quad\quad\quad\quad\quad sq : Int\ \ \ \ \rightarrow Nat$
$\quad\quad\quad\quad\quad\quad\quad\ \ s : Nat \rightarrow Nat \quad\quad\quad\quad |\ |: Int\ \ \ \ \rightarrow Nat$
$\quad\quad\quad\quad\quad\quad\quad\quad\quad\quad\quad\quad\quad\quad\quad\quad\quad\quad *:\ Int, Int \rightarrow Int$

Rules: $\pi_1 : |\ x_{Nat}\ | \rightarrow x_{Nat} \quad\quad \pi_2 : sq(y_{Int}) \rightarrow y_{Int} * y_{Int}$

$$\begin{array}{ccc} & |s^n(sq(y_{Int}))| & \\ \swarrow \pi_1 & & \searrow \pi_2 \\ s^n(sq(y_{Int})) & & |s^n(y_{Int} * y_{Int})| \\ \searrow \pi_2 & & \swarrow \pi_1 \\ & s^n(y_{Int} * y_{Int}) & \end{array}$$

Since[2] $s^n(y_{Int} * y_{Int})$ is not a (syntactically) well-sorted term of sort Nat we are not able to reduce $|s^n(y_{Int} * y_{Int})|$ to $s^n(y_{Int} * y_{Int})$. Thus, the terms $s^n(sq(y_{Int}))$ and $|s^n(y_{Int} * y_{Int})|$ do not converge (for all $n \in \mathbf{N}$). In order to decide the local confluence of $\rightarrow$, we are looking for a *finite* set CP of pairs of terms such that $\rightarrow$ is locally confluent iff every pair of CP converges. Obviously, there does not exist such a set CP for the last example. Note that same happens if we consider only well-sorted terms but add $s : Int \rightarrow Int$.

---

[2] $s^n(t)$ is an abbreviation: $s^0(t) := t$ and $s^{n+1}(t) := s^n(s(t))$.

In [Smolka & al 87], therefore, a critical pair lemma is proven for sort decreasing rules. However, this approach is too restrictive: we are not able to express that $y_{Int} * y_{Int}$ is a (syntactically) well-sorted term of sort Nat, hence $sq(y_{Int}) \to y_{Int} * y_{Int}$ is not sort decreasing.

The problem discussed in Example 3 originates in the fact that $s^n(y_{Int} * y_{Int})$ is not a well-sorted term of sort Nat, therefore rule π_2 cannot be applied to $|s^n(y_{Int} * y_{Int})|$. But $s^n(y_{Int} * y_{Int})$ is a semantically well-sorted term of sort Nat, since it is equivalent to $s^n(sq(y_{Int}))$. This observation allows us to prove a critical pair lemma without assuming the rules to be sort decreasing. In order to keep rewriting decidable, we restrict it to so-called range unique signatures. We show that each non-range-unique signature can be transformed equivalently to a range-unique one.

The paper is organized as follows: Section 2 contains some basic notations. In Section 3 we demonstrate the benefits of semantically well-sorted terms for sort decreasing rules. In Section 4 and Section 5 we show how the requirement of sort decreasingness can be dropped. Finally, we compare our approach with some others.

2 Foundations

2.1 Well-sorted terms

An *order-sorted signature* Σ is a triple $(\mathcal{S}, \leq, \mathcal{F})$, where $\mathcal{S}$ is a set of sorts, $\leq$ is a partial order on $\mathcal{S}$, and $\mathcal{F} := \bigcup_{w,s} \mathcal{F}_{w,s}$ for some (not necessarily disjoint) sets of function symbols $\mathcal{F}_{w,s}$ with $w \in \mathcal{S}^*$ and $s \in \mathcal{S}$. We call $f(w) : s$ or $f : w \to s$ *function declaration* and use it as an abbreviation for $f \in \mathcal{F}_{w,s}$. If w is the empty word, we write $f : s$. In both cases s is called *range sort* of f.

A *set of variables* $\mathcal{V}$ for an order-sorted signature Σ is a union of pairwise disjoint, infinite, but enumerable sets $\mathcal{V}_s$ with $s \in \mathcal{S}$ and $\mathcal{V}_s \cap \mathcal{F} = \emptyset$. x_s always denotes a variable of sort s (i.e., $x_s \in \mathcal{V}_s$).

We define the set $\mathcal{T}_\Sigma(\mathcal{V})_s$ of all Σ-*terms (well-sorted terms)* of sort s as follows:

(1) if $\bar{s} \leq s$ and $x_{\bar{s}} \in \mathcal{V}_{\bar{s}}$, then $x_{\bar{s}} \in \mathcal{T}_\Sigma(\mathcal{V})_s$,

(2) if $f(s_1, \ldots, s_n) : \bar{s}, t_i \in \mathcal{T}_\Sigma(\mathcal{V})_{s_i}$ and $\bar{s} \leq s$, then $f(t_1, \ldots, t_n) \in \mathcal{T}_\Sigma(\mathcal{V})_s$.

The set $\mathcal{T}_\Sigma(\mathcal{V})$ of all Σ-terms is defined by $\mathcal{T}_\Sigma(\mathcal{V}) := \bigcup_{s \in \mathcal{S}} \mathcal{T}_\Sigma(\mathcal{V})_s$. A signature Σ is said to be *regular* if every term $t \in \mathcal{T}_\Sigma(\mathcal{V})$ has a least sort.

If we neglect the sort constraints, we get the set of *extended terms* $ET_\Sigma(\mathcal{V})$:

(1) $\mathcal{V} \subseteq ET_\Sigma(\mathcal{V})$,

(2) if $f(s_1, \ldots, s_n) : \bar{s}, t_i \in ET_\Sigma(\mathcal{V})$ then $f(t_1, \ldots, t_n) \in ET_\Sigma(\mathcal{V})$.

Note that $\mathcal{T}_\Sigma(\mathcal{V}) \subseteq ET_\Sigma(\mathcal{V})$. We call t *ill-formed*, if $t \in ET_\Sigma(\mathcal{V}) \setminus \mathcal{T}_\Sigma(\mathcal{V})$.

$Occ(t)$ denotes the set of all occurrences of the (extended) term t. t/u is the subterm of t at occurrence u, $t[u \leftarrow t']$ the term obtained from t by replacing the subterm t/u by t'. Note that for well-sorted terms t, t' and an occurrence $u \in Occ(t)$, $t[u \leftarrow t']$ is not necessarily well-sorted. $Var(t)$ denotes the set of all variables of t.

2.2 Substitutions

An *extended substitution* is a mapping $\sigma : \mathcal{V} \to ET_\Sigma(\mathcal{V})$ such that $Dom(\sigma) := \{x \mid \sigma(x) \neq x\}$ is finite. We extend an extended substitution σ by $\sigma(f(t_1, \ldots, t_n)) = f(\sigma(t_1), \ldots, \sigma(t_n))$ to a mapping on $ET_\Sigma(\mathcal{V})$.

Let the set of *S-substitutions* be an arbitrary subset of the set of extended substitutions.[3] An *S*-substitution σ is called *S-unifier* of t and t' if $\sigma(t) = \sigma(t')$. A set of *S*-unifiers $CSU_S(t, t')$ of t and t' is said to be a *complete set of S-unifiers of t and t'* on a set of variables V with $Var(t) \cup Var(t') \subseteq V \subseteq \mathcal{V}$ if for every *S*-unifier σ of t and t' there are a $\mu \in CSU_S(t, t')$ and a *S*-substitution λ such that $\sigma(x) = \lambda(\mu(x))$ for all $x \in V$.

A *Σ-substitution* is an extended substitution such that if x_s is a variable of sort s then $\sigma(x_s)$ is a Σ-term of sort s.

2.3 Rewriting

A *term rewriting system* $\mathcal{R}$ is a set of rules $l \to r$ with $l, r \in ET_\Sigma(\mathcal{V})$, $Var(r) \subseteq Var(l)$ and $l \notin \mathcal{V}$.

Definition 1 ($\Rightarrow_\mathcal{R}^{Ext}$). Given a rewriting system $\mathcal{R}$ and an extended term t, we write $t \Rightarrow_{\mathcal{R}\;[l \to r, u, \sigma]}^{Ext} t'$ (possibly omitting indices) if there are a rule $l \to r \in \mathcal{R}$, an occurrence $u \in Occ(t)$ and an extended substitution σ with $t/u = \sigma(l)$ and $t' = t[u \leftarrow \sigma(r)]$.

Note that t' always is an extended term.

In this paper, for all relations defined on extended terms we use the symbol $\Rightarrow$. The superscript will always indicate which kind of substitution is used, for example $\Rightarrow_\mathcal{R}^{Ext}$ is defined using extended substitutions. If σ is a Σ-substitution, then we also use $\Rightarrow_\mathcal{R}^{\Sigma}$ instead of $\Rightarrow_\mathcal{R}^{Ext}$.

Remark. If $l, r \in T_\Sigma(\mathcal{V})$ for each rule $l \to r \in \mathcal{R}$, then $t \Rightarrow_{\mathcal{R}[u]}^{\Sigma} t'$ implies $t/u, t'/u \in T_\Sigma(\mathcal{V})$.

Given any binary relation $\to$, we define $\leftrightarrow$ to be its symmetric closure, $\xrightarrow{+}$ to be its transitive closure, $\xrightarrow{*}$ to be its reflexive and transitive closure, and $\xleftrightarrow{*}$ to be its reflexive, symmetric and transitive closure. We say that t_1 and t_2 *converge* if there exists a t' with $t_1 \xrightarrow{*} t'$ and $t_2 \xrightarrow{*} t'$. If $\to$ is noetherian and (locally) confluent, then for every t there exists a unique normal form $t\downarrow$ (with respect to $\to$). For more details see [Huet & Oppen 80].

---

[3] For example, we define below Σ-substitutions, $(\Sigma, \mathcal{R})$-substitutions, and *T*-substitutions.

2.4 Semantics

The following notation is based on the semantics used in [Smolka & al 87][4]. We write $t \doteq_E t'$ if every model of E satisfies the equation $t \doteq t'$. To avoid problems, we assume that for every sort s there is at least one ground term t_s of sort s (i.e., $\mathcal{T}_\Sigma(\emptyset)_s \neq \emptyset$).

2.5 Semantically well-sorted terms

Definition 2. $\mathcal{T}_{\Sigma,\mathcal{R}}(\mathcal{V})_s$ is the set of all $(\Sigma, \mathcal{R})$-terms (semantically well-sorted terms) of sort s:

$$t \in \mathcal{T}_{\Sigma,\mathcal{R}}(\mathcal{V})_s \quad \text{iff} \quad \text{there exists a } t' \in \mathcal{T}_\Sigma(\mathcal{V})_s \text{ with } t \stackrel{*\Sigma}{\Leftrightarrow}_\mathcal{R} t'$$

$\mathcal{T}_{\Sigma,\mathcal{R}}(\mathcal{V})$ denotes the set of all $(\Sigma,\mathcal{R})$-terms: $\mathcal{T}_{\Sigma,\mathcal{R}}(\mathcal{V}) := \bigcup_{s \in \mathcal{S}} \mathcal{T}_{\Sigma,\mathcal{R}}(\mathcal{V})_s$.

For instance, $f(b)$ is a $(\Sigma,\mathcal{R})$-term of sort A in Example 1. Since $\mathcal{T}_\Sigma(\mathcal{V}) \subseteq \mathcal{T}_{\Sigma,\mathcal{R}}(\mathcal{V})$ holds, $(\Sigma,\mathcal{R})$-terms generalize Σ-terms. It can be shown that each semantically well-sorted term denotes a well-defined element in every model of $\mathcal{R}$ if $l, r \in \mathcal{T}_\Sigma(\mathcal{V})$ for each rule of $\mathcal{R}$ [Werner 93].

For all relations defined on $(\Sigma, \mathcal{R})$-terms, we will use $\rightarrow$. The superscript will always indicate which kind of substitution is used.

2.6 Assumption

From now on (except in Section 4 and in the Subsections 5.1 and 5.2) we assume $l, r \in \mathcal{T}_\Sigma(\mathcal{V})$ for each rule $l \rightarrow r$ of the given rewriting system $\mathcal{R}$.

3 Using Σ-Substitutions

In this section we extend rewriting with sort decreasing rules to semantically well-sorted terms and show the advantages of this extension.

Definition 3 ($\rightarrow_\mathcal{R}^\Sigma$). We write $t \rightarrow_\mathcal{R}^\Sigma t'$ if $t, t' \in \mathcal{T}_{\Sigma,\mathcal{R}}(\mathcal{V})$ and $t \Rightarrow_{\mathcal{R}\,[\sigma]}^{Ext} t'$ for some Σ-substitution σ.

Note that $t \rightarrow_\mathcal{R}^\Sigma t'$ iff $t, t' \in \mathcal{T}_{\Sigma,\mathcal{R}}(\mathcal{V})$ and $t \Rightarrow_\mathcal{R}^\Sigma t'$. Furthermore, if t or t' is in $\mathcal{T}_{\Sigma,\mathcal{R}}(\mathcal{V})$ and $t \Rightarrow_\mathcal{R}^\Sigma t'$ then both terms belong to $\mathcal{T}_{\Sigma,\mathcal{R}}(\mathcal{V})$ (by Definition 2) and $t \rightarrow_\mathcal{R}^\Sigma t'$.

Definition 4. A rewriting rule $l \rightarrow r$ is called *sort decreasing*, if for all sorts $s \in \mathcal{S}$ and all Σ-substitutions σ, $\sigma(l) \in \mathcal{T}_\Sigma(\mathcal{V})_s$ implies $\sigma(r) \in \mathcal{T}_\Sigma(\mathcal{V})_s$. We call a rewriting system $\mathcal{R}$ (or $\rightarrow_\mathcal{R}^\Sigma$) sort decreasing if all its rules are sort decreasing.

---

[4] In other words, we use the *non-overloaded* semantics. A function symbol may still have more than one declaration.

For instance, rule π_1 in Example 3 is sort decreasing, whereas rule π_2 of the same example is not sort decreasing.

Definition 5. Let $l_1 \to r_1$, $l_2 \to r_2 \in \mathcal{R}$ be two variable disjoint rules, u in $Occ(l_1)$ an occurrence such that l_1/u is not a variable, and $CSU_S(l_1/u, l_2)$ a complete set of S-unifiers of l_1/u and l_2 on $Var(l_1/u) \cup Var(l_2)$. Then the pairs $\langle \sigma(r_1), \sigma(l_1[u \leftarrow r_2]) \rangle$ with $\sigma \in CSU_S(l_1/u, l_2)$ are called *S-overlaps* (of $\mathcal{R}$).

Proposition 6. Let $t, t' \in \mathcal{T}_{\Sigma,\mathcal{R}}(\mathcal{V})$, $u \in Occ(t)$ and σ be a Σ-substitution.
(Stability) $\quad t \xrightarrow[\mathcal{R}]{*\Sigma} t'$ implies $\sigma(t) \xrightarrow[\mathcal{R}]{*\Sigma} \sigma(t')$.
(Monotonicity) $t/u \xrightarrow[\mathcal{R}]{*\Sigma} t'$ implies $t \xrightarrow[\mathcal{R}]{*\Sigma} t[u \leftarrow t']$.

Theorem 7 (Critical pair lemma for sort decreasing rules). *Let $\mathcal{R}$ be sort decreasing. $\xrightarrow[\mathcal{R}]{\Sigma}$ is locally confluent iff all Σ-overlaps of $\mathcal{R}$ converge.*

Note that the lemma differs from the lemma in [Smolka & al 87] in the fact that we do not restrict rewriting to Σ-terms. However, the proofs are similar (see [Werner 91]). $\xrightarrow[\mathcal{R}]{\Sigma}$ is called *weakly sort decreasing*, if $t \in \mathcal{T}_\Sigma(\mathcal{V})_s$ and $t \xrightarrow[\mathcal{R}]{\Sigma} t'$ implies there exists a $t_s \in \mathcal{T}_\Sigma(\mathcal{V})_s$ such that $t' \xrightarrow[\mathcal{R}]{*\Sigma} t_s$. The lemma remains true, if we replace *sort decreasing* by *weakly sort decreasing* [With 92]. Furthermore, $\Rightarrow_\mathcal{R}^\Sigma$ is locally confluent (or noetherian respectively) iff $\xrightarrow[\mathcal{R}]{\Sigma}$ is [Werner 93].

In Example 3, we have seen that (weak) sort decreasingness is needed for a critical pair lemma, if we use $\xrightarrow[\mathcal{R}]{\Sigma}$. One advantage of using $\xrightarrow[\mathcal{R}]{\Sigma}$ on $\mathcal{T}_{\Sigma,\mathcal{R}}(\mathcal{V})$ instead of $\mathcal{T}_\Sigma(\mathcal{V})$ is that Birkhoff's completeness theorem always holds which will be shown in the next section. On the other hand, $\mathcal{T}_{\Sigma,\mathcal{R}}(\mathcal{V})$ allows to handle meaningful but ill-formed terms. This was an open problem for the semantics we use [Goguen & Diaconescu 92].

Theorem 8. *Let $\xrightarrow[\mathcal{R}]{\Sigma}$ be (weakly) sort decreasing, locally confluent and noetherian. Then for all $t \in ET_\Sigma(\mathcal{V})$ and for all $s \in \mathcal{S}$: $t \in \mathcal{T}_{\Sigma,\mathcal{R}}(\mathcal{V})_s$ iff $t \Downarrow_\mathcal{R}^\Sigma \in \mathcal{T}_\Sigma(\mathcal{V})_s$.*

Proof. If $t \Downarrow_\mathcal{R}^\Sigma \in \mathcal{T}_\Sigma(\mathcal{V})_s$, then $t \in \mathcal{T}_{\Sigma,\mathcal{R}}(\mathcal{V})_s$ by Definition 2. Conversely, if $t \in \mathcal{T}_{\Sigma,\mathcal{R}}(\mathcal{V})_s$ then there exists a $t_s \in \mathcal{T}_\Sigma(\mathcal{V})_s$ such that $t \Leftrightarrow_\mathcal{R}^\Sigma t_s$. Hence, $t \xleftrightarrow[\mathcal{R}]{*\Sigma} t_s$. Since $\xrightarrow[\mathcal{R}]{\Sigma}$ is locally confluent and noetherian, we have $t_s \xrightarrow[\mathcal{R}]{*\Sigma} t\downarrow_\mathcal{R}^\Sigma$. We prove that $t_s \in \mathcal{T}_\Sigma(\mathcal{V})_s$ implies $t\downarrow_\mathcal{R}^\Sigma \in \mathcal{T}_\Sigma(\mathcal{V})_s$ by noetherian induction. If t_s is in normal form, then $t\downarrow_\mathcal{R}^\Sigma = t_s \in \mathcal{T}_\Sigma(\mathcal{V})_s$. Otherwise we have $t_s \xrightarrow[\mathcal{R}]{+\Sigma} t'_s$ for some $t'_s \in \mathcal{T}_\Sigma(\mathcal{V})_s$, since $\xrightarrow[\mathcal{R}]{\Sigma}$ is weakly sort decreasing. By induction hypothesis, we have $t\downarrow_\mathcal{R}^\Sigma = t'_s\downarrow_\mathcal{R}^\Sigma \in \mathcal{T}_\Sigma(\mathcal{V})_s$. Thus, $t\Downarrow_\mathcal{R}^\Sigma = t\downarrow_\mathcal{R}^\Sigma \in \mathcal{T}_\Sigma(\mathcal{V})_s$. □

Therefore, under the conditions of Theorem 8, a given expression is a $(\Sigma, \mathcal{R})$-term if its normal form with respect to $\Rightarrow_\mathcal{R}^\Sigma$ is a Σ-term. Note that for each $(\Sigma, \mathcal{R})$-term the normal form with respect to $\Rightarrow_\mathcal{R}^\Sigma$ and the normal form with respect to $\xrightarrow[\mathcal{R}]{\Sigma}$ coincide.

Example 4 [Goguen & Diaconescu 92].
Sorts: $\qquad\qquad\qquad$ Nat. $\qquad\qquad$ NEStack $<$ Stack.
Function symbols:

$$empty : Stack \qquad\qquad top : NEStack \to Nat$$
$$push : \quad Nat, Stack \to NEStack \qquad pop : NEStack \to Stack$$

Rules: $\quad top(push(n_{Nat}, s_{Stack})) \to n_{Nat} \qquad pop(push(n_{Nat}, s_{Stack})) \to s_{Stack}$

For example, $top(pop(push(2, push(1, empty))))$ is not a Σ-term. But it can be reduced to a Σ-term: $top(pop(push(2, push(1, empty)))) \overset{*}{\Rightarrow}{}_{\mathcal{R}}^{\Sigma} 1$. Hence, it is a meaningful expression and a $(\Sigma, \mathcal{R})$-term by definition. Furthermore, we can replace $\overset{*}{\Rightarrow}{}_{\mathcal{R}}^{\Sigma}$ by $\overset{*}{\to}{}_{\mathcal{R}}^{\Sigma}$ by the remarks to the definition of $\to_{\mathcal{R}}^{\Sigma}$. On the other hand, we have $top(pop(push(2, empty))) \overset{*}{\Rightarrow}{}_{\mathcal{R}}^{\Sigma} top(empty)$, hence the first term is neither a meaningful expression nor a $(\Sigma, \mathcal{R})$-term. We do not need any retracts [Goguen & al 85].

4 Using $(\Sigma, \mathcal{R})$-Substitutions

Now, we explain how a critical pair lemma that does not require the rules to be sort decreasing can be obtained on principle. The problem in Example 3 arises from the fact that $s^n(sq(y_{Int}))$ belongs to the set of Σ-terms of sort Nat, but if we apply rule π_2 to it then the result $s^n(y_{Int} * y_{Int})$ does not belong to this set. Since both terms are equivalent, both are $(\Sigma, \mathcal{R})$-terms of sort Nat. Hence, we can solve the problem on principle if we use more suitable substitutions:

Definition 9. A $(\Sigma, \mathcal{R})$-*substitution* is an extended substitution such that if x_s is a variable of sort s then $\sigma(x_s)$ is a $(\Sigma, \mathcal{R})$-term of sort s.

In Subsection 5.2 we will see that the rewriting systems obtained by completion of a non-sort-decreasing rewriting system may contain terms not belonging to the set of Σ-terms. Therefore, in this section and Subsection 5.1, we assume that $\mathcal{R}$ is a rewriting system such that there exists a rewriting system $\mathcal{R}^\Sigma$ with $l^\Sigma, r^\Sigma \in \mathcal{T}_\Sigma(\mathcal{V})$ for each rule $l^\Sigma \to r^\Sigma \in \mathcal{R}^\Sigma$, $l, r \in \mathcal{T}_{\Sigma, \mathcal{R}^\Sigma}(\mathcal{V})$ for each rule $l \to r \in \mathcal{R}$, and $\overset{*}{\Leftrightarrow}{}_{\mathcal{R}}^{\Sigma} = \overset{*}{\Leftrightarrow}{}_{\mathcal{R}^\Sigma}^{\Sigma}$. We will make this point clearer in Subsection 5.2. Thus, $\mathcal{T}_{\Sigma, \mathcal{R}}(\mathcal{V})_s = \mathcal{T}_{\Sigma, \mathcal{R}^\Sigma}(\mathcal{V})_s$ for all $s \in \mathcal{S}$ and $\mathcal{T}_{\Sigma, \mathcal{R}}(\mathcal{V}) = \mathcal{T}_{\Sigma, \mathcal{R}^\Sigma}(\mathcal{V})$.

Definition 10 ($\to_{\mathcal{R}}^{\Sigma, \mathcal{R}}$). We write $t \to_{\mathcal{R}}^{\Sigma, \mathcal{R}} t'$ if $t, t' \in \mathcal{T}_{\Sigma, \mathcal{R}}(\mathcal{V})$ and $t \Rightarrow_{\mathcal{R}[\sigma]}^{Ext} t'$ for some $(\Sigma, \mathcal{R})$-substitution σ.

Theorem 11. *Given* $t, t', t'' \in \mathcal{T}_{\Sigma, \mathcal{R}}(\mathcal{V})$, $u \in Occ(t)$ *and a binary relation* $\to_A$ *such that* $\to_{\mathcal{R}}^{\Sigma} \subseteq \to_A \subseteq \to_{\mathcal{R}}^{\Sigma, \mathcal{R}}$, *the following holds:*
(Strong compat.) $t' \to_A t''$ impl. $(t[u \leftarrow t'] \in \mathcal{T}_{\Sigma, \mathcal{R}}(\mathcal{V})$ iff $t[u \leftarrow t''] \in \mathcal{T}_{\Sigma, \mathcal{R}}(\mathcal{V}))$
(Completeness) $\quad t' \overset{*}{\leftrightarrow}_A t''$ iff $\quad t' \doteq_{\mathcal{R}} t''$ iff $\quad t' \overset{*}{\Leftrightarrow}{}_{\mathcal{R}}^{\Sigma} t''$

The *strong compatibility* of $\to_A$ implies we do not need to check that both terms in definition of $\to_{\mathcal{R}}^{\Sigma}, \to_{\mathcal{R}}^{\Sigma, \mathcal{R}}$ (and later $\to_{\mathcal{R}}^{T}$) belong to $\mathcal{T}_{\Sigma, \mathcal{R}}(\mathcal{V})$; it suffices to perform the check on one of them. From the *completeness* it follows that rewriting using $\to_{\mathcal{R}}^{\Sigma}$ or $\to_{\mathcal{R}}^{\Sigma,\mathcal{R}}$ is complete (Birkhoff's completeness theorem). Furthermore, all relations $\to_A$ induce the same congruence relation on $(\Sigma, \mathcal{R})$-terms.

Proposition 12. *Let $\to_a$ and $\to_b$ be two binary relations such that $\to_a \subseteq \to_b$ and $\stackrel{*}{\leftrightarrow}_a = \stackrel{*}{\leftrightarrow}_b$.*

- *If $\to_a$ is confluent then $\to_b$ is confluent.*
- *If $\to_b$ is noetherian then $\to_a$ is noetherian.*

Proposition 13. *Let $t, t' \in \mathcal{T}_{\Sigma,\mathcal{R}}(\mathcal{V})$, $u \in Occ(t)$ and σ be a $(\Sigma, \mathcal{R})$-substitution:*
(Stability) $\quad t \stackrel{*}{\to}_{\mathcal{R}}^{\Sigma,\mathcal{R}} t'$ *implies* $\sigma(t) \stackrel{*}{\to}_{\mathcal{R}}^{\Sigma,\mathcal{R}} \sigma(t')$.
(Monotonicity) $t/u \stackrel{*}{\to}_{\mathcal{R}}^{\Sigma,\mathcal{R}} t'$ *implies* $t \stackrel{*}{\to}_{\mathcal{R}}^{\Sigma,\mathcal{R}} t[u \leftarrow t']$.

Theorem 14 (Critical pair lemma for $\to_{\mathcal{R}}^{\Sigma,\mathcal{R}}$). $\to_{\mathcal{R}}^{\Sigma,\mathcal{R}}$ *is locally confluent iff all $(\Sigma, \mathcal{R})$-overlaps of $\mathcal{R}$ converge.*

Sketch of Proof. The proof is similar to the unsorted one given in [Huet 80], see [Werner 91] for details. The fact that $t \in \mathcal{T}_{\Sigma,\mathcal{R}}(\mathcal{V})_s$ and $t \stackrel{*}{\leftrightarrow}_{\mathcal{R}}^{\Sigma,\mathcal{R}} t'$ implies $t' \in \mathcal{T}_{\Sigma,\mathcal{R}}(\mathcal{V})_s$ is important. □

There are two remaining problems:

- In general, there need not to be a *finite* complete set of $(\Sigma, \mathcal{R})$-unifiers for two $(\Sigma, \mathcal{R})$-unifiable terms. See Example 5.
- In general, it is not decidable whether a given mapping is a $(\Sigma, \mathcal{R})$-substitution as we will prove in Theorem 26. Therefore, rewriting and the computation of overlaps are not decidable.

The following example shows that $(\Sigma, \mathcal{R})$-unification is infinitary:

Example 5. $Pos < Nat$. $s(Nat) : Pos$, $0 : Nat$, $mod_2(Nat) : Nat$.
$$\text{Rules}: \quad \pi_1 : mod_2(0) \to 0,$$
$$\pi_2 : mod_2(s(0)) \to s(0),$$
$$\pi_3 : mod_2(s(s(x_{Nat}))) \to mod_2(x_{Nat}).$$

Consider the following unification problem: $y_{Pos} \doteq mod_2(z_{Nat})$. $\{y_{Pos} \mapsto mod_2(s^{2n+1}(0)), z_{Nat} \mapsto s^{2n+1}(0)\}$ is for each n a $(\Sigma, \mathcal{R})$-unifiers since $mod_2(s^{2n+1}(0)) \stackrel{*}{\Rightarrow}_{\mathcal{R}}^{\Sigma} s(0)$. But $\{y_{Pos} \mapsto mod_2(z_{Nat})\}$ is not a $(\Sigma, \mathcal{R})$-unifier.

5 Using T-Substitutions

5.1 Basic Results

The problem of unifying two $(\Sigma, \mathcal{R})$-terms with $(\Sigma, \mathcal{R})$-substitutions is closely related to E-unification. Therefore, we now define substitutions which are of more syntactical nature.

The idea is to use the range sort of the top symbol of a term to decide whether it may be assigned to a variable. Consider Example 3 again. If $n > 0$ then the range sort of the top symbol of $s^n(y_{Int} * y_{Int})$ is Nat. Hence, using the idea above $|s^n(y_{Int} * y_{Int})|$ can be reduced to $s^n(y_{Int} * y_{Int})$. In the remaining case $n = 0$, we add $\langle sq(y_{Int}), |y_{Int} * y_{Int}| \rangle$ to the set of critical pairs.

To keep rewriting decidable and unification finitary (see [Werner 93] for details), we require that every function symbol has for every arity n a uniquely determined range sort:

Definition 15. An order-sorted signature Σ is called *range unique* if $f(s_1, \ldots, s_n) : s$ and $f(s'_1, \ldots, s'_n) : s'$ implies $s = s'$.

In Subsection 5.3 we show how to transform arbitrary signatures to range-unique ones. Note that every range unique signature is regular.

Definition 16. Let Σ be a range unique signature. A *T-substitution* is an extended substitution $\sigma : \mathcal{V} \to \mathcal{T}_{\Sigma,\mathcal{R}}(\mathcal{V})$ such that:
- $\sigma(x_s) = f(t_1, \ldots, t_n)$ implies there is a s' with $f(\ldots) : s'$ and $s' \leq s$,
- $\sigma(x_s) = y_{s'}$ implies $s' \leq s$.

For instance, in Example 3 $\{x_{Nat} \mapsto s^{n+1}(y_{Int} * y_{Int})\}$ is a T-substitution, whereas $\{x_{Nat} \mapsto y_{Int} * y_{Int}\}$ is not a T-substitution.

Lemma 17. *Let Σ be a range unique signature. Every T-substitution is a $(\Sigma, \mathcal{R})$-substitution.*

Proof. If $\sigma(x_s)$ is a variable, then the theorem holds trivially. Therefore, let $t = \sigma(x_s) = f(t_1, \ldots, t_n)$, and let s' be the range sort of f (for arity n). We have to show $t \in \mathcal{T}_{\Sigma,\mathcal{R}}(\mathcal{V})_s$.

By assumption $t \in \mathcal{T}_{\Sigma,\mathcal{R}}(\mathcal{V})$. Hence, there is a $\bar{t} \in \mathcal{T}_\Sigma(\mathcal{V})$ such that $t \overset{*}{\underset{\mathcal{R}_\Sigma}{\Leftrightarrow}}{}^{\Sigma} \bar{t}$ (cf. remark to Definition 10). We show by induction on the length of $\overset{*}{\underset{\mathcal{R}_\Sigma}{\Leftrightarrow}}{}^{\Sigma}$ that $t \in \mathcal{T}_{\Sigma,\mathcal{R}}(\mathcal{V})_s$. If $t \in \mathcal{T}_\Sigma(\mathcal{V})$ then $t = f(t_1, \ldots, t_n) \in \mathcal{T}_\Sigma(\mathcal{V})_s$, since Σ is range unique.

Therefore, let $t \notin \mathcal{T}_\Sigma(\mathcal{V})$. By Definition 2, there exist $\bar{t} \in \mathcal{T}_\Sigma(\mathcal{V})$ and $t' \in \mathcal{T}_{\Sigma,\mathcal{R}}(\mathcal{V})$ such that $t \underset{\mathcal{R}_\Sigma}{\Leftrightarrow}{}^{\Sigma} t' \overset{*}{\underset{\mathcal{R}_\Sigma}{\Leftrightarrow}}{}^{\Sigma} \bar{t}$. Since $t \notin \mathcal{T}_\Sigma(\mathcal{V})$, the reduction step $t \underset{\mathcal{R}_\Sigma}{\Rightarrow}{}^{\Sigma} t'$ or $t' \underset{\mathcal{R}_\Sigma}{\Rightarrow}{}^{\Sigma} t$ respectively takes place at an occurrence $u \neq \epsilon$ by the remark in Subsection 2.3. Thus, t' has the form $f(t'_1, \ldots, t'_n)$. By induction hypothesis $t' \in \mathcal{T}_{\Sigma,\mathcal{R}}(\mathcal{V})_s$. Hence, $t \in \mathcal{T}_{\Sigma,\mathcal{R}}(\mathcal{V})_s$. □

Theorem 18. *Let Σ be a finite, range unique signature, the sets $\mathcal{V}_s$ be decidable, $t, t' \in \mathcal{T}_{\Sigma,\mathcal{R}}(\mathcal{V})$ and V be a finite set of variables such that $Var(t) \cup Var(t') \subseteq V \subseteq \mathcal{V}$. Then it is decidable whether there exists a T-unifier of t and t', and there exists a finite, computable, complete set of T-unifiers $CSU_T(t, t')$ on V.*

Sketch of Proof.

- If $\mathcal{C}$ is the set of maximal lower bounds of the sorts A and B then the set $\{\{x_A \mapsto z_C, y_B \mapsto z_C\} \mid C \in \mathcal{C}\}$ is a complete set of T-unifiers of x_A and y_B on $\mathcal{V} \setminus \{z_C \mid C \in \mathcal{C}\}$.
- On the other side, if we have to T-unify a variable x_s and a $(\Sigma, \mathcal{R})$-term t that is not a variable, then
  - either $\{x_s \mapsto t\}$ is a T-unifier on $\mathcal{V}$
  - or there doesn't exist a T-unifier of x_s and t (since Σ is range unique).

This discovery enables us to prove the theorem using the proofs of Lemma 8.2 and Lemma 8.3 in [Waldmann 89]. □

Definition 19 ($\to_{\mathcal{R}}^{T}$). We write $t \to_{\mathcal{R}}^{T} t'$ if $t, t' \in \mathcal{T}_{\Sigma, \mathcal{R}}(\mathcal{V})$ and $t \to_{\mathcal{R}}^{Ext}{}_{[\sigma]} t'$ for some T-substitution σ.

Recall that we determined in the last section what kind of rewriting systems we consider in this subsection. Note that $\to_{\mathcal{R}}^{\Sigma} \subseteq \to_{\mathcal{R}}^{T} \subseteq \to_{\mathcal{R}}^{\Sigma,\mathcal{R}}$. By Theorem 11, Birkhoff's completeness theorem holds for rewriting with $\to_{\mathcal{R}}^{T}$.

Proposition 20. *Let $t, t' \in \mathcal{T}_{\Sigma, \mathcal{R}}(\mathcal{V})$, $u \in Occ(t)$ and σ be a T-substitution:*
(Stability) $t \overset{*}{\to}_{\mathcal{R}}^{T} t'$ *implies* $\sigma(t) \overset{*}{\to}_{\mathcal{R}}^{T} \sigma(t')$.
(Monotonicity) $t/u \overset{*}{\to}_{\mathcal{R}}^{T} t'$ *implies* $t \overset{*}{\to}_{\mathcal{R}}^{T} t[u \leftarrow t']$.

Definition 21. Let $l_1 \to r_1$, $l_2 \to r_2 \in \mathcal{R}$ be two variable disjoint rules, u in $Occ(l_1)$ an occurrence such that l_1/u is a variable, $\sigma := \{l_1/u \mapsto l_2\}$ is a T-substitution, but $\{l_1/u \mapsto r_2\}$ is not a T-substitution. Then the pair $\langle \sigma(r_1), \sigma(l_1[u \leftarrow r_2]) \rangle$ is called T-contact (of $\mathcal{R}$).

Note, if $\mathcal{R}$ is sort decreasing then the set of T-contacts of $\mathcal{R}$ is empty.

Theorem 22 (Critical pair lemma for $\to_{\mathcal{R}}^{T}$). *Let Σ be a range unique signature. $\to_{\mathcal{R}}^{T}$ is locally confluent iff all T-overlaps and all T-contacts of $\mathcal{R}$ converge.*

Sketch of Proof. We only consider the non-trivial if-case. Let $t \to_{\mathcal{R}[l_1 \to r_1, u_1, \sigma_1]}^{T}$ $t_1 := t[u_1 \leftarrow \sigma_1(r_1)]$ and $t \to_{\mathcal{R}[l_2 \to r_2, u_2, \sigma_2]}^{T} t_2 := t[u_2 \leftarrow \sigma_2(r_2)]$.

Case 1: u_1 and u_2 are disjoint occurrences. Similar to the unsorted lemma, one uses $(t[u_1 \leftarrow \sigma_1(r_1)])[u_2 \leftarrow \sigma_2(r_2)] = (t[u_2 \leftarrow \sigma_2(r_2)])[u_1 \leftarrow \sigma_1(r_1)]$ to prove the convergence of t_1 and t_2.

Case 2: w.l.o.g. $u_2 = u_1 v$. **Case 2a:** $v \in Occ(l_1)$ and l_1/v is not a variable. In this case, to show the convergence of t_1 and t_2 one uses the convergence of the T-overlaps of $\mathcal{R}$ in the same way as in the unsorted lemma.

Case 2b: $u_2 = u_1 v_1 v_2$ and l_1/v_1 is a variable. In the unsorted lemma it can be shown that t_1 and t_2 converge without using the convergence of the overlaps of R. In the order-sorted lemma, we have to distinguish three cases:

- If $v_2 \neq \epsilon$, then t_1 and t_2 converge in the same way as in the unsorted lemma. To prove this, one uses the fact that if $\{x \mapsto t\}$ is a T-substitution and $t/u \overset{*}{\leftrightarrow}_{\mathcal{R}}^{\Sigma,\mathcal{R}} t'$ for $u \neq \epsilon$ then $\{x \mapsto t[u \leftarrow t']\}$ also is a T-substitution.

- If $v_2 = \epsilon$ and $\{l_1/v_1 \mapsto r_2\}$ is a T-substitution, then we can show the convergence of t_1 and t_2 in the same way as in the unsorted lemma.
- If $v_2 = \epsilon$ and $\{l_1/v_1 \mapsto r_2\}$ is not a T-substitution, then we use (in the same way as in *case 2a*) the convergence of the T-contacts of $\mathcal{R}$ to prove the convergence of t_1 and t_2. □

5.2 Completion

If $\to_{\mathcal{R}}^{T}$ is not locally confluent, we can try to transform it into a locally confluent relation by completion. In this subsection we discuss the particularities of completion procedures for rewriting with the relation $\to_{\mathcal{R}}^{T}$ and the differences to classical completion [Bachmair & al 86].

Given a set of equations $\mathcal{E}$ and a rewriting system $\mathcal{R}$, we consider $\mathcal{E} \cup \mathcal{R}$ to be a rewriting system (the direction of the orientation of the equations in $\mathcal{E}$ is unimportant for our purposes) for the reason of simplicity.

We start with a set of equations $\mathcal{E}_0$ and a rewriting system $\mathcal{R}_0$ both containing only Σ-terms[5]. The inference rules for completion differ mainly from the classical ones in the fact that we have to add both kind of critical pairs (T-overlaps and T-contacts) as equational consequences. The rules obtained by completion may contain ill-formed terms. But the following holds:

Theorem 23. *Let* $(\mathcal{E}_0, \mathcal{R}_0) \vdash (\mathcal{E}_1, \mathcal{R}_1) \vdash \ldots \vdash (\mathcal{E}_i, \mathcal{R}_i) \vdash \ldots$ *be a sequence of completion steps. Then all terms occurring in $\mathcal{E}_i$ and $\mathcal{R}_i$ are $(\Sigma, \mathcal{E}_0 \cup \mathcal{R}_0)$-terms. Furthermore,* $\overset{*}{\leftrightarrow}{}_{\mathcal{E}_0 \cup \mathcal{R}_0}^{\Sigma} = \overset{*}{\leftrightarrow}{}_{\mathcal{E}_i \cup \mathcal{R}_i}^{\Sigma}$.

In particular, each pair $(\mathcal{E}_i, \mathcal{R}_i)$ defines the same set of semantically well-sorted terms for each sort $s \in \mathcal{S}$. Furthermore, $(\mathcal{E}_i, \mathcal{R}_i) \vdash (\mathcal{E}_{i+1}, \mathcal{R}_{i+1})$ implies $\overset{*}{\leftrightarrow}{}_{\mathcal{E}_i \cup \mathcal{R}_i}^{T} = \overset{*}{\leftrightarrow}{}_{\mathcal{E}_{i+1} \cup \mathcal{R}_{i+1}}^{T}$.

The same results can be shown as for classical completion [Bachmair & al 86]. This is not surprising since in [Comon 91] variable-overlaps are also added as equational consequences and similar results are obtained.

Example 6. Consider Example 3 again. We have

$$|s^{n+1}(x_{Int} * x_{Int})| \quad \to_{\mathcal{R}}^{T} \quad s^{n+1}(x_{Int} * x_{Int}).$$

Furthermore, $\langle\, sq(x_{Int}), |x_{Int} * x_{Int}| \,\rangle$ is a T-contact of $\mathcal{R}$. If we add the rule $|x_{Int} * x_{Int}| \to x_{Int} * x_{Int}$ to $\mathcal{R}$ then $\to_{\mathcal{R}}^{T}$ is noetherian and confluent.

5.3 How to Transform Signatures to Range Unique Ones

An arbitrary signature $\Sigma := (\mathcal{S}, \leq, \mathcal{F})$ can be transformed to a range unique one Σ^{ru} in the following way:

---

[5] In Section 6, we explain why we use this assumption.

Step 1: If Σ is regular then let $\Sigma^{re} := \Sigma$, otherwise we transform Σ into the regular signature Σ^{re} by the method given in [With 92] (where function declarations are added on intersection sorts).

Step 2: If there is a greatest sort $\Omega \in S^{re}$, then let $\Sigma^{\Omega} := \Sigma^{re}$, otherwise let Σ^{Ω} be the signature obtained by adding the sort $\Omega \notin S^{re}$ to S^{re} with $s \leq^{\Omega} \Omega$ for all sorts s. In both cases, if for $f \in \mathcal{F}^{re}$ there is no $s' \in S^{re}$ such that for all declarations $f(s_1, \ldots, s_n) : s$ there exists a declaration $f(s'_1, \ldots, s'_n) : s'$ with $s_1 \leq^{\Omega} s'_1, \ldots, s_n \leq^{\Omega} s'_n$, and $s \leq^{\Omega} s'$ then we add for each declaration $f(s_1, \ldots, s_n) : s$ the declaration $f(s_1, \ldots, s_n) : \Omega$ to $\mathcal{F}^{re}$.

Step 3: Let Σ^{ru} be the signature obtained from Σ^{Ω} by replacing the function declaration $f(s_1, \ldots, s_n) : s$ by $f_s(s_1, \ldots, s_n) : s$ for each function symbol $f \in \mathcal{F}^{\Omega}$.

We define a mapping $T : \mathcal{T}_{\Sigma}(\mathcal{V}) \to \mathcal{T}_{\Sigma^{ru}}(\mathcal{V})$ by $T(x) = x$ for $x \in \mathcal{V}$ and $T(f(t_1, \ldots, t_n)) = f_s(T(t_1), \ldots, T(t_n))$ where s is the least sort of $f(t_1, \ldots, t_n)$ and a second mapping T^{-1} by $T^{-1}(x) = x$ for $x \in \mathcal{V}$ and $T^{-1}(f_s(t_1, \ldots, t_n)) = f(T^{-1}(t_1), \ldots, T^{-1}(t_n))$.

Furthermore, let $\mathcal{R}^{ru}$ consist of the rules $T(l) \to T(r)$ for $l \to r \in \mathcal{R}$ and the rules $f_s(x_{s''_1}, \ldots, x_{s''_n}) \to f_{s'}(x_{s''_1}, \ldots, x_{s''_n})$ for each function symbol $f \in \mathcal{F}^{\Omega}$ with declarations $f(s_1, \ldots, s_n) : s$ and $f(s'_1, \ldots, s'_n) : s'$ such that $s' \leq^{\Omega} s$ but $s' \neq s$, and s''_i is a maximal lower bound of s_i and s'_i (with respect to $\leq^{\Omega}$).

Example 7. The signature Σ defined by $S = \{Nat, Int\}$ with $Nat < Int$ and the function declarations

$||: Int \to Nat \qquad *: Int, Int \to Int, \qquad *: Nat, Nat \to Nat$

and the rule $|x_{Nat}| \to x_{Nat}$ are transformed into the signature Σ^{ru} defined by $S^{ru} = \{Nat, Int\}$ with $Nat <^{ru} Int$ and the function declarations

$||_{Int}: Int \to Nat \qquad *_{Int} : Int, Int \to Int, \qquad *_{Nat} : Nat, Nat \to Nat$

and the rules $|x_{Nat}|_{Int} \to x_{Nat}$ and $x_{Nat} *_{Int} y_{Nat} \to x_{Nat} *_{Nat} y_{Nat}$.

Note that $|x_{Nat} * x_{Nat}| \to_{\mathcal{R}}^{\Sigma} x_{Nat} * x_{Nat}$ but *not* $|x_{Int} * x_{Int}| \to_{\mathcal{R}}^{\Sigma} x_{Int} * x_{Int}$ and $|x_{Nat} *_{Int} x_{Nat}|_{Int} \to_{\mathcal{R}^{ru}}^{\Sigma^{ru}} x_{Nat} *_{Nat} x_{Nat}$ but *not* $|x_{Int} *_{Int} x_{Int}|_{Int} \to_{\mathcal{R}^{ru}}^{\Sigma^{ru}} x_{Int} *_{Int} x_{Int}$.

Proposition 24. *If there is at least one ground term $t_s \in \mathcal{T}_{\Sigma}(\mathcal{V})_s$ for every sort $s \in S$ then there is at least one ground term $t_s^{ru} \in \mathcal{T}_{\Sigma^{ru}}(\mathcal{V})_s$ for every $s \in S^{ru}$.*

Theorem 25. *Let $t_1, t_2 \in \mathcal{T}_{\Sigma}(\mathcal{V})$ and $t_1^{ru}, t_2^{ru} \in \mathcal{T}_{\Sigma^{ru}}(\mathcal{V})$ such that $T^{-1}(t_1^{ru})$, $T^{-1}(t_2^{ru}) \in \mathcal{T}_{\Sigma}(\mathcal{V})$.*

- *If $t_1 \stackrel{*}{\leftrightarrow}_{\mathcal{R}}^{\Sigma} t_2$ then $T(t_1) \stackrel{*}{\leftrightarrow}_{\mathcal{R}^{ru}}^{\Sigma^{ru}} T(t_2)$.*
- *If $t_1^{ru} \stackrel{*}{\leftrightarrow}_{\mathcal{R}^{ru}}^{\Sigma^{ru}} t_2^{ru}$ then $T^{-1}(t_1^{ru}) \stackrel{*}{\leftrightarrow}_{\mathcal{R}}^{\Sigma} T^{-1}(t_2^{ru})$*
- *$t_1^{ru} \stackrel{*}{\to}_{\mathcal{R}^{ru}}^{\Sigma^{ru}} T(T^{-1}(t_1^{ru}))$.*

In [Ganzinger 91] and [Goguen & al 85] similar techniques are used to transform order-sorted signatures and rewriting systems into many-sorted ones. Note that neither Σ^{ru} nor $\mathcal{R}^{ru}$ are many-sorted. The transformation presented above can be used if the non-overloaded semantics is considered, whereas the other ones are based on the overloaded semantics (cf. [Waldmann 92]). In the Section 7 we will discuss the differences of our approach to other ones.

6 Undecidability Results

In Section 3, we have seen that it is decidable whether a given expression belongs to $\mathcal{T}_{\Sigma,\mathcal{R}}(\mathcal{V})$ if $\rightarrow_{\mathcal{R}}^{\Sigma}$ is sort decreasing, locally confluent and noetherian, and for example $\mathcal{S}$ and $\mathcal{R}$ are finite and the sets $\mathcal{T}_{\Sigma}(\mathcal{V})_s$ are decidable. However, in general, the set $\mathcal{T}_{\Sigma,\mathcal{R}}(\mathcal{V})$ is undecidable:

Theorem 26. *There is no algorithm which, when presented*

- *a (finite) signature Σ (such that every function symbol has only one declaration)*
- *a rewriting system $\mathcal{R}$ (such that the equational theory defined by $\mathcal{R}$ is decidable)*
- *$t \in ET_{\Sigma}(\mathcal{V})$ and $s, s' \in \mathcal{S}$*

that can decide whether

- *$t \in \mathcal{T}_{\Sigma,\mathcal{R}}(\mathcal{V})_s$,*
- *$t \in \mathcal{T}_{\Sigma,\mathcal{R}}(\mathcal{V})$,*
- *$\mathcal{T}_{\Sigma,\mathcal{R}}(\mathcal{V})_s \cap \mathcal{T}_{\Sigma,\mathcal{R}}(\mathcal{V})_{s'} = \emptyset$.*

Proof. The idea is the same as the one used in [Bockmayr 87] to show that unification and matching are undecidable for canonical theories. Consider the sorts R, $Zero$ and Int: $R \leq Int$ and $Zero \leq Int$. We use the following function symbols:

$$
\begin{array}{llll}
0: & Zero & *: Int, Int & \rightarrow Int \\
su: Int \rightarrow Int & & +: Int, Int & \rightarrow Int \\
pr: Int \rightarrow Int & & -: Int & \rightarrow Int \\
f: R \rightarrow Int & & py: Int, \ldots, Int \rightarrow R
\end{array}
$$

Let $\mathcal{R}$ contain the 24 rules for integer arithmetic in [Réty & al 85] and the rule

$$py(X_1, \ldots, X_n) \rightarrow \langle \text{multivariate integer polynomial P} \rangle.$$

$\rightarrow_{\mathcal{R}}^{T}$ is locally confluent (by Theorem 22) and noetherian. Hence, $\leftrightarrow_{\mathcal{R}}^{*T}$ is decidable. P has a zero in $t_1, \ldots, t_n$ iff $py(t_1, \ldots, t_n) \xrightarrow{*T}_{\mathcal{R}} 0$. On the other hand, it is undecidable whether a multivariate integer polynomial has a zero (Hilbert's tenth problem). Hence the theorem is valid, since $0 \in \mathcal{T}_{\Sigma,\mathcal{R}}(\mathcal{V})_R$, $f(0) \in \mathcal{T}_{\Sigma,\mathcal{R}}(\mathcal{V})$ and $\mathcal{T}_{\Sigma,\mathcal{R}}(\mathcal{V})_R \cap \mathcal{T}_{\Sigma,\mathcal{R}}(\mathcal{V})_{Zero} \neq \emptyset$ iff there are $t_1, \ldots, t_n$ with $py(t_1, \ldots, t_n) \xrightarrow{*T}_{\mathcal{R}} 0$. □

The problems that arise by these undecidability results for rewriting with $\to_{\mathcal{R}}^{T}$ can be solved in the following way: Starting from well-defined $(\Sigma, \mathcal{R})$-terms, e.g. Σ-terms, all terms obtained by rewriting and completion are also $(\Sigma, \mathcal{R})$-terms by Theorem 11 and 23. In the previous sections, we showed that if we assume that all given expressions are $(\Sigma, \mathcal{R})$-terms then rewriting and computation of critical pairs are essentially the same as if we use sort decreasing rules and Σ-terms.

7 Related Works

In [Chen & Hsiang 91], new function symbols and function declarations are added to the given signature in order to get weakly sort decreasing rules. In [With 92], term declarations are used to obtain a weakly sort decreasing rewriting system. Each rewriting system of the class of rewriting systems used in the proof of Theorem 26 is noetherian and locally confluent due to Critical Pair Lemma 22. But it is not possible to transform all of them equivalently into a weakly sort decreasing one by the Theorems 8 and 26 (note that $\overset{*}{\leftrightarrow}_{\mathcal{R}}^{\Sigma} = \overset{*}{\leftrightarrow}_{\mathcal{R}}^{T}$ by Theorem 11).

C. and H. Kirchner presented a different approach. To each term a type expression is associated expressing possible types of the term which are deduced during computation. Unfortunately, the criteria given in [Kirchner & Kirchner 91] for the completeness of their unification is undecidable by Theorem 26.

In [Comon 91], sort constraints based on a fragment of second-order logic are used in order to get a critical pair lemma. Unfortunately, H. Comon's completion procedure does not terminate even in cases of termination of the other approaches.

In [Ganzinger 91], order-sorted term rewriting systems are translated into many-sorted ones. The rewriting system of Example 2 is locally confluent in our approach but it is not locally confluent if we transform it to a many-sorted one.

8 Conclusions

We demonstrated the benefits of $(\Sigma, \mathcal{R})$-terms: Birkhoff's completeness theorem always holds. Meaningful but ill-formed terms are contained in the set of $(\Sigma, \mathcal{R})$-terms. We showed some decidability and undecidability results and demonstrated how to overcome the problems.

Basically, it is advisable to use sort decreasing rewriting systems (and semantically well-sorted terms) as far as possible. Natural examples, however, show the necessity to generalize the critical pair lemma to non-sort-decreasing rewriting systems. Alternatively, the relation $\to_{\mathcal{R}}^{T}$ can be used.

The difference of our approach to other ones was shown in the last section. In [Werner 91], we carried over the completeness results for many narrowing strategies from the unsorted to the order-sorted case for both approaches.

Acknowledgments: I am grateful to Christoph Brzoska, Hubert Comon, Uwe Waldmann and Lars With for discussion and comments. I thank Michael Gollner and Steve Heracleous for reading earlier versions of this paper and the referees for their remarks.

References

[Bachmair & al 86] L. Bachmair, N. Derschowitz and J. Hsiang, Orderings for Equational Proofs, Proc. IEEE Symposium Logic in Computer Science, Cambridge, MA, 1986, pages 346 - 357.

[Bockmayr 87] A. Bockmayr: A Note on a Canonical Theory with Undecidable Unification And Matching Problem, Journal of Automated Reasoning 3, 1987, pages 379 - 381.

[Chen & Hsiang 91] H. Chen, J. Hsiang: Order-Sorted Specification and Completion, Technical Report, Department of Computer Science, SUNY at Stony Brook, 1991.

[Comon 91] H. Comon: Completion of Rewrite Systems with Membership Constraints. Research Report. CNRS & LRI, Paris. 1991.

[Ganzinger 91] H. Ganzinger: Order-Sorted Completion: The Many-Sorted way, Theoretical Computer Science 89, 1991, pages 3 - 32.

[Goguen & al 85] J.A. Goguen, J.P. Jouannaud and J. Meseguer: Operational Semantics for Order-Sorted Algebra, Proc. of the 12th ICALP, LNCS 194, Springer-Verlag, 1985, pages 221-231.

[Goguen & Diaconescu 92] J.A. Goguen, R. Diaconescu: A Short Survey of Order Sorted Algebra, ETACS Bulletin, Number 49, 1992, pages 121 - 133.

[Huet 80] G. Huet: Confluent Reductions: Abstract Properties and Application to Term Rewrite Systems, Journal of ACM 27 (1980), pages 797 - 821.

[Huet & Oppen 80] G. Huet and D.C. Oppen: Equations and Rewrite Rules: A Survey. In R. Book, editor, Formal Languages: Perspectives and Open Problems, Academic Press, 1980, pages 349 - 405.

[Kfoury & al 82] A.J. Kfoury, R.N. Moll, M.A. Arbib: A Programming Approach to Computability, Springer-Verlag, 1982.

[Kirchner & Kirchner 91] C. Kirchner and H. Kirchner: Order-Sorted Rewriting Computation in G-algebra. Technical Report. INRIA-Lorraine & CRIN, Nancy. 1992.

[Réty & al 85] P. Réty, C. Kirchner, H. Kirchner, P. Lescanne: NARROWER: A New Algorithm for Unification and its Application to Logic Programming, Proc. of 1st RTA, Dijon, France, LNCS 202, Springer-Verlag, pages 141 - 157.

[Smolka & al 87] G. Smolka, W. Nutt, J.A. Goguen and J. Meseguer: Order-Sorted Equational Computation, in: H. Ait-Kaci, M. Nivat: Resolution Of Equations In Algebraic Structure, volume 2, pages 297 - 367, Academic Press 1989.

[Waldmann 89] U. Waldmann: Unification in Order-Sorted Signatures, Forschungsbericht Nr. 298, Universität Dortmund, 1989.

[Waldmann 92] U. Waldmann: Semantics in Order-Sorted Specifications, Theoretical Computer Science 94, 1992, pages 1 - 33.

[Werner 91] A. Werner: Termersetzung und Narrowing mit geordneten Sorten, Diplomarbeit, Fakultät für Informatik, Universität Karlsruhe, July 1991.

[Werner 93] A. Werner: A Semantic Approach to Order-Sorted Rewriting, Interner Bericht 5/93 , Fakultät für Informatik, Universität Karlsruhe, 1993.

[With 92] L. With: Completeness and Confluence of Order-Sorted Term Rewriting, Proc. of 3rd CTRS, 1992, LNCS 656, Springer-Verlag.

Distributing equational theorem proving

J. Avenhaus · J. Denzinger
Fachbereich Informatik, Universität Kaiserslautern
6750 Kaiserslautern
{avenhaus , denzinge}@informatik.uni-kl.de

Abstract

In this paper we show that distributing the theorem proving task to several experts is a promising idea. We describe the team work method which allows the experts to compete for a while and then to cooperate. In the cooperation phase the best results derived in the competition phase are collected and the less important results are forgotten. We describe some useful experts and explain in detail how they work together. We establish fairness criteria and so prove the distributed system to be both complete and correct. We have implemented our system and show by non-trivial examples that drastical time speed-ups are possible for a cooperating team of experts compared to the time needed by the best expert in the team.

1 Introduction

The success of general theorem provers is limited by the fact that even for relatively simple problems the search space for finding a proof becomes too large. There are several possibilities to deal with this problem. On the syntactic level one may restrict the search space by designing powerful inference rules or by imposing order restrictions. On the semantic level one may incorporate domain specific knowledge and proof plans. On the machine level one may use parallelism. In this paper we propose the team work method, it easily allows one to combine these ideas.

The basic idea of the team work method is as follows: There is a supervisor that activates severals experts – running on different processors – to work on the given problem. Each expert is a prover by itself, it may focus on a subproblem (for example on a special part of the database) and follow its own heuristics. So for a given amount of time the experts work independently and compete. When this time is elapsed the supervisor stops the experts and calls the referees to judge the work done by the experts. Based on the referee reports one of the experts is declared to be the winner. The referees determine the best results of the losers and send these results to the winner. Now the supervisor creates a new team of experts and starts a new round. The whole process stops as soon as one expert has found a proof. This approach seems to be very flexible: If the supervisor has some knowledge on the problem he can decompose the problem into subproblems, select domain specific experts and combine their results. This allows one to execute a given proof plan efficiently. On the other side, if only little is known about the given problem, the supervisor may start a standard team of experts and after each round exchange those experts by others that did not contribute to solving the problem. In this case it is indispensible to forget those results that will not help to find the proof. Otherwise, the database would explode and the "wrong" experts would prevent the system to find a proof. It is one task of referees to extract the useful results derived by the experts.

It is the aim of this paper to make these ideas precise and to study the problems that come along with the approach. We believe that the approach is useful for a wide class of theorem provers based on inference rules that generate new and simplify old facts (e.g. resolution based provers), but we restrict in this paper to pure equational reasoning. To be precise, the underlying prover is the unfailing Knuth-Bendix completion procedure [BDP89]. In this case we are able to present different useful experts. We analyse the tasks of the supervisor and the referees and we discuss communication problems. We give simple criteria to guarantee fairness, they imply that the whole distributed system is both correct and complete. Experiments show that the approach is promising. For example, we prove the ring example of Stickel [St84] (a ring satisfying $x^3 = x$ is commutative) without AC-unification and AC-rewriting with a team consisting of two experts in 308 seconds. Here the best expert alone needs 5153 seconds. This speed-up factor much greater than two for a team of two experts is not unusual as other examples show.

The paper is organized as follows: In section 2 we review the proof method unfailing completion. In section 3 we describe the team work method in detail and we give conditions to guarantee fairness in section 4. We discuss several experts and the tasks of the referees and the supervisor in section 5. In section 6 we discuss some examples and prove that remarkable speed-ups are possible for team experts working together. Finally, in section 7 we relate our approach to some of those known in the literature. Because of lack of space we omit all proofs. For details see [De93] and [AD93].

2 Unfailing completion as the basic proof procedure

We apply the team work method outlined in the introduction to purely equational reasoning. So we are interested in the following problem:

Input: E, a set of equations over a fixed signature sig; $s = t$, an equation over sig
Question: Does $s = t$ hold in every model of E ?

Let $Th(E)$ denote the set of equations over sig that hold in every model of E. By Birkhoff's theorem we have $s = t \in Th(E)$ iff s can be transformed into t by *replacing equals by equals*. It is well-known that provers based on rewriting and completion techniques developed by Knuth and Bendix [KB70] are efficient for this problem. In order to avoid abortion of the completion procedure due to the fact that equations may not be orientable into rules we use the *unfailing completion procedure* of Bachmair, Dershowitz and Plaisted [BDP89] as our basic proof procedure.

We assume the reader to be familiar with rewriting and completion techniques. For an overview see [AM90] and [DJ90]. We use the standard notations. A signature $sig = (S, F, \tau)$ consists of a set S of sorts, a set F of operators and a function $\tau : F \to S^+$ that fixes the input and output sorts of the operators. Let $\mathcal{T}(F, V)$ denote the set of terms over F and a set V of variables. We write

$t[s]_p$ to denote that $s \equiv t/p$, i.e. s is the subterm of t at position p. By $\mathcal{T}(F) = \mathcal{T}(F, \emptyset)$ we denote a set of *ground terms* over F. Let K be a set of new constants. A *reduction ordering* $\succ$ is a well-founded ordering on $\mathcal{T}(F \cup K, V)$ that is compatible with substitutions and the term structure, i.e. $t_1 \succ t_2$ implies $\sigma(t_1) \succ \sigma(t_2)$ and $t[t_1]_p \succ t[t_2]_p$. If $\succ$ is total on $\mathcal{T}(F \cup K)$ then $\succ$ is called a *ground reduction ordering*.

A *rule* is an oriented equation, written $l \to r$ such that $Var(r) \subseteq Var(l)$. A set R of rules is *compatible* with $\succ$ if $l \succ r$ for every $l \to r$ in R. If E is a set of equations then $R_E = \{\sigma(u) \to \sigma(v) \mid u \doteq v \text{ in } E, \sigma \text{ a substitution}, \sigma(u) \succ \sigma(v)\}$ is the set of orientable instances of equations in E. (We use $u \doteq v$ to denote $u = v$ or $v = u$.) Finally, we have $R(E) = R \cup R_E$.

Let $u \doteq v$ and $s \doteq t$ be equations in $E \cup R$. Let u/p be a non-variable subterm of u that is unifiable with s, say with most general unifier $\sigma = mgu(u/p, s)$. Then $\sigma(u[t]_p) = \sigma(v)$ is in $Th(R \cup E)$. If $\sigma(u[t]_p) \not\succ \sigma(u)$ and and $\sigma(v) \not\succ \sigma(u)$ then $\sigma(u[t]_p) = \sigma(v)$ is a critical pair of R, E. We denote by $CP(R, E)$ the set of all critical pairs of R, E.

We are now ready to define the unfailing completion procedure. It works on triples of the form (E, R, g) and is parameterized by a ground reduction ordering $\succ$. Here E is a set of equations (originally the input), R a set of rules compatible with $\succ$ (originally empty) and g a ground equation over $F \cup K$ (originally the skolemized input goal $s = t$). The completion procedure is given by a set of inference rules and a set of fairness conditions that restrict the application of the inference rules.

Definition 2.1 (Inference system $\mathcal{U}$, see [BDP89])

(U1)   Orient an equation $(E \cup \{s \doteq t\}, R, g) \vdash_\mathcal{U} (E, R \cup \{s \to t\}, g)$ if $s \succ t$

(U2)   Deduce an equation $(E, R, g) \vdash_\mathcal{U} (E \cup \{s = t\}, R, g)$ if $s = t \in CP(R, E)$

(S1)   Delete an equation $(E \cup \{s = t\}, R, g) \vdash_\mathcal{U} (E, R, g)$ if $s \equiv t$

(S2)   Simplify an equation $(E \cup \{s \doteq t\}, R, g) \vdash_\mathcal{U} (E \cup \{u = t\}, R, g)$ if $s \longrightarrow_{R(E)} u$

(S3)   Subsume an equation $(E \cup \{s \doteq t, u \doteq v\}, R, g) \vdash_\mathcal{U} (E \cup \{s = t\}, R, g)$ if $u/p \equiv \sigma(s), v \equiv u[\sigma(t)]_p$ for some σ and position p and $u \triangleright s$

(S4)   Simplify a rule, right $(E, R \cup \{s \to t\}, g) \vdash_\mathcal{U} (E, R \cup \{s \to u\}, g)$ if $t \longrightarrow_{R(E)} u$

(S5)   Simplify a rule, left $(E, R \cup \{s \to t\}, g) \vdash_\mathcal{U} (E \cup \{s = u\}, R, g)$ if $s \longrightarrow_{R(E)} u$ using $l \to r$ and $s \triangleright l$

(G1)   Simplify the goal $(E, R, s = t) \vdash_\mathcal{U} (E, R, u = t)$ if $s \longrightarrow_{R(E)} u$

(G2)   Success $(E, R, s = t) \vdash_\mathcal{U} SUCCESS$ if $s \equiv t$

In this definition $\triangleright$ denotes the encompassment ordering. It is the strict part of the quasi-ordering defined by $s \trianglerighteq t$ iff $\sigma(t) \equiv s/p$ for some substitution σ and some position p. Notice that we have added subsumption rule (S3) that is missing in [BDP89]. This rule is indispensible for efficiency reasons. For instance, if commutative and associative operators are present it prevents an explosion of the set E.

Using the orderings $\succ$ and $\triangleright$ a proof ordering $\succ_P$ can be constructed such that the following holds (see [BDP89]): If $(E, R, g) \vdash_\mathcal{U} (E', R', g')$ and B is proof for $s = t$ in (E, R) then there is a proof B' for $s = t$ in (E', R') with $B \succeq_P B'$. In particular, if $s = t$ is in E' then $B \succeq_P B_{s,t}$ where $B_{s,t}$ is the one step proof consisting of applying the equation $s = t$.

Definition 2.2 (Fairness)
A $\mathcal{U}$-derivation is a sequence $(E_i, R_i, g_i)_{i \geq 0}$ with $(E_i, R_i, g_i) \vdash_\mathcal{U} (E_{i+1}, R_{i+1}, g_{i+1})$ for all i. It defines the sets R^∞ and E^∞ of persistent rules and equations by
$$R^\infty = \bigcup_{j \geq 0} \bigcap_{i \geq j} R_i \qquad E^\infty = \bigcup_{j \geq 0} \bigcap_{i \geq j} E_i$$
The derivation is fair if either it ends with SUCCESS or else for every critical pair $u = v$ of E^∞, R^∞ there is an $i \geq 0$ and a proof B_i for $u = v$ in (E_i, R_i) with $B_{u,v} \succeq_P B_i$.

The main theorem on unfailing completion now is ([BDP89])

Theorem 2.1 Let $(E_i, R_i, g_i)_{i \geq 0}$ be a fair $\mathcal{U}$-derivation with $(E_0, R_0, g_0) = (E, \emptyset, \overline{s} = \overline{t})$ where $\overline{s} = \overline{t}$ is the skolemized version of $s = t$. We have $s = t \in Th(E)$ iff the derivation is finite and ends with SUCCESS.

This theorem directly gives raise to a theorem prover for the problem "$s = t \in Th(E)$?" that is both, correct and complete.

Definition 2.3 (Basic prover)
A basic prover is any algorithm that with input $(E, s = t, \succ)$ produces only $\mathcal{U}$-derivations. The basic prover is fair, if it produces only fair $\mathcal{U}$-derivations.

3 Team work completion

The team work method was mainly designed to use distributed computation in situations where almost nothing is known of how to find a proof for the problem instance $(E, s = t)$. In this case the supervisor activates a team of probably good experts (basic provers) and lets them try to solve the problem independently. It hopes that at least one of the experts is well suited for the problem instance and some of the other experts deliver valuable subresults at the right time. So after a while it stops the competition phase and starts a team meeting for cooperation. Now the work of the experts has to be judged and this is the task of the referees. So the supervisor really selects a team of expert/referee pairs. Each referee gives a report on the overall behavior of its expert and selects the most important results. On the basis of this information the supervisor declares one of the experts as the winner and the selected results of the losers are sent to the winner. Using this extended database of the winner the supervisor now starts a new round of competition and cooperation. It stops all computations as soon as one proof has been found.

So the computation time is splitted into rounds. The $k - th$ round has the following form:

Cooperation: The supervisor accepts the referee reports from round $k-1$. Based on this information it determines the winner and accepts the selected results of the losers. Then it selects a new n-tuple of expert/referee pairs.

Competition: The experts work independently.

Judgement: The referees prepare their reports.

This concept sounds simple. In section 6 we will demonstrate by examples that it works and that it makes remarkable speed-ups possible. Even more, the concept seems to be very flexible. Very different sorts of knowledge can be implemented either in the supervisor (e.g. proof plans) or in the experts (e.g. domain knowledge).

Clearly, the concept can only work if

- a) the tasks of the supervisor and the referees are carefully examined
- b) useful experts are created
- c) reasonable criteria for referees to judge the work of experts are developed
- d) communication time is reduced to a minimum.

We will discuss these problems in section 5. Here we describe in more detail the general form of an expert and how the cooperation of the experts is organized by the supervisor. This will allow us to develop fairness criteria for the distributed system. This discussion is on a conceptual level to simplify proofs. For implementation aspects see section 5.

An expert is a basic prover P, so on input $(E_0, s = t)$ it produces only $\mathcal{U}$-derivations. It uses a fixed reduction ordering $\succ$ to transform an equation into a rule according to inference rule (U1). For efficiency reasons it applies the simplification rules (S1) to (S5) with highest priority. Then it computes the critical pairs as soon as possible and stores them in the set CP. This is done in order to give the heuristic of P – coded as *choose-CP* function – the chance to find a good one to be processed next. So the state of P is described by a quadruple (R, E, g, CP). Here R and E are the current sets of rules and equations, respectively, g is the current goal $\overline{s} = \overline{t}$ and CP is the set of critical pairs not processed so far. P performs a while loop with the following loop invariant: R is compatible with $\succ$ and for any equation $u = v$ in E the terms u and v are uncomparable by $\succ$. All critical pairs in $CP(R, E)$ are already computed, fully reduced by $R(E)$ and stored in CP. Within the loop the next element of CP is selected according to the *choose-CP* function and processed: It is integrated in the R- or E-component and all critical pairs with the new equation are computed. So the quadruple (R, E, g, CP) of a basic prover corresponds to the triple (E, R, g) of the inference system $\mathcal{U}$. The CP-component in the quadruple is used to keep track of the critical pairs not processed so far and for choosing a good one (according to the heuristic used) to be processed next. So the behaviour of a basic prover mainly depends on its *choose-CP* function.

A basic prover is fair if its *choose-CP* function rejects no element in CP infinitely often. The supervisor may interrupt a basic prover only at the end of a loop as described above.

The cooperation during a team meeting is organized as follows:
(1) The supervisor determines the winner of the latest round.
(2) He accepts the selected rules/equations from the losers and integrates them into the quadruple (R, E, g, CP) of the winner by processing them.
(3) The supervisor determines an n-tuple of new expert/referee pairs for the new round, including the winner. He starts the $n-1$ experts (besides the winner) with the quadruple $(\emptyset, \emptyset, g, R \cup E \cup CP)$, where (R, E, g, CP) is the quadruple of the winner.

Note that simplification, also backward simplification, is a fundamental part of our basic provers and so of the distributed system also. So we do not have the bottleneck backward subsumption as, for example, the approach of Slaney and Lusk [SL90] has.

4 Fairness

The computation in the distributed system with input $(E, s = t)$ is controlled by a team strategy S. A *team strategy* determines in a team meeting from the referee reports the winner and the n-tuple of expert/referee pairs for the next round. A team strategy is *complete* if for any input $(E, s = t)$ with $s = t \in Th(E)$ the result YES is produced. (By construction, if YES is produced then $s = t \in Th(E)$ holds. So every team strategy S is correct.) We are going to develop criteria for the completeness of a team strategy.

To do so we first extend the inference system $\mathcal{U}$ for describing sequential provers to an inference system $\mathcal{DU}$ for describing our distributed prover. Then we express completeness criteria for a team strategy by fairness criteria in $\mathcal{DU}$.

We extend $\mathcal{U}$ to $\mathcal{DU}$ by adding rules for describing the integration of the selected rules/equations of the losers into the database of the winner.

Definition 4.1 (Inference system $\mathcal{DU}$)
The inference system $\mathcal{DU}$ consists of the inference rules in $\mathcal{U}$ and the two rules

(D1)   Introduce rule $(E, R, g) \vdash_{\mathcal{DU}} (E, R \cup \{l \to r\}, g)$ *if* $l =_{E \cup R} r$ *and* $l \succ r$

(D2)   Introduce equation $(E, R, g) \vdash_{\mathcal{DU}} (E \cup \{u = v\}, R, g)$ *if* $u =_{E \cup R} v$ *and* u, v *are* $\succ$*-incomparable.*

Lemma 4.1 *Suppose the distributed system is started with input quadruple $(\emptyset, \emptyset, g, E)$ and in every round the winner uses a given reduction ordering $\succ$. Let (R_0, E_0, g_0, CP_0) be the actual quadruple of an active winner. Then we have $(E, \emptyset, g) \vdash^*_{\mathcal{DU}} (E_0 \cup CP_0, R_0, g_0)$.*

The main idea of the proof is to use the inference rules D1 and D2 to add the selected results of the losers to the system of the winner during the cooperation part of a round.

Lemma 4.1 indicates that the distributed computation can be described as a sequential computation according to the inference system $\mathcal{DU}$. The definition of fairness of a $\mathcal{DU}$-derivation is as in Definition 2.2. Now Theorem 2.1 can be carried over.

Theorem 4.1 *Let $(E_i, R_i, g_i)_{i \geq 0}$ be a fair $\mathcal{DU}$-derivation with $(E_0, R_0, g_0) = (E, \emptyset, \overline{s} = \overline{t})$. We have $s = t \in Th(E)$ iff the derivation is finite and ends with SUCCESS.*

Now we have to find fairness criteria for a team strategy $\mathcal{S}$ such that using $\mathcal{S}$ will lead to fair $\mathcal{DU}$-derivations. For an input $(E, s = t)$ the team strategy $\mathcal{S}$ may determine the basic prover P_0 as winner several times, say for the rounds $i_0, i_1, i_2, \ldots$. Let (R_j, E_j, g_j, CP_j) be the starting quadruple of P_0 in round i_j. We call $(R_j, E_j, g_j, CP_j)_{j \geq 0}$ the P_0-sequence for $\mathcal{S}$ and $(E, s = t)$. This leads us to the following fairness criteria that also weakens the restriction on the reduction ordering used by the winners.

Definition 4.2 (Fairness of a team strategy)
A team strategy $\mathcal{S}$ is fair if there is a reduction ordering $\succ$ such that (1) for the reduction ordering $\succ_i$ of the winner of the i-th round $\succ_i \subseteq \succ_{i+1} \subseteq \succ$ holds and (2) either the computation stops or there is a basic prover P_0 with an infinite P_0-sequence $(R_j, E_j, g_j, CP_j)_{j \geq 0}$ for $\mathcal{S}$ and $(E, s = t)$ such that for every critical pair $u = v$ of E^∞, R^∞ there is a j and a proof B_j for $u = v$ in (E_j, R_j) with $B_{u,v} \succeq_P B_j$.

Theorem 4.2 *Every fair team strategy is complete.*

According to Theorem 4.2 a team strategy $\mathcal{S}$ is complete if for every input $(E, s = t)$ either the computation stops or an expert P_0 becomes the winner infinitely often and for P_0 condition (2) of Definition 4.2 holds.
Note that fairness of P_0 alone is not sufficienced for condition (2). It is possible that the integration of the results of the losers leads always to critical pairs that are better rated by the *choose-CP* function of P_0 than already existing ones. Then these equations will eventually never be selected thus leading to a contradiction to condition (2). The next definition gives us conditions for P_0 that guarantee the condition (2) of Definition 4.2. Here we identify equations that are equal up to a variable renaming.

Definition 4.3 (strongly fair)
An expert P is strongly fair if there is a quasi-ordering $\leq$ on the equations such that

- *$\{e' \mid e' \leq e\}$ is finite for every equation e*
- *choose-CP(E) is a $\leq$-minimal element in E for every set E of equations*

Lemma 4.2 *Let $\mathcal{S}$ be a team strategy and $(E, s = t)$ an input. If expert P is strongly fair and appears infinitely often in the sequence of winners for $\mathcal{S}$ and $(E, s = t)$ then condition (2) of Definition 4.2 holds.*

Corollary 4.1 *Let $\mathcal{S}$ be a team strategy such that for every input $(E, s = t)$ the sequence of winners is either finite or it contains a strongly fair expert infinitely often. Then $\mathcal{S}$ is complete.*

It is easy to construct strongly fair experts. For example, the experts ADD-WEIGHT and MAX-WEIGHT discussed in the next section are strongly fair. To guarantee fairness of the team strategy, such an expert should periodically become the winner. In the meantime unfair experts may become the winner.

We can relax the condition "strongly fair" a little bit. What we really need is that the *choose-CP* function for the distinguished strongly fair expert P never rejects an equation in the CP-component infinitely often, even if not all the equations in the CP-component are generated by P itself but may be added from outside during a team meeting. There are several possibilities to guarantee this. One is indicated in Definition 4.4. Another one would be to use time stamps and let the *choose-CP* function always select the oldest equation.

Condition (1) of Definition 4.2 restricts only the reduction ordering of the winners. All the other experts in the team may use an arbitrary reduction ordering. So completeness of a team strategy is easy to achieve.

There are also ways to weaken condition (1). For further details see [AD93] and [De93]. Note that our way of proving the team work completion to be complete can easily be adapted to prove the completeness of the team work method for other theorem proving methods based on generation and simplification of facts.

5 Experts, referees, the supervisor and implementation aspects

5.1 Experts

Every expert is a basic prover P, its behavior is mainly determined by its *choose-CP* function. In this function the heuristic of P for traversing the search space is encoded. We have implemented generic experts according to the following classification

- using syntactic arguments
- focusing on subproblems by focusing on a subset of function symbols
- focusing on special aspects of the (completion) method
- focusing on goal-oriented deduction

We discuss some of them.

Syntactic arguments: Experiments show that it is often advantagous to process short critical pairs first (see [Hu80]). Generalizing this idea we define a numerical weight for each term. This leads to two very useful experts called ADD-WEIGHT and MAX-WEIGHT. They give precedence to those critical pairs that have a small sum (a small maximum) of the two terms in the pair. It turns out that these experts in general perform very differently. These experts can be created without any knowledge of the problem instance, so they can be used as a member of the standard team.

Focusing on function symbols: The expert POLYNOM-WEIGHT associates to every function symbol a polynomial and a constant to all variables and so it defines a weight for each term. To focus on the operators in $F_0 \subseteq F$, one associates small polynomials to the $f \in F_0$ and large polynomials to the $f \in$

$F - F_0$. Experience shows that this method allows a fine tuning of the search for a proof.

Focusing on the method: Sometimes it is known that a result of a subproblem is needed for the rest of the proof. In order to get that result early an unfair expert may be needed. We have implemented FORCED-DIV and PREFER-RULE. The first of these experts concentrates on a subset of the database even if there is a high risk of divergence (i.e. generating an infinite set of equations). The second expert only selects critical pairs that are orientable by its reduction ordering. We discuss the use of these experts more deeply in section 6 in combination with the examples *div* and *ring*.

Focusing on the goal: Experience shows that near the end of the proof often all needed results are already deduced but the prover can not find the final steps of the proof at this moment. To solve this problem we have created the expert GOAL-SIM. This expert defines a measure for the similarity between the goal and a critical pair. We have implemented several measures, they depend on the facts whether subterms of the goal and the whole critical pair or subterms of the pair and the whole goal are unifiable. This expert has proven to be very useful in the situation lined out above. It is comparable to the terminator in resolution based theorem provers using connection graphs (see e.g. [AO83]).

There is a wide variety to define other experts that use special knowledge to focus on parts of critical pairs. It seems also possible to learn heuristics from analogous successful proofs. The team work method provides a good basis to activate such an expert even if the risk is high that it will be unsuccessful. In this case its results are just forgotten – provided the situation is correctly analysed by the corresponding referee.

5.2 Referees

A referee has to judge the work of its expert: It has to determine the appropriateness of its expert to the given situation and it has to extract the best results derived by his expert. Without special information on the given problem instance this seems to be hard and much work is to be done in this direction. Up to now we have experimented with referees that base their judgement on statistical information.

To determine the appropriateness of an expert to the given situation the referee computes a weighted sum of the following components:

- the number of rules, equations and critical pairs generated during the latest round
- the number of reductions of the goal
- the number of reductions of rules, equations, critical pairs
- the average weight of all processed critical pairs in the latest round in relation to the last k critical pairs

The reasons for introducing the first three of these components seem to be clear. The fourth component is used to indicate whether the expert became better during the latest round.

To determine the value of a given rule/equation one can restrict the first three components to this rule/equation. So the referee computes a weighted sum for every new rule/equation it generates and delivers the best ones according to this measure.

The referee has to be fair to the expert: Experts (for example ADD-WEIGHT and GOAL-SIM) are created for totally different purposes and this has to be taken into account by the referee. This can be done by adjusting the weights for the components in the weighted sum mentioned above.

5.3 The supervisor

The supervisor is responsible for the team meetings. It
- determines the winner for the next round
- integrates the selected results from the losers into the winner's database
- determines the new n-tuple of expert/referee pairs
- determines the time for the next team meeting.

The first task is based on the referee reports about the appropriateness of the experts in the latest round. For the integration of the results of the losers see section 3.

We give some hints to create the team for the next round in the case where almost nothing is known about the problem instance. For the first rounds a standard team should be activated, including an expert of type ADD-WEIGHT or MAX-WEIGHT. Later on every expert should be activated periodically, it should replace the expert with the lowest rating. Additionally, if during a team meeting an expert gets a rating far below the others it should be replaced by another one. The details have to be fixed by the user.

To determine the length of a round the following rules have turned out to be useful. For the first rounds the length should be kept fixed. Next, since the database grows and henceforth it costs some time to find new useful results, the length of the rounds should grow linearly. Finally even faster growing is recommended, i.e. an exponential growth.

5.4 Implementation aspects

A crucial point with distributed systems is the need to reduce the communication overhead and the idle times of processors to a minimum. From the conceptual point of view the team work method takes this into account by limiting the communication to fixed events, the team meetings. We now discuss implementation aspects.

We have implemented the conceptual units expert, referee and supervisor as "quasi-processes" (see below). In order to minimize the transport of data on the net we in general do not send data to the quasi-processes but run the quasi-processes on that processor that has the data. So we always run an expert/referee pair on the same processor. The supervisor is active only during the team meetings. At the beginning the supervisor is run on the processor of the old winner. Here it determines the new winner for the next round. After that the supervisor is run on the processor of the new winner, here it integrates the results from the losers, determines the new team and sends the starting information to the processors of the other team members. Technically, we have implemented a single process with the three modes *expert*, *referee*, and *supervisor*. Now a quasi-process for an expert is just a process in mode *expert*. This trick allows one to realize the ideas developed above. We call this concept *floating control*.

To reduce idle times we interleave the tasks of the supervisor with the preprocessing of the team members: If an expert uses the same ordering as the new winner then it can accept the starting quadruple (R, E, g, CP) separated into these components. Otherwise it has to accept this information in the form $(\emptyset, \emptyset, g, R \cup E \cup CP)$. In any case it has to sort the CP-component according to its *choose-CP* function and that costs more time than sending data. So the supervisor first sends the CP-component of the winner without the results of the losers, it then processes the results of the losers and then sends this information to the other team members. So the time for processing the results of the old losers can be used by the new experts to preprocess their input data.

We have implemented our team work completion in C under UNIX on a cluster of SUN ELC machines. Unfortunately, up to now we have implemented the communication by message passing for a cluster of two machines only. This is the basis for the results reported in section 6. An implementation of broadcasting allowing for bigger clusters is under way.

6 Results

We will demonstrate the usefulness of the team work method on five examples from different areas of equational reasoning. Each team consists of two experts that work together. In Table 1 we compare the run time needed by the team with the sequential run time of each member of the team. The speed-up factor is the time needed by the best of the two experts divided by the time needed by the team.

The run times given in the table include the communication overhead and the idle times. So it is the time the user has to wait for the proof. For the sequential prover this is very close to the CPU-time.

example	team	1st expert	2nd expert	speed-up
Z22	5.032	16.241	39.760	3.2
div	2.813	34.698	–	12.3
luka1	15.044	95.407	40.908	2.7
luka2	13.518	23730.000	81.383	6.0
ring	307.962	–	5153.000	16.7

Table 1: run-time comparison team *vs* sequential experts (in seconds)

Before we comment on these results we will give brief descriptions of the examples and the teams used.

Example Z22:

Input:
$$\begin{aligned}
a(b(c(x))) &= d(x) & b(c(d(x))) &= e(x) \\
c(d(e(x))) &= a(x) & d(e(a(x))) &= b(x) \\
e(a(b(x))) &= c(x) & a(a1(x)) &= x
\end{aligned}$$

$a1(a(x))$	=	x	$b(b1(x))$	=	x	$b1(b(x))$	=	x
$c(c1(x))$	=	x	$c1(c(x))$	=	x	$d(d1(x))$	=	x
$d1(d(x))$	=	x	$e(e1(x))$	=	x	$e1(e(x))$	=	x

Ordering: LPO with precedence $e1 > e > d1 > d > c1 > c > b1 > b > a1 > a$

Task: Complete system

Team: expert1: POLYNOM-WEIGHT
expert2: MAX-WEIGHT

The example Z22 was brought to our attention by J. A. Kalman during the CADE-10 conference. The completion of the equational system shows that the equations represent the cyclic group of order 22 (therefore the name Z22).

The system is completed by our team in two rounds. The winner of the first round is MAX-WEIGHT. POLYNOM-WEIGHT (that assigns in this example to all function symbols polynomials of the form $x + c_f$ with c_f a positive number) finishes the completion in the second round. The speed-up is due to the change of heuristic for choosing critical pairs because all rules selected from the results of POLYNOM-WEIGHT after the first round were already in the set of rules of MAX-WEIGHT.

Example div:

Input:
$f(g(f(x))) = g(f(x))$
$h(f(g(x))) = c(e)$
$b(c(d(a^4(x)))) = a^2(b(c(a^2(x))))$
$c(d(b(a^3(x)))) = a^3(x)$
$a^8(x) = c(x)$
$b^7(x) = a(x)$

Ordering: Knuth-Bendix ordering KBO with weight 1 for all symbols and precedence $h > f > g > a > b > c > d > e$

Task: Prove $c(d(b(c(e)))) = h(g^{20}(f(e)))$

Team: expert1: POLYNOM-WEIGHT
expert2: FORCED-DIV

This example shows the advantages of focusing on different parts of the set of equations. Only using the first two equations of the input $h(g^{20}(f(e))) = c(e)$ can be proved. Only using the last 4 equations $c(d(b(c(e)))) = c(e)$ can be proved. The expert FORCED-DIV can prove the right side of the goal in approx. 2 seconds and POLYNOM-WEIGHT, again only using polynomials of the form $x + c_f$ as interpretations, with big c_f values for the symbols f, g and h needs the same time to prove the left side. So, after a round of 2 seconds, the expert POLYNOM-WEIGHT is the winner and gets from FORCED-DIV the rule $h(g^{20}(f(e))) \to c(e)$, which is considered very good by its referee, because it can reduce the goal. As POLYNOM-WEIGHT has already found the rule $c(d(b(c(e)))) \to c(e)$ the proof is finished.

All experts, except FORCED-DIV, generate the rule $h(g^{20}(f(e))) \to c(e)$ very late, because it is big. They concentrate mainly on the consequences of the last four input equations. Therefore they need much time until they can complete the proof (ADD-WEIGHT, MAX-WEIGHT or GOAL-SIM need the same or more time as POLYNOM-WEIGHT). On the other hand, FORCED-DIV concentrates on the divergence $f(g^i(f(x))) \to g^i(f(x))$ and therefore neglects the other equations. The cooperation forced by the team work method leads to an enormous speed-up by combining the strengths of both experts.

Example luka1 and luka2:

Input: $C(T,x) = x \quad C(x,C(y,x)) = T \quad C(x,N(N(x))) = T$
$C(C(x,y),C(N(y),N(x))) = T \quad C(C(x,C(y,z)),C(C(x,y),C(x,z))) = T$
$C(N(N(x)),x) = T \quad C(C(x,C(y,z)),C(y,C(x,z))) = T$

Ordering: LPO with precedence $C > N > T > p > q > r$

Task: luka1: Prove $C(C(p,q),C(C(q,r),C(p,r))) = T$
luka2: Prove $C(C(N(p),p),p) = T$

Team: luka1: expert1: ADD-WEIGHT
expert2: GOAL-SIM
luka2: expert1: POLYNOM-WEIGHT
expert2: MAX-WEIGHT

The examples luka1 and luka2 are taken from [Ta56]. The input equations are an equational axiomatization for propositional calculus by Frege. Lukasiewicz gave another set of axioms of which luka1 and luka2 are the first two.

Fair sequential basic provers have problems with these examples in so far as they simply try to complete the set of input equations. The goals do not influence the computation. This is also one of the major critisims on completion based equational theorem proving. But in our team work approach there are many concepts that force the team to concentrate on the given goal. For example, the referees take into account in their judgements reductions of the goal. Further we can include heuristics that concentrate on the goal. They are not fair, but a team strategy using them can be fair. For luka1 the winner of the first round is ADD-WEIGHT. No result of GOAL-SIM is integrated in the winning system. But in the second round GOAL-SIM completes the proof. Again, the change of the heuristic is responsible for the speed-up. GOAL-SIM is not able to generate the facts it needs for appropriate use of its heuristic. This is done by ADD-WEIGHT. For luka2 we have the same situation. POLYNOM-WEIGHT wins the first round, while MAX-WEIGHT finds no good results. But in the second round MAX-WEIGHT finishes the proof.

Example ring:

Input:

$j(0,x) = x \quad j(x,0) = x$
$j(x,g(x)) = 0 \quad j(j(x,y),z) = j(x,j(y,z))$
$f(f(x,y),z) = f(x,f(y,z)) \quad f(x,j(y,z)) = j(f(x,y),f(x,z))$
$j(g(x),x) = 0$
$j(x,y) = j(y,x)$
$f(j(x,y),z) = j(f(x,z),f(y,z))$

Ordering: KBO with weights
$\varphi(f) = 5, \quad \varphi(j) = 4, \quad \varphi(g) = 3$
$\varphi(0) = 1, \quad \varphi(b) = 1, \quad \varphi(a) = 1$
and precedence $f > j > g > 0 > b > a$

Task: Prove $f(a,b) = f(b,a)$

Team: expert1: PREFER-RULE
expert2: ADD-WEIGHT

This example is mentioned as a challenging problem in [St84]. Reported automated proofs were obtained by using completion prover with build-in theory AC. However, our team does not use build-in theories. It needs 5 rounds to find the proof. The winner of each round is PREFER-RULE, but the proof is completed by ADD-WEIGHT. After the first round the referee of ADD-WEIGHT selects the two equations $j(x, j(y, z)) = j(y, j(z, x))$ and $j(x, j(y, z)) = j(z, j(y, x))$ that are added to the system of PREFER-RULE. Although PREFER-RULE selects no critical pairs that can not be oriented, results of other experts are considered. These equations are necessary, because they introduce the commutativity of j in the system of PREFER-RULE. (Note that although $j(x, y) = j(y, x)$ is an input equation, it will not be selected by PREFER-RULE !) The proof can only be found by ADD-WEIGHT, because the commutativity of f is needed, which will never be selected by PREFER-RULE.

So we have achieved speed-ups in very different areas – semi-thue systems (coded here by monadic function symbols), ring theory, equational propositional logic – for both completion and proving tasks. The combination of different heuristics in a competitive but also cooperative way leads to speed-ups that are more than linear in comparison to the sequential heuristics used in the team. The example ring suggests that there may be other and better ways to deal with theories than using the expensive theory completion.

Note that the sequential run times are fast. Our sequential prover can compete with such systems as OTTER or REVEAL on these examples. Therefore the speed-ups of the teams are not due to weakness of our sequential prover.

7 Related work

In the literature several attempts are reported to use parallel or distributed computing to enhance the power of theorem proving. They differ in the granularity of the parallelisation and in the degree of cooperation.

Yelick and Garland [GY92] use a very fine granularity of parallelism. Their approach is based on the inference system of Bachmair, Dershowitz and Hsiang [BDH86] for the Knuth-Bendix completion procedure and the parallelisation takes place on the level of these inference rules. According to our experience this granularity is too fine. There is no aspect of cooperation and competition discussed in the paper.

On the contrary, Ertel [Er90] uses a very coarse granularity of parallelism. He uses a tableaux-based theorem prover and observes that for a fixed input problem and a fixed strategy the running time may heavily depend on the order the input data are given to the prover. So he starts parallel computations with several, randomly generated, permutations of the input data and stops as soon as one processor has found a proof. Also every decision a prover has to make is done randomly. In this approach there is no cooperation between the provers, they only compete.

Slaney and Lusk [SL90] have proposed to use parallelism to compute the clousure of a set of facts (i.e. clauses) under some inference rules. The processors share a common memory and the inferences are distributed among the processors such that each processor gets assigned facts. Then the processor generates all inferences of its facts with all other facts. The drawback with this approach is that backward subsumption (and simplification) is costly. Here only parallelism is used, there is no cooperation or competition between the processors.

The DARES system ([CMM90]) does not only distribute the process of generating new facts, but also the initial facts are distributed among the processors. Then requests have to be started to get new facts from other processors. For starting and answering such requests DARES uses heuristics and no central control is needed. In DARES the different behavior of the problem solving nodes is only achieved by different facts. No further control knowledge, like in our team work method, is used.

In a recent paper of Bonacina and Hsiang [BH92] the problem of how to guarantee fairness in distributed automatic deduction is studied. Each processor p_k has stored in its own memory at time i the set S_i^k of facts. So at time i the whole data basis consists of $S_i^1 \cup \cdots \cup S_i^n$. Each processor works on its own data and simultaneously sends messages to draw inferences between his and foreign data. Criteria are developed that guarantee fairness (and so completeness) of the whole system. The problems here arise from the fact that the data base is distributed. In our approach there is a common data base for all experts whenever a new round is started. This simplifies the problem to guarantee fairness. We believe that it also saves unproductive time for communication.

References

[AD93] Avenhaus, J., Denzinger, J.: Distributing equational theorem proving, *to appear as SEKI-Report, Universität Kaiserslautern, 1993*.

[AM90] Avenhaus, J., Madlener, K.: Term Rewriting and Equational Reasoning, in R.B. Banerji (ed): *Formal Techniques in Artificial Intelligence, Elsevier, 1990, pp. 1-43*.

[AO83] Antoniou, G., Ohlbach, H.J.: Terminator, *Proc. 8th IJCAI, Karlsruhe, 1983*.

[BDH86] Bachmair, L., Dershowitz, N., Hsiang, J.: Orderings for equational proofs, *Proc. Symposium on Logic in Computer Science, 1986, pp. 346-357*.

[BDP89] Bachmair, L., Dershowitz, N., Plaisted, D.A.: Completion without Failure, *Coll. on the Resolution of Equations in Algebraic Structures, Austin (1987), Academic Press, 1989*.

[BH92] Bonacina, M.P., Hsiang, J.: On fairness in distributed automated deduction, *to be published*.

[CMM90] Conry, S.E.; MacIntosh, D.J., Meyer, R.A.: DARES: A Distributed Automated REasoning System, *Proc. AAAI-90, 1990, pp. 78-85*.

[De93] Denzinger, J.: TEAMWORK: A method to design distributed knowledge based equational theorem provers, *forthcoming Ph.D. thesis, University of Kaiserslautern, 1993*.

[DJ90] Dershowitz, N., Jouannaud, J.P.: Rewriting systems, in J. van Leeuwen (Ed.): *Handbook of theoretical computer science, Vol. B., Elsevier, 1990, pp. 241-320*.

[Er90] Ertel, W.: Random Competition: A Simple, but Efficient Method for Parallelizing Inference Systems, *Int. Report TUM-19050, Technical University of Munich, 1990*.

[GY92] Garland, S.J., Yelick, K.A.: A Parallel Completion procedure for Term Rewriting Systems, *Proc. 11th CADE, 1992, pp. 109-123*.

[Hu80] Huet, G.: Confluent Reductions: Abstract Properties and Applications to Term Rewriting Systems, *Journal of ACM, Vol. 27, No. 4, 1980, pp. 798-821*.

[KB70] Knuth, D.E., Bendix, P.B.: Simple Word Problems in Universal Algebra, *Computational Algebra, J. Leech, Pergamon Press, 1970, pp. 263-297*.

[SL90] Slaney, J.K., Lusk, E.L.: Parallelizing the Closure Computation in Automated Deduction, *Proc. 10th CADE, LNAI 449, Springer, Kaiserslautern, 1990, pp. 28-39*.

[St84] Stickel, M.E.: A Case Study of Theorem Proving by the Knuth-Bendix Method: Discovering that $x^3 = x$ implies Ring Commutativity, *Proc. CADE-7, LNCS 170, Springer, 1984, pp. 248-258*.

[Ta56] Tarski, A.: Logic, Semantics, Meta mathematics, *Oxford University Press, 1956*.

On the Correctness of a Distributed Memory Gröbner Basis Algorithm*

Soumen Chakrabarti and Katherine Yelick

University of California, Berkeley, CA 94720, USA

Abstract. We present an asynchronous MIMD algorithm for Gröbner basis computation. The algorithm is based on the well-known sequential algorithm of Buchberger. Two factors make the correctness of our algorithm nontrivial: the *nondeterminism* that is inherent with asynchronous parallelism, and the *distribution* of data structures which leads to inconsistent views of the global state of the system. We demonstrate that by describing the algorithm as a nondeterministic sequential algorithm, and presenting the optimized parallel algorithm through a series of refinements to that algorithm, the algorithm is easier to understand and the correctness proof becomes manageable. The proof does, however, rely on algebraic properties of the polynomials in the computation, and does not follow directly from the proof of Buchberger's algorithm.

1 Introduction

Buchberger introduced the notion of a *Gröbner basis* of a set of polynomials and presented an algorithm for computing it [4]. We present an algorithm based on his for computing Gröbner bases on a MIMD distributed memory multiprocessor.

Although somewhat controversial [12], Buchberger and others believe that interreduction (keeping the basis reduced with respect to itself) is essential to performance. Our algorithm executes interreduction steps concurrently with the standard critical pair and reduction steps. We believe this is the first attempt at computing Gröbner bases in parallel in an asynchronous message passing framework while performing interreduction. The completion method used in the Gröbner basis computation is typical of other completion procedures, so we expect our design techniques to have wider application.

In this paper, we focus on the question of correctness of the algorithm and how it is affected by parallelization. We identify some of the key points here.

- The proofs rely on algebraic properties of polynomials, rather than being a direct proof that the parallel program is equivalent to the sequential one. The parallelization is not a simple semantics-preserving transformation on the sequential program.

---

* This work was supported in part by the Advanced Research Projects Agency of the Department of Defense monitored by the Office of Naval Research under contract DABT63-92-C-0026, by AT&T, and by a National Science Foundation through an Infrastructure Grant (number CDA-8722788) and Research Initiation Award (number CCR-9210260). The information presented here does not necessarily reflect the position or the policy of the Government and no official endorsement should be inferred.

- The proofs are structured around distributed data structures. While a single data structure may be quite complicated internally, its value is abstracted in the proof to a single shared object which does not necessarily exist in the computation.
- Interreduction complicates the parallel algorithm and its proof. Without interreduction, the basis grows monotonically, but with interreduction, elements may be modified and deleted. Thus the algorithm has to ensure correctness in the presence of multiple, *inconsistent* copies of the basis.
- Finally, the design extends the transition-based approach [21] to distributed memory machines.

We have reported on engineering issues and more extensively on performance elsewhere [8]. The algorithm has been implemented on a CM-5 multiprocessor. It outperforms previous parallel algorithms on shared memory machines.

This paper is organized as follows. §2 gives background definitions. §3 presents the parallel algorithm and correctness proof, using a succession of refinements. §4 gives performance numbers, and §5 discusses the relation to the Knuth-Bendix procedure. We finish with some concluding remarks in §6.

2 Notation

In this section we briefly introduce some notation. A more detailed treatment can be found in [14].

Let K be a field and $x_1, ..., x_n$ be variables, arbitrarily ordered as $x_1 > x_2 > \cdots > x_n$. Then $\mathcal{K} = K[x_1, ..., x_n]$ defines a ring of polynomials under standard polynomial arithmetic. A total order $\succ$ on monomials is *admissible* if for all monomials a, p, q it satisfies (1) $p \succeq 1$ (note that $1 = x_1^0 \cdots x_n^0$) and (2) $p \succeq q \Rightarrow ap \succeq aq$. Also, $p \succ q$ iff $p \succeq q$ and $p \neq q$.

When written in decreasing order of monomials, TERM(p, i) denotes the i-th term of polynomial p ($i \geq 1$). A term contains the *coefficient* and the *monomial*: TERM(p, i) = COEF$(p, i) \times$ MONO(p, i). The *head term* of a polynomial p is the leading term: HTERM(p) = TERM$(p, 1)$. Similarly, HCOEF(p) = COEF$(p, 1)$ and HMONO(p) = MONO$(p, 1)$. HMONO, HCOEF and HTERM are naturally extended to sets of polynomials: HMONO(S) = {HMONO(p) : $p \in S$}, etc. The admissible ordering $\succ$ is extended to polynomials by defining $p \succ q$ iff HMONO$(p) \succ$ HMONO(q), and $p \succeq q$ iff HMONO$(p) \succeq$ HMONO(q).

Given polynomials p and r such that HMONO(r) divides MONO(p, i) for some i, *reduction* of p by r is defined as:

$$p' = p - \frac{\text{TERM}(p, i)}{\text{HTERM}(r)} \times r. \tag{1}$$

Note that TERM(p, i) vanishes out of p'. Reduction by a *set* S of polynomials is done by repeatedly reducing p by some element of S. When no element of S can reduce p, it is *irreducible* or in *normal form*, also written NORMAL(p, S). The collection of all possible normal forms of p when reduced by S is denoted

$\mathrm{NF}_S(p)$. The zero polynomial, 0, is in normal form with respect to any S. The *highest common factor* of two monomials is denoted

$$\mathrm{HCF}(x_1^{i_1}\cdots x_n^{i_n}, x_1^{j_1}\cdots x_n^{j_n}) = x_1^{\min(i_1,j_1)}\cdots x_n^{\min(i_n,j_n)}. \tag{2}$$

Given polynomials p_1 and p_2, with head terms $k_1 m_1$ and $k_2 m_2$ respectively, their *s-polynomial* is given by

$$\mathrm{SPOL}(p_1, p_2) = p_1 \frac{k_2 m_2}{\mathrm{HCF}(m_1, m_2)} - p_2 \frac{k_1 m_1}{\mathrm{HCF}(m_1, m_2)}. \tag{3}$$

The *ideal* generated by a set S of polynomials is denoted by $\mathrm{IDEAL}(S)$.
Given a set P of polynomials, a *Gröbner basis* of P is a set G of polynomials satisfying the following:
- $\mathrm{IDEAL}(G) = \mathrm{IDEAL}(P)$ and
- For each $p \in \mathrm{IDEAL}(P)$, $\mathrm{NF}_G(p) = \{0\}$.

A survey of the theory can be found in Mishra [14]. Parallel implementations have been surveyed by Vidal [19]. Earlier network implementations have been reported by Siegl [17], Attardi *et al* [1] and Hawley [13].

3 Algorithm Design

In this section we develop the parallel algorithm, starting from the sequential algorithm. Correctness is proved at each step as part of the design process. §3.1 reviews Buchberger's algorithm without interreduction; §3.2 gives a nondeterministic version, which is essentially a parallel algorithm with atomic operations on shared data structures; §3.3 extends the nondeterministic algorithm to handle interreduction; §3.4 presents the distributed algorithm.

3.1 Sequential Algorithm

Figure 1(a) shows Buchberger's sequential algorithm, called S. The two main data structures are G (the basis) and gpq (the set of pairs for SPOL computation). For simplicity, this version is without interreduction: polynomials entering G are completely reduced with respect to all previous elements in G, but old basis elements are not checked for reducibility by new entrants. The effect is that polynomials that have entered the basis once are never modified or deleted. We enhance the algorithm with interreduction after we refine S to a transition axiom form.

A correctness proof of S is given by Mishra and Yap [14]; we sketch their proof of partial correctness and give a different proof of termination. Our proof generalizes better to the interreducing algorithm to be described later.

Theorem 1 (Buchberger). *G is a Gröbner basis iff for all $f, g \in G$, $0 \in \mathrm{NF}_G(\mathrm{SPOL}(f, g))$ ([14], Theorem 5.8).*

Definition 2. *A ring R is defined to be Noetherian iff it has no infinite ascending sequence $R_1 \subset R_2 \subset R_3 \subset \ldots$ of ideals of R.*

(a)	(b)

Input: F, a finite set of polynomials.
Initially:
$\quad G = F$
$\quad gpq = \{ \{f,g\} : f,g \in G \}$
while $gpq \neq \emptyset$ {
$\quad$ let $\{f,g\}$ be any pair in gpq
$\quad gpq = gpq \setminus \{\{f,g\}\}$
$\quad h = \text{SPOL}(f,g)$
$\quad h' = \text{REDUCE}(h,G)$
$\quad$ if $h' \neq 0$ {
$\quad\quad gpq = gpq \cup \{\{f,h'\} : f \in G\}$
$\quad\quad G = G \cup h'$
$\quad$ }
}
	Input: F, a finite set of polynomials.
	Initially:
	$\quad grq = \emptyset$, $G = F$,
	$\quad gpq = \{ \{f,g\} : f,g \in G \}$.
	S-POLYNOMIAL
	$\quad \exists \{p,q\} \in gpq \Rightarrow$
	$\quad\quad gpq = gpq \setminus \{p,q\}$
	$\quad\quad grq = grq \cup \{\langle p,q,\text{SPOL}(p,q)\rangle\}$
	AUGMENT BASIS
	$\quad \exists \langle p,q,r\rangle \in grq : \text{NORMAL}(r,G),\ r \neq 0 \Rightarrow$
	$\quad\quad grq = grq \setminus \{\langle p,q,r\rangle\}$
	$\quad\quad gpq = gpq \cup \{\ \{s,r\},\ s \in G\ \}$
	$\quad\quad G = G \cup \{r\}$
	REDUCE
	$\quad \exists \langle p,q,r\rangle \in grq : \neg \text{NORMAL}(r,G) \Rightarrow$
	$\quad\quad r = \text{REDUCE}(r,G)$

Fig. 1. (a) Sequential Algorithm S [Buchberger]. G is initialized to the input set F and grows to become a Gröbner basis. Elements in G are never modified. gpq is the set of pairs of polynomials. The function REDUCE(h,G) returns some element $h' \in \text{NF}_G(h)$, i.e., it reduces h completely to normal form. (b) G-1: Transition Axiom formulation with one copy of G. Data structures G and gpq as before. Unlike in Algorithm S, REDUCE(r,G) need not return a normal form; a partially reduced form will do. Note that p,q are not needed in grq; they just help write cleaner invariants in the correctness proof.

Theorem 3 (Hilbert's Basis Theorem). *If R is a Noetherian ring then so is $R[x_1, x_2, \ldots, x_n]$ ([15], Pages 420–425).*

Lemma 4. *Algorithm S terminates with G a Gröbner basis of F.*

Proof. For partial correctness, observe the loop invariant

$$\forall\, p,q \in G,\ \{p,q\} \notin gpq \ \Rightarrow\ 0 \in \text{NF}_G(\text{SPOL}(p,q)). \tag{4}$$

If S terminates, $gpq = \emptyset$, so G is Gröbner by theorem 1. For termination, note that REDUCE is a terminating computation [14] and consider tuples of the form $\langle M, p\rangle$, built of an ideal M over $\mathcal{K}$ and integer $p \geq 0$, ordered lexicographically as $\langle M_1, p_1\rangle \sqsupset \langle M_2, p_2\rangle$ iff

$$M_1 \subset M_2 \ \text{or}\ \left(M_1 = M_2 \ \text{and}\ p_1 > p_2 \right). \tag{5}$$

Each loop iteration of S reduces the tuple $\left\langle \text{IDEAL}(\text{HMONO}(G)), |gpq| \right\rangle$, as can be verified easily by examining each axiom. (See Dershowitz and Manna [10] for similar termination proving techniques.)

Algebraic optimizations to the basic algorithm have been developed that test s-polynomials to quickly detect reduction to zero, without actually performing the reduction [5]. Although our implementation includes such improvements, we omit them from the proofs for simplicity.

3.2 Transition Axiom Specification

Transition axioms are a means to exploit non-determinism in a sequential algorithm description. Inspired by guarded command languages [9, 11], and augmented by linearizable data types [21], this style was used to implement a shared-memory Knuth-Bendix procedure [22]. Transition axioms help break the computation into independently schedulable chunks, so the scheduling decisions are deferred until late in the design process. They are written in the form $C \Rightarrow A$ where C is the *enabling condition* (a guard predicate) and A is the action. An execution proceeds by repeatedly *firing* enabled axioms nondeterministically. Termination occurs when none of the axioms can be fired. Parallelism results from being able to overlap axioms in time on multiple processors.

There are two sources of non-determinism in S.

- Reduction has many degrees of freedom, since the choice of a reducer is not specified. Also, it is not required to reduce the argument polynomial completely to normal form with respect to the reducing set; any positive number of reduction steps will do.
- The choice of a pair from *gpq* to compute the SPOL is not specified (although selection heuristics affect performance). Thus one can work on several pairs simultaneously.

Algorithm G-1 in figure 1(b) is the result of rewriting S as a transition axiom specification. There are three data structures: G is the growing basis, *gpq* is the pair set as before and *grq* is a temporary set of polynomials in some stage of being reduced[2]. We now prove that G-1 correctly computes a Gröbner basis.

Definition 5. Let S, T be finite multisets of polynomials with $|S| = |T|$. Define the irreflexive, non-symmetric and transitive ordering $\triangleright$ between such multisets as $S \triangleright T$ iff there is a bijection $\sigma : S \to T$ such that $\forall s \in S : s \succeq \sigma(s)$ and $\exists s \in S : s \succ \sigma(s)$.

Clearly, $\triangleright$ is Noetherian. We use it to show that G-1 terminates.

Lemma 6. *Algorithm G-1 terminates with G a Gröbner basis of F.*

Proof. For partial correctness, we give an "axiom invariant" that is true *between* any two successive rules in the firing sequence, specifically, $\forall f, g \in G$:

$$\Big(\{f,g\} \in \mathit{gpq}\Big) \text{ or } \Big(\exists r \neq 0 : \langle f, g, r \rangle \in \mathit{grq}\Big) \text{ or } \Big(0 \in \mathrm{NF}_G(\mathrm{SPOL}(f,g))\Big). \quad (6)$$

---

[2] An explanation of names: *g* for global, meaning they are shared by all processors; *p* for pairs and *r* for reducts; *q* for queues because of the heuristic ordering on monomials in *gpq* and *grq*.

The result follows from theorem 1 and the observation that when all the guards are false, $gpq = \emptyset$ and $\langle f, g, r\rangle \in grq \Rightarrow r = 0$. For termination, consider tuples of the form $\langle M, p, R\rangle$ constructed as in the proof of S, but with the additional field R, a multiset of polynomials. Tuples are ordered lexicographically as before, but with R ordered by $\triangleright$. Then the tuple $\langle \text{IDEAL}(\text{HMONO}(G)), |gpq|, grq\rangle$ is reduced upon firing any axiom.

Input: F, a finite set of polynomials.
Initially:
  $grq = \emptyset$, $G = F$,
  $gpq = \{\ \{f, g\} : f, g \in G\}$.
S-POLYNOMIAL
  $\exists \{p, q\} \in gpq \Rightarrow$
    $gpq = gpq \setminus \{p, q\}$
    $grq = grq \cup \{\langle p, q, \text{SPOL}(p, q)\rangle\}$
REDUCE
  $\exists \langle p, q, r\rangle \in grq : \neg\text{NORMAL}(r, G) \Rightarrow$
    $r = \text{REDUCE}(r, G)$

AUGMENT BASIS
  $\exists \langle p, q, r\rangle \in grq : \text{NORMAL}(r, G), r \neq 0 \Rightarrow$
    $grq = grq \setminus \{\langle p, q, r\rangle\}$
    $gpq = gpq \cup \{\ \{s, r\},\ s \in G\ \}$
    $G = G \cup \{r\}$
INTERREDUCE
  $\exists p, q \in G :\ q$ reduces $p \Rightarrow$
    $p' = \text{REDUCE}(p, \{q\})$
    $G = (G \setminus \{p\}) \cup \{p'\}$
    $gpq = gpq \cup \{\ \{p', g\} : g \in G,\ g \neq p'\}$

Fig. 2. IG-1: Transition Axiom formulation for interreduction with one copy of G. INTERREDUCE might reduce a basis element to zero; we assume for simplicity that zero elements are left around in G but are never considered as reducers.

3.3 Interreduction

We have parallelized and distributed S without interreduction [7]. The proof of that algorithm is a special case of the algorithm with interreduction: interreduction introduces mutation of polynomials in the basis. We present only the more general case here, and therefore proceed by introducing interreduction into our nondeterministic algorithm.

Buchberger describes an elaborate way to keep track of polynomials that become reducible each time the basis grows, so that after each addition the basis is *interreduced*, i.e., basis polynomials are sequentially reduced by each other until nothing more can be reduced [4]. In a parallel algorithm, this global interreduction could potentially change a polynomial used by any other transition axiom, yet we cannot afford to stop other work while interreduction proceeds. We therefore introduce a single interreduction step as a separate transition axiom.

Figure 2 shows the transition axioms IG-1 for interreducing Gröbner basis computation. The only modification is the addition of INTERREDUCE. As in G-1, there is one shared copy of G. While the extent of reduction done in REDUCE is not specified, in the proof we assume only a single reduction step occurs in INTERREDUCE. It follows that correctness is preserved if INTERREDUCE were to reduce multiple steps, which is done in the implementation. The additional properties to be proved for IG-1 are that interreduction maintains the axiom invariant, and does not destroy progress.

Lemma 7. *Let* INTERREDUCE, *be invoked on* G_1, *resulting in* G_2. *Then for any polynomial* p, *if* $0 \in \mathrm{NF}_{G_1}(p)$ *then* $0 \in \mathrm{NF}_{G_2}(p)$.

Proof. There must be $g, h \in G_1$ such that h reduces g for INTERREDUCE to be enabled. Suppose $g \xrightarrow{h} g'$. Let $0 \in \mathrm{NF}_{G_1}(p)$. We need to show that $0 \in \mathrm{NF}_{G_2}(p)$. Consider the reduction sequence $p \to \cdots p_i \to p_{i+1} \cdots \to 0$ in G_1. If there is no reduction by g there is nothing to prove, so suppose the $p_i \to p_{i+1}$ reduction is by g, denoted $p_i \xrightarrow{g} p_{i+1}$. In G_2 we can achieve the same reduction of p_i to p_{i+1} in two steps:

$$p_i \xrightarrow{h} p_{\mathrm{new}} \xrightarrow{g'} p_{i+1}. \tag{7}$$

We can do this for all steps that used g as a reducer to get a reduction sequence using only reducers from G_2. Thus, $0 \in \mathrm{NF}_{G_2}(p)$.

Lemma 8. *Let* INTERREDUCE *be invoked on* G_1, *resulting in* G_2. *Then*

$$\mathrm{IDEAL}(\mathrm{HMONO}(G_1)) \subseteq \mathrm{IDEAL}(\mathrm{HMONO}(G_2)).$$

Proof. For INTERREDUCE to be enabled, $\exists g, h \in G_1$ such that h reduces g to g'. If $\mathrm{HMONO}(g)$ is unaffected there is nothing to prove, so suppose $\mathrm{HMONO}(h)$ divides $\mathrm{HMONO}(g)$. So, $\mathrm{IDEAL}(\mathrm{HMONO}(G_1 \setminus \{g\})) = \mathrm{IDEAL}(\mathrm{HMONO}(G_1))$. Thus, $\mathrm{IDEAL}(\mathrm{HMONO}(G_2)) = \mathrm{IDEAL}\left(\mathrm{HMONO}(G_1 \setminus \{g\} \cup \{g'\})\right) \supseteq \mathrm{IDEAL}(\mathrm{HMONO}(G_1 \setminus \{g\}))$.

Lemma 9. *Algorithm IG-1 terminates with* G *a Gröbner basis of* F.

Proof. Partial correctness is direct from lemma 6 and lemma 7. For termination, we augment our proof for G-1. Consider tuples of the form $\langle M, S, p, R \rangle$ as in the proof of Lemma 6, but with the additional field S, a set of polynomials. Tuples are lexicographically ordered as before, with S ordered by $\triangleright$. We can show that firing any axiom in IG-1 reduces the tuple

$$\left\langle \mathrm{IDEAL}\left(\mathrm{HMONO}(G)\right), G, |gpq|, grq \right\rangle. \tag{8}$$

3.4 Replicating the Basis

For a distributed memory algorithm, it is not realistic to assume that processors always have consistent copies of shared data. Replication may occur either on a large scale by replicating an entire data structure, or on a small scale by keeping temporary copies of individual pointers and values. A consistency problem arises when any of the replicated values may be mutated.

In the Gröbner basis computation, the most important data structure in question is the basis, since it is shared most extensively. The basis could be distributed by partitioning or replication, but a pragmatic analysis of load balance, granularity and communication requirements [8] favor replication. Given a replicated basis, we have to address the problem of maintaining consistency without introducing excessive overhead. Fortunately, the consistency requirement on the basis is rather lax: a processor can do significant amounts of useful work while having an incomplete or even inconsistent copy of the basis.

Allowing Inconsistent Copies

An example of a consistency problem that may occur is the following "race condition" [16]. Suppose processors P_1 and P_2 both have copies of polynomials g, h, which happen to be equal. INTERREDUCE fires on P_1 and P_2. Say g is reduced by h to 0 on P_1. Processor P_2 does not modify its copy of g, instead it reduces h by g to 0. Subsequent invalidation messages lead both processors to discard their copies of g and h, possibly destroying the correctness of the solution. A solution to this special case is to impose a total order AGE on polynomials such that if $f = g$, f is allowed to reduce g to 0 only if the order is favorable.

In general, a stronger check is needed, namely, the total order should be used whenever the *head monomials* of the reducer and the reduced are equal, even if they are not completely equal. It is easy to verify that this check prevents the particular error indicated, but it is still non-trivial to show correctness in general.

Version Sequences

To keep track of mutable basis polynomials we introduce the notion of the *version sequence* of a polynomial. Suppose a polynomial $p(0)$ enters the basis, and is successively reduced by $r_1, r_2, \ldots, r_t$ to $p(1), p(2), \ldots, p(t)$. We represent this life history by the notation

$$p = \left[p(0) \xrightarrow{r_1} p(1) \xrightarrow{r_2} p(2) \xrightarrow{r_3} \ldots \xrightarrow{r_t} p(t) \right], \qquad (9)$$

where p represents the version sequence and $p(t)$ represents the t-th version of p. In our implementation, version sequences are identified by unique ID's.

The Model

Let each processor i have access to a (possibly inconsistent) local copy G_i of the basis, $1 \leq i \leq P$. In addition, we have a global *shadow set* G' which contains version sequences of all polynomials that ever entered the basis. Also, for ease of both implementation and proof, we assume each polynomial is *owned* by its creator processor which thereafter is the *only* processor authorized to mutate the polynomial[3].

When processor i creates a new polynomial f and value $f(0)$, it creates a new version sequence (which, by abuse of notation, we also call f) $f = [f(0)]$ in G'. When processor i modifies an owned polynomial $f(t)$ to $f(t+1)$ (as a result of reducing by $h_{t+1}(e_{t+1})$: version e_{t+1} of polynomial h_{t+1}) it appends the new value to the version sequence f in G', changing it to

$$\left[f(0) \xrightarrow{h_1(e_1)} \ldots \xrightarrow{h_t(e_t)} f(t) \xrightarrow{h_{t+1}(e_{t+1})} f(t+1) \right]. \qquad (10)$$

---

[3] This is not a serious limitation. The alternative is to associate modify locks with each polynomial.

> **Input:** F, a finite set of polynomials.
> **Initially:**
> $grq = \emptyset$,
> $gpq = \{ \{f,g\} : f(0), g(0) \in F \}$.
> $\forall i : 1 \leq i \leq P, G_i = F$
> $G' = \{[f(0)] : f(0) \in F\}$
> Processor i, $1 \leq i \leq P$.
> VALIDATE
> $(gpq \neq \emptyset$ or $grq \neq \emptyset)$ and $\exists g(t) \in G' : \forall g(\ell) \in G_i, \ell < t \Rightarrow$
> $G_i = \left(G_i \setminus \bigcup_{0 \leq e < t} g(e)\right) \cup g(t)$
> S-POLYNOMIAL
> $\exists \{f,g\} \in gpq : f^*, g^* \in G_i \Rightarrow$
> $gpq = gpq \setminus \{ \{f,g\} \}$
> $grq = grq \cup \{\langle f^*, g^*, \text{SPOL}(f^*, g^*)\rangle\}$
> AUGMENT BASIS AND INVALIDATE
> $G_i = \{g^* : g \in G'\}, \exists \langle p,q,r \rangle \in grq : r \neq 0$, and NORMAL$(r, G_i) \Rightarrow$
> $grq = grq \setminus \{\langle p,q,r\rangle\}$
>   Create unique ID h for r, so that $h(0) = r$
> $G' = G' \cup \{[h(0)]\}$ /* create new version sequence */
> $gpq = gpq \cup \{ \{g,h\}, g^* \in G_i \}$
> $G_i = G_i \cup \{r\}$
> REDUCE
> $\exists \langle p,q,r\rangle \in grq : \neg\text{NORMAL}(r, G_i) \Rightarrow$
> $r = \text{REDUCE}(r, G_i)$
> INTERREDUCE AND INVALIDATE
> $\exists f^* = f(t), h(e) \in G_i : h(e)$ reduces f^*, f^* owned,
> $(\text{HMONO}(f^*) \neq \text{HMONO}(h(e))$ or $\text{AGE}(f^*) > \text{AGE}(h(e))) \Rightarrow$
> $f(t+1) = \text{REDUCE}(f(t), \{h(e)\})$
> $G' = (G' \setminus \{f\}) \cup \{[f \xrightarrow{h(e)} f(t+1)]\}$ /* append latest version */
> $gpq = gpq \cup \{ \{f,g\} : g \in G' \}$

Fig. 3. IG-P: Transition axioms for interreduction using P copies of G. Note that since gpq now contains ID's, not polynomial values, we effectively generate new polynomial pairs in INTERREDUCE by requiring in the guard of S-POLYNOMIAL that the polynomials in G_i are the latest.

The local copy G_i consists of a selection of versions from a subset of version sequences in G'. A validation operation either puts the first element of a new version sequence in G_i or replaces version $g(t)$ from a version sequence g by $g(t+\ell)$, $\ell > 0$. The latest element in a version sequence f at a given time is special; we call it f^*.

As before, we will need to define an abstract basis $\mathcal{G}$ in terms of the physical data structures. The following definition will serve our purpose.

$$\mathcal{G} = \left\{ g^* : g \in G' \right\}. \qquad (11)$$

Using this model, we now write the transition axioms IG-P in figure 3. As mentioned before, VALIDATE picks some polynomial in the system of which processor i has no copy or a stale copy, and gets a copy or advances to a later version. Invalidation has two forms. When AUGMENT BASIS fires, processor i adds a new version sequence in G'; when INTERREDUCE fires, processor i appends the new version to the extant sequence. REDUCE and S-POLYNOMIAL are as before. Each polynomial (alias version sequence) has a unique ID from a totally ordered set (integer in our implementation) which will be used by AGE to break reduction loops as mentioned before.

The act of copying a version from a version sequence in G' to G_i models the communication step to update processor i's copy of the basis. The value of a polynomial cannot be used unless this is performed. However, we can still manipulate the ID, like putting it into gpq as in INTERREDUCE. ID's are very lightweight (8 bytes) compared to the polynomials they represent (hundreds to thousands of bytes). Hence communicating ID's is faster and cheaper than transporting polynomials. This has guided the formulation of the model.

Lemma 10. *If f reduces g then $g \succeq f$.*

Lemma 11. *Let an invocation of INTERREDUCE modify the basis from $\mathcal{G}_1$ to $\mathcal{G}_2$. For any polynomial y, if $0 \in \mathrm{NF}_{\mathcal{G}_1}(y)$ then $0 \in \mathrm{NF}_{\mathcal{G}_2}(y)$.*

Proof. Suppose $0 \in \mathrm{NF}_{\mathcal{G}_1}(y)$. We need to show that $0 \in \mathrm{NF}_{\mathcal{G}_2}(y)$. Say the invocation of INTERREDUCE reduces $f_1(t_1)$ to $f_1(t_1 + 1)$. Suppose $y_0 = y \to \cdots y_i \to y_{i+1} \cdots \to 0 = y_j$. The problem is that all reducers in the above reduction chain, even though elements in $\mathcal{G}_1$, may not be in $\mathcal{G}_2$ (using $f_1(t_1)$ as a reducer, for example).

We demonstrate how to replace *one* occurrence of a non-latest element in the reducers by latest elements alone. Since the reduction sequence is finite, we can replace such occurrences one by one.

We can use lemma 7 to replace old versions of reducers by new ones. Suppose G' contains the interreduction step

$$f_i(j) \xrightarrow{h_{ij}(e_{ij})} f_i(j+1). \tag{12}$$

Then any reduction step $y_{k-1} \xrightarrow{f_i(j)} y_k$ can be replaced as in the proof of lemma 7 by the equivalent computation

$$y_{k-1} \xrightarrow{h_{ij}(e_{ij})} y_{\mathrm{new}} \xrightarrow{f_i(j+1)} y_k. \tag{13}$$

This seems closer to our goal: $f_i(j+1)$ is closer to f^* than $f_i(j)$, and we have brought in a different element from a finite set. We can continue this until all reducers that take x to x' are in $\mathcal{G}_2$. This gives rise to a transformation tree: latest version polynomials are leaf nodes. A non-latest reducer $f_i(j)$ has children $h_{ij}(e_{ij})$ and $f_i(j+1)$. How are we guaranteed that the tree is finite? Define the lexicographic extension of ordering $\succeq$ and the ordering imposed by AGE on polynomials as $p > q$ if $p \succ q$ or $\mathrm{HMONO}(p) = \mathrm{HMONO}(q)$ and $\mathrm{AGE}(p) > \mathrm{AGE}(q)$. Without loss of generality, let the root of some subtree be $f_{i_1}(j_1)$. Since the tree is heap ordered with respect to $>$ defined above (with the root as the greatest element), $f_{i_1}(j_1)$ cannot occur anywhere in the subtree. Since the total number of versions of all polynomials in the system is finite, the result follows.

Lemma 12. *Let an invocation of* INTERREDUCE *modify the basis from* $\mathcal{G}_1$ *to* $\mathcal{G}_2$. *Then* IDEAL(HMONO($\mathcal{G}_1$)) $\subseteq$ IDEAL(HMONO($\mathcal{G}_2$)).

Proof. Similar to the proof of lemma 11. Suppose INTERREDUCE performed the reduction
$$f_1(t_1) \longrightarrow f_1(t_1 + 1). \tag{14}$$
If the reducer is in $\mathcal{G}_2$ there is nothing more to prove. Suppose the reducer is not a final element, i.e., there is a reduction
$$f_i(j) \stackrel{h_{ij}(e_{ij})}{\longrightarrow} f_i(j+1) \tag{15}$$
so that HMONO($h_{ij}(e_{ij})$) divides HMONO($f_i(j)$). In that case
$$\text{IDEAL}\left(M \cup \{\text{HMONO}(f_i(j))\}\right) \subseteq \text{IDEAL}\left(M \cup \{\text{HMONO}(h_{ij}(e_{ij}))\}\right) \tag{16}$$
for any set of monomials M. Continue till a final version reducer is encountered. Since versions are drawn from a finite set and AGE prevents repeating reducers, we must reach a final version reducer.

Lemma 13. *IG-P terminates, computing a Gröbner basis of* F.

Proof. Partial correctness follows from lemma 9 and 11. For termination, we adapt the proof of lemma 9, using lemma 12 and replacing G by $\mathcal{G}$ in the tuple in lemma 9.

4 Implementation and Performance

Our prototype runs on the CM-5 multiprocessor [6]. Each processor is a 33 MHz (15–20 MIPS) Sparc with about 8 MB of memory. The network is a fat-tree supporting at most 20 MB/s point-to-point data transfer. The prototype is in C, with the *active message* layer [20] for communication. For the benchmarks we used, there was no remarkable difference in performance with and without interreduction. The results are quoted without interreduction. Also, we reduced only head terms in REDUCE, not all terms. This also produces a Gröbner basis, but not necessarily a unique one. In Figure 4 some of the speedups on the CM-5 multiprocessor are given. The first set (a) is done on a small number of processors using some standards benchmarks [19].

Scalability

The standard benchmarks complete in a few seconds. Scalability is better than some previous shared memory implementations, but is still limited by the small total number of tasks (pairs added to the pair queue). To see if this is a fundamental limiting factor, we synthesized problems that lead to a large

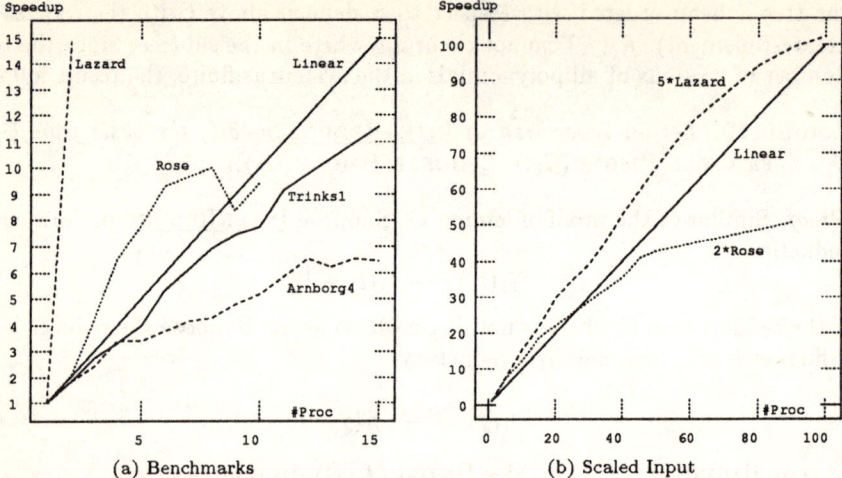

Fig. 4. Speedups for (a) standard benchmarks on a few processors and (b) Inputs created with multiple copies of a benchmark on many processors. X-axis: number of processors; Y-axis: ratio of 1-processor running time and P-processor running time. Superlinear behavior is seen in some cases.

number of tasks, using multiple copies (two and five, respectively) of the standard benchmarks, with variables renamed between the copies. The results are shown in figure 4(b). We also proved a geometry theorem using Gröbner basis that was too large to be run sequentially.

Even though the algorithm has good time scalability for long running problems, the indiscriminate replication makes it scale poorly in space. We came across a few examples that are extremely long-running, but replication exhausts memory. It is clear from our analysis [8] that time scalability is favored by replication. Solving large real problems seems to need a compromise with partitioning. We are designing a general object library that permits replication as far as memory capacity permits, thus making the compromise on a continuum.

5 Applications to Term Rewriting

Many of the parallelization techniques and correctness results in this paper could be applied to Knuth-Bendix and other completion procedures. A precursor to this work was a shared memory implementation of the Knuth-Bendix procedure, which was also based on a transition axiom style [22]. See also [18] for a generic parallel completion procedure and [3] for some related correctness results for distributed computations. The pragmatic question of whether other completion procedures would perform well on distributed memory machines is beyond the scope of this paper. However, in this section we discuss the ways in which the algorithms and proofs could be extended to a distributed memory Knuth-Bendix procedure.

Proving partial correctness of a distributed Knuth-Bendix procedure follows roughly the same lines as for Gröbner basis. However, the completion problem for term rewriting systems is undecidable. For some sets of rewrite rules, no finite complete system exists. The termination requirement for Gröbner basis is therefore replaced by a *liveness* condition, stating that the procedure must continually make progress towards a complete system. The procedure is usable as a semi-decision procedure, in that any equational theorem must eventually be provable by rewriting. In addition, the Knuth-Bendix procedure may fail. Failure also impacts the correctness criteria for the procedure, although incorporating it into the proofs should be straightforward. Interreduction is a performance improvement in both procedures, and is not essential for correctness. Whereas its value is arguable in the Gröbner basis computation, it is considered essential in Knuth-Bendix.

The basic outline of the liveness proofs could also be extended. Note that the nontermination property of Knuth-Bendix creates a subtle distinction in distributing the two computations. Hilbert's Basis Theorem guarantees that *all* increasing chains of ideals were finite, so we could have relaxed the guard of the AUGMENT BASIS AND INVALIDATE axiom in Figure 3 by not requiring the local copy G_i to be up to date. This would still terminate, but would not necessarily be practical. A Knuth-Bendix procedure must have regularly scheduled validations, since there is no analog to Hilbert's Theorem for term rewriting systems. Completing a subset of rewrite rules could lead to nonterminating executing, forever missing some critical pair. Requiring validations at all add points, as our Gröbner basis algorithm does, is sufficient in either domain, and a less stringent policy might also be possible.

In spite of these differences, the similarity between the two sequential procedures carries over to the distributed case. The race condition mentioned in §3.4, that comes with parallel interreduction, also exists for rewrite rules: two copies of the same rule can be used to reduce one another so that both disappear. It also has a similar solution, in that rewrite rules can be time stamped to prevent reduction cycles. Informally, the analog to Lemma 11 says that one can reduce using out-of-date copies of the rewrite rule set, since any reductions done there could have been performed with the latest set. Similarly, the liveness argument is analogous to termination for Gröbner basis. However, Hilbert's Basis Theorem would be replaced by the proof ordering notion of Bachmair *et al* as the basic measure of progress [2]. The proof ordering results are already quite general, giving the correctness of nondeterministic algorithm with interreduction, similar to IG-1 here.

6 Future work and Conclusion

In this paper, we have described the design and implementation of a parallel Gröbner basis procedure. We believe that current performance can be further improved in the following ways.

- As mentioned in §4, a replicated basis favors scalability in terms of achievable speedup. For large problems, it is not practical to maintain complete copies at all processors. We are implementing a generic library for application

level caching of data structures with some weak consistency models that are profitable for the application.
- After each interreduction step reducing p to p', the algorithm has to add pairs involving p' (to maintain the correctness invariant). In the sequential algorithm (where a random access on the pair "queue" is assumed), one can also remove all pairs involving p, for efficiency reasons. This is not feasible in a distributed memory setting with high communication expense. Are there efficient techniques to reorganize the distributed pair queue? Are there algebraic properties that obviate adding all new pairs with p'?
- In our design, the limit to granularity is a reduction step. This appeared reasonable for the target architecture. A general parallelization recipe for a variety of architectures will be useful. In particular, vectorizing the infinite precision coefficient computations should improve absolute performance.

In conclusion, we have presented a distributed memory MIMD algorithm for computing Gröbner basis. Our implementation out-performs the shared memory implementation of Vidal [19] fairly consistently, and has the additional advantage that shared memory hardware is not assumed. The transition-based approach, previously used for shared memory [21], is extended here for distributed memory. The key idea is to replace the shared data structures with distributed data structures, for which replication and partitioning of the data is hidden. The encapsulation of distributed objects and the structure provided by the transition axioms helps in both the algorithm presentation and in the correctness proof, and we believe it will be useful in other problems that have irregular patterns of communication and control.

Acknowledgements

Steve Schwab provided the packages for *bignum* and polynomial arithmetic and a shared memory Gröbner basis program developed at CMU. Chih-Po Wen contributed a distributed memory task queue package. We are also grateful to the referees for their review and comments.

References

1. G. Attardi and C. Traverso. A Network Implementation of Buchberger Algorithm. Technical Report 1177, University di Pisa, January 1991.
2. L. Bachmair, N. Dershowitz, and J. Hsiang. Orderings for Equational Proofs. In *Proceedings of the Symposium on Logic in Computer Science*, pages 346–357. IEEE, 1986.
3. M. P. Bonacina. *Distributed Automated Deduction*. PhD thesis, Department of Computer Science, SUNY at Stony Brook, December 1992.
4. B. Buchberger. Gröbner basis: an algorithmic method in polynomial ideal theory. In N. K. Bose, editor, *Multidimensional Systems Theory*, chapter 6, pages 184–232. D. Reidel Publishing Company, 1985.
5. B. Buchberger. A Criterion for detecting Unnecessary Reductions in the construction of Gröbner Bases. In *Proceedings of the EUROSAM '79, An International Symposium on Symbolic and Algebraic Manipulation*, pages 3–21, Marseille, France, June 1979.

6. N. J. Burnett. The Architecture of the CM-5. In *IEEE Colloquium on 'Medium Grain Distributed Computing' (Digest 070)*, pages 1–2, London, 26 March 1992.
7. S. Chakrabarti. A distributed memory Gröbner basis algorithm. Master's thesis, University of California, Berkeley, December 1992.
8. S. Chakrabarti and K. Yelick. Implementing an Irregular Application on a Distributed Memory Multiprocessor. In *Principles and Practices of Parallel Programming*, May 1993.
9. K. M. Chandy and J. Misra. *Parallel Program Design : a Foundation*. Addison-Wesley Publishing Company, Reading, Mass., 1988.
10. N. Dershowitz and Z. Manna. Proving Termination with Multiset Orderings. *Communications of the ACM*, 22:465–476, 1979.
11. E. W. Dijkstra. *A Discipline of Programming*. Prentice-Hall, 1976.
12. A. Giovini, T. Mora, G. Niesi, L. Robbiano, and C. Traverso. "One sugar cube, please" OR Selection strategies in the Buchberger algorithm. In *Proceedings of the 1991 International Symposium on Symbolic and Algebraic Computation*, pages 49–54, Bonn, Germany, 15–17 July 1992.
13. D. J. Hawley. A Buchberger algorithm for Distributed Memory Multi-processors. In *Proceedings of the 1st International ACPC Conference on Parallel Computation*, pages 385–390, Salzburg, Austria, 30 September – 2 October 1991. Springer-Verlag.
14. B. Mishra and C. Yap. Notes on Gröbner basis. In *Information Sciences 48*, pages 219–252. Elsevier Science Publishing Company, 1989.
15. Nathan Jacobson. *Basic Algebra — Volume 2*. W. H. Freeman and Company, New York, 1989.
16. C. G. Ponder. Evaluation of "performance enhancements" in algebraic manipulation systems. Technical Report UCB/CSD 88/438, University of California, Berkeley, 1988. Chapter 7, Parallel Algorithms for Gröbner Basis Reduction.
17. K. Siegl. Parallel Gröbner basis computation in ||MAPLE||. Technical Report 92-11, Research Institute for Symbolic Computation, Linz, Austria, 1992.
18. J. K. Slaney and E. W. Lusk. Parallelizing the Closure Computation in Automated Deduction. In *Proceedings of the 10th International Conference on Automated Deduction*, pages 28–29. Springer-Verlag, LNCS 449, 1990.
19. J.-P. Vidal. The computation of Gröbner bases on a shared memory multiprocessor. Technical Report CMU-CS-90-163, School of Computer Science, Carnegie Mellon University, Pittsburgh, PA 15213, 1990.
20. T. von Eicken, D. E. Culler, S. C. Goldstein, and K. E. Schauser. Active messages: A mechanism for integrated communication and computation. In *Proceedings of the 19th Annual International Symposium on Computer Architecture*, pages 256–266, 1992.
21. K. Yelick. Using abstraction in explicitly parallel programs. Technical Report MIT/LCS/TR-507, Massachusetts Institute of Technology, 545 Technology Square, Cambridge, MA 02139, July 1991.
22. K. A. Yelick and S. J. Garland. A parallel completion procedure for term rewriting systems. In *Conference on Automated Deduction*, Saratoga Springs, NY, 1992.

Improving Transformation Systems for General E-Unification

Max Moser

Institut für Informatik, Technische Universität München
Arcisstr. 21, D-8000 München 2
moser@informatik.tu-muenchen.de

Abstract. In this paper we motivate and present a new and improved transformation system for general E-unification. It can be seen as a modification of the original transformation system by Gallier and Snyder refined by ordinary unification and basic paramodulation. We present a short proof of completeness. Besides completeness we can also show an important property of the transformation system which is not known for the original system: independence of the selection rule. This motivates the abstraction of transformation sequences to equational proof trees thus obtaining static proof objects which facilitates finding further refinements of the procedure.

1 Introduction

Transformation systems are a rather recent approach to general E-unification (e.g. [MMR86], [GS89], [Sny91]). Unlike completion based approaches to E-unification (e.g. [HR87], [BDP89]), which (for the general case) have to proceed in a bottom-up fashion, transformation systems operate in a goal-oriented manner. They solve E-unification problems by repeated transformation of 'complex' problems to 'less complex' ones. The price to be payed for the goal-orientedness first of all is that it is not possible to impose ordering constraints on the application of equations. Secondly, in order to be complete without the need to paramodulate into variables, the application of equations must be done lazily. This means, that instead of immediately unifying the subterm to be rewritten and the term on one side of an equation, a new E-unification problem to be solved is generated. Unfortunately, this leads to a system of transformation rules with a considerably large search space.

Despite the fact that in the meantime some effort has been made to improve transformation systems (e.g. [DJ90b]), up to now they were not attractive enough to be applied for example in automated theorem proving. In addition to the huge search space involved, a reason could be that it has not yet been proven, whether transformation systems are independent of the selection rule. Although this is commonly assumed, in order to guarantee completeness during the search for solutions one would have to examine all possible orderings of selections of goals, since transformation systems were just shown to be nondeterministically complete.

In this paper we want to attack both the inherent inefficiency of transformation systems and the unsolved question of independence of the selection rule. We motivate and present a refined transformation system which is characterized by ordinary unification and the restriction of applications of equations to *basic* terms. The *basic* restriction (for details cf. e.g. [Hul80], [NRS89], [BGLS92]) serves to abstract whether a variable is instantiated or not, and ordinary unification replaces three rules for unification in the original transformation system by Gallier and Snyder. We can show that we are still complete and have a substantially reduced search space. Moreover, we can prove the important property that the completeness of the proposed transformation system is independent of the selection rule. We not only show that for different selection rules (variants of) the same E-unifiers can be obtained, but moreover, that we can get the same structure of the derivation of an E-unifier. This allows us to abstract transformation sequences to proof objects: basic equational proof trees. A basic equational proof tree will be defined as a representative for a class of transformation sequences (disregarding their selection rules) and it represents the structure of the actual proof. This abstract view facilitates the recognition and proof of properties and refinements of the transformation system more easily than in the original sequence setting. We will demonstrate its significance on a first proposition about equivalent transformations which leads to a further refinement of the system.

Our intention with this paper is twofold. First, we want to present an improved general and complete transformation system which satisfies certain important properties for its application in automated theorem proving (like the independence of the selection rule). The second point is to show that the established properties are the basis for a variety of straightforward proofs of improvements. The transformation system and its refinements in this paper are a first step in this direction.

2 Preliminaries

This section provides a short sketch of some of the basic concepts and definitions used in this paper. A more detailed description can be found e.g. in [DJ90a] or [JK91]. The notational conventions we use were adopted from [DJ89] and [BGLS92].

An *equation* is a pair of terms related by the special (symmetric) predicate symbol '$\simeq$'. A *closure* $e \cdot \sigma$ is a pair consisting of an *skeleton* e and a substitution σ. The skeleton can be an arbitrary structure, e.g. an equation or a multiset of equations. Closures will serve to express the notion that no equation may be applied to a term or subterm introduced by a substitution (known as *basic* restriction, e.g. [Hul80]; for more details on closures cf. [BGLS92]).

A *position* p in a term t is represented by a sequence of positive natural numbers. The set of all positions in a term t is denoted by $\mathcal{P}os(t)$, the set of non-variable positions by $\mathcal{FP}os(t)$. The top-most position in a term is Λ. $t_{|p}$ represents the *subterm* of t at the position p and $t[s]_p$ the result of replacing the subterm in t at position p by the term s. By $p \parallel q$ we denote that the positions p

and q are disjoint, i.e. no one is above the other. The definition of positions can analogously be extended from terms to equations. For a term t, $\mathcal{H}ead(t)$ denotes the function symbol heading it.

An *E-unification problem* $s =_E^? t$ is the question whether two terms s and t are *E-unifiable*. In our refutational setting, $\neg(s \simeq t)$ (or for convenience $s \not\simeq t$) is called an *E-unification goal*. An *E-unifier* for an *E-unification problem* $s =_E^? t$ is a substitution σ s.t. $s\sigma =_E t\sigma$. A *complete set of E-unifiers* $\mathcal{CSU}_E(s,t)$ for terms s,t is a set of substitutions which is *correct* (i.e. for all $\sigma \in \mathcal{CSU}_E(s,t) : s\sigma =_E t\sigma$) and *complete* (i.e. for each *E*-unifier θ of s and t there is a $\sigma \in \mathcal{CSU}_E(s,t)$ such that $\sigma \leq_E \theta$). An *E-unification procedure* is a procedure which takes an equational theory E and an *E*-unification problem $s =_E^? t$ and generates a set of *E*-unifiers $\mathcal{U}(s,t)$ such that $\forall \sigma \in \mathcal{U}(s,t) : s\sigma =_E t\sigma$. An *E*-unification procedure is called *complete* if it generates a complete set of *E*-unifiers for all *E*-unification problems, and *general* if it is complete for arbitrary equational theories E.

3 A Refined Transformation System

In this section we present the new transformation system $\mathcal{T}_{BP}$ and compare it with the original transformation system as described in [GS89].

$\mathcal{T}_{BP}$ consists of two transformation rules: $\Downarrow$unify and $\Downarrow$lazy-param. They are presented as state-transition rules which describe how a given multiset of equational goals can be transformed into a new multiset of goals whenever certain conditions are satisfied.

Unification ($\Downarrow$unify)

$$\frac{(R \cup \{s \not\simeq t\}) \cdot \sigma}{R \cdot \sigma\theta} \left\langle \theta = mgu(s\sigma, t\sigma) \right\rangle$$

Lazy Basic Paramodulation ($\Downarrow$lazy-param)

$$\frac{(R \cup \{s \not\simeq t\}) \cdot \sigma}{(R \cup \{s_{|p} \not\simeq l, s[r]_p \not\simeq t\}) \cdot \sigma} \left\langle \begin{array}{l} p \in \mathcal{FP}os(s) \\ (l \simeq r) \in E \\ \text{if } l \notin \mathcal{V}ar \text{ then } \mathcal{H}ead(s_{|p}) = \mathcal{H}ead(l) \end{array} \right\rangle$$

A *transformation sequence* is a sequence of multiset closures where each member of the sequence can be obtained by applying a transformation rule to the preceding member of the sequence. A transformation sequence for an *E*-unification problem $s =_E^? t$ starts with $\{s \not\simeq t\} \cdot \varepsilon$ (where ε is the empty substitution). It is called *terminating* if the empty closure $\{\} \cdot \sigma$ is derived. Then σ is the *E*-unifier resulting from the transformation. The equational goal $s_{|p} \not\simeq l$ as the result of a $\Downarrow$lazy-param step is called *witness pair*, the goal $s[r]_p \not\simeq t$ is called *result pair*. The witness pairs represent the task to justify the application of the equation.

Example 1. Applying $\Downarrow$unify to $\{g(x,a) \not\simeq z\} \cdot \sigma$ with $\sigma = \{z \mapsto g(b,y)\}$ yields an empty multiset of goals $\{\} \cdot \sigma\theta$ where $\theta = mgu(g(x,a)\sigma, z\sigma) = \{x \mapsto b, y \mapsto a\}$. For an equational theory E containing an equation $g(u,v) \simeq c$ also a $\Downarrow$lazy-param step can be applied to $g(x,a)$ since it is not a variable and the root function symbols are the same. This leads to the new goal multiset $\{g(x,a) \not\simeq g(u,v),\ c \not\simeq z\} \cdot \sigma$. Please note that to $\{g(x,a) \not\simeq z\} \cdot \sigma$ no lazy-paramodulation step is applicable to the right side since z is a variable.

Transformation systems for E-unification are a means for transforming complex E-unification problems into (hopefully) less complex E-unification problems. Here, however, we present them in a refutational setting, i.e. the transformation is a search for a contradiction starting from equational goals. Although this is just a notational detail, we adopt this notation since we want to emphasize the connection with common automated theorem proving approaches. Another difference to the transformation system presented e.g. by Gallier and Snyder is that their system aims at constructing an equational unifier out of a system in solved form, i.e. each equation must have the form $x \simeq t$ where $x \notin Var(t)$. By having replaced the three transformation rules for syntactic unification by just one rule we no longer construct solved forms. However, the solved form is implicitly contained in the substitution part of the closure.

$\mathcal{T}_{\text{BP}}$ gains a considerable reduction of the search space for two reasons. First, many applications of equations are no longer possible since instantiated variables are 'blocked' by the basic restriction (realized by closures). Second, it is no longer possible to interleave the syntactic unification and the application of equations as it was with the original transformation system and thus eliminates many redundant transformation sequences.

4 Completeness of $\mathcal{T}_{\text{BP}}$

We now want to show that the transformation system $\mathcal{T}_{\text{BP}}$ can be used for a complete E-unification procedure, i.e. it enumerates complete sets of E-unifiers for E-unification problems. This is done by proving that $\mathcal{T}_{\text{BP}}$ can simulate inferences of a variant of another calculus: *basic superposition* $\mathcal{S}$.

Basic superposition is proven to be a refutationally complete calculus for clausal form formulae ([BGLS92]). For the purpose of simulating it by $\mathcal{T}_{\text{BP}}$, we have to make some adaptions to the inference rules due to the restricted logic consisting just of an equational theory and one equational goal (see Subsection 6.1). Furthermore, we have to replace the ordering constraints by a weaker restriction in order to facilitate the simulation. The resulting variant of $\mathcal{S}$ will be called $\mathcal{S}^\star$ and consists of the following three inference rules.

Negative Basic Paramodulation

$$\frac{(s \not\simeq t) \cdot \rho \qquad (u \simeq v) \cdot \rho}{(s[v]_p \not\simeq t) \cdot \rho\theta} \left\langle \begin{array}{l} p \in \mathcal{FPos}(s) \\ \theta = mgu(s|_p\rho, u\rho) \end{array} \right\rangle$$

Positive Constrained Basic Paramodulation

$$\frac{(s \simeq t) \cdot \rho \qquad (u \simeq v) \cdot \rho}{(s[v]_p \simeq t) \cdot \rho\theta} \left\langle \begin{array}{l} p \in \mathcal{FPos}(s) \\ \theta = mgu(s_{|p}\rho, u\rho) \\ \text{if } p = \Lambda \text{ then } u \notin \mathit{Var} \end{array} \right\rangle$$

Unification

$$\frac{(s \not\simeq t) \cdot \rho}{\bot \cdot \rho\theta} \langle \theta = mgu(s\rho, t\rho) \rangle$$

$\mathcal{S}^\star$ is a new calculus for which we first have to show that it generates complete sets of unifiers for E-unification problems (Theorem 14). This will be done in Section 6 and will be based on a new completeness result for $\mathcal{S}$. The completeness result for $\mathcal{S}^\star$ is already stated in the following theorem.

Theorem 2. *$\mathcal{S}^\star$ is a set of inference rules for general and complete E-unification.*

This result allows us to derive the completeness proof of $\mathcal{T}_{\text{BP}}$ in a quite straightforward manner. The overall idea will be to simulate refutations by $\mathcal{S}^\star$ in a goal-oriented setting. For this we will separate the negative basic paramodulation and positive constrained basic paramodulation inferences of $\mathcal{S}^\star$-refutations: before any negative basic paramodulation step is carried out all necessary positive constrained basic paramodulations must have been done.[1]

For the following let $\hat{E}$ be the set of all equations which are eventually generated in this 'preprocessing' phase and which are necessary for the refutation. In the first step we show that we can simulate each negative basic paramodulation step using an equation in $\hat{E}$ by a corresponding step in $\mathcal{T}_{\text{BP}}$. We then have to show that positive constrained basic paramodulation can be simulated also. For this we replace all inferences where an equation resulting from some positive constrained basic paramodulation step has been applied by applications of the equations involved in this step. Due to the use of closures it is ensured that the basic restriction is not violated in the course of the simulation.

Lemma 3 (Simulation of Negative Basic Paramodulation). *Let σ be an E-unifier of the goal $(s \not\simeq t)$ generated by a sequence of $\mathcal{S}^\star$ inferences in the equational theory E. Let $\hat{E}$ be the set of equations used for negative basic paramodulation. Then there is also a sequence of transformation steps in $\mathcal{T}_{\text{BP}}$ based on $\hat{E}$ terminating with σ.*

Proof. Straightforward construction. For each negative basic paramodulation step

$$\frac{(s \not\simeq t) \cdot \rho \qquad (l \simeq r) \cdot \rho}{(s[r]_p \not\simeq t) \cdot \rho\theta} \left\langle \begin{array}{l} p \in \mathcal{FPos}(s) \\ \theta = mgu(s_{|p}\rho, l\rho) \end{array} \right\rangle$$

there is a corresponding transformation sequence of a ⇓lazy-param step and an immediate ⇓unify step. (Please note that the equational goal printed in boldface is an intermediate result of the simulation.)

---
[1] We want to indicate that already the proof in [GS89] allows to derive that what we call negative basic paramodulation can be simulated. However, this is not the case for positive basic paramodulation.

$$\Downarrow \mathsf{lazy\text{-}param}_p(l \simeq r) \frac{\{s \not\simeq t\} \cdot \rho}{\Downarrow \mathsf{unify} \frac{\{s_{|p} \not\simeq l \,,\, s[r]_p \not\simeq t\} \cdot \rho}{\{s[r]_p \not\simeq t\} \cdot \rho\theta}}$$

Since $s_{|p}$ is not a variable and $s_{|p}\rho$ and $l\rho$ are unifiable it follows that if $l \notin \mathcal{V}ar$ then $\mathcal{H}ead(s_{|p}) = \mathcal{H}ead(l)$. □

Lemma 4 (Simulation of Positive Constrained Basic Paramodulation).
Let σ be an E-unifier of the goal $(s \not\simeq t)$ generated by an arbitrary sequence of $\mathcal{S}^*$ inferences in the equational theory E. Let $\hat{E}$ be the set of equations used for negative basic paramodulation (and which are either from E or are generated by positive constrained basic paramodulation) and Q a corresponding sequence of transformation steps in T_{BP} based on $\hat{E}$. Then there exists a sequence Q' where just equations from E are used and which terminates with a variant of σ.

Proof. The proof is in essence similar to the one given in [Sny91].

For the following we define the level of an equation to be the number of positive constrained basic paramodulations which were necessary to obtain it. We do multiset induction on μ, where μ is a multiset containing the levels k of the equations in a transformation sequence based on $\hat{E}$. For an equation in $\hat{E}$ with level $k > 0$ there must be some positive constrained basic paramodulation inference

$$\frac{(u \simeq v) \cdot \rho \qquad (u' \simeq v') \cdot \rho}{(u[v']_q \simeq v) \cdot \rho\beta} \left\langle \begin{array}{l} q \in \mathcal{FP}os(u) \\ \beta = mgu(u_{|q}\rho, u'\rho) \\ \text{if } q = \Lambda \text{ then } u' \notin \mathcal{V}ar \end{array} \right\rangle$$

where the input equations have levels $k', k'' < k$.

For the induction step we have to distinguish among the direction of the application of the equation $(u[v']_q \simeq v) \cdot \rho\beta$.
Case (a): left-to-right: In this case, we furthermore have to distinguish among whether $q = \Lambda$ or not.
   Case (a.1): left-to-right and $q \neq \Lambda$: We replace the transformation step

$$\Downarrow \mathsf{lazy\text{-}param}_p(u[v']_q \simeq v) \frac{\{s \not\simeq t\} \cdot \rho\beta}{\{s_{|p} \not\simeq u[v']_q \,,\, s[v]_p \not\simeq t\} \cdot \rho\beta}$$

by two successive applications of $(u \simeq v) \cdot \rho$ and $(u' \simeq v') \cdot \rho$ and an additional unification step applied to $(u_{|q} \not\simeq u') \cdot \rho\beta$ $(= (u_{|q} \not\simeq u') \cdot \beta\rho)$ as follows:

$$\Downarrow \mathsf{lazy\text{-}param}_p(u \simeq v) \frac{\{s \not\simeq t\} \cdot \rho\beta}{\Downarrow \mathsf{lazy\text{-}param}_q(u' \simeq v') \frac{\{s_{|p} \not\simeq u \,,\, s[v]_p \not\simeq t\} \cdot \rho\beta}{\Downarrow \mathsf{unify} \frac{\{u_{|q} \not\simeq u' \,,\, s_{|p} \not\simeq u[v']_q \,,\, s[v]_p \not\simeq t\} \cdot \rho\beta}{\{s_{|p} \not\simeq u[v']_q \,,\, s[v]_p \not\simeq t\} \cdot \rho\beta}}}$$

Case (a.2): left-to-right and $q = \Lambda$: In this case the respective equation is $(v' \simeq v) \cdot \rho\beta$ and we replace the transformation step

$$\Downarrow \text{lazy-param}_p(v' \simeq v) \frac{\{s \not\simeq t\} \cdot \rho\beta}{\{s_{|p} \not\simeq v', \, s[v]_p \not\simeq t\} \cdot \rho\beta}$$

by two successive applications of $(v' \simeq u') \cdot \rho$ and $(u \simeq v) \cdot \rho$ and an additional unification step. Note that in this case the simulation is just possible because of $u' \notin \mathit{Var}$! This is the reason for the use of $\mathcal{S}^\star$ instead of simply $\mathcal{S}$ here.

$$\Downarrow \text{lazy-param}_p(v' \simeq u') \frac{\{s \not\simeq t\} \cdot \rho\beta}{\Downarrow \text{lazy-param}_p(u \simeq v) \frac{\{s_{|p} \not\simeq v', \, s[u']_p \not\simeq t\} \cdot \rho\beta}{\Downarrow \text{unify} \frac{\{s_{|p} \not\simeq v', \, u' \not\simeq u, \, s[v]_p \not\simeq t\} \cdot \rho\beta}{\{s_{|p} \not\simeq v', \, s[v]_p \not\simeq t\} \cdot \rho\beta}}}$$

It is important to verify that in the cases above $\Downarrow$lazy-param is just applied to non-variable positions and obeys the condition that the root symbols must be the same whenever the equation is not applied with a variable.

Case (b): right-to-left: In this case the simulation is done similarly to the case (a.2). However, we do not need to distinguish among the position q. □

A direct consequence of the combination of the two lemmata above together with the theorem stating that $\mathcal{S}^\star$ generates complete sets of unifiers is the following.

Theorem 5 (Completeness of the System $\mathcal{T}_{\text{BP}}$). $\mathcal{T}_{\text{BP}}$ *is a transformation system for general and complete E-unification.*

5 Independence of $\mathcal{T}_{\text{BP}}$

We now turn to the second central result of our paper: $\mathcal{T}_{\text{BP}}$ is independent of the selection rule, i.e. the completeness of the calculus is independent of the ordering in which the subgoals to be solved are selected. But we can even prove a more drastic result: also the 'structure' of the proof is independent of the selection rule, i.e. no matter when an equational goal is selected, it is always possible to apply the same transformation steps to it. This property is sometimes called *strong independence* of the selection rule and *cannot* be obtained for the original transformation system in [GS89].[2]

The following switching lemma can be seen as an extension of the one presented in [Llo87] for SLD-resolution. (A more detailed discussion of selection rules can also be found there.) It will be the basis for the strong independence theorem.

---

[2] We want to indicate that it is foreseeable that the *weak independence* (i.e. variants of the same substitutions can be found for arbitrary selection rules but NOT necessarily the same structure of the derivation) of the original transformation system by Gallier and Snyder can be proven based on the strong independence of $\mathcal{T}_{\text{BP}}$. Although assumed to be true, this property was not proven before.

Lemma 6 (Switching Lemma). *Let E be an equational theory and R a selection rule. Suppose that there is a transformational sequence controlled by R which contains the subsequence* $S_{k-1} \Rightarrow_{T_{\mathrm{BP}}} S_k \Rightarrow_{T_{\mathrm{BP}}} S_{k+1}$ *where*

$$S_{k-1} \text{ is } \{\ldots\ldots, e_i, \ldots\ldots\ldots, e_j, \ldots\ldots\} \cdot \sigma$$
$$S_k \text{ is } \{\ldots \overbrace{f_1, \ldots, f_r}, \ldots\ldots, e_j, \ldots\ldots\} \cdot \sigma\theta$$
$$S_{k+1} \text{ is } \{\ldots, f_1, \ldots, f_r, \ldots, \overbrace{g_1, \ldots, g_p}, \ldots\} \cdot \sigma\theta\phi \;.$$

Then there exists a corresponding transformational sequence via the selection rule R', which is the same as R except that e_j is selected in S_{k-1} instead of e_i and e_i is selected in S_k instead of e_j. Furthermore, the R'-computed E-unifier for the whole sequence is just a variant of the R-computed E-unifier.

Proof. We have to distinguish four cases corresponding to the four combinations of applying ⇓unify and ⇓lazy-param to e_i and e_j. For the following, let $e_i = (s \simeq t)$ and $e_j = (s' \simeq t')$.

- $S_{k-1} \Rightarrow_{\text{unify}} S_k \Rightarrow_{\text{unify}} S_{k+1}$: That is, ⇓unify is applied first to $e_i \cdot \sigma$ with the result $\theta = mgu(s\sigma, t\sigma)$, and then to $e_j \cdot \sigma\theta$ with $\phi = mgu(s'\sigma\theta, t'\sigma\theta)$. It is commonly known (and is proven in the context of resolution for example in [Llo87]) that we can switch (i.e. first select e_j and then e_i) and obtain a variant of $\sigma\theta\phi$.

- $S_{k-1} \Rightarrow_{\text{unify}} S_k \Rightarrow_{\text{lazy-param}} S_{k+1}$: That is, after applying ⇓unify to $e_i \cdot \sigma$ with the result $\theta = mgu(s\sigma, t\sigma)$, we apply ⇓lazy-param to $e_j \cdot \sigma\theta = (s' \not\simeq t') \cdot \sigma\theta$, which results in two new subgoals $g_1 \cdot \sigma\theta = (s'_{|p} \not\simeq l) \cdot \sigma\theta$ and $g_2 \cdot \sigma\theta = (s'[r]_p \not\simeq t') \cdot \sigma\theta$. This is the most important case where we make implicit usage of the basic restriction: since to $e_j \cdot \sigma$ already ⇓lazy-param is applicable and has no impact on the following application of ⇓unify we can switch.

- For the cases $(S_{k-1} \Rightarrow_{\text{lazy-param}} S_k \Rightarrow_{\text{unify}} S_{k+1})$ and $(S_{k-1} \Rightarrow_{\text{lazy-param}} S_k \Rightarrow_{\text{lazy-param}} S_{k+1})$ it suffices to see that ⇓lazy-param does not instantiate any variables and is independent of the actual instance of a term. Thus it can be shown by a similar argumentation as above that they also can be switched. □

This allows us to state almost immediately the desired independence theorem.

Theorem 7 (Strong Independence of the System T_{BP}). *The system T_{BP} is strongly independent of the selection rule.*

Proof. By the completeness of T_{BP} (Theorem 5) there exists a terminating sequence of transformation steps for E-unifiable terms. By iterated application of the switching lemma (Lemma 6) we can obtain a terminating transformational sequence for any selection rule. □

Since the actual proof by T_{BP} is independent of the selection rule used, we can abstract a sequence of transformation steps to a 'proof object': *basic equational proof trees*. They will just represent to which goal which operation has been applied and what its results are. It will allow us to make propositions about properties of T_{BP} without needing to take into account which selection rule is used in the individual situation.

Definition 8 (Basic Equational Proof Tree). A *basic equational proof tree* T is a binary tree, the nodes of which are labelled with equational goals. Furthermore, each label of an inner node together with the labels of its immediate successor nodes must constitute a lazy paramodulation step. A basic equational proof tree T is *closed* if there is a substitution σ such that $s\sigma = t\sigma$ for each leaf node label $s \not\simeq t$. The most general unifier σ such that a basic equational proof tree T is closed is denoted by $\mathcal{U}nif(T)$. A basic equational proof tree T *corresponds* to a transformation sequence if the transformation sequence can be obtained by a tree traversal controlled by some selection rule.

Example 9. The two trees T_1, T_2 below represent the structures of two derivations of E-unifiers for the goal $\{f(x) \not\simeq g(y)\} \cdot \varepsilon$ in the equational theory $E = \{f(a) \simeq c, g(b) \simeq c\}$.

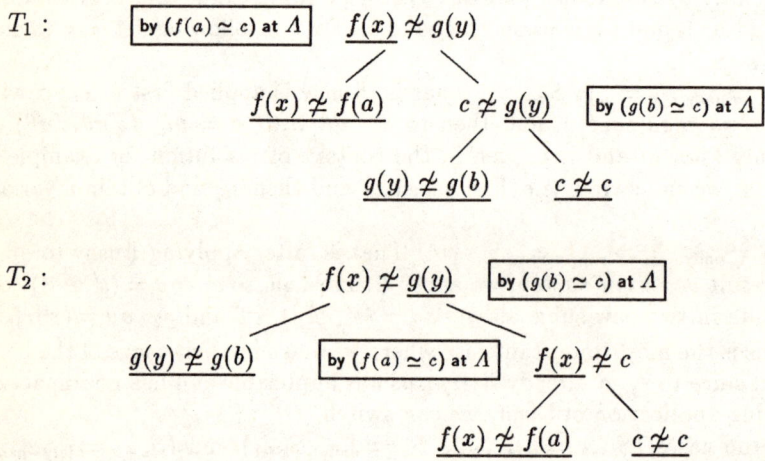

Although the two derivations have a different structure, we have that $\mathcal{U}nif(T_1) = \mathcal{U}nif(T_2) = \{x \mapsto a, y \mapsto b\}$. The reason for this is that lazy paramodulation was applied at disjoint positions. This phenomenon is, however, just an instance of a more general source of redundancy and thus inefficiency. In the following lemma we will therefore utilize the new definition of basic equational proof trees and describe this property.

Lemma 10 (Commutation Lemma). *Let E be an equational theory and T a basic equational proof tree containing the subtree T' with the root goal g. Suppose that to g first an equation e_1 at position p_1 is applied and then an equation e_2 at position p_2 with $p_1 \parallel p_2$. Then, in T we can replace T' by a corresponding equational proof tree S' to obtain S where the ordering of the applications of e_1 and e_2 has been switched. Furthermore, $\mathcal{U}nif(S) = \mathcal{U}nif(T)$.*

Proof. Since $p_1 \parallel p_2$, the two applications of $\Downarrow$lazy-param do not interfere and thus the ordering of their application can be switched without impact on the global substitution. □

The commutation property above has tremendous impact on the search space of $\mathcal{T}_{\text{BP}}$. A direct outcome for example is the completeness of a strategy which allows no application to the left side of a goal after an application at its right side has taken place. Furthermore, similar strategies can also be found for the term level.

6 Basic Superposition and Complete Sets of Unifiers

In this section we want to present a short sketch of the proof that the inference system $\mathcal{S}^\star$ generates complete sets of unifiers. This property is the basis for the proof of completeness of the transformation system $\mathcal{T}_{\text{BP}}$ in Section 4. Since the details in this proof are not relevant for our paper, the reader may skip this section without lack of understanding. A more detailed proof however can be found in [Mos93]. For the basics on orderings we refer to [Bac91] or [Sny91].

The actual proof of completeness of $\mathcal{S}^\star$ will be made in two steps. The first is to show in Subsection 6.1 that basic superposition $\mathcal{S}$ generates complete sets of unifiers of equational goals in an equational theory. This is actually a new result and allows to use $\mathcal{S}$ as a general and complete E-unification procedure (in a different setting similar work has been done in [NR92]).

From this, we will derive in Subsection 6.2 the completeness of $\mathcal{S}^\star$ which is in essence a version of $\mathcal{S}$ without ordering constraints but with an additional variable constraint. This variable constraint of $\mathcal{S}^\star$ is important for the simulation proof of $\mathcal{T}_{\text{BP}}$ since without it, the simulation in a certain case would not have been possible.[3]

6.1 Complete Sets of Unifiers by $\mathcal{S}$

For the restricted logic of an equational theory and one equational goal the inference system $\mathcal{S}$ presented in [BGLS92] consists of the following three rules.

Basic Left Superposition

$$\frac{(s \not\simeq t) \cdot \rho \qquad (u \simeq v) \cdot \rho}{(s[v]_p \not\simeq t) \cdot \rho\theta} \left\langle \begin{array}{l} p \in \mathcal{FP}os(s) \\ \theta = mgu(s|_p \rho, u\rho) \\ t\rho\theta \not\succeq s\rho\theta \text{ and } v\rho\theta \not\succeq u\rho\theta \end{array} \right\rangle$$

Basic Right Superposition

$$\frac{(s \simeq t) \cdot \rho \qquad (u \simeq v) \cdot \rho}{(s[v]_p \simeq t) \cdot \rho\theta} \left\langle \begin{array}{l} p \in \mathcal{FP}os(s) \\ \theta = mgu(s|_p \rho, u\rho) \\ t\rho\theta \not\succeq s\rho\theta \text{ and } v\rho\theta \not\succeq u\rho\theta \end{array} \right\rangle$$

---

[3] It is conjectured that it is possible to prove the completeness of $\mathcal{S}$ even in presence of the constraint "if $p = \Lambda$ then $u \notin \mathcal{V}ar$" of positive constrained basic paramodulation. Obviously, this would eliminate the need to consider $\mathcal{S}^\star$ at all and thus lead to a simplified proof of completeness of $\mathcal{T}_{\text{BP}}$.

Equality Resolution

$$\frac{(s \not\simeq t) \cdot \rho}{\bot \cdot \rho\theta} \langle \theta = mgu(s\rho, t\rho) \rangle$$

Comparing $\mathcal{S}$ and $\mathcal{S}^\star$, *basic left superposition* corresponds to *negative basic paramodulation*, *basic right superposition* to *positive constrained basic paramodulation* and *equality resolution* is the same as *unification*.

The most important property which we use in our proof is that $\mathcal{S}$ is a refutationally complete calculus for first order logic with equality (cf. [BGLS92]). Based on this theorem we are able to prove the following lemma. It states that $\mathcal{S}$ is able to find contradictions for certain E-unifiers. In the following, $Sat(M)$ represents the saturation of a set M using inference rules of $\mathcal{S}$ i.e., it represents the fixpoint of M under $\mathcal{S}$.

Lemma 11. *Let E be an equational theory and σ an E-unifier of s and t. Then there exists an E-unifier $\theta \leq_E \sigma[\text{Var}(s,t)]$ of s and t s.t. $Sat(E \cup \{(s \not\simeq t) \cdot \theta\})$ contains $\bot \cdot \theta$.*

Proof. Without loss of generality we may first assume that the theory E is not trivial, i.e. $x \simeq y$ should not be valid in E for $x, y \in \text{Var}$. In this case the E-unification problem is trivial.

We introduce new Skolem constants for all variables occurring in s and t and identify the instantiation of the variables by the corresponding Skolem constants with a substitution κ. The ordering $\prec$ is homogeneously extended so that the constants introduced by κ are smaller than any other function symbol.

Since $\sigma\kappa$ is ground it can be normalized to a $\nu = \sigma'\kappa$ for some σ' with $\sigma' \leq_E \sigma$ and $s\sigma' =_E t\sigma'$.

Due to the refutational completeness of $\mathcal{S}$, $Sat(E \cup \{(s\sigma'\kappa \not\simeq t\sigma'\kappa)\})$ must contain a contradiction. This tells us that there is a refutation starting with the E-unifier σ' which does not specialize the variables in s and t. Furthermore, since $x \simeq y$ must not be valid in E and the Skolem constants are new and smaller than any other function symbol, no rewrite step is applied to a Skolem constant. Thus, we can infer that $Sat(E \cup \{(s\sigma' \not\simeq t\sigma') \cdot \kappa\})$ contains a contradiction as well. Since beyond that no rewrite step in the refutation was applied to a term or subterm introduced by σ' ($\sigma'\kappa$ is normal!), we can "lift" away σ' and obtain that even $Sat(E \cup \{(s \not\simeq t) \cdot \sigma'\})$ contains some contradiction $\bot \cdot \sigma'\beta$ for some β where $\mathcal{D}om(\beta) \cap \text{Var}(s,t) = \emptyset$. □

From this lemma it is quite straightforward to infer the desired completeness theorem.

Theorem 12 (Completeness of the System $\mathcal{S}$). *Let E be a theory, and s and t two E-unifiable terms. Then $\mathcal{S}$ generates a $\mathcal{CSU}_E(s,t)$.*

Proof. By the lemma above we know that if we already start with a normalized E-unifier which is 'blocked' by a closure, $\mathcal{S}$ is able to derive a contradiction without applying ordered paramodulation to a term or subterm introduced by the substitution and without specializing it. This implies that $\mathcal{S}$ is able to find a more E-general unifier for an arbitrary E-unifier and is thus complete. □

6.2 Complete Sets of Unifiers by $\mathcal{S}^\star$

In the second step we will base the proof of completeness of $\mathcal{S}^\star$ on a variant of $\mathcal{S}$. Actually, this variant called $\mathcal{S}^+$ is a slightly stronger calculus than $\mathcal{S}$, where the ordering constraints of basic left and right superposition are weakened: by applying an equation $(u \simeq v) \cdot \rho$ the ordering constraints are just applied whenever $u\rho \in \mathit{Var}$. Obviously, $\mathcal{S}^+$ is complete since $\mathcal{S}$ is complete and any refutation with $\mathcal{S}$ is also possible with $\mathcal{S}^+$.

The actual proof shows that any $\mathcal{S}^+$-refutation can be 'rearranged' to a $\mathcal{S}^\star$-refutation such that the resulting E-unifier is a variant of the original E-unifier. This will be presented in the following lemma.

Lemma 13. *For every $\mathcal{S}^+$-refutation yielding σ there exists a $\mathcal{S}^\star$-refutation yielding θ such that $\theta \leq_E \sigma$.*

Proof. Since the detailed proof of this lemma does not reveal new and interesting aspects we will present just a sketch of it here. The proof consists of two phases.

In the first phase we eliminate successive basic left superposition steps of $\mathcal{S}^+$ taking place at positions p_1, p_2 with $p_1 \leq p_2$. This is necessary for the second phase where we obtain equations with possibly more positions blocked by the closure and where without this first phase some negative basic paramodulations would no longer be possible. This is done by first applying basic right superposition between the two equations of the basic left superpositions at p_1 and p_2 and then basic left superposition with the result. It can be assured very easily that this procedure yields a refutation, the resulting substitution of which is a variant. It is important to verify that this manipulations do not sacrifice the relaxed ordering constraints of $\mathcal{S}^+$ which are necessary for the following manipulations.

The aim of the second phase is to rearrange basic right superposition steps between closures $(u \simeq v) \cdot \rho$ and $(x \simeq t) \cdot \rho$ taking place at the root position of u with $x\rho \notin \mathit{Var}$. But since the term $x\rho$ must have been introduced by unification of some basic right superposition step(s) in the generation of $(x \simeq t) \cdot \rho$, we can apply $(u \simeq v) \cdot \rho$ immediately to the position(s) of the equations involved. It is important to note that it is possible to rearrange the positive basic superposition steps such that no new violations of the rules of $\mathcal{S}^\star$ are created. Termination of this procedure is assured by an inductive argument. □

This lemma entails the desired completeness result for $\mathcal{S}^\star$.

Theorem 14 (Completeness of the System $\mathcal{S}^\star$). *Let E be a theory, and s and t two E-unifiable terms. Then $\mathcal{S}^\star$ generates a $\mathcal{CSU}_E(s,t)$.*

Proof. Since $\mathcal{S}$ is complete (Theorem 12) and $\mathcal{S}^+$ is a stronger calculus than $\mathcal{S}$, $\mathcal{S}^+$ is also complete. Since for any $\mathcal{S}^+$-refutation there exists an as general $\mathcal{S}^\star$-refutation (Lemma 13), $\mathcal{S}^\star$ must be complete and generates complete sets of unifiers. □

7 Conclusion

We have presented a refined transformation system for general E-unification. Our motivation was to obtain an E-unification procedure which is better suited for automated theorem proving. The original transformation system not only lacks a proof of the independence of the selection rule but also suffers from some inherent sources of inefficiency.

The new transformation system is characterized by its ordinary unification and the restriction of the application of equations to basic terms. This leads to a significant restriction of the search space. We have given a proof of completeness which is based on the new proof that the basic superposition calculus can generate complete sets of unifiers, and show its simulation by T_{BP}. Furthermore, we have presented a proof of the strong independence of the selection rule which motivates the definition of basic equational proof trees as proof objects, disregarding the selection rule. We have demonstrated that this abstracted view allows to realize and prove important properties of the transformation system in an elegant way.

Although properties established for the transformation system by Gallier and Snyder (like the immediate decomposition of the witness pair) have not been proven for T_{BP} in this paper, it is foreseeable that they can (at least partially) also be established for T_{BP}. But apart from them, T_{BP} holds a significant potential for further major refinements which we plan to investigate in the future.

Acknowledgements The author wants thank Klaus Mayr, Christian Suttner and other members of the group for helping to improve the presentation of the paper.

References

[Bac91]   L. Bachmair. *Canonical Equational Proofs*. Birkhäuser, Boston, 1991.

[BDP89]   L. Bachmair, N. Dershowitz, and D. Plaisted. Completion Without Failure. In H. Ait-Kaci and M. Nivat, editors, *Resolution of Equations in Algebraic Structures, Vol. 2*, chapter 1, pages 1 – 30. Academic Press, Boston, 1989.

[BGLS92]   L. Bachmair, H. Ganzinger, Ch. Lynch, and W. Snyder. Basic Paramodulation and Superposition. In D. Kapur, editor, *CADE'92, 11th International Conference on Automated Deduction*, number 607 in LNCS, pages 462 – 476. Springer, 1992.

[DJ89]   N. Dershowitz and J.-P. Jouannaud. Notations for Rewriting. In *BECAT-1991*, pages 162 – 172, Berlin, 1989. Springer.

[DJ90a]   N. Dershowitz and J.-P. Jouannaud. Rewrite Systems. In J.van Leeuwen, editor, *Handbook of Theoretical Computer Science*, pages 245 – 321. Elsevier Science Publishers, Amsterdam, 1990.

[DJ90b]   D.J. Dougherty and P. Johann. An Improved General E-Unification Method. In M.E. Stickel, editor, *CADE'90, 10th International Conference on Automated Deduction*, number 449 in LNCS, pages 261 – 275, Berlin, 1990. Springer.

[GS89]   J. Gallier and W. Snyder. Complete Sets of Transformations for General E-Unification. *Theoretical Computer Science*, 67:203 – 260, 1989.

[HR87]   J. Hsiang and M. Rusinowitch. On Word Problems in Equational Theories. In *14th International Colloquium of Automata, Languages and Programming*, number 267 in LNCS, pages 54 – 71, Berlin, 1987. Springer.

[Hul80]   J.-M. Hullot. Canonical Forms and Unification. In *CADE'80, 5th International Conference on Automated Deduction*, number 87 in LNCS, pages 318 – 334, Berlin, 1980. Springer.

[JK91]   J.-P. Jouannaud and C. Kirchner. Solving Equations in Abstract Algebras: A Rule-Based Survey of Unification. In *Computational Logic. Essays in honor of Alan Robinson*, chapter 8, pages 257–321. MIT Press, 1991.

[Llo87]   J.W. Lloyd. *Foundations of Logic Programming*. Springer, Berlin, 2 edition, 1987.

[MMR86]   A. Martelli, C. Moiso, and C.F. Rossi. An Algorithm for Unification in Equational Theories. In *International Symposium on Logic Programming*, pages 180–186, 1986.

[Mos93]   M. Moser. Complete Sets of Unifiers by Basic Superposition. Technical report, Technische Universität München, 1993.

[NR92]   R. Nieuwenhuis and A. Rubio. Basic Superposition is Complete. In B. Krieg-Brückner, editor, *ESOP'92, 4th European Symposium on Programming*, number 582 in LNCS, pages 371 – 389. Springer, 1992.

[NRS89]   W. Nutt, P. Réty, and G. Smolka. Basic Narrowing Revisited. *Journal of Symbolic Computation*, 7:295 – 317, 1989.

[Sny91]   W. Snyder. *A Proof Theory for General Unification*. Birkhäuser, Boston, 1991.

Equational and Membership Constraints for Infinite Trees

Joachim Niehren[1]* Andreas Podelski[2] and Ralf Treinen[1]**

[1] Deutsches Forschungszentrum für Künstliche Intelligenz (DFKI), Stuhlsatzenhausweg 3, D-6600 Saarbrücken 11, Germany, (niehren/treinen)@dfki.uni-sb.de
[2] Digital Equipment Corporation, Paris Research Laboratory, 85, Avenue Victor Hugo, F-92563 Rueil-Malmaison, France, podelski@prl.dec.com

Abstract. We present a new constraint system with equational and membership constraints over infinite trees. It provides for complete and correct satisfiability and entailment tests and is therefore suitable for the use in concurrent constraint programming systems which are based on cyclic data structures.

Our set defining devices are greatest *fixpoint solutions* of regular systems of equations with a deterministic form of union. As the main technical particularity of the algorithms we present a novel memorization technique. We believe that both satisfiability and entailment tests can be implemented in an efficient and incremental manner.

1 Introduction

Concurrent constraint programming (CCP) systems factorize into a constraint system, which may be seen as a parameter to the system, and an extension facility to compute with relations or processes. The constraint system consists of a universal data structure and a set of logical formulae, called constraints, that express relations between the data objects.

There are several computation models for different CCP systems and paradigms, such as AKL [8], ALPS [13], cc-languages [17], constraint logic programing (CLP) [11, 10], LIFE [1] and Oz [18, 9]. They all require the constraints to be closed under conjunction and raise the need for an efficient and incremental constraint simplification algorithm that yields a test for satisfiability of constraints. All of them use existential quantification of constraints implicitly or explicitly, and most of them require an efficient and incremental entailment test (*i.e.*, a test of of the implication between two constraints). In particular this test is necessary for committed choice mechanisms depending on the satisfaction of guards as in Oz, AKL, LIFE and ALPS.

---

* Supported by the Graduierten-Kolleg Informatik der Universität des Saarlandes and by the Hydra project at DFKI.
** Supported by the Bundesminister für Forschung und Technology, contract ITW 9105, and by the Esprit working group CCL, contract EP 6028.

In many programming languages, *memberships* come in the form of static type assertions. In the CCP context however, it is natural to have memberships as relations. Having definitions for the two sets *Nat* and *NatList* like

$$Nat = 0 \cup succ(Nat)$$
$$NatList = nil \cup cons(Nat, NatList)$$

we could of course define according unary predicates *Nat* and *NatList* in the extension facility (for instance as a logic program). The problem is that the extension facility is by design decision in general *in*complete for disjunctive information, while the sort definitions are inherently disjunctive. For instance the conjunction of the atoms $Nat(x) \wedge NatList(x)$ will not be reduced to $\bot$ unless the language provides some kind of backtracking, which often is not the case in CCP systems. Even worse, in the context of the set definitions

$$\begin{array}{ll} Even = 0 \cup succ(Odd) & Nat = 0 \cup succ(Nat) \\ Odd = succ(Even) & Inf = succ(Inf) \end{array}$$

the computation rules of the extension facility will not detect that the denotation of *Even* is a subset of the denotation of *Nat*, since this requires an inductive argument. Hence, a rule like `if Nat(x) then` $\cdots$ will not fire in a context where $Even(x)$ is given. The third reason why we can not employ the extension facility for dealing with memberships is founded in the use of *infinite trees* as the basic data structure. Infinite trees have been introduced in Prolog II [5] in order to model cyclic data structures. With the definition of *Nat* as above, the conjunction $x \doteq succ(x) \wedge Nat(x)$ will forever unfold *Nat*. Again, an inductive reasoning is missing here.

Consequently, we claim that CCP systems will benefit from the incorporation of memberships of some restricted form into the constraint part.[3] This allows to delegate some computation from the extension facility to a possibly complete constraint solver. Hence, our constraint system comprises equational constraints *and* membership constraints. The syntax in BNF style of our constraints is as follows:

$$\gamma ::= x \dot{\in} p \mid x \doteq y \mid x \doteq f(y_1, \cdots, y_n) \mid \gamma \wedge \gamma' \mid \exists x\, \gamma \mid \bot \mid \top\,.$$

As defining device we use regular systems of equations with *deterministic* union and its *greatest* fixpoint solution. These equations are not part of the constraint system. Nevertheless, it is possible to extended the system by new equations in the course of computation.

For instance in the definition of *Nat* given above, $x \dot{\in} Nat$ holds exactly if x is a natural number including ∞. This conforms with the fact that ∞ has an equational representation as the unique solution of $x \doteq sux.x(x)$.

The union is used in a *deterministic* manner, since the constructors in the different possibilities of an equation are distinct. We use the name determinism for this concept, since the components of the *least* fixpoint solutions of our

---

[3] This idea is due to Gert Smolka.

deterministic regular systems are exactly the sets recognized by deterministic top down tree automata. Without an appropriate restriction of the union like determinism we could not hope for any efficient algorithm. Furthermore, our entailment test relies on the determinism condition.

Our algorithms for testing the satisfiability and entailment are based on a novel technique that we call *memorization*. The correctness of memorization depends mainly on the *greatest* fixpoint solution. We illustrate this technique by proving the entailment:

$$x \doteq succ(x) \models x \dot\in Nat .$$

By unfolding the definition of *Nat*, we obtain a constraint which simplifies to $x \dot\in Nat$ relatively to $x \doteq succ(x)$. [4] Now, instead of reducing to the same subproblem infinitely often, we memorize all constraints once unfolded, and throw them away when they reappear. In this way $x \dot\in Nat$ is simplified to $\top$, and entailment is proven.

We prove that the step of deleting once unfolded constraints is correct in the greatest fixpoint solution, while it can be wrong in other fixpoints. More technically, we use the fact that the greatest fixpoint solution is obtained by ω iteration steps from 'top'. Note that for arbitrary logic programs this is in general not the case [12].

In order to check the satisfiability of the conjunction of membership constraints, we need to be able to compute the *intersection* of sets. Furthermore, the entailment problem for two membership constraints amounts to the computation of the *subset relation* for the two corresponding sets. Our constraint system provides both computations. Note that we can *not* decide the subset relation $p \subseteq q$ with an emptiness test of $p \cap q^c$, since the family of sets defined by deterministic equation systems is closed under intersection but neither under union nor under complement (either would lead to inefficiency by combinatorial explosion). Instead, we will give a system of transformation rules on conjunctions of subset formulas $p \subseteq q$ according to the equation system, and again apply the memorization technique.

Entailment tests for feature constraints, which refine equational constraints for infinite trees, have been treated in [19, 2]. In most of those contexts rational and infinite trees can not be distinguished by means of logical formulae [3, 14].

Membership constraints over sets of finite trees have been considered in [6, 22]. The case of finite feature trees is discussed in [15]. In these works (generalized) tree automata or regular equation systems with least fixpoint solutions are used. The proposed simplification algorithms are *not* efficient, since the union in the set defining devices is not restricted such that combinatoric explosion is possible.

As an alternative to the approach chosen here, we could have taken Rabin automata to define sets of infinite trees (*cf.*, [21]). In a constraint system for

---

[4] Relative simplification [19, 2] of a constraint ϕ relatively to a constraint ψ means that we transform ϕ into a constraint ϕ' which is equivalent to ϕ modulo ψ (*i.e.*, $\phi \wedge \psi$ is equivalent to $\phi' \wedge \psi$).

CCP, however, it would be irrealistic to hope that one could use this theory. The complexity of the algorithms involved is far too high. Clearly, we don't need the expressiveness of the corresponding second-order logic.

A full version of this paper with complete proofs is published as [16].

2 Equational and Membership Constraints

We assume a non empty, finite or infinite, one-sorted signature Σ of function symbols $f, g, \ldots$. $\mathcal{IT}$ denotes the set of all finite and infinite trees over Σ. We also assume an infinite alphabet of variables ranged over by x, y, z and a possibly infinite collection $\mathcal{Q}$ of set expressions ranged over by p, q. We will be more specific about the set expressions in Section 3.

Finite sequences of set expressions and variables are abbreviated as $\bar{p}$ and $\bar{x}$. We will also use similar notions like $\bar{x} \doteq \bar{y}$ or $\bar{x} \dot{\in} \bar{p}$ for finite sequences of formulae.

As *atomic constraints* we take *equational constraints* of the form $x \doteq y$ or $x \doteq f(\bar{y})$, *membership constraints* $x \dot{\in} p$ and $\bot$. The set of *constraints* is the closure of the set of atomic constraints under conjunction and existential quantification. $\top$ is a constraint standing for the empty conjunction. Note that, without loss of generality, we consider only flat terms $f(\bar{y})$. The symbols for constraints of several restricted forms are given in Figure 1. A membership constraint $x \dot{\in} p$

$$\theta ::= x \doteq y \mid \top \mid \theta \wedge \theta'$$
$$\eta ::= x \doteq f(\bar{y}) \mid \top \mid \eta \wedge \eta'$$
$$\mu ::= x \dot{\in} q \mid \top \mid \mu \wedge \mu'$$

$$\phi ::= \theta \wedge \eta \wedge \mu \mid \bot$$

Fig. 1. The fragments of constraints without $\exists$.

can be seen as a convenient notion for the application $p(x)$ of a unary predicate p to the variable x.

As semantics of this first order language we consider $\mathcal{IT}$-structures. These are structures with the domain $\mathcal{IT}$ that interpret the function symbols f of Σ as the pertaining tree constructor $f^{\mathcal{IT}}$. The possible interpretations of the unary predicate symbols of $\mathcal{Q}$ will be restricted in Section 3 by the choice of special $\mathcal{IT}$-structures. It is understood that $\bot$ and $\doteq$ get their standard meaning. As usual, we use the notions of existential (resp. universal) closure, $\tilde{\exists}w$ (resp. $\tilde{\forall}w$), and the set of free variables $\mathcal{V}(w)$ occurring in w.

The notion of a structure $\mathcal{A}$ being a model of a closed formula w ($\models_\mathcal{A} w$) is defined as usual. An arbitrary formula w is *satisfiable* in a structure $\mathcal{A}$ if $\models_\mathcal{A} \tilde{\exists} w$, otherwise it is *unsatisfiable* in $\mathcal{A}$. A formula v *entails* a formula w in a structure $\mathcal{A}$ ($v \models_\mathcal{A} w$) if $\models_\mathcal{A} \tilde{\forall}(v \rightarrow w)$. Two formulae v, w are *equivalent* in a structure $\mathcal{A}$ ($v \models\!\!\!\dashv_\mathcal{A} w$) if $\models_\mathcal{A} \tilde{\forall}(v \leftrightarrow w)$. The notions of entailment and equivalence can be extended to classes of structures. Sometimes, we furthermore use the notion $v \models_\mathcal{A}^\phi w$ for $\phi \models_\mathcal{A} \tilde{\forall}(v \rightarrow w)$ and $v \models\!\!\!\dashv_\mathcal{A}^\phi w$ for $\phi \models_\mathcal{A} \tilde{\forall}(v \leftrightarrow w)$.

3 Set Definitions

When simplifying membership constraints such as $x \mathbin{\dot\in} p \land x \mathbin{\dot\in} q$, we need set expressions representing intersections. Therefore, we require that the set $\mathcal{Q}$ of set expressions is closed under $\cap$, which is taken to be an associative, commutative and idempotent constructor for set expressions. For instance, $q \cap (p \cap q)$ is identified with $p \cap q$.

The possible interpretations of the unary predicates are described by a given *regular system of equations* $\mathcal{E}$. This is a set of equations of one of the two following forms:

$$q = f_1(\overline{q_1}) \cup \ldots \cup f_n(\overline{q_n}) \quad \text{or} \quad q = \top. \tag{1}$$

We restrict the union in the equations to be *deterministic*, which means that the constructors on the right hand side of an equation have to be pairwise distinct. In particular the empty disjunction, denoted as $\bot$, is allowed.

We say that a set expression is *defined in* $\mathcal{E}$, if it appears on the left hand side of an equation in $\mathcal{E}$. We require that no set expression is defined twice and that each set expression appearing on the right hand side of $\mathcal{E}$ is defined. In the following sections we will often consider a constraint together with an equation system $\mathcal{E}$ and assume that all the set expressions used in the constraint are defined in $\mathcal{E}$.

A structure $\mathcal{A}$ is a *model of* $\mathcal{E}$ if the statement

$$x \mathbin{\dot\in} q \models_{\mathcal{A}} \exists \overline{y_1} \ldots \exists \overline{y_n} ((x \mathbin{\dot=} f(\overline{y_1}) \land \overline{y_1} \mathbin{\dot\in} \overline{q_1}) \lor \ldots \lor (x \mathbin{\dot=} f(\overline{y_n}) \land \overline{y_n} \mathbin{\dot\in} \overline{q_n}))$$

holds for all equations in $\mathcal{E}$ of the first form of (1), and if $x \mathbin{\dot\in} q \models_{\mathcal{A}} \top$ holds in the second case of (1).

An equation system $\mathcal{E}$ can be considered as a syntactic characterization of its $\mathcal{IT}$-models. Therefore, we identify $\mathcal{E}$ with its $\mathcal{IT}$-models in notions like $v \models_{\mathcal{E}} w$ and $v \models_{\mathcal{E}} w$.

We restrict ourselves to equation systems $\mathcal{E}$ with *appropriate definitions of compound set expressions*. If p, q and $p \cap q$ are defined in $\mathcal{E}$, then we require:

$$x \mathbin{\dot\in} p \cap q \models_{\mathcal{E}} x \mathbin{\dot\in} p \land x \mathbin{\dot\in} q.$$

We will often make use of the following observation. If η contains the equation $x \mathbin{\dot=} f_i(\overline{y})$, then we get by the determinism condition of $\mathcal{E}$ and the restriction to tree structures:

$$x \mathbin{\dot\in} q \models_{\mathcal{E}}^{\eta} \overline{y} \mathbin{\dot\in} \overline{q_i}.$$

In the rest of this section we discuss computational properties of the greatest $\mathcal{IT}$-model $\mathfrak{M}$ and the least $\mathcal{IT}$-model $\mathfrak{m}$ of an equation system $\mathcal{E}$.

The set of $\mathcal{IT}$-structures over the defined set expressions of $\mathcal{E}$ is a complete lattice in its canonical order. We denote its greatest and least elements by $\mathcal{A}_\top$ and $\mathcal{A}_\bot$. The equation system $\mathcal{E}$ defines a monotone operator, also called $\mathcal{E}$, on this lattice. If $q^{\mathcal{A}}$ denotes the interpretation of the unary relation q in the $\mathcal{IT}$-structure $\mathcal{A}$, then the definition of the $\mathcal{IT}$-structure $\mathcal{E}(\mathcal{A})$ is given by:

$$q^{\mathcal{E}(\mathcal{A})} := \bigcup_{i=1}^{n} f_i^{\mathcal{IT}}(\overline{q_i^{\mathcal{A}}}) \quad \text{if } q = f_1(\overline{q_1}) \cup \ldots \cup f_n(\overline{q_n}) \text{ in } \mathcal{E}$$
$$q^{\mathcal{E}(\mathcal{A})} := \mathcal{IT} \quad \text{if } q = \top \text{ in } \mathcal{E}.$$

Hence, the $\mathcal{IT}$-models $\mathcal{A}$ of $\mathcal{E}$ are exactly the fixpoints of the operator $\mathcal{E}$. By monotonicity and Tarski's fixed point theorem (see for instance [7]) the operator $\mathcal{E}$ has a least and a greatest fixpoint. This proves the existence of $\mathfrak{m}$ and $\mathfrak{M}$.

The operator $\mathcal{E}$ is upward and downward continuous, as the reader easily verifies. This means that $\mathcal{E}$ preserves least upper (greatest lower) bounds of every upward (downward) directed chain $\mathcal{A}_\alpha$, i.e. $\mathcal{E}(\sup \mathcal{A}_\alpha) = \sup \mathcal{E}(\mathcal{A}_\alpha)$ ($\mathcal{E}(\inf \mathcal{A}_\alpha) = \inf \mathcal{E}(\mathcal{A}_\alpha)$). With an application of Kleene's fixed point theorem to the complete lattice of $\mathcal{IT}$ structures and to its dual lattice, we get that the least (resp. greatest) fixed points of $\mathcal{E}$ can be reached in ω iteration steps from bottom (resp. top). This is well known for the least fixed point of $\mathcal{E}$, since $\mathcal{E}$ considered as a logic program defines an upward continuous operator. For the greatest fixed point of $\mathcal{E}$ it is surprising, since it takes in general more than ω steps to iterate the greatest fixed point of a logic program from top [12].

Lemma 1. $$\mathfrak{m} = \bigcup_{m=0}^{\infty} \mathcal{E}^m(\mathcal{A}_\bot) \quad \text{and} \quad \mathfrak{M} = \bigcap_{m=0}^{\infty} \mathcal{E}^m(\mathcal{A}_\top).$$

We intend to interpret set expressions in greatest $\mathcal{IT}$-models $\mathfrak{M}$. Therefore we call a subset of $\mathcal{IT}$ *definable*, if it is a component of the greatest $\mathcal{IT}$-model of some deterministic equation system. An example of a non-definable set is $\{f(a,a), f(b,b)\}$, since our equation systems are deterministic. In general, the restrictions of $\mathcal{IT}$ definable sets to finite trees are exactly those that are recognizable by a deterministic tree automaton.

4 Normal Forms of Constraints

In order to decide the satisfiability of constraints, we present a transformation of constraints into either $\bot$ or a satisfiable normal form. Since $\exists x \phi$ is satisfiable iff ϕ is, we will restrict ourselves to constraints without existential quantification. These are considered as multisets of atomic constraints. In other words the conjunction is seen to be associative and commutative, but not idempotent.

Since we use membership constraints, all our normal forms are calculated with respect to the maximal model $\mathfrak{M}$ of an equation system $\mathcal{E}$.

A variable is called *constrained* (in ϕ) if it appears on the left hand side of an atomic constraint in ϕ which is not equivalent to $\top$. With $\mathcal{C}(\phi)$ we denote the set of all constrained variables in the multiset ϕ. The problem $\phi \models_{\mathfrak{M}} \top$ can be decided syntactically. This is trivial for infinite, and a little bit more complicated for finite signatures. For example, let $\mathcal{E}$ contain the definition of *Nat* from the introduction and let ϕ be the constraint $x \doteq x \wedge y \dot\in Nat$. Then $\mathcal{C}(\phi) = \emptyset$, if the signature consists of $\{succ, 0\}$ only, and $\mathcal{C}(\phi) = \{y\}$ otherwise.

For the case of an infinite signature, x is always constrained in $x \doteq f(\bar{y})$ and constrained in $x \dot\in p$ iff $p = \top$ is not in $\mathcal{E}$ (both statement can be wrong for finite signatures). Note that $x \doteq y$ constrains x if $x \not\equiv y$, but not y.

Definition 2. A constraint ϕ is in *normal form*, iff $\phi = \theta \wedge \eta \wedge \mu$ with

1. every variables of ϕ is constrained at most once.
2. every variable constrained in θ does not occur in $\eta \wedge \mu$.
3. μ is satisfiable in $\mathfrak{M}$.

ψ is a *normal form of* ϕ, if ψ is in normal form and $\phi \models_\mathfrak{M} \psi$.

A normal form θ can be considered as an idempotent substitution with domain $\mathcal{C}(\theta)$. The application of this substitution to a formula w is denoted by θw and corresponds exactly to the elimination of the constrained variables of θ in w.

The following proposition implies in particular the satisfiability of normal forms. We will exploit this proposition again for the entailment check.

Proposition 3. *If ϕ is in normal form, then every assignment of the non constrained variables of ϕ can be extended to a solution of ϕ in $\mathfrak{M}$:*

$$\models_\mathfrak{M} \tilde\forall \exists \mathcal{C}(\phi) \phi .$$

The proof reduces immediately to the case of equational constraints only, which has been solved in [14].

Normal forms of equational constraints can be obtained by the well known unification rules. We obtain normal forms of arbitrary constraints in four steps. First we calculate a normal form $\theta \wedge \eta$ of the equational part. Second we apply θ to the membership part and call the result μ. Third we simplify μ relative to η and $\mathcal{E}$ by memorization. In the last step we calculate intersections and detect unsatisfiable membership constraints.

The memorization technique is described by schemes of rewrite rules which depend on η and $\mathcal{E}$. It transforms expressions of the form $\mu \square \mu'$, where $\square$ is a new symbol. We say that μ_0 *simplifies to* μ_1 *relative to* η *and* $\mathcal{E}$, if there is a μ'_1 such that $\mu_0 \square \top$ rewrites to $\mu_1 \square \mu'_1$ relative to η and $\mathcal{E}$. In this case we will prove μ_0 and μ_1 to be equivalent relative to η in $\mathfrak{M}$ (correctness of memorization).

On the right hand side of $\square$ we memorize the constraints which have already been unfolded. The rules are presented in Figure 2. They forbid multiple unfolding of the same constraint and delete those that have been unfolded before.

The termination of memorization is obvious. The main problem is the correctness of the *memo*-rule, which is proven in Section 5.

$$\frac{x \dot\in Inf \; \square \; \top}{x \dot\in Inf \; \square \; x \dot\in Inf} \textit{unfold}$$
$$\frac{x \dot\in Inf \; \square \; x \dot\in Inf}{\top \; \square \; x \dot\in Inf} \textit{memo}$$

A typical example is the simplification of the constraint $x \dot\in Inf$ relative to $x \doteq succ(x)$ and $\mathcal{E}$ containing $Inf = succ(Inf)$. in $\mathfrak{M}$).

The last set of rules handles empty sets in membership constraints and conjunctions of membership constraints for the same variable. It is given in Figure 3. During its execution we want to maintain two invariants. First, each occurring

unfold	$\dfrac{x \dot{\in} q \wedge \mu \Box \mu'}{\bar{y} \dot{\in} \bar{p} \wedge \mu \Box x \dot{\in} q \wedge \mu'}$	if $x \dot{\in} q$ is not in μ', $q = \ldots \cup f(\bar{p}) \cup \ldots$ is in $\mathcal{E}$, and $x \doteq f(\bar{y})$ is in η.
memo	$\dfrac{x \dot{\in} q \wedge \mu \Box \mu'}{\mu \Box \mu'}$	if $x \dot{\in} q$ is in μ'
clash 2	$\dfrac{x \dot{\in} q \wedge \mu \Box \mu'}{\bot \Box \mu'}$	if $x \doteq f(\bar{y})$ is in η the definition of q does not contain f, and $q = \top$ is not in $\mathcal{E}$.

Fig. 2. Simplification of Memberships relative to Equational Constraints with Memorization.

intersect	$\dfrac{x \dot{\in} p \wedge x \dot{\in} q}{x \dot{\in} p \cap q}$	
empty	$\dfrac{x \dot{\in} p \wedge \mu}{\bot}$	if $p = \bot$ is in $\mathcal{E}$

Fig. 3. Simplification of Empty Sets and Conjunctions of Membership.

set expression should be defined in $\mathcal{E}$. This means that we have to calculate equations for intersections and to extend the equation system by need without changing the interpretations of previously defined set expressions in $\mathfrak{M}$. This will be done in Section 7.

Second, a set expression p should be defined by $p = \bot$ iff $p^{\mathfrak{M}} = \emptyset$. This can easily be done by propagating $\bot$ in $\mathcal{E}$.

Theorem 4. *When started with the constraint ϕ, the above algorithm terminates with $\bot$ if ϕ is unsatisfiable in $\mathfrak{M}$, and in a normal form of ϕ otherwise.*

Here is an example that illustrates our algorithm in action. $\mathcal{E}$ contains the equations for *Nat*, *NatList*, *Even* and *Odd* from the introduction. We compute a normal form of

$$x \doteq cons(y, x) \wedge x \doteq cons(z, x) \wedge x \dot{\in} NatList \wedge y \dot{\in} Even \wedge z \dot{\in} Odd .$$

The equational part simplifies to $\theta \wedge \eta$ with $\theta = y \doteq z \wedge x \doteq x$, $\eta = x \doteq cons(z, x)$. By applying θ to the membership part we get $\mu = x \dot{\in} NatList \wedge z \dot{\in} Even \wedge z \dot{\in} Odd$ The memorization algorithm simplifies μ relative to η and $\mathcal{E}$ to $\mu_1 = z \dot{\in} Nat \wedge z \dot{\in} Even \wedge z \dot{\in} Odd$. This is transformed with the intersection rule to $z \dot{\in} Nat \cap Even \cap Odd$. The intersection algorithm of Section 7 adds the following equation to $\mathcal{E}$: $Nat \cap Even \cap Odd = succ(Nat \cap Even \cap Odd)$. To be precise it also adds an equation for $Even \cap Odd$, $Nat \cap Even$ or $Nat \cap Odd$ depending on which intersection is calculated first. We get the normal form

$$y \doteq z \wedge x \doteq x \wedge x \doteq cons(z, x) \wedge z \dot{\in} Nat \cap Even \cap Odd .$$

This normal form is satisfiable since $\infty \in (Nat \cap Even \cap Odd)^{\mathfrak{M}}$. Note that we could replace $z \dot{\in} Nat \cap Even \cap Odd$ by the $\mathfrak{M}$-equivalent constraint $z \doteq succ(z)$. This will be necessary in the entailment check.

5 Correctness of Relative Simplification with Memorization

Since the *clash* rule terminates the rewriting, we can restrict ourselves to the *memo* and *unfold* rule. The relation 'rewrites to in one *memo* or *unfold* step' on expressions of the form $\mu \square \mu'$ will be denoted by $\triangleright_{\eta,\mathcal{E}}$ and its reflexive transitive closure by $\triangleright_{\eta,\mathcal{E}}^*$.

Roughly speaking, the following theorem states that the symbol $\square$ can be interpreted as the logical connective $\wedge$ with respect to all $\mathcal{IT}$-models of $\mathcal{E}$, and also as $\rightarrow$ with respect to the greatest $\mathcal{IT}$-model $\mathfrak{M}$.

Theorem 5 Correctness. *For each computation* $\mu_0 \square \mu_0' \triangleright_{\eta,\mathcal{E}}^* \mu_1 \square \mu_1'$ *the following two statements are invariant:*

$$\mu_0 \wedge \mu_0' \models_\mathcal{E}^\eta \mu_1 \wedge \mu_1' \quad \text{and} \quad \mu_0' \models_\mathfrak{M}^\eta \mu_1 \rightarrow \mu_1'.$$

If μ_0 simplifies to μ_1 by memorization relative to η and $\mathcal{E}$ then $\mu_0 \models_\mathfrak{M}^\mathcal{E} \mu_1$ holds.

Only the statement about $\rightarrow$ is interesting to prove, since the assumption of the implication is weakened by the *memo* rule.

It can be proven with the help of Lemma 6, which reflects an important property of the greatest $\mathcal{IT}$-model $\mathfrak{M}$.

In order to be general enough we need the concept of *derivable constructors*. If $f \in \Sigma$ is a constructor with arity n, m a natural number and $\sigma : \{1, \cdots, n\} \rightarrow \{1, \cdots, m\}$ an arbitrary mapping, then the pair f_σ is called derivable constructor with arity m. The interpretation of f_σ in a Σ-structure $\mathcal{I}$ is defined by

$$f_\sigma^\mathcal{I}(d_1, \ldots, d_m) = f^\mathcal{I}(d_{\sigma(1)}, \ldots, d_{\sigma(n)})$$

for all elements d_i of the domain of $\mathcal{I}$. Each constructor is itself a derivable constructor, since we may chose σ to be the identity. We will freely use derivable constructors as abbreviations in terms. For example $f_\sigma(x,y)$ stands for $f(y,y,y)$ if σ is the mapping $\sigma(1) = \sigma(2) = \sigma(3) = 2$. In the sequel we will not distinguish between constructors and derivable constructors.

Using this notion in the rest of this section we will always assume finite sequence of objects to have the form $\bar{o} = (o_i)_i$.

Lemma 6 Main. *Let $\mathcal{E}$ be an equation system, $\bar{p}$ and $\bar{q}$ finite sequences of set expressions, $\bar{x}$ and $\bar{y}$ finite sequences of variables, $\bar{f}$ a finite sequence of derivable constructors and η a constraint. Under the assumptions*

$$p_j = \ldots \cup f_j(\bar{p}, \bar{q}) \cup \ldots \text{ in } \mathcal{E} \quad \text{and} \quad \models_\mathfrak{M}^\eta x_j \doteq f_j(\bar{x}, \bar{y})$$

the following implication relative to η and the greatest model $\mathfrak{M}$ of $\mathcal{E}$ is valid:

$$\models_\mathfrak{M}^\eta \bar{y} \dot{\in} \bar{q} \rightarrow \bar{x} \dot{\in} \bar{p}.$$

In order to illustrate the contents of the Main Lemma, let η be $x_1 \doteq f(x_1, x_2) \wedge x_2 \doteq f(x_1, x_2)$ and let $\mathcal{E}$ contain the equations $p_1 \doteq f(p_1, p_2)$ and $p_2 \doteq f(p_1, p_2)$. The Main Lemma implies $\top \models^\eta_{\mathfrak{M}} x_1 \dot\in p_1 \wedge x_2 \dot\in p_2$. Note that this does not hold in any other solution of $\mathcal{E}$, i.e. for $p_1^{\mathfrak{m}} = \emptyset$ and $p_2^{\mathfrak{m}} = \emptyset$.

Proof of Theorem 5. For simplicity we assume $\mu'_0 = \top$. We call the expression $\bar{y} \dot\in \bar{q} \square \bar{x} \dot\in \bar{p}$ *appropriate with respect to η and $\mathcal{E}$* iff there is a finite sequence of derived constructors $\bar{f}$ and a finite sequence of variables $\bar{y}$ such that for all j the equation $p_j = \ldots \cup f_j(\bar{p}, \bar{q}) \cup \ldots$ is in $\mathcal{E}$ and $\models^\eta x_j \doteq f_j(\bar{x}, \bar{y})$ holds. $\mu_0 \square \top$ is appropriate even for arbitrary η and $\mathcal{E}$. We can show that *unfold* and *memo* steps relative to η and $\mathcal{E}$ maintain appropriateness relative to η and $\mathcal{E}$. Therefore $\mu_1 \square \mu'_1$ is appropriate relative to η and $\mathcal{E}$. The Main Lemma yields $\models^\eta_{\mathfrak{M}} \mu_1 \to \mu'_1$.

6. The Entailment Check

In this section, we show how to decide entailment between existentially quantified constraints in the greatest model $\mathfrak{M}$. *For the purpose of this section we assume that the signature contains at least two elements, since otherwise the domain of the models under consideration will be singleton, and hence* every *equation will hold.* Initially, we are given the question whether

$$\exists X'\phi' \models_{\mathfrak{M}} \exists X \phi \qquad (2)$$

holds, where X, X' are finite sets of variables. We may assume without loss of generality that ϕ' is satisfiable in $\mathfrak{M}$, since otherwise (2) holds vacuously. Hence, we can by Theorem 4 assume ϕ' to be in normal form. For the purpose of entailment checks it is convenient to exclude certain forms of degenerate membership constraints. Hence, we furthermore require that for all membership constraints $x \dot\in p$ of ϕ' the definition of p is disjunctive. A normal form meeting this additional condition is called a *branching normal form*. Note that a branching normal form contains only membership constraints $x \dot\in p$ for which $p^{\mathfrak{M}}$ is not singleton. We can always transform a normal form into an equivalent branching normal form by introducing new existentially quantified variables:

- A membership constraint $x \dot\in p$, where $p^{\mathfrak{M}}$ is singleton, is equivalent to a corresponding equation. For example, if $\mathcal{E}$ contains $p = f(q)$ and $q = g(p)$, then $x \dot\in p$ is in $\mathfrak{M}$ equivalent to $\exists y (x \doteq f(y) \wedge y \doteq g(x))$.
- A membership constraint $x \dot\in p$, where the definition of p is of the form $p = f(\bar{q})$ and where $p^{\mathfrak{M}}$ is not singleton, is replaced by $\exists \bar{y}(x \doteq f(\bar{y}) \wedge \bar{y} \dot\in \bar{q})$.

Both rules maintain normal forms (modulo existential quantification) and terminate, since the second rule applies only when $p^{\mathfrak{M}}$ is not a singleton. Note that by Lemma 6, $p^{\mathfrak{M}}$ is non-singleton iff the definition of p depends on a definition which is disjunctive, or which is $\top$.

Taking the definition of *Inf*, *Even* and *Odd* as given in the introduction, we transform the existentially quantified normal form $\exists y(x \doteq Inf \land y \dot\in Odd)$ into the existentially quantified branching normal form

$$\exists y \exists z (x \doteq succ(x) \land y \doteq succ(z) \land z \dot\in Even) .$$

The next lemma states that membership constraints in a branching normal form can not contribute to equalities:

Lemma 7. *Let $\eta' \land \mu'$ be in branching normal form. Then $\eta' \land \mu' \models_{\mathfrak{M}} \theta$ iff $\eta' \models_{\mathfrak{M}} \theta$.*

Hence, we assume without loss of generality in (2) that ϕ' is in branching normal form. Since we may in (2) assume without loss of generality that X' is disjoint to $\mathcal{V}(\phi)$, we may drop the existential quantifier on the left hand side. The normal form ϕ' can be written as $\theta' \land \eta' \land \mu'$. Since θ' is an idempotent substitution and since we may assume X to be disjoint to $\mathcal{V}\theta'$, we arrive at the problem

$$\eta' \land \mu' \models_{\mathfrak{M}} \exists X \theta' \phi$$

where $\eta' \land \mu'$ is in branching normal form.

Before we state the entailment theorem we consider the special case of a right hand side consisting of equations only. We say that some θ is *complete* for some η if

$$\theta \models x \doteq y \quad \text{and} \quad x \doteq f(\bar{x}) \in \eta \quad \text{and} \quad y \doteq f(\bar{y}) \in \eta \quad \text{implies} \quad \theta \models \bar{x} \doteq \bar{y} .$$

For instance, $x \doteq v \land y \doteq v$ is complete for $x \doteq f(x,y) \land y \doteq f(v,x) \land v \doteq f(y,x)$.

Lemma 8 Determined Equations. *Let θ be complete for η', let θ contain no trivial equation $x \doteq x$ and let $\theta \land \eta'$ be satisfiable in $\mathfrak{M}$. Then $\eta' \models_{\mathfrak{M}} \theta$ iff $\mathcal{V}(\theta) \subseteq \mathcal{C}(\eta')$.*

A proof of a more general lemma (in the context of feature constraints) has been given in [20].

Before we state the entailment theorem we have to introduce some more notation. We call θ *X-directed* if θ contains no equation $x \doteq y$ with $x \notin X$ and $y \in X$. For a constraint ϕ we define ϕ^X to be the subset of atomic constraints which constrain only variables from X, and ϕ^{-X} to be the subset of atomic constraints which constrain only variables alien to X. Since every constraint is is either equivalent to $\top$ or constrains a variable, we have $\phi \models_{\mathfrak{M}} \phi^X \land \phi^{-X}$.

Definition 9. *Let $\eta' \land \mu'$ be in branching normal form. The constraint $\exists X(\theta \land \eta \land \mu)$ is in normal form relative to $\eta' \land \mu'$ if*

1. *θ is X-directed,*
2. *θ is complete for $\eta' \land \eta$, and $\theta \land \eta'$ is satisfiable in $\mathfrak{M}$,*
3. *$\mathcal{C}(\eta')$ and $\mathcal{C}(\mu)$ are disjoint,*
4. *$\theta \land \eta \land \mu$ is in normal form.*

For instance,
$$\exists v(v \doteq z \land x \doteq y \land y \doteq f(y) \land z \dot{\in} p) \tag{3}$$
is in normal form relative to
$$x \doteq f(y) \land y \doteq f(x) \land w \doteq h(z) \land z \dot{\in} q . \tag{4}$$

This does not hold if we drop $x \doteq y$ from (3), since then clause 2 of Definition 9 is violated.

Theorem 10 Entailment. *Let $\eta' \land \mu'$ be in branching normal form, let X be disjoint to $\mathcal{V}(\eta' \land \mu')$ and let $\exists X(\theta \land \eta \land \mu)$ be in normal form relative to $\eta' \land \mu'$. Then $\eta' \land \mu' \models_{\mathfrak{M}} \exists X(\theta \land \eta \land \mu)$ iff the three following statements hold:*

1. $\mathcal{V}(\theta^{-X}) \subseteq \mathcal{C}(\eta')$,
2. *for every $x \dot{\in} p$ in μ^{-X} there is an $x \dot{\in} q$ in μ' with $q^{\mathfrak{M}} \subseteq p^{\mathfrak{M}}$,*
3. $\eta^{-X} \subseteq \theta\eta'$.

For instance (4) $\models_{\mathfrak{M}}$ (3) holds provided that $q^{\mathfrak{M}} \subseteq p^{\mathfrak{M}}$.

Next we show how to transform a constraint $\exists X \phi$ into normal form relative to $\eta' \land \mu'$. First, we transform the equational part of ϕ using the rules of Figure 4. This rules are equivalence transformations relative to η' in all $\mathcal{IT}$ structures. The rules terminate with either $\bot$ or with $\theta \land \eta$, such that the clauses 1 and 2 of Definition 9 hold. Let μ be the membership part of ϕ. Now we simplify $\theta\mu$

$\dfrac{x \doteq y \land \theta \land \eta}{x \doteq y \land (\theta \land \eta)[x \leftarrow y]}$	$x \neq y, x \in \mathcal{V}(\theta \land \eta)$
$\dfrac{x \doteq y \land \theta \land \eta}{y \doteq x \land \theta \land \eta}$	$x \notin X, y \in X$
$\dfrac{\theta \land \eta}{\bot}$	θ is substitution, $x \doteq f(\bar{y}) \land x \doteq g(\bar{z}) \subseteq \eta \land \theta\eta', f \neq g$
$\dfrac{\theta \land \eta}{\bar{y} \doteq \bar{z} \land \theta \land \eta}$	θ is substitution, $x \doteq f(\bar{y}) \land x \doteq f(\bar{z}) \subseteq \eta \land \theta\eta'$

Fig. 4. Relative Simplification of Equations.

relative to $\eta \land \eta'$ as explained in Section 4. Finally, we simplify constraints of the form $x \dot{\in} p \land x \dot{\in} q$ and check for membership in empty sets, as explained in Section 4. If this does not lead to $\bot$, we arrive at a relative normal form.

As an example of equational simplification, the constraint
$$\exists v(x \doteq v \land v \doteq f(v))$$
simplifies relative to $x \doteq f(y) \land y \doteq f(z) \land z \doteq f(x)$ to
$$\exists v(v \doteq z \land x \doteq z \land y \doteq z \land z \doteq f(z)) .$$

7 Equations for Intersections

We need an algorithm that extends a deterministic equation system $\mathcal{E}$ containing definitions of p and q by an appropriate definition for $p \cap q$.

In the terminology of model theory, we will extend the formula $\mathcal{E}$ to a formula $\mathcal{E}'$, such that every $\mathcal{IT}$-model of $\mathcal{E}$ extends conservatively to a $\mathcal{IT}$-model of $\mathcal{E}'$, and such that $x \in p \cap q \models_{\mathcal{E}'} x \in p \land x \in q$ for all new set expressions.

This extension can be achieved by iterated applications of the non-deterministic rewrite rules in Figure 5 that are easily proven correct in the above sense. Note that the rules maintain determinism of equation systems. This completion

$$int\ 1 \quad \frac{\mathcal{E}}{\mathcal{E} \cup \{e\}} \quad \begin{array}{l} p = f_1(\overline{p_1}) \cup \ldots \cup f_n(\overline{p_n}) \cup f_{n+1}(\overline{p_{n+1}}) \cup \ldots \text{ in } \mathcal{E} \\ q = f_1(\overline{q_1}) \cup \ldots \cup f_n(\overline{q_n}) \cup g_{n+1}(\overline{q_{n+1}}) \cup \ldots \text{ in } \mathcal{E} \\ \text{with } f_j \neq g_k \text{ for all } j, k \geq n+1. \\ e \text{ is } p \cap q = f_1(\overline{p_1} \cap \overline{q_1}) \cup \ldots \cup f_n(\overline{p_n} \cap \overline{q_n}) \end{array}$$

$$int\ 2 \quad \frac{\mathcal{E}}{\mathcal{E} \cup \{e\}} \quad \begin{array}{l} \text{if } q = \top \text{ and } p = def_p \text{ are contained in } \mathcal{E} \\ \text{and } e \text{ is the equation } p \cap q = def_p. \end{array}$$

Fig. 5. Computation of Intersections

process can be organised in a terminating manner, by adding $p \cap q$ to $\mathcal{E}$ only under the assumption that p, q and all set expressions on the right hand sides of $\mathcal{E}$ are defined in $\mathcal{E}$. For example we can extend an equations system $\mathcal{E}$ containing the above definitions for *Even* and *Nat* with an equation for $Even \cap Nat$. First the first rule adds the equation $Even \cap Nat = 0 \cup succ(Odd \cap Nat)$. A further application the same rule adds $Odd \cap Nat = succ(Even \cap Nat)$.

8 Deciding the Subset Relation

We will decide the subset relation $p^{\mathcal{A}} \subset q^{\mathcal{A}}$ for $\mathcal{A} = \mathfrak{m}$ or $\mathcal{A} = \mathfrak{M}$ using the memorization technique. Note that this includes a subset check for sets recognized by deterministic top down tree automata as well as for $\mathcal{IT}$ definable sets.

Therefore we define the following fragment of new constraints:

$$\Gamma ::= p \subset q \mid \top \mid \Gamma \land \Gamma'.$$

The memorization technique is carried out by rewriting expressions of the form $\Gamma \square \Gamma'$. We say that Γ_0 simplifies to Γ_1 relative to $\mathcal{E}$ if there is a Γ_1' such that $\Gamma_0 \square \top$ rewrites to $\Gamma_1 \square \Gamma_1'$ relative to $\mathcal{E}$.

Without loss of generality we make two assumptions on E. First we assume that $p^{\mathcal{A}} = \emptyset$ iff $p = \bot$ in $\mathcal{E}$, and that set expressions which are used on the right hand side of $\mathcal{E}$ do not denote the empty set in $\mathcal{A}$. Second, we assume that $p^{\mathcal{A}} = \mathcal{IT}$ iff $p = \top$ in $\mathcal{E}$. Both conditions can be assured for $\mathfrak{M}$ as well as for $\mathfrak{m}$,

$$unfold1 \quad \frac{p \,\dot{\subset}\, q \wedge \Gamma \square \Gamma'}{\Gamma_1 \wedge \Gamma \square p \,\dot{\subset}\, q \wedge \Gamma'} \quad \begin{array}{l} \text{if } p \,\dot{\subset}\, q \text{ is not in } \Gamma', \text{ the equations} \\ p = f_1(\overline{p_1}) \cup \ldots \cup f_n(\overline{p_n}) \\ q = f_1(\overline{q_1}) \cup \ldots \cup f_n(\overline{q_n}) \cup \ldots \\ \text{are in } \mathcal{E} \text{ and } \Gamma_1 = \overline{p_1} \,\dot{\subset}\, \overline{q_1} \wedge \ldots \wedge \overline{p_n} \,\dot{\subset}\, \overline{q_n}. \end{array}$$

$$memo1 \quad \frac{p \,\dot{\subset}\, q \wedge \Gamma \square \Gamma'}{\Gamma \square \Gamma'} \quad \text{if } p \,\dot{\subset}\, q \text{ is in } \Gamma'.$$

$$clash3 \quad \frac{p \,\dot{\subset}\, q \wedge \Gamma \square \Gamma'}{\bot \square \Gamma} \quad \begin{array}{l} \text{if } p = \ldots \cup f(\bar{p}) \cup \ldots \text{ is in } \mathcal{E}, \\ \text{but the definition of } q \text{ in } \mathcal{E} \text{ is not of form} \\ q = \ldots \cup f(\bar{q}) \cup \ldots \text{ or } q = \top. \end{array}$$

$$clash4 \quad \frac{p \,\dot{\subset}\, q \wedge \Gamma \square \Gamma'}{\bot \square \Gamma} \quad \begin{array}{l} \text{if } p = \top \text{ is in } \mathcal{E}, \text{ but the definition of } q \\ \text{is not } q = \top. \end{array}$$

Fig. 6. Deciding the Subset Relation with Memorization

for finite as well as for infinite signatures. The rules of the rewrite system are presented in Figure 6.

Theorem 11 Correctness and Completeness. *The rewrite system of Figure 6 terminates. If Γ_0 simplifies to Γ_1 relative to $\mathcal{E}$ then $\Gamma_0 \models_{\mathcal{A}} \Gamma_1$ holds. A constraint $\Gamma_1 \neq \bot$ that can not be simplified is valid.*

Termination and the last statement are trivial. The clash rules are correct by the assumptions on $\mathcal{E}$. It remains to show that the rules $unfold1$ and $memo1$ are correct. This can be done in analogy to Section 5.

9 Future Work

There are several direction of future research. Most important is the complexity analysis of the algorithms presented here. An immediate question is the decidability of the full first-order theory of the constraint system. We conjecture a positive answer, encouraged by the decidability result of [6] for the case of a minimal fixpoint solution of the equation system. Another important extension is the relaxation of the determinism condition. Finally, it will be interesting to apply the methods developed here to the other formalism modeling cyclic data structures: features trees [19, 3, 2].

Acknowledgments. We are grateful to Gert Smolka. He inspired this work and contributed ideas during the whole development. We would like to thank Hassan Aït-Kaci and Hubert Comon for stimulating questions and fruitful discussions.

References

1. H. Aït-Kaci and A. Podelski. Towards a meaning of LIFE. In *Third PLIPS*, pages 255–274. Springer-Verlag, LNCS 528, August 1991.

2. H. Aït-Kaci, A. Podelski, and G. Smolka. A feature-based constraint system for logic programming with entailment. In *5th FGCS*, pages 1012–1022, June 1992.
3. R. Backofen and G. Smolka. A complete and recursive feature theory. Research Report RR-92-30, German Research Center for Artificial Intelligence (DFKI), Stuhlsatzenhausweg 3, 6600 Saarbrücken 11, Germany, September 1992.
4. A. Colmerauer. Equations and inequations on finite and infinite trees. In *2nd FGCS*, pages 85–99, 1984.
5. A. Colmerauer, H. Kanoui, and M. V. Caneghem. Prolog, theoretical principles and current trends. *Technology and Science of Informatics*, 2(4):255–292, 1983.
6. H. Comon and C. Delor. Equational formulae with membership constraints. Rapport de Recherche 649, LRI, Université de Paris Sud, Orsay,France, Mar. 1991.
7. I. Guessarian. Some fixpoint techniques in algebraic structures and applications to computer science. In Aït-Kaci and M. Nivat, editors, *Resolution of Equations in Algebraic Structures*, volume 1, pages 263–292. Academic Press, 1989.
8. S. Haridi and S. Janson. Kernel andorra prolog and its computation model. In *7th ICLP*, pages 31–48, Cambridge, June 1990. MIT Press.
9. M. Henz, G. Smolka, and J. Würtz. Oz - a programming language for multi-agent systems. In *13th IJCAI*, Chambéry, France, Aug. 1993.
10. M. Höhfeld and G. Smolka. Definite relations over constraint languages. LILOG Report 53, IWBS, IBM Deutschland, Postfach 80 08 80, 7000 Stuttgart 80, Germany, Oct. 1988.
11. J. Jaffar and J.-L. Lassez. Constraint logic programming. In *14th POPL*, pages 111–119, Munich, Germany, Jan. 1987.
12. J. W. Lloyd. *Foundations of Logic Programming*. Springer-Verlag, 1984.
13. M. J. Maher. Logic semantics for a class of committed-choice programs. In *Fourth ICLP*, pages 858–876. MIT Press, 1987.
14. M. J. Maher. Complete axiomatizations of the algebras of finite, rational and infinite trees. In *Third LICS*, pages 348–357. IEEE Computer Society, 1988.
15. J. Niehren and A. Podelski. Feature automata and recognizable sets of feature trees. In *Tapssoft*, Apr. 1993.
16. J. Niehren, A. Podelski, and R. Treinen. Equational and membership constraints for infinite trees. Research Report RR-93-14, Deutsches Forschungszentrum für Künstliche Intelligenz, Stuhlsatzenhausweg 3, D-W-6600 Saarbrücken, Germany, 1993.
17. V. Saraswat and M. Rinard. Semantic foundations of concurrent constraint programming. In *18th POPL*, pages 333–351, Jan. 1991.
18. G. Smolka. A calculus for higher-order concurrent constraint programming. Research report, Deutsches Forschungszentrum für Künstliche Intelligenz, Stuhlsatzenhausweg 3, D-W-6600 Saarbrücken, Germany, 1993. Forthcomming.
19. G. Smolka and R. Treinen. Records for logic programming. In *Proceedings of the Joint International Conference and Symposium on Logic Programming*, pages 240–254, Washington, USA, 1992. The MIT Press.
20. G. Smolka and R. Treinen. Records for logic programming. Research Report RR-92-23, Deutsches Forschungszentrum für Künstliche Intelligenz, Stuhlsatzenhausweg 3, D-W-6600 Saarbrücken, Germany, Aug. 1992.
21. W. Thomas. Automata on infinite objects. In J. van Leeuwen, editor, *Handbook of Theoretical Computer Science*, volume B - Formal Models and Semantics, chapter 4, pages 133–191. Elsevier Science Publishers and The MIT Press, 1990.
22. T. E. Uribe. Sorted unification using set constraints. In *11th CADE* LNCS vol. 607, pages 163–177, June 1992. Springer-Verlag.

Regular Path Expressions in Feature Logic*

Rolf Backofen

Deutsches Forschungszentrum für Künstliche Intelligenz (DFKI)
W-6600 Saarbrücken, Germany
`backofen@dfki.uni-sb.de`

Abstract. We examine the existential fragment of a feature logic, which is extended by regular path expressions. A regular path expression is a subterm relation, where the allowed paths for the subterms are restricted by a regular language. We will prove that satisfiability is decidable. This is achieved by setting up a quasi-terminating rewrite system.

1 Introduction

Feature descriptions are used as the main data structure of so-called unification grammars, which have become the predominant paradigm for natural language processing (for a good introduction see [Shi86]). More recently, feature descriptions have been proposed as a constraint system for logic programming (e.g. see [ST92]). They provide for a typical partial description of abstract objects by means of functional attributes called features. As an example consider the feature description (in matrix notation)

$$x : \exists y \begin{bmatrix} woman \\ father : \begin{bmatrix} engineer \\ age : y \end{bmatrix} \\ husband : \begin{bmatrix} painter \\ age : y \end{bmatrix} \end{bmatrix},$$

which may be read as saying that x is a woman whose father is an engineer, whose husband is a painter and whose father and husband are both of the same age.

Feature descriptions have been proposed in various forms with various formalizations. We will follow the logical approach introduced by Smolka [Smo88, Smo92], where feature descriptions are standard first order formulae interpreted in first order structures. In this formalization features are considered as functional relations. Atomic formula (which we will call atomic constraints) are of the

---

* This work was supported by a research grant, ITW 9002 0, from the German Bundesministerium für Forschung und Technologie to the DFKI project DISCO. I would like to thank Jochen Dörre, Joachim Niehren and Ralf Treinen for reading draft version of this paper. An extended version of this paper containing the complete proofs will be published as a technical report [Bac93].

form $A(x)$ or xfy, where x, y are first order variables, A is some sort predicate and f is a feature (written infix notation). Using this notation, we can express the above feature description by the (less suggestive) formula

$$\exists y, x_1, x_2 \ (\ woman(x) \ \land$$
$$x \ father \ x_1 \ \land \ engineer(x_1) \ \land \ x_1 \ age \ y \ \land$$
$$x \ husband \ x_2 \ \land \ painter(x_2) \ \land \ x_2 \ age \ y \ ).$$

This feature logic has been investigated in detail. A complete axiomatization of the standard model (the so-called feature graphs) is given in [BS92]. There it was shown, that the standard model is elementarily equivalent to a tree model. Additionally, some connection to first order constructor terms has been examined [ST92].

In this paper we will be concerned with an extension to feature descriptions, which has been introduced under the notion of "functional uncertainty" by Kaplan and Maxwell [KM88]. This extension is done by adding a subterm relation, where the allowed paths for the subterms are required to be in a given regular language. It was invented for handling so-called long-distance dependencies in the grammar formalism LFG [KB82]. For a detailed description the reader is referred to [KZ88]. Further applications can be found in [Kel91].

For this extension we first have to generalize the constraints of the form xfy to constraints of the form xwy, where $w = f_1 \cdot \ldots \cdot f_n$ is a string of features (called *feature path*). The feature paths are interpreted using simple relational composition.

As Smolka [Smo88] shows, this generalization is just syntactic sugar. This changes if we add functional uncertainty in form of constraints xLy, where L is a regular language of feature paths. A constraint xLy holds if there is a path $w \in L$ such that xwy holds. By this existential interpretation a constraint xLy can be seen as the disjunction

$$xLy = \bigvee \{xwy \mid w \in L\}.$$

Because this disjunction can be infinite, functional uncertainty yields additional expressivity. Note that the constraint xwy can also be expressed by $x\{w\}y$.

Kaplan and Maxwell [KM88] have shown that the satisfiability problem of the pure existential fragment (i.e. the satisfiability of formulae built with $A(x)$, xLy and equations $x \doteq y$) is decidable, provided that a certain acyclicity condition is met. Baader et al. [BBN+91] have shown, that satisfiability is undecidable if negation is added. But it is an open problem whether satisfiability of the pure existential fragment without any additional conditions (such as acyclicity) is decidable. In this work we show that this is indeed decidable, thus filling this gap.

2   The Method

At first we will briefly describe the main part of solving standard feature descriptions and then turn over to the extension by functional uncertainty. To get a good intuition note that there is some sort of tree model which is canonical for satisfiability. This means that a pure existential formula is satisfiable if it is satisfiable in this tree model. Hence, the feature paths used in the language can directly be compared with paths in trees.

For instance, consider a clause $\phi = xp_1y_1 \wedge xp_2y_2$ (in the rest of the paper we will call pure conjunctive formulae *clauses*). Although only subterm relations for x, y_1 and x, y_2 are contained in this clause, some additional subterm or equality relation can be deduced depending on the paths p_1 and p_2. If p_1 equals p_2, then this is also the case for y_1 and y_2, which implies that ϕ is equivalent to $xpy_1 \wedge y_1 \doteq y_2$. If p_1 is a prefix of p_2 (i.e. $p_2 = p_1p'$), we can transform ϕ equivalently into the formulae $xp_1y_1 \wedge y_1p'y_2$, from which we can conclude that y_2 is a subterm of y_1. The reverse case is treated analogous. If neither prefix or equality relation holds between the paths, there is nothing to do. By and large, clauses where this holds for every x and every pair of different constraints $xp_1y \in \phi$ and $xp_2z \in \phi$ are the solved forms in Smolka [Smo88], which are satisfiable.

For a clause of the form $\phi = xL_1y_1 \wedge xL_2y_2$ the relation between y_1 and y_2 has to be checked, too. But now there is in general no unique relation determined by ϕ, since this depends on which paths p_1 and p_2 we choose as elements from L_1 and L_2. Hence, we have to guess the relation between p_1 and p_2 before we can calculate the relation between y_1 and y_2. But there is a problem with the original syntax, namely that it allows not to express any relation between the chosen paths[2]. Therefore we extend the syntax by introducing so-called *path variables* (written as $\alpha, \beta, \alpha', \ldots$), which are interpreted as feature paths. If we use in addition the modified subterm relation $x\alpha y$ and a restriction constraint $\alpha \dot\in L$, a path expression xLy can equivalently be expressed as $x\alpha y \wedge \alpha \dot\in L$ (α being new).

By using this extended (two-sorted) syntax we are now able to reason about the relations between different path variables. To do this we introduce the additional constraints $\alpha \doteq \beta$ (equality), $\alpha \prec\!\!\cdot\, \beta$ (prefix) and $\alpha \ddot{\mathrm{u}}\, \beta$ (divergence). Divergence holds if neither equality nor prefix holds. Now we can describe an equivalent to the solved clauses in Smolka's work, which we will call *pre-solved* clauses. A clause ϕ is pre-solved iff for each pair of different constraint $x\alpha y_1$ and $x\beta y_2$ in ϕ there is a constraint $\alpha \ddot{\mathrm{u}}\, \beta$ in ϕ. We call such clauses pre-solved since they are not necessarily consistent. It may happen that the divergence constraints together with restrictions of the form $\alpha \dot\in L$ are inconsistent (e.g. consider the clause $\alpha \dot\in f^+ \wedge \beta \dot\in ff^+ \wedge \alpha \ddot{\mathrm{u}}\, \beta$). But pre-solved clauses have the property, that if we find a valuation for the path variables, then the clause is satisfiable.

In a first phase our algorithm transforms a clause into a set of pre-solved clauses, which is (seen as a disjunction) equivalent to the initial clause. In a

---

[2] Kaplan and Maxwell solved this problem by directly using operation on regular languages such as intersection and calculation of prefix languages. By using this method they were forced to introduce a new variable each time a transformation rule was applied. For a feature description that contains a cycle of the form $xL_1y_1 \wedge \ldots y_{n-1}L_nx$ this caused the introduction of an infinite number of variables.

second phase the pre-solved clauses are checked for satisfiability with respect to the path variables. In this paper we will mainly concentrate on the first phase, since this is the more difficult one. The second phase is briefly described in section 5.

Before starting with the technical part we will illustrate the first phase. For the rest of the paper we will write clauses as sets of atomic constraints. Now consider the clause $\gamma = \{x\alpha y,\ \alpha_1 \dot\in L_1,\ x\beta z,\ \beta \dot\in L_2\}$. At the beginning one guesses the relation between the path variables α and β. In our example there are four different possibilities. Therefore γ can equivalently be expressed by the set of clauses

$$\gamma_1 = \{\alpha \between \beta,\ x\alpha y,\ \alpha \dot\in L_1,\ x\beta z,\ \beta \dot\in L_2\}$$
$$\gamma_2 = \{\alpha \doteq \beta,\ x\alpha y,\ \alpha \dot\in L_1,\ x\beta z,\ \beta \dot\in L_2\}$$
$$\gamma_3 = \{\alpha \dot\prec \beta,\ x\alpha y,\ \alpha \dot\in L_1,\ x\beta z,\ \beta \dot\in L_2\}$$
$$\gamma_4 = \{\beta \dot\prec \alpha,\ x\alpha y,\ \alpha \dot\in L_1,\ x\beta z,\ \beta \dot\in L_2\}.$$

The clause γ_1 is pre-solved. For the others we have to evaluate the relation between α and β, which is done as follows. For γ_2 we substitute β by α and z by y, which yields

$$\{y \doteq z,\ x\alpha y,\ \alpha \dot\in L_1,\ \alpha \dot\in L_2\}.$$

We keep only the equality constraint for the first order variables since we are only interested in their valuation. Combining $\{\alpha \dot\in L_1, \alpha \dot\in L_2\}$ to $\{\alpha \dot\in (L_1 \cap L_2)\}$ then will give us an equivalent pre-solved clause. For γ_3 we know that the variable β can be split up into two parts, one of them covered by α. We can use concatenation of path variables to express this, that means we can replace β by the term $\alpha \cdot \beta'$ where β' is new. This would lead to the clause

$$\{\alpha \dot\prec \alpha \cdot \beta',\ x\alpha y,\ \alpha \dot\in L_1,\ x\alpha \cdot \beta' z,\ \alpha \cdot \beta' \dot\in L_2\}.$$

However, this could easily be expressed in a more simpler way. First, the constraint $\alpha \dot\prec \alpha \cdot \beta'$ is superfluous. Second, the constraint $x\alpha \cdot \beta' z$ in combination with $x\alpha y$ can also be expressed by $\{x\alpha y,\ y\beta' z\}$. Then we get the clause

$$\gamma'_3 = \{x\alpha y,\ \alpha \dot\in L_1,\ y\beta' z,\ \alpha \cdot \beta' \dot\in L_2\},$$

This shows that we do not need concatenation of path variables within subterm agreements, and we will avoid them for simplicity.

The only thing that we have to do additionally in order to achieve a pre-solved clause is to resolve the constraint $\alpha \cdot \beta' \dot\in L_2$. To do this we have to guess a so-called decomposition P, S of L_2 with $P \cdot S \subseteq L_2$ such that $\alpha \dot\in P$ and $\beta' \dot\in S$. In general, there can be an infinite number of decompositions (think of the possible decompositions of the language f^*g). But as we use regular languages, there is a finite set of regular decomposition which covers all possibilities. Finally, reducing $\{\alpha \dot\in L_1,\ \alpha \dot\in P\}$ to $\{\alpha \dot\in (L_1 \cap P)\}$ will yield a pre-solved clause.

Note that the evaluation of the prefix relation in γ_3 has the additional effect that we introduce a new constraint $y\beta' z$. In general this implies that after the evaluation of prefix constraints there again may be some path variables the relation of which is unknown. Hence, after reducing the terms of form $\alpha \doteq \beta$ or $\alpha \mathrel{\dot\prec} \beta$ we may have to repeat the non-deterministic choice of relation between path variables. At the end, the only remaining constraints between path variables are of form $\alpha \mathrel{\dot{\text{п}}} \beta$.

Now let's turn to some additional point we have to consider, namely that the rules we present will (naturally) loop in some cases. Roughly speaking, one can say that this always occurs if a cycle in the graph coincides with a cycle in the regular language. To see this let us vary the above example and let γ now be the clause

$$\{x\alpha x, \ \alpha \mathrel{\dot\in} f, \ x\beta z, \ \beta \mathrel{\dot\in} f^*g\}$$

A possibly looping derivation could be

$\{\alpha \mathrel{\dot\prec} \beta, \ x\alpha x, \ \alpha \mathrel{\dot\in} f, \ x\beta z, \ \beta \mathrel{\dot\in} f^*g\}$	adding relation $\alpha \mathrel{\dot\prec} \beta$
$\{x\alpha x, \ \alpha \mathrel{\dot\in} f, \ x\beta' z, \ \alpha\cdot\beta' \mathrel{\dot\in} f^*g\}$	splitting β into $\alpha\cdot\beta'$
$\{x\alpha x, \ \alpha \mathrel{\dot\in} f, \ x\beta' z, \ \alpha \mathrel{\dot\in} f^*, \ \beta' \mathrel{\dot\in} f^*g\}$	decomposing $\alpha\cdot\beta' \mathrel{\dot\in} f^*g$
$\{x\alpha x, \ \alpha \mathrel{\dot\in} f, \ x\beta z, \ \beta' \mathrel{\dot\in} f^*g\}$	joining α-restrictions

But we will show that our rule system is quasi-terminating, which means that the rule system may cycle, but produces only finitely many different clauses (see [Der87]). This is achieved by the following measures: First we guarantee that the rules do not introduce additional variables; second we restrict concatenation to length two; and third we show that the rule system produces only finitely many regular languages. In order to prove that our rewrite system is complete, we have additionally to show that every solution can be found in a pre-solved clause.

3 Preliminaries

Our *signature* consist of a set of *sorts* $\mathcal{S}$ $(A, B, \ldots)$, *features* $\mathcal{F}$ $(f, g \ldots)$, *first order variables* $\mathcal{X}$ $(x, y, \ldots)$ and *path variables* $\mathcal{P}$ $(\alpha, \beta, \ldots)$. We assume a finite set of features and infinite sets of variables and sorts. A *path* is a finite string of features. We say that a path u is a *prefix* of a path v (written $u \prec v$) if there is a non-empty path w such that $v = uw$. Note that $\prec$ is neither symmetric nor reflexive. We say that two paths u, v *diverge* (written $u \mathrel{\text{п}} v$,) if there are features f, g with $f \neq g$ and possibly empty paths w, w_1, w_2 such that $u = wfw_1 \land v = wgw_2$. It is clear that п is a symmetric relation.

Proposition 1. *Given two paths u and v, then exactly one of the relations $u = v$, $u \prec v$, $u \succ v$ or $u \mathrel{\text{п}} v$ holds.*

A *path term* $(p, q, \ldots)$ is either a path variable α or concatenation of path variables $\alpha \cdot \beta$. We will allow path terms only in divergence constraint and not in prefix or equality constraints. The set of atomic constraints is given by

$$
\begin{array}{lll}
c \to & Ax & \text{sort restriction} \\
& x \doteq y & \text{agreement} \\
& x\, f_1 \cdot \ldots \cdot f_n\, y & \text{subterm agreement 1} \\
& x\alpha y & \text{subterm agreement 2} \\
& p \dot\epsilon\, L & \text{path restriction} \\
& p \mathbin{\dot{\text{ü}}} q & \text{divergence} \\
& \alpha \mathrel{\dot\prec} \beta & \text{prefix} \\
& \alpha \doteq \beta & \text{path equality}
\end{array}
$$

We exclude empty paths in subterm agreements, since $x\epsilon y$ is equivalent to $x \doteq y$. Therefore we require $f_1 \cdot \ldots \cdot f_n \in \mathcal{F}^+$ and $L \subseteq \mathcal{F}^+$.

A *clause* is a finite set of atomic constraint denoting their conjunction. We will say that a path term $\alpha \cdot \beta$ is *contained* (or *used*) in some clause ϕ if ϕ contains either a constraint $\alpha \cdot \beta \,\dot\epsilon\, L$ or a constraint $\alpha \cdot \beta \mathbin{\dot{\text{ü}}} q$.[3] Constraints of the form $p \dot\epsilon\, L$, $p \mathbin{\dot{\text{ü}}} q$, $\alpha \mathrel{\dot\prec} \beta$ and $\alpha \doteq \beta$ will be called *path constraints*.

An *interpretation* $\mathcal{I}$ is a standard first order structure, where every feature $f \in \mathcal{F}$ is interpreted as a binary, functional relation $F^{\mathcal{I}}$ and where sort symbols are interpreted as unary, disjoint predicates (hence $A^{\mathcal{I}} \cap B^{\mathcal{I}} = \emptyset$ for $A \neq B$). A *valuation* is a pair $(V_{\mathcal{X}}, V_{\mathcal{P}})$, where $V_{\mathcal{X}}$ is a standard first order valuation of the variables in X and $V_{\mathcal{P}}$ is a function $V_{\mathcal{P}} : \mathcal{P} \to \mathcal{F}^+$. We define $V_{\mathcal{P}}(\alpha \cdot \beta)$ to be $V_{\mathcal{P}}(\alpha) V_{\mathcal{P}}(\beta)$,

The *validity* of an atomic constraint in an interpretation $\mathcal{I}$ under a valuation $(V_{\mathcal{X}}, V_{\mathcal{P}})$ is defined as follows:

$$
\begin{array}{lll}
(V_{\mathcal{X}}, V_{\mathcal{P}}) \models_{\mathcal{I}} Ax & :\Longleftrightarrow & V_{\mathcal{X}}(x) \in A^{\mathcal{I}} \\
(V_{\mathcal{X}}, V_{\mathcal{P}}) \models_{\mathcal{I}} x \doteq y & :\Longleftrightarrow & V_{\mathcal{X}}(x) = V_{\mathcal{X}}(y) \\
(V_{\mathcal{X}}, V_{\mathcal{P}}) \models_{\mathcal{I}} x\, f_1 \cdot \ldots \cdot f_n\, y & :\Longleftrightarrow & V_{\mathcal{X}}(x)\; F_1^{\mathcal{I}} \circ \ldots \circ F_n^{\mathcal{I}}\; V_{\mathcal{X}}(y) \\
(V_{\mathcal{X}}, V_{\mathcal{P}}) \models_{\mathcal{I}} x\alpha y & :\Longleftrightarrow & (V_{\mathcal{X}}, V_{\mathcal{P}}) \models_{\mathcal{I}} x\, V_{\mathcal{P}}(\alpha)\, y \\
(V_{\mathcal{X}}, V_{\mathcal{P}}) \models_{\mathcal{I}} p \dot\epsilon\, L & :\Longleftrightarrow & V_{\mathcal{P}}(p) \in L \\
(V_{\mathcal{X}}, V_{\mathcal{P}}) \models_{\mathcal{I}} p \mathbin{\dot\diamond} q & :\Longleftrightarrow & V_{\mathcal{P}}(p) \diamond V_{\mathcal{P}}(q) \text{ for } \dot\diamond \in \{\dot{\text{ü}}, \dot\prec, \doteq\}
\end{array}
$$

Note that subterm agreement 2 is the only constraints where an interaction between $V_{\mathcal{X}}$ and $V_{\mathcal{P}}$ happens. The validity of sort restriction, agreement and subterm agreement 1 depends only on $V_{\mathcal{X}}$ and $\mathcal{I}$. Hence, we will sometimes omit the path valuation $V_{\mathcal{P}}$ and write $V_{\mathcal{X}} \models_{\mathcal{I}} \phi$ if ϕ consists only of these constraints and ϕ is valid under $\mathcal{I}$ and $V_{\mathcal{X}}$. Similar, validity of path constraints depends only on the path valuation $V_{\mathcal{P}}$. We will write $V_{\mathcal{P}} \models \phi$ if ϕ consists of path constraints which are valid under $V_{\mathcal{P}}$.

For a set $\xi \subseteq \mathcal{X}$ we define $=_\xi$ to be the following relation on first order valuation:

$$V_{\mathcal{X}} =_\xi V'_{\mathcal{X}} \quad \text{iff} \quad \text{for all } x \in \xi \text{ the equation } V_{\mathcal{X}}(x) = V'_{\mathcal{X}}(x) \text{ holds.}$$

---

[3] We will not make a distinction between $p \mathbin{\dot{\text{ü}}} q$ and $q \mathbin{\dot{\text{ü}}} p$.

Similarly, we define $=_\pi$ with $\pi \subseteq \mathcal{P}$ for path valuations. Let $\vartheta \subseteq \mathcal{X} \cup \mathcal{P}$ be a set of variables. For a given interpretation $\mathcal{I}$ we say that a valuation $(V_\mathcal{X}, V_\mathcal{P})$ is a ϑ-*solution* of a clause ϕ if there is a valuation $(V'_\mathcal{X}, V'_\mathcal{P})$ in $\mathcal{I}$ such that $V_\mathcal{X} =_{\mathcal{X} \cap \vartheta} V'_\mathcal{X}$, $V_\mathcal{P} =_{\mathcal{P} \cap \vartheta} V'_\mathcal{P}$ and $(V'_\mathcal{X}, V'_\mathcal{P}) \models_\mathcal{I} \phi$. The set of all ϑ-solutions of ϕ in $\mathcal{I}$ is denoted by $[\![\phi]\!]^\mathcal{I}_\vartheta$. We will call X-solutions just solutions and write $[\![\phi]\!]^\mathcal{I}$ instead of $[\![\phi]\!]^\mathcal{I}_\mathcal{X}$.

For checking satisfiability we use transformation rules. A rule R is ϑ-*sound* if $\phi \to_R \gamma \Rightarrow [\![\phi]\!]^\mathcal{I}_\vartheta \supseteq [\![\gamma]\!]^\mathcal{I}_\vartheta$ for every interpretation $\mathcal{I}$. R is called ϑ-*preserving* if $\phi \to_R \gamma \Rightarrow [\![\phi]\!]^\mathcal{I}_\vartheta \subseteq [\![\gamma]\!]^\mathcal{I}_\vartheta$. R is *globally ϑ-preserving* if

$$[\![\phi]\!]^\mathcal{I}_\vartheta \subseteq \bigcup_{\phi \to_R \gamma} [\![\gamma]\!]^\mathcal{I}_\vartheta.$$

4 The First Phase

4.1 A Set of Rules

Recall that we have switched from the original syntax to a (two-sorted) syntax by translating constraints xLy into $\{x\alpha y, \alpha \dot{\in} L\}$, where α is new. This implies, that we have to consider only a restricted set of clauses.

Let ϕ be some clause and x, y be different variables. We say that ϕ *binds* y *to* x if $x \doteq y \in \phi$ and y occurs only once in ϕ. Here it is important that we consider equations as directed, that is, we assume that $x \doteq y$ is different from $y \doteq x$. We say that ϕ *eliminates* y if ϕ binds y to some variable x. A clause is called *basic* if

1. an equation $x \doteq y$ appears in ϕ if and only if ϕ eliminates y,
2. For every path variable α used in ϕ there is *at most* one constraint $x\alpha y \in \phi$.

A clause ϕ is called *prime* if ϕ is basic and ϕ does not contain an atomic constraint of form $p \ddot{\shortparallel} q$, $\alpha \dot{\prec} \beta$ or $\alpha \doteq \beta$.

Kaplan and Maxwell stated the satisfiability problem for functional uncertainty in an unsorted syntax. By and large, this syntax consists of the atomic constraints Ax, $x f_1 \cdot \ldots \cdot f_n y$ and $x \doteq y$ together with the additional constraint xLy. As we have mentioned the constraint xLy is interpreted as

$$xLy = \bigvee \{xwy \mid w \in L\}.$$

It is easy to show that every clause in this syntax can be transformed in an equivalent prime clause.

Proposition 2. *Every clause ϕ in the Kaplan/Maxwell syntax can be translated into a prime clause γ such that for every interpretation $\mathcal{I}$ and for every first order valuation $V_\mathcal{X}$ $V_\mathcal{X} \models_\mathcal{I} \phi$ iff there is a $V_\mathcal{P}$ with $(V_\mathcal{X}, V_\mathcal{P}) \models_\mathcal{I} \gamma$.*

This implies that it suffices to check satisfiability of prime clauses in order to check satisfiability of clauses in the Kaplan/Maxwell syntax. Hence, prime clauses are the input clauses for the first phase.

Now let's turn to the output clauses of the first step. A basic clause is said to be *pre-solved* if the following holds:

1. $Ax \in \phi$ and $Bx \in \phi$ implies $A = B$.
2. $\alpha \mathbin{\dot{\in}} L \in \phi$ and $\alpha \mathbin{\dot{\in}} L' \in \phi$ implies $L = L'$. Furthermore, $\alpha \mathbin{\dot{\in}} \emptyset$ is not in ϕ.
3. $\alpha \cdot \beta$, $\alpha \mathbin{\dot{=}} \beta$ or $\alpha \mathbin{\dot{\prec}} \beta$ is not contained in ϕ.
4. $\alpha \mathbin{\dot{\sqcup}} \beta \in \phi$ if and only if $\alpha \neq \beta$, $x\alpha y \in \phi$ and $x\beta z \in \phi$.

Lemma 3. *Let ϕ be a pre-solved clause. Then ϕ is consistent iff there is a path valuation $V_{\mathcal{P}}$ with $V_{\mathcal{P}} \models \phi_p$, where ϕ_p is the set of path constraints in ϕ.*

This lemma is proven by constructing an appropriate valuation in the tree model. The domain of the tree model consists of so-called *feature trees*. A feature tree is a partial function $\sigma : \mathcal{F}^* \to \mathcal{S}$, the domain of which is prefix-closed (a domain $D \subseteq \mathcal{F}^*$ is prefix closed if $wu \in D$ implies $w \in D$). Note that this implies that the tree model is canonical for satisfiability, since our algorithm transforms each prime clause into an equivalent set of pre-solved clauses.

Now let's turn to the rule system. The first rule is the non-deterministic addition of relational constraints between path variables. We will add the relations between one fixed variable α and all other path variables β which are used under the same node x as α in one step. Furthermore, we will consider only the constraints $\alpha \mathbin{\dot{=}} \beta$, $\alpha \mathbin{\dot{\sqcup}} \beta$ and $\alpha \mathbin{\dot{\prec}} \beta$ and not additionally the constraint $\alpha \mathbin{\dot{\succ}} \beta$. Thus the rule can be described by the following pseudo code:

    Choose $x \in \mathcal{V}ars_{\mathcal{X}}(\phi)$ (don't care)
    Choose $x\alpha y \in \phi$ (don't know)
        For each $x\beta z \in \phi$ **with** $\alpha \neq \beta$ and $\alpha \mathbin{\dot{\sqcup}} \beta \notin \phi$
        add $\alpha \mathbin{\dot{\circ}_\beta} \beta$ with $\dot{\circ}_\beta \in \{\dot{=}, \dot{\prec}, \dot{\sqcup}\}$ (don't know)

Formally, this rule is written as

$$\text{(PathRel)} \ \frac{\{x\alpha y\} \cup \psi}{\{\alpha \mathbin{\dot{\circ}_\beta} \beta \mid x\beta z \in \psi \land \alpha \neq \beta \land \alpha \mathbin{\dot{\sqcup}} \beta \notin \psi\} \cup \{x\alpha y\} \cup \psi}$$

where $\dot{\circ}_\beta \in \{\dot{=}, \dot{\prec}, \dot{\sqcup}\}$.

> This rule will only by applied if ϕ contains no prefix constraints, path equality constraints or path concatenation. Furthermore, the application must add at least one constraint.

Although we have restricted the relations $\dot{\circ}_\beta$ to $\{\dot{=}, \dot{\prec}, \dot{\sqcup}\}$, this rule is globally preserving since we have non-deterministically chosen $x\alpha y$. To see this let ϕ be a clause, $\mathcal{I}$ be an interpretation and $(V_\mathcal{X}, V_\mathcal{P})$ be a valuation in $\mathcal{I}$ with $(V_\mathcal{X}, V_\mathcal{P}) \models_\mathcal{I} \phi$. To find an instance of (PathRel) such that $(V_\mathcal{X}, V_\mathcal{P}) \models_\mathcal{I} \gamma$ where γ is the result of applying this instance, we choose $x\alpha y \in \phi$ with $V_\mathcal{P}(\alpha)$ is prefix minimal in

$$\{V_\mathcal{P}(\beta) \mid x\beta z \in \phi\}.$$

Then for each $x\beta z \in \phi$ with $\alpha \neq \beta$ and $\alpha \mathbin{\dot{\sqcup}} \beta \notin \phi$ we add $\alpha \mathbin{\dot{\circ}_\beta} \beta$ where $V_\mathcal{P}(\alpha) \mathbin{\dot{\circ}_\beta} V_\mathcal{P}(\beta)$ holds. Note that $\dot{\circ}_\beta$ equals $\dot{\succ}$ will not occur since we have chosen a path variable α the interpretation of which is prefix minimal. Therefore the restriction $\dot{\circ}_\beta \in \{\dot{=}, \dot{\prec}, \dot{\sqcup}\}$ is satisfied.

We have defined (PathRel) in a very special way. The reason for this is that only by using this special definition we can maintain the condition that concatenation of path variables is restricted to binary concatenation. To see this assume that we would have added both $\beta_1 \mathrel{\dot\prec} \alpha$ and $\alpha \mathrel{\dot\prec} \beta_2$ to a clause γ. Then first splitting up the variable β_2 into $\alpha \cdot \beta_2'$ and then α into $\beta_1 \cdot \alpha'$ will result in a substitution of β_2 in γ by $\beta_1 \cdot \alpha' \cdot \beta_2'$. By the definition of (PathRel) we have ensured that this does not occur.

The second non-deterministic rule is used in the decomposition of regular languages. For decomposition we have the following rules:

(DecClash) $\dfrac{\{\alpha \cdot \beta \mathrel{\dot\in} L\} \cup \psi}{\bot}$ if $\{w \in L \mid |w| > 1\} = \emptyset$

(LangDec$_\Lambda$) $\dfrac{\{\alpha \cdot \beta \mathrel{\dot\in} L\} \cup \psi}{\{\alpha \mathrel{\dot\in} P\} \cup \{\beta \mathrel{\dot\in} S\} \cup \psi}$ $P \cdot S \subseteq L$

where $P, S, L \subseteq F^+$ and Λ is a finite set of reg. languages with $L, P, S \in \Lambda$. L must contain a word w with $|w| > 1$.

The clash rule is needed since we require regular languages not to contain the empty word. The remaining rules are listed in figure 1. Note that we have not considered subterm agreement 1 constraints since $x\, f_1 \cdot \ldots \cdot f_n\, y$ is equivalent to $x\, \{f_1 \cdot \ldots \cdot f_n\}\, y$.

(Eq)	$\dfrac{\{\alpha \mathrel{\dot=} \beta,\ x\alpha y,\ x\beta z\} \cup \psi}{\{y \mathrel{\dot=} z,\ x\alpha y\} \cup \psi[\beta \leftarrow \alpha,\ z \leftarrow y]}$	(Join)	$\dfrac{\{\alpha \mathrel{\dot\in} L,\ \alpha \mathrel{\dot\in} L'\} \cup \psi}{\{\alpha \mathrel{\dot\in} (L \cap L')\} \cup \psi}$	$L \neq L'$
(Div1)	$\dfrac{\{\alpha \mathbin{\dot{\sqcup}} \beta'\} \cup \{\alpha \cdot \beta \mathbin{\dot{\sqcup}} \beta'\} \cup \psi}{\{\alpha \mathbin{\dot{\sqcup}} \beta'\} \cup \psi}$	(Div2)	$\dfrac{\{\alpha \cdot \beta \mathbin{\dot{\sqcup}} \alpha \cdot \beta'\} \cup \psi}{\{\beta \mathbin{\dot{\sqcup}} \beta'\} \cup \psi}$	
(DClash1)	$\dfrac{\{\alpha \cdot \beta \mathbin{\dot{\sqcup}} \alpha\} \cup \psi}{\bot}$	(DClash2)	$\dfrac{\{\alpha \mathbin{\dot{\sqcup}} \alpha\} \cup \psi}{\bot}$	
(Empty)	$\dfrac{\{\alpha \mathrel{\dot\in} \emptyset\} \cup \psi}{\bot}$	(SClash)	$\dfrac{\{Ax,\ Bx\} \cup \psi}{\bot}$	$A \neq B$
	(Pre)	$\dfrac{\{\alpha \mathrel{\dot\prec} \beta,\ x\alpha y,\ x\beta z\} \cup \psi}{\{x\alpha y\} \cup \{y\beta z\} \cup \psi[\beta \leftarrow \alpha \cdot \beta]}$	$\alpha \neq \beta$	

Fig. 1. Simplification rules

We use Λ in (LangDec$_\Lambda$) as a global restriction, i.e. for every Λ we obtain a different rule (LangDec$_\Lambda$) (and hence a different rule system $\mathcal{R}_\Lambda$). This is done because the rule system is quasi-terminating. By restricting (LangDec$_\Lambda$) we can guarantee that only finitely many regular languages are produced.

For (LangDec$_\Lambda$) to be globally preserving we need to find for every possible valuation of α and β a suitable pair P, S in Λ. Therefore we require Λ to satisfy

$$\forall L \in \Lambda, \forall w_1, w_2 \neq \epsilon :$$
$$[w_1 w_2 \in L \Rightarrow \exists P, S \in \Lambda : (P \cdot S \subseteq L \wedge w_1 \in P \wedge w_2 \in S)].$$

We call Λ *closed under decomposition* if it satisfies this condition. Additionally, we have to ensure that $L \in \Lambda$ for every L that is contained in some clause ϕ. We will call a such set Λ ϕ-*closed*.

Lemma 4.

1. If Λ is ϕ-closed and closed under intersection, then Λ is γ-closed for all $(\phi, \mathcal{R}_\Lambda)$-derivatives γ.
2. For every prime clause ϕ there is a finite Λ such that Λ is ϕ-closed, closed under intersection and decomposition.

Proof. The first claim is easy to prove. For the second claim let $\{L_1, \ldots, L_n\} \subseteq P(\mathcal{F}^+)$ be the set of regular languages used in ϕ and let $\mathcal{A}_i = (Q_{\mathcal{A}_i}, i_{\mathcal{A}_i}, \sigma_{\mathcal{A}_i}, Fin_{\mathcal{A}_i})$ be finite, deterministic automatons such that $\mathcal{A}_i$ recognizes L_i. For each $\mathcal{A}_i$ we define dec($\mathcal{A}_i$) to be the set

$$\text{dec}(\mathcal{A}_i) = \{\overline{L_p^q} \mid p, q \in Q_{\mathcal{A}_i}\},$$

where $\overline{L_p^q} = \{w \in \mathcal{F}^+ \mid \sigma_{\mathcal{A}_i}^*(p, w) = q\}$. It is easy to show that dec($\mathcal{A}_i$) is a set of regular languages that contains L_i and is closed under decomposition. Hence, the set $\Lambda_0 = \bigcup_{i=1}^n \text{dec}(\mathcal{A}_i)$ contains each L_i and is closed under decomposition. Now let $\Lambda = \text{fi}(\Lambda_0)$ be the least set that contains Λ_0 and is closed under intersection. Then Λ is finite and ϕ-closed, since it contains each L_i.

We will prove that Λ is also closed under decomposition. Given some $L \in \Lambda$ and a word $w = w_1 w_2 \in L$, we have to find an appropriate decomposition P, S in Λ. Since each L in Λ can be written as a finite intersection $L = \bigcap_{k=1}^m L_{i_k}$ where L_{i_k} is in Λ_0, we know that $w = w_1 w_2$ is in L_{i_k} for $1..m$. As Λ_0 is closed under decomposition, there are languages P_{i_k} and S_{i_k} for $k = 1..m$ with $w_1 \in P_{i_k}$, $w_2 \in S_{i_k}$ and $P_{i_k} \cdot S_{i_k} \subseteq L_{i_k}$. Let $P = \bigcap_{k=1}^m P_{i_k}$ and $S = \bigcap_{k=1}^m S_{i_k}$. Clearly, $w_1 \in P$, $w_2 \in S$ and $P \cdot S \subseteq L$. Furthermore, $P, S \in \Lambda$ as Λ is closed under intersection. This implies that P, S is an appropriate decomposition for $w_1 w_2$.

4.2 Soundness, Completeness and Quasi-Termination

Proposition 5. *The rule* (PathRel) *is* $\mathcal{X} \cup \mathcal{V}$-*sound and globally* $\mathcal{X} \cup \mathcal{V}$-*preserving. If Λ is closed under decomposition, then* (LangDec$_\Lambda$) *is* $\mathcal{X} \cup \mathcal{V}$-*sound and globally* $\mathcal{X} \cup \mathcal{V}$-*preserving. The* (Pre) *rule is* $\mathcal{X}$-*sound and* $\mathcal{X}$-*preserving. All other rules are* $\mathcal{X} \cup \mathcal{V}$-*sound and* $\mathcal{X} \cup \mathcal{V}$-*preserving.*

Next we will prove some syntactic properties of the clauses derivable by the rule system. For the rest of the paper we will call clauses that are derivable from prime clauses *admissible*.

Proposition 6.

1. *Every admissible clause is basic.*
2. *If $\alpha \precdot \beta$, $\alpha \doteq \beta$ or $\alpha \mathbin{\ddot{\sqcup}} \beta$ is contained in some admissible clause ϕ, then there is a variable x such that $x\alpha y$ and $x\beta z$ is in ϕ.*

Note that by proposition 6 (Pre) (resp. (Eq)) can always be applied if a constraint $\alpha \precdot \beta$ (resp. $\alpha \doteq \beta$) is contained in some admissible clause. The next lemma will show that different applications of (Pre) or (Eq) will not interact. This means the application of one of these rules to some prefix or path equality constraint will not change any other prefix or path equality constraint contained in the same clause. This is a direct consequence of the way (PathRel) was defined.

Lemma 7. *Given two admissible clauses γ, γ' with $\gamma \rightarrow_r \gamma'$ and r different from (PathRel). Then $\alpha \doteq \beta \in \gamma'$ (resp. $\alpha \precdot \beta \in \gamma'$) implies $\alpha \doteq \beta \in \gamma$ (resp. $\alpha \precdot \beta \in \gamma$). Furthermore, if $\alpha \cdot \beta$ is contained in γ', then either $\alpha \cdot \beta$ or $\alpha \precdot \beta$ is contained in γ.*

Note that this lemma implies that new path equality or prefix constraints are only introduced by (PathRel). We can derive from this lemma some syntactic properties of admissible clauses which are needed for proving completeness and quasi-termination.

Lemma 8. *If ϕ is an admissible clause, then*

1. *If $\alpha \precdot \beta$ is contained in ϕ, then there is no other prefix or equality constraint in ϕ involving β. Furthermore, neither $\beta \cdot \beta'$ nor $\beta' \cdot \beta$ is in ϕ.*
2. *if $\alpha \cdot \beta \mathbin{\ddot{\sqcup}} \beta'$ is in ϕ, then either β' equals α or ϕ contains a constraint of form $\alpha \mathbin{\ddot{\sqcup}} \beta'$, $\alpha \doteq \beta'$ or $\alpha \precdot \beta'$.*

The first property will guarantee that concatenation does not occur in prefix or equality constraints and that length of path concatenation is restricted to 2. The second property ensures that a constraint $\alpha \cdot \beta \mathbin{\ddot{\sqcup}} \beta'$ is always reducible. If β' equals α, then we could apply (DClash1). If $\alpha \mathbin{\ddot{\sqcup}} \beta'$ is in ϕ, we can apply (Div1). If $\alpha \doteq \beta'$ is in ϕ we can apply (Eq) followed by (DClash1). And finally, if $\phi = \{\alpha \precdot \beta', \alpha \cdot \beta \mathbin{\ddot{\sqcup}} \beta'\} \cup \psi$, then we can apply (Pre) yielding $\{\alpha \cdot \beta \mathbin{\ddot{\sqcup}} \alpha \cdot \beta'\} \cup \psi'$, where we can apply (Div2).

Theorem 9. *For every finite Λ the rule system $\mathcal{R}_\Lambda$ is quasi-terminating.*

Proof. The rule system produces only finitely many different clauses since the rules introduce no additional variables or sort symbols and the set of used languages is finite. Additionally length of concatenation is restricted to 2.

Lemma 10. *There are no infinite derivations using only finitely many instances of (Pre).*

Since the rule system is quasi-terminating, the completeness proof consists of two parts. In the first part we prove that pre-solved clauses are just the irreducible clauses. In the second part we will show that one can find for each solution $V_\mathcal{X}$ of a prime clause ϕ a pre-solved ϕ-derivative γ such that $V_\mathcal{X}$ is also a solution of γ.

Theorem 11 Completeness I. *Given an admissible clause $\phi \neq \bot$ such that ϕ is not in pre-solved form. If Λ is ϕ-closed and closed under decomposition, then ϕ is $\mathcal{R}_\Lambda$-reducible.*

Theorem 12 Completeness II. *For every prime clause ϕ and for every Λ that is ϕ-closed, closed under decomposition and intersection we have*

$$[\![\phi]\!]^\mathcal{I} \subseteq \bigcup_{\gamma \in \text{pre-solved}(\phi, R_\Lambda)} [\![\gamma]\!]^\mathcal{I},$$

where $\text{pre-solved}(\phi, R_\Lambda)$ *is the set of pre-solved (ϕ, R_Λ)-derivatives.*

Proof (Sketch) We have to show, that for each prime clause ϕ and each $V_\mathcal{X}, V_\mathcal{P}, \mathcal{I}$ with $(V_\mathcal{X}, V_\mathcal{P}) \models_\mathcal{I} \phi$ there is a pre-solved $(\phi, \mathcal{R}_\Lambda)$-derivative γ such that $V_\mathcal{X} \in [\![\gamma]\!]^\mathcal{I}$. To do this we will control the derivations using the valuation $(V_\mathcal{X}, V_\mathcal{P})$. The control will guarantee finiteness of derivations, but will keep the first completeness property, namely that the irreducible clauses are exactly the pre-solved clauses.

We allow only those instances of the non-deterministic rules (PathRel) and (LangDec$_\Lambda$), which preserve exactly the valuation $(V_\mathcal{X}, V_\mathcal{P})$. That means if $(V_\mathcal{X}, V_\mathcal{P}) \models_\mathcal{I} \phi$ and $\phi \to_r \gamma$ for one of these rules, then $(V_\mathcal{X}, V_\mathcal{P}) \models_\mathcal{I} \gamma$ must hold. Note that the control depends only on $V_\mathcal{P}$. E.g. given the clause $\phi = \{x\alpha y, \alpha \in L_1, x\beta z, \beta \in L_2\}$, an arbitrary $\mathcal{I}, V_\mathcal{X}$ and a path valuation $V_\mathcal{P}$ with $V_\mathcal{P}(\alpha) = f$, $V_\mathcal{P}(\beta) = g$ and $(V_\mathcal{X}, V_\mathcal{P}) \models_\mathcal{I} \phi$, then the rule (PathRel) may transform ϕ only into $\{\alpha \mathbin{\ddot{\sqcup}} \beta\} \cup \phi$.

If $V_\mathcal{P}$ satisfies $V_\mathcal{P}(\alpha) \not\prec V_\mathcal{P}(\beta)$ for $x\alpha y \in \phi$, $x\beta z \in \phi$ and α different from β, then we cannot add any prefix constraint using this control. Hence, (Pre) cannot be applied. By lemma 10 this implies that there are no infinite controlled derivations in this case. We will call such path valuations prefix-free with respect to ϕ.

If $V_\mathcal{P}$ is not prefix-free, then (Pre) will be applied during the derivations. In this case we have to change the path valuation, since (Pre) is not $\mathcal{P}$-preserving. If $(V_\mathcal{X}, V_\mathcal{P}) \models_\mathcal{I} \{\alpha \prec \beta\} \cup \psi$ and (Pre) has been applied on $\alpha \prec \beta$ yielding γ, then the valuation $V_\mathcal{P}'$ with $V_\mathcal{P}(\beta) = V_\mathcal{P}(\alpha)V_\mathcal{P}'(\beta)$ and $V_\mathcal{P}(\alpha) = V_\mathcal{P}'(\alpha)$ for $\alpha \neq \beta$ will satisfy $(V_\mathcal{X}, V_\mathcal{P}') \models_\mathcal{I} \gamma$. We will use $V_\mathcal{P}'$ for controlling the further derivations.

If we change the path valuation in this way, there are again only finite derivations. To see this note that every time (Pre) is applied and the path valuation is changed, the valuation of one variable is shortened by a non-empty path. As the number of variables used in derivatives does not increase, this shortening can only be done finitely many times. This implies, that (Pre) can only finitely often be applied under this control. But lemma 10 states that an infinite number of (Pre)-applications is a necessary prerequisite for an infinite derivation. □

5 Satisfiability of Pre-Solved Clauses

In the second phase we will check satisfiability of pre-solved clauses. By proposition 3 a pre-solved clause ϕ is satisfiable if the set of path constraints ϕ_P of ϕ is. Hence, we assume for simplicity that the transformations we present below apply only to ϕ_P. Note that by definition of a pre-solved clause ϕ_P consists only of divergence constraints and path restrictions.

We will first do a minor redefinition of divergence. We say that two paths u, v are *directly diverging* (written $u \, \text{ii}_0 \, v$) if there are features $f \neq g$ such that $u \in f\mathcal{F}^*$ and $v \in g\mathcal{F}^*$. Then $u \, \text{ii} \, v$ holds if there are a possible empty prefix w and paths u', v' such that $u = wu'$ and $v = wv'$ and $u' \, \text{ii}_0 \, v'$.

Using this definition of divergence, we can (non-deterministicly) transforming $\phi_P = \{\alpha_1 \, \text{ii} \, \alpha_2\} \cup \psi$ into either $\{\alpha_1 \, \text{ii}_0 \, \alpha_2\} \cup \psi$ or $\{\alpha_1 \doteq \beta \cdot \alpha'_1, \alpha_2 \doteq \beta \cdot \alpha'_2, \alpha'_1 \, \text{ii}_0 \, \alpha'_2\} \cup \psi$, where $\alpha_1 \, \text{ii}_0 \, \alpha_2$ is a new atomic constraint that corresponds to the ii_0 relation.[4] By the definition of ii_0 we can reduce (non-deterministicly) the constraints of form $\alpha_1 \, \text{ii}_0 \, \alpha_2$ into $\{\alpha_1 \, \dot\in \, f\mathcal{F}^*, \alpha_2 \, \dot\in \, g\mathcal{F}^*\}$ with $f \neq g$. The aim is to process all divergence constraints this way in order to yield a clause that consists only of non-empty path restrictions, since such a clause is necessarily satisfiable.

But we have to reformulate the reduction of divergence constraints as we have to evaluate the constraints of form $\alpha_1 \doteq \beta \cdot \alpha'_1$. This can produce constraints of the form $\alpha \cdot \beta \, \dot\in \, L$ and $\alpha \cdot \beta \, \text{ii} \, \beta'$. The second one is problematic, since we have to guess the relation between α and β'. This complicates the termination proof.

We will avoid this problem by using a special property of pre-solved clauses, namely that $\alpha \, \text{ii} \, \beta$ is contained in a pre-solved clause ϕ iff $x\alpha y$ and $x\beta z$ is in ϕ. Hence, if $\alpha \, \text{ii} \, \beta$ and $\beta \, \text{ii} \, \delta$ is in ϕ, then $\alpha \, \text{ii} \, \delta$ is also in ϕ. This implies, that we can write ϕ_P as $\text{ii}(A_1) \uplus \ldots \uplus \text{ii}(A_n) \uplus \psi$, where $\text{ii}(A)$ is a syntactic sugar for

$$\text{ii}(A) = \{\alpha \, \text{ii} \, \alpha' \mid \alpha \neq \alpha' \land \alpha, \alpha' \in A\},$$

$A_1, \ldots, A_n$ are disjoint sets of path variables and ψ consists only of path restrictions.[5] Now given such a constraint $\text{ii}(A)$, we assume that a whole set of path variables $A_1 \subseteq A$ diverge with the same prefix. That means we can replace $\text{ii}(A_1) \subseteq \text{ii}(A)$ by

$$A_1 \doteq \beta \cdot A'_1 \cup \text{ii}_0(A'_1),$$

where β is new, $A'_1 = \{\alpha'_1, \ldots, \alpha'_n\}$ is a disjoint copy of $A_1 = \{\alpha_1, \ldots, \alpha_n\}$ and $A \doteq \beta \cdot A'_1$ is a shortcut for the clause $\{\alpha_1 \doteq \beta \cdot \alpha'_1, \ldots, \alpha_n \doteq \beta \cdot \alpha'_n\}$. $\text{ii}_0(A)$ is defined similar as $\text{ii}(A)$. Assuming additionally that the common prefix β is maximal implies that $\beta \, \text{ii} \, \alpha$ holds for $\alpha \in (A - A_1)$. Finally we get the rule

---

[4] The first case is needed as we do not allow valuations of path variables to be empty paths. In contrast to the first phase we use here also path equality constraints of the form $\alpha \doteq \beta \cdot \beta'$.

[5] Initially, the A_i are exactly the sets of path variables that are used under the same first order variable. This means that for every A_i there is a variable x such that $A_i = \{\alpha \mid \exists y : x\alpha y \in \phi\}$

$$(\text{Reduce}_1) \quad \frac{\dot{\mathrm{u}}(A) \cup \psi}{A_1 \doteq \beta \cdot A_1' \;\cup\; \dot{\mathrm{u}}_0(A_1') \;\cup\; \dot{\mathrm{u}}(\{\beta\} \cup A_2) \;\cup\; \psi[A_1 \leftarrow \beta \cdot A_1']}$$

where β new, $A_1 \uplus A_2 = A$ and $|A_1| > 1$. A_1' is a disjoint copy of A_1. ψ may not contain constraints of form $\delta \cdot \delta \,\dot\in\, L$ in ψ.

Here $\psi[A \leftarrow \beta \cdot A']$ is a shortcut for $\psi[\alpha_1 \leftarrow \beta \cdot \alpha_1', \ldots, \alpha_n \leftarrow \beta \cdot \alpha_n']$. Note that we have avoided constraints of the form $\alpha \cdot \beta \;\dot{\mathrm{u}}\; \beta'$. The use of constraints $\alpha \doteq \beta \cdot \alpha'$ is not crucial, since α is not contained in the rest of the clause. The rules

$$(\text{Reduce}_2) \quad \frac{\dot{\mathrm{u}}(A) \cup \psi}{\dot{\mathrm{u}}_0(A) \cup \psi}$$

$$(\text{Solv}) \quad \frac{\dot{\mathrm{u}}_0(A) \cup \psi}{\{\alpha \in f_\alpha \cdot F^* \mid \alpha \in A\} \cup \psi} \quad f_\alpha \neq f_{\alpha'} \text{ for } \alpha \neq \alpha'$$

together with the rules (LangDec_Λ), (Join) and (Empty) completes the rule system $\mathcal{R}_\Lambda^{\text{Solv}}$. ($\text{Reduce}_2$) is needed as path variables always denote non-empty paths. We will see (Reduce_1) and (Reduce_2) as one single rule (Reduce).

A clause ϕ is said to be *solved* if (1) $\alpha \cdot \beta \,\dot\in\, L$ and $\alpha \,\dot\in\, \emptyset$ is not in ϕ_P, (2) $\alpha \,\dot\in\, L_1$ in ϕ_P and $\alpha \,\dot\in\, L_2$ in ϕ_P implies $L_1 = L_2$ and (3) ϕ does not contain divergence constraints $\alpha \;\dot{\mathrm{u}}\; \beta$ or $\alpha \;\dot{\mathrm{u}}_0\; \beta$. Clearly, every solved clause is consistent. Recall that we assume above transformations to work only on the set of path constraints ϕ_P of a clause ϕ.

Lemma 13. *The rules* (Reduce) = (Reduce_1) + (Reduce_2) *and* (Solv) *are* $\mathcal{X} \cup \mathcal{P}$-*sound and globally* $\mathcal{X} \cup \mathcal{P}$-*preserving. Furthermore,* $\mathcal{R}_\Lambda^{\text{Solv}}$ *is terminating.*

Lemma 14. *Let ϕ be a pre-solved clause, and let ϕ_P be the set of path constraints of ϕ. If Λ is ϕ-closed, closed under intersection and decomposition, then a $(\phi_P, \mathcal{R}_\Lambda^{\text{Solv}})$-derivate different from $\bot$ is irreducible if and only if it is solved.*

Finally we can combine both phases of the algorithm.

Theorem 15. *Satisfiability of prime clauses is decidable.*

This implies (by proposition 2) that satisfiability of clauses in the Kaplan/Maxwell syntax is decidable.

6 Conclusion

We have shown, that the pure existential fragment of feature logic extended by regular path expressions is decidable. The main prerequisite for achieving this result was to switch from the original, unsorted syntax to a two-sorted syntax. For each clause in the original syntax we get an equivalent clause in the new syntax by translating a regular path expression xLy into $\{x\alpha y, \alpha \,\dot\in\, L\}$ with α new.

The result of the translation constitutes a special class of clauses, namely the class of prime clauses. The main restriction imposed on prime clauses is that for each path variable α there is *at most* one constraint $x\alpha y$ contained in a clause. For prime clauses we have presented an algorithm, that first transforms a clause into an equivalent set of pre-solved clauses. In a second phase pre-solved clauses are checked for satisfiability by transforming them into an equivalent set of solved clauses.

Our syntax is more expressive than the original one. Although restricting to prime clauses was sufficient for our purposes, it may therefore be interesting to examine whether decidability can be preserved if we skip the restriction.

References

[Bac93]   Rolf Backofen. Regular path expressions in feature logic. Research report, DFKI, 1993. To appear.

[BBN+91] Franz Baader, Hans-Jürgen Bürckert, Berhard Nebel, Werner Nutt, and Gert Smolka. On the expressivity of feature logics with negation, functional uncertainity, and sort equations. Research Report RR-91-01, DFKI, 1991.

[BS92]   Rolf Backofen and Gert Smolka. A complete and recursive feature theory. Research Report RR-92-30, DFKI, 1992.

[Der87]   Nachum Dershowitz. Termination of rewriting. *Journal of Symbolic Computation*, 3:69–116, 1987.

[KB82]   Ronald M. Kaplan and Joan Bresnan. Lexical-Functional Grammar: A formal system for grammatical representation. In J. Bresnan, editor, *The Mental Representation of Grammatical Relations*, pages 173–381. MIT Press, Cambridge (MA), 1982.

[Kel91]   Bill Keller. Feature logics, infinitary descriptions and the logical treatment of grammar. Cognitive Science Research Report 205, Univerity of Sussex, School of Cognitive and Computing Sciences, 1991.

[KM88]   R. M. Kaplan and J. T. Maxwell III. An algorithm for functional uncertainty. In *Proceedings of the 12th International Conference on Computational Linguistics*, pages 297–302, Budapest, Hungary, 1988.

[KZ88]   Ronald M. Kaplan and Annie Zaenen. Long-distance dependencies, constituent structure, and functional uncertainty. In M. Baltin and A. Kroch, editors, *Alternative Conceptions of Phrase Structure*. University of Chicago Press, Chicago, 1988.

[Shi86]   Stuart M. Shieber. *An Introduction to Unification-Based Approaches to Grammar*, volume 4 of *CSLI Lecture Notes*. Stanford University, Stanford (CA), 1986.

[Smo88]   Gert Smolka. A feature logic with subsorts. LILOG-Report 33, IWBS, IBM Deutschland, Stuttgart, May 1988.

[Smo92]   Gert Smolka. Feature constraint logics for unification grammars. *Journal of Logic Programming*, 12:51–87, 1992.

[ST92]   Gert Smolka and Ralf Treinen. Records for logic programming. In *Proceedings of the 1992 Joint International Conference and Symposium on Logic Programming*, Washington, DC, November 1992. The MIT Press.

Proving Properties of Typed Lambda Terms: Realizability, Covers, and Sheaves

Jean Gallier[1]

University of Pennsylvania at Philadelphia
e-mail: jean@cis.upenn.edu

Abstract

We present a general method for proving properties of typed lambda terms. This method is obtained by introducing a semantic notion of realizability which uses the notion of a cover algebra (as in abstract sheaf theory). For this, we introduce a new class of semantic structures equipped with preorders, called pre-applicative structures. In this framework, a general realizability theorem can be shown. Applying this theorem to the special case of the term model, yields a general theorem for proving properties of typed lambda terms, in particular, strong normalization and confluence. This approach clarifies the reducibility method by showing that the closure conditions on candidates of reducibility can be viewed as sheaf conditions.

Some lambda calculi with categorical sums and products

Daniel J. Dougherty

Dept. of Mathematics,
Wesleyan University,
Middletown, CT, 06459 USA.
ddougherty@eagle.wesleyan.edu

Abstract. We consider the simply typed λ-calculus with primitive recursion operators and types corresponding to categorical products and coproducts.. The standard equations corresponding to extensionality and to surjectivity of pairing and its dual are oriented as expansion rules. Strong normalization and ground (base-type) confluence is proved for the full calculus; full confluence is proved for the calculus omitting the rule for strong sums. In the latter case, fixed-point constructors may be added while retaining confluence.

1 Introduction

The systems investigated here are simply typed λ-caluli whose types include pairs, unit, sums, an empty type, and a type of natural numbers supporting constructions by primitive recursion. In the core system the types behave as categorical product and coproducts, so the subject at hand is equivalently ([LS86]) the equational theory of the free bicartesian closed category (generated by objects for the base types) with weak natural numbers object. Such λ-calculi play a role in modeling several aspects of programming languages, and we are further led to investigate the consequences of adding fixed-point operators to the systems.

An important aspect of the present treatment is that we orient certain of the standard axioms (for example, (η)) as expansion rules. The resulting calculus enjoys certain pleasant properties lacking in the traditional reduction systems.

We prove strong normalization for the core calculus; it is then easy to conclude that ground (i.e., closed, base-type) terms reduce to numerals, and from that to derive ground confluence. The core system fails to be fully confluent. When the equation characterizing sums as categorical coproducts is dropped, together with the equation for the empty type, confluence is recovered. Furthermore, when fixed-point operators are added to this system, it remains confluent. As a consequence we conclude that the equational theory involving the fixed-points is a conservative extension of the theory with primitive recursion only. It is well-known that adding fixed-point operators to the theory of coproducts is inconsistent [Law69]; we think the analysis here sheds some additional light on that result (see the remarks at the end of the paper).

Normalization implies, as usual, that closed normal form terms are numerals, abstractions, pairs, etc. (depending on their type). In retrospect, this makes the use of expansions seem even more natural — closed terms will reduce to expanded terms in any event, so the calculus does it explicitly.

The ground confluence result for the theory with categorical coproducts implies that the reduction system can serve as an evaluator for a programming language (with primitive recursion only) whose observational equivalences are true in the theory. The existence of (weak) sums in ordinary programming languages provides a mechanism for building data supporting computation by cases; the use of true coproducts allows the programmer to *reason about the code* by cases.

The confluence-with-fixed-points result serves as a contrast to some work of J. W. Klop. Klop has shown [Klo80] that adding surjective pairing to the *untyped* λ-calculus — with the uniqueness axiom oriented in the traditional way — spoils confluence (although unique normal forms are retained [KV89]). The terms other than the fixed-point combinator used in Klop's argument can be simply-typed, so that the argument shows that adding fixed points to the typed system with products results in a non-confluent calculus when the pairing axiom is oriented as a contraction.

The λ-calculus with pair types was shown to be strongly normalizing (SN) by deVrier in 1982 [deV87] adapting Tait's method [Tai67]. The presence of a unit type (terminal object) spoils confluence for the traditional reduction. Say that a system has *products* if the types include both pairs and a terminal object. Poigné and Voss [PV87] explored a rich calculus including products, and gave proofs of termination and confluence, but the proofs contained errors. Curien and Di Cosmo [CD91] showed confluence and termination for a second-order calculus with products.

Meanwhile Mints [Min80] considered the use of expansion rules for products and function types, and gave proofs of weak normalization and confluence for the system. Jay [Jay91], motivated by category theory, showed strong normalization and confluence for a calculus with products and a natural numbers object (supporting iteration), assuming that all types are inhabited. Independently Cubric [Cub92] repaired some errors in Mint's proof and gave applications in category theory. Akama [Aka93] has recently shown strong normalization and confluence for the general expansion lambda calculus with products.

Let us say that a calculus has *sums* if there are types behaving as weak sums (in the category-theoretic sense), but not necessarily an empty type (initial object). Gandy [Gan80] proved termination for a typed lambda calculus with sums. The present author [Dou90] and independently Okada and Scott [OS91] showed strong normalization for a calculus with products, coproducts and primitive recursion (with contracting reductions). Most recently, and independently of the present work, Di Cosmo and Kesner [DK93b] have investigated the calculus with products, sums, and full recursion (using fixed-point operators), orienting the reductions as expansions, and proved strong normalization and confluence. It is remarkable that so much (independent) recent work has investigated expansions;

note that the only difference between the subject of present paper and that of Di Cosmo and Kesner is our use of true coproducts. The techniques are quite different, however. We recommend [DK93a] for a very careful analysis of some of the subtleties involved in these systems, with all details provided.

In the Girard/Reynolds polymorphic lambda calculus standard data types such as pairs, sums, and lists are definable implicitly, and of course that calculus is known to be terminating. But the "uniqueness" equations which characterize pairs and sums as categorical are *not* theorems of that calculus, so those results will not apply here.

We will assume familiarity with standard notation and results in the λ-calculus [Bar84] and rewriting [DJ91].

2 Preliminaries

Definition 2.1 Fix a set of *base types*, including at least the types $\mathbf{1}$, $\mathbf{0}$, and N. The set of *types* is the closure of the set of base types under the constructions $(A \times B)$, $(A + B)$, and $(A \to B)$.

For each type T, fix an infinite set of explicitly typed *variables* and an arbitrary set of *free constants*.

The set Λ of *terms* is the closure of the variables and constants under the following constructions (write $t : T$ to assert that t is a term of type T):

I. Introduction terms

$\lambda x.b : A \to B$ when $x : a$ and $b : B$
$[f_1 \ f_2] : (A_1 + A_2) \to B$ when $f_1 : A_1 \to B$ and $f_2 : A_2 \to B$
$\Box : 0 \to A$
$\langle a_1, a_2 \rangle : A_1 \times A_2$ when $a : A_1$ and $b : A_2$
$* : \mathbf{1}$
$\sigma_i d : A_1 + A_2$ when $d : A_i$, $\quad i \in \{1, 2\}$
$0 : N$
$\mathsf{succ}\, n : N$ when $n : N$

II. Elimination terms

$fa : B$ when $f : A \to B$ and $a : A$
$\pi_i p : A_i$ when $p : A_1 \times A_2$, $\quad i \in \{1, 2\}$
$\mathsf{rec}\, fan : A$ when $f : N \to A \to A$, $a : A$, and $n : N$

The rec constructor builds primitive recursive functions.

We have abused notation here by not decorating the constructors with type information. For example, the type of $\sigma_i d$ cannot be inferred from the type of d with our notation, but this should never cause confusion.

Introduction terms (so designated because they correspond to logical introduction rules under the Curry-Howard isomorphism) will be referred to as *I-terms*. A *numeral* is either 0 or ($\mathsf{succ}\, n$) where n is a numeral. The *size* of a type or term is the number of operations used in constructing it. The substitution of term a for variable x in b is denoted $b[x := a]$.

In the concrete syntax parentheses will suppressed whenever possible (under the usual conventions that the function-space constructor associates right and term application associates left), terms will be considered identical if they differ only by renaming of bound variables, and type information will be omitted if it can be easily inferred.

When h is a function with domain $A_1 + A_2$ we will often have occasion to consider its respective "summands" with domains A_i. We intend to expand abstractions $h: (A_1 + A_2) \to B$ to an explicit sum of their summands. To avoid creating loops, we define the following notions.

Let $h: (A_1 + A_2) \to B$. For $i \in \{1, 2\}$ we define $h \cdot \sigma_i : A_i \to B$ by:

1. if $h \equiv [h_1\ h_2]$ then $h \cdot \sigma_i \equiv h_i$,
2. if $h \equiv \lambda x.b$ then $h \cdot \sigma_i \equiv \lambda x_i.b[x := (\sigma_i x_i)]$,
3. else $h \cdot \sigma_i \equiv \lambda x_i.h(\sigma_i x_i)]$.

For function f and argument a we define $f^!a$ by:

1. If $f \equiv \lambda x.b$ then $f^!a \equiv b[x := a]$,
2. otherwise $f^!a \equiv (fa)$.

Definition 2.2 The relation E is generated by the substitution instances of the following axioms and rules of inference. An equation is, by definition, given by a pair of terms of the same type.

I. Computational axioms:

(β) $\quad (\lambda x.b)a = b[x := a]$
(σ) $\quad [f_1\ f_2](\sigma_i a) = f_i^! a, \quad i \in \{1, 2\}$
(π) $\quad \pi_i \langle t_1, t_2 \rangle = t_i, \quad i \in \{1, 2\}$
(rec) $\quad$ rec $fa0 = a$,
$\quad\quad\quad$ rec $fa(\text{succ } s) = fs(\text{rec } fas)$

I. Uniqueness axioms:

(η) $\quad f = \lambda x.fx$ (x not free in f)
$(+!)$ $\quad h = [h \cdot \sigma_1\ h \cdot \sigma_2]$
$(\times!)$ $\quad p = \langle pr_1 p, pr_2 p \rangle$
$(1!)$ $\quad u = *$
$(0!)$ $\quad f = \square$

The equations provable in this theory are precisely those obtained by omitting the first three uniqueness axioms and adding the following rules of inference to the primary axioms:

- From $(fx) = (gx)$ infer $f = g$, when x is not free in f or h.
- From $g(\sigma_1 x_1) = h(\sigma_1 x_1)$ and $g(\sigma_2 x_2) = h(\sigma_2 x_2)$ infer $g = h$.
- From $\pi_1 p = \pi_1 q$ and $\pi_2 p = \pi_2 q$ infer $p = q$.

The rule for the sum ensures that the meaning of a term g of type $(A_1 + A_2) \to B$ is determined by its action on terms from A and from B. This is the sense in which the sum type is "categorical"; a similar remark applies to for product type.

Definition 2.3 The reduction relation R^∞ is obtained by orienting each equation in Definition 2.2 from left to right.

Write $a \Longrightarrow b$ to indicate that a reduces to b under R^∞.

Certainly R^∞ generates E, in the sense that E is the least equivalence relation containing R^∞. But R^∞ is clearly not strongly normalizing. For example, the "expanding" reductions corresponding to the uniqueness axioms can be applied indefinitely, the the terms fa and $(\lambda x.fx)a$ reduce to each other, and (η) and $(+!)$ can alternate to produce an infinite reduction sequence. The reduction we will be most interested in is obtained by imposing the obvious (local) restrictions to prohibit these chains and loops:

Definition 2.4 The reduction relation R is obtained by orienting each equation in Definition 2.2 from left to right and adding the following constraints:
  in (η): f is neither of the form $\lambda y.b$ nor of the form $[f_1\ f_2]$ and
      the reduced occurrence of f is not applied to an argument,
  in $(+!)$: h is an abstraction,
  in $(\times!)$: p is not of the form $\langle p_1, p_2 \rangle$ and
      the reduced occurrence of p is not the subject of a projection,
  in $(1!)$: t is not $*$,
  in $(0!)$: f is not $\square$.

Write $a \longrightarrow b$ to indicate that a reduces to b under R. We abuse notation by using the same notation to refer to an equation and to the R-rule which it induces.

Proposition 2.5 *The relation R generates E.*

Proof. It suffices to show that if $a \Longrightarrow b$ while $a \longrightarrow b$ fails then $a \longleftrightarrow b$, by induction on the size of a.

If the reduction is one of $(\times!)$, $(1!)$, $(0!)$, or an (η) whose redex is not a $[f_1\ f_2]$-term, the argument is easy. Suppose $a \equiv C[[f_1\ f_2]] \Longrightarrow C[\lambda x.[f_1\ f_2]\,x] \equiv b$. Then b reduces to $C[[\lambda x_1.f_1{}^!x_1\ \lambda x_2.f_2{}^!x_2]]$, and it is easy to check that each $\lambda x_i.f_i{}^!x_i \longleftrightarrow f_i$, possibly using the induction hypothesis at term f_i.

Finally suppose that b is obtained by (η)-expanding a non-abstraction h. If h is an explicit sum of functions, $a \equiv b$, so consider $a \equiv C[h]$ reducing to $b \equiv C[[\lambda x_1.h(\sigma_1 x_1)\ \lambda x_2.h(\sigma_2 x_2)]]$. Then $C[\lambda x.hx] \longleftrightarrow a$ by induction the hypothesis, and $C[\lambda x.hx]$ reduces to b directly. □

The restrictions on the expansion rules have the unfortunate consequence that the reduction is not closed under substitution. We will have to be careful about this in certain places below.

Lawvere [Law69] has pointed that one can do propositional logic in a bicartesian closed category and so cannot postulate fixed points for all maps. It will be useful to sketch the argument here. Fix any type A; the type $B = A + A$ will play the role of a boolean type, in which elements of the form $\sigma_1 u$ and $\sigma_2 v$ play the roles of "true" and "false" values, respectively. For any T define

$$\text{not} : B \to B \equiv [\lambda x_2.(\sigma_2 x_2)\ \lambda x_1.(\sigma_2 x_1)]$$

and

$$\text{if} : B \to T \to T \to T \equiv \lambda zxy.[\mathsf{K}x\ \mathsf{K}y]\,z,$$

where K abbreviates $\lambda uv.u$. Then

$$\text{not}\ \sigma_1 u \longrightarrow \sigma_2 u,\quad \text{not}\ \sigma_2 u \longrightarrow \sigma_1 u,\quad \text{if}\ (\sigma_1 u)xy \longrightarrow\!\!\!\!\!\ast\ x,\text{ and if }(\sigma_2 v)xy \longrightarrow\!\!\!\!\!\ast\ y.$$

A weak form of equality-testing is provided by the term

$$\text{tst} : B \to B \to T \to T \to T \equiv \lambda uvxy.\text{if}\ u(\text{if}\ vxy)(\text{if}\ vyx).$$

Then

$$\text{tst}\ (\sigma_i a)(\sigma_i b)xy \longrightarrow\!\!\!\!\!\ast\ x\ \text{and}\ \text{tst}\ (\sigma_i a)(\sigma_j b)xy \longrightarrow\!\!\!\!\!\ast\ y\ \text{when } i \neq j.$$

None of the above constructions relied on the uniqueness equations for the sum. But in the presence of +! the following holds of the function tst :

$$\text{tst}\ zzxy = x\ \text{and tst}\ z(\text{not}\ z)xy = y.$$

To prove this observe that $\lambda z.\text{tst}\ zzxy$ and $\mathsf{K}x$ have a common reduct, as do $\lambda z.\text{tst}\ z(\text{not}\ z)xy$ and $\mathsf{K}y$.

If we now postulate the existence of fixed-point operators, so that every term has a fixed point, let w be a fixed point of not. Then for any x and y,

$$\text{tst}\ wwxy = x = \text{tst}\ w(\text{not}\ w)xy = y$$

and the theory is inconsistent.

The technique developed in section 4 will shed some additional light on this situation.

3 Strong Normalization

In this section we show that every sequence of R-reductions terminates.

An individual term t will be said to be strongly normalizing (in any of the reduction systems we consider) if every sequence of reductions out of t terminates. If a term is not strongly normalizing, we will say it is *infinite*.

Some preliminary observations will be helpful. It is not obvious even that simple variables are SN, since non-base-type variables can undergo expansions. On the other hand, if t is an I-term other than an abstraction whose immediate subterms are SN, then t is SN — it is straightforward to prove this using the fact that root expansions never apply to such terms. The situation with λ-abstractions is more delicate, since they can undergo $(+!)$-reductions.

To handle abstractions we require the following notion.

Definition 3.1 Given a type T, the *pseudo-variables* $PV(T)$ of T are the variables of type T, together with, in case $T \equiv T_1 + T_2$, the set $\{\sigma_i p_i \mid p_i \in PV(T_i),\ i \in \{1,2\}\}$.

Lemma 3.2 *Let U be a type such that $PV(U) \subseteq SN$. Then $\lambda x.b : U \to V$ is SN provided $\{b[x := p] \mid p \in PV(U)\} \subseteq SN$.*

Proof. Easy (use induction on U). □

Definition 3.3 The set of *computable* terms of type T is defined by induction on T. The term $t : T$ is computable if t is strongly normalizing, and if furthermore

1. if $t \longrightarrow\!\!\!\!\!\twoheadrightarrow \lambda x.b$ then for all computable a, $b[x := a]$ is computable,
2. if $t \longrightarrow\!\!\!\!\!\twoheadrightarrow [f_1\ f_2]$ then each f_i is computable,
3. if $t \longrightarrow\!\!\!\!\!\twoheadrightarrow \langle t_1, t_2 \rangle$ then t_1 and t_2 are each computable, and
4. if $t \longrightarrow\!\!\!\!\!\twoheadrightarrow \sigma_i a$ then A is computable.

Let $\mathcal{C}^T$ denote the computable terms of type T, and set $\mathcal{C}$ to be $\bigcup \{\mathcal{C}^T \mid T \text{ a type}\}$.

Tait [Tai67] originated the strategy of using an inductively defined predicate such as computability to prove termination in the λ-calculus. Prawitz [Pra70] pointed out the possibility of basing a notion of computability, there termed *validity*, based on I-terms rather than on E-terms (as Tait's method is), having observed that the latter approach breaks down when sum types are involved. He proves termination of a certain calculus by a method based on I-terms but slightly different from the definition above. The reductions in that calculus include "commuting reductions" for sum terms, inspired directly by proof theory rather than computation, and do not include reductions corresponding to "uniqueness" equations.

We will use the following observations about computability below. A term of base type is in $\mathcal{C}$ precisely when it is SN. A non-abstraction I-term is in $\mathcal{C}$ if all of its immediate subterms are in $\mathcal{C}$ — this submits to same sort of argument as the corresponding remark about SN. The set $\mathcal{C}$ is closed under reduction, so (since every computable term is SN) we will be justified in proofs by induction over maximal-reduction-length for terms in $\mathcal{C}$.

When t is SN, define $\sharp t$ to be the maximum length of a reduction path out of t.

Notation 3.4 Let us call the rules arising from the uniqueness axioms *expansion* rules. (Note that an "expansion" is still a reduction, as a rule in the calculus). A *root* reduction of a is a reduction whose redex is a itself; a *proper* reduction is one which is not a root expansion.

The following lemma establishes the key facts we need for the termination proof. The importance of the second assertion in the lemma was highlighted by Girard in his proof of termination for the λ-calculus with polymorphic types. The crucial point here is the restriction to proper reductions.

Lemma 3.5 *For each T:*

1. *$\lambda x.b : T$ is computable if for all computable a, the term $b[x := a]$ is computable,*
2. *if $t : T$ is not an I-term and if each proper one-step reduct of t is computable, then t is computable,*
3. *$PV(T) \subseteq \mathcal{C}$,*
4. *$\pi_i p : T$ is computable if p is computable, and*
5. *if $f : T$ is computable then for all computable a, (fa) is computable.*

Proof.

The proof is by induction on T, and it importanmt that we prove the clauses in the order stated.

Clause 1. Write $T \equiv U \to V$, so that $x : U$ and $b : V$. First note that $\lambda x.b$ is SN by Lemma 3.2, using (3) at type U.

Next, suppose $U \equiv U_1 + U_2$ and $\lambda x.b \longrightarrow\!\!\!\twoheadrightarrow [\lambda x_1.b_1 \ \lambda x_2.b_2]$; we want to show that each $\lambda x_i.b_i$ is computable at type $U_i \to V$. Without loss of generality the reduction must have begun with a $(+!)$-reduction, and so, since $\mathcal{C}$ is closed under reduction it suffices to consider the case where each b_i is $b[x := \sigma_i x_i]$. By induction, it suffices to show that for each $d \in \mathcal{C}^{U_i}$, $b_i[x_i := d]$ is computable; but this is simply $b[x := \sigma_i d]$, computable by assumption on b.

Finally, we wish to show that if $\lambda x.b \longrightarrow\!\!\!\twoheadrightarrow \lambda x.c$ then for any computable a, $c[x := a]$ is computable.

Let us introduce the notation $\mathcal{C}^V_x$ for the set of terms t such that $t[x := a] \in \mathcal{C}^V$ whenever $a \in \mathcal{C}^U$ (remember that $x : U$).

Now suppose we can show that for $a \in \mathcal{C}^U$, $(\lambda x.b)a \in \mathcal{C}V$. Then we can simply observe that $(\lambda x.b)a \longrightarrow\!\!\!\twoheadrightarrow (\lambda x.c)a \longrightarrow\!\!\!\twoheadrightarrow c[x := a]$. We concentrate, then, on showing $(\lambda x.b)a \in \mathcal{C}V$.

We may use (2) at type V, and be content with showing that all proper reducts are computable. Furthermore b and a are SN, and so we may argue by induction over $\sharp b + \sharp a$, *provided* we know that $\mathcal{C}^V_x$ is closed under reduction. The reducts are of the form $(\lambda x.b')a$, $(\lambda x.b)a'$, or $b[x := a]$; the first two are computable by induction hypothesis, the latter by assumption on b. But it is not clear that $\mathcal{C}^V_x$ is in fact closed under reduction. The remainder of the argument is devoted to verifying this.

Consider a reduction $b \equiv C[u] \longrightarrow C[u'] \equiv c$ in which u is the redex. To argue that $c \in \mathcal{C}^V_x$, choose a computable term a to replace x in c; we want $c[x := a] \in \mathcal{C}^V$. Now $b[x := a] \in \mathcal{C}^V$, and if $b[x := a] \longrightarrow c[x := a]$ we can simply use the fact that $\mathcal{C}$ is closed under reduction. Of course, $b[x := a] \longrightarrow c[x := a]$ will not hold if one of the expansion-restrictions is violated. So we restrict attention to this situation.

Now (since $b \equiv C[u] \longrightarrow C[u'] \equiv c$) the way in which $b[x := a] \longrightarrow c[x := a]$ can fail is for a restriction as to the form of the redex to be violated . Indeed, it must be the case that u is in fact the variable x, the original $b \longrightarrow c$ was an expansion of x, and the term a was either $*$, $\square$, $\langle a_1, a_2 \rangle$, $\lambda y.d$, or $[f_1 \ f_2]$, according to the expansion rule.

We claim that the expansions a' of a are themselves computable and that $a' \longrightarrow\!\!\!\!\!\rightarrow a$. If so, then $b[x := a']$ is computable by assumption on b, and $c[x := a]$ is either $b[x := a']$ or (in case there are occurrences of x other than as the redex in b) obtained by reduction from $b[x := a']$. Then we are done.

Actually, in case $a \equiv *$, $a \equiv \square$, or $a \equiv [f_1\ f_2]$ we have $a \equiv a'$ and there is nothing to prove.

Taking the other cases in turn (note that a has type U and so we may invoke the current Lemma there): if $a \equiv \langle a_1, a_2 \rangle$ the corresponding expansion is $\langle \pi_1 \langle a_1, a_2 \rangle, \pi_2 \langle a_1, a_2 \rangle \rangle$. But this is computable by part (4) and the remarks following Definition 3.3. If $a \equiv \lambda y.d$ the corresponding expansion is $\lambda x.(\lambda y.d)x$, and this is computable by (1) and then (5). The fact that $a' \longrightarrow\!\!\!\!\!\rightarrow a$ is easily verified.

Clause 2. Since t is neutral, it is clear that if *every* one-step reduction of t were computable, then t would be computable. So it suffices here to show that each 1-step root expansion of t is computable. When T is either a base type or a sum, there is then nothing to prove.

When $T \equiv T_1 \times T_2$ we consider the term $\langle \pi_1 t, \pi_2 t \rangle$. It suffices to show that each $\pi_i t$ is computable. By the induction hypothesis it suffices to show that an arbitrary proper reduct of $\pi_i t$ is computable. Since t is not an I-term such a reduction is of the form $\pi_i t'$, where $t \longrightarrow t'$ via a proper reduction by the context restriction on expansions. Thus t' is computable by hypothesis on t. Part (4) at type T_i then tells us $\pi_i t'$ is computable, as desired.

When $T \equiv U \rightarrow V$ we consider $\lambda x.tx$. By (1) it suffices to see that for $a \in \mathcal{C}^U$, $ta \in \mathcal{C}^V$. By (2) at type V we may examine the proper reducts and induct over $\sharp a$. But since t is neutral the argument is easy.

Clause 3. When T is a base type, use (2) together with the fact that there are no improper reductions out of a base-type variable. Otherwise use the fact that the pseudo-variables are generated by the σ_i contexts, which preserve computability.

Clause 4. Induct on $\sharp p$. By (2) it suffices to show that each proper reduct of $\pi_i p$ is in $\mathcal{C}$. There are two forms such a reduction can take. One is a proper reduction $p \longrightarrow p'$ inside of p, which yields $\pi_i p'$, in $\mathcal{C}$ by the induction hypothesis. The other is a root reduction, in case $p \equiv \langle p_1, p_2 \rangle$, yielding p_i, which is in $\mathcal{C}$ since p was.

Clause 5. Induct on $\sharp f + \sharp a$. Invoking (2), we examine the proper reducts of fa; non-root reductions submit to the induction hypothesis. There are two possible root reductions, in the cases $f \equiv (\lambda x.b)$ and $f \equiv [f_1\ f_2]$. In each case the result is computable by hypothesis on f.

□

Corollary 3.6 *If f, a, and n are each computable then* rec fan *is computable.*

Proof. It suffices to show that all proper reducts are computable; we show this by induction on $\sharp f + \sharp a + \sharp n$, with a secondary induction on the length of the normal form of n. The only interesting reduction is rec $fa(\text{succ } n) \longrightarrow fn(\text{rec } fan)$. The last argument is computable since the normal form of n is smaller than the normal form of succ n, now use (5) of the previous Lemma. □

Theorem 3.7 R *is SN.*

Proof. It suffices to show that each term is computable. Using the standard trick, define C^{Vars} to be the set of all substitutions θ such that for every x, $\theta x \in C$; and let C^* be the set of terms t such that for all $\theta \in C^{Vars}$, $\theta t \in C$. We show that all terms t are in C^* by induction on t. This suffices since variables are computable and thus the identity substitution is in C^{Vars}.

Choose $\theta \in C^{Vars}$. When t is a variable $\theta t \in C$ by definition of C^*; when t is a constant $\theta t \equiv t$, and so is in C by Lemma 3.5 (2).

When $t \equiv \lambda x.b$ then (since we may assume that x is not in the domain of θ) $\theta t \equiv \lambda x.\theta b$, and by Lemma 3.5 it suffices to show that for any $a \in C$, $b[x := a] \in C$. But $b[x := a]$ is $\theta' b$ where $\theta' \equiv \theta \cup \{x \mapsto a\}$, and this is a substitution in C^{Vars}. So by the induction hypothesis for b, $b[x := a] \in C$ as desired.

In every other case, the substitution θ filters down to the immediate subterms of t and the result can be seen to be computable by invoking the induction hypothesis and applying Lemma 3.5 or the remarks following Definition 3.3. □

Corollary 3.8 *When there are no constants other than 0, every closed term reduces to an I-term.*

Proof. By induction on terms; by strong normalization it suffices to show that no closed non-I term is irreducible. The argument is straightforward. □

4 Confluence

Proposition 4.1 *When there are no constants other than 0, R is confluent on closed terms of base type.*

Proof. It suffices to show that closed base-type normal forms are equal only if they are identical. But the terms in question are the numerals, and the full set-theoretic type hierarchy generated by the standard natural numbers provides a model for the equality theory, so distinct numerals can never be proved equal. □

Remark 4.2 R is not confluent.

Proof. Let $x : A + A$ and observe that

$$x = [\lambda x_1.(\sigma_1 x_1) \ \lambda x_2.(\sigma_2 x_2)] \, x$$

is provable in E, but each term is irreducible (if A is not a sum).

Abstracting over x gives a closed example. A similar example shows that the presence of rule (0!) also blocks confluence. □

We are led to consider the theory of sums, obtained by deleting equations (+!) and (0!). The natural step in constructing a corresponding reduction system is to use R without (+!) and (0!) rules. We can further and relax the restriction on (η) by allowing the redex to be be an explict sum of functions.

In fact, rule (+!) was used in demonstrating that

$$[f_1 \ f_2] = \lambda x . [f_1 \ f_2] x$$

was admissible in the theory generated by R, so we *must* relax the restriction on (η).

Definition 4.3 The equational theory E^- is generated by the axioms of Definition 2.2 with (+!) and (0!) omitted.

The reduction relation $R^{\bowtie}$ is obtained by orienting each equation in E^- from left to right. Write $a \Longrightarrow b$ for this relation.

The reduction relation R^- is generated by the rules of Definition 2.4 with (+!) and (0!) omitted, and with the first restriction on rule (η) revised to read simply "f is not of the form $\lambda y.b$". Write $a \longrightarrow b$ for this relation.

It is not hard to see that R^- generates E^-. The main results of this section are that R^- is confluent and furthermore remains confluent when fixed-point operators are added.

The technique makes use of the strong normalization of R^-, but this does not follow from the results of the previous section since R^- is not a sub-system of R. The proof is a slight modification of the arguments of the last section, so we will just give an outline:

Theorem 4.4 R^- *is SN*.

Proof (outline). The analogue of Lemma 3.2 is the easy observation that $\lambda x.b$ is SN if b is — abstractions undergo no expansions in the present system. The definition of *computable* terms is precisely the same as in Definition 3.3, and most of the remarks about the computability of I-terms still hold, with the exception that it is no longer clear that $[f_1 \ f_2]$ is computable when the f_i are (such terms may now enjoy root (η) expansions).

So we add this as a final assertion in the main Lemma. Clearly $[f_1 \ f_2]$ is SN. The argument will reduce to showing (since $[f_1 \ f_2] \longrightarrow \lambda x. [f_1 \ f_2] x$) that for computable a, $[f_1 \ f_2] a$ is computable. A routine application of (2) suffices. □

The system R^- has better substitutivity properties than R.

Proposition 4.5 1. *If $a \Longrightarrow b$ then either $a \longrightarrow b$ or $b \longrightarrow\!\!\!\twoheadrightarrow a$.*

2. *If $a \longrightarrow a'$ then for all t $t[x := a]$ and $t[x := a']$ have a common $\longrightarrow\!\!\!\twoheadrightarrow$-reduct.*

3. *If $t \longrightarrow t'$ then for all a, either*
    (a) $t[x := a] \longrightarrow t'[x := a]$ *or*
    (b) $t'[x := a] \longrightarrow\!\!\!\twoheadrightarrow t[x := a]$.

Proof. Part (1) is an easy examination of cases. For (2) first note that if $a \xrightarrow{} a'$ then $a \Longrightarrow a'$. Since there are no restrictions on R^{M} it is easy to see that $t[x := a] \Longrightarrow t[x := a']$, by rewriting the various occurrences of a to a'. Now use (1), but note that the direction of the $\twoheadrightarrow$-reductions may be different for different occurrences of x. Part (3) is similar, and easier. □

Theorem 4.6 R^{-} *is confluent.*

Proof. It suffices to establish local confluence. Since R^{-} must obey context constraints we cannot, *a priori*, restrict attention to finding common reducts for the critical pairs. We proceed by induction on terms t to show that

If $t \xrightarrow{} u$ and $t \xrightarrow{} v$ then there exists w such that $u \twoheadrightarrow w$ and $v \twoheadrightarrow w$.

First consider the case in which $t \xrightarrow{} u$ is a root expansion. If $t \xrightarrow{} u$ is $t \xrightarrow{} \lambda x.tx$ then, writing t' for v, note that $t' \xrightarrow{} \lambda x.t'x$ and that $\lambda x.tx \Longrightarrow \lambda x.t'x$. Proposition 4.5 yields the desired w. The case of a $(\times!)$-expansion is similar.

We may assume then that neither $t \xrightarrow{} u$ nor $t \xrightarrow{} v$ is a root expansion. If t is an I-term it is easy to construct a common reduct of u and v. The other cases are straightforward unless $t \equiv fa$, u is $f'a$ and v is fa'. Here consider the term $f'a'$. We can be sure that $f'a \xrightarrow{} f'a'$ since there are no context restriction on rewriting an argument; the fact that $fa' \Longrightarrow f'a'$ and an application of Proposition 4.5 yield our w.

We are left with the case in which u is obtained by a root non-expansion from fa. We discuss only the case $fa \equiv (\lambda x.b)a \xrightarrow{} b[x := a] \equiv u$. The two possibilities for v are $(\lambda x.b')a$ and $(\lambda x.b)a'$; these reduce, respectively, to $b'[x := a]$ and $b[x := a']$. A final application of Proposition 4.5 completes the proof. □

We conclude that the unrestricted system is confluent as well:

Corollary 4.7 R^{M} *is confluent.*

Proof. R^{M} and R^{-} generate the same equational theory, and R^{-} is a subsytem of R^{M}. □

Motivated by the interpretation of λ-calculi as theories of programming languages, we next consider the consequences of adding fixed-point operators to Λ. As described in section 2, adding fixed-point operators to E yields an inconsistent theory, so we certainly cannot expect confluence there. We again turn to E^{-}.

It is easy to see that adding fixed-point operators to E^{-} yields a consistent theory — indeed the familiar category of complete partial orders (with either separated or coalesced sums) is cartesian closed (but not bicartesian closed) and so provides models. But a confluence result gives more information, of course. As described in the introduction, the traditional system for products and fixed point fails to be confluent. We show now in contrast that R^{-} with fixed points added remains confluent.

Definition 4.8 Assume that for each type T other than 0 there is a distinguished constant $\varphi_T : (T \to T) \to T$. The reduction system R^φ is obtained from R^- by adding the following rules:

$$(\varphi) \quad \varphi_T f \xrightarrow{\varphi} f(\varphi_T f).$$

For emphasis, when fixed-point constants have been distiguished we designate the set of terms by Λ^φ. We will henceforth supress the type subscript on the φ_T.

We will analyze the behavior of this new calculus by simulating the fixed-point operators by operators providing *bounded* recursion. This technique apparently originates with Lévy, and was used by Poigné and Voss [PV87] (I am indebted to Roberto Di Cosmo and Delia Kesner for this reference).

Definition 4.9 Assume that for each type T other than 0 and each natural number n there is a distinguished constant $\varphi_T^n : (T \to T) \to T$. The reduction system $\mathsf{R}^\#$ is obtained from R^- by adding the following rules (omitting, as usual, type tags):

$$(\varphi^n) \quad \varphi^{n+1} f \xrightarrow{\#} f(\varphi^n f).$$

In this setting we will designate the set of terms by $\Lambda^\#$.

Lemma 4.10 $\mathsf{R}^\#$ *is SN and confluent.*

Proof. We can code the φ^n as terms $\overline{\varphi^n}$ in the ordinary R^- system by letting $\overline{\varphi^0}$ be an arbitrary free constant and $\overline{\varphi^{n+1}}$ be $\lambda f.f(\overline{\varphi^n} f)$. Then a reduction in $\mathsf{R}^\#$ induces a corresponding reduction in R^-, hence is finite. Note that there were no restrictions on the set of free constants in the SN proof.

For confluence, it suffices to check local confluence (observe that a naive argument based on the coding above will not suffice). The argument is just as for the proof of Theorem 4.6; the additional cases due to the φ^n cause no difficulty. □

We now show how to lift the previous result to obtain CR for the pure R^φ system.

Definition 4.11 Given t from $\Lambda^\#$, the *erasure* $|t|$ of t is obtained by replacing each occurrence of a φ^n by φ.

It is easy to prove an "erasing lemma" showing that if $t \xrightarrow{\#}_* s$ then $|t| \xrightarrow{\varphi}_* |s|$.

The following "lifting" lemma shows that any computation can be simulated by the bounded recursors. Say that the *index* of a term in $\Lambda^\#$ is the least n such that some φ^n occurs in the term (or 0, if no φ^n appear).

Lemma 4.12 *Let* $t \in \Lambda^\#$ *have index k and suppose that*

$$|t| \xrightarrow{\varphi}_* u \text{ in no more than } k \text{ steps.}$$

Then there exists $s \in \Lambda^\varphi$ *such that* $t \xrightarrow{\#}_* s$ *and* $|s| \equiv u$.

Proof. It suffices to consider a 1-step reduction under the hypothesis that k is at least 1, provided we show that in this case the index of the resulting s is at least $k - 1$.

But s can be determined by applying the same reduction (in the obvious sense) out of t as the given reduction out of $|t|$. Specifically, (and assuming for the sake of ease of notation that the given reduction is at the root of t) the relevant fact is that if $|t|$ is of the form θl where $l \xrightarrow{\varphi} r$ is one of the rules of R^φ, then t will be of the form $\theta' l'$ where the erasures of l' and θ' (pointwise) are l and θ. The term s will then be $\theta' r$. It is clear that the index is either unchanged or diminished by one.

It is just here that we use the fact that all of our reduction rules are left-linear. For example, if the rule for surjective pairing were oriented as $\langle \pi_1 p, \pi_2 p \rangle \longrightarrow p$ then the left-hand side could arise as the erasure of a term t without t being a redex. $\square$

Theorem 4.13 R^φ *is confluent.*

Proof. Suppose $a \xrightarrow{\varphi}\!\!\!\twoheadrightarrow b_1$ and $a \xrightarrow{\varphi}\!\!\!\twoheadrightarrow b_2$; we seek c such that $b_1 \xrightarrow{\varphi}\!\!\!\twoheadrightarrow c$ and $b_2 \xrightarrow{\varphi}\!\!\!\twoheadrightarrow c$. Let n be the maximum of the lengths of the given reductions, and construct $a^\# \in \Lambda^\#$ from a by replacing each occurrence of φ by φ^n.

By the lifting lemma there are $b_1^\#$ and $b_2^\#$ such that $a^\# \xrightarrow{\#}\!\!\!\twoheadrightarrow b_1^\#$ and $a^\# \xrightarrow{\#}\!\!\!\twoheadrightarrow b_2^\#$ and $|b_i^\#| \equiv b_i$.

By the confluence of $\mathsf{R}^\#$ there is a $c^\#$ such that for each i, $b_i^\# \xrightarrow{\#}\!\!\!\twoheadrightarrow c^\#$. Apply the erasing lemma to produce $|c^\#|$ as the desired c. $\square$

Corollary 4.14 *Adding fixed-point operators yields a conservative extension of* E^-.

$\square$

As a final note we reconsider the original theory E. The construction above strengthens Remark 4.2 by showing that no *ground confluent* left-linear reduction system can be constructed for a functionally complete theory of coproducts, without some restriction on the free constants allowed. If such a system existed, we could build in the bounded recursors, show preservation of confluence in the presence of true fixed-point operators, and derive the above Corollary for the full theory. But the argument in section 2 shows that this is a contradiction.

Acknowledgements.

Roberto Di Cosmo and Delia Kesner pointed out an error in an early version of Proposition 4.5. I am grateful to Ramesh Subrahmanyam for several enlightening discussions.

References

[Aka93] Y. Akama. On Mint's Reduction for ccc-Calculus. *Proc. Typed Lambda Calculus and Applications* 1993.

[Bar84] H. P. Barendregt. *The Lambda Calculus: Its Syntax and Semantics*. Volume 103 of *Studies in Logic and the Foundations of Mathematics*, North-Holland, Amsterdam, 1981. Revised edition, 1984.

[CD91] P-L Curien, R. Di Cosmo. A confluent reduction system for the λ-calculus with surjective pairing and terminal object, in Leach et. al. (eds.), *Int'l. Conf. on Automata, Languages, and Programming*, LNCS 510, 291–302, Springer Verlag 1991.

[Cub92] D. Cubric. Embedding of a free cartesian closed category into the category of Sets. Manuscript, McGill University 1992.

[DK93a] R. Di Cosmo, D. Kesner. Simulating expansions without expansions. Tech Rep, INRIA 1993.

[DK93b] R. Di Cosmo, D. Kesner. A confluent reduction system for the extensional λ-calculus with pairs, sums, recursion and terminal object. Proc ICALP 1993.

[DJ91] N. Dershowitz, J.-P. Jounnaud. Term Rewriting Systems, in *Handbook of Theoretical Computer Science*, 243–320, North-Holland, Amsterdam, 1991.

[Dou90] D. J. Dougherty. Some reduction properties of a λ-calculus with categorical sums and products. Manuscript, Wesleyan University 1990.

[Gan80] R. O. Gandy. Proofs of strong normalization, in J. P. Seldin and J. R. Hindley (eds.) *To H. B. Curry, Essays on Combinatory Logic, Lambda Calculus and Formalism*. Academic Press, New York, 1980.

[Jay91] C. Barry Jay. Long β-normal forms and confluence. Manuscript, University of Edinburgh, 1991.

[Klo80] J. W. Klop. *Combinatory Reduction Systems*. Mathematical Centre Tracts, v. 127. Centre for Mathematics and Computer Science, Amsterdam 1980.

[KV89] J. W. Klop, R.C. deVrijer. Unique normal forms for lambda calculus with surjective pairing. *Information and Computation*, v.80 no. 2, 97–113, 1989.

[LS86] J. Lambek, P. Scott. *Introduction to Higher-order Categorical Logic*. Cambridge Studies in Advanced Mathematics 7. Cambridge University Press 1986.

[Law69] F. W. Lawvere. Diagonal arguments and cartesian closed categories, in *Category Theory, Homology Theory, and Their Applications II*, LNM 92, Springer-Verlag, 1969.

[Min80] G. E. Mints. Category Theory and Proof Theory. In *Aktualnie Voprosi Logiki i Metodologiinauki* (Russian) 252–278. Kiev, 1980.

[OS91] M. Okada, P. Scott. Rewriting theory for uniqueness conditions: coproducts. Talk presented First Montreal Workshop on Programming Language Theory, April 1991.

[Pra70] D. Prawitz. Ideas and Results in Proof Theory, in J. E. Fenstad, ed., *Proceedings of the Second Scandinavian Logic Symposium*. North-Holland, Amsterdam, 1971.

[PV87] A. Poigné, J. Voss. On the implementation of Abstract Data Types by Programming Language Constructs. *Journal of Computer and System Science* **34** (1987), 340–376.

[Tai67] W. W. Tait. Intensional interpretation of functionals of finite type I, *J. Symbolic Logic* **32**, pp. 198-212, 1967.

[deV87] R. C. de Vrijer, Strong Normalization in $N - HA^{\omega}$. *Proc. Koninklijke Nederlandse Akademie van Wetenschappen* Series A, 90/4, pp. 473-478, 1987.

Paths, Computations and Labels in the λ-calculus*

Andrea Asperti
Dip. di Matematica, Bologna

Cosimo Laneve
INRIA Sophia-Antipolis

Abstract

We provide a new characterization of Lévy's redex-families in the λ-calculus [11] as suitable paths in the initial term of the derivation. The idea is that redexes in a same family are created by "contraction" (via β-reduction) of a unique common path in the initial term. This fact gives new evidence about the "common nature" of redexes in a same family, and about the possibility of sharing their reduction. From this point of view, our characterization underlies all recent works on optimal graph reduction techniques for the λ-calculus [9,6,7,1], providing an original and intuitive understanding of optimal implementations.

As an easy by-product, we prove that neither overlining nor underlining are required in Lévy's labelling.

1 Introduction

Take the (labelled) λ-term $\Delta\Delta$, represented in the following picture:

Consider the application node labelled by c. Our question is: "is it possible that this application node will be ever involved in a β-reduction?". In order to solve the

---

*Partially supported by the ESPRIT Basic Research Project 6454 - CONFER. The work was mostly carried out while the first author was at INRIA Rocquencourt and the second one was at the Dipartimento di Informatica, Universitá di Pisa.

question we must look for a λ-node to match against the application. We eventually start our search towards the left daughter of the @-node (this is the *principal port* of the application, in Lafont's terminology [8,2], that is the only port where we may have interaction with a dual operator). Thus, we pass the edge labelled by d and we find a variable. If the variable was free we had finished: no redex involving our application could possibly exist. Since the variable is bound, the "control" is passed back to its binder, that is the λ-node labelled by b in the picture. Indeed, a β-reduction involving the binder will replace the variable with the argument of the application, and we must continue our search inside this argument. Hence we must pose a symmetrical question about this λ-node, namely: "is it possible that this λ-node will be ever involved in a β-reduction?". Since the question concerns a λ-node, we must now start travelling towards the root (note that this is still the principal port of the λ-node, according to Lafont). This time, we find an application. Moreover we enter the application at its principal port, thus we have found a redex. The last question is solved positively, and we must resume the previous one, looking into the argument of the application. Thus, we pass f, and we find a λ. This λ-node is reached at its principal port, so we have finally found a (virtual) redex for the original @-node.

During this process, we have described a "path" dbf in t. If we suppose that bound variables are explicitly connected to (positive, auxiliary ports of) their respective binders (an idea going back to Bourbaki), the "path" is indeed connected. In the following, we shall always make this hypothesis, even if we shall not explicitly draw the connections in the pictures. Note also that, by firing the redex b, we "contract" the path dbf into a single redex-edge. This is the "actual" redex, corresponding to its virtual path-description dbf.

Let us consider another example, starting with the @-node labelled by g, in the picture above. We travel along h, then back to the binder for y, and up along f. We have found a @-node, but we entered this node at its negative, auxiliary port (not its principal port!). So we did not find a redex, and we must open a new session, looking for a redex involving this @-node. This is immediately found: it is the redex b. Hence we resume the previous search. Since the "control" was coming from the argument of the @-node labelled by a, we must pass it to some of the variables bound by the λ-node in the redex b, that is d or e (note the non-determinism of the search algorithm, at this point). Suppose to follow the first possibility. We have found the same redex dbf of the previous computation (travelling in the opposite direction), and we may resume the first question. We must go down to e, up to the binder for x, up again along the redex b and finally down along f. We have described the path $hfbdebf$ beginning at the principal port of a @-node and ending to the principal port of a λ-node. The reader may convince himself that this path corresponds to the unique redex created after two reductions of $\Delta\Delta$ (see also example 4.2). Indeed, by firing first the redex b and then the (newly created) redex dbf, we contract the path $hfbdebf$ into a single redex-edge.

Actually, the search algorithm is a bit more contrived of what we have described. In particular, the non deterministic choice at the level of bound variables (when travelling down through a λ) is not always free.

Our aim, for the moment, was just to provide the main intuitions behind the path-description of "virtual" redexes in a λ-term t (i.e. redexes which can be created along some reduction from t). Every "virtual" redex t defines a suitable "legal" path in t from an application node to a λ-node.

Do different "virtual" redexes define different paths? The answer is no, due to a duplication problem. Consider a term $t = (\lambda x.M)N$, and suppose to have a redex r inside N. This redex may have several residuals in $M[N/x]$, and all these different residuals define the same path in t. This fact suggests that we can reasonably hope to have an injection from "virtual" redexes to paths, up to some notion of "sharing".

This is actually the case. Formally, we have a *bijective correspondence* between families of redexes in the sense of Lévy [11] and (a suitable class of *legal*) paths. In other words, two redexes are in a same family, if and only if their associated paths in the initial term of the derivation coincide.

The first hints behind the notion of legal path were already in Lamping's work [9]. In particular, every *prerequisite chain* is always a prefix or a suffix of our legal paths. However, in order to characterize every "virtual" redex as a legal path, we must suitably compose prerequisite chains, that is the not trivial part of the work (and one of our original contributions). Moreover, Lamping did never establish a clear correspondence between prerequisite chains and Lévy's labelling. An (implicit) account of this correspondence is already in [6]. However, the notion of legal path is based on Lamping's control operators, and it has pretty operational flavour that makes it difficult to understand.

Up to our knowledge, the first ones to pursue the program of describing "virtual" redexes (computations) by means of paths have been Danos and Regnier [13,4]. Anyway, they do not seem to have established the correspondence between paths and redex families (optimal reductions). Fixing the relationship between Danos and Regnier's paths and our legal paths is an open problem.

An (implicit) account of this correspondence is already in [6]. Unfortunately the paper is rather cryptic (you will have to deal with the *bus notation*). Moreover, the notion of legal path is based on Lamping's control operators, and it has pretty operational flavour that makes it difficult to understand.

The structure of the paper is the following. We shall start with introducing the notion of redex-family in the λ-calculus, devoting a particular attention to Lévy's labelling [11] (section 3). In Section 4 we shall define the notion of *path associated with a redex*, by relying on its label. Obviously, not every path in the initial term of a derivation does correspond to a redex-family. The final part of the paper aims to provide a complete characterization of these paths (*legal* paths), independent from labels.

2 The family relation

In 1978, Lévy introduced the notion of redex family in the λ-calculus, with the aim to formally capture an *intuitive* idea of *optimal sharing* between "copies" of a same redex. For a long time, no λ-calculus implementation has been able to achieve the theoretical performance fixed by Lévy (see [5]), and it is only in recent years [9,6] that this problem has been finally solved (see also [3,10] for a generalization of these results to a wider class of higher order term rewriting systems).

In order to comfort his notion of family, Lévy proposed several alternative definitions, inspired by different perspectives, and proved their equivalence.

The most abstract approach to the notion of family [11] is the so called *zig-zag*. In this case, duplication of redexes is formalized as residuals modulo permutations. In particular, a redex u with history σ (notation σu) is a *copy* of a redex v with history ρ iff $\rho v \leq \sigma u$ (i.e., there exists τ such that $\sigma = \rho\tau$ up to permutation equivalence, and u is a residual of v after τ). The family relation $\simeq$ is then the symmetric and transitive closure of the copy-relation (pictorially, this gives rise to the "zig-zag"). Note that, intuitively, the family relation has to be an equivalence relation.

Another approach is that of considering the *causal history* of the redex. Intuitively, two redexes can be "shared" if and only if they have been created "in the same way" (or, better, their *causes* are the same). This is formalized by defining an *extraction relation* over redexes (with history) σu, which throws away all the redexes in σ that have not been relevant for the creation of u. The canonical form we obtain at the end of this process essentially expresses the causal dependencies of u along the derivation (we may deal with causal *chains* instead of *partial orders* since *only standard derivations* are considered).

The most "operational" approach to the family relation is based on a suitable labelled variant of the λ-calculus [11], described in the next section. The idea of labels is essentially that of marking the "points of contact" created by reductions. In particular, labels grow along the reduction, keeping a trace of its history. Two redexes are in a same family if and only if their labels are identical.

The equivalence between zig-zag and extraction is not particularly problematic [12]. On the contrary, the proof of the equivalence between extraction (or zig-zag) and labelling is much more difficult, and it forms the core of Lévy's Thesis [11]. The relation between labels and "paths" described in this paper allow us to provide a simpler proof of this fact [2,10]. Actually, in our experience, the approach to the family relation based on "paths" provides the most friendly and intuitive tool for reasoning about redex-families, as well in theory as in practice (implementation).

3 Labelling

The *labelled λ-calculus* is an extension of the λ-calculus proposed by Lévy in [11]. In order to avoid some annoying problems concerning the associativity of the concatenation operator over labels, our presentation will be slightly different (but equivalent) w.r.t. Lévy's one. In particular, we shall delay the concatenation of labels until it is required for firing a redex.

Definition 3.1 Let $L = \{a, b, \cdots\}$ be a denumerable set of *atomic labels*. The set **L** of *labels*, ranged over by $\ell, \ell_1, \cdots$, is defined by the following rules:
$$L \mid \ell_1\ell_2 \mid \underline{\ell} \mid \overline{\ell}$$
The operation of concatenation $\ell_1\ell_2$ will be assumed to be associative. The set $\mathbf{L}_p$ of *proper labels* contains exactly all those labels $\ell \in \mathbf{L}$ such that ℓ is atomic or $\ell = \underline{\ell'}$ or $\ell = \overline{\ell'}$. $\mathbf{L}_p$ will be ranged over by $\alpha, \beta, \cdots$. ∎

For instance $\overline{ab}\underline{ab}c$ is a label. Let us come to the formal definition. Labelled λ-terms are those obtained by the following syntax:
$$t ::= x \mid \lambda x.t \mid @(t, t') \mid \alpha(t)$$
where $\alpha \in \mathbf{L}_p$. The β-reduction is now a set of rules: for every n-tuple $\alpha_1, \cdots, \alpha_n$, we have the following rule:
$$@(\alpha_1(\cdots(\alpha_n(\lambda x.X)\cdots), Y) \longrightarrow \overline{\alpha_1\cdots\alpha_n}(X[\underline{\alpha_1\cdots\alpha_n}(Y)/x])$$

When a redex is fired, a label ℓ is captured between @ and λ: this is the *degree* of the redex. Then, every possible interaction created in the rhs of the rewriting rule must be suitably "marked" with ℓ, in order to keep a trace of the history of the creation. In the case of λ-calculus, there are two ways to create redexes: "towards the top", i.e. a redex in which the outermost symbol of the functional part of the rule is involved, and "towards the bottom", i.e. a redex in which the outermost symbol of the argument part is involved. Overlinings and underlining essentially express this double possibility.

We will say that a given subexpression $t' = \lambda x.t''$ or $t' = @(t_1, t_2)$ occurring in t has label $\alpha_1\cdots\alpha_n$ if $t = \alpha_1(\cdots\alpha_n(t')\cdots)$ or $\alpha_1(\cdots\alpha_n(t')\cdots)$ is the body of an abstraction or an argument of an application.

Definition 3.2 Let t be a labelled λ-expression. The predicate **INIT**(t) is true if and only if the labels of all subterms of t are atomic and pairwise different. ∎

4 Labels as paths

Labels provide a very simple approach to the notion of computation as a travel along a path. In particular, every label trivially define a path in the initial term of the computation. The interesting problem will be to provide an independent characterization of these paths (see Section 5).

Let us assume here the graph representation of terms mentioned in the introduction (i.e. bound variables are supposed connected to the respective binders). Consider an expression t such that $\text{INIT}(t)$. Every edge in t is labelled with a different atomic symbol, so we may call each edge by its label.

Definition 4.1 If ℓ is a label of an edge generated along some reduction from t, the *path* of ℓ in t is inductively defined as follows:

$$\begin{aligned}
\text{path}(a) &= a \\
\text{path}(\ell_1 \ell_2) &= \text{path}(\ell_1) \cdot \text{path}(\ell_2) \\
\text{path}(\overline{\ell}) &= \text{path}(\ell) \\
\text{path}(\underline{\ell}) &= (\text{path}(\ell))^r
\end{aligned}$$

where "·" means concatenation (it will be omitted in the following) and $(\varphi)^r$ is as p but reverted. ■

It is easy to prove by induction on the length of the derivation generating the label ℓ that the previous definition is sound, i.e. that $\text{path}(\ell)$ is indeed a path in t (it simply follows by the labelled β-contraction).

Example 4.2 Consider again the labelled term $\Delta\Delta$ of the introduction. After two labelled β-reduction, you obtain a redex with degree $\ell = hd\underline{b}f\underline{e}bf$. Thus $\text{path}(\ell) = hfbdebf$, that is the path we obtained in the introduction for the application labelled with g. ■

The interesting fact is that different degrees define different paths in the initial term. A preliminary proposition is required, concerning the structure of labels.

Proposition 4.3 Let $\alpha_1 \cdots \alpha_n$ be a label generated along some derivation σ starting at t (every α_i is a proper label). Then

1. n is odd; for every i odd, α_i is atomic; for every i even, $\alpha_i = \overline{\ell}$ or $\alpha_i = \underline{\ell}$;

2. if $\alpha_{2i} = \overline{\ell}$ then α_{2i-1} marks the output edge of the node @ (in t) determinated by the leftmost label in ℓ. α_{2i+1} marks the input port of the abstraction (in t) individuated by the rightmost label in ℓ.

3. if $\alpha_{2i} = \underline{\ell}$ then α_{2i-1} labels an edge in t incoming into the bound port of the abstraction individuated by the rightmost atomic label in ℓ. α_{2i+1} marks the edge of the second argument of the application determinated by the leftmost label in ℓ.

Proof: Easy induction on the length of the derivation. ∎

Remark 4.4 Observe that, if ℓ is a label relative to a redex, it may appear inside another label only if it is overlined or underlined, and surrounded by atomic labels. ∎

An easy consequence of the above proposition is that both underlinings and overlinings can be safely omitted in Lévy's labelling (a property that was already known to Gonthier and Lévy, but never published).

Corollary 4.5 Neither overlining nor underlining are needed in Lévy's labelling.

Proof: (Hint) Remove all underlinings and overlinings from a label. The initial structure can be retrieved by working "inside out", using Proposition 4.3. That is, starting from atomic labels relative to redexes, we overline or underline them according to the surrounding labels, and then repeat the process with the structured labels yielded so far. ∎

Even if overlinings and underlinings are needless, they are convenient. For instance, it is possible to determinate the canonical derivation (in the sense of the extraction process) given a Lévy's label or its path characterization. In order to obtain the same result from the flattened representation of Corollary 4.5, we have to resort underlinings and overlinings.

Proposition 4.6 The function **path** is injective over labels generated along derivations starting from an expression owning INIT.

Proof: Let ℓ_1 and ℓ_2 be two different labels. The proof is by induction on the structure of ℓ_1. The case ℓ_1 atomic is easy. Thus, assume that $\ell_1 = \alpha_1 \ldots \alpha_n$. If $\ell_1 = \ell_2 \ell$ (or vice versa) the thesis follows trivially, since the two paths have different lengths. Note that this case is not possible when one of the two labels ℓ_1 or ℓ_2 is relative to a redex, by Remark 4.4. Suppose then $\ell_2 = \beta_1 \ldots \beta_m$, and let k be the first index such that $\alpha_k \neq \beta_k$. Three subcases are possible.

- α_k is atomic. By Proposition 4.3, also β_k must be atomic. Since they are different, the two paths diverge here.

- $\alpha_k = \overline{\ell'_1}$. Then $\beta_k = \overline{\ell'_2}$ by Proposition 4.3 (note that β_k cannot be underlined, since otherwise, $\alpha_{k-1} \neq \beta_{k-1}$). Then we use the inductive hypothesis over ℓ'_1 and ℓ'_2 (note in particular that the two paths must really diverge and they cannot be one an initial subpath of the other).

- $\alpha_k = \underline{\ell'_1}$. Analogous to the previous one. ∎

5 Legal paths

We have proved in the previous section that different degrees correspond to different paths in the original term. However, there are paths that are not associated with degrees. In this section we shall provide a complete characterization of paths yielded by degrees (*legal paths*), independently from the notion of labelling.

We start with providing the notion of *well balanced paths* (wbp's), that is a superclass of *legal paths*. We shall obtain legal paths by suitably constraining wbp's.

Definition 5.1 Let t be an expression. Any edge connecting the principal port of an application to an arbitrary operator, whose "type" ? may be λ, @ or v (variable), is a *well balanced path* (shortened into wbp) of type @-?.

Well balanced paths are then composed in the following way:

(λ-composition) Let ψ be a wbp of type @-v whose ending variable is bound by a λ-node c and φ be a wbp of type @-λ coming into c. Then $\psi \cdot (\varphi)^r \cdot u$ is a wbp, where u is the edge outgoing the second argument of the initial node of φ. The type of $\psi \cdot (\varphi)^r \cdot u$ depends on the node connected to u;

(@-composition) Let ψ be a wbp of type @-@ ending into a node d and φ be of type @-λ leading from d to some λ-node c. Then $\psi \cdot \varphi \cdot u$ is a wbp, where u outgoes c towards its body. The type of $\psi \cdot \varphi \cdot u$ depends on the connection of the edge u. ∎

The idea is the following. Every wbp's of type @-λ corresponds to a "session" in the terminology of the introduction. Composition with a "session" explains how to resume a previous search according to the way we "entered" the session. Another way to understand wbp's is in relation with proposition 4.3: every well balanced (sub-)path of type @-λ corresponds to an underlined or overlined label (see section 6), which must be surrounded by suitables atomic edges.

Example 5.2 Consider again the λ-term $\Delta\Delta$ (see the picture in the introduction). The path b is a wbp of type @-λ. d is a wbp of type @-v. By rule 1, $d(b)^r f = dbf$ is a wbp of type @-λ. It corresponds to the redex created after one step, along the unique derivation for $\Delta\Delta$. Now we can proceed. h is a wbp of type @-v, and we have already built the wbp dbf leading from an application to the binder of the ending variable of h. So we can apply again rule 1, obtaining the wbp $h(dbf)^r e = hfbde$ of type @-v. By a further application of rule 1, we finally get the wbp $hfbde(b)^r f = hfbdebf$ of type @-λ. This is the path associated to the unique redex created after two β-reductions.

Up to now, we have a complete correspondence between "virtual" redexes and wbp's. Unfortunately, this is lost at the next step. Note first that we have now two @-λ paths leading to the same application marked with f in the original term, namely $\varphi = dbf$

and $\psi = hfbdebf$. By rule 1, we may build a @-v-path, $h(\psi)^r k$. The final variable is bound by the λ marked with f in the original term. So we may now proceed in two ways, according to the previous paths. If we follow φ, we end up with $h(\psi)^r k(\varphi)^r ebf$, that correctly correspond to the redex created after three β-reductions. But we may also repeat ψ an arbitrary number of times, building wbp's of the kind

$$h(\psi)^r k(\psi)^r k \cdots (\psi)^r k(\varphi)^r ebf$$

and no one of these paths is associated with a redex. ∎

Let us provide a simpler example where the correspondence between wbp's and (virtual) redexes fails.

Example 5.3 Consider the λ-term $(\lambda x.(xM)(xN))(\lambda y.y)$, represented in the following picture:

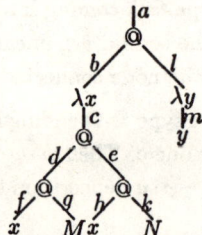

We have two wbp's $\varphi = fbl$ and $\psi = hbl$ leading to the same λ. The two paths $d\varphi m(\varphi)^r g$ and $d\varphi m(\psi)^r k$ are both well balanced, but only the first one is "legal". ∎

So, wbp's are a superset of the set of paths yielded by degrees. This means that wbp's must be constrained by some proviso. Let us try to understand the problem. As we remarked in the introduction, the only non-determinism in the search algorithm (in the definition of wbp) is at the level of bound variables. If a bound variable v appears inside a wbp, it is eventually followed by a wbp φ of type @-λ (reversed) and an access to the argument N of the application. Now, suppose φ describes a cycle internal to N. Intuitively, we are working inside the instance of N individuated by the particular bound variable v. So, when we exit from N we "cannot jump" inside another instance, but we are forced to follow back the same path we used to access the argument.

We shall now formalize the previous intuition. The main problem is to capture the "right" notion of cycle, that will require an inductive definition. The basic case is when the cycle is *physically* internal to the argument N of the application. The problem is that we can exit from some free variable of N (free in N, but eventually bound in the initial term), make a cycle somewhere else, and come back again inside N.

Definition 5.4 Let φ be a wbp. An *elementary @-cycle* of φ is a subpath ψ starting from and ending to the argument edge p of a @-node (the negative auxiliary port), and internal to the argument N of the application (i.e. not traversing variables which are free in N). ∎

Let us come to the definition of cycle. For this purpose we shall also inductively define the notion of *v-cycle* (a cycle over a variable!).

Definition 5.5 Let φ be a wbp.

basic case Every elementary @-cycle is a @-cycle.

induction (*v-cycles*) Every cyclic path of the form $v\lambda(\varphi)^r @\psi @\varphi \lambda v$ where φ is a wbp and ψ is a @-cycle, is a *v*-cycle.

(@-cycles) Every path ψ starting from and ending to the argument edge p of a @-node (the negative auxiliary port), and composed of subpaths internal to the argument N with *v*-cycles over free variables of N is a @-cycle. ∎

The next step is to require that @-cycles must be surrounded by a unique common path to the "associated" λ. The unicity of such a λ is stated by the following proposition.

Proposition 5.6 Let φ be a wbp.

1. For every λ-node traversed by φ at its principal port there exists a unique decomposition of φ as $\zeta_1 @\zeta_2 \lambda \zeta_3$ or $\zeta_1 \lambda (\zeta_2)^r @\zeta_3$, where $@\zeta_2 \lambda$ is a wbp.

2. For every @-node traversed by φ, except the initial one, there exists a unique decomposition of φ as $\zeta_1 @\zeta_2 \lambda \zeta_3$ or $\zeta_1 \lambda (\zeta_2)^r @\zeta_3$, where $@\zeta_2 \lambda$ is a wbp. The initial @-node is joined with a λ-node if and only if the type of φ is @-λ.

Proof: Easy induction on the definition of wbp. ∎

Corollary 5.7 Let φ be a wbp with a @-cycle $@\psi@$. Then φ can be uniquely decomposed as

$$\zeta_1 \lambda \zeta_2 @ \psi @ (\zeta_3)^r \lambda \zeta_4$$

where both ζ_2 and ζ_3 are wbp's. ∎

In the situation of Corollary 5.7, we will say that ζ_2 and ζ_3 are the *call* and *return* paths of the @-cycle ψ. The last label of ζ_1 and the first label of ζ_4 will be named the *discriminants* of the call and return paths, respectively (note that discriminants are eventually edges relative to variables bound by λ).

Now, we are ready to state our *legality* condition for wbp's.

Definition 5.8 A wbp is a *legal path* if and only if the call and return paths of *any* @-cycle are one the reverse of the other and their discriminants are equal. ∎

Example 5.9 Consider the two wbp's $dfblmlbfg$ and $dfblmlbhk$ of Example 5.3. In both cases we have an (elementary) @-cycle lml over the @-node labelled with @. In both cases the call and return path is b, but their discriminants are different in the second case (they are f and h). ∎

6 Legal paths and redex families

In this section, we shall prove the bijective correspondence between legal paths and paths yielded by degrees.

Some preliminary definitions and results are required.

Definition 6.1 Let φ be a wbp over a labelled term t (**INIT**(t) is assumed). The label of φ is defined inductively as follows:

(basic case) the label of the edge marked by φ;

(@-composition) if $\varphi = @\varphi_1 @\varphi_2 \lambda u$ then its label is $\ell_1 \overline{\ell_2} a$, where ℓ_1 is the label of φ_1, ℓ_2 is the label of φ_2 and a is the label of the edge u;

(λ-composition) if $\varphi = @\varphi_1 \lambda \varphi_2 @u$ then its label is $\ell_1 \underline{\ell_2} a$, where ℓ_1 is the label of φ_1, ℓ_2 is the label of φ_2 and a is the label of the edge u. ∎

According to the above definition, it is immediate to verify that, if ℓ is the label of a wbp φ, then $\varphi = \text{path}(\ell)$.

Let $t \xrightarrow{u} t'$. Every (arbitrary) path in t' has an *ancestor* in t. Its definition is pretty intuitive, so we shall not be pedantic, here. In particular, we shall just define the notion of ancestor for single edges in t'. This is extended to paths by composition, in the obvious way.

Suppose that $t = C[(\lambda x.M)N]$ and $t' = C[M[^N/_x]]$, where $C[\ ]$ is some context. Note first that some edges in t' are "residuals" of edges in t. This is the case for every edge v' internal to M, to some instance of N, or belonging to the context. If v' is a residual of v in the previous sense, then v is the origin of v'. The problem is when v' is a new connection created by firing the redex u. Let @ and λ be the two nodes connected by u in t. Three cases are possible:

- v' is a connection between the context and the body M. The ancestor of v' is $a \cdot u \cdot b$, where a is the edge leading from the context to the positive port of @, and b is the edge leading from λ to M.

- v' is a connection between M and the i-th instance of N. The ancestor of v' is $b \cdot (u)^r \cdot c$, where b is the edge leading from M to the i-th instance of the variable bound by λ, and c is the edge leading from @ to N.

- v' is a connection between the context and N. This is only possible when $M = x$, and it is an obvious combination of the previous cases. In particular, the ancestor of v' is $a \cdot u \cdot b \cdot (u)^r \cdot c$, with a, b and c as above.

We leave to the reader the care to generalize the definition of ancestor from edges to arbitrary paths (we have to prove that we preserve the connections, but this is easy).

The definition of ancestor is then extended to derivations in the obvious way.

Proposition 6.2 Every ancestor of a wbp is still a wbp. ∎

Lemma 6.3 Let $t \xrightarrow{u} t'$, and φ be the ancestor of a *legal* path φ' in t'. Then

1. Every @-cycle in φ relative to an application **d** not fired by u is ancestor of an @-cycle in φ' over a residual of **d**.

2. Every v-cycle in φ relative to an instance x of a variable not bound by the λ in u is ancestor of an v-cycle in φ' over a residual of x.

Proof: The proof is by induction on the definition of cycles.

- In case the @-cycle is elementary, property 1 is trivial.

- Let us consider the inductive case.

  (**v-cycles**) Let $v\lambda(\varphi_1)^r @\psi @\varphi_1 \lambda v$ be the cycle. Since v is not bound by the λ in u, the @-cycle ψ cannot be relative to the application fired by u (indeed, u is the unique wbp in t relative to this application). So we may apply the induction hypothesis, and we know that there exists a cycle ψ' inside φ' whose ancestor is ψ. Recall that φ' is legal, so we must found some subpath of the shape $v'\lambda(\varphi'_1)^r @\psi' @\varphi'_1 \lambda v'$ containing ψ'. Since the ancestor of a wbp is still a wbp, φ_1 must eventually be the ancestor of φ'_1, and similarly for v' and v.

  (**@-cycles**) Consider a @-cycle ψ inside φ and relative to an application **d** in t. Let ψ' be the sub-path of φ' whose ancestor is ψ. This eventually exists by the edge-wise definition of ancestors. Consider all the "maximal" v-cycles (i.e. v-cycles not contained in other v-cycles) in ψ not relative to variables bound by the λ in u. By induction, these are ancestor of v-cycles in ψ'. Moreover these are all the possible maximal v-cycles in ψ' since variables bound by the lambda in u disappear after the reduction. The remaining portion of paths in ψ' is eventually inside the argument of the unique residual **d**$'$ of **d** (by the connection of ψ' and a simple case inspection on the position of **d** w.r.t. u). So ψ' is indeed a @-cycle over **d**$'$. ∎

Lemma 6.4 Let $t \xrightarrow{u} t'$, and φ be the ancestor of a *legal* path φ' in t'. Then every @-cycle ψ in φ relative to an application **d** not fired by u is legal.

Proof: By lemma 6.3, ψ is ancestor of a @-cycle ψ' in φ'. Since φ' is legal we must have in φ' some subpath of the kind $v'\lambda(\varphi'_1)^r@\psi'@\varphi'_1\lambda v'$ containing ψ'. So the structure of φ around the @-cycle ψ has the shape $v\lambda(\varphi_1)^r@\psi@\varphi_1\lambda v$, where φ_1 is the ancestor of φ'_1, and v' is the ancestor of v. ∎

Lemma 6.5 Let $t \xrightarrow{u} t'$, and φ be the ancestor of a *legal* path φ' in t'. Then every @-cycle ψ in φ relative to the application **d** fired by u is legal.

Proof: The call and return paths are obviously equal (they are u). We must only prove that also the discriminants are equal. The proof is by induction on the definition of @-cycle. The basic case is easy. Let us consider the inductive case. The path is then composed by subpaths internal to the argument N of the redex u and v-cycles over free variables of N. By lemma 6.3 these v-cycles are ancestors of v-cycles in φ'. Consider such a cycle. It will be relative to some free variable of some instance N_i of N in t. Note now that all instances of N are disjoint in t'. Since the path φ' is connected, it must eventually define the "cycle" originating ψ inside this instance N_i, that ensures the equality of discriminants.

Proposition 6.6 The ancestor of a legal path φ' in t' along $t \xrightarrow{u} t'$ is a legal path φ in t.

Proof: Easy consequence of Lemmas 6.3 and 6.4. ∎

Proposition 6.7 Labelling of paths is unchanged in ancestors. ∎

Theorem 6.8 Every path yielded by degrees of redexes is a legal path.

Proof: By Proposition 6.6, and the fact that if ℓ is the label of a path φ, then $\varphi = \mathbf{path}(\ell)$. ∎

The strategy to show the vice versa of Theorem 6.8 is based on the proof that legal paths can be uniquely "contracted" to legal paths by firing the leftmost outermost redex. A preliminary immediate result is required.

Proposition 6.9 Let $u = (\lambda x.M)N$ be a redex in t, and let φ, be a legal path internal to N. Let $t \xrightarrow{u} t'$. Then there exists a legal path φ_i in each instance of N_i whose ancestor is φ. ∎

Proposition 6.10 Let φ, $|\varphi| > 1$, be a legal path in t and let $u = (\lambda x.M)N$ be the leftmost outermost redex traversed by φ in t. Let $t \xrightarrow{u} t'$. Then there exists a *unique* legal path φ' in t' whose ancestor is φ.

Proof: (Sketch) We start proving that there exist a *unique* wbp φ' in t' whose ancestor is φ (the only problem is the connection of φ'). We work by induction on the structure of wbp of φ.

The basic case is vacuous. For simplicity we shall only consider the case of @-composition. Thus, let then $\varphi = @_1\varphi_1@_2\varphi_2\lambda_1$. If φ_1 (φ_2) is internal to N, by Proposition 6.9, we have a "copy" of φ_1 (φ_2) inside every instance of N, and we are free to choose the only one that matches the "contractum" of φ_2 (φ_1). Suppose otherwise. By induction we know that there are two paths $@_1'\varphi_1'@_2'$ and $@_2''\varphi_2'\lambda_1'$ whose ancestors are φ_1 and φ_2, respectively. $@_2'$ and $@_2''$ are both residuals of $@_2$, but, a priori, nothing ensures that $@_2' = @_2''$. The only problematic case is when the composition is internal to the argument N. Recall that both paths φ_1 and φ_2 exit from N. Since u is the leftmost outermost redex, they must eventually exit from the "top", i.e. we have an access to N. In particular, we have a final sub-path $u\zeta_1@_1$ of φ_1, where ζ_1 is internal to N. Similarly we must have an initial sub-path $@_1\zeta_2 u$ of φ_2 with ζ_2 internal to N. So the connected path $\zeta_1\zeta_2$ is a (elementary) cycle for the application in u. Since φ is legal, the discriminants for the two occurrences of u must be the same. So $@_2' = @_2''$, since they will be in the same instance of N. Therefore φ' is a wbp.

We must still prove that φ' is legal. This is done by following the same ideas as for proposition 6.6. Namely, we have that the unique contractum in φ' of a @-cycle in φ not relative to the application fired by u is still a @-cycle, and similarly for a v-cycle not relative to a variable bound by the λ fired by u. Legality easily follows, by the unicity of the contractum.

Theorem 6.11 If φ is a legal path, there exists a degree ℓ such that $\text{path}(\ell) = \varphi$.

Proof: By Proposition 6.10 we can "contract" a legal path into another legal path by firing the leftmost outermost redex. The length of the contractum is strictly shorter than that of its ancestor, and we eventually end up with a single redex. Since labels are preserved by contraction, the path yielded by the ending redex is exactly φ. ∎

7 Legal paths and optimal reductions

The notion of legal path helps both in the simplification of the theory of optimality and in the understanding of an optimal evaluator for the λ calculus. For instance, we have provided a new, simple proof of the coincidence of labelling and extraction [2,10]. As far as the optimal evaluator is concerned, due to Theorem 6.11, the implementation must always have a unique representation of every legal path. In particular, legal paths must be *physically* shared. The evaluation proceeds by contracting legal paths, provided that every contractum of a legal subpath has a unique representation.

More precisely, suppose to have an explicit operator of duplication (a *fan*, in Lamping's terminology). Consider a legal path φ. An essential condition to get optimality is that a *fan* external to φ can never enter the path. On the other side, a *fan* already internal to φ can be freely moved along φ (provided it does not enter subpaths of type @-λ). That is, we may pursue the duplication *inside* a path @-λ, since the portion of paths we are duplicating eventually belong to different legal paths.

There is a strong relation between our legal paths and Lamping-Gonthier's *consistent paths*. These are used to read back λ-expressions form sharing graphs. In the (forthcoming) full paper we prove that legal paths are exactly wbp's that are consistent w.r.t. the *context semantics* [9,6] yielding a further, alternative characterization of legal paths.

Note also that a prerequisite chain [9] for @ is always a prefix of a legal path, up to the first atomic edge representing a redex. The interesting fact is that this prefix is always *unique* for every legal path starting from a same node. In a sense, prerequisite chains are the deterministic part of legal paths. Similarly, a prerequisite chain of a λ-node is the *unique* common suffix of every legal path starting form the last edge representing a redex.

8 Conclusions

We have provided a new characterization of the family relation in the λ calculus in terms of suitable paths in the initial term of the derivation. In, particular, we have proved that two redexes in a same family comes from the contraction of a unique common path in the initial term, yielding new evidence to their "sharable" nature. From this respect, our work sheds a new light over graph reduction techniques for optimal implementations, elaborating Danos and Regnier's idea of a computation as a *travel along a path* [13].

From the theoretical point of view, legal paths help to understand Levy's labels, providing a new insight in the algebraic structure of degrees. Moreover, in our experience, legal paths are the most friendly and intuitive tool for reasoning about redex-families, leading to a significant simplification of many results concerning them.

Studying the relationships between Danos and Regnier's paths [4] and legal paths is surely an exciting task. The former ones have a strong geometric motivation (in the sense of Girard's "Geometry of Interaction") and a more algebraic flavour. Legal paths have been mainly motivated by an operational viewpoint, exploiting properties of Lévy's labels.

Acknowledgements We have enjoyed discussions with Gerard Boudol, Georges Gonthier and Jean Jacques Lévy.

References

[1] A. Asperti. Linear logic, comonads, and optimal reductions. Draft, INRIA-Rocquencourt, 1991.

[2] A. Asperti and C. Laneve. Interaction systems 1: The theory of optimal reductions. Technical Report 1748, INRIA-Rocquencourt, September 1992.

[3] A. Asperti and C. Laneve. Optimal reductions in interaction systems. In *TapSoft '93*, Lecture Notes in Computer Science. Springer-Verlag, 1993.

[4] V. Danos and L. Regnier. Local and asynchronous beta-reduction. In *Proceedings 8th Annual Symposium on Logic in Com puter Science*, Montreal, 1993.

[5] J. Field. On laziness and optimality in lambda interpreters: tools for specification and analysis. In *Proceedings 17th ACM Symposium on Principles of Programmining Languages*, pages 1 – 15, 1990.

[6] G. Gonthier, M. Abadi, and J.J. Lévy. The geometry of optimal lambda reduction. In *Proceedings 19th ACM Symposium on Principles of Programmining Languages*, pages 15 – 26, 1992.

[7] G. Gonthier, M. Abadi, and J.J. Lévy. Linear logic without boxes. In *Proceedings 7th Annual Symposium on Logic in Computer Science*, 1992.

[8] Y. Lafont. Interaction nets. In *Proceedings 17th ACM Symposium on Principles of Programmining Languages*, pages 95 – 108, 1990.

[9] J. Lamping. An algorithm for optimal lambda calculus reductions. In *Proceedings 17th ACM Symposium on Principles of Programmining Languages*, pages 16 – 30, 1990.

[10] C. Laneve. *Optimality and Concurrency in Interaction Systems*. PhD thesis, Technical Report TD – 8/93, Dip. Informatica, Università di Pisa, March 1993.

[11] J.J. Lévy. *Réductions correctes et optimales dans le lambda calcul*. PhD thesis, Université Paris VII, 1978.

[12] J.J. Lévy. Optimal reductions in the lambda-calculus. In J.P. Seldin and J.R. Hindley, editors, *To H.B. Curry, Essays on Combinatory Logic, Lambda Calculus and Formalism*, pages 159 – 191. Academic Press, 1980.

[13] L. Regnier. Lambda Calcul et Réseaux. Thèse de doctorat, Université Paris VII. 1992.

Confluence and Superdevelopments

Femke van Raamsdonk*

CWI, P.O. Box 4079, 1009 AB Amsterdam, The Netherlands

Abstract. In this paper a short proof is presented for confluence of a quite general class of reduction systems, containing λ-calculus and term rewrite systems: the orthogonal combinatory reduction systems. Combinatory reduction systems (CRSs for short) were introduced by Klop generalizing an idea of Aczel. In CRSs, the usual first-order term rewriting format is extended with binding structures for variables. This permits to express besides first order term rewriting also λ-calculus, extensions of λ-calculus and proof normalizations. Confluence will be proved for orthogonal CRSs, that is, for those CRSs having left-linear rules and no critical pairs. The proof proceeds along the lines of the proof of Tait and Martin-Löf for confluence of λ-calculus, but uses a different notion of 'parallel reduction' as employed by Aczel. It gives rise to an extended notion of development, called 'superdevelopment'. A superdevelopment is a reduction sequence in which besides redexes that descend from the initial term also some redexes that are created during reduction may be contracted. For the case of λ-calculus, all superdevelopments are proved to be finite. A link with the confluence proof is provided by proving that superdevelopments characterize exactly the Aczel's notion of 'parallel reduction' used in order to obtain confluence.

1 Introduction

The study of λ-calculus and of the foundations of functional programming has led to a rich variety of classes of reduction systems, an important one being first-order term rewriting. For a lot of different although related reduction systems, different although related proofs of syntactic properties have been given.

A first attempt to provide a uniform framework for a number of extensions of λ-calculus was given by Hindley's $\lambda(a)$-reductions [3]. A much more extensive format in the form of contraction schemes, was developed by Aczel [1]. However, contraction schemes do not contain first-order term rewriting. Inspired by Aczel, Klop defined the Combinatory Reduction Systems (CRSs) [6]. The set-up of CRSs originates from term rewriting. Term rewriting systems (TRSs for short) handle pattern-matching, but they lack a notion of binding, like for instance in

$$\int_a^b x^2 dx$$
$$\forall x(P(x) \Rightarrow Q(x))$$
$$\lambda x.x$$

---

* supported by NWO/SION project 612-316-606 *Extensions of orthogonal rewrite systems - syntactic properties*.

So the first thing to be done is adding binding structures for variables. If it comes to 'using' binding structures, a notion of substitution is needed, like in

$$1/3\ b^3 - 1/3\ a^3$$
$$P(t) \Rightarrow Q(t)$$
$$x[x := z]$$

This leads to introducing metavariables, that indicate places where an arbitrary term can be plugged in, with the possibility to express substitutions. In the case of CRSs, this is done by assigning to all metavariables a fixed arity. If it is not specified in the example which function, which predicates and which term is meant, the expressions can be written as

$$\int_a^b Z(x)dx$$
$$\forall x(Z(x) \Rightarrow Z'(x))$$
$$\lambda x.Z(x)$$

So CRSs can be viewed as TRSs with binding structures for variables and with metavariables having a fixed arity. Extending term rewriting considerably in this manner yields a general framework in which proofs of syntactic properties can be obtained in a uniform way.

In this paper, the property 'confluence' is a matter of concern. It means that for every two coinitial reduction sequences $s \twoheadrightarrow t$ and $s \twoheadrightarrow u$ a term v can be found such that $t \twoheadrightarrow v$ and $u \twoheadrightarrow v$. The term v is called a common reduct of t and u. An equivalent notion is the Church-Rosser property, stating that every pair of convertible terms has a common reduct.

In this paper confluence is proved for all orthogonal CRSs. Confluence for orthogonal CRSs has been proved by Klop [6], by proving first that all developments are finite. The strategy of the proof presented in this paper is very similar to the way confluence for λ-calculus has been proved by Tait and Martin-Löf. In a similar way, Aczel proves confluence for his contraction schemes.

In the proof a relation $\geq$ on terms is used which reflexive-transitive closure equals reduction. When characterizing this relation in terms of reduction, it turns out that the reduction sequences exactly corresponding to this relation $\geq$ form a generalization of developments; they are therefore called *superdevelopments*. We prove that superdevelopments in λ-calculus are always finite.

Another new proof of confluence of a large class of reduction systems with bound variables is given by Khasidashvili [5]. This proof proceeds along the lines of the one by Klop, but a slightly stronger version of the Church-Rosser property is proved. Takahashi [11] proves confluence of λ-calculi with conditional rules in a way similar to the method of Tait and Martin-Löf. Nipkow [8] proves confluence for orthogonal higher-order rewrite systems (HRSs) in the same manner. HRSs are close to CRSs but the starting point is λ-calculus rather than term rewriting. Kahrs investigates an extension of λ-calculus with first-order rewriting, and proves several syntactic properties [4]. Confluence results for another class of general reduction systems, the so-called subtree replacement systems, are studied by Rosen and O'Donnell in respectively [10] and [9].

2 Combinatory Reduction Systems

Metavariables and variables are distinguished in the following way:
- Metavariables indicate places in rules where an arbitrary term can be plugged in, like x in the TRS rule $F(x) \to G(x)$ and like M and N in the β-reduction rule for λ-calculus, $(\lambda x.M)N \to M[x := N]$.
- Variables are on the one hand used to build up terms, like in the term $F(x)$ in some TRS and in xz in λ-calculus, and on the other hand they are used to serve as placeholders indicating a place where a substitution can be carried out, like in the term $\lambda x.x$ in λ-calculus.

Metavariables occur only in left- and right-hand side of rules. A nullary metavariable Z occurring in the left-hand side of some rule, can be instantiated by an arbitrary term. A unary metavariable occurring in the form $Z(x)$ in the left-hand side of a rule can be instantiated by a term t possibly but not necessarily containing free occurrences of the variable x. If this metavariable occurs in the right-hand side of the rule in the form $Z(s)$ then it is instantiated by the term t in which all free occurrences of x are replaced by s. For example, the β-reduction rule of λ-calculus, with metavariables M and N is usually written as $(\lambda x.M)N \to M[x := N]$ but is given as well as $(\lambda x.M(x))N \to M(N)$. In the latter formulation, the metavariables M and N have arity 1 and 0 respectively.

The alphabet of a CRS consists of
- variables, written as $x\ y\ z\ \ldots$,
- metavariables with a fixed arity, written as $Z\ Z_0\ Z_1\ \ldots$,
- function symbols with a fixed arity, written as $F\ G\ H\ \ldots$,
- an abstraction operator, written as $[\_]\_$,
- parentheses and commas.

A term t from which some variable x has been abstracted is written as $[x]t$. Function symbols of arity 0 are called *constant symbols*. Metaterms and terms are built up from the alphabet given above.

Definition 1. The set MTerms of metaterms is the smallest set satisfying
(1) $x \in$ MTerms for every variable x,
(2) if $t \in$ MTerms then $[x]t \in$ MTerms for every variable x,
(3) if F is a function symbol of arity n and $t_1, \ldots, t_n \in$ MTerms, then $F(t_1, \ldots, t_n) \in$ MTerms,
(4) if Z is a metavariable of arity n and $t_1, \ldots, t_n \in$ MTerms, then $Z(t_1, \ldots, t_n) \in$ MTerms.

The set Terms of terms consists of all metaterms without any metavariable.

If C is a constant symbol then we write C instead of $C()$. Identity on MTerms and on Terms is denoted by $=$.

In a metaterm or term of the form $[x]t$, we call t the *scope* of the abstraction $[x]$. A variable x occurs *free* in a metaterm or term if it is not in the scope of a $[x]$. It occurs *bound* otherwise. A metaterm or term is called *closed* if all variables occur bound. Like in λ-calculus, bound variables can be renamed. (Meta)terms that are identical up to a renaming of bound variables are considered equal. This permits to adopt the convention that in a term no variable is abstracted over twice or more. Instead of $[x_1] \ldots [x_n]t$ we write $[x_1 \ldots x_n]t$.

Note that the abstraction in a (meta)term of the form $[x]t$, is purely syntactic. The actual (operational) meaning of this abstraction has to be expressed by a function symbol and its reduction rules.

Let $\Box$ be a fresh symbol. A *context* is a (meta)term with one or more occurrences of $\Box$. A context with exactly one occurrence of $\Box$ is written as $C[\ ]$, and one with n occurrences of $\Box$ as $C[,\ldots,]$. If $C[,\ldots,]$ is a context with n occurrences of $\Box$ and $t_1, \ldots, t_n$ are (meta)terms, then $C[t_1, \ldots, t_n]$ denotes the result of replacing from left to right the occurrences of $\Box$ by $t_1, \ldots, t_n$. A (meta)term s is said to be a *sub(meta)term* of of a (meta)term t if a context $C[\ ]$ exist such that $t = C[s]$.

Example 1. The alphabet of the untyped λ-calculus consists of
- a unary function symbol λ for λ-abstraction
- a binary function symbol Ap for application

Then we write for instance $\lambda([x]x)$ for $\lambda x.x$, $\lambda([x]\mathsf{Ap}(x,y))$ for $\lambda x.xy$, and $\mathsf{Ap}(\lambda([x]y), z)$ for $(\lambda x.y)z$.

Note that according to the definition of Terms from this alphabet a lot of terms can be built that don't correspond to λ-terms. One reason is that it cannot be specified that in order to form a λ-term, the argument of the symbol λ must be an abstraction term. This can be done by extending the notion of arity. The presence of 'junk' terms doesn't yield a problem when applying for instance the confluence result, since a term corresponding to a λ-term reduces always to terms corresponding to λ-terms.

A reduction relation on the terms of a CRS is generated by a set of reduction rules.

Definition 2. A *reduction rule* is a pair of metaterms (α, β) written as $\alpha \rightarrow \beta$, satisfying the following constraints:
- α and β are closed,
- α is of the form $F(\alpha_1, \ldots, \alpha_n)$,
- metavariables occurring in β occur in α as well,
- metavariables in α occur only in the form $Z(x_1, \ldots, x_n)$ with $x_1, \ldots, x_n$ distinct variables.

A reduction rule acts as a scheme from which actual reduction steps can be obtained. Metavariables in the left-hand side of a rule indicate places where an arbitrary term can be substituted. Variables occur only bound and serve to indicate places where substitutions are carried out. A left-hand side of a rule is not allowed to be a metavariable nor an abstraction term; the former because this would permit to rewrite an arbitrary term and the latter because the abstraction is considered to be purely syntactic without any operational meaning. Allowing to have metavariables in β that do not occur in α would permit to introduce terms out of the blue by reducing. Substitution mechanisms are expressed by the reduction rules as follows: $Z(x_1, \ldots, x_n)$ in the left-hand side of some rule is instantiated by a term s possibly but not necessarily containing the variables $x_1, \ldots, x_n$. An occurrence of Z in the right-hand side in the form $Z(t_1, \ldots, t_n)$ is then instantiated by s in which the free occurrences of $x_1, \ldots, x_n$ are replaced by $t_1, \ldots, t_n$ respectively.

Example 2. The β-reduction rule of λ-calculus is in CRS format written as $\mathsf{Ap}(\lambda([x]Z(x)), Z') \to Z(Z')$.

Now it will be described how the reduction rules generate an actual reduction relation on Terms. Therefore the concept of 'valuation' is introduced. Valuations express how metavariables are instantiated by terms. Before defining valuations we introduce as a notational device the n-ary *substitute* (a name due to Kahrs [4]).

Definition 3. Let t be a term is some CRS R.
(1) For an n-tuple of distinct variables $x_1, \ldots, x_n$, the expression $\underline{\lambda}(x_1, \ldots .x_n).t$ is an n-ary substitute.
(2) An n-ary substitute $\underline{\lambda}(x_1, \ldots .x_n).t$ can be applied to an n-tuple of terms $(s_1, \ldots, s_n)$, yielding as result the term t with $s_1, \ldots, s_n$ simultaneously substituted for $x_1, \ldots, x_n$ respectively:

$$(\underline{\lambda}(x_1, \ldots .x_n).t)(s_1, \ldots, s_n) = t[x_1 := s_1, \ldots, x_n := s_n]$$

So an n-ary substitute can be considered as a function $\mathsf{Terms}^n \to \mathsf{Terms}$. For an n-ary substitute $\underline{\lambda}(x_1, \ldots .x_n).t$ the variables $x_1, \ldots, x_n$ are considered to be bound in t and may be renamed so that no name clashes occur. The variables in t that don't occur in $(x_1, \ldots, x_n)$ and that are not bound in t are called the free variables of the substitute $\underline{\lambda}(x_1, \ldots .x_n).t$.

Now the definition of a valuation can be given.

Definition 4. A valuation is a map σ that assigns to an n-ary metavariable Z an n-ary substitute:

$$\sigma(Z) = \underline{\lambda}(x_1, \ldots, x_n).s$$

The map σ is extended to a homomorphism on metaterms (denoted as σ as well) in the following way:
(1) $\sigma(x) = x$,
(2) $\sigma([x]t) = [x]\sigma(t)$ for a variable x and a metaterm t,
(3) $\sigma(F(t_1, \ldots, t_n)) = F(\sigma(t_1), \ldots, \sigma(t_n))$ for a function symbol F of arity n and metaterms $t_1, \ldots, t_n$,
(4) $\sigma(Z(t_1, \ldots, t_n)) = \sigma(Z)(\sigma(t_1), \ldots, \sigma(t_n))$ for an metavariable Z of arity n and metaterms $t_1, \ldots, t_n$.

Without any conditions on the valuations, in instantiated reduction rules variables can be bound unintendedly. Two kinds of problems can occur.

First, variables can be captured by abstractors by plugging in terms for metavariables. This is for instance the case if $F([x]Z)$ is instantiated by a valuation that assigns x to Z. So we have to require that bound variables in rules are renamed such that they differ from free variables in substitutes.

Second, in the instance of a right-hand side, substitution can yield unintended bindings. This is for example the case if $Z(Z')$ is instantiated by the valuation σ with $\sigma(Z(u)) = \underline{\lambda}(u).F([y]u)$ and $\sigma(Z') = y$. Then $\sigma(Z(Z')) = F([y]y)$. Situations like this can be avoided by requiring for every two different metavariables $Z_1(x_1, \ldots, x_{k_1})$ and $Z_2(x_1, \ldots, x_{k_2})$ occurring in the same reduction rule, the free variables in $\sigma(Z_1(x_1, \ldots, x_{k_1}))$ to be different from the bound variables in $\sigma(Z_2(x_1, \ldots, x_{k_2}))$. In the following we will suppose that these requirements are met. Informally, they can be thought of as 'rename bound variables as much as possible, in order to avoid free occurrences of x to be captured unintentionally by abstractors $[x]$.'

Finally the actual reduction relation $\to$ on Terms can be defined.

Definition 5. Let $\alpha \to \beta$ be a reduction rule and σ a valuation. An instance $\sigma(\alpha)$ of the left-hand side of a reduction rule is called a *redex*. The associated right-hand side $\sigma(\beta)$ is called its *contractum*. Replacing a redex by its contractum in a context is called a *reduction step* and is written as $C[\sigma(\alpha)] \to C[\sigma(\beta)]$. A sequence of zero or more reduction steps is called a *reduction sequence* or *reduction*. If a reduction from s to t exists we write $s \twoheadrightarrow t$ and say that s *reduces* to t, and t is called a *reduct* of s.

Example 3. The reduction step $(\lambda x.x)y \to y$ in λ-calculus is obtained by instantiating the β-reduction rule $\mathsf{Ap}(\lambda([x]Z(x)), Z') \to Z(Z')$ in the following way: $\sigma(Z) = \underline{\lambda}(z).z$ and $\sigma(Z') = y$. Then $\sigma(\lambda[x].Z(x)) = \lambda([x]x)$ so the left-hand side of the rule instantiated by σ is $\mathsf{Ap}(\lambda[x]x, y)$. This term reduces to $\sigma(Z(Z')) = (\underline{\lambda}(z).z)(y) = y$.

A CRS is called *left-linear* if in none of its reduction rules the same metavariable occurs twice or more in the left-hand-side. Two reduction rules $\alpha \to \beta$ and $\alpha' \to \beta'$ are said to *overlap* if there exist valuations σ and σ' such that $\sigma(\alpha) = \sigma'(\alpha'_1)$ for a submetaterm α'_1 of α' that is not of the form $Z(x_1, \ldots, x_n)$. Then a context $C[\ ]$ exists such that $C[\sigma(\alpha)] = \sigma'(\alpha')$. In the case that the reduction rules $\alpha \to \beta$ and $\alpha' \to \beta'$ are the same, we require additionally the context $C[\ ]$ to be non-trivial. The term $C[\sigma(\alpha)] = \sigma'(\alpha')$ can be reduced in two different ways, yielding as a result $C[\sigma(\beta)]$ or $\sigma'(\beta')$ respectively.

If a CRS doesn't contain overlapping rules then it is called *non-ambiguous*. If for every two overlapping rules $\alpha \to \beta$ and $\alpha' \to \beta'$ with $C[\sigma(\alpha)] = \sigma'(\alpha')$ it holds that $C[\sigma(\beta)] = \sigma'(\beta')$, then the CRS is said to be *weakly non-ambiguous*.

A CRS that is left-linear and non-ambiguous is called *orthogonal*. A CRS that is left-linear and weakly non-ambiguous is called *weakly orthogonal*.

Example 4. An orthogonal CRS is λ-calculus with β-reduction; if η-reduction is added it is a weakly orthogonal one. The η-rule is in CRS format written as $\lambda([x]\mathsf{Ap}(Z, x)) \to Z$. Each time that a term contains a β-and an η-redex such that they share a λ, both possible reduction steps yield the same result. For instance, the term $@(\lambda([x]@(y, x), z))$ is itself a β-redex and contains non-trivially the η-redex $\lambda([x]@(y, x))$. Reducing the β-redex yields the same result as reducing the η-redex, namely $@(y, z)$.

CRSs themselves are untyped systems. Nevertheless various typed systems like simply typed λ-calculus and system F can be written in the CRS framework.

In the original definition of CRSs by Klop, all function symbols are nullary and a distinguished symbol for application is used. Metavariables have, exactly like in this set-up, a fixed arity.

It is easily seen that every applicative system can be written as a functional one by writing all applications explicitly. A functional system can be written in applicative format by turning all function symbols into constants and adding a binary operator for application that is not written explicitly. Then the functional CRS correspond to a sub-CRS of its applicative version. A *sub-CRS* of a CRS R is defined as a CRS obtained from R by restricting the set of terms to a subset that is closed under reduction. So the functional and applicative formats have the same expressive power.

The confluence result, like most syntactic results for CRSs, carries over directly to sub-CRSs. This explains why it is no problem when the CRS representation of a system contains 'junk' terms.

3 Confluence for Orthogonal CRSs

In this section all orthogonal CRSs will be proved to be confluent. The strategy of the proof is essentially the same as the one of the proof of confluence for λ-calculus with β-reduction by Tait and Martin-Löf [2]. This proof method is employed by several others [1], [8], [11], mostly in order to prove confluence.

A relation $\geq$ on Terms is defined such that its reflexive-transitive closure equals reduction. For this relation $\geq$ the diamond property, given in the next definition, will be proved.

Definition 6. A binary relation $\triangleright$ satisfies the *diamond property* if for every a, b and c such that $a \triangleright b$ and $a \triangleright c$ there exists a d such that $b \triangleright d$ and $c \triangleright d$.

Having proved the diamond property for $\geq$, confluence of the reduction relation follows immediately.

Definition 7. The relation $\geq$ on Terms is defined as follows:
(1) $x \geq x$ for every variable x,
(2) if $s \geq t$ then $[x]s \geq [x]t$ for every variable x,
(3) if $s_1 \geq t_1, \ldots, s_n \geq t_n$ then $F(s_1, \ldots, s_n) \geq F(t_1, \ldots, t_n)$ for every n-ary function symbol F,
(4) if $s_1 \geq t_1, \ldots, s_n \geq t_n$ and $F(t_1, \ldots, t_n) = \sigma(\alpha)$ for some reduction rule $\alpha \to \beta$ and valuation σ, then $F(s_1, \ldots, s_n) \geq \sigma(\beta)$.

The fourth clause can be depicted as follows:

$$F(s_1, \ldots, s_n)$$
$$\mathbin{\rotatebox[origin=c]{90}{$\geq$}} \quad \mathbin{\rotatebox[origin=c]{90}{$\geq$}} \quad \searrow$$
$$\sigma(\alpha) = \quad F(t_1, \ldots, t_n) \quad \to \quad \sigma(\beta)$$

The first three clauses of the definition state that $\geq$ is a reflexive relation that is closed under term formation. The fourth clause expresses that $s \geq t$ if s reduces to t by a parallel 'inside-out' reduction, in which possibly redexes that are 'created upwards' are contracted. Note that in this clause $F(s_1, \ldots, s_n)$ is not necessarily a redex.

The next proposition states that $\geq$ is indeed a useful relation to prove the diamond property for.

Proposition 8. *The transitive closure of $\geq$ equals reduction.*

The crucial step in proving the diamond property for $\geq$ is proving that $\geq$ satisfies a property named 'coherence'. This notion is originally introduced by Aczel [6].

Definition 9. A binary relation ▷ on Terms is said to be *coherent* with respect to reduction if the following holds: if $F(a_1, \ldots, a_n) = \sigma(\alpha)$ for some reduction rule $\alpha \rightarrow \beta$ and valuation σ, and $a_1 \vartriangleright b_1, \ldots, a_n \vartriangleright b_n$, then we have for some valuation τ that $F(b_1, \ldots, b_n) = \tau(\alpha)$ with $\sigma(\beta) \vartriangleright \tau(\beta)$.

Coherence can be depicted as follows:

$$\begin{array}{ccc} F(a_1, \ldots, a_n) & \rightarrow & a \\ \triangledown & \triangledown & \triangledown \\ F(b_1, \ldots, b_n) & \rightarrow & b \end{array}$$

Two technical propositions are needed in order to prove coherence of $\geq$ with respect to reduction.

Proposition 10. *If $a \geq b$ and $s_1 \geq t_1, \ldots, s_n \geq t_n$ then*

$$a[x_1 := s_1, \ldots, x_n := s_n] \geq b[x_1 := t_1, \ldots x_n := t_n]$$

PROOF. Induction on the derivation of $a \geq b$. □

Proposition 11. *Let t be a metaterm containing only the metavariables $Z_1, \ldots, Z_k$. Let σ and τ be valuations. If $\sigma(Z_i(x_1, \ldots, x_{n_i})) \geq \tau(Z_i(x_1, \ldots, x_{n_i}))$ for $i = 1, \ldots, k$, then $\sigma(t) \geq \tau(t)$.*

PROOF. Induction on the structure of t. □

Lemma 12. *The relation $\geq$ is coherent with respect to reduction.*

PROOF. Suppose $F(a_1, \ldots, a_n) = \sigma(\alpha) \rightarrow \sigma(\beta)$ and $a_1 \geq b_1, \ldots, a_n \geq b_n$. By Proposition 8 we have $a_1 \twoheadrightarrow b_1, \ldots, a_n \twoheadrightarrow b_n$. By non-ambiguity, we know that all reduction steps take place in instances of metavariables in α. Together with left-linearity this yields that $F(b_1, \ldots, b_n)$ is still an instance of α, say $F(b_1, \ldots, b_n) = \tau(\alpha)$ with contractum $\tau(\beta)$. Now it has to be proved that $\sigma(\beta) \geq \tau(\beta)$. Appropriate first parts of derivations of $a_1 \geq b_1, \ldots, a_n \geq b_n$ form a derivation for $\sigma(Z_i(x_1, \ldots, x_{k_i})) \geq \tau(Z_i(x_1, \ldots, x_{k_i}))$ for every k_i-ary metavariable Z_i occurring in α. Note that metavariables in α only occur in this form. In β occur only metavariables that occur in α as well, so by Proposition 11 we can conclude $\sigma(\beta) \geq \tau(\beta)$. □

Theorem 13. *The relation $\geq$ satisfies the diamond property.*

PROOF. We shall prove that for any a, b and c such that $a \geq b$ and $a \geq c$ there exists a d such that $b \geq d$ and $c \geq d$. The proof proceeds by induction on the derivation of $a \geq b$.
- If $a \geq b$ is $x \geq x$ then necessarily $c = x$. Take $d := x$.
- If $a \geq b$ is $[x]a' \geq [x]b'$ with $a' \geq b'$, then c is necessarily of the form $[x]c'$. By induction hypothesis, a d' exists such that $b' \geq d'$ and $c' \geq d'$. By defining $d := [x]d'$, both $b \geq d$ and $c \geq d$ are satisfied.
- If $a \geq b$ is $F(a_1, \ldots, a_n) \geq F(b_1, \ldots, b_n)$ with $a_1 \geq b_1, \ldots, a_n \geq b_n$, then $a \geq c$ can be due to the third or to the fourth clause of the definition of $\geq$.

  First we consider the case that $a \geq c$ is $F(a_1, \ldots, a_n) \geq F(c_1, \ldots, c_n)$ with $a_1 \geq c_1, \ldots, a_n \geq c_n$. By induction hypothesis $d_1, \ldots, d_n$ exist such that $b_i \geq d_i$ and $c_i \geq d_i$ for $i = 1, \ldots, n$. Define $d := F(d_1, \ldots, d_n)$. Then $b \geq d$ and $c \geq d$ hold.

  Second we consider the case that $a \geq c$ is due to the fourth clause of the definition of $\geq$. In that case $a = F(a_1, \ldots, a_n)$ and there exist $c_1, \ldots, c_n$ such that $a_1 \geq c_1, \ldots, a_n \geq c_n$ and $F(c_1, \ldots, c_n) = \sigma(\alpha) \to \sigma(\beta) = c$ for some reduction rule $\alpha \to \beta$ and valuation σ. By induction hypothesis $d_1, \ldots, d_n$ exist such that $b_i \geq d_i$ and $c_i \geq d_i$ for $i = 1, \ldots, n$. By coherence of $\geq$ we have $F(d_1, \ldots, d_n) = \tau(\alpha)$ with $\sigma(\beta) \geq \tau(\beta)$. Define $d := \tau(\beta)$. Then $b \geq d$ holds by the fourth clause of the definition of $\geq$ and $c \geq d$ holds by coherence of $\geq$.

- The last case to be considered is when $a \geq b$ is due to the fourth clause of the definition of $\geq$. Then $a = F(a_1, \ldots, a_n)$ and there exist $b_1, \ldots, b_n$ such that $a_1 \geq b_1, \ldots, a_n \geq b_n$ and $F(b_1, \ldots, b_n)$ is a redex with contractum b, say $F(b_1, \ldots, b_n) = \sigma(\alpha) \to \sigma(\beta) = b$ for a reduction rule $\alpha \to \beta$. Again, $a \geq c$ can be due to either the third or the fourth clause of the definition of $\geq$.

  The case that $a \geq c$ is due to the third clause of the definition is similar to the second case in the previous step in the proof.

  Second we consider the case that $a \geq c$ is a consequence of the fourth clause of the definition of $\geq$. Then there exist $c_1, \ldots, c_n$ such that $a_1 \geq c_1, \ldots, a_n \geq c_n$ and we have $F(c_1, \ldots, c_n) = \sigma'(\alpha') \to \sigma'(\beta') = c$ for some reduction rule $\alpha' \to \beta'$ and valuation σ'. By induction hypothesis, $d_1, \ldots, d_n$ exist such that $b_i \geq d_i$ and $c_i \geq d_i$ for $i = 1, \ldots, n$. Coherence of $\geq$ yields that one has $F(d_1, \ldots, d_n) = \tau(\alpha)$ with $\sigma(\beta) = b \geq \tau(\beta)$ and $F(d_1, \ldots, d_n) = \tau'(\alpha')$ with $\sigma'(\beta') = c \geq \tau'(\beta')$. So $\tau(\alpha) = \tau'(\alpha')$ and by orthogonality we have $\tau = \tau'$ and $\alpha = \alpha'$. Define $d := \tau(\beta)$. By coherence we have $b = \sigma(\beta) \geq d$ and $c = \sigma'(\beta') \geq d$.

□

The main result of this section is a direct result of this theorem.

Corollary 14. *All orthogonal CRSs are confluent.*

4 Superdevelopments for λ-calculus

In this section we fix attention to λ-calculus. Confluence for λ-calculus is proved by Tait and Martin-Löf using a relation $\mapsto_1$ whose transitive closure equals reduction. Another proof of confluence can be given by first proving that all developments are finite (see [2]). A development is a reduction sequence in which only descendents of redexes that are present in the initial term may be contracted. It is not allowed to contract redexes that are created along the way. The crucial notions in both proofs are related in the following way: $M \mapsto_1 N$ if and only if a (complete) development $M \twoheadrightarrow N$ exists (see [2]).

The relation between $\geq$ and $\mapsto_1$ is as follows: $M \mapsto_1 N$ implies $M \geq N$ but not necessarily vice versa. Questions arising are: can the reduction sequences corresponding exactly to $\geq$ be characterized, and, if so, are these reduction sequences always finite?

In this section we shall characterize the reduction sequences corresponding exactly to the relation $\geq$ on λ-terms. In order to do so, a set of labelled λ-terms Λ_l and labelled β-reduction $\rightarrow_{\beta_l}$ on them will be defined. Lambda's will be labelled by a label from a countably infinite set of labels I, and application nodes will be labelled by a subset of I. If the labelling of a λ-term M satisfies certain conditions, then its β_l-reduction to normal form is, after having erased all labels, a superdevelopment. All superdevelopments are proved to be finite.

In [7] Lévy analyses the different ways in which β-redexes can be created. The following possibilities are distinguished (written in the usual notation for λ-calculus):
(1) $((\lambda x.\lambda y.M)N)P \rightarrow_\beta (\lambda y.M[x := N])P$
(2) $(\lambda x.x)(\lambda y.M)N \rightarrow_\beta (\lambda y.M)N$
(3) $(\lambda x.C[xM])(\lambda y.N) \rightarrow_\beta C'[(\lambda y.N)M']$ where C' and M' stand for C respectively M in which all free occurrences of x have been replaced by $\lambda y.N$.

The first two created redexes are 'innocent' and may be contracted in a superdevelopment. Note that, if we think of a λ-term as a tree built from application- and λ-nodes, the redexes in the first two cases are 'created upwards'. In the last case, on the other hand, the redex isn't created upwards, and may not be contracted in a superdevelopment. The result that all superdevelopments are finite illustrates that all infinite β-reduction sequences in λ-calculus are due to the third way of redex creation; indeed redex creation e.g. in the reduction sequence of $(\lambda x.xx)(\lambda x.xx)$ happens in this way.

In the following, we shall write the application nodes explicitly, but abstraction terms as usual. Further, the relation $\geq$ when only used on λ-terms can be simplified a bit.

Definition 15. The relation $\geq$ on λ-terms is defined in the following way:
(1) $x \geq x$ for each variable x,
(2) if $M \geq M'$ then $\lambda x.M \geq \lambda x.M'$ for a λ-term M,
(3) if $M \geq M'$ and $N \geq N'$ then $\mathsf{Ap}(M, N) \geq \mathsf{Ap}(M', N')$ for λ-terms M and N,

(4) if $M \geq \lambda x.M'$ and $N \geq N'$, then $\mathsf{Ap}(M,N) \geq M'[x := N']$ for λ-terms M and N.

We proceed by defining the set of labelled λ-terms.

Definition 16. The set Λ_l of labelled λ-terms is defined as the smallest set such that
(1) $x \in \Lambda_l$ for every variable x,
(2) if $M \in \Lambda_l$ and $i \in I$, then $\lambda_i x.M \in \Lambda_l$,
(3) if $M, N \in \Lambda_l$ and $X \subset I$, then $\mathsf{Ap}^X(M,N) \in \Lambda_l$.

Erasing all labels of a term $M \in \Lambda_l$ is done by a function $E : \Lambda_l \to \Lambda$ that is defined inductively as follows:

$$E(x) = x$$
$$E(\lambda_i x.M) = \lambda x.E(M)$$
$$E(\mathsf{Ap}^X(M_1, M_2)) = \mathsf{Ap}(E(M_1), E(M_2))$$

Let $\wp(I)$ denote the powerset of I, i.e. the set of all subsets of I. A *labelling* L for a λ-term M is a partial function from the symbols of M to $I \cup \wp(I)$, that is only defined on symbols Ap and λ, to which a subset of I respectively an element of I is assigned. The result of applying L to M is written as M^L. So $E(M^L) = M$.

The reduction rule β_l on Λ_l is defined as

$$\mathsf{Ap}^X(\lambda_i x.M, N) \to_{\beta_l} M[x := N] \quad \text{if } i \in X$$

where the substitution $[x := N]$ is defined as usual. Like usually in λ-calculus, we adopt the variable convention, i.e. all bound variables in a statement are supposed to be different from the free ones.

Definition 17.
- A term $M \in \Lambda_l$ is called *good* if no label X of an application node contains the index i of a λ occurring outside the scope of this application node.
- A labelling L is said to be *good for* a λ-term M if M^L is a good term.
- A labelling L is an *initial labelling* for a λ-term M if it is good for M and all λ's have a different label.
- Two labelling are said to be *disjoint* if no element of I occurs in both labellings.
- Two terms M and N of Λ_l are said to be *disjointly labelled* if there exist λ-terms M' and N' and disjoint labellings L_1 and L_2 such that $M'^{L_1} = M$ and $N'^{L_2} = N$.

For example, a good term is $\mathsf{Ap}^{\{2\}}(\mathsf{Ap}^{\{1\}}(\lambda_1 x.\lambda_2 y.xy, z), u)$, but, on the other hand, the term $\mathsf{Ap}^{\{1\}}(\lambda_1 x.\mathsf{Ap}^{\{2\}}(x,y), \lambda_2 y.y)$ isn't good. It is clear that all subterms of a good term are good. The property 'good' is preserved under reduction, i.e. β_l-reduction cannot push a λ outside the scope of an application node in which it occurred originally. This is proved in the following proposition, that will be used implicitly.

Proposition 18. *If $M \in \Lambda_l$ is a good term and $M \to_{\beta_l} N$, then N is a good term.*

PROOF. Let $\mathsf{Ap}^X(\lambda_i x.P, Q)$ be a good term with $i \in X$ (so it is a β_l-redex). It is proved by induction on the structure of P that $P[x := Q]$ is a good term. □

Definition 19. A reduction sequence $M \twoheadrightarrow_\beta N$ is a *superdevelopment* if there exists an initial labelling L such that $M^L \twoheadrightarrow_{\beta_l} N^{L'}$ is a β_l-reduction sequence to normal form (with L' some labelling).

The following proposition states that no β_l-redexes are created by substitution. With the 'pattern' of a redex, the application and lambda symbol on top are meant.

Proposition 20. *If $\mathsf{Ap}^X(\lambda_i x.P, Q) \in \Lambda_l$ is a good term and a β_l-redex, then all patterns of β_l-redexes in $P[x := Q]$ descend either totally from P or totally from Q.*

PROOF. Suppose $\mathsf{Ap}^X(\lambda_i x.P, Q)$ is a good term with $i \in X$ and we have in $P[x := Q]$ a subterm of the form $\mathsf{Ap}^Y(\lambda_j y.R, S)$. If the symbol Ap with label Y originates from P and λ_j from Q, then $j \notin Y$, because $\mathsf{Ap}^X(\lambda_i x.P, Q)$ is a good term. So in that case $\mathsf{Ap}^Y(\lambda_j y.R, S)$ is not a β_l-redex. It is impossible to have in $P[x := Q]$ a subterm $\mathsf{Ap}^Y(\lambda_j y.R, S)$ with Ap^Y originating from Q and λ_j from P. So if $\mathsf{Ap}^Y(\lambda_j y.R, S)$ is a β_l-redex in $P[x := Q]$, then Ap^Y and λ_j originate either both from P or both from Q. □

Theorem 21. (FINITE SUPERDEVELOPMENTS) *If a λ-term M is labelled by an initial labelling L then all its β_l-reductions are finite.*

PROOF. Suppose infinite β_l-reduction sequences exist, and let M be a minimal (with respect to the number of symbols) λ-term, labelled by an initial labelling, that admits an infinite β_l-reduction sequence. By minimality M has to be an application, so M is of the form $\mathsf{Ap}^X(M_1, M_2)$. The infinite β_l-reduction sequence starting with M then must be of the form

$$\mathsf{Ap}^X(M_1, M_2) \twoheadrightarrow_{\beta_l} \mathsf{Ap}^X(\lambda_i x.M_1', M_2') \to_{\beta_l} M_1'[x := M_2'] \to_{\beta_l} \ldots$$

In this reduction sequence, we have $M_1 \twoheadrightarrow_{\beta_l} \lambda_i x.M_1'$ and $M_2 \twoheadrightarrow_{\beta_l} M_2'$, and moreover $i \in X$. Turn M_1' into a context $C[, \ldots,]$ by replacing all free occurrences of x by $\square$. So $C[x, \ldots, x] = M_1'$. The last step in the reduction sequence above can now be written as $\mathsf{Ap}^X(\lambda_i x.C[x, \ldots, x], M_2') \to_{\beta_l} C[M_2', \ldots, M_2']$ Now we claim that all reducts of this reduction sequence are of the form $C'[M_{21}'', \ldots, M_{2n}'']$ with $C[, \ldots,] \twoheadrightarrow_{\beta_l} C'[, \ldots,]$ and $M_2' \twoheadrightarrow_{\beta_l} M_{2i}''$ for $i = 1, \ldots, n$. So all reductions take place either in descendants of $C[, \ldots,]$ or in descendants of M_2'. The claim follows from proposition 20 and the observation that nothing can be substituted into a descendant of M_2'. From the claim it follows immediately that either M_1 or M_2 admits an infinite reduction sequence, contradicting the minimality of M. □

Definition 22. A reduction relation $\to$ is called *weakly confluent* if for every two coinitial one-step reductions a common reduct can be found. So if $s \to t$ and $s \to u$ then a term v exists such that $t \twoheadrightarrow v$ and $u \twoheadrightarrow v$.

In exactly the same way as weak confluence for λ-calculus with β-reduction is obtained one obtains (by checking that the labels match) the same result for β_l-reduction on labelled λ-terms. Together with the property that all β_l-reduction sequences are finite we can conclude by Newman's Lemma that β_l-reduction is confluent. So each term of Λ_l has a unique β_l-normal form. Confluence for Λ_l with β_l-reduction can also be obtained by noting that it is an orthogonal CRS.

Proposition 23. *Let P and Q be good terms of Λ_l that are disjointly labelled. If $P \twoheadrightarrow_{\beta_l} P'$ and $Q \twoheadrightarrow_{\beta_l} Q'$ are β_l-reductions to normal form, then $P[x := Q] \twoheadrightarrow_{\beta_l} P'[x := Q']$ and $P'[x := Q']$ is a β_l-normal form.*

PROOF. The proof proceeds by induction on the structure of P.
- If P is a variable then the statement follows trivially.
- If $P = \lambda_i y.P_1$, then a β_l-reduction sequence of P to its normal form P' is of the form $\lambda_i y.P_1 \twoheadrightarrow_{\beta_l} \lambda_i y.P_1'$ with $P_1 \twoheadrightarrow_{\beta_l} P_1'$ a reduction to normal form. By induction hypothesis, $P_1[x := Q] \twoheadrightarrow_{\beta_l} P_1'[x := Q']$ is a reduction to normal form. Since $(\lambda_i y.P_1)[x := Q] = \lambda_i y.P_1[x := Q]$ it follows that $P[x := Q] \twoheadrightarrow_{\beta_l} P'[x := Q']$ is a reduction sequence to normal form.
- The last case to be considered is when $P = \mathsf{Ap}^X(P_1, P_2)$. Let $P_1 \twoheadrightarrow_{\beta_l} P_1'$ and $P_2 \twoheadrightarrow_{\beta_l} P_2'$ be reduction sequences to normal form. We distinguish two cases. First we consider the case that the reduction of P to normal form is of the form $\mathsf{Ap}^X(P_1, P_2) \twoheadrightarrow_{\beta_l} \mathsf{Ap}^X(P_1', P_2')$. By induction hypothesis we have that $P_1[x := Q] \twoheadrightarrow_{\beta_l} P_1'[x := Q']$ and $P_2[x := Q] \twoheadrightarrow_{\beta_l} P_2'[x := Q']$ are reduction sequences to normal form. We have $\mathsf{Ap}^X(P_1, P_2)[x := Q] \twoheadrightarrow_{\beta_l} \mathsf{Ap}^X(P_1', P_2')[x := Q']$. The result is in β_l-normal form, because even if $P'[x := Q']$ has a λ as head symbol this doesn't yield a β_l-redex. Namely, if this λ originates from P_1', its label is not contained in X (otherwise $\mathsf{Ap}^X(P_1', P_2')$ wouldn't be a β_l-normal form), and if the λ originates from Q' it doesn't yield a β_l-redex either since P and Q are disjointly labelled.

In the second case $\mathsf{Ap}^X(P_1', P_2') = \mathsf{Ap}^X(\lambda_i y.P_{11}', P_2')$ is a β_l-redex with contractum $P_{11}'[y := P_2']$. Since $\lambda_i y.P_{11}'$ and P_2' are in β_l-normal form, the term $P_{11}'[y := P_2']$ is by proposition 20 in β_l-normal form. By induction hypothesis we have that $P_1[x := Q] \twoheadrightarrow_{\beta_l} \lambda_i y.P_{11}'[x := Q']$ and $P_2[x := Q] \twoheadrightarrow_{\beta_l} P_2'[x := Q']$ are reduction sequences to normal form. We have $\mathsf{Ap}^X(P_1, P_2)[x := Q] \twoheadrightarrow_{\beta_l} \mathsf{Ap}^X(\lambda_i y.P_{11}'[x := Q'], P_2'[x := Q']) \twoheadrightarrow_{\beta_l} (P_{11}'[x := Q'])[y := P_2'[x := Q']]$. By the substitution lemma of λ-calculus, which holds as well for the labelled case, this term equals $(P_{11}'[y := P_2'])[x := Q']$ which is by Proposition 20 a β_l-normal form.

□

This proposition yields, together with the property of unique normal forms, that if the term $\mathsf{Ap}^X(\lambda_i x.P, Q)$ is good and a β_l-redex, and its reduct $P[x := Q]$ reduces to a β_l-normal form M, then M is of the form $P'[x := Q']$, with P' and Q' the normal forms of P and Q respectively.

Theorem 24. *If $M \geq M'$, then there exists an initial labelling L such that $M^L \twoheadrightarrow_{\beta_l} M'^{L'}$ for some labelling L' and $M'^{L'}$ is a β_l-normal form.*

PROOF. The proof proceeds by induction on the derivation of $M \geq M'$. The first two easy steps are omitted.
- If $M \geq M'$ is $\mathsf{Ap}^X(M_1, M_2) \geq \mathsf{Ap}^X(M'_1, M'_2)$ with $M_1 \geq M'_1$ and $M_2 \geq M'_2$, then by induction hypothesis labellings L_1 and L_2 exist such that $M_1^{L_1} \twoheadrightarrow_{\beta_l} M_1'^{L'_1}$ and $M_2^{L_2} \twoheadrightarrow_{\beta_l} M_2'^{L'_2}$ are reductions to β_l-normal form. Without loss of generality we can suppose L_1 and L_2 to be disjoint. Take as labelling L for $\mathsf{Ap}^X(M_1, M_2)$ the union of L_1 and L_2 extended by assigning $\emptyset$ to the head-symbol Ap. Then $\mathsf{Ap}^\emptyset(M_1^{L_1}, M_2^{L_2}) \twoheadrightarrow_{\beta_l} \mathsf{Ap}^\emptyset(M_1'^{L'_1}, M_2'^{L'_2})$ is a reduction sequence to normal form.
- If $M \geq M'$ is due to the fourth clause of the definition of $\geq$, then $M = \mathsf{Ap}(M_1, M_2)$ and $M' = M'_1[x := M'_2]$ with $M_1 \geq \lambda x.M'_1$ and $M_2 \geq M'_2$. By induction hypothesis, labelling L_1 and L_2 exist such that $M_1^{L_1} \twoheadrightarrow_{\beta_l} (\lambda x.M'_1)^{L'_1}$ and $M_2^{L_2} \twoheadrightarrow_{\beta_l} M_2'^{L'_2}$ are reduction sequences to normal form. Again, L_1 and L_2 are supposed to be disjoint. Define an initial labelling L as the union of L_1 and L_2 extended by assigning $\{i\}$ to the head-symbol Ap if i is the label assigned by L'_1 to the head-symbol λ of $\lambda x.M'_1$. Then

$$M^L = \mathsf{Ap}^{\{i\}}(M_1^{L_1}, M_2^{L_2}) \twoheadrightarrow_{\beta_l} \mathsf{Ap}^{\{i\}}((\lambda x.M'_1)^{L'_1}, M_2'^{L'_2}) \twoheadrightarrow_{\beta_l} M_1'^{L''_1}[x := M_2'^{L'_2}]$$

The result of this reduction sequence is in β_l-normal form by Proposition 20. □

Theorem 25. *If $M \in \Lambda_l$ is a good term and $M \twoheadrightarrow_{\beta_l} M'$ is a β_l-reduction sequence to normal form, then $E(M) \geq E(M')$.*

PROOF. The proof proceeds by induction on the structure of M. We omit the first two steps which are trivial. If $M = \mathsf{Ap}^X(M_1, M_2)$, two possibilities have to be distinguished. First the case is considered that the reduction of M to its normal form M' is of the form $\mathsf{Ap}^X(M_1, M_2) \twoheadrightarrow_{\beta_l} \mathsf{Ap}^X(M'_1, M'_2)$ with $M_1 \twoheadrightarrow_{\beta_l} M'_1$ and $M_2 \twoheadrightarrow_{\beta_l} M'_2$ β_l-reductions to normal form. By induction hypothesis, we have that $E(M_1) \geq E(M'_1)$ and $E(M_2) \geq E(M'_2)$. This yields, applying the third clause of the definition of $\geq$, that $\mathsf{Ap}(E(M_1), E(M_2)) \geq \mathsf{Ap}(E(M'_1), E(M'_2))$, or $E(M) \geq E(M')$.

Second we consider the case that the reduction sequence of M is of the form $\mathsf{Ap}^X(M_1, M_2) \twoheadrightarrow_{\beta_l} \mathsf{Ap}^X(\lambda_i x.M'_1, M'_2) \twoheadrightarrow_{\beta_l} M'_1[x := M'_2] \twoheadrightarrow_{\beta_l} M'$. By Proposition 23, M' is of the form $M''_1[x := M''_2]$ with M''_1 and M''_2 the normal forms of M'_1 and M'_2 respectively. Then $M_1 \twoheadrightarrow_{\beta_l} \lambda_i x.M''_1$ and $M_2 \twoheadrightarrow M''_2$ are β_l-reduction sequences to normal form. By induction hypothesis, we have $E(M_1) \geq E(\lambda_i x.M''_1)$ and $E(M_2) \geq E(M''_2)$. By the fourth clause of the definition of $\geq$, we have $E(M) \geq E(M''_1)[x := E(M''_2)] = E(M''_1[x := M''_2]) = E(M')$ □

Corollary 26. *$M \geq N$ if and only if there exists a superdevelopment $M \twoheadrightarrow N$.*

Acknowledgements

I am very grateful to my supervisor Jan Willem Klop for introducing me to the subject of higher-order rewriting. Special thanks to Vincent van Oostrom for several examples and counterexamples. I would like to thank Aart Middeldorp, Tobias Nipkow and Fer-Jan de Vries for comments on earlier versions of this paper. The paper benefitted from the remarks of the anonymous referees.

References

1. P. Aczel. A general Church-Rosser theorem. Technical report, University of Manchester, 1978.
2. H.P. Barendregt. *The Lambda Calculus, its Syntax and Semantics*. North Holland, second edition, 1984.
3. R. Hindley. The equivalence of complete reductions. *Transactions of the American Mathematical Society*, 229:227–248, 1977.
4. S. Kahrs. λ-*rewriting*. PhD thesis, Universität Bremen, 1991.
5. Z. Khasidashvili. Church-Rosser theorem in orthogonal combinatory reduction systems. Technical Report 1824, INRIA Rocquencourt, 1992.
6. J.W. Klop. *Combinatory Reduction Systems*. Mathematical Centre Tracts Nr. 127. CWI, Amsterdam, 1980. PhD Thesis.
7. J.-J. Lévy. *Réductions correctes et optimales dans le lambda-calcul*. PhD thesis, Université de Paris VII, 1978.
8. T. Nipkow. Orthogonal higher-order rewrite systems are confluent. In M. Bezem and J.F. Groote, editors, *Proceedings of the International Conference on Typed Lambda Calculi and Applications*, pages 306–317, Utrecht, 1993. Springer LNCS 664.
9. M.J. O'Donnell. *Computing in Systems described by Equations*. Lecture Notes in Computer Science 58. Springer-Verlag, 1977.
10. B.K. Rosen. Tree-manipulating systems and Church-Rosser theorems. *JACM*, 20(1):160–187, 1973.
11. M. Takahashi. λ-calculi with conditional rules. In M. Bezem and J.F. Groote, editors, *Proceedings of the International Conference on Typed Lambda Calculi and Applications*, pages 306–317, Utrecht, 1993. Springer LNCS 664.

Relating Graph and Term Rewriting via Böhm Models

Zena M. Ariola

Computer & Information Science Department, University of Oregon
Eugene, OR 97403-1202

Abstract. Dealing properly with sharing is important for expressing some of the common compiler optimizations, such as common subexpressions elimination, lifting of free expressions and removal of invariants from a loop, as source-to-source transformations. Graph rewriting is a suitable vehicle to accommodate these concerns. In [4] we have presented a term model for graph rewriting systems (GRSs) without interfering rules, and shown the partial correctness of the aforementioned optimizations. In this paper we define a different model for GRSs, which allows us to prove total correctness of those optimizations. Differently from [4] we will discard sharing from our observations and introduce more restrictions on the rules. We will introduce the notion of Böhm tree for GRSs, and show that in a system without interfering and non-left linear rules (orthogonal GRSs), Böhm tree equivalence defines a congruence. Total correctness then follows in a straightforward way from showing that if a program M contains less sharing than a program N, then both M and N have the same Böhm tree.

We will also show that orthogonal GRSs are a correct implementation of orthogonal TRSs. The basic idea of the proof is to show that the behavior of a graph can be deduced from its finite approximations, that is, graph rewriting is a continuous operation. Our approach differs from that of other researchers [6, 9], which is based on infinite rewriting.

1 Introduction

Dealing properly with sharing is important in a framework to reason about some compiler optimizations, such as common subexpressions elimination, lifting of free expressions and removal of invariants from a loop. All these optimizations can be characterized as merely increasing the sharing of sub-computations in a program. If we want to express these optimizations as source-to-source transformations, we need a formalism which can distinguish between, for example, the following two programs:

$$x = e;\ y = x + x \qquad y = e + e$$

Graph rewriting is a computational model to accommodate these concerns. Moreover, due to its implicit parallelism it is also suitable as an intermediate language for compilation on parallel machines. We have successfully described the operational semantics and the compilation of Id [11] using two different graph rewriting systems, Kid and P-TAC [1, 2, 3].

Graph rewriting has usually been described in terms of labeled graphs and homomorphisms [5]. Instead, in [4] we have introduced a novel approach to graph rewriting systems (GRSs), where the graph is represented as a set of mutually recursive equations, *i.e.*, a *letrec* expression, and reduction is explained in a way similar to term rewriting. In [4] we have also presented a term model for GRSs without interfering (*i.e.*, overlapping) rules, where the terms in the model are graphs and showed that a *less-sharing* relation corresponds to the pre-congruence relation induced by the model. Using the *less-sharing* relation we are able to prove the *partial correctness* of the aforementioned optimizations.

In this paper we are interested in defining a different model for GRSs, which will allow us to show *total correctness* of those optimizations. We also use this model to compare GRSs and TRSs. As in the earlier paper, our emphasis is on what a graph computes from an operational point of view, *i.e.*, its *answer*, and on a suitable notion of equality on the set of graph terms. As an example, consider the following terms M and N, respectively.

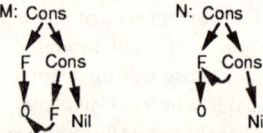

Notice that $M < N$, where $<$ is the sharing relation, and as such, according to the model in [4], $M \sqsubseteq N$, where $\sqsubseteq$ is the relation induced by the model. In other words, we would say that M will produce less observable information than N in any context. However, we would like to equate them because, if the internal representation of lists is ignored by an observer, they both represent the same unfolded list, $\mathsf{F}(0) : \mathsf{F}(0) : \mathsf{Nil}$.

We are also interested in equating terms *without* normal forms to terms *with* normal forms. Consider the following two terms M_1 and N_1,

$$M_1 \equiv x = \mathsf{Cons}(1, x) \qquad N_1 \equiv \mathsf{F}(1) \quad (\text{where } \mathsf{F}(y) \to \mathsf{Cons}(y, \mathsf{F}(y)))$$

Suppose Cons does not have any rule associated with it, then, M_1 is in normal form, while N_1 does not have one. However, intuitively it can be said that both M_1 and N_1 represent the infinite list of one's. M_1 has a finite representation of it, while N_1 does not. (This representation issue is similar to the representation issue for real numbers, where the number with decimal expansion $.999999\cdots$ is the same as 1).

We will define the *answer* of a graph in terms of its Böhm tree, and discuss the restrictions we have to impose on the rules to turn the Böhm tree equivalence into a *congruence*. Correctness of optimizations can then be formalized in terms of Böhm tree equivalence, that is, if a term M gets optimized into a term N then we say that the optimization is correct if M and N have the same Böhm tree. Moreover, we will show that the relation of *less-sharing* introduced in [4] guarantees the Böhm tree equivalence, and thus showing the total correctness of the optimizations that only affect sharing.

The Böhm tree model also constitutes an adequate notion for comparing GRSs and TRSs. Barendregt et al. [5] base the relation between GRSs and TRSs on the notion of normal form, which leads to some undesirable conclusions. We believe that the notion of normal form is inadequate for such a comparison. This point was already stressed by Wadsworth in his analysis of the relation between the syntactic and the semantic aspects of the λ-calculus [13, 14]. On the other hand, our approach is based on showing that the behavior of a graph can be deduced from its finite approximations. In other words, we show that graph rewriting is a *continuous* operation. This approach is different from that of other researchers [6, 9]. For example, Farmer and Watro [6] have shown the soundness of the cyclic Y-rule based on infinite rewriting. Similarly the approach taken by Klop et al. [9] is also based on rewriting infinite terms.

In Section 2 we will briefly introduce the reader to graph rewriting systems (GRSs). We restrict our attention to GRSs which are adequate to describe sharing in combinatory systems. In Section 3 we introduce a function $\mathcal{P}rint$ which given a term M returns the *finite stable* information associated with M, and we collect the information gathered by reducing M in a set called $\mathcal{P}rint^*(M)$, which represents the answer computed by M. For example, considering the terms M_1 and N_1 given earlier, $\mathcal{P}rint^*(N_1)$ is the infinite list of 1's. What term does M_1 compute? It is in normal form, but by performing its expansion, we can say that $\mathcal{P}rint^*(M_1)$ is also the infinite list of 1's. Thus, by taking the answer as our criterion for equating terms, we can say that terms M_1 and N_1 are the same. We will show that $\mathcal{P}rint^*$ equality is a congruence for a subclass of GRSs, namely those without interfering, and non-left linear rules (*orthogonal* GRSs).

Finally, using the notion of Böhm tree, in Section 4 we will show that orthogonal GRSs are a correct implementation of orthogonal term rewriting systems (TRSs). The notion of Böhm tree allows us to consider cyclic graphs, differently from [5]. We conclude the paper with our thoughts on future work.

This paper is self contained and only requires some understanding of TRSs. In particular, the model presented can be understood without reading [4].

2 Overview of a GRS

We express graph rewriting through a novel system, called a GRS. The objects of a GRS are simply a set of mutually recursive equations, *i.e.*, letrec expressions, instead of being complicated objects such as labeled graphs. For example, the following graph:

is expressed in a GRS system as $\{t_1 = +(1,2); t_2 = \mathsf{G}(t_1,t_1) \mathsf{\ In\ } t_2\}$, where the name following the keyword In denotes the root of the graph. In other words, the terms of a GRS are such that each subexpression has a name. The syntax of GRS terms is given in Figure 1. The variable names on the left hand side of bindings in a block are required to be pairwise distinct. Furthermore, the order of bindings in a block does not matter. Substitution is allowed for constants and variables only, thus no duplication of work occurs during reduction. For example, the expression $+(1,2)$ will be substituted for each occurrence of t_1 only when $+(1,2)$ becomes a value, *i.e.*, 3. The term obtained after all the substitutions are performed is called a *canonical term*. In that respect notice that if a term contains a binding of the kind "$x = x$" then we will substitute a constant, $\circ$, for each occurrence of x; $\circ$ stands for a circular binding. As shown in [4] this guarantees the confluence of a certain subclass of GRSs. Furthermore, the internal block structure of a term is merely syntactic sugar, that is, all internal blocks are flattened to compute the canonical form. Thus, analogously to the λ-calculus, we have a notion of α-equivalence, which consists in equating, up to renaming, all terms having the same canonical form. GRS rules are

$$
\begin{array}{ll}
SE & \in \text{Simple Expression} \\
E & \in \text{Expression} \\
\mathsf{F}^k & \in \mathcal{F}^k \\
Constant & \in \mathcal{F}^0 \\
SE & ::= Variable \mid Constant \\
E & ::= SE \mid \mathsf{F}^k(SE_1, \cdots, SE_k) \mid Block \\
Block & ::= \{\,[Binding;]^*\ \mathsf{In}\ SE\,\} \\
Binding & ::= Variable = E \\
Term & ::= E
\end{array}
$$

Fig. 1. Syntax of terms of a GRS with signature $\mathcal{F}$

not expressed in terms of multi-rooted graphs but as familiar TRS rules plus a precondition. The presence of a precondition is based on the observation that, differently from term rewriting, the root of a term plays a special role during its reduction. For example, consider the rule $\mathsf{F}(\mathsf{G}(x)) \longrightarrow 0$ and the term $\mathsf{F}(\mathsf{G}(1))$. During graph reduction only the pointers to F (*i.e.*, the root) are redirected to 0, that is, the subterm $\mathsf{G}(1)$ remains unaffected. Thus, we call the subterm $\mathsf{G}(x)$ the *precondition* of the above rule, which will be written in GRS notation as follows: $\dfrac{x_1 = \mathsf{G}(y)}{x = \mathsf{F}(x_1) \longrightarrow x = 0}$. We will call x, x_1 and y the variables and metavariables, respectively, of the above rule. In general, a GRS rule is a set of preconditions, $x_1 = e_1, \cdots, x_n = e_n$, and a left-hand-side, l, and a right-hand-side, r, and is written as: $\dfrac{x_1 = e_1 \mid \cdots \mid x_n = e_n}{x = l \longrightarrow x = r}$, where $e_i, 1 \leq i \leq n$, and l are terms of the form $\mathsf{F}^k(y_1, \cdots, y_k)$, where each y_i is either a variable or a constant and $k > 0$; r is a term such that its free variables appear either in the left hand side or in the precondition. The term $\{x_1 = e_1; \cdots x_n = e_n\ \mathsf{In}\ x\}$ is referred to as the pattern of the rule. A redex is then defined in terms of a substitution function σ from all the variables and metavariables of a rule to the set of variables and constants of a term. Take, for example, the following rule τ and term M, respectively:

$$\tau : \dfrac{x_1 = \mathsf{F}(y)}{x = \mathsf{G}(x_1, x_1) \longrightarrow x = y} \qquad M : \begin{array}{l} \{t_1 = \mathsf{F}(0); \\ t_2 = \mathsf{G}(t_1, t_1) \\ \mathsf{In}\ t_2\} \end{array}$$

the subterm $G(t_1,t_1)$ is a redex because $G(t_1,t_1) \equiv G(x_1,x_1)^\sigma$, where σ is: $x = t_2, x_1 = t_1, y = 0$. Reduction consists in replacing the redex by an instantiation of the right hand side of rule τ; that is, we will say that $M \xrightarrow[t_2]{\tau} M[t_2 \leftarrow r^\sigma] \equiv M[t_2 \leftarrow 0] \equiv_\alpha \{t_1 = F(0) \text{ In } 0\}$. We will also make use of the notation $M[\mathcal{F} \leftarrow e]$, where $\mathcal{F}$ is a set of redexes, indicating that all redexes in $\mathcal{F}$ are replaced by the expression e.

GRSs are formally introduced in [4], where GRSs without interfering rules are proved to be confluent. The notion of *interference* extends the familiar overlapping notion of TRSs; we will say that rule τ_1 interferes with rules τ_2 if there exists an instance of a non-variable subterm of the pattern of rule τ_1 which is the same, up to α-equivalence, to an instance of rule τ_2. A rule is said to be *non-left linear* if the pattern of the rule contains two distinct paths to a node; notice that left linearity implies that the left hand side of rules has to be acyclic. GRSs without interfering and non-left linear rules are called orthogonal GRSs. Hereafter, we will only consider orthogonal GRSs.

3 Böhm Model for Acyclic Orthogonal GRSs

3.1 Printable Value of a Term

Given a term we define its printable value as the stable part of the term, that is, the part which will never change during its reduction. The printable value corresponds to the notions of instant semantics [15] and direct approximation [10, 12] introduced for the λ-calculus. As for the λ-calculus, all redexes must first be substituted by a new constant Ω, which stands for no information. However, this is not enough to guarantee the monotonicity of the information content with respect to reduction. The problem is due to the upward creation of redexes; that is, even though a term M is not a redex, it can become so when some redexes under it are performed. To cope with this phenomenon in the λ-calculus, both Wadsworth [12] and Lévy [10] have introduced the notion of ω-rule, which states $\Omega P \longrightarrow \Omega$. Huet and Lévy have applied the same concept to orthogonal TRSs [8]. Analogously, for GRSs we introduce a set of ω-rules. For example, given the GRS rules: $x = F(0) \longrightarrow x = 1$ and $x = G(y) \longrightarrow x = 0$, we will generate the ω-rule: $x = F(\Omega) \longrightarrow x = \Omega$, thus, reducing to Ω the term $\{t_1 = F(t_2); t_2 = G(y) \text{ In } t_1\}$. Notice that the above term will also be reduced to Ω if we replace the second rule by $x = G(y) \longrightarrow x = 1$, that is, the ω-rules are generated without analyzing the right hand side of rules.

In [4] the stable part of a term was computed by applying the ω-reductions on the term obtained by replacing all its redexes by Ω, that is, sharing was part of our observations. Instead, in this paper we go one step further by considering the *expansions* or *finite information*. For example, consider the rule: $\dfrac{x_1 = A(y)}{x = F(x_1) \longrightarrow x = 0}$, and the term $M \equiv \{t = F(t) \text{ In } t\}$. Since M does not contain a redex, following [4], we will consider M as stable information. However, since we want to guarantee the continuity of graph rewriting with respect to graph expansions, we are forced to infer that the printable value of M is Ω, because none of the expansions, i.e., $\{\Omega, \mathsf{F}(\Omega), \mathsf{F}(\mathsf{F}(\Omega)), \cdots\}$, constitute stable information. For example, $\mathsf{F}(\Omega)$ is not stable because it becomes a redex by replacing Ω by the term $\mathsf{A}(x)$. As another example, the printable value of the term $M \equiv \{x = +(y,y); y = +(x,x) \text{ In } x\}$ will also be Ω. In other words, in both the previous two examples the terms contain only infinite information but they do not contain any finite information.

To obtain the printable value of a term, we first replace all its redexes by Ω and then perform ω-reductions on each expansion of the term so obtained. The expansions are obtained by applying Kleene recursion. In Kleene recursion the kth expansion, M^k, of a term $M \equiv \{x_1 = e_1; \cdots; x_n = e_n \text{ In } x\}$ is defined as follows:

M^0 is the expression bound to x^0 in $\{x_1^0 = \Omega; \cdots; x_n^0 = \Omega \text{ In } x^0\}$
$\vdots$

M^{k+1} is the expression bound to x^{k+1} in
$\{x_1^{k+1} = e_1[x_1^k/x_1, \cdots, x_n^k/x_n]; \cdots; x_n^{k+1} = e_n[x_1^k/x_1, \cdots, x_n^k/x_n] \text{ In } x^{k+1}\}$

where x_i^k stands for the kth approximation of the value associated with the variable x_i, $(1 \leq i \leq n)$. The printable value of M is then the chain of *stable trees* corresponding to variable x; all the other trees, i.e., x_j^k ($x_j \neq x$), are needed only to compute x^k.

Notice that the expansions are TRS terms defined over a signature extended with the constant Ω. Let us call T_Ω the set of such terms, and let us define the following prefix ordering:

Definition 1 Prefix Order on T_Ω. $(T_\Omega, \leq_t)$ is a partial order, where $\leq_t$ is defined as follows:
(i) $\Omega \leq_t M, \forall M \in T_\Omega$;
(ii) $x \leq_t x, \forall x \in Variable$;
(iii) $c \leq_t c, \forall c \in Constant$;
(iv) $F^n(M_1, \cdots, M_n) \leq_t F^n(N_1, \cdots, N_n)$, if $M_i \leq_t N_i$ $1 \leq i \leq n$.

Remark 2 Monotonicity of Expansions. *Given a GRS term M and its two expansions M^k and $M^{k'}$, if $k \leq k'$ then $M^k \leq_t M^{k'}$.*

In order to generate the ω-rules let us first introduce the following function Fl, which given an acyclic term M produces its flattened version.
$\text{Fl}[\![x]\!] = x$
$\text{Fl}[\![Constant]\!] = Constant$
$\text{Fl}[\![\mathsf{F}^\mathsf{n}(y_1, \cdots, y_n)]\!] = \mathsf{F}^\mathsf{n}(y_1, \cdots, y_n)$
$\text{Fl}[\![\{x_1 = e_1; \cdots x_n = e_n \text{ In } x_i\}]\!] =$
$e_i[\text{Fl}(\{x_1 = e_1; \cdots x_n = e_n \text{ In } x_1'\})/x_1', \cdots, \text{Fl}(\{x_1 = e_1; \cdots x_n = e_n \text{ In } x_m'\})/x_m']$
$\qquad$ where $x_1' \cdots x_m'$ are free variables of e_i
$\text{Fl}[\![\{x_1 = e_1; \cdots x_n = e_n \text{ In } x\}]\!] = x$
$\qquad$ where x is either a constant or $x \notin \{x_1, \cdots, x_n\}$

Definition 3 ω-rule. Given a GRS rule $\tau : \dfrac{x_1 = e_1 \mid \cdots \mid x_n = e_n}{x = l \longrightarrow x = r}$, a rule τ' : $l' \longrightarrow \Omega$ is said to be a ω-rule iff $l' <_t \mathtt{Fl}(\{x_1 = e_1;\ \cdots x_n = e_n;\ x = l \ \mathsf{In}\ x\})$, and $l' \not\equiv \Omega$.

Definition 4 ω-TRS. Given a GRS=$(A(\mathcal{F}), R)$, we define its corresponding ω-TRS to be $(A(\mathcal{F} \cup \Omega), R_\omega)$, where:
(i) $A(\mathcal{F} \cup \Omega)$ is obtained by first substituting Ω to each redex occurring in a term of $A(\mathcal{F})$, and then performing its expansions;
(ii) R_ω is obtained by generating all ω-rules for each τ in R.

Proposition 5. *A ω-TRS is strongly normalizing and confluent.*

A normal form in a ω-TRS will be referred to as a ω-normal form. Thus, the stable part of each expansion of a term M will be computed by the following ω-function, where M_Ω is the term obtained by substituting all its redexes by Ω.

Definition 6 ω-function. Given a GRS term M, for any k, $\omega(M_\Omega^k)$ is the ω-normal form of M_Ω^k in the corresponding ω-TRS.

Proposition 7 Monotonicity of ω with respect to $\leq_t$. *Given ω-TRS terms M and N, if $M \leq_t N$ then $\omega(M) \leq_t \omega(N)$.*

We thus collect all the finite information contained in a term in a set called ω-$\mathcal{T}$rees.

Definition 8 ω-Trees: Set of Observations. Given a GRS, the set of all finite observations is ω-$\mathcal{T}rees = \bigcup\{\omega(M^k) \mid \forall \text{ GRS terms } M, k \geq 0\}$.

Since we want to guarantee that infinite chains have a limit, we turn ω-$\mathcal{T}rees$ into a complete partial order using the ideal completion method [7].

Definition 9 Böhm Domain: ω-$\mathcal{T}rees^\infty$. Given a GRS, the domain of finite observations is the ideal completion of ω-$\mathcal{T}rees$, that is, ω-$\mathcal{T}rees^\infty = \{S \mid S \subseteq \omega\text{-}\mathcal{T}rees,\ S \text{ is an ideal }\}$.

Proposition 10. *ω-$\mathcal{T}rees^\infty$ is a complete algebraic partial order.*

Definition 11 Printable Value of a GRS Term. Given a GRS term M, the printable value of M is $\mathcal{P}rint(M) = \{a \mid a \leq_t b,\ a \in \omega\text{-}\mathcal{T}rees,\ b \in \{\omega(M^k) \mid k \geq 0\}\}$.

Notice that in defining what is observable of a term we get rid off all nodes not accessible from the root. Moreover, the monotonicity of the expansions (Remark 2) and the monotonicity of the ω-function with respect to the $\leq_t$ ordering (Proposition 7) guarantee that the printable value defines an element of ω-$\mathcal{T}rees$.

3.2 Answer of a Term

We define the answer of a GRS term as the collection of printable information obtained by reducing that term.

Definition 12 Answer of a GRS Term. Given a GRS term M, the answer of M is $\mathcal{P}rint^*(M) = \bigcup \{\mathcal{P}rint(M') \mid M \longrightarrow M'\}$.

Since orthogonal GRSs are confluent [4], in order to guarantee that the answer is well defined, it is sufficient to show that the printable value is monotonic with respect to reduction. To that end, let us introduce an ordering, $\leq_\omega$, on GRS terms, which captures both the sharing and the prefix ordering. Intuitively, $M \leq_\omega N$ if N can be obtained from M by replacing Ω with any other term or by increasing the sharing in M. For example, in the following, the term on the left is $\leq_\omega$ to the term on the right, but the term on the right is not $\leq_\omega$ to the term on the left:

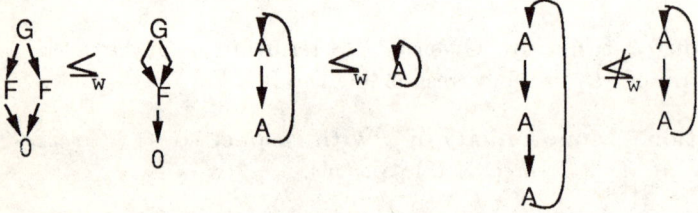

We will then show that if $M \leq_\omega N$ then $\mathcal{P}rint(M) \subseteq \mathcal{P}rint(N)$.

In the following, $\mathsf{FV}(M)$ and $\mathsf{BV}(M)$ will denote the free and bound variables of term M, respectively. Moreover, if $M \equiv \{x_1 = e_1; \cdots x_n = e_n \text{ In } x\}$ and x is a variable we will say that M is rooted at x.

Definition 13 ω-ordering: $\leq_\omega$. Given GRS terms M and N, rooted at z_1 and z_2, respectively, $M \leq_\omega N$ iff $\exists$ a function $\sigma : (\mathsf{BV}(M) \cup \mathsf{FV}(M) \cup Constants) \rightarrow (\mathsf{BV}(N) \cup \mathsf{FV}(N) \cup Constants)$ such that:
- $\forall c \in \mathcal{F}^0, \sigma(c) = c$;
- $\forall x \in \mathsf{FV}(M), \sigma(x) = x$;
- $\forall x \in \mathsf{BV}(M)$, if x is bound to $\mathsf{F}^k(y_1, \cdots, y_k)$ in M then $\exists z, z = \sigma(x)$ such that z is bound to $\mathsf{F}^k(\sigma(y_1), \cdots, \sigma(y_k))$ in N;
- $\sigma(z_1) = z_2$.

Proposition 14 Monotonicity of $\mathcal{P}rint$ with respect to $\leq_\omega$. Given GRS terms M and N, if $M \leq_\omega N$ then $\mathcal{P}rint(M) \subseteq \mathcal{P}rint(N)$.

Proof. Follows from the monotonicity of the ω-function with respect to $\leq_t$ (Proposition 7). □

Proposition 15 Monotonicity of $\mathcal{P}rint$ with respect to $\longrightarrow$. Given GRS terms M and N, if $M \longrightarrow N$ then $\mathcal{P}rint(M) \subseteq \mathcal{P}rint(N)$.

Proof. Follows from the monotonicity of the $\mathcal{P}rint$ function with respect to $\leq_\omega$ (Proposition 14). □

Using the notion of answer, we can define an ordering on the set of terms as follows:

Definition 16 $\sqsubseteq_{BT}$: **Ordering on finite observable information.** Given GRS terms M and N, $M \sqsubseteq_{BT} N$ iff $\mathcal{P}rint^*(M) \subseteq \mathcal{P}rint^*(N)$.

The symmetric closure of $\sqsubseteq_{BT}$ will be denoted by $\equiv_{BT}$.

If we want $\mathcal{P}rint^*$ to be our interpretation function, $\mathcal{P}rint^*$ will have to satisfy some properties; that is, the meaning will have to be preserved by reduction (*soundness*), and it will have to be compositional (*congruence*). Soundness follows trivially from the confluence of the system and from the monotonicity of $\mathcal{P}rint$ with respect to reduction (Proposition 15), while congruence will require some more machinery.

Since $\mathcal{P}rint^*$ equality deals with expansions of term, *i.e.*, trees, it is natural that if a rule in a GRS can distinguish between different sharing of subterms, then $\mathcal{P}rint^*$ would not be a congruence. A way of assuring that $\equiv_{BT}$ is a congruence is to show that for any context $C[\Box]$ the behavior of $C[M]$ can be inferred from the observations about M; that is, $\forall C[\Box], C[M] \equiv_{BT} \bigsqcup \{C[\tilde{a}] \mid a \in \mathcal{P}rint^*(M)\}$ where $\mathcal{P}rint^*(\bigsqcup S) = \bigsqcup\{\mathcal{P}rint^*(s) \mid s \in S\}$, and $\tilde{a}$ represents the GRS form of a TRS term. This can be considered as a syntactic version of the continuity of the context operation. In other words, to get a *finite approximation* of the answer of $C[M]$ a *finite approximation* of the answer of M suffices. In the following, we show through examples that in that respect *non-left linear rules* are discontinuous rules [1], because they rely on an infinite amount of information.

Consider the rule: $\dfrac{x_1 = \mathsf{A}(x_1)}{x = \mathsf{G}(x_1) \longrightarrow x = 1}$. Let $M \equiv \{t' = \mathsf{A}(t') \text{ In } t'\}$ and $C[\Box] \equiv \{t_1 = \mathsf{G}(t_2); t_2 = \Box \text{ In } t_1\}$, then we have $C[M] \not\equiv_{BT} \bigsqcup\{C[\tilde{a}] \mid a \in \mathcal{P}rint^*(M)\}$ because, $\mathcal{P}rint^*(C[M]) = \{\Omega, 1\}$ while $\mathcal{P}rint^*(C[\tilde{a}]) = \{\Omega\}$, for any observation a of M. As another example, consider the rule: $\dfrac{x_1 = \mathsf{A}(y, y)}{x = \mathsf{G}(x_1) \longrightarrow x = 1}$. Let $M \equiv \{t = \mathsf{A}(t_1, t_1); t_1 = \mathsf{B}(0); \text{ In } t\}$ and $C[\Box] \equiv \{t_1 = \mathsf{G}(t_2); t_2 = \Box \text{ In } t_1\}$. Then we have $C[M] \not\equiv_{BT} \bigsqcup\{C[\tilde{a}] \mid a \in \mathcal{P}rint^*(M)\}$, because, $\mathcal{P}rint^*(C[M]) = \{\Omega, 1\}$ while $\mathcal{P}rint^*(C[\tilde{a}]) = \{\Omega\}$, for any observation a of M.

The direction $\forall C[\Box], \bigsqcup\{C[\tilde{a}] \mid a \in \mathcal{P}rint^*(M)\} \sqsubseteq_{BT} C[M]$ trivially follows from the monotonicity of $\sqsubseteq_{BT}$ with respect to the $\leq_\omega$ ordering.

Lemma 17 Monotonicity of $\sqsubseteq_{BT}$ with respect to $\leq_\omega$. *Given GRS terms M and N, if $M \leq_\omega N$ then $M \sqsubseteq_{BT} N$.*

Proof. We prove, by induction on the number of reduction steps, that if $M \longrightarrow M_1$, then $\exists N_1$ such that $N \longrightarrow N_1$ and $\mathcal{P}rint(M_1) \subseteq \mathcal{P}rint(N_1)$. Suppose

---

[1] Private communications with Georges Gonthier and Jean-Jacques Lévy.

$M \xrightarrow{z} M_1$, by reducing redex z in M. Let z_1 be the corresponding redex in N, and let N_1 be such that $N \xrightarrow{z_1} N_1$. Let $\mathcal{F}$ be the set of all redexes, distinct from z, occurring in M which are mapped into z_1. Then, $M_1[\mathcal{F} \leftarrow \Omega] \leq_\omega N_1$. Since $\mathcal{P}rint(M_1) = \mathcal{P}rint(M_1[\mathcal{F} \leftarrow \Omega])$, from the monotonicity of the $\mathcal{P}rint$ function with respect to $\leq_\omega$ (Proposition 14) we have $\mathcal{P}rint(M_1) \subseteq \mathcal{P}rint(N_1)$. The result then follows from the observation that in the absence of interfering rules any residual of a redex $z' \notin \mathcal{F}$ occurs also in $M_1[\mathcal{F} \leftarrow \Omega]$ and therefore there will exists a corresponding redex in N_1. □

In order to prove the other direction we show that if $M \twoheadrightarrow N$ then each expansion of N can be obtained by reducing an expansion of M.

Lemma 18. *Given a GRS term M, if $M \twoheadrightarrow M_1$ then $\forall k, \exists i, M^i \twoheadrightarrow M'$ and $\tilde{M}_1^k \leq_\omega M'$.*

Proof. The proof is by induction on the number of reduction steps of $M \twoheadrightarrow M_1$. Suppose $M \xrightarrow{\tau}_z M_1 \equiv_\alpha M[z \leftarrow r^\sigma]$, where r is the right hand side of rule τ. By definition of expansions we have that $\forall k, \exists j$, such that $(M[z \leftarrow r^\sigma])^k \leq_\omega M^j[z_1 \leftarrow r^{\sigma_1}] \cdots [z_n \leftarrow r^{\sigma_n}]$, where:
(i) if we let σ' be the substitution induced by $M^j \leq_\omega M$, we have $\{z_1, \cdots, z_n\} = \{z' \mid z' \in \text{BV}(M^j) \text{ such that } z'^{\sigma'} = z\}$;
(ii) $\forall i, 1 \leq i \leq n, \sigma_i$ is a mapping from $\text{FV}(r)$ to M^j.
Since rules are left-linear and non-interferent $\exists i \geq j$ such that $\forall l, 1 \leq l \leq n$ z_l is a τ-redex occurring in M^i. If we let $\mathcal{F} = \{z_1, \cdots, z_n\}$ then $M^i \xrightarrow[cdv(\mathcal{F})]{} M'$, where $\tilde{M}_1^k \leq_\omega M'$ ($\xrightarrow[cdv(\mathcal{F})]{}$ denotes a a complete development with respect to $\mathcal{F}$).
Suppose $M \twoheadrightarrow M_{n-1} \xrightarrow{\tau}_z M_n$, then we have that $\forall k, \exists i$ such that $M_{n-1}^i \xrightarrow[cdv(\mathcal{F}_1)]{} M'_n$, where $\mathcal{F}_1 = \{z_1, \cdots, z_n\}$ and $z_1, \cdots z_n$ being τ-redexes, and $\tilde{M}_n^k \leq_\omega M'_n$. Moreover, by induction hypothesis, $\exists j$, such that $M^j \twoheadrightarrow M'_{n-1}$, where $\tilde{M}_{n-1}^i \leq_\omega M'_{n-1}$, and let σ' be the induced substitution function. Notice that $\forall z_l \in \mathcal{F}_1$, the corresponding node in M'_{n-1} is a τ-redex. Let $\mathcal{F}'_1 = \{z' \mid z' \in \text{BV}(M'_{n-1})$ such that $\exists s \in \mathcal{F}_1$ with $\sigma'(s) = z'\}$. Notice that each node in M'_{n-1} which corresponds to a node in $\mathcal{F}'_1$ either belongs to $\mathcal{F}'$ or is Ω. Therefore, $M'_{n-1} \xrightarrow[cdv(\mathcal{F}'_1)]{} M''_n$ and $M'_n \leq_\omega M''_n$. Subsequently, $\tilde{M}_n^k \leq_\omega M''_n$. □

Theorem 19. *Given a GRS term M, $C[M] \equiv_{\mathsf{BT}} \bigsqcup \{C[\tilde{a}] \mid a \in \mathcal{P}rint^*(M)\}$.*

Proof. $- \bigsqcup\{C[\tilde{a}] \mid a \in \mathcal{P}rint^*(M)\} \sqsubseteq_{\mathsf{BT}} C[M]$. Since $C[\tilde{a}] \leq_\omega C[M]$ the result follows from the monotonicity of $\sqsubseteq_{\mathsf{BT}}$ with respect to the $\leq_\omega$ ordering (Lemma 17);
$- C[M] \sqsubseteq_{\mathsf{BT}} \bigsqcup\{C[\tilde{a}] \mid a \in \mathcal{P}rint^*(M)\}$. Suppose $C[M] \twoheadrightarrow N$. By the commutativity of a GRS [4] we have that $\exists M', C[M] \twoheadrightarrow C[M']$ and $C[M']$

reduces to N without performing any redexes occurring in M', i.e., $C[M']$ $\xrightarrow{}_{M'} N$. Let $a \in \mathcal{P}rint(N)$, that is, $\exists k, a \leq_t \omega(N^k)$. By Lemma 18 we have that each expansion of N can be obtained by reducing an expansion of $C[M']$, that is, $\exists i$ such that: $(C[M'])^i \longrightarrow N'$ and $\tilde{N}^k \leq_\omega N'$. By the monotonicity of the $\mathcal{P}rint$ function with respect to the $\leq_\omega$ ordering (Proposition 14) we have that $a \in \mathcal{P}rint(N')$ and, consequently, $a \in \mathcal{P}rint^*(C[M']^i)$. Since $\exists j, C[\tilde{M}']^i \leq_\omega C[\tilde{M}'^j]$, we have that $a \in \mathcal{P}rint^*(C[M'^j])$. Moreover, since a is obtained without executing any redex inside M', and, due to the non-interference of the rules, ω-reductions do not change the meaning of a term, we have that $a \in \mathcal{P}rint^*(C[\omega(M'^j)])$.

□

Corollary 20 *Congruence of* $\equiv_{BT}$. *Given a GRS and terms M and N, if $M \equiv_{BT} N$ then $C[M] \equiv_{BT} C[N]$.*

The restriction of the ordering $\leq_\omega$ to Ω free terms will be called $\leq_{sharing}$.

Theorem 21. *Given GRS terms M and N, if $M \leq_{sharing} N$ then $M \equiv_{BT} N$.*

Proof. We only have to prove that $N \sqsubseteq_{BT} M$. To that end, if $N \longrightarrow N_1$, then it is possible to show that $\exists M_1, M \longrightarrow M_1$, where $M_1 \leq_{sharing} N_1$. The result then follows from the observation that $\mathcal{P}rint(M_1) = \mathcal{P}rint(N_1)$. □

We can express the common subexpression elimination in terms of a GRS rule as : $\dfrac{y = +(a,b)}{x = +(a,b) \longrightarrow x = y}$.

Proposition 22. *The common subexpression elimination is totally correct.*

Proof. Note that if $M \longrightarrow M_1$ by applying the common subexpression elimination rule then $M \leq_{sharing} M_1$. Therefore, the result follows from Theorem 21.

4 Relation between GRSs and TRSs

We want to explore if GRSs are a correct implementation of TRSs. A GRS is correct with respect to a TRS if the following two conditions are satisfied:
(i) each information obtained by reducing a term in a GRS can be obtained by reducing an expansion to the term in the corresponding TRS;
(ii) each information obtained by reducing a term in a TRS can be obtained by reducing the term in the corresponding GRS.

We do not impose any restriction on the set of GRS terms; that is, the terms may contain cycles. However, their expansions will be TRS terms. To that end, we introduce the function BT which returns the answer of a TRS term.

Notice that, as in Section 3, we can associate to each TRS a ω-TRS, with the only difference that no expansions of terms is required. Thus, the definition of the ω_t function follows trivially.

Definition 23 Answer of a TRS term or its Böhm Tree. Given a TRS term M, the answer of M is $\mathsf{BT}(M) = \{a \mid a \in \omega\text{-}Trees, a \leq_t \omega_t(N), M \longrightarrow\!\!\!\!\!\rightarrow N\}$.

Analogous to GRSs, we can show that the answer defines an interpretation function. The proof methodology is the same as the one developed in Section 3, and most of the proofs carry over to the TRS case. The only difference shows up in the proof of the continuity of the context operation. In particular, we remind the reader that in proving $C[M] \sqsubseteq_g \bigsqcup \{C[\tilde{a}] \mid a \in Print^*(M)\}$ we made use of a property of reductions, namely that a reduction of the form $C[M] \longrightarrow\!\!\!\!\!\rightarrow N$ can be re-ordered in a way consisting of first reducing all redexes occurring in M and then performing the rest of the reduction. This property came out directly from the commutativity property of GRSs without interfering rules. For orthogonal TRSs, due to duplication of redexes, commutativity is lost. However, it is not difficult to show that for TRSs reductions can be re-ordered in an inside-out manner.

4.1 Soundness and Completeness of GRS

Due to the presence of cycles, the translation of a GRS term will not always produce a TRS term, but a possible infinite sequence of them. To define an infinite term we perform the ideal completion of $\langle T_\Omega, \leq_t \rangle$.

Definition 24 Infinite Term. The set of infinite terms, called T_Ω^∞, is the ideal completion of T_Ω.

Definition 25 Answer of an infinite term. Given an infinite term M, the answer of M is $\mathcal{P}rint_\infty^*(M) = \bigcup \{\mathsf{BT}(t) \mid t \in M\}$.

Proposition 26. *Given an infinite term M, $\mathcal{P}rint_\infty^*(M)$ is in $\omega\text{-}Trees^\infty$.*

Definition 27 Expansion of a GRS Term.
Given a GRS term M, the corresponding infinite term is $\mathsf{Exp}_\infty(M) = \{a \mid a \leq_t b, a \in T_\Omega, b \in \{M^k \mid k \geq 0\}\}$.

Proposition 28. *Given a GRS term M, $\mathsf{Exp}_\infty(M)$ is in T_Ω^∞.*

The translation between a GRS rule and a TRS rule requires also attention because the rhs of a GRS rule may contain cycles. As such, the expansion of a GRS rule may result in a set of TRS rules. Instead, we remind the reader that the translation of the lhs of a GRS rule is merely obtained by flattening the pattern of the rule. For example, the expansion of the following GRS rule:

$$x = \mathsf{C}(y) \longrightarrow x = \{t_1 = \mathsf{B}(t_2);\\ t_2 = \mathsf{A}(t_1, y)\\ \mathsf{In}\ t_2\}$$

is the set of TRS rules: $C(y) \longrightarrow \Omega_{t_2}(y)$, $\Omega_{t_2} \longrightarrow A(\Omega_{t_1}(y), y)$, and $\Omega_{t_1}(y) \longrightarrow B(\Omega_{t_2}(y))$, where Ω_{t_i} are new function symbols. Notice that the second and third rule generate all the expansions of the rhs of the GRS rule. The expansion of the cyclic Y-rule, which in GRS notation is expressed as $x = \mathsf{Ap}(Y, F) \longrightarrow x = \mathsf{Ap}(F, x)$, is the set of TRS rules: $\mathsf{Ap}(Y, F) \longrightarrow \Omega_x$, and $\Omega_x \longrightarrow \mathsf{Ap}(F, \Omega_x)$. In order to define our translation procedure, called TR, we associate with a GRS term a set of rules, called $R_{exp}(M)$, which will generate all its expansions.

Rules to generate the expansions of a GRS term M

Given a term $M \equiv \{x_1 = e_1; \cdots x_n = e_n \text{ In } x_i\}$, let $\{y_1, \cdots, y_k\}$ be the set of free variables of M. Then, $\forall x_i, 1 \leq i \leq n$, $R_{exp}(M)$ will contain the following rules:

$$\Omega_{x_i}(y_1, \cdots, y_k) \longrightarrow e_i[\Omega_{x_1}(y_1, \cdots, y_k)/x_1, \cdots, \Omega_{x_n}(y_1, \cdots, y_k)/x_n]$$

The expansions of M are then obtained by rewriting the term $\Omega_x(y_1, \cdots, y_k)$ following $R_{exp}(M)$.

For example, given the term $M \equiv \{x = \mathsf{Cons}(1, y); \; y = \mathsf{Cons}(1, x) \text{ In } x\}$, then the expansions of M are computed by reducing Ω_x following the rules $\Omega_x \longrightarrow \mathsf{Cons}(1, \Omega_y)$ and $\Omega_y \longrightarrow \mathsf{Cons}(1, \Omega_x)$.

Translation of GRS rules

Given a GRS rule $\tau : \dfrac{x_1 = e_1 \mid \cdots \mid x_n = e_n}{x = l \rightarrow x = r}$, let $M \equiv \{x_1 = e_1; \cdots x_n = e_n; \; x = l \text{ In } x\}$, $N \equiv \{x_1 = e_1; \cdots x_n = e_n; \; x = r \text{ In } x\}$, and $\{y_1, \cdots, y_k\} = \mathsf{FV}(N)$. Then the translation of GRS rule τ is: $\mathsf{TR}(\tau) = \{\mathsf{Fl}(M) \longrightarrow \Omega_x(y_1, \cdots, y_k)\} \cup R_{exp}(N)$.

Notice that $\mathsf{Fl}(M)$ is a finite term since rules are left-cyclic.

Given a set of GRS rules R, let us denote by $\longrightarrow_*$ the reduction relation induced by the set of rules $R^* = \cup\{\mathsf{TR}(r) \mid r \in R\}$. Notice that all rules in R^* are right-acyclic; that is, they are TRS rules. Moreover, if M is acyclic and $M \longrightarrow_* M'$, then M' is also acyclic. Let $\mathcal{P}rint_*(M)$ be the printable value of a term computed with respect to R^*, and let $\mathcal{P}rint_*^*$ be its answer; that is, $\mathcal{P}rint_*^*(M) = \bigcup \{\mathcal{P}rint_*(M') \mid M \longrightarrow_* M'\}$. In the following we will show that this translation does not change the meaning of a term.

Proposition 29. *Given a GRS, $\mathcal{P}rint^*(M) = \mathcal{P}rint_*^*(M)$.*

Proof. Follows from the observation that $M \equiv_{\mathsf{BT}} \bigsqcup \{M^i \mid i \geq 0\}$ (from Lemma 18). □

Definition 30 GTRS. Given a GRS $=(A_c, R_c)$ its corresponding TRS, called GTRS, is (A, R), where:
(1) $A = \{a \mid a \in \mathsf{Exp}_\infty(M), \forall M \in A_c\}$;
(2) $R = \{r \mid r \in \mathsf{TR}(r_c), \forall r_c \in R_c\}$.

Notice that the terms of a GTRS is the set of all expansions of graph terms.

Definition 31 Soundness. Given a GRS $\mathcal{A} = (A_c, R_c)$ and its corresponding GTRS $\mathcal{T} = (A, R)$, then $\mathcal{A}$ is sound with respect to $\mathcal{T}$, if $\forall\ M \in A_c$, $\mathcal{P}rint^*(M) \subseteq \mathcal{P}rint^*_\infty(\text{Exp}_\infty(M))$.

Definition 32 Completeness. Given a GRS $\mathcal{A} = (A_c, R_c)$ and its corresponding GTRS $\mathcal{T} = (A, R)$, then $\mathcal{A}$ is complete with respect to $\mathcal{T}$, if $\forall\ t \in A$, $\text{BT}(t) \subseteq \mathcal{P}rint^*(\tilde{t})$.

Theorem 33. *A GRS $\mathcal{A} = (A_c, R_c)$ is sound with respect to its corresponding GTRS.*

Proof. By Proposition 29 we have: $\mathcal{P}rint^*(M) = \mathcal{P}rint^*_*(M)$. By Lemma 18 we have $\mathcal{P}rint^*_*(M) = \bigcup \{\mathcal{P}rint^*_*(\tilde{M^i}) \mid M^i \in \text{Exp}_\infty(M)\}$. Since rules are both right acyclic, and left-linear we have $\mathcal{P}rint^*_*(\tilde{M^i}) \subseteq \text{BT}(M^i)$. Therefore,
$\mathcal{P}rint^*(M) = \bigcup \{\mathcal{P}rint^*_*(\tilde{M^i}) \mid M^i \in \text{Exp}_\infty(M)\}$
$\subseteq \bigcup \{\text{BT}(M^i) \mid M^i \in \text{Exp}_\infty(M)\} = \mathcal{P}rint^*_\infty(\text{Exp}_\infty(M))$. □

Let GRS^Y be the GRS containing the cyclic Y-rule, and TRS^Y be the TRS containing the common Y-rule. Since the cyclic Y-rule is left-linear, then from the soundness theorem we have:

Corollary 34. *GRS^Y is sound with respect to the TRS^Y.*

Theorem 35. *A GRS $\mathcal{A} = (A_c, R_c)$ is complete with respect to its corresponding GTRS.*

Conclusions

GRSs have already proven very useful in expressing the operational semantics and the compilation process of Id. In particular, the notion of $\mathcal{P}rint^*$ defines what an Id program computes. GRSs allow also the formalization of optimizations in terms of source-to-source transformations, and the term model developed provide a criterion for their correctness.

We would like to extend GRSs with λ-abstraction and to provide a term model that covers multi-rooted rules to express side-effect operations. This would provide a sound mathematical basis for the Id language. It would also be interesting to investigate the suitability of GRSs as an intermediate language for other classes of languages, such as imperative languages.

5 Acknowledgements

This work is part of the author's PhD thesis, and was done at the Laboratory for Computer Science at MIT. Funding for this work has been provided in part by the Advanced Research Projects Agency of the U.S. Department of Defense under the Office of Naval Research contracts N00014-84-K-0099 (MIT) and N0039-88-C-0163 (Harvard).

Many thanks to Arvind, J.W. Klop, Art Farley, Evan Tick and Will Clinger for reading the current draft of the paper.

References

1. Z. Ariola and Arvind. P-TAC: A Parallel Intermediate Language. In *Proc. ACM Conference on Functional Programming Languages and Computer Architecture, London*, September 1989.
2. Z. Ariola and Arvind. A Syntactic Approach to Program Transformations. In *Proc. ACM SIGPLAN Symposium on Partial Evaluation and Semantics Based Program Manipulation, Yale University, New Haven, CT*, June 1991.
3. Z. Ariola and Arvind. Compilation of Id. In *Proc. of the Fourth Worskshop on Languages and Compilers for Parallel Computing, Santa Clara, California, Springer-Verlag LNCS 589*, August 1991.
4. Z. Ariola and Arvind. *Graph Rewriting Systems For Efficient Compilation*, chapter 6. John Wiley & Sons, 1993.
5. H. Barendregt, M. van Eekelen, J. Glauert, J. Kennaway, M. Plasmeijer, and M. Sleep. Term Graph Rewriting. In *Proceedings of the PARLE Conference, Eindhoven, The Netherlands, Springer-Verlag LNCS 259*, pages 141–158, June 1987.
6. W. Farmer and R. Watro. Redex Capturing in Term Graph Rewriting. Technical Report M89-59, MITRE corporation, Massachusetts, 1989.
7. M. Hennessy. *Algebraic Theory of Processes*. MIT Press, 1988.
8. G. Huet and J.-J. Lévy. Computations in Orthogonal Rewriting Systems 1 and 2. In *Computational logic. Essays in Honor of Alan Robinson. Ed. J.-L. Lassez & G.D. Plotkin*, 1991.
9. J. Kennaway, J. Klop, M. Sleep, and F. de Vries. Transfinite Reductions in Orthogonal Term Rewriting Systems. In *Proc. RTA '91, Springer-Verlag LNCS*, 1991.
10. J.-J. Lévy. *Réductions Correctes et Optimales dans le Lambda-Calcul*. PhD thesis, Universite Paris VII, October 1978.
11. R. S. Nikhil. Id (Version 90.0) Reference Manual. Technical Report CSG Memo 284-a, MIT Laboratory for Computer Science, 545 Technology Square, Cambridge, MA 02139, USA, July 1990.
12. C. Wadsworth. *Semantics And Pragmatics Of The Lambda-Calculus*. PhD thesis, University of Oxford, September 1971.
13. C. Wadsworth. The Relation between Computational and Denotational Properties for Scott's D_∞-Models of the Lambda-Calculus. *Theoretical Computer Science*, 5, 1976.
14. C. Wadsworth. Approximate Reduction and Lambda Calculus Models. *Theoretical Computer Science*, 7, 1978.
15. P. Welch. Continuous Semantics and Inside-out Reductions. In *λ-Calculus and Computer Science Theory, Italy (Springer-Verlag LNCS 37)*, March 1975.

Topics in Termination*

Nachum Dershowitz and Charles Hoot

Department of Computer Science, University of Illinois, Urbana, IL 61801, U.S.A.
nachum,hoot@cs.uiuc.edu

Abstract. We generalize the various path orderings and the conditions under which they work, and describe an implementation of this general ordering. We look at methods for proving termination of orthogonal systems and give a new solution to a problem of Zantema's.

1 Introduction

If no infinite sequences of rewrites are possible, a rewrite system is said to have the *termination* property. In practice, one usually guarantees termination by devising a well-founded (strict partial) ordering $\succ$, such that $s \succ t$ whenever s rewrites to t. As suggested in [Manna and Ness, 1970], it is often convenient to separate reduction orderings into a homomorphism from terms to an algebra with a well-founded ordering. The use, in particular, of *polynomial interpretations* which map terms into the natural numbers, was developed by Lankford [1979]. For a survey of termination methods, see [Dershowitz, 1987].

Virtually all orderings used in practice are *simplification orderings* [Dershowitz, 1982], satisfying the *replacement* property, that $s \succ t$ implies that any term containing s is not less (under $\succ$) than the same term with that occurrence of s replaced by t, and the *subterm* property, that any term containing s is greater or equal to s. Simplification orderings cannot be used to prove termination of "self-embedding" systems, that is, when a term t can be derived in one or more steps from a term t', and t' can be obtained by repeatedly replacing subterms of t with subterms of those subterms.

Knuth and Bendix [1970] designed a particular class of well-orderings which assigns a weight to a term which is the sum of the weights of its constituent function symbols. Terms of equal weight and headed by the same symbol have their subterms compared lexicographically. Another class of simplification orderings, the *path orderings* [Dershowitz, 1982], is based on the idea that a term u should be bigger than any term that is built from smaller terms, all held together by a structure of function symbols that are smaller in some precedence ordering than the root symbol of u. The notion of path ordering was extended by Kamin and Lévy [1980] to compare subterms lexicographically and to allow for a semantic component; see [Dershowitz, 1987]. Here, we generalize these orderings and the conditions under which they work. In the appendix, we describe an implementation of the general ordering.

---

* This research was supported in part by the U. S. National Science Foundation under Grants CCR-90-07195 and CCR-90-24271. The first author was also supported by a Lady Davis fellowship at the Hebrew University of Jerusalem.

We also look at methods of proving termination of *orthogonal* (left-linear non-overlapping) systems and related issues. These may be compared with ordinary structural induction proofs used for recursively-defined functions; see [Burstall, 1969; Manna, 1974]. In particular, we give a solution to a problem posed by Zantema [personal communication].

2 Path orderings

We use quasi-orderings (reflexive-transitive binary relations) to prove termination of rewrite systems. A quasi-ordering is *well-founded* if it has no infinite strictly descending sequences of elements. A *precedence* is a well-founded quasi-ordering of function symbols. An ordering might be called *syntactic* if it is based on a precedence and is invariant under shifts of symbols. In other words, we require that consistently replacing function symbols in two terms with others of the same arity and with the same relative ordering has no effect on the ordering of the two. The recursive path orderings [Dershowitz, 1982; Kamin and Lévy, 1980; Lescanne, 1990] are syntactic; the Knuth-Bendix and polynomial orderings are not.

The rule
$$x \times (y + z) \quad \to \quad (x \times y) + (x \times z) \tag{1}$$
is terminating. This can be shown by considering the multiset of "natural" interpretations of all products in a term, letting + and × stand for addition and multiplication, and assigning some fixed value to constants; see [Dershowitz and Manna, 1979] for similar examples. Syntactic "path" orderings (see [Dershowitz, 1987]) work in this case, too. Lipton and Snyder [1977] gave a method for proving termination with interpretations (order-isomorphic to ω) for which rules are "value-preserving", as this example is for the natural interpretation.

Consider the following contrived system for computing factorial in unary arithmetic (expanding on one in [Kamin and Lévy, 1980]):

$$\begin{aligned}
p(s(x)) &\to x \\
fact(0) &\to s(0) \\
fact(s(x)) &\to s(x) \times fact(p(s(x))) \\
0 \times y &\to 0 \\
s(x) \times y &\to (x \times y) + y \\
x + 0 &\to x \\
x + s(y) &\to s(x + y) \, .
\end{aligned} \tag{2}$$

It would be nice were we able to use a natural interpretation, but that does not prove termination, since the rules preserve the value of the interpretation, rather than cause a decrease. Nor can we use multisets of the values of the argument of *fact*, since some rules can multiply occurrences of that symbol. Though path orderings [Dershowitz, 1987] have been successfully applied to many termination proofs, they suffer from the same limitation as do all simplification orderings: they are not useful when a rule embeds as does $fact(s(x)) \to s(x) \times fact(p(s(x)))$.

What is needed is a way of combining the semantics given by a natural interpretation with a non-simplification ordering that takes the structure of terms into account.

Definition 1 (Termination Function). A *termination function* τ takes a term as argument and is of one of the following types:

a. a function that returns the outermost function symbol of a term to be compared using a precedence;
b. a homomorphism from terms to some well-founded set of values (that is, $\tau(f(s_1,\ldots,s_n)) = f_\tau(\tau(s_1),\ldots,\tau(s_n))$, for each function symbol f);
c. a monotonic homomorphism from terms to some well-founded set with the strict subterm property $(f_\tau(\ldots x \ldots)) > x)$ (a homomorphism is *monotonic* with respect to the given ordering $\geq$ if $f_\tau(\ldots x \ldots) \geq f_\tau(\ldots y \ldots)$ whenever $x > y$; it has the *strict subterm property* if $f_\tau(\ldots x \ldots) > x$);
d. a strictly monotonic homomorphism from terms to some well-founded set which has the strict subterm property (it is *strictly monotonic* if $f_\tau(\ldots x \ldots) > f_\tau(\ldots y \ldots)$) whenever $x > y$);
e. a function that extracts the immediate subterm at a specified position (which position can depend on the outermost function symbol of the term);
f. a function that extracts the immediate subterm of a specified rank (the kth largest in the path ordering defined recursively below); or
g. a constant function.

Simple examples of homomorphisms from terms to the natural numbers are size (number of function symbols, including constants), depth (maximum nesting of function symbols), and weight (sum of weights of function symbols). Size and weight are strictly monotonic; depth is monotonic. (The subterm property is guaranteed for strictly monotonic homomorphisms into well-ordered sets [Dershowitz, 1982].)

Definition 2 (General Path Ordering). Let $\tau_0,\ldots,\tau_n$ be termination functions. The induced *path ordering* $\succ$ is as follows:

$$s = f(s_1,\ldots,s_m) \succeq g(t_1,\ldots,t_n) = t$$

if either of the following cases (1 or 2) hold:

(1) $s_i \succeq t$ for some s_i, $i = 1,\ldots,m$; or
(2) $s \succ t_1,\ldots,t_n$ and $\langle \tau_1(s),\ldots,\tau_k(s) \rangle$ is lexicographically greater than or equal to $\langle \tau_1(t),\ldots,\tau_k(t) \rangle$, where function symbols are compared according to their precedence, homomorphic images are compared in the corresponding well-founded ordering, and subterms are compared recursively in $\succ$.

As usual, $s \succ t$ if $s \succeq t$, but $s \not\preceq t$.

Lemma 3. *The path ordering satisfies the strict subterm property* $f(\ldots,s_i,\ldots) \succ s_i$, *for all* i.

Proof. By (1) $f(\ldots,s_i,\ldots) \succeq s_i$, but $s_i \not\succeq f(\ldots,s_i,\ldots)$, since the first part of (2) cannot hold for the ith subterm on the right. □

Thus, the ordering is strict when Case (1) applies, or, for Case (2), if the lexicographic comparison is strictly greater.

Lemma 4. *For the path ordering, $s \succeq t$ implies $s \succ t|_i$ for each proper subterm $t|_i$ of t and implies $u[s] \succ t$ for each immediately enclosing context $u[\cdot]$ of s.*

Lemma 5. *The path ordering is a quasi-ordering.*

Proof. Reflexivity is an easy induction. For transitivity, we show that $s \succeq t \succeq u$ implies $s \succeq u$ and that $s \succeq t \succ u$ or $s \succ t \succeq u$ implies $s \succ u$, simultaneously, by induction on the size of the three terms and a case analysis. This requires the preceding lemma. □

Theorem 6. *Let $\tau_0, \ldots, \tau_{i-1}$ ($i \geq 0$) be monotonic homomorphisms, all but possibly the last strict, and let $\tau_i, \ldots, \tau_k$ be any other kinds of termination functions. A rewrite system terminates if $l\sigma \succ r\sigma$ in the corresponding path ordering $\succ$ for all rules $l \rightarrow r$ and ground substitutions σ, and also $\tau(l\sigma) = \tau(r\sigma)$ for each of the non-monotonic homomorphisms among the τ_i.*

The proof of this theorem is akin to [Kamin and Lévy, 1980] and uses a minimal counter-example argument.

Proof. First we show that

$$s \rightarrow t \text{ and } s \succ t \text{ imply } f(\ldots, s, \ldots) \succeq f(\ldots, t, \ldots),$$

for all terms $s, t, \ldots$ and function symbols f. Then, $l\sigma \succ r\sigma$ will imply a decrease with each rewrite.

Other than for the monotonic homomorphisms, we have $\tau_i(f(\ldots, s, \ldots)) \geq \tau_i(f(\ldots, t, \ldots))$: For τ, a precedence, value-preserving homomorphism, specified subterm, or constant, $s \rightarrow t$ clearly implies $\tau(f(\ldots, s, \ldots)) \geq \tau(f(\ldots, t, \ldots))$ in the relevant ordering. For a τ that extracts the kth largest subterm u of $f(\ldots, s, \ldots)$: if $u \succ s$ or $t \succ u$, then replacing s by t has no impact on rank k and $\tau(f(\ldots, s, \ldots)) = u = \tau(f(\ldots, t, \ldots))$; if $s \succeq u \succeq t$, then $\tau(f(\ldots, s, \ldots)) \succeq \tau(f(\ldots, t, \ldots))$.

Let $s \succ t$ because the τ_i for some subterm $s|_p$ of s are lexicographically greater than for t. If the first point of difference between the τ_i is a strict homomorphism, then this (with the subterm property) implies a strict decrease $\tau_i(f(\ldots, s, \ldots)) \succ \tau_i(f(\ldots, t, \ldots))$ and, therefore, $f(\ldots, s, \ldots) \succ f(\ldots, t, \ldots)$. If it's at the non-strict homomorphism, then $\tau_i(f(\ldots, s, \ldots)) \succeq \tau_i(f(\ldots, t, \ldots))$ and $f(\ldots, s, \ldots) \succeq f(\ldots, t, \ldots)$.

To prove well-foundedness of $\succ$, consider a minimal infinite descending sequence $t_1 \succ t_2 \succ \cdots$, minimal in the sense that from all proper subterms of each term in the example there are only finite descending sequences. (By the subterm property, if $t_j \succ t_{j+1}$ then t_j is also greater than the subterms of t_j.) Case (1) of the definition of $\succ$ could not be the justification for any pair $t_j \succ t_{j+1}$, since then we would have $t_{j-1} \succ t_j|_i \succ t_{j+2}$, for some proper subterm $t_j|_i$ of the jth term in the example, and the example would not be minimal. Since Case (2) uses a lexicographic combination of well-founded orderings (including $\succ$ on proper subterms), it, too, is well-founded, and the descending sequence could not be infinite. □

For System (2), let τ_0 interpret everything naturally: *fact* as factorial, s as successor, p as predecessor, × as multiplication, + as addition, and 0 as zero. Let all constants be interpreted as natural numbers, making all terms non-negative. Let the precedence τ_1 be $\mathit{fact} \succ \times \succ + \succ s$. Each rule causes a strict decrease with respect to $\succ$.

One must also make sure that all terms and subterms in any derivation are interpretable as natural numbers; otherwise a rule like $\mathit{fact}(x) \rightarrow \mathit{fact}(p(x))$ would give pretense of being terminating.

The following orderings are special cases of the general path ordering. For all but one, the conditions of the theorem hold:

Knuth-Bendix ordering [Knuth and Bendix, 1970]. τ_0 gives the sum of (non-negative integer) "weights" of the function symbols appearing in a term; τ_1 gives a (total) precedence; $\tau_2, \ldots, \tau_{n+1}$ give a permutation of the subterms.

Polynomial path ordering [Lankford, 1979]. τ_0 is a strictly monotonic homomorphism (each f_τ is a polynomial with positive coefficients); τ_1 gives a precedence; $\tau_2, \ldots, \tau_{n+1}$ give a permutation of the subterms.

Multiset path ordering [Dershowitz, 1982]. τ_0 is a total precedence; $\tau_1, \ldots, \tau_n$ give the subterms in non-increasing order. (The multiset path ordering is also defined for partial precedences; that would require comparing the τ_i as a multiset, rather than lexicographically .)

Lexicographic path ordering [Kamin and Lévy, 1980]. τ_0 is a precedence; $\tau_1, \ldots, \tau_n$ give a permutation of the subterms.

Semantic path ordering [Kamin and Lévy, 1980; Plaisted, 1979]. τ_0 is the identity function (a non-monotonic homomorphism), with terms compared in some well-founded ordering; τ_1 gives a precedence; $\tau_2, \ldots, \tau_{n+1}$ give a permutation of the subterms. *(For this ordering, one must separately insure that $s \rightarrow t$ implies $\tau_0(s) \geq \tau_0(t)$.)*

Recursive path ordering [Lescanne, 1990]. τ_0 is a total precedence; $\tau_1, \ldots, \tau_n$ give a permutation of the subterms or give the subterms in non-increasing order, depending on the function symbol.

Extended Knuth-Bendix ordering [Dershowitz, 1982; Steinbach and Zehnter, 1990]. τ_0 is a monotonic interpretation; τ_1 gives a precedence; $\tau_2, \ldots, \tau_{n+1}$ give the subterms in order, permuted, or sorted, depending on the function symbol.

For a system like

$$\begin{aligned} f(s(x)) &\rightarrow s(h(d(f(x)))) \\ f(0) &\rightarrow 0 \\ d(0) &\rightarrow 0 \\ d(s(x)) &\rightarrow s(s(d(x))) \\ h(s(s(x))) &\rightarrow s(h(x)) \, , \end{aligned} \qquad (3)$$

a precedence ($f > h > d > s$) ought to be considered first, before looking at subterms, as with a lexicographic path ordering. In a system like

$$\begin{aligned} f(s(x)) &\to p(s(f(f(x)))) \\ f(0) &\to 0 \\ p(s(x)) &\to x \, , \end{aligned} \qquad (4)$$

with nested defined symbols on the right, an interpretation ($f_\tau(x) = 0, s_\tau(x) = x+1, p_\tau(x) = x-1$) could be considered first, followed by a precedence ($f > s, p$), as with an extended Knuth-Bendix ordering. (With $f(0) \to s(0)$, instead of 0, the system would be nonterminating.)

In the appendix, we describe how an implementation of this ordering performs on a sorting example.

3 Orthogonal systems

Consider a recursive definition like

$$f(x) = \text{if } x > 0 \text{ then } f(f(x-1)) + 1 \text{ else } 0 \, .$$

By a straightforward use of structural induction, one can prove that the least fixpoint (over the natural numbers) is the always-defined identity function. This definition translates into the rewrite system:

$$\begin{aligned} f(s(x)) &\to s(f(f(p(s(x))))) \\ f(0) &\to 0 \\ p(s(x)) &\to x \, . \end{aligned} \qquad (5)$$

It would be nice to be able to mimic the proof for the recursive function definition in the rewriting context, but several issues arise:

1. One cannot use a syntactic simplification ordering like the simple path ordering [Plaisted, 1978], since the first rule is embedding. In fact, we must combine termination with the semantics ($f(x) = x$), as one must for the functional proof.
2. In the functional case, one can show that call-by-value terminates, which implies that all fixpoint computation rules also terminate. We will see under what conditions the same holds for rewriting.
3. For rewriting in general, one must consider the possibility that the x in the definition of $f(x)$ us itself a term containing occurrences of the defined function f (or of mutually recursive defined functions), something usually ignored in the functional case.

   Consider the system:

$$\begin{aligned} f(s(x)) &\to s(f(p(s(x)))) \\ f(0) &\to 0 \\ p(s(x)) &\to x \, . \end{aligned} \qquad (6)$$

The general path ordering works with a natural interpretation of the argument of f and a precedence $f > s, p$.

Alternatively, one can employ the following result:

Proposition 7 [O'Donnell, 1977]. *A non-erasing orthogonal system is terminating if and only if every term has a normal form.*

Therefore, the offending rule may be immediately followed by an application of the last rule, effectively replacing the former with $f(s(x)) \to s(f(x))$. Now termination can be shown with a standard recursive path ordering, demonstrating that the orginal system is normalizing, and, hence, terminating.

This method does not apply to a system like

$$\begin{aligned} x \times 0 &\to 0 \\ x \times s(y) &\to (x \times y) + x \\ x + 0 &\to x \\ x + s(y) &\to s(x+y) \, , \end{aligned} \quad (7)$$

with its erasing rule (the first one).

Still, we can employ the following:

Proposition 8 [Gramlich, 1992]. *A locally confluent overlaying system is terminating if and only if innermost rewriting always leads to a normal form.*

An *overlaying* system is one whose only critical pairs are obtained from an overlap at the topmost position. In particular, orthogonal systems are locally confluent and have no (non-trivial) critical pairs; the proposition for this case was shown in [O'Donnell, 1977].

We turn now to the question of when termination of ground constructor instances of left-hand sides suffices for establishing termination in all cases.

Definition 9 [Dershowitz, 1981]. The set of *forward closures* for a given rewrite system is inductively defined as follows:

- Every rule $l \to r$ is a forward closure.
- If $c \to c'$ and $d \to d'$ are forward closures such that $c' = u[s]$ for nonvariable s and $s\mu = d\mu$ for most general unifier μ, then $c\mu \to u\mu[d'\mu]$ is also a forward closure.

The idea is to restrict application of rules to that part of a term created by previous rewrites. In the same way, we can define *innermost* and *outermost* forward closures—restricting the position at which unification is performed so that the derivations captured by closure are of the desired type.

Proposition 10 [Geupel, 1989]. *A non-overlapping rewrite system is terminating if, and only if, no right-hand side of a forward closure initiates an infinite derivation.*

In general, though, a term-rewriting system need not terminate even if all its forward closures do [Dershowitz, 1981].

Consider the following system for symbolic differentiation with respect to t (proving termination of the first five of these rules was one of the problems on a qualifying exam given at Carnegie-Mellon University in 1967):

$$\begin{aligned}
D_t\, t &\rightarrow 1 \\
D_t\, a &\rightarrow 0 \\
D_t\, (x+y) &\rightarrow D_t\, x + D_t\, y \\
D_t\, (x \cdot y) &\rightarrow y \cdot D_t\, x + x \cdot D_t\, y \\
D_t\, (x-y) &\rightarrow D_t\, x - D_t\, y \\
D_t\, (-x) &\rightarrow -D_t\, x \\
D_t\, (x/y) &\rightarrow D_t\, x/y - x \cdot D_t\, y/y^2 \\
D_t\, (\ln x) &\rightarrow D_t\, x/x \\
D_t\, (x^y) &\rightarrow y \cdot x^{y-1} \cdot D_t\, x + x^y \cdot (\ln x) \cdot D_t\, y \, ,
\end{aligned} \qquad (8)$$

where a is any constant symbol other than t. It is orthogonal (hence, non-overlapping), so the above method applies. Since D's are not nested on the right, forward closures cannot have nested D's. Since the arguments to D on the left are always longer than those on the right, all forward closures must lead to terminating derivations; hence, regardless of the rewriting strategy and initial term, rewriting terminates.

For a system like

$$\begin{aligned}
f(s(x)) &\rightarrow s(s(f(p(s(x))))) \\
f(0) &\rightarrow 0 \\
p(s(x)) &\rightarrow x \, ,
\end{aligned} \qquad (9)$$

we can also restrict our attention to forward closures. Since f's won't nest, termination can be shown by comparing the argument on the left, $s(x)$, with the one on the right, $p(s(x))$. This time we need to use a semantic comparison, making the left argument always larger.

Theorem 11. *A locally-confluent overlaying rewrite system is terminating if, and only if, no right-hand side of an innermost forward closure initiates an infinite derivation.*

In particular, orthogonal systems satisfy the prerequisites for application of this termination test; one need only prove termination of such innermost derivations.

The proof is similar to [Geupel, 1989]:

Proof. Consider a minimal example of nontermination $t_1 \rightarrow t_2 \rightarrow \cdots$, minimal in the sense that at each point any rewrite lower down in the term than the redex in the example would have to lead to a normal form. Replace the largest terminating subterms of each t_i with their unique normal form (which they have by local confluence). The fact that all overlaps occur at the top ensures that none of these replacements prevents application of a rule above the replaced terms. Hence, the result is an infinite derivation with the desired characteristics. □

This method applies to Systems 2 and 5: Since we need only consider innermost derivations, we can assume that the problematic $p(s(x))$ on the right rewrites immediately to x (and that the x is in normal form).

Suppose an orthogonal system is constructor-based, that is, all proper subterms of left-hand sides have only free constructors and variables. All its forward closures begin with constructor-based instances of left-hand sides. Thus, termination proofs need not consider initial terms containing nested defined function symbols (even when the symbol is not completely defined). That makes proving termination of such systems no more difficult than proving termination of ordinary recursive functions: the instances of rule variables can be presumed to be in normal form and the context can be ignored.

For System 5, say, we can compare the multiset of right-hand side arguments of the (mutually-)recursive function symbols $\{f(p(s(x))), p(s(x))\}$ with that of left-hand side, $\{s(x)\}$. Semantics are necessary for this comparison. If we let $p(s(x)) \to x$ and $f(x) \to x$, we have $\{s(x)\}$ greater (in the multiset ordering) than $\{x, x\}$. But one must ensure that the semantics are consistent with the rules (which is analogous to showing that $f(x) = x$ is a fixpoint of the definition). This can be done using standard rewriting technique ("proof by consistency").

It is instructive to compare the above examples with the following nonterminating rewrite system:

$$\begin{aligned} f(s(x)) &\to s(s(f(f(p(s(x)))))) \\ f(0) &\to 0 \\ p(s(x)) &\to x \,. \end{aligned} \qquad (10)$$

It is the rewriting analogue of the recursively-defined function

$$f(x) = \text{if } x > 0 \text{ then } f(f(x-1)) + 2 \text{ else } 0 \,,$$

which does not terminate for 2. Indeed, $f(x) = x$ would be inconsistent with the rules.

The above results can be used to prove termination of systems that can be decomposed into two terminating systems that do not share defined symbols.

Proposition 12 [Dershowitz, 1993]. *Let R contain defined symbols and free constructors, and S contain defined symbols from a disjoint set of defined symbols and from the same set of constructors. If R and S are each non-overlapping and terminating, then so is their union.*

4 String rewriting

Proposition 13 [Dershowitz, 1981]. *A right-linear rewrite system is terminating if, and only if, no right-hand side of a forward closure initiates an infinite derivation.*

In particular, forward closures suffice for string-rewriting systems. String systems are also non-erasing.

Zantema's Problem (circulated via electronic mail) is to prove termination of the following one-rule string-rewriting system:

$$1100 \to 000111 \,. \qquad (11)$$

It provides a nice example of termination proofs based on an analysis of restricted derivations.

Suppose it is nonterminating. Consider a minimal infinite derivation

$$t_1 \to t_2 \to \cdots ,$$

minimal in the sense that no substring of t_1 is nonterminating and there is no infinite derivation from t_1 taking place at higher positions (further left). More specifically, $t_1 \to t_2$ takes place at the top (leftmost symbol) and among all infinite derivations beginning $t_1 \to t_2 \to \cdots \to t_i$, none starts higher than does $t_i \to t_{i+1} \to \cdots$.

Divide each string t_i into three parts (from left to right): dead, active, and passive. The dead part never develops a redex; the passive part is a residual substring of the initial string which has not yet been touched; the active part contains letters introduced by right-hand sides. The dead part is in normal form and for (11) always ends in 000.

To start off, t_1 is all passive, except for its first letter. This minimal derivation must be leftmost (outermost). Suppose this were not the case. Either the outer redex is eventually rewritten, or it never is. In the former case, the derivation

$$t_1 \to \cdots \to u1100v \to u1100v' \to \cdots \to u1100v'' \to u000111v'' \to \cdots,$$

where $v \to v' \to \cdots \to v''$, can be rearranged to

$$t_1 \to \cdots \to u1100v \to u000111v \to u000111v' \to \cdots \to u000111v'' \to \cdots,$$

and, therefore, is not minimal. In the latter case, rewriting the outer redex doesn't preclude nontermination, and the smaller alternative is also nonterminating.

Similarly, redexes are always in the active part. For suppose the minimal derivation did have some steps in the passive part. There would have to be a subsequent step in the active part (or else that passive proper substring of t_1 would be nonterminating), which is perhaps enabled by the step in the passive part:

$$t_1 = sw \to \cdots \to uvw \to uvw' \to \cdots \to uvw'' \to uv'w'' \to \cdots,$$

where u is dead, v is active, and w is passive. Since the alternate derivation

$$sw'' \to \cdots \to uvw'' \to uv'w'' \to \cdots$$

(starting out after the rewriting of the passive part) is smaller (the v redex is higher up than the w one), the given derivation can not be minimal.

More generally:

Proposition 14. *A non-erasing orthogonal system terminates if and only if no right-hand side of an outermost forward closure initiates an infinite derivation.*

System (9) is of this form (all its forward closures are outermost anyway.)

For this specific system, we need only consider three active parts: 111, 1110111, or 1111110111, since it takes only finitely many steps to get from one of these to another. Call these states A, B, and C, respectively.

For there to be a redex in the active part, the passive part must begin with 00 or with 100. The leftmost derivations (with redex underlined, and dead parts bracketed) of the six cases are shown in Fig. 1. In each case, termination follows from the fact that the passive part decreases in size.

The same approach works for other examples of the form $1^i 0^j \to 0^k 1^l$.

$A00 = 11\underline{1100} \to [1000]111 \in A$

$B00 = 11101\underline{1100} \to [11101000]111 \in A$

$C00 = 1111110\underline{1100} \to [11111101000]111 \in A$

$A100 = 11\underline{1100} \to \underline{11000}111 \to [000]1110111 \in B$

$B100 = 1110\underline{11100} \to 11101\underline{1000}111 \to 11\underline{1100}001110111 \to [1000]\underline{111001}110111 \to [10001000]1111110111 \in C$

$C100 = 111111011\underline{1100} \to 111111011\underline{000}111 \to 11111\underline{1100}001110111 \to 111\underline{1100}0111001110111 \to 1\underline{1100}01110111001110111 \to [000]1110111011001110111 \to [00011101110100]1111110111 \in C$

Fig. 1. Derivations for Zantema's problem

Appendix

Our general path ordering termination code (GPOTC) is implemented in Common Lisp on a Macintosh. (No special features of Macintosh Common Lisp were used, so the code should be capable of running under any Common Lisp with just a few minor changes.)[2] The implementation supports termination functions for precedence, term extraction (given, minimum, and maximum), and homomorphisms.

Interpretations involving addition, multiplication, negation, and exponentiation are expressible. Currently, the burden of proving that functions are either value-preserving or monotonic is placed on the user. As is usual for such functions, one often ends up needing to know if a given function is positive over some range. When the functions are rational polynomials, this is decidable, but time consuming. Our code does not attempt a full solution, but merely applies some quick and dirty heuristics, such as testing the function at endpoints and checking coefficients of polynomials. In cases where the code cannot make a determination, it will query the user for an authoritative answer. The part of the code that does this testing could be upgraded to provide heuristics such as those described in [Lankford, 1979; Ben Cherifa and Lescanne, 1987; Steinbach and Zehnter, 1990]. We are also in the process of implementing Paul Cohen's decision procedure [Cohen, 1969] for the first-order theory of real polynomials within Mathematica®.

The following brief example shows the use of GPOTC. The rewrite rules in Fig. 2 are an implementation of insertion sort over the natural numbers. The function choose is used to determine whether X should be inserted before or after the first element of the list which is the second argument to insert. Rule 2, for example, would be defined for the system as follows:

---

[2] Those interested in obtaining a copy of GPOTC should send electronic mail to hoot@cs.uiuc.edu.

```
(setf ins2 (make-production :lhs '(!Sort (!Cons ?X ?Y))
                            :rhs '(!Insert ?X (!Sort ?Y)))) .
```

The characters "!" and "?" are macro symbols indicating symbols and variables, respectively.

```
RULE 1:  sort(nil)  -->  nil
RULE 2:  sort(cons(X, Y))  -->  insert(X, sort(Y))
RULE 3:  insert(X, nil)  -->  cons(X, nil)
RULE 4:  insert(X, cons(V, W))  -->  choose(X, cons(V, W), X, V)
RULE 5:  choose(X, cons(V, W), 0, 0)  -->  cons(X, cons(V, W))
RULE 6:  choose(X, cons(V, W), s(P), 0)  -->  cons(X, cons(V, W))
RULE 7:  choose(X, cons(V, W), 0, s(Q))  -->  cons(V, insert(X, W))
RULE 8:  choose(X, cons(V, W), s(P), s(Q))  -->  choose(X, cons(V, W), P, Q)
```

Fig. 2. Rules for insertion sort.

The code for creating the ordering is

```
(setf SymOrd1 '(!Sort !Insert !Choose !Cons))
(setf SymOrd2 '(!Sort (!Insert !Choose) !Cons))
(makeorder ord1
  (list
    (make_prec_tau SymOrd2)
    (make_subterm_tau ((!Sort 1) (!Choose 2) (!Insert 2)) ord1)
    (make_prec_tau SymOrd1)
    (make_subterm_tau ((!Sort 1) (!Choose 3) (!Insert 2)) ord1)
  ))
```

Three termination functions are used; they are lexicographically compared from first to last. The macro **make_prec_tau** creates a precedence ordering based on its argument; **make_subterm_tau** ((f n) ...) ord1 extracts the nth subterm for function symbol f and compares it using the ordering **ord1**. The **makeorder** macro creates a function with the name of the first argument which accepts two terms (s and t) and may return one of three values: Ge ($s \succeq t$), Gr ($s \succ t$), or Un (unknown).

If one uses a precedence ordering based on **SymOrd1**, all of the rules except for Rule 7 would be oriented in the appropriate direction. Unfortunately, Rules 4 and 7 interact with each other. In particular, there is a **choose** and an **insert** on opposite sides of each rule. The precedence order **SymOrd2** with (sort $\succ$ insert = choose $\succ$ cons) is chosen to guarantee that the lexicographical ordering of the terms in Rule 7 is from left to right, while leaving Rule 4 equal. This means that the left-hand side of Rule 7 is compared with each of the two subterms on the right. The comparison of interest is choose(X, cons(V, W), 0, s(Q)) with insert(X, W). These terms are equal under the precedence ordering **SymOrd2**, but by selecting the second subterm, the subterms cons(V, W) and W are recursively compared giving the necessary decrease. Fortunately, the second subterm on both sides of Rule 4 is identical, leaving the lexicographical ordering unaffected. The precedence ordering SymOrd1 with (sort ≻ insert ≻ choose ≻ cons) breaks the tie, and all that remains is to verify that the left-hand side of Rule 4 is greater than the subterms on the right.

The code in Fig. 3 shows an example of a monotonic homomorphism where $F_f(X) = 2X + 4$, $F_g(X, Y) = 3Y + 6$, $F_a = 0$ and $F_b = 1$. The macro make-fn accepts a list of symbols and their associated functions. Notice that the expressions are essentially the equivalent Lisp expressions with (arg n) giving the nth argument.

```
(setq example-FNtau
     (make-fn ((!f (+ (* 2 (arg 1)) 4))
      (!g (+ (* 3 (arg 2)) 6))
      (!a 0)
      (!b 1))))
```

Fig. 3. Example code for creating a function τ.

To apply the ordering function ord to each of the rules in the list InsSort (containing the six rules in Fig. 2), one issues the command

(term-cond InsSort #'ord1) ,

with the result

(:GR :GR :GR :GR :GR :GR :GR :GR) .

Figure 4 displays the justification for Rule 4. The system is able to determine that insert(X, cons(V, W)) is greater than choose(X, cons(V, W), X, V) by first showing that insert(X, cons(V, W)) is strictly greater than each of the subterms of the right-hand side. These sub-proofs (for X, cons(V, W), and V) are all similar: a sub-term of the left-hand side is found to be syntactically equal to the right-hand side, and Case (1) of the path ordering applies. Showing the lexicographic part of the ordering comes next: one of the termination functions must show a strict increase. The first two do not result in a strict decrease (they are equal). The third, however, compares insert with compare in the precedence given by SymOrd1 where there is the desired strict decrease. That concludes Case (2) of Definition 2, showing that insert(X, cons(V, W)) is strictly greater than choose(X, cons(V, W), X, V).

References

[Ben Cherifa and Lescanne, 1987] Ahlem Ben Cherifa and Pierre Lescanne. Termination of rewriting systems by polynomial interpretations and its implementation. *Science of Computer Programming*, 9:137–159, 1987.

[Burstall, 1969] Robert M. Burstall. Proving properties of programs by structural induction. *Computing J.*, 12(1):41–48, February 1969.

```
(term-cond (list ins4) #'ord1 :keep-causes t)
((:GR
 insert(X, cons(V, W)) > choose(X, cons(V, W), X, V) by case (2)
 Case 2a: Check that the LHS > all subterms of the RHS:
 | insert(X, cons(V, W)) > X by case (1)
 | | X is syntactically equal to term X
 | |
 | insert(X, cons(V, W)) > cons(V, W) by case (1)
 | | cons(V, W) is syntactically equal to term cons(V, W)
 | |
 | insert(X, cons(V, W)) > X by case (1)
 | | X is syntactically equal to term X
 | |
 | insert(X, cons(V, W)) > V by case (1)
 | | cons(V, W) >= V by case (1)
 | | | V is syntactically equal to term V
 Case 2b: Check that the LHS > RHS via lexicographic comparison:
 | 1:insert(X, cons(V, W)) >= choose(X, cons(V, W), X, V) by basic ordering
                                                            of a precedence tau
 | |
 | 2:immediate subterms insert|2 with choose|2: cons(V, W) >= cons(V, W)
 | | cons(V, W) is syntactically equal to term cons(V, W)
 | |
 | 3:insert(X, cons(V, W)) > choose(X, cons(V, W), X, V) by basic ordering
                                                            of a precedence tau
))
```

Fig. 4. Proof for a single rule.

[Cohen, 1969] Paul J. Cohen. Decision procedures for real and p-adic fields. *Comm. Pure and Applied Math*, 22(2): 279–301, March 1969.

[Dershowitz, 1981] Nachum Dershowitz. Termination of linear rewriting systems. In *Proceedings of the Eighth International Colloquium on Automata, Languages and Programming*, pages 448–458, Acre, Israel, July 1981. European Association of Theoretical Computer Science. Vol. 115 of *Lecture Notes in Computer Science*, Springer-Verlag, Berlin.

[Dershowitz, 1982] Nachum Dershowitz. Orderings for term-rewriting systems. *Theoretical Computer Science*, 17(3):279–301, March 1982.

[Dershowitz, 1987] Nachum Dershowitz. Termination of rewriting. *J. of Symbolic Computation*, 3(1&2):69–115, February/April 1987. Corrigendum: *4*, 3 (December 1987), 409–410.

[Dershowitz, 1993] Nachum Dershowitz. Hierarchical termination. Technical Report, Leibnitz Center for Research in Computer Science, Hebrew University, Jerusalem, Israel.

[Dershowitz and Manna, 1979] Nachum Dershowitz and Zohar Manna. Proving termination with multiset orderings. *Communications of the ACM*, 22(8):465–476, August 1979.

[Geupel, 1989] Oliver Geupel. Overlap closures and termination of term rewriting systems. Report MIP-8922, Universität Passau, Passau, West Germany, July 1989.

[Gramlich, 1992] Bernhard Gramlich. Relating innermost, weak, uniform and modular termination of term rewriting systems. *Proceedings of the Conference on Logic Programming and Automated Reasoning*, pp. 285–296, St. Petersburg, Russia, July 1992. Vol. 624 of *Lecture Notes in Computer Science*, Springer-Verlag, Berlin.

[Kamin and Lévy, 1980] Sam Kamin and Jean-Jacques Lévy. Two generalizations of the recursive path ordering. Unpublished note, Department of Computer Science, University of Illinois, Urbana, IL, February 1980.

[Knuth and Bendix, 1970] Donald E. Knuth and P. B. Bendix. Simple word problems in universal algebras. In J. Leech, editor, *Computational Problems in Abstract Algebra*, pp. 263–297. Pergamon Press, Oxford, U. K., 1970.

[Lankford, 1979] Dallas S. Lankford. On proving term rewriting systems are Noetherian. Memo MTP-3, Mathematics Department, Louisiana Tech. University, Ruston, LA, October 1979.

[Lescanne, 1990] Pierre Lescanne. On the recursive decomposition ordering with lexicographical status and other related orderings. *J. Automated Reasoning*, 6:39–49, 1990.

[Lipton and Snyder, 1977] R. Lipton and L. Snyder. On the halting of tree replacement systems. In *Proceedings of the Conference on Theoretical Computer Science*, pp. 43–46, Waterloo, Canada, August 1977.

[Manna, 1974] Zohar Manna. *Mathematical Theory of Computation*. McGraw-Hill, New York, 1974.

[Manna and Ness, 1970] Zohar Manna and Steven Ness. On the termination of Markov algorithms. In *Proceedings of the Third Hawaii International Conference on System Science*, pp. 789–792, Honolulu, HI, January 1970.

[O'Donnell, 1977] Michael J. O'Donnell. *Computing in systems described by equations*, volume 58 of *Lecture Notes in Computer Science*. Springer, Berlin, West Germany, 1977.

[Plaisted, 1978] David A. Plaisted. Well-founded orderings for proving termination of systems of rewrite rules. Report R-78-932, Department of Computer Science, University of Illinois, Urbana, IL, July 1978.

[Plaisted, 1979] David A. Plaisted. Personal communication, 1979.

[Steinbach and Zehnter, 1990] Joachim Steinbach and Michael Zehnter. Vade-mecum of polynomial orderings. Report SR-90-03, Fachbereich Informatik, Universität Kaiserslautern, Kaiserslautern, West Germany, 1990.

Total Termination of Term Rewriting

M. C. F. Ferreira* and H. Zantema**

Utrecht University, Department of Computer Science
P.O. box 80.089, 3508 TB Utrecht, The Netherlands

Abstract. We investigate proving termination of term rewriting systems by interpretation of terms in a compositional way in a total well-founded order. This kind of termination is called *total termination*. On one hand it is more restrictive than simple termination, on the other it generalizes most of the usual techniques for proving termination. For total termination it turns out that below ϵ_0 the only orders of interest are built from the natural numbers by lexicographic product and the multiset construction. By examples we show that both constructions are essential. For a wide class of term rewriting systems we prove that total termination is a modular property. Most of our techniques are based on ordinal arithmetic.

1 Introduction

One of the main problems in the theory of term rewriting systems (TRS) is the detection of termination: for a fixed system of rewrite rules, detect whether there exist infinite rewrite chains or not. In general this problem is undecidable ([6]). However, there are several methods for deciding termination that are successful for many special cases. Roughly these methods can be divided into two main types: *syntactical* methods and *semantical* methods. In a syntactical method terms are ordered by a careful analysis of the term structure. A well-known representative of this type is the *recursive path order* ([2]). All of these orderings are simplification orderings, i.e., a term is always greater than its proper subterms. An overview and comparison of simplification orderings is given in [13].

Here we focus on a semantical method: terms are interpreted compositionally in some well-founded ordered set. This is done in such a way that each rewrite chain will map to a descending chain, and hence will terminate. The general framework has been introduced in [14]. One problem is how to choose a suitable well-founded ordered set. The variation among well-founded ordered sets is so unwieldy that some restriction is reasonable. A natural way is the restriction to total orders: then the ordered sets correspond to ordinal numbers, having a very elegant structure that has been studied extensively in the past. This kind of termination of term rewriting systems is called *total termination*.

Total termination turns out to be a slightly stronger restriction than simple termination. However, most of the general techniques of proving termination like

---

* phone +31-30-532249, e-mail maria@cs.ruu.nl
** phone +31-30-534116, e-mail hansz@cs.ruu.nl

polynomial interpretations ([9, 1]), *elementary interpretations* ([10]), recursive path order (RPO) with status and Knuth-Bendix order (KBO) with status, all fit in the notion of total termination.

This paper is an investigation of total termination, in particular of which totally ordered sets are useful. One of the main conclusions is that apart from some minor exceptions only ordinals of the shape ω^α are of interest. The basic observation leading to this result is the following. The existence of a binary operation in a total well-founded order that is strictly monotonic in both coordinates implies that the order type is ω^α. Stated without ordinals this means that the order is isomorphic to the finite multisets over another order. Below the ordinal ϵ_0 this implies that all totally ordered sets of interest can be constructed from the natural numbers in finitely many steps using only lexicographic product and the multiset construction. We show that these constructions are essential by presenting examples of TRS's for which a termination proof can be given (by an interpretation) in ω^η, for any fixed $\eta \leq \omega$, but not in a totally ordered set of a smaller order type.

Another main topic of this paper is the modularity of total termination. Surprisingly the tree structure of mixed terms that is essential in other modularity questions ([12]) does not play a role here. The essential problem is how to lift an interpretation in an ordinal to an interpretation in a greater ordinal without affecting monotonicity and compatibility. We did not succeed in proving modularity of total termination in full generality. However, we found some interesting partial results. For example, if two systems are totally terminating and not both of them contain duplicating rules, then the direct sum is also totally terminating.

2 Monotone Algebras

Let $\mathcal{F}$ be a set of operation symbols each having a fixed arity. We define a *well-founded monotone $\mathcal{F}$-algebra* $(A, >)$ to be an $\mathcal{F}$-algebra A for which the underlying set is provided with a well-founded order $>$ and each algebra operation is monotone[3] in all of its coordinates, more precisely: for each operation symbol $f \in \mathcal{F}$ and all $a_1, \ldots, a_n, b_1, \ldots, b_n \in A$ for which $a_i > b_i$ for some i, and $a_j = b_j$ for all $j \neq i$, we have $f_A(a_1, \ldots, a_n) > f_A(b_1, \ldots, b_n)$.

Let $(A, >)$ be a well-founded monotone $\mathcal{F}$-algebra. Let $A^{\mathcal{X}} = \{\sigma : \mathcal{X} \to A\}$. We define $\phi_A : \mathcal{T}(\mathcal{F}, \mathcal{X}) \times A^{\mathcal{X}} \to A$ inductively by

$$\phi_A(x, \sigma) = \sigma(x),$$
$$\phi_A(f(t_1, \ldots, t_n), \sigma) = f_A(\phi_A(t_1, \sigma), \ldots, \phi_A(t_n, \sigma))$$

for $x \in \mathcal{X}, \sigma : \mathcal{X} \to A, f \in \mathcal{F}, t_1, \ldots, t_n \in \mathcal{T}(\mathcal{F}, \mathcal{X})$. This function induces a partial order $>_A$ on $\mathcal{T}(\mathcal{F}, \mathcal{X})$ as follows:

$$t >_A t' \iff (\forall \sigma \in A^{\mathcal{X}} : \phi_A(t, \sigma) > \phi_A(t', \sigma)).$$

---

[3] By monotone we mean *strictly increasing*.

Intuitively $t >_A t'$ means that for each interpretation of the variables in A the interpreted value of t is greater than that of t'.

We say that a non-empty well-founded monotone algebra $(A, >)$ *normalizes* a TRS if $l >_A r$ for every rule $l \to r$ of the TRS. This terminology is motivated by the following proposition.

Theorem 1. *A TRS is terminating if and only if it is normalized by a non-empty well-founded monotone algebra.*

For the proof we refer to [14]. The way of proving termination of a TRS is now as follows: choose a well-founded poset A, define for each operation symbol a corresponding operation that is strictly monotone in all of its coordinates, and for which $\phi_A(l, \sigma) >_A \phi_A(r, \sigma)$ for all rewrite rules $l \to r$ and all $\sigma : \mathcal{X} \to A$. Then according to the above proposition the TRS is terminating. A typical example is the system

$$f(f(x,y), z) \to f(x, f(y, z)).$$

Choose $(A, >) = (\mathbb{N}, >)$ where $\mathbb{N}$ is defined to be the set of strictly positive integers, and choose $f_A(x, y) = 2x + y$. Clearly f_A is strictly monotone in both coordinates, and

$$f_A(f_A(x, y), z) = 4x + 2y + z > 2x + 2y + z = f_A(x, f_A(y, z))$$

for all $x, y, z \in A$. Hence $f(f(x,y), z) >_A f(x, f(y, z))$, proving termination.

Definition 2. A TRS is called *totally terminating* if it is normalized by a non-empty well-founded monotone algebra in which the underlying order is total.

Every totally terminating TRS allows a simplification order (as defined in [3] and many other texts); in fact this follows from lemma 3 presented below. The converse does not hold, for example, termination of the system

$$f(a) \to f(b) \; ; \; g(b) \to g(a)$$

is easily proved by a simplification order, but the system is not totally terminating since the interpretations of a and b have to be incomparable.

However, most of the existing methods of proving termination of TRS also prove total termination. By definition the methods of polynomial interpretations ([9, 1]) and elementary interpretations ([10]) are nothing else than our approach in which A is chosen to be the naturals and the operations have a particular shape. Hence a termination proof by these interpretations implies total termination. The same can be said for recursive path order and Knuth-Bendix order, both with status. Here we choose A to be the set of ground terms modulo some congruence. If there are no constants, one constant can be added to force the existence of ground terms. The congruence is generated by interchanging the arguments of the operations that have multiset status. The order on these congruence classes is defined by the RPO or KBO itself, where the precedence is extended to a total precedence. For both RPO and KBO it can be proved by induction on the size of the terms that any two terms, modulo this congruence, are comparable. As a consequence, the orders are total and prove total termination. For a finite TRS proved terminating by recursive path order with only multiset status, Hofbauer ([5]) proved that a proof of total termination can be given in the natural numbers with primitively recursive operations.

A main topic of this paper is the investigation of useful total orders for total termination. The main tool is the arithmetic of ordinals, i.e., of total well-founded orders modulo order-isomorphism. We say that a proof of total termination is in an ordinal α if the underlying order of the monotone algebra has order type α. Since in this algebra we allow all possible monotone functions this does not mean that the proof can be given in α in the proof-theoretical sense. For example, the term rewriting system describing the Ackermann function can be proven terminating by a monotone algebra of which the underlying order corresponds to the natural numbers, so in our sense its termination proof is in ω. Another approach connecting termination orderings and ordinals is given by [11].

In the next section we summarize notions and results from ordinal arithmetic we need. For many of the proofs we refer to [7].

3 Tools from Ordinal Theory

A total well-founded order is called a *well-order*. In a well-order every non-empty subset has a minimal element. A simple but useful lemma is the following.

Lemma 3. *Let $\mathcal{A} = (A, >)$ be well-ordered and let $f : A \to A$ be any monotone function. Then $f(x) \geq x$ for every $x \in A$.*

Proof. Suppose there is $x \in A$ such that $x > f(x)$. Monotonicity of f leads to an infinite decreasing sequence: $x > f(x) > f(f(x)) > f(f(f(x))) > \ldots$; contradicting well-foundedness. $\square$

Two ordered sets are called *similar* if they are *order-isomorphic*, i. e. there is a monotone bijection between them. Since monotonicity implies injectivity we have:

Lemma 4. *Let $\mathcal{A}$ and $\mathcal{B}$ be totally ordered sets and $f : \mathcal{A} \to \mathcal{B}$. Then f is monotone and surjective $\iff$ f is an order-isomorphism.*

Similarity classes of well-orders are called *ordinal numbers* (or for short *ordinals*). For finite well-ordered sets their ordinals coincide with their cardinality and are denoted by natural numbers. The ordinal corresponding to a well-ordered set is called its *order type* or *type*.

A proper subset X of a well-order $\mathcal{A} = (A, >)$ is called an *initial segment* of $\mathcal{A}$ if $\forall x \in X \, \forall y \in A \, (y < x \Rightarrow y \in X)$. Equivalently X is an initial segment of $\mathcal{A}$ if and only if $X = \{y \mid y < x\}$ for some $x \in \mathcal{A}$.

Theorem 5. *Let $\mathcal{A}$ and $\mathcal{B}$ be well-ordered sets. Then either $\mathcal{A}$ is similar to $\mathcal{B}$ or $\mathcal{A}$ is similar to an initial segment of $\mathcal{B}$ or $\mathcal{B}$ is similar to an initial segment of $\mathcal{A}$.*

Let Ord denote the class of ordinal numbers and define a relation $<$ on Ord by: $\alpha < \beta \iff$ any set of type α is similar to an initial segment of a set of type β. From theorem 5 follows that $<$ totally orders Ord.

An ordinal is an equivalence class and it is convenient to describe it by a canonical representative of this class. If $(\mathcal{A}, >)$ has type α, it can be seen that $\mathcal{A}$ is similar to the set $\{\beta \in Ord \mid \beta < \alpha\}$. We choose this set to be the canonical representative. As a consequence we have: $\beta < \alpha \iff \beta \in \alpha \iff \beta \subset \alpha$. We shall freely switch between the class and the canonical representative. We list below some basic properties of Ord.

I. $<$ well-orders the class Ord, that is:
  - $<$ is a total ordering in Ord.
  - Every non-empty class $B \subseteq Ord$ has a minimal element in B.
  - For every $\alpha \in Ord$, $\{\xi \in Ord | \xi < \alpha\}$ is a set.

II. For every set of ordinals U there is an ordinal α such that $\alpha = sup(U) = \bigvee U$. If $U = \{f(\xi) | p(\xi)\}$ (for any predicate p) we sometimes use the notation $\bigvee_{p(\xi)} f(\xi)$.

III. $W(\alpha) = \{\xi \mid \xi < \alpha\}$ is well-ordered and has type α.

The second condition in **I** above is equivalent to the principle of transfinite induction that we will use in some proofs.

The ordinal 0 is defined to be the minimal element of Ord; it is the type of the empty set. For every ordinal ξ, its *successor* ξ' is defined by $\xi' = min\{\alpha \mid \xi < \alpha\}$. We use the notation $0' = 1$, $1' = 2$, and so forth. We will sometimes denote the successor ordinal by $\xi + 1$. Clearly $\xi < \xi'$ and there is no ordinal α such that $\xi < \alpha < \xi'$.

An ordinal ξ is defined to be a limit ordinal if

$$(\exists \alpha < \xi) \wedge (\forall \alpha < \xi \ \exists \eta < \xi : \alpha < \eta)$$

The first condition states that a limit ordinal is non-empty, and the second condition says that it has no maximal element. An ordinal ξ is a limit ordinal if and only if $\alpha < \xi \Rightarrow \alpha' < \xi$, if and only if $\xi = \bigvee_{\alpha<\xi} \alpha$. The class of limit ordinals is denoted by Lim. The ordinal ω is defined to be the minimum of Lim; it is the type of the natural numbers.

Every ordinal is either 0, a *successor* ordinal or a *limit* ordinal. These three kinds often appear in inductive proofs and definitions.

The operations of addition, multiplication and exponentiation are inductively defined in Ord as follows:

	$\alpha + \beta$	$\alpha.\beta$	α^β
$\beta = 0$	α	0	1
$\beta = \beta_0'$	$(\alpha + \beta_0)'$	$\alpha.\beta_0 + \alpha$	$\alpha^{\beta_0}.\alpha$
$\beta \in Lim$	$\bigvee_{\xi<\beta}(\alpha+\xi)$	$\bigvee_{\xi<\beta}(\alpha.\xi)$	$\bigvee_{\xi<\beta}(\alpha^\xi)$

We remark that:
- $+$ and $.$ are both associative and non-commutative and $.$ left-distributes over $+$.
- $+$ is (strictly) monotone in the right argument and weakly monotone in the left argument. Consequently there is a left-cancellation law. The same holds for $.$ whenever the left argument is not zero, for strict monotonicity in the right argument and left cancellation, or the right argument is not zero, for weak monotonicity in the left argument.
- for a fixed base greater than 1, exponentiation is strictly monotone in the exponent; also for any α, β, γ, $(\alpha^\beta)^\gamma = \alpha^{\beta.\gamma}$.
- $0.\alpha = 0$, for any α. Also $\alpha.\beta = 0 \iff \alpha = 0 \vee \beta = 0$.
- for any α, if $\beta \in Lim$, then $\alpha + \beta \in Lim$. Additionally if $\alpha \neq 0$ then $\alpha.\beta$, $\beta.\alpha \in Lim$.

Lemma 6. $\lambda \in Lim \iff \lambda = \omega.\beta$, for some $\beta \neq 0$.

Some ordinals are closed under the operations of addition and/or multiplication; they are crucial in this paper.

Definition 7. An ordinal $\alpha \neq 0$ is additive principal if it satisfies $\xi, \eta < \alpha \Rightarrow \xi + \eta < \alpha$. An ordinal $\alpha > 1$ is multiplicative principal if it satisfies $\xi, \eta < \alpha \Rightarrow \xi.\eta < \alpha$.

Lemma 8. Let $\alpha \in Ord$. Then the following conditions are equivalent:
- α is additive (respectively multiplicative) principal;
- $\alpha = \omega^\eta$ (respectively $\alpha = \omega^{\omega^\eta}$ or $\alpha = 2$), for some $\eta \geq 0$;
- $\forall \beta < \alpha : \beta + \alpha = \alpha$ (respectively $\beta.\alpha = \alpha$).

We conclude this section with some useful standard results.

Lemma 9. If $\alpha \leq \beta$ then there is a unique ordinal δ such that $\beta = \alpha + \delta$.

Lemma 10. Let $f : \alpha \to \beta$ be monotone. Then $\alpha \leq \beta$.

Proof. Suppose that $\beta < \alpha$. Then there is a unique δ such that $\alpha = \beta + \delta$ and since $\delta > 0$, $\beta \in \beta+\delta$. We remark that in particular f is also a monotone function from $\beta + \delta$ to $\beta + \delta$ and therefore, by lemma 3, we have that $x \leq f(x)$, for any $x \in \beta + \delta$. Consequently $\beta \leq f(\beta) < \beta$ (since $f(\beta) \in \beta$), giving a contradiction. □

If $\alpha \leq \beta$ then the ordinal $\beta - \alpha$ is defined by the property $\alpha + (\beta - \alpha) = \beta$. Its existence and uniqueness is guaranteed by lemma 9.

Lemma 11. If $\tau \leq \alpha$ then $(\alpha + \delta) - \tau = (\alpha - \tau) + \delta$.

Lemma 12. If $\alpha < \beta$ and $\beta = \omega^\eta$, for some ordinal η, then $\beta - \alpha = \beta$.

Lemma 13.
1. $\forall \alpha, \beta \in Ord\ \exists! \gamma, \delta \in Ord : \beta = \alpha.\gamma + \delta,\ \delta < \alpha$
2. $\forall \beta \geq 1 \forall \alpha \geq 2\ \exists! \eta :\ \alpha^\eta \leq \beta < \alpha^{\eta+1}$.
3. If $\alpha < \beta.\gamma$ then $\exists! \beta_1, \gamma_1 : \beta_1 < \beta\ \wedge\ \gamma_1 < \gamma\ \wedge\ \alpha = \beta.\gamma_1 + \beta_1$.

4 Multisets and Binary Functions

We give a constructive description of ordinal exponentiation. Let

$$Exp(\alpha, \eta) = \{\sigma : \eta \to \alpha \mid \{y \in \eta \mid \sigma(y) \neq 0\} \text{ is finite}\},$$

for any $\alpha, \eta \in Ord$. In $Exp(\alpha, \eta)$ we define the relation $>$ by

$$\sigma > \sigma' \iff \exists x \in \eta : (\sigma(x) >_\alpha \sigma'(x)) \wedge (\forall y \in \eta : y >_\eta x \Rightarrow \sigma(y) = \sigma'(y))$$

for any $\sigma, \sigma' \in Exp(\alpha, \eta)$. One easily verifies that $>$ is a total order.

Theorem 14. *Let $\alpha, \eta \in Ord$. Then $(Exp(\alpha, \eta), >)$ is order-isomorphic to ordinal exponentiation α^η.*

We present only a sketch of the proof. If $\alpha = 0$ or $\eta = 0$, the result is trivially verified. Suppose then that $\alpha, \eta \geq 1$. Any $x \in \alpha^\eta$ admits a unique finite decomposition in base α (see [7]), i. e. we can write $x = \alpha^{\eta_1}.\gamma_1 + \ldots + \alpha^{\eta_k}.\gamma_k$, with $1 \leq k < \omega$, $\eta > \eta_1 > \ldots > \eta_k$ and $\alpha > \gamma_i$, for $1 \leq i \leq k$. Given such a decomposition, the function $\phi : \alpha^\eta \to Exp(\alpha, \eta)$ is defined as

$$\forall x \in \alpha^\eta \ \forall \xi \in \eta : \ \phi(x)(\xi) = \begin{cases} \gamma_i & \text{if } \xi = \eta_i \text{ for any } 1 \leq i \leq k \\ 0 & \text{otherwise} \end{cases}$$

It can be checked that ϕ is an order-isomorphism, yielding the result.

Note that in the case of $\alpha = \omega$ the definition of $Exp(\alpha, \eta)$ coincides with that of the set $M(\eta)$ of finite multisets over η, together with its multiset order as described in [4]. So the order type of $M(\eta)$ is ω^η. In the sequel we shall freely switch between $M(\eta)$ and ω^η. For example, considering multisets in $M(\eta)$ as functions from η to ω multiset union is pointwise addition. This corresponds exactly to natural addition of ordinals below ω^η.

We shall prove that the existence of a monotone operation of arity greater than one in some ordinal implies that the ordinal has the form ω^η. As a consequence, for a TRS containing operation symbols of arity > 1 the only monotone algebras of interest are those whose underlying order is a multiset order. First we need two lemmas.

Lemma 15. *Let λ be an ordinal for which $\exists \alpha < \lambda : \lambda - \alpha \leq \alpha$. Then no function from $\lambda \times \cdots \times \lambda$ to λ exists which has more than one argument and is monotone in all arguments.*[4]

Proof. Suppose such a function exists. Then by fixing all arguments but two we obtain a binary function f that is monotone in both arguments. Define $\varphi : \lambda \to \lambda$ by $\varphi(x) = f(x, \alpha) - \alpha$. We have to see that φ is well-defined. If we fix the first argument of f to 0_λ, the minimum of λ, we have, since $f(0_\lambda, x)$ is strictly monotone and by lemma 3, that $f(0_\lambda, \alpha) \geq \alpha$. So $f(x, \alpha) \geq \alpha$ for any x, hence φ is well-defined. Actually φ is a function from λ to $\lambda - \alpha$. If $x > y$ then $\alpha + \varphi(x) = f(x, \alpha) > f(y, \alpha) = \alpha + \varphi(y)$. Due to the left cancellation law, we conclude that φ is strictly monotone. By lemma 10 we conclude that $\lambda \leq \lambda - \alpha$. Since $\alpha < \lambda$ we get $\alpha < \lambda - \alpha$ contradicting the hypothesis. □

---
[4] $\times$ denotes cartesian product.

Lemma 16. *For $\lambda \neq 0$, $\lambda = \omega^\gamma$ for some γ $\iff$ $\forall \alpha < \lambda : \lambda - \alpha > \alpha$.*

Proof. We will prove that $\forall \alpha, \beta < \lambda : \alpha + \beta < \lambda$ if and only if $\forall \alpha < \lambda : \lambda - \alpha > \alpha$; then the result follows from lemma 8.

For the only-if part, let $\alpha < \lambda$. We always have $\lambda - \alpha \leq \lambda$. If $\lambda - \alpha < \lambda$, by hypothesis we get $\alpha + (\lambda - \alpha) < \lambda$, a contradiction. Therefore $\lambda - \alpha = \lambda$, so $\lambda - \alpha > \alpha$.

For the if part, take $\alpha, \beta < \lambda$. The hypothesis implies $\alpha < \lambda - \alpha$ and $\beta < \lambda - \beta$. If $\beta \leq \alpha$ then $\alpha + \beta \leq \alpha + \alpha < \alpha + (\lambda - \alpha) = \lambda$. If $\alpha < \beta$ then $\alpha + \beta \leq \beta + \beta < \beta + (\lambda - \beta) = \lambda$. In both cases we conclude $\alpha + \beta < \lambda$, which we had to prove. □

Theorem 17. *Let $\mathcal{A} = (A, >)$ be a well-ordered set such that $A \neq \emptyset$. Then there is a function from $A \times \cdots \times A$ to A with more than one argument, monotone in all arguments if and only if $\mathcal{A}$ is order-isomorphic to $M(\mathcal{B})$, for some well-ordered set $\mathcal{B}$.*

Proof. Assume $\mathcal{A} \cong M(\mathcal{B})$. The multiset union from $M(\mathcal{B}) \times \cdots \times M(\mathcal{B})$ to $M(\mathcal{B})$ is monotone in all arguments. The isomorphism gives us a similar function in A.

On the other hand assume there is a function that is monotone in several arguments. According to lemmas 15 and 16 the order type of $\mathcal{A}$ is ω^γ, so $\mathcal{A}$ is order-isomorphic to $M(\gamma)$. □

Stated in different words, the previous result says that if we have a TRS R containing at least a function symbol of arity $n \geq 2$ and totally terminating in an algebra $\mathcal{A}$, then $\mathcal{A}$ has type ω^γ, for some $\gamma \geq 0$.

5 Extension to Higher Ordinals and Modularity

In this section we look at modularity of total termination. If two TRS's are totally terminating, what can be said about their disjoint union? From [8] follows that the disjoint union is simply terminating, but is it also totally terminating? This is not clear if the proofs of total termination are given in distinct ordinals. That arises the question whether a total termination proof in some ordinal can be lifted to a similar proof in another ordinal.

Definition 18. *For a TRS R we define $U(R)$ to be the class of ordinals in which a proof of total termination of R can be given. The minimum of $U(R)$ is denoted by u_R.*

By definition $U(R)$ is non-empty for every totally terminating TRS R. For example, if R consists of one rule involving two different constants then $U(R)$ is the class of all ordinals > 1. Note that the disjoint union $R_1 \oplus R_2$ of two TRS's R_1 and R_2 is totally terminating if and only if $U(R_1 \oplus R_2) = U(R_1) \cap U(R_2) \neq \emptyset$.

The next lemmas state some basic properties of $U(R)$.

Lemma 19. *Let $\alpha \in U(R)$ and let β be an arbitrary non-zero ordinal. Let either all function symbols in R have arity ≤ 1 or $\beta = \omega^\gamma$ for some ordinal γ. Then $\beta.\alpha \in U(R)$.*

Proof. Remember that $\beta.\alpha$ is the lexicographic product with weight on α. Its elements will be denoted by pairs (b, a), with $a \in \alpha$ and $b \in \beta$. Since $\alpha \in U(R)$, we have an interpretation f_α of every function symbol f of R in α, strictly monotone in each argument, such that for every rule $l \to r$ in R and every substitution $\tau : X \to \alpha$, it holds $\phi_\alpha(l, \tau) >_\alpha \phi_\alpha(r, \tau)$. For every function symbol f we introduce an interpretation f_β in β: for constants c we choose $c_\beta = 0$ and for unary f we choose f_β to be the identity on β. If there are symbols of arity > 1 we assumed β to be the finite multisets over γ, in this case we define f_β to be the multiset union of all of its arguments. For every f define

$$f_{\beta.\alpha}((b_1, a_1), \ldots, (b_n, a_n)) = (f_\beta(b_1, \ldots, b_n), f_\alpha(a_1, \ldots, a_n)).$$

Monotonicity of $f_{\beta.\alpha}$ in all arguments is easily verified. We still have to check that $\phi_{\beta.\alpha}(l, \tau) >_{\beta.\alpha} \phi_{\beta.\alpha}(r, \tau)$ for every rule $l \to r$ in R and every $\tau : X \to \beta.\alpha$. For this we need a lemma, which is easily proven by induction on terms.

Lemma 20. *Let t be any term, $\tau : X \to \beta.\alpha$, and let π_j be the projection on the jth coordinate for $j = 1, 2$. Then $\phi_{\beta.\alpha}(t, \tau) = (\phi_\beta(t, \pi_1 \circ \tau), \phi_\alpha(t, \pi_2 \circ \tau))$.*

Since $\phi_\alpha(l, \sigma) >_\alpha \phi_\alpha(r, \sigma)$ for any $\sigma : X \to \alpha$, we conclude that $\phi_{\beta.\alpha}(l, \tau) = (\phi_\beta(l, \pi_1 \circ \tau), \phi_\alpha(l, \pi_2 \circ \tau)) >_{\beta.\alpha} (\phi_\beta(r, \pi_1 \circ \tau), \phi_\alpha(r, \pi_2 \circ \tau)) = \phi_{\beta.\alpha}(r, \tau)$. □

Theorem 21. *If $\alpha \in U(R)$ then $\omega^\alpha \in U(R)$.*

Proof. Again f_α will denote the interpretation of the function symbols f of R in α. In this proof we identify ω^α with the finite non-empty multisets over α instead of all finite multisets. In terms of ordinals this does not make any difference since for $\alpha \geq 1$, $\omega^\alpha - 1 = \omega^\alpha$. Write $[a]$ for the multiset containing only one element a and $\bigsqcup$ for multiset union. Multiset union indexed over finite multisets is defined as follows:

$$\bigsqcup_{x \in [a]} \Phi(x) = \Phi(a); \quad \bigsqcup_{x \in X \sqcup Y} \Phi(x) = (\bigsqcup_{x \in X} \Phi(x)) \bigsqcup (\bigsqcup_{x \in Y} \Phi(x)),$$

for any function $\Phi : \alpha \to M(\alpha)$. For constants c and function symbols f of arity $n \geq 1$, we define:

- $c_{\omega^\alpha} = [c_\alpha]$.
- $f_{\omega^\alpha}(X_1, \ldots, X_n) = \bigsqcup_{x_1 \in X_1} \cdots \bigsqcup_{x_n \in X_n} [f_\alpha(x_1, \ldots, x_n)]$.

It can be verified that f_{ω^α} is strictly monotone in each argument for all function symbols f; for functions with arity > 1 the non-emptiness restriction is essential.

Let $l \to r$ be an arbitrary rewrite rule and let $\tau : X \to \omega^\alpha$. We still have to prove that $\phi_{\omega^\alpha}(l, \tau) > \phi_{\omega^\alpha}(r, \tau)$. For any such τ, we define a substitution $\sigma_{max} : X \to \alpha$ by $\sigma_{max}(x) = \max(\tau(x))$ (recall that for any $x \in X$, $\tau(x) \neq \emptyset$).

Using the definition of f_{ω^α}, it can be easily proven by induction that, for any term t, $\max(\phi_{\omega^\alpha}(t,\tau)) = \phi_\alpha(t,\sigma_{max})$. For all $a \in \phi_{\omega^\alpha}(r,\tau)$ we have

$$a \leq \max(\phi_{\omega^\alpha}(r,\tau)) = \phi_\alpha(r,\sigma_{max}) < \phi_\alpha(l,\sigma_{max}) \in \phi_{\omega^\alpha}(l,\tau)).$$

Consequently we obtain $\phi_{\omega^\alpha}(l,\tau) > \phi_{\omega^\alpha}(r,\tau)$. We have proven that R is totally terminating in ω^α, so $\omega^\alpha \in U(R)$. □

Now we are ready to prove modularity of total termination under certain conditions.

Theorem 22. *Let R_1 and R_2 be totally terminating TRS's, at least one of them not containing duplicating rules. Then $R_1 \oplus R_2$ is totally terminating.*

Proof. Let α and β be ordinals in which the proofs of total termination of R_1 and R_2 can respectively be given. Due to theorem 21 we may, and shall, assume that $\alpha = \omega^\gamma$ and $\beta = \omega^\eta$, for some $\gamma, \eta \geq 1$. Suppose that R_1 has no duplicating rules (the other case is symmetric). Identify $\beta = \omega^\eta$ with finite multisets over η and define interpretations in β for the functions symbols of R_1 in the following way:

- $c_\beta = [\,]$, for any constant c, where $[\,] = 0_\beta$ represents the empty multiset.
- $f_\beta(x_1,\ldots,x_n) = \bigsqcup_{i=1}^{n} x_i$, where $\bigsqcup$ represents multiset union.

For a term t let X_t be the multiset of variables occurring in t. For any $\tau: X \to \beta$ we obtain $\phi_\beta(t,\tau) = \bigsqcup_{x \in X_t} \tau(x)$; here the multiset union over an empty index is defined to be $[\,]$. Since there are no duplicating rules the multiset X_r is contained in X_l for all rewrites rules $l \to r$. Consequently,

$$\phi_\beta(l,\tau) = \bigsqcup_{x \in X_l} \tau(x) \geq \bigsqcup_{x \in X_r} \tau(x) = \phi_\beta(r,\tau).$$

Note that the inequality is not strict in general.

Now in $\alpha.\beta$ (the lexicographic product with weight on β) we define for any n-ary, $n \geq 0$, function symbol f of R_1:

$$f_{\alpha.\beta}((a_1,b_1),\ldots,(a_n,b_n)) = (f_\alpha(a_1,\ldots,a_n), f_\beta(b_1,\ldots,b_n)),$$

where f_α comes from the total termination proof of R_1 in α. Since f_α and f_β are strictly monotone in all coordinates, the same holds for $f_{\alpha.\beta}$.

Let $l \to r$ be a rule in R_1 and let $\tau: X \to \alpha.\beta$. Applying lemma 20 and using $\phi_\beta(l,\pi_2 \circ \tau) \geq \phi_\beta(r,\pi_2 \circ \tau)$ and $\phi_\alpha(l,\pi_1 \circ \tau) > \phi_\alpha(r,\pi_1 \circ \tau)$, we conclude $\phi_{\alpha.\beta}(l,\tau) = (\phi_\alpha(l,\pi_1 \circ \tau), \phi_\beta(l,\pi_2 \circ \tau)) > (\phi_\alpha(r,\pi_1 \circ \tau), \phi_\beta(r,\pi_2 \circ \tau)) = \phi_{\alpha.\beta}(r,\tau)$.

So we have a proof of total termination of R_1 in $\alpha.\beta$, hence $\alpha.\beta \in U(R_1)$. On the other hand, since $\alpha = \omega^\gamma$, we can apply lemma 19 to conclude that $\alpha.\beta \in U(R_2)$. Hence $\alpha.\beta \in U(R_1) \cap U(R_2)$, so $R_1 \oplus R_2$ is totally terminating. □

Note that if both R_1 and R_2 contain duplicating rules, there are particular cases in which we can prove the union is totally terminating.[5] For example, let R_1 and R_2 be totally terminating in α, β, respectively, and assume there are ordinals γ, δ such that $\gamma + \omega^{\cdot^{\cdot^{\omega^\alpha}}} = \delta + \omega^{\cdot^{\cdot^{\omega^\beta}}} = A$, for finite exponentiations on both right summands. Then it easily follows from lemma 19 and theorem 21 that $\omega^A \in U(R_1 \oplus R_2)$, so $R_1 \oplus R_2$ is totally terminating. However, not all α, β satisfy this property; for example $\alpha = \omega^2$ and $\beta = \omega^\omega$. The problem boils down to extending functions (of any arity) defined on a certain ordinal, to a given higher one, in such a way that the requirements of total termination are met, that is, in the new ordinal the functions are strictly monotone in all coordinates and for every rule the interpretation of the left hand side is greater than that of the right hand side.

6 String Rewriting Systems

In the previous sections we saw that when trying to prove total termination of TRS's containing at least a function symbol of arity $n \geq 2$, only ordinals of the form ω^η were relevant. In this section, we discuss whether the same holds for string rewriting systems (SRS's), i.e., rewriting systems containing only unary function symbols. First we need a lemma.

Lemma 23. *Let $\alpha \neq 0$ and $f : \alpha \to \alpha$ be (strictly) monotone. Then there is a unique ordinal η such that $\omega^\eta \leq \alpha < \omega^{\eta+1}$ and $f(\omega^\eta) \subseteq \omega^\eta$.*

Proof. Lemma 13 guarantees the existence of a unique $\eta \in Ord$ such that $\omega^\eta \leq \alpha < \omega^{\eta+1}$. If $\alpha = \omega^\eta$ we are done, otherwise we can write $\alpha = \omega^\eta + \delta$, with $\delta > 0$.

We suppose $f(\omega^\eta) \not\subseteq \omega^\eta$ and will derive a contradiction. That means there is $b \in \omega^\eta$ such that $f(b) \geq \omega^\eta$. We now define a function $g : (\omega^\eta + \delta) - b \to \delta$ by $g(x) = f(b + x) - \omega^\eta$. We see that

- since $f(b+x) \geq f(b) \geq \omega^\eta$, g is well-defined.
- g is (strictly) monotone since for $x' > x$ also $b + x' > b + x$ and then $\omega^\eta + g(x') = f(b + x') > f(b + x) = \omega^\eta + g(x)$; by left cancellation we get $g(x') > g(x)$.

By lemma 10 we obtain $(\omega^\eta + \delta) - b \leq \delta$. Since $b < \omega^\eta$, by lemmas 11 and 12, we get $(\omega^\eta + \delta) - b = (\omega^\eta - b) + \delta = \omega^\eta + \delta$. So we have $\omega^\eta + \delta \leq \delta$, hence $\omega^\eta + \delta = \delta$. Since $\delta \leq \alpha < \omega^{\eta+1} = \omega^\eta.\omega$, by lemma 13 there are uniquely determined ordinals β, γ such that $\beta < \omega$, $\gamma < \omega^\eta$ and $\delta = \omega^\eta.\beta + \gamma$. However, $\delta = \omega^\eta + \delta$, so also $\delta = \omega^\eta.(1 + \beta) + \gamma$. From the uniqueness now follows $\beta = 1 + \beta$, contradicting $\beta < \omega$. □

Remember that for a totally terminating TRS R the ordinal u_R is defined to be the minimal ordinal in which the total termination proof can be given.

---
[5] The obvious case is when the proof of termination is given in the same ordinal for both TRS's.

Theorem 24. Let R be a totally terminating SRS. Then $u_R = \omega^\eta$ for some $\eta \geq 1$.

Proof. From lemma 23 we obtain a unique ordinal η such that $\omega^\eta \leq u_R < \omega^{\eta+1}$ and $f(\omega^\eta) \subseteq \omega^\eta$ for all operation symbols f. By restricting $f : u_R \to u_R$ to ω^η for all operation symbols f, we see that we also have a proof of total termination of R in ω^η, so $\omega^\eta \in U(R)$. Since u_R is the minimum of $U(R)$ and $\omega^\eta \leq u_R$ we obtain $u_R = \omega^\eta$. □

Note that this result is essentially weaker than theorem 17 for the case of arity > 1. The fact that $u_R = \omega^\eta$ does not imply that every ordinal in $U(R)$ is of that shape. For example, every proof of total termination of a SRS in ω is easily extended to a similar proof in $\omega + \omega$, which is not of the required shape.

A natural operation on SRS's is reversing: all left hand sides and right hand sides are reversed, considered as strings. For example, the reverse of $f(f(g(x)))$ is $g(f(f(x)))$. Clearly there is a bijective correspondence between reductions in the original system and reductions in the reversed system. As a consequence, a SRS is terminating if and only if the reversed system is terminating. However, a similar observation does not hold for total termination. For example, the system

$$f(f(x)) \to f(g(x)), \quad g(g(x)) \to g(f(x))$$

is not totally terminating since $f(a)$ and $g(a)$ are incomparable for any a in any corresponding monotone algebra. On the other hand, the reversed system

$$f(f(x)) \to g(f(x)), \quad g(g(x)) \to f(g(x))$$

is totally terminating in the natural numbers; a possible interpretation is $f(x) = 4x + 2, g(x) = 4x + 1$ for x even, and $f(x) = 4x, g(x) = 4x + 3$ for x odd. Further, if for a totally terminating system the reversed system is totally terminating too, the corresponding ordinal may change. An example is $f(g(x)) \to g(f(f(x)))$; in the next section we shall see that the minimal ordinal of this totally terminating system is ω^2, while termination of the reversed system $g(f(x)) \to f(f(g(x)))$ is proven in the natural numbers by choosing $f(x) = x + 1, g(x) = 3x$.

We conclude this section with some remarks about TRS's that also contain constants, and no function symbols of arity > 1. From theorems 17 and 24 we know that otherwise total termination implies that $u_R = \omega^\eta$ for some $\eta \geq 0$. However, if there are constants then the proof of theorem 24 does not hold any more since the interpretation of the constants can be too great. The simplest example is the TRS R consisting of the rule $a \to b$, where a and b are constants. It is totally terminating and $u_R = 2$. If we allow infinitely many constants and rewrite rules then for any ordinal α a TRS R can be given with $u_R = \alpha$.

The infinite TRS R consisting of the rules $c \to f^i(d)$, for each $i < \omega$, and the rule $f(x) \to x$, satisfies $u_R = \omega + \omega$. An interpretation in $\omega + \omega$ is given by $d = 0$, $c = \omega$ and $f(x) = x + 1$. It can not be done in a smaller ordinal since the interpretation of c is at least ω. In a similar way for every ordinal $\alpha < \omega^\omega$ an infinite TRS R with finitely many unary symbols and constants can be constructed satisfying $u_R = \alpha$.

We conjecture that for any finite totally terminating TRS R without function symbols of arity > 1 and containing at least one variable, the ordinal u_R is of the form ω^η. In the next section we investigate which ordinals actually occur as u_R of such a finite TRS R.

However, even if u_R is not of the form ω^η, due to theorem 21 we need only consider those ordinals for proving total termination.

7 Minimal Ordinals

As we have seen previously, when trying to establish total termination of (finite or infinite) SRS's or TRS's containing symbols of arity > 0, we only need to consider algebras with type ω^η for some $\eta > 0$. Is it the case that all ordinals of that form are important or can we restrict the class even further? Partially answering this question, we have the following theorem.

Theorem 25. *For any ordinal η with $1 \leq \eta \leq \omega$, there is a SRS R such that $u_R = \omega^\eta$.*

Due to lack of space we do not give the technical proof here. We only give the SRS's for which u_R is of the required shape. For $\eta = 1$, the SRS $f(x) \to x$ fulfills the requirements. For $1 < \eta < \omega$, let R_η consist of the $\eta - 1$ rules

$$f_i(f_{i+1}(x)) \to f_{i+1}(f_i(f_i(x)))$$

for $i = 1, \ldots, \eta - 1$. Then $u_{R_\eta} = \omega^\eta$ (for $\eta = 2$ this was already shown in [14], report version). Proving $u_{R_\eta} \geq \omega^\eta$ is difficult; $u_{R_\eta} \leq \omega^\eta$ follows from the following interpretation in ω^η. Identify ω with strictly positive integers and define in ω^η:

$$f_i(x_1, \ldots, x_\eta) = (x_1, \ldots, x_{i-1}, x_i + 2^{x_{i+1}}, x_{i+1}, \ldots, x_\eta)$$

for $i = 1, \ldots, \eta$, where $x_{\eta+1}$ is defined to be 1. With this interpretation, we can easily see that all the requirements of total termination are fulfilled.

For the ordinal ω^ω we consider the TRS R

$$f(g(x)) \to g(f(f(x)))$$
$$f(h(x)) \to h(g(x))$$

Proving $u_R \geq \omega^\omega$ is again difficult; $u_R \leq \omega^\omega$ is a consequence of the following interpretation. Identify ω with natural numbers, including 0. Recall from Theorem 14 that we can identify an element $X \in \omega^\omega$ with a certain function $X : \omega \to \omega$, therefore we will denote such an element by the sequence $(p_0, \ldots, p_k)$ where:

- $X(i) = p_i$, if $0 \leq i \leq k$.
- $X(k) \neq 0$ and $X(i) = 0$ for $i > k$.

We restrict to the part of ω^ω for which $k \geq 1$ in this notation. This means that we skip the first ω elements of ω^ω; since $\omega^\omega - \omega = \omega^\omega$ this does not affect the ordinal. We now define $f, g, h : \mathcal{A} \to \mathcal{A}$ by:

- $f(p_0, \ldots, p_{k-1}, p_k) = (p_0 + p_k, \ldots, p_{k-1} + p_k, p_k)$
- $g(p_0, \ldots, p_{k-1}, p_k) = (p_0, \ldots, p_{k-1}, 2.p_k + 1)$
- $h(p_0, \ldots, p_{k-1}, p_k) = (p_0, \ldots, p_{k-1}, p_k, 0, 1)$

With some easy calculations, it can be shown that the functions are indeed strictly monotonic and that for both rules the interpretation of the left hand side is greater than the interpretation of the right hand side.

8 Further Work and Conclusions

Total termination is a semantic notion and we would like to have an equivalent syntactical notion. We define the truncation closure $TC(R)$ of a TRS R to be the TRS consisting of all rules $t \to t'$ for which there is a context C such that $C[t] \to_R^+ C[t']$. It is not difficult to see that R is totally terminating if and only if $TC(R) \cup Emb(\mathcal{F})$ is totally terminating, where $Emb(\mathcal{F})$ consists of all embedding rules for R.[6] As a consequence if $TC(R) \cup Emb(\mathcal{F})$ allows an infinite reduction, R is not totally terminating. This is a useful tool for proving that a TRS is not totally terminating. It is not clear whether total termination of R is equivalent to termination of $TC(R) \cup Emb(\mathcal{F})$.

Proving termination of term rewriting systems by interpretation is not easy. We focussed on interpretation in monotone algebras in which the underlying order is total. We have shown that the existence of a function symbol of arity > 1 implies that the underlying order has type ω^η, i. e. is equivalent to finite multisets over some well-order. Furthermore, for any TRS the class of total orders in which it can be shown totally terminating, is closed under multiset construction and lexicographic product. However, it is not clear how to extend a total termination proof in a particular well-order to well-orders that can not be finitely obtained from the original one by these constructions. This problem is closely connected to modularity of total termination, on which we obtained some interesting partial results.

We found examples of TRS's showing that proofs of total termination cannot always be given in well-orders of type smaller than ω^ω. Most of our techniques are based upon ordinal arithmetic; ordinal arithmetic appears to be a strong and useful tool for proving termination of TRS's. We believe that our framework is a step towards generalizing and combining existing techniques like recursive path order and Knuth-Bendix order.

References

1. BEN-CHERIFA, A., AND LESCANNE, P. Termination of rewriting systems by polynomial interpretations and its implementation. *Science of Computing Programming 9*, 2 (1987), 137–159.

---

[6] For any function symbol f of arity n, its n embedding rules are of the form $f(x_1, \ldots, x_n) \to x_i$, for $1 \leq i \leq n$.

2. DERSHOWITZ, N. Termination of rewriting. *Journal of Symbolic Computation 3*, 1 and 2 (1987), 69–116.
3. DERSHOWITZ, N., AND JOUANNAUD, J.-P. Rewrite systems. In *Handbook of Theoretical Computer Science*, J. van Leeuwen, Ed., vol. B. Elsevier, 1990, ch. 6, pp. 243–320.
4. DERSHOWITZ, N., AND MANNA, Z. Proving termination with multiset orderings. *Communications ACM 22*, 8 (1979), 465–476.
5. HOFBAUER, D. Termination proofs by multiset path orderings imply primitive recursive derivation lengths. *Theoretical Computer Science 105*, 1 (1992), 129–140.
6. HUET, G., AND LANKFORD, D. S. On the uniform halting problem for term rewriting systems. Rapport Laboria 283, INRIA, 1978.
7. KURATOWSKI, K., AND MOSTOWSKI, A. *Set Theory.* North-Holland Publishing Company, 1968.
8. KURIHARA, M., AND OHUCHI, A. Modularity of simple termination of term rewriting systems. *Journal of IPS Japan 31*, 5 (1990), 633–642.
9. LANKFORD, D. S. On proving term rewriting systems are noetherian. Tech. Rep. MTP-3, Louisiana Technical University, Ruston, 1979.
10. LESCANNE, P. Termination of rewrite systems by elementary interpretations. In *Algebraic and Logic Programming* (1992), H. Kirchner and G. Levi, Eds., vol. 632 of *Lecture Notes in Computer Science*, Springer, pp. 21 – 36.
11. MARTIN, U., AND SCOTT, E. The order types of termination orderings on terms, strings and multisets. In *Proceedings of the 8th Annual IEEE Symposium on Logic in Computer Science* (1993).
12. MIDDELDORP, A. *Modular Properties of Term Rewriting Systems.* PhD thesis, Free University Amsterdam, 1990.
13. STEINBACH, J. Extensions and comparison of simplification orderings. In *Proceedings of the 3rd Conference on Rewriting Techniques an Applications* (1989), N. Dershowitz, Ed., vol. 355 of *Lecture Notes in Computer Science*, Springer, pp. 434–448.
14. ZANTEMA, H. Termination of term rewriting by interpretation. In *Conditional Term Rewriting Systems, Proceedings Third International Workshop CTRS-92* (1993), M. Rusinowitch and J. Rémy, Eds., vol. 656 of *Lecture Notes in Computer Science*, Springer, pp. 155–167. Full version appeared as report RUU-CS-92-14, Utrecht University.

Simple Termination is Difficult

Aart Middeldorp

Institute of Information Sciences and Electronics
University of Tsukuba, Tsukuba 305, Japan
ami@softlab.is.tsukuba.ac.jp

Bernhard Gramlich

Fachbereich Informatik, Universität Kaiserslautern
Postfach 3049, D-6750 Kaiserslautern, Germany
gramlich@informatik.uni-kl.de

ABSTRACT

A terminating term rewriting system is called simply terminating if its termination can be shown by means of a simplification ordering, an ordering with the property that a term is always bigger than its proper subterms. Almost all methods for proving termination yield, when applicable, simple termination. We show that simple termination is an undecidable property, even for one-rule systems. This contradicts a result by Jouannaud and Kirchner. The proof is based on the ingenious construction of Dauchet who showed the undecidability of termination for one-rule systems.

1. Introduction

It is well-known that termination is an undecidable property of term rewriting systems. This result was obtained by Huet and Lankford [8] in 1978. They showed that every Turing machine can be coded as a string rewriting system—a term rewriting system with only unary function symbols—such that termination of the resulting string rewriting system is equivalent to the uniform halting problem for the originating Turing machine. The number of rules in their construction depends on the number of Turing machine instructions. Later, Dershowitz [3] showed that every Turing machine can be simulated by means of a two-rule term rewriting system. This result was improved by Dauchet [2], who showed that termination remains undecidable even if we restrict our attention to one-rule term rewriting systems that are orthogonal and variable preserving. His skillful construction will be explained in detail later in this paper. On the other hand, Caron [1] recently showed that termination is an undecidable property of length-preserving string rewriting systems—systems in which the left-hand side and the right-hand side of each rule have the same length—by a reduction to the uniform halting problem for linear bounded automata—a restricted kind of Turing machines.

From this last result one easily obtains the undecidability of *simple termination* for the same class of term rewriting systems. Simple termination is a stronger notion than termination. A term rewriting system is simply terminating if the addition of all rewrite rules of the form $f(x_1,\ldots,x_n) \to x_i$ results in a terminating system. Virtually all methods for proving termination yield, when applicable, simple termination. Simple termination is closely related to the *non-self-embedding* property, since every simply terminating term rewriting system is non-self-embedding. Plaisted [15] showed that the non-self-embedding property is undecidable. From this result we cannot infer the undecidability of simple termination, however. As a matter of fact, it is known that negative results for the class of non-self-embedding systems do not always carry over to the class of simply terminating systems, see [6]. In this paper we show the undecidability of simple termination for one-rule term rewriting systems. This contradicts a result of Jouannaud and Kirchner [9]. The undecidability proof is based on the ingenious construction of Dauchet. He showed in [2] that with every Turing machine M one can associate a term rewriting system $\mathcal{R}_M$ consisting of a single rewrite rule such that

M halts for all configurations
$\iff$
$\mathcal{R}_M$ is terminating.

From this we cannot immediately infer the undecidability of simple termination for one-rule systems, since the implication "$\mathcal{R}_M$ is terminating $\Rightarrow$ $\mathcal{R}_M$ is simply terminating" does not hold for every Turing machine M. However, we will show that if we start the construction of Dauchet from a linear bounded automaton M instead of a Turing machine, termination and simple termination of $\mathcal{R}_M$ coincide.

The paper is organized as follows. The next section contains a brief introduction to term rewriting, including a discussion of the property simple termination. In Section 3 we define linear bounded automata. Section 4 describes Dauchet's construction. Actually, we present a somewhat simpler construction. We show that the equivalence

M halts for all configurations
$\iff$
$\mathcal{R}_M$ is terminating

is easily obtained for all linear bounded automata M by using a recent result of Zantema [16] on type removal. In Section 5 we show that the equivalence

$$\mathcal{R}_M \text{ is terminating}$$
$$\Longleftrightarrow$$
$$\mathcal{R}_M \text{ is simply terminating}$$

follows for all linear bounded automata M by using a powerful method of Zantema [17] for establishing (simple) termination.

2. Simple Termination

We start with a brief introduction to term rewriting. Term rewriting is surveyed in Dershowitz and Jouannaud [4] and Klop [10].

A *signature* is a set $\mathcal{F}$ of *function symbols*. Associated with every $f \in \mathcal{F}$ is a natural number denoting its arity. Function symbols of arity 0 are called *constants*. Let $\mathcal{T}(\mathcal{F}, \mathcal{V})$ be the set of all terms built from $\mathcal{F}$ and a countably infinite set $\mathcal{V}$ of *variables*, disjoint from $\mathcal{F}$. If t is a term then $Var(t)$ denotes the set of variables occurring in t. A term t is called *ground* if $Var(t) = \emptyset$. The set of all ground terms is denoted by $\mathcal{T}(\mathcal{F})$. A term t is called *linear* if it does not contain multiple occurrences of the same variable. The *root symbol* of a term t is defined as follows: $root(t) = t$ if t is a variable and $root(t) = f$ if $t = f(t_1, \ldots, t_n)$. The *size* $|t|$ of a term t is the number of variables and function symbols occurring in t.

We introduce a fresh constant symbol $\square$, named *hole*. A *context* C is a term in $\mathcal{T}(\mathcal{F} \cup \{\square\}, \mathcal{V})$. The designation *term* is restricted to members of $\mathcal{T}(\mathcal{F}, \mathcal{V})$. A context may contain zero, one or more holes. If C is a context with n holes and $t_1, \ldots, t_n$ are terms then $C[t_1, \ldots, t_n]$ denotes the result of replacing from left to right the holes in C by $t_1, \ldots, t_n$. A term s is a *subterm* of a term t if there exists a context C such that $t = C[s]$. A subterm s of t is *proper*, denoted by $t \triangleright s$, if $s \neq t$. A *substitution* is a map σ from $\mathcal{V}$ to $\mathcal{T}(\mathcal{F}, \mathcal{V})$. If σ is a substitution and t a term then $t\sigma$ denotes the result of applying σ to t. We call $t\sigma$ an *instance* of t. A binary relation $\succ$ on terms is a *rewrite relation* if it is closed under contexts and substitutions, i.e. if $s \succ t$ then $C[s\sigma] \succ C[t\sigma]$ for all contexts C (with precisely one hole) and substitutions σ.

A *rewrite rule* is a pair (l, r) of terms such that the left-hand side l is not a variable and variables which occur in the right-hand side r occur also in l, i.e. $Var(r) \subseteq Var(l)$. Rewrite rules (l, r) will henceforth be written as $l \to r$. A rewrite rule is *collapsing* if its right-hand side is a single variable. A rewrite rule is *duplicating* if its right-hand side contains more occurrences of some variable than its left-hand side. A rewrite rule is *left-linear* (*right-linear*) if its left-hand (right-hand) side is a linear term,

A *term rewriting system* (TRS for short) is a pair $(\mathcal{F}, \mathcal{R})$ consisting of a signature $\mathcal{F}$ and a set $\mathcal{R}$ of rewrite rules between terms in $\mathcal{T}(\mathcal{F}, \mathcal{V})$. We often present a TRS as a set of rewrite rules, without making explicit its signature, assuming that the signature consists of the function symbols occurring in the rewrite rules.

If $(\mathcal{F}, \mathcal{R})$ is a TRS then $\to_\mathcal{R}$ denotes the smallest rewrite relation on $\mathcal{T}(\mathcal{F}, \mathcal{V})$ containing $\mathcal{R}$. So $s \to_\mathcal{R} t$ if there exists a rewrite rule $l \to r$ in $\mathcal{R}$, a substitution σ and a context C such that $s = C[l\sigma]$ and $t = C[r\sigma]$. The subterm $l\sigma$ of s is called a *redex* and we say that s rewrites to t by *contracting* redex $l\sigma$. We call $s \to_\mathcal{R} t$ a *rewrite* or *reduction step*. If $C = \square$ then we speak of a *root* reduction. The transitive closure of $\to_\mathcal{R}$ is denoted by $\to_\mathcal{R}^+$ and $\to_\mathcal{R}^*$ denotes the transitive-reflexive closure of $\mathcal{R}$. If $s \to_\mathcal{R}^* t$ we say that s *reduces* to t. A TRS $(\mathcal{F}, \mathcal{R})$ is called *terminating* if there are no infinite reduction sequences $t_1 \to_\mathcal{R} t_2 \to_\mathcal{R} t_3 \to_\mathcal{R} \cdots$ of terms in $\mathcal{T}(\mathcal{F}, \mathcal{V})$.

A rewrite relation that is also a (strict) partial order is called a *rewrite order*. A TRS $(\mathcal{F}, \mathcal{R})$ is *compatible* with a rewrite order $\succ$ on $\mathcal{T}(\mathcal{F}, \mathcal{V})$ if $l \succ r$ for every rewrite rule $l \to r$ of $\mathcal{R}$. It is easy to show that a TRS is terminating if and only if it is compatible with a well-founded rewrite order.

DEFINITION 2.1.
- A *simplification order* is a rewrite order $\succ$ with the *subterm property*, i.e. $C[t] \succ t$ for all contexts $C \neq \square$ (with precisely one hole) and terms t.
- A TRS is called *simplifying* if it is compatible with a simplification order.
- A TRS is called *simply terminating* if it is compatible with a well-founded simplification order.

Clearly every simply terminating TRS is both simplifying and terminating. A simplifying TRS $(\mathcal{F}, \mathcal{R})$ with $\mathcal{F}$ or $\mathcal{R}$ finite is simply terminating, as a consequence of Kruskal's Tree Theorem. There exists (infinite) simplifying and terminating TRSs that are not simply terminating, see Ohlebusch [14]. This does not concern us too much as we will deal with decidability issues in the sequel, in which one considers only finite (both with respect to signature and set of rewrite rules) TRSs. Next we present a useful characterization of simple termination.

DEFINITION 2.2. Let $\mathcal{F}$ be a signature. The TRS $\mathcal{E}mb(\mathcal{F})$ consists of all rewrite rules

$$f(x_1, \ldots, x_n) \to x_i$$

with $f \in \mathcal{F}$ a function symbol of arity $n \geqslant 1$ and $i \in \{1, \ldots, n\}$. We write $s \lesssim t$ for terms $s, t \in \mathcal{T}(\mathcal{F}, \mathcal{V})$ if $t \to_{\mathcal{E}mb(\mathcal{F})}^* s$. The relation $\lesssim$ is called *embedding*.

LEMMA 2.3. Let $(\mathcal{F}, \mathcal{R})$ be a TRS. The following statements are equivalent.
- *The TRS $(\mathcal{F}, \mathcal{R})$ is simply terminating.*
- *The TRS $(\mathcal{F}, \mathcal{R}) \cup \mathcal{E}mb(\mathcal{F})$ is simply terminating.*
- *The TRS $(\mathcal{F}, \mathcal{R}) \cup \mathcal{E}mb(\mathcal{F})$ is terminating.*

$\square$

The proof is not difficult. This lemma appeared for the first time in Zantema [17], although it is implicit in many earlier works on termination, see Dershowitz [3] for a survey. Kurihara and Ohuchi [11, 12] proved the related equivalence "a TRS $(\mathcal{F}, \mathcal{R})$ is simplifying $\iff$ the transitive closure of the rewrite relation associated to the TRS $(\mathcal{F}, \mathcal{R}) \cup \mathcal{E}mb(\mathcal{F})$ is irreflexive".

The above lemma facilitates an easy proof of the undecidability of simple termination. For that matter we need some background on string rewriting systems. A *string rewriting system* (SRS) is a TRS $(\mathcal{F}, \mathcal{R})$ whose signature $\mathcal{F}$ contains only unary function symbols. A SRS $(\mathcal{F}, \mathcal{R})$ is called *non-length-increasing* if every rewrite rule $l \to r$ of $\mathcal{R}$ satisfies $|l| \geq |r|$. We call $\mathcal{R}$ *length-preserving* if $|l| = |r|$ for every rewrite rule $l \to r \in \mathcal{R}$. In the introduction we already mentioned that Caron [1] showed the undecidability of termination for length-preserving SRSs. Combining this result with Lemma 2.3 yields the undecidability of simple terminating for the same class of TRSs since it is very easy to show that a length-preserving SRS $(\mathcal{F}, \mathcal{R})$ is terminating if and only if the non-length-increasing SRS $(\mathcal{F}, \mathcal{R}) \cup \{f(x) \to x \mid f \in \mathcal{F}\}$ is terminating.

In the following sections we show that simple termination is an undecidable property of one-rule TRSs. This contradicts a result by Jouannaud and Kirchner [9]. They claimed that a one-rule TRS $\{l \to r\}$ is simply terminating if and only if l does not unify with any non-variable term embedded in r. This decision procedure is wrong as can be seen from the one-rule TRS $\mathcal{R} = \{f(a,b,x) \to f(x,x,x)\}$. The only non-variable term embedded in the right-hand side $f(x,x,x)$ is $f(x,x,x)$ itself, which clearly does not unify with the left-hand side $f(a,b,x)$. On the other hand, the term $f(a,b,f(a,b,b))$ has an infinite reduction with respect to the TRS

$$\left\{ \begin{array}{rcl} f(a,b,x) & \to & f(x,x,x) \\ f(x,y,z) & \to & x \\ f(x,y,z) & \to & y \\ f(x,y,z) & \to & z \end{array} \right\}$$

and hence $\mathcal{R}$ is not simply terminating, as a consequence of Lemma 2.3.

We would like to conclude this section with mentioning a (famous) open problem: the decidability of termination for one-rule SRSs. Partial results were obtained by Kurth [13]. He showed that termination is decidable in case the number of function symbols in the right-hand side of the single rewrite rule does not exceed six. Deciding the termination of one-rule non-length-increasing SRSs is much easier: a non-length-increasing SRS $\{l \to r\}$ is terminating if and only if $l \neq r$. Another open problem is whether termination is decidable for TRSs having only one left and right-linear rewrite rule (problem 21 in [5]).

3. Linear Bounded Automata

In this section we introduce linear bounded automata. Before presenting formal definitions, we give an intuitive description.

A linear bounded automaton consists of a tape which is divided into cells, a tape head that scans one cell at a time, and a finite control, see Figure 1(i). Each cell of the tape contains one symbol of a finite alphabet. A linear bounded

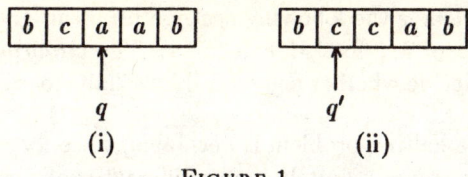

FIGURE 1.

automaton operates as follows. Depending on the state of the finite control and the symbol scanned by the tape head, a linear bounded automaton
- changes state,
- replaces the symbol scanned by the tape head by another symbol, and
- moves the tape head one cell to the left or to the right.

It is not required that the new state or the new tape symbol differ from the previous ones. On certain combinations of state and tape symbol, the linear bounded automaton stops operating. Moves to the left are not allowed if the tape head is positioned at the leftmost cell of the tape. Likewise, a right-move is forbidden if the tape head points to the rightmost cell of the tape. So a linear bounded automaton is like a Turing machine operating on a finite tape.

DEFINITION 3.1.
- A (deterministic) *linear bounded automaton* (LBA for short) is a triple $M = (Q, \Gamma, \delta)$ consisting of a finite set Q of *states*, a finite set Γ of *tape symbols*, disjoint from Q, and a *transition function* δ, which is a partial mapping from $Q \times \Gamma$ to $Q \times \Gamma \times \{L, R\}$.
- Let $M = (Q, \Gamma, \delta)$ be an LBA. A *configuration* is an element of $\Gamma^* Q \Gamma^+$, i.e. a string $w_1 q w_2$ with q a state, w_1 a string of tape symbols and w_2 a non-empty string of tape symbols. The idea is that the LBA scans the leftmost symbol of w_2. If $w_1 = \varepsilon$ then the tape head is positioned at the leftmost cell of the tape. The transition function δ determines a relation $\vdash_M$ on configurations as follows:

transition step	provided
$w_1 q a b w_2 \vdash_M w_1 a' q' b w_2$	$\delta(q, a) = (q', a', R)$
$w_1 b q a w_2 \vdash_M w_1 q' b a' w_2$	$\delta(q, a) = (q', a', L)$

Here $q, q' \in Q$, $a, a', b, b' \in \Gamma$ and $w_1, w_2 \in \Gamma^*$. Observe that for every configuration α there is at most one configuration β such that $\alpha \vdash_M \beta$. In other words, the transition relation $\vdash_M$ is deterministic.

The situation of Figure 1(i) can be described by the configuration $bcqaab$. If $\delta(q, a) = (q', c, L)$ then $bcqaab \vdash bq'ccab$, i.e. the situation of Figure 1(ii) is obtained.

DEFINITION 3.2.
- Let M be an LBA and α a configuration of M. We say that M *halts* for α if there is no infinite sequence $\alpha \vdash_M \alpha' \vdash_M \alpha'' \vdash_M \cdots$

- The *halting problem* is the following decision problem: given an LBA M and a configuration α of M, does M halt for α? The *uniform halting problem* is the problem to decide whether a given LBA M halts for all its configurations.

Observe that the halting problem is decidable, since for any configuration α of an LBA M there are only finitely many different configurations α' reachable from α (i.e. $\alpha \vdash_M^* \alpha'$). Hence halting can be decided by enumerating the (unique) sequence $\alpha \vdash_M \alpha' \vdash_M \alpha'' \vdash_M \cdots$. If M does not halt for α then at some stage we will reach a configuration that occurred earlier in the sequence. The uniform halting problem is undecidable though.

THEOREM 3.3. *Let M be an arbitrary LBA. It is undecidable whether M halts for all its configurations.* □

A proof of this statement can be found in Caron [1], where a reduction to Post's Correspondence Problem is given. Caron ascribes the above result to Hooper [7], but she obtained it independently. Moreover, [7] is very hard to read—there is for instance no notion of LBA—and it is not clear at all whether we may assume the simple definition of LBA given above (in order to conclude the undecidability of the uniform halting problem). By coding every LBA as a length-preserving SRS, similar to the construction described in Huet and Lankford [8], Caron reduced the undecidability of termination for length-preserving SRSs to the uniform halting problem for LBAs.

We conclude this section with a concrete example of an LBA, which will be used to illustrate subsequent developments.

EXAMPLE 3.4. Consider the LBA $M = (Q, \Gamma, \delta)$ with $Q = \{p, q\}$, $\Gamma = \{a, b\}$ and δ defined by the following table:

	a	b
p	(p, b, R)	(q, a, L)
q		(p, a, R)

The LBA M halts for configuration pab since $pab \vdash_M bpb \vdash_M qba \vdash_M apa$ and there is no transition step possible from configuration apa since abp is not a configuration.

4. Dauchet's Construction

In this section we associate with every LBA M a one-rule TRS $\mathcal{R}_M$ such that M halts for all its configurations if and only if $\mathcal{R}_M$ is terminating. In the next section we show that simple termination of $\mathcal{R}_M$ coincides with termination. Our construction is somewhat simpler than the one by Dauchet [2]—we use for instance only five variables as opposed to the six used by Dauchet—but the essence is the same.

DEFINITION 4.1. Let $M = (Q, \Gamma, \delta)$ be an arbitrary LBA. Suppose $Q = \{q_1, \ldots, q_m\}$, $\Gamma = \{a_1, \ldots, a_n\}$ and the number of pairs in $Q \times \Gamma$ for which δ is defined equals p. (So M contains p instructions.) The signature $\mathcal{F}_M$ of $\mathcal{R}_M$ consists of the following symbols:
- constants $q_1, \ldots, q_m$ and $a_1, \ldots, a_n$,
- a binary function symbol c and two constants $\#$ and NIL,
- a function symbol L of arity $m + n + 3$ and a function symbol R of arity p.

The use of the same characters for function symbols on the one hand, and states and tape symbols on the other hand, will cause no confusion. Next we define the single rewrite $l_M \to r_M$ of the TRS $\mathcal{R}_M$. The left-hand side l_M is the term

$$L(\text{c}(x_1, x_2), x_3, \text{c}(x_4, x_5), q_1, \ldots, q_m, a_1, \ldots, a_n).$$

Here $x_1, \ldots, x_5$ are (pairwise different) variables. The right-hand side r_M is the term

$$R(r_1, \ldots, r_p)$$

with r_k $(1 \leqslant k \leqslant p)$ defined as follows:

$$r_k = L(\text{c}(a', \text{c}(x_1, x_2)), q', x_5, Q_1, \ldots, Q_m, A_1, \ldots, A_n)$$

if the k-th instruction of M is a right-moving instruction $\delta(q_i, a_j) = (q', a', R)$, and

$$r_k = L(x_2, q', \text{c}(x_1, \text{c}(a', x_5)), Q_1, \ldots, Q_m, A_1, \ldots, A_n)$$

if the k-th instruction of M is a left-moving instruction $\delta(q_i, a_j) = (q', a', L)$. Here the terms $Q_1, \ldots, Q_m, A_1, \ldots, A_n$ are defined by

$$Q_l = \begin{cases} x_3 & \text{if } i = l, \\ q_l & \text{if } i \neq l \end{cases}$$

for $1 \leqslant l \leqslant m$ and

$$A_l = \begin{cases} x_4 & \text{if } j = l, \\ a_l & \text{if } j \neq l \end{cases}$$

for $1 \leqslant l \leqslant n$.

Let us try to explain the construction. The idea is that every configuration corresponds to an instance of the left-hand side. The first argument will contain the contents of the tape to the left of the tape head, the second argument will contain the state of the configuration, and the third argument will contain the contents of the tape cell scanned by the tape head as well as the contents of the tape to the right of the tape head. Tape parts are represented as terms by using the constructors c and NIL. For instance, aab will correspond to the term $\text{c}(a, \text{c}(a, \text{c}(b, \text{NIL})))$. However, the contents of the tape to the left of the tape head should be represented in reverse order since the rightmost symbol will be accessed first. So the instance of the left-hand side representing configuration *abqab* would have $c(b,c(a,\text{NIL})), q$ and $c(a,c(b,\text{NIL})))$ as first three arguments. There is only one problem with this approach: if the tape head is positioned at the leftmost tape cell then the term representing the empty tape part to its left would simply be NIL which is not an instance of $c(x_1, x_2)$, the first argument of the left-hand side. For that reason we introduced the special constant ♯, the trick being to represent a configuration like *abqab* by the three terms $c(b, c(a, c(♯, \text{NIL}))), q$ and $c(a, c(b, \text{NIL})))$. So the configuration *qa* will be represented by $c(♯, \text{NIL}), q$ and $c(a, \text{NIL})$. Observe that there is no need to add the symbol ♯ to the third term since the string w_2 in a configuration $w_1 q w_2$ is always non-empty. After this informal discussion, the following definition is easy.

DEFINITION 4.2. Let $M = (Q, \Gamma, \delta)$ be an LBA. We define two translations ϕ_1 and ϕ_2 from Γ^* to $\mathcal{T}(\Gamma \cup \{c, ♯, \text{NIL}\})$ as follows:

$$\phi_1(w) = \begin{cases} c(♯, \text{NIL}) & \text{if } w = \varepsilon, \\ c(a, \phi_1(w')) & \text{if } w = w'a \end{cases}$$

and

$$\phi_2(w) = \begin{cases} \text{NIL} & \text{if } w = \varepsilon, \\ c(a, \phi_2(w')) & \text{if } w = aw'. \end{cases}$$

These mappings are used to define a mapping ϕ from configurations of M to instances of the left-hand side of the single rewrite rule of $\mathcal{R}_M$ by means of the equation

$$\phi(w_1 q w_2) = L(\phi_1(w_1), q, \phi_2(w_2), q_1, \ldots, q_m, a_1, \ldots, a_n).$$

We still have to explain the remaining $m + n$ arguments of the L-terms occurring in the single rewrite rule, which really is the ingenious part of Dauchet's construction. This can best be done by means of a concrete example.

EXAMPLE 4.3. Consider the LBA M of Example 3.4. Its associated TRS $\mathcal{R}_M$ has the rewrite rule

$$L(c(x_1, x_2), x_3, c(x_4, x_5), p, q, a, b)$$

$$\to_R \begin{pmatrix} L(c(b, c(x_1, x_2)), p, x_5, x_3, q, x_4, b) \\ L(x_2, q, c(x_1, c(a, x_5)), x_3, q, a, x_4) \\ L(c(a, c(x_1, x_2)), p, x_5, p, x_3, a, x_4) \end{pmatrix}.$$

We have $pab \vdash_M bpb$. How is this transition step reflected at the rewrite level? In $\mathcal{R}_M$ we have the rewrite step

$$L(\text{C}(\sharp,\text{NIL}),p,\text{C}(a,\text{C}(b,\text{NIL})),p,q,a,b)$$

$$\to R \begin{pmatrix} L(\text{C}(b,\text{C}(\sharp,\text{NIL})),p,\text{C}(b,\text{NIL}),p,q,a,b) \\ L(\text{NIL},q,\text{C}(\sharp,\text{C}(a,\text{C}(b,\text{NIL}))),p,q,a,a) \\ L(\text{C}(a,\text{C}(\sharp,\text{NIL})),p,\text{C}(b,\text{NIL}),p,p,a,a) \end{pmatrix}$$

starting from $\phi(pab)$. The first argument

$$t_1 = L(\text{C}(b,\text{C}(\sharp,\text{NIL})),p,\text{C}(b,\text{NIL}),p,q,a,b)$$

of the resulting term corresponds to performing the instruction $\delta(p,a) = (p,b,R)$. This step is allowed since in configuration pab the state is p and the tape cell scanned by the tape head contains the symbol a. Notice that $t_1 = \phi(bpb)$. The second argument

$$t_2 = L(\text{NIL},q,\text{C}(\sharp,\text{C}(a,\text{C}(b,\text{NIL}))),p,q,a,a)$$

corresponds to performing the instruction $\delta(p,b) = (q,a,L)$. This step is of course not allowed as the symbol scanned by the tape head in configuration pab is a, not b. Observe that t_2 is no longer reducible since its last argument is an a instead of a b. In addition, t_2 is not reducible because its first argument is NIL instead of an instance of $\text{C}(x_1,x_2)$, signaling the fact than an illegal left-move has been attempted. Finally, the third argument

$$t_3 = L(\text{C}(a,\text{C}(\sharp,\text{NIL})),p,\text{C}(b,\text{NIL}),p,p,a,a)$$

is not reducible since its last four arguments are p,p,a,a instead of p,q,a,b. This means that an instruction of the form "$\delta(q,b) = \ldots$" has been attempted where "$\delta(p,a) = \ldots$" was required.

The easy implication in the desired equivalence "an LBA M halts for all configurations if and only if the TRS $\mathcal{R}_M$ is terminating" is stated in the following lemma.

LEMMA 4.4. *Let M be a LBA. If M does not halt for configuration α then $\mathcal{R}_M$ has an infinite reduction starting from the term $\phi(\alpha)$.*

PROOF. By construction, every transition step $\alpha \vdash_M \beta$ translates to $\phi(\alpha) = l_M\sigma \to r_M\sigma$ with one of the arguments of the resulting term $r_M\sigma$ equal to $\phi(\beta)$. Thus $\phi(\alpha) \to_{\mathcal{R}_M} C[\phi(\beta)]$ for some context C. Hence an infinite transition sequence $\alpha \vdash_M \alpha' \vdash_M \alpha'' \vdash_M \cdots$ corresponds to an infinite rewrite sequence $\phi(\alpha) \to_{\mathcal{R}_M} C[\phi(\alpha')] \to_{\mathcal{R}_M} C[C'[\phi(\alpha'')]] \to_{\mathcal{R}_M} \cdots$ □

The validity of the implication "M halts for all configurations $\Longrightarrow$ $\mathcal{R}_M$ is terminating" remains to be shown. This is less easy since there are many reducible terms in $\mathcal{R}_M$ that do not correspond to a configuration. However, since $\mathcal{R}_M$ contains no collapsing rules, we can use a recent result of Zantema. In [16] he showed that the termination behaviour of a TRS is not affected if we restrict our attention to well-typed terms according to some many-sorted type discipline, provided the system contains not both collapsing and duplicating rules. (See [16] for a precise formulation.) For $\mathcal{R}_M$ we take the following type discipline:

symbol	sort declaration
q_i $(1 \leq i \leq m)$, x_3	S_Q
a_i $(1 \leq i \leq n)$, $\#$, x_1, x_4	S_Γ
NIL, x_2, x_5	S_{LIST}
C	$S_\Gamma \times S_{\text{LIST}} \to S_{\text{LIST}}$
L	$S_{\text{LIST}} \times S_\Gamma \times S_{\text{LIST}} \times S_Q^m \times S_\Gamma^n \to S$
R	$S^p \to S$

Observe that both the left-hand side and right-hand side of the rewrite rules of $\mathcal{R}_M$ type-check and have the same sort S. Clearly only terms of sort S are reducible and hence the theorem of Zantema amounts to the equivalence of "$\mathcal{R}_M$ is terminating" and "$\mathcal{R}_M$ is terminating for all terms of sort S". It is not difficult to show that this last statement can be strengthened to "$\mathcal{R}_M$ is terminating for all *ground redexes* of sort S". So the problem remains how to extract an infinite transition sequence $\alpha_1 \vdash_M \alpha_2 \vdash_M \alpha_3 \vdash_M \cdots$ from an infinite rewrite sequence $t_1 \to_{\mathcal{R}_M} t_2 \to_{\mathcal{R}_M} t_3 \to_{\mathcal{R}_M} \cdots$ of ground terms of sort S with t_1 a redex.

DEFINITION 4.5. Let $M = (Q, \Gamma, \delta)$ be an LBA. Let $\Gamma_\# = \Gamma \cup \{\#\}$. We define two translations ψ_1 and ψ_2 from the set of ground terms of sort S_{LIST} to $\Gamma_\#^*$ as follows:

$$\psi_1(t) = \begin{cases} \varepsilon & \text{if } t = \text{NIL}, \\ \psi_1(t_2)\, t_1 & \text{if } t = \text{C}(t_1, t_2) \end{cases}$$

and

$$\psi_2(t) = \begin{cases} \varepsilon & \text{if } t = \text{NIL}, \\ t_1 \psi_2(t_2) & \text{if } t = \text{C}(t_1, t_2). \end{cases}$$

Observe that $\psi_2(t)$ is simply the reverse of $\psi_1(t)$. These mappings induce a mapping ψ from ground redexes of sort S to elements of $\Gamma_\#^* Q \Gamma_\#^+$ by means of the equation

$$\psi(L(t_1, t_2, t_3, q_1, \ldots, q_m, a_1, \ldots, a_n)) = \psi_1(t_1)\, t_2\, \psi_2(t_3).$$

Because of the presence of $\#$, elements of $\Gamma_\#^* Q \Gamma_\#^+$ are not configurations in the sense of Definition 3.1. The transition relation $\vdash_M$ however easily extends to elements of $\Gamma_\#^* Q \Gamma_\#^+$ by relaxing $b \in \Gamma$ and $w_1, w_2 \in \Gamma^*$ in Definition 3.1 to $b \in \Gamma_\#$ and $w_1, w_2 \in \Gamma_\#^*$. In the proof of Lemma 4.7 below, we will extract an infinite $\vdash_M$-sequence of elements of $\Gamma_\#^* Q \Gamma_\#^+$ from a presupposed infinite reduction sequence (of ground terms of sort S) in $\mathcal{R}_M$. In order to obtain an infinite $\vdash_M$-sequence of configurations, we have to get rid of the $\#$'s. The next definition provides an easy solution.

DEFINITION 4.6. We define a mapping χ from $\Gamma_\#^*$ to Γ^* inductively as follows:

$$\chi(w) = \begin{cases} \varepsilon & \text{if } w = \varepsilon, \\ a\chi(w') & \text{if } w = aw' \text{ with } a \in \Gamma, \\ \chi(w') & \text{if } w = \#w'. \end{cases}$$

This mapping is extended to elements of $\Gamma_\#^* Q \Gamma_\#^+$ by putting

$$\chi(w_1 q w_2) = \chi(w_1) q \chi(w_2).$$

Observe that $\chi(w_1 q w_2)$ is not necessarily a configuration, since $\chi(w_2)$ may be the empty string. However, it is not difficult to see that if $\alpha_1 \vdash_M \alpha_2 \vdash_M \alpha_3$ with $\alpha_1, \alpha_2, \alpha_3 \in \Gamma_\#^* Q \Gamma_\#^+$, then $\chi(\alpha_1)$ and $\chi(\alpha_2)$ are configurations such that $\chi(\alpha_1) \vdash_M \chi(\alpha_2)$. This implies that an infinite $\vdash_M$-sequence of elements of $\Gamma_\#^* Q \Gamma_\#^+$ is transformed by χ into an infinite $\vdash_M$-sequence of configurations.

LEMMA 4.7. *Let M be a LBA. If $\mathcal{R}_M$ is not terminating then M does not halt for all configurations.*

PROOF. Suppose $\mathcal{R}_M$ is not terminating. From the preceding discussion we know that there exists an infinite reduction sequence $t_1 \to_{\mathcal{R}_M} t_2 \to_{\mathcal{R}_M} t_3 \to_{\mathcal{R}_M} \cdots$ of ground terms of sort S with t_1 a redex. Consider the first step $t_1 \to_{\mathcal{R}_M} t_2$. Because M is deterministic, at most one of the arguments of t_2 is a ground redex (of sort S). From the reducibility of t_2 we infer that precisely one of its arguments is reducible. Let us call this argument t_2'. We have $\psi(t_1) \vdash_M \psi(t_2')$ by construction of $\mathcal{R}_M$. Let C be the context such that $t_2 = C[t_2']$. There exist terms t_i' for $i \geq 3$ such that $t_i = C[t_i']$ $(i \geq 3)$ and $t_2' \to_{\mathcal{R}_M} t_3' \to_{\mathcal{R}_M} \cdots$ is an infinite reduction sequence of ground terms of sort S with t_2' being a redex. Repeating the above argument yields an infinite sequence $\psi(t_1) \vdash_M \psi(t_2') \vdash_M \psi(t_3'') \to_{\mathcal{R}_M} \cdots$ of elements of $\Gamma_\#^* Q \Gamma_\#^+$. Applying the transformation χ to this sequence yields an infinite sequence of configurations $\chi(\psi(t_1)) \vdash_M \chi(\psi(t_2')) \vdash_M \chi(\psi(t_3'')) \to_{\mathcal{R}_M} \cdots$. Hence M does not halt for all configurations. □

5. Simple Termination is Undecidable for One-Rule Systems

Our main result follows if we can show that for the one-rule TRSs $\mathcal{R}_M$ introduced in the previous section, termination and simple termination coincide. It suffices to show that every terminating $\mathcal{R}_M$ is simply terminating. So consider an LBA M such that $\mathcal{R}_M$ is terminating. In the previous version of this paper we constructed a complicated well-founded order on $\mathcal{T}(\mathcal{F}_M)$ which extends the rewrite relation associated to the TRS $\mathcal{R}_M \cup \mathcal{E}mb(\mathcal{F}_M)$. This implies that $\mathcal{R}_M \cup \mathcal{E}mb(\mathcal{F}_M)$ terminates for all ground terms, which in turn implies the termination of $\mathcal{R}_M \cup \mathcal{E}mb(\mathcal{F}_M)$ (since every infinite reduction sequence can be transformed into an infinite reduction sequence involving only ground terms by simply substituting some constant for all variables). According to Lemma 2.3 this is equivalent to the simple termination of $\mathcal{R}_M$. In this paper we show that the powerful *distribution elimination* technique of Zantema [17] gives rise to a much simpler proof.

We start with a brief description of Zantema's technique, specialized to the present situation. Let $(\mathcal{F}, \mathcal{R})$ be a TRS and $f \in \mathcal{F}$ a function symbol of arity $n \geqslant 1$ that does not occur in the left-hand sides of the rewrite rules in $\mathcal{R}$. We inductively define a mapping E_f that assigns to every term $t \in \mathcal{T}(\mathcal{F}, \mathcal{V})$ a subset of $\mathcal{T}(\mathcal{F} \setminus \{f\}, \mathcal{V})$ as follows:

$$E_f(t) = \begin{cases} \{t\} & \text{if } t \in \mathcal{V}, \\ \bigcup_{i=1}^{n} E_f(t_i) & \text{if } t = f(t_1, \ldots, t_n), \\ \{g(u_1, \ldots, u_m) \mid \forall i \, u_i \in E_f(t_i)\} & \text{if } t = g(t_1, \ldots, t_m) \text{ and } f \neq g. \end{cases}$$

The set of rewrite rules $\{l \to u \mid l \to r \in \mathcal{R} \text{ and } u \in E_f(r)\}$ is denoted by $E_f(\mathcal{R})$.

THEOREM 5.1 (Zantema [17]). *If $E_f(\mathcal{R})$ is simply terminating and right-linear then $\mathcal{R}$ is simply terminating.* □

We would like to stress that Theorem 5.1 is only a very special case of the results in [17]. The idea is now to apply Theorem 5.1 to the TRS $\mathcal{R}_M$ with respect to the function symbol R, which only occurs in the right-hand side of the single rewrite rule of $\mathcal{R}_M$. If R happens to be a constant, i.e. if the LBA M contains no instructions, then $\mathcal{R}_M$ is immediately seen to be simply terminating. So we may assume that R is not a constant. Recall that the single rewrite rule of $\mathcal{R}_M$ has the form $l_M \to R(r_1, \ldots, r_p)$. One easily verifies that

$$E_R(\mathcal{R}_M) = \left\{ \begin{array}{ccc} l_M & \to & r_1 \\ & \vdots & \\ l_M & \to & r_p \end{array} \right\}.$$

Before we can apply Theorem 5.1 we have to check that $E_R(\mathcal{R}_M)$ is right-linear and simply terminating. Right-linearity is obvious. Simple termination follows from the termination of $\mathcal{R}_M$. First we show that $E_R(\mathcal{R}_M)$ is terminating.

LEMMA 5.2. *The TRS $E_R(\mathcal{R}_M)$ is terminating.*

PROOF. We use again the result of Zantema [16] on type removal. This is allowed since $E_R(\mathcal{R}_M)$ lacks collapsing rules. Consider the type discipline of Section 4. If $E_R(\mathcal{R}_M)$ is not terminating then there exists an infinite reduction sequence in which all terms have sort S. This implies that such an infinite reduction sequence contains only root reductions. However, if $s \to_{E_R(\mathcal{R}_M)} t$ is a root reduction then there exists a context C such that $s \to_{\mathcal{R}_M} C[t]$, and hence any infinite $E_R(\mathcal{R}_M)$-reduction sequence containing only root reductions can trivially be embedded into an infinite $\mathcal{R}_M$-reduction sequence. This contradicts the termination of $\mathcal{R}_M$. □

LEMMA 5.3. *The TRS $E_R(\mathcal{R}_M)$ is simply terminating.*

PROOF. One easily verifies that $|s| = |t|$ whenever $s \to_{E_R(\mathcal{R}_M)} t$. Since clearly $|s| > |t|$ whenever $s \to_{\mathcal{E}mb(\mathcal{F}_M \setminus \{R\})} t$, termination of $E_R(\mathcal{R}_M) \cup \mathcal{E}mb(\mathcal{F}_M \setminus \{R\})$ follows from the termination of $E_R(\mathcal{R}_M)$ (Lemma 5.2). Lemma 2.3 yields the simple termination of $E_R(\mathcal{R}_M)$. □

It should be stressed that the above lemma does not hold if M is an arbitrary Turing machine instead of an LBA. Actually this is the only place where we use a property of $\mathcal{R}_M$ which does not hold for the single rewrite rule of Dauchet. Theorem 5.1 now yields the desired result.

THEOREM 5.4. *The TRS $\mathcal{R}_M$ is simply terminating.* □

COROLLARY 5.5. *Simple termination is an undecidable property of one-rule TRSs.* □

Since every TRS $\mathcal{R}_M$ is *orthogonal* (left-linear and no critical pairs), *variable preserving* ($\mathcal{V}ar(l_M) = \mathcal{V}ar(r_M)$) and a *constructor system* (proper subterms of l_M do not contain the symbol $root(l_M)$), we can state that simple termination is an undecidable property of orthogonal, variable preserving, one-rule constructor systems.

References

1. A.-C. Caron, *Linear Bounded Automata and Rewrite Systems: Influence of Initial Configuration on Decision Properties*, Proceedings of the Colloquium on Trees in Algebra and Programming, Brighton, Lecture Notes in Computer Science **493**, pp. 74–89, 1991.

2. M. Dauchet, *Simulation of Turing Machines by a Regular Rewrite Rule*, Theoretical Computer Science **103**, pp. 409–420, 1992. Previous version in the Proceedings of the 3rd International Conference on Rewriting Techniques and Applications, Chapel Hill, Lecture Notes in Computer Science **355**, pp. 109–120, 1989.

3. N. Dershowitz, *Termination of Rewriting*, Journal of Symbolic Computation **3**(1), pp. 69–116, 1987.

4. N. Dershowitz and J.-P. Jouannaud, *Rewrite Systems*, in: Handbook of Theoretical Computer Science, Vol. B (ed. J. van Leeuwen), North-Holland, pp. 243–320, 1990.

5. N. Dershowitz, J.-P. Jouannaud, and J.W. Klop, *Open Problems in Rewriting*, Proceedings of the 4th International Conference on Rewriting Techniques and Applications, Como, Lecture Notes in Computer Science **488**, pp. 445–456, 1991.

6. B. Gramlich, *A Structural Analysis of Modular Termination of Term Rewriting Systems*, SEKI report SR-91-15, Universität Kaiserslautern, 1991.

7. P.K. Hooper, *The Undecidability of the Turing Machine Immortality Problem*, Journal of Symbolic Logic **31**(2), pp. 219–234, 1966.

8. G. Huet and D. Lankford, *On the Uniform Halting Problem for Term Rewriting Systems*, report 283, INRIA, 1978.

9. J.-P. Jouannaud and H. Kirchner, *Construction d'un Plus Petit Ordre de Simplification*, RAIRO Informatique Théorique **18**(3), pp. 191–207, 1984 (in French).

10. J.W. Klop, *Term Rewriting Systems*, in: Handbook of Logic in Computer Science, Vol. II (eds. S. Abramsky, D. Gabbay and T. Maibaum), Oxford University Press, pp. 1–112, 1992.

11. M. Kurihara and A. Ohuchi, *Modularity of Simple Termination of Term Rewriting Systems*, Journal of the Information Processing Society Japan **31**(5), pp. 633–642, 1990.

12. M. Kurihara and A. Ohuchi, *Modularity of Simple Termination of Term Rewriting Systems with Shared Constructors*, Theoretical Computer Science **103**, pp. 273–282, 1992.

13. W. Kurth, *Termination und Konfluenz von Semi-Thue-Systems mit nur einer Regel*, Ph.D. thesis, Technische Universität Clausthal, 1990 (in German).

14. E. Ohlebusch, *A Note on Simple Termination of Infinite Term Rewriting Systems*, report nr. 7, Universität Bielefeld, 1992.

15. D.A. Plaisted, *The Undecidability of Self-Embedding for Term Rewriting Systems*, Information Processing Letters **20**, pp. 61–64, 1985.

16. H. Zantema, *Type Removal in Term Rewriting*, Proceedings of the 3rd International Workshop on Conditional Term Rewriting Systems, Pont-à-Mousson, Lecture Notes in Computer Science **656**, pp. 148–154, 1993.

17. H. Zantema, *Termination of Term Rewriting by Interpretation*, Proceedings of the 3rd International Workshop on Conditional Term Rewriting Systems, Pont-à-Mousson, Lecture Notes in Computer Science **656**, pp. 155–167, 1993.

Optimal Normalization in Orthogonal Term Rewriting Systems

Zurab Khasidashvili

INRIA-Rocquencourt, 78153 Le Chesnay, France
Zurab.Khasidashvili@inria.fr

Abstract. We design a normalizing strategy for orthogonal term rewriting systems (OTRSs), which is a generalization of the call-by-need strategy of Huet-Lévy [4]. The redexes contracted in our strategy are essential in the sense that they have "descendants" under any reduction of a given term. There is an essential redex in any term not in normal form. We further show that contraction of the innermost essential redexes gives an optimal reduction to normal form, if it exists. We classify OTRSs depending on possible kinds of redex creation as non-creating, persistent, inside-creating, non-left-absorbing, etc. All these classes are decidable. TRSs in these classes are sequential, but they do not need to be strongly sequential. For non-creating and persistent OTRSs, we show that our optimal strategy is efficient as well.

1 Introduction

In this paper, we study correct and optimal computations in Orthogonal Term Rewriting Systems (OTRSs). We only consider one-step rewriting strategies, which select a single redex to contract at each rewriting step. A strategy is correct when it succeeds in computing the normal form of any term, whenever such a normal form exists. A correct strategy is optimal when the computation of normal form is performed for a minimal cost. The choice of one-step rewriting provides us with a very simple and natural cost measure: the cost of a reduction is just the number of elementary steps it comprises, which is also the number of redexes it contracts. Clearly, such a cost notion assumes a term implementation of term rewriting. By contrast, the more sophisticated notions of cost of [1, 12, 13] take the contraction of one set of similar redexes, a so-called redex family, as the unitary cost. Thus, in a realistic implementation, redex families must be implemented as a single shared object, so that optimal reduction are indeed optimal, when real (i.e., implemented) rewriting steps are considered.

The seminal paper on strategies in Orthogonal Term Rewriting Systems is [4] by Huet and Lévy. They designed the call-by-need strategy and proved its correctness for computation of normal forms. Following Huet and Lévy's definition, a redex u in a term t is needed when at least one residual of u is contracted by all normalizing reductions starting from t. Call-by-need is then defined as any strategy that contracts needed redexes only. Huet and Lévy did not limit themselves to the existence of correct strategies; they were also concerned with the feasibility of such strategies. They thus introduced the class of strongly sequential systems. In a strongly sequential system, it is possible to give an algorithm to find a needed redex in any term not in normal form. They also proved that strong sequentiality is a decidable property.

This fundamental work was later extended by many authors. Kennaway and Sleep [6] gave a proof of correctness of the call-by-need strategy for all orthogonal combinatory reduction systems using labeling. Maranget [13] studied the optimality of the graph implementation of call-by-need. Kennaway [5] showed that every OTRS admits a computable correct strategy, and not only strongly sequential systems. In Klop and Middeldorp [10, 11], simpler proofs of decidability of strong sequentiality are given. In Thatte [18] the notion of left-sequentiality is introduced. In Oyamaguchi [15] a class of sufficiently sequential OTRSs, which properly contains the class of strongly sequential OTRSs, is shown to be decidable. In Kesner [7] sequentiality is studied in OTRSs with order-sorted alphabets. Toyama [19] generalized some results from [4] to left-linear overlapping TRSs.

The essential strategy, as defined in this paper, is based on a notion of descendant, which allows the tracing of all subterms, including contracted redexes, during reductions. Essential subterms of a term t have descendants under any reduction starting from t. So they contribute to the finite or infinite result of a computation. For the case of redexes, essentiality coincides with Maranget's improved notion of neededness (developed independently) [13]. By contrast with the notion of neededness of Huet and Lévy, the latter notion also makes sense in the case when the term does not have a normal form.

The existence of an essential redex in any term t not in normal form follows from two observations. First, one can easily construct an "erasing reduction" starting from t in which inessential redexes have no descendants. Second, for any reduction from t to some term s there is at least one redex in t that has a descendant in s. Applying this property to the erasing reduction for t, we obtain a proof that there is at least one redex in t that is essential. (In the body of paper we prove a somewhat stronger theorem.)

The correctness proof of call-by-need in [4] is not very direct, as pointed out by Klop in [10]. Our proof is simple and short. The only thing we need is that essential redexes have essential residuals under any reduction that does not contract them. Therefore, if a term t has a reduction P that contracts infinitely many essential redexes and Q is a reduction from t to a term s, then s possesses a reduction P' contracting infinitely many essential redexes. Essential redexes contracted in P' are residuals of essential redexes contracted in P that are not contracted in Q; thus t is not normalizable. Hence, essential reductions always succeed in finding a normal form, whenever it exists.

Further, we show that reductions are optimal if and only if they contract non-duplicating essential redexes, i.e., redexes that do not duplicate other essential redexes. In particular, innermost essential reductions are optimal. This optimality result relies on the crucial fact that the essential residuals of non-duplicating essential redexes remain non-duplicating. Our strategy is also optimal in the sense of [1, 12, 13], because non-duplicating essential redexes are alone in their families.

We note that complexity of an OTRS depends crucially on the possible kinds of redex creation. We define non-creating, persistent, inside-creating, and outside-creating OTRSs. In non-creating systems, there is no creation of redexes. In persistent systems, creation depends on rewrite rules only; the actual arguments and the context in which reduction takes place do not matter. In particular, Recursive (Applicative) Program Schemes [2] are persistent. In inside-creating systems, the arguments, but not the context, can also take part in creation; any created redex is inside the contractum of a creating redex. In outside-creating systems, the context always takes part in redex creation.

Since before erasure a redex must be absorbed in an argument of a redex above it, it is also relevant to consider a classification of outside creation depending on what kind of absorption is possible. For example, the leftmost outermost redex in a term is essential in left-normal TRSs (where function symbols precede variables in the left-hand sides of rules [14]), because these TRSs are non-left-absorbing. So we distinguish non-absorbing, non-left-absorbing, and non-right-absorbing systems.

All these classes have different normalization behavior. For example, in outside-creating systems, all reductions are terminating [17]. In persistent systems, normalization and termination are decidable. In non-absorbing systems, all outermost redexes are essential, etc. Furthermore, these classes are easily decidable. In fact, the above classification is a classification depending on critical pairs between left-hand sides and right-hand sides of rewrite rules. In our opinion, a similar classification of TRSs will be useful for further development of TRS theory. Here we study the Persistent TRSs in more depth. In particular, we show that in Persistent TRSs the innermost essential strategy is computable and that weak normalization is decidable. Note that some of the Persistent TRSs may not be strongly sequential. Such a system can be obtained from any non-strongly-sequential system by replacing in its rules right-hand sides with fresh constants.

The rest of the paper is organized as follows. In section 2 we introduce the notion of strict equivalence of reductions, based on the descendant notion, and establish the "strict" Church-Rosser theorem. In section 3 we prove the correctness of the essential strategy. Section 4 is devoted to the classification of OTRSs. In section 5 we obtain the optimality result. In section 6 we study properties of inessential and independent subterms. In section 7 we establish the relation between neededness and essentiality, and in section 8 we study optimal reductions in Persistent TRSs.

2 Strict Confluence for OTRSs

We refer to [3] and [10] for introductions in the TRS theory. Below we consider only OTRSs.

Notation We use R for OTRSs, t, s, e, o for terms, r for rewrite rules, u, v, w for redexes, and P, Q for reductions. We write $s \in t$ if s is a subterm of t. A one-step reduction in which a redex u in a term t is contracted is written as $t \xrightarrow{u} s$ or $t \to s$. We write $P : t \twoheadrightarrow s$ if P denotes a reduction of t to s comprising 0 or more steps. $\emptyset_t$, or simply $\emptyset$, denotes the empty reduction of a term t; the symbol $\emptyset$ is also used to denote the empty set. A normal form (nf for short) of t is denoted by $!t$. If the last term of P coincides with the initial term of Q, then $P + Q$ denotes the concatenation of P and Q. $|P|$ denotes the length, i.e., the number of steps, of P. WN stands for (weak) normalization (i.e., existence of a nf) and SN stands for strong normalization (i.e., termination of all reductions).

Definition 2.1 (1) Let $r : t \to s$ be a rewrite rule and v be an r-redex. Subterms of v that correspond to variables of t are the *arguments* of v and the rest is the *pattern* of v. Subterms of v rooted at the pattern are called the *pattern-subterms* of v. The *arguments, pattern,* and *pattern-subterms* are defined analogously in the contractum of v.

(2) Let $t' \xrightarrow{u} s'$ and let e be the contractum of u in s'. For each argument o of u there are 0 or more arguments of e. We call them $(u\text{-})descendants$ of o. Correspondingly, subterms of o have 0 or more *descendants*. By definition, the *descendant* of each pattern-subterm of u is e. Descendants of all redexes of t' except u are also called *residuals*. By definition, u does not have *residuals* in s. A redex of s' is said to be *created* by contracting u or to be an (u)-*new* redex if it is not a residual of a redex of t'. It is clear what is to be meant by *descendants* of subterms that are not in u. The notions of *descendant* and *residual* extend naturally to arbitrary reductions.

Example 2.1 The notion of descendant is illustrated by the picture below of a reduction in the OTRS $R = \{f(g(x,y), b) \to h(h(b,y), y), h(x,c) \to d\}$.

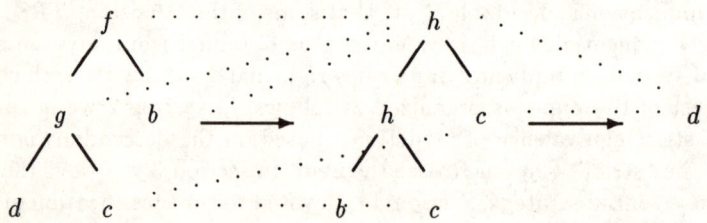

Notation If F is a set of redexes in t and $P : t \twoheadrightarrow s$, then F/P denotes the set of all residuals of redexes from F in s. If $F = \{u\}$, then we write u/P for $\{u\}/P$. In the following, F will also denote a complete F-development, where the residuals of redexes from F are contracted as long as possible. Similarly, if $u \in t$, then u will also denote the reduction $t \xrightarrow{u} s$.

Definition 2.2 Let $Q : t \twoheadrightarrow s$ and $t \xrightarrow{u} e$. Then the *residual* Q/u *of* Q *by* u is defined by induction on $|Q|$ as follows. If $Q = \emptyset_t$, then $Q/u = \emptyset_e$. If $Q = Q_1 + v$, then $Q/u = Q_1/u + v/(u/Q_1)$.

Definition 2.3 Let $P : t \twoheadrightarrow s$ and $Q : t \twoheadrightarrow e$. Then the *residual* P/Q *of* P *by* Q and the *residual* Q/P *of* Q *by* P are defined by induction on $|P|$ as follows.

(1) If $P = \emptyset_t$, then $P/Q = \emptyset_e$ and $Q/P = Q$.
(2) If $P = P_1 + u$, then $P/Q = P_1/Q + u/(Q/P_1)$ and $Q/P = (Q/P_1)/u$.
We write $P \sqcup Q$ for $P + Q/P$.

Definition 2.4 We call the reductions $P : t \twoheadrightarrow t'$ and $Q : s \twoheadrightarrow s'$ *strictly equivalent* (written $P \approx Q$) if $t = s$, $t' = s'$, and P-descendants and Q-descendants of any subterm of t are the same occurrences in t' and s'.

Strict equivalence is weaker then strong equivalence [4], but it is enough for our purposes.

Theorem 2.1 (Strict Church-Rosser Theorem) Let P and Q be any co-initial reductions in an OTRS R. Then $P \sqcup Q \approx Q \sqcup P$.

Proof. Similar to the ordinary case. One has only to check that reduction steps u and v in a term "strictly commute": $u + v/u \approx v + u/v$.

3 Correctness of the Essential Strategy

Definition 3.1 We call a subterm s in t *essential* (written $ES(s,t)$) if s has at least one descendant under any reduction starting from t and *inessential* (written $IE(s,t)$) otherwise.

Lemma 3.1 Let $s_0, \ldots, s_k \in t$ be such that $IE(s_i, t)$ for all $i = 0, \ldots, k$. Then there exists a reduction P starting from t such that none of the subterms $s_0, \ldots, s_k$ have P-descendants.

Proof. Let P_i be a reduction starting from t such that s_i does not have P_i-descendants (P_i exists since $IE(s_i, t)$). Then, by Theorem 2.1, one can take $P = (\ldots (P_1 \sqcup P_2) \sqcup \ldots \sqcup P_n)$.

Lemma 3.2 Let $t \xrightarrow{u} t'$ and $s \in t$. Then $IE(s,t)$ iff, for every u-descendant s' of s, $IE(s', t')$.

Proof. ($\Rightarrow$) Let $IE(s,t)$. Then there is some reduction P starting from t such that s does not have P-descendants. By Theorem 2.1, $P+u/P \approx u+P/u$. Hence, s' does not have P/u-descendants, i.e., $IE(s', t')$. ($\Leftarrow$) If all u-descendants of s are inessential in t', then, by Lemma 3.1, there is some reduction P starting from t' under which none of them have descendants. Thus s does not have $u + P$-descendants, i.e., $IE(s,t)$.

Corollary 3.1 Let $P : t \twoheadrightarrow t'$ and $s \in t$. Then $IE(s,t)$ iff all P-descendants of s are inessential in t'. In particular, if t' is a normal form, then $ES(s,t)$ iff s has a P-descendant.

Lemma 3.3 (1) Let $t \xrightarrow{u} t'$ and $e \in s \in t$, and let e' be a u-descendant of e. Then s has a u-descendant that contains e'.
   (2) Let $e \in s \in t$ and $ES(e,t)$. Then $ES(s,t)$.
   (3) Let s be a pattern-subterm of a redex $u \in t$. Then $ES(u,t)$ iff $ES(s,t)$.

Proof. (1) By considering all relative positions of u and e, using Definition 2.1.

(2) By (1).

(3) Let $t \xrightarrow{u} t'$. The contractum of u is the unique descendant of u, as well as of s. Hence the lemma follows from Corollary 3.1.

Definition 3.2 We call a subterm s of a term t *unabsorbed in a reduction* $P : t \twoheadrightarrow e$ if none of the descendants of s appear in redex-arguments of terms in P, and call s *absorbed in* P otherwise. We call s *unabsorbed in* t if it is unabsorbed in any reduction starting from t and *absorbed in* t otherwise.

Remark 3.1 It is easy to see that a redex $u \in t$ is unabsorbed iff u is *external* [4] in t. Note that if $s \in t$ is unabsorbed in $P : t \twoheadrightarrow e$, then s has exactly one descendant in each term of P.

Lemma 3.4 Let $P : t_0 \xrightarrow{u_0} t_1 \xrightarrow{u_1} \ldots \to t_n$. Then for any outermost redex $v \in t_n$ there is an unabsorbed in P redex $w \in t_0$ that has a descendant $s \in t_n$ containing a pattern-subterm e of v.

Proof. By induction on $|P|$. If $|P| = 0$, then the result is obvious. So suppose that $|P| > 0$.

(a) Let v be a residual of a redex v' of t_{n-1}. Further, let $v^* = v'$ if $v' \not\in u_{n-1}$ and $v^* = u_{n-1}$ otherwise. Since v is outermost, v^* is outermost as well. By the induction assumption, the descendant $s' \in t_{n-1}$ of $w \in t_0$ contains a pattern-subterm of v^*. Therefore, by Lemma 3.3.(1), s', and hence also w, have a descendant that contains a pattern-subterm of v, and w is unabsorbed in P.

(b) Assume that v is a new redex. Then u_{n-1} is outermost and, by the induction assumption, a pattern-subterm e' of u_{n-1} is inside a descendant s' of w. Let e be the maximal common subterm of v and the contractum of u_{n-1}. Then e is a pattern-subterm of v. Since e' is a pattern-subterm of u_{n-1}, its descendant contains e. By Lemma 3.3.(1), s', and hence also w, have a descendant s that contains the pattern-subterm e of v. So w is unabsorbed in P as well.

Lemma 3.5 For any reduction $P : t \twoheadrightarrow s$ starting from a term t not in nf there is an outermost redex u in t that is unabsorbed in P.

Proof. If $|P| = 0$ or $|P| > 0$ and s is not in nf, then the lemma follows immediately from Lemma 3.4. Otherwise, let $P : t \twoheadrightarrow s' \xrightarrow{v} s$. By Lemma 3.4 there is an unabsorbed in $t \twoheadrightarrow s'$ redex $u \in t$ the only descendant o of which contains a pattern-subterm of v. Thus u has a descendant in s that is not in a redex-argument, since s is in nf. Hence, u is unabsorbed in P.

Lemma 3.6 Let u be absorbed in $P : t \twoheadrightarrow s$ and let $Q : t \twoheadrightarrow e$. Then u is absorbed in $Q \sqcup P$.

Proof. It suffices to prove the lemma for $|Q| = 1$, i.e., $Q = w$ for a redex w in t. The rest follows by induction on $|Q|$. Furthermore, we can assume that u is absorbed in the last step of P, i.e., there is a redex v created in the last step of P

and containing the only descendant o of u in its argument. If o does not have a w/P-descendant, then, by Theorem 2.1, u does not have $(w \sqcup P)$–descendants, so u is absorbed in $w \sqcup P$ (otherwise its descendants cannot be erased). If o has a w/P-descendant o', then, by Lemma 3.3.(1), v also has a w/P-descendant that contains o'. Since in w/P only residuals of w are contracted and v is a new redex, v has a residual v' that contains o'. But, by Theorem 2.1, o' also is a $(w \sqcup P)$–descendant of u. Hence, u is absorbed in $w \sqcup P$.

Theorem 3.1 Let t be a term not in normal form in an OTRS. Then there is an unabsorbed outermost — hence essential — redex in t.

Proof. If, for any outermost redex u_i of t, there is a reduction P_i that absorbs u_i $(i = 1, \ldots, k)$, then, by Lemma 3.6, $P_1 \sqcup \ldots \sqcup P_k$ absorbs all redexes u_i, contrary to Lemma 3.5.

Lemma 3.7 Let $t \xrightarrow{u} t'$, $IE(u,t)$, and let u' be a u-new redex. Then $IE(u',t')$.

Proof. Let e' be the contractum of u. By Corollary 3.1, $IE(e',t')$. So if $u' \in e'$, then by Lemma 3.3.(2), $IE(u',t')$, and if $e' \in u'$, then e' is a pattern-subterm of u' and, by Lemma 3.3.(3), $IE(u',t')$.

Lemma 3.8 Let $P : t_0 \xrightarrow{u_0} t_1 \xrightarrow{u_1} \ldots$ contain infinitely many essential steps and let $t_0 \xrightarrow{u} s_0$. Then s_0 also has a reduction with infinitely many essential steps.

Proof. Let $F_i = u/(u_0 + \ldots + u_{i-1})$ for $i = 0, 1, \ldots$ Using Theorem 2.1, we can construct the diagram

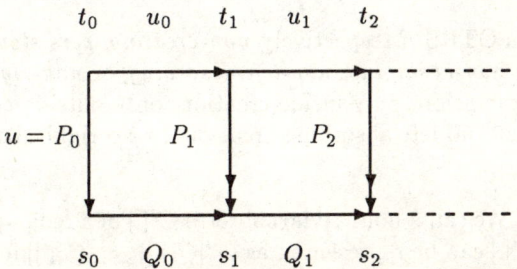

where $P_i : t_i \twoheadrightarrow s_i$ is a complete F_i-development and $Q_i : s_i \twoheadrightarrow s_{i+1}$ is a complete u_i/P_i-development. It is easy to see that there are infinitely many k such that $ES(u_k, t_k)$ and $u_k \notin F_k$, since otherwise there would be some m such that in $t_m \xrightarrow{u_m} t_{m+1} \xrightarrow{u_{m+1}} \ldots$ all contracted essential redexes were residuals of u. But this is impossible, since by Lemmas 3.3.(2) and 3.7 inessential steps cannot duplicate or create essential redexes, and residuals of u are disjoint. By Corollary 3.1, $u_k \notin F_k$ and $ES(u_k, t_k)$ imply that u_k has at least one essential P_k-residual in s_k, i.e., Q_k contains at least one essential step. Hence $Q_0 + Q_1 + \ldots$ contains infinitely many essential steps.

Theorem 3.2 Let t be a term in an OTRS R. Then t has a normal form iff it does not possess a reduction contracting infinitely many essential redexes.

Proof. From Lemma 3.8 and Theorem 3.1.

4 A Classification of OTRSs

Definition 4.1 (1) Let $t \xrightarrow{u} s$; let v be a new redex in s; let o be the contractum of u, $C[\,]$ the pattern of o, and $C'[\,]$ the pattern of v.

(a) We call v a *generated* redex if C' is inside C.

(b) We call v an *inside-created* redex if the top of C' is in C, but C' may contain a symbol outside C. In the latter case, we call v an *argument-dependent* inside-created redex.

(c) We call v an *argument-dependent outside-created* redex if the top of C is in C' and is strictly below the top of C', and if there is at least one edge symbol of C that is strictly inside C' (i.e. that is not an edge symbol of C').

(d) We call v an *argument-independent outside-created* redex if the top of C is in C' and is strictly below the top of C', and if there is no edge symbol of C strictly inside C'.

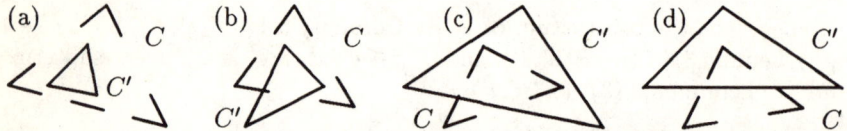

(e) We call v a *collapsing outside-created* redex if C is empty and call v *non-collapsing* otherwise.

(f) We call v *non-absorbing* if all its arguments are in o. We call v *left-absorbing* (resp. *right-absorbing*) if v has an argument to the left (resp. to the right) of o.

(2) We call an OTRS R respectively *non-creating, persistent, inside-creating, outside-creating, non-absorbing, non-left-absorbing*, or *non-right-absorbing*, if no creation, only generation, only inside-creation, only outside-creation, only non-absorbing creation, no left-absorbing creation, or no right-absorbing-creation is possible in R.

Definition 4.2 We call a non-trivial subterm $C'[\,]$ of a redex-pattern $C[\,]$ a *sub-redex-pattern*. $C'[\,]$ can be represented as $C'''[C_1[\,],\ldots,C_n[\,]]$ for some non-trivial context C'' (in general, in various ways). We call C'' a *pre-sub-redex-pattern*. C'' is *proper* if at least one C_i is non-trivial. *Sub-contractum-pattern* and *pre-sub-contractum-pattern* are defined analogously. We call the pre-sub-redex-pattern C'' *non-left* (resp. *non-right*) if there is a hole in the redex-pattern $C[\,]$ to the left (resp. to the right) of the sub-redex-pattern C'.

Lemma 4.1 Generation of a redex is possible in an OTRS R iff a redex-pattern is a pre-sub-contractum pattern.

Proof. From Definitions 4.1 and 4.2 (see the picture).

Lemma 4.2 (1) Inside-creation is possible in an OTRS R iff a pre-redex-pattern C^* coincides with a pre-sub-contractum-pattern C'' and, for any edge symbol of C^* that is not an edge symbol in the redex-pattern, the corresponding symbol in the contractum-pattern is an edge symbol.

(2) Moreover, the inside-creation is argument-dependent iff the pre-redex-pattern C^* is proper.

Proof. (1) ($\Rightarrow$) By Definition 4.1. ($\Leftarrow$) If $r : t \to s$ is the rule corresponding to the contractum-pattern, then, to create a redex inside contractum, it suffices to contract an appropriate r-redex. The variables of t that correspond to holes of C'' must be instances of the sub-redex-patterns, obtained from the redex-pattern by removing the pre-redex-pattern C^*.
(2) By (1) and Definition 4.2.

Remark 4.1 Consider the OTRS $R = \{f(g(x)) \to f(h(x))\}$. The pre-redex-pattern $f([\ ])$ is also a pre-(sub)-contractum-pattern, but there is no inside-creation in R, since f is an edge symbol of $f([\ ])$ but is not an edge symbol of the contractum-pattern $f(h[\ ])$.

Lemma 4.3 A collapsing outside-creation is possible in an OTRS R iff R contains a collapsing rule (whose right-hand side is a variable) and a rule with at least two function symbols in the redex-pattern.

Proof. ($\Rightarrow$) By Definition 4.1. ($\Leftarrow$) A redex-pattern that has at least two function symbols can be represented as $C_1[C_2[\ ]]$ with non-trivial contexts C_1 and C_2. Now, if $r : t \to x$ is a collapsing rule, then there is a collapsing outside-creation after contraction, in the context $C_1[\ ]$, of any r-redex in which x is instantiated by a term $C_2[e]$.

Lemma 4.4 A non-collapsing outside-creation is possible in an OTRS R iff a pre-proper-sub-redex-pattern C^* coincides with a pre-contractum-pattern C'' and, for any edge symbol of C'' that is not an edge symbol of the contractum-pattern, the corresponding symbol in the redex-pattern is an edge symbol. The non-collapsing outside creation is argument-dependent iff there is at least one edge symbol in C'' whose corresponding symbol in C^* is not an edge symbol in the redex-pattern.

Proof. ($\Rightarrow$) By Definition 4.1. ($\Leftarrow$) The redex-pattern can be represented as $C_1[C_2[\ ]]$ where the top of C_2 coincides with the top of C''. If the contractum-pattern corresponds to the rule $r : t \to s$, then there is an argument-dependent outside-creation after contraction of an appropriate r-redex u in the context C_1. The arguments of u that correspond to edge symbols of the contractum-pattern that are strictly inside the redex-pattern must be instances of the sub-redex-patterns rooted at these edge symbols.

Obviously, in (b)-(d) all possible cases of redex creation are considered, and (e)-(f) is a further classification of outside-creation. Note, that the collapsing outside-creation in Lemma 4.3 is left-absorbing (resp. right-absorbing) iff C_2 is non-left (resp. non-right). The non-collapsing outside creation in Lemma 4.4 is left-absorbing (resp. right-absorbing) iff C^* is non-left (resp. non-right). Hence, we have the following proposition.

Proposition 4.1 It is decidable whether an OTRS is non-creating, persistent, inside-creating, outside-creating, non-absorbing, non-left-absorbing, or non-right-absorbing.

Theorem 4.1 (1) A term t in a non-left-absorbing OTRS has a normal form iff it does not have a reduction contracting infinitely many leftmost outermost redexes.

(2) A term t in a non-absorbing OTRS has a normal form iff it does not have a reduction contracting infinitely many outermost redexes.

(3) A term t in a non-right-absorbing OTRS has a normal form iff it does not have a reduction contracting infinitely many rightmost outermost redexes.

Proof. (1) For any reduction P starting from t, one can show by induction on $|P|$ that the leftmost outermost redex u of t is unabsorbed in P. Hence u is essential and the theorem follows from Theorem 3.2. Parts (2) and (3) are proved analogously.

5 Optimal Normalization

Definition 5.1 We call a reduction P *quasi-innermost-essential* if each essential redex contracted in P is innermost among essential redexes. $|P|_I$ denotes the number of innermost essential steps in P. We call a redex $v \in t$ *non-duplicating essential* if it is essential and any essential redex u of t other than v has exactly one essential v-residual (u may have inessential residuals as well).

Lemma 5.1 Let $t \xrightarrow{u} s$, where u is an innermost essential redex or an inessential redex in t. Further, let v be any redex in t different from u and let $t \xrightarrow{v} e$.

(1) If v is essential, then it has exactly one u-residual v', which is essential.

(2) If u is innermost essential, then u/v is quasi-innermost-essential, $|u/v|_I \geq 1$, and if v is non-duplicating essential, then $|u/v|_I = 1$. If u is inessential, then each step in u/v is inessential.

(3) If v is non-duplicating essential, then v' is non-duplicating essential.

Proof. (1) By Corollary 3.1 and Lemma 3.3.(2).

(2) Again by Corollary 3.1 and Lemma 3.3.(2), since the residuals of u are disjoint.

(3) Let v be non-duplicating essential. Further, let w' be an essential redex in s, different from v'. If w' is a residual of a redex $w \in t$, then $w \neq v$ and by Corollary 3.1, w is essential. Since, by (2), u/v is quasi-innermost-essential, it follows from (1) that w' is the only essential residual of w, and the only essential v-residual of w has exactly one essential u/v-residual. Now it follows from Theorem 2.1 that w' has exactly one essential v'-residual. If w' is a new redex, then the contractum s' of u is a pattern-subterm of w' or $w' \in s'$. By Lemma 3.7, u is essential. It follows from (2) that u has exactly one essential $(v \sqcup u)$-descendant. Hence, by Theorem 2.1, s' has exactly one essential v'-descendent. So if s' is

a pattern-subterm of w', then it follows from Lemma 3.3.(3) that the essential v'-residual of w' is the one for which the essential v'-descendant of s' is a pattern-subterm. If $w' \in s'$, then each residual of w' that is inside an inessential v'-descendent of s' is inessential by Lemma 3.3.(2). Thus in this case w' has at most one essential residual. Hence, v' is non-duplicating essential.

Lemma 5.2 Let $P : t_0 \twoheadrightarrow t_n$ be a normalizing quasi-innermost-essential reduction and let $t_0 \xrightarrow{v_0} s_0$. Then P/v_0 also is a normalizing quasi-innermost-essential reduction and $|P|_I \leq |P/v_0|_I + 1$. Moreover, $|P|_I = |P/v_0|_I + 1$ iff v_0 is non-duplicating essential.

Proof. Let $P : t_0 \xrightarrow{u_0} t_1 \xrightarrow{u_1} \ldots \to t_n$. Using Theorem 2.1, we can construct the diagram

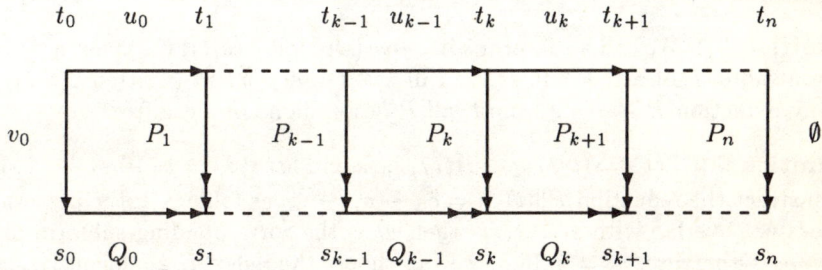

where $P_i : t_i \twoheadrightarrow s_i$ is a complete $v_0/(u_0 + \ldots + u_{i-1})$-development and $Q_j : s_j \twoheadrightarrow s_{j+1}$ is a complete u_j/P_j-development, $i = 1, \ldots, n; j = 0, \ldots, n-1$. If u_j is inessential, then, by Corollary 3.1, Q_j is also inessential. If v_0 is inessential, then, by Corollary 3.1, the reductions P_i are inessential. Hence, it follows from Lemma 5.1 that P/v_0 is quasi-innermost-essential as well, and $|P/v_0|_I = |P|_I$. Now let us consider the case $ES(v_0, t_0)$. Since t_n is a nf, we have $P_n = \emptyset$. Let k be the minimal number such that $P_k = \emptyset$. Then, for any $j \geq k$, $P_j = \emptyset$ and $Q_j = u_j$. By Lemma 5.1.(1), for all $l = 1, \ldots, k-1$, P_l contracts only one redex — say, v_l — which is the only essential residual of v_0. Hence, $P_k = \emptyset$ implies that $v_{k-1} = u_{k-1}$ and $Q_{k-1} = \emptyset$. By Lemma 5.1.(2), for all m such that $1 \leq m \leq k-2$ and u_m is innermost essential, Q_m is quasi-innermost-essential and $|Q_m|_I \geq 1$. It remains to show that $|Q_m|_I = 1$ iff v_0 is non-duplicating essential. Indeed, if v_0 is non-duplicating essential, then, by Lemma 5.1.(3), $v_1, \ldots, v_{k-1}$ are non-duplicating essential and, by Lemma 5.1.(2), $|Q_m|_I = 1$. If v_0 is not non-duplicating essential, then there is an essential redex $w \in t_0$ that has at least two essential v_0-residuals. By Lemma 5.1.(1), w has exactly one essential residual in each term of P until it is contracted, say in $t_{m'}$. Since the residuals of w are not contracted in $Q_0, \ldots, Q_{m-1}$, it follows from Corollary 3.1 that there are at least two essential residuals of w in $s_{m'}$. So, by Theorem 2.1, $u_{m'}$ has at least two essential residuals in $s_{m'}$. Therefore, $m' < k-1$ and, by Lemma 5.1.(2), $|Q_{m'}|_I > 1$.

Theorem 5.1 Let t be a normalizable term in an OTRS. A normalizing reduction P starting from t is optimal iff in each term it contracts a non-duplicating essential redex. In particular, innermost essential normalizing reductions are optimal.

Proof. Let $P : t \twoheadrightarrow s$ be an optimal normalizing reduction and Q be an innermost essential normalizing reduction (Q exists by **Theorem 3.2**). It follows from Lemma 5.2 that $|Q|_I \leq |Q/P|_I + |P| = |P|$ and $|Q|_I = |P|$ iff each step in P is non-duplicating essential. But P is optimal, so $|P| \leq |Q| = |Q|_I \leq |P|$. Hence each step in P is non-duplicating essential and Q is optimal.

6 Properties of Inessential and Independent Subterms

Notation We write $t = (t_1//s_1, \ldots, t_n//s_n)s$, if $s_1, \ldots, s_n$ are disjoint subterms in s and t is obtained from s by replacing them with $t_1, \ldots, t_n$, respectively.

Definition 6.1 We call a subterm s in t *free* (written $F(s,t)$) if s is not a proper pattern-subterm of a redex in t. We call $s \in t$ *independent* (written $I(s,t)$) if, for any reduction P starting from t, all P-descendants of s are free.

Definition 6.2 Let $t = (t_1//s_1, \ldots, t_k//s_n)s$ and let $P : s = e_0 \xrightarrow{v_0} e_1 \xrightarrow{v_1} \ldots$ Let us construct the reduction $P||(t) : t = o_0 \xrightarrow{u_0} o_1 \xrightarrow{u_1} \ldots$ as follows. If the pattern of v_0 does not overlap with $s_1, \ldots, s_n$, then u_0 is the corresponding subterm of v_0 in $t = o_0$. Otherwise, $u_0 = \emptyset$. In o_1 we can choose the redex u_1 analogously, since o_1 is obtained from e_1 by replacing the descendants of $s_1, \ldots, s_n$ with $t_1, \ldots, t_n$ respectively, and so on.

Remark 6.1 The reduction $P||(t)$ depends not only on P and t, but also on the choice of $s_1, \ldots, s_k$, but the notation does not give rise to ambiguity.

Lemma 6.1 Let $t = (t_1//s_1, \ldots, t_n//s_n)s$, let the subterms $s_1, \ldots, s_k$ be independent and essential in s, and let $s_{k+1}, \ldots, s_n$ be inessential in s. Further, let $P : s \twoheadrightarrow s'$ and $Q = P||(t) : t \twoheadrightarrow t'$. Then

(1) The term t' can be obtained from s' by replacing the descendants of $s_1, \ldots, s_k$ with $t_1, \ldots, t_k$ and replacing the descendants of $s_{k+1}, \ldots, s_n$ with some terms. The replaced occurrences of $t_1, \ldots, t_k$ in t' are the only descendants of $t_1, \ldots, t_k \in t$.

(2) If e and o are corresponding subterms in s and t, then e has P-descendants iff o has Q-descendants; each Q-descendant of o is in the subterm that corresponds to some P-descendant of e.

Proof. By induction on $|P|$.

Corollary 6.1 Let $t = (t_1//s_1, \ldots, t_n//s_n)s$, where $s_1, \ldots, s_k$ are essential and independent in s, subterms $s_{k+1}, \ldots, s_n$ are inessential in s, and $t_1, \ldots, t_k$ are independent in t. If s' and t' are any corresponding subterms in s and t, then $ES(s', s)$ iff $ES(t', t)$.

Lemma 6.2 Let $I(s,t)$, $P : t \twoheadrightarrow t'$, and let s have a P-descendant $s' \in t'$. Then there exists a reduction $Q : s \twoheadrightarrow s'$ such that any subterm $e \in s$ has a P-descendant in $s' \in t'$ iff e has a Q-descendant in s'.

Proof. Easy induction on $|P|$.

Corollary 6.2 Let $ES(e,s)$, $ES(s,t)$, and $I(s,t)$. Then $ES(e,t)$.

7 Neededness and Essentiality

Definition 7.1 A redex u in t is called *needed* [4] if in each reduction of t to normal form (if any) at least one residual of u is contracted; u is called *needed** [13] if u has at least one residual under any reduction starting from t that does not contract residuals of u.

Theorem 7.1 (1) Let t be a normalizable term in an OTRS R and let u be a redex in t. Then u is essential in t iff it is needed in t.

(2) Let t be a term in an OTRS R and let u be a redex in t. Then u is essential in t iff it is needed* in t.

Proof. (1) ($\Rightarrow$) Let $IE(u,t)$. By Theorem 3.2, there is a normalizing essential reduction $P : t \twoheadrightarrow !t$. By Corollary 3.1, the residuals of u are not contracted in P. Hence u is not needed in t. ($\Leftarrow$) From Definitions 3.1 and 7.1.

(2) ($\Rightarrow$) Let $IE(u,t)$. Further, let x be a fresh variable and let $s = (x//u)t$. By Corollary 6.1, $IE(x,s)$, i.e., there is some reduction P starting from s such that x does not have P-descendants. Hence, by Lemma 6.1.(2), s does not have descendants under $P\|(t)$, which does not contract residuals of u. ($\Leftarrow$) From Definitions 3.1 and 7.1.

8 Optimal Reductions in Persistent TRSs

Lemma 8.1 Let t be a term in a PTRS. Then $F(e,t)$ iff $I(e,t)$.

Proof. By considering all relative positions of e and u.

Definition 8.1 Let r be a rule in a PTRS. We call the sequence of numbers of essential arguments of the left-hand side of r the *essentiality indicator* of r. By definition, all r-redexes have the same *essentiality indicator*.

Lemma 8.2 The i-th argument of a redex u is essential iff i belongs to the essentiality indicator of u.

Proof. From Corollary 6.1.

Lemma 8.3 Let t be a term in a PTRS R.
 (1) Each outermost redex in t is essential.
 (2) If $F(s,t)$, $ES(s,t)$, and $ES(e,s)$, then $ES(e,t)$.
 (3) $IE(e,t)$ iff e is in an inessential argument of an essential redex of t.

Proof. (1) By Definition 4.1.

(2) By Lemma 8.1 and Corollary 6.2.

(3) ($\Rightarrow$) Since $IE(e,t)$, it follows from the persistency of R that e is inside an outermost, hence essential, redex. Let u be the innermost essential redex containing e. By Lemma 3.3.(3), e is in an argument s of u. It follows from (2) that there is no redex in s that contains e. Thus $ES(e,s)$. But $F(u,t)$ and $F(s,u)$. Hence, by (2), $IE(e,u)$ and $IE(s,u)$. ($\Leftarrow$) By Lemma 3.3.(2) and (2).

Corollary 8.1 Let t be a term in a PTRS and let essentiality indicators of redexes in t be known. Then one can find all essential redexes in t as follows: Mark outermost redexes in t; in essential arguments of marked redexes mark outermost redexes; and so on, as long as possible. The marked redexes are all the essential redexes in t.

Definition 8.2 Let R be a PTRS and $r, r' \in R$. We call r' an *essential successor* of r if an r'-redex has an essential occurrence in the right-hand side of r. We call a sequence of rules $r_0, r_1, \ldots$, where r_{i+1} is an essential successor of r_i, an *essential chain* or an *essential r_0-chain*. For any r_0-redex u, we also call an essential r_0-chain an *essential u-chain*.

Lemma 8.4 Let r' be an essential successor of r. If u is an essential r-redex in t and $t \xrightarrow{u} s$, then s contains an essential r'-redex created by u.

Proof. By Definition 8.2, the contractum e of u contains an essential r'-redex v created by u. By Corollary 3.1, $ES(e,s)$. Thus, by Lemmas 8.1 and 8.3.(2), $ES(v,s)$.

Lemma 8.5 Let t be a term in a PTRS. Then t is WN iff any essential chain of an essential redex in t is finite.

Proof. ($\Rightarrow$) Consider an infinite essential chain $r_0, r_1, \ldots$ of an essential redex in t. We show by induction on i that $(\alpha)_i$: for any reduction $P : t = t_0 \xrightarrow{u_0} t_1 \xrightarrow{u_1} \ldots$ and for each $i < |P|$ there is a k_i such that t_i contains an essential r_{k_i}-redex. $(\alpha)_0$ is obvious. If u_{k_i} is an essential r_{k_i}-redex, then, by Lemma 8.4, t_{i+1} contains an essential r_{k_i+1}-redex. Otherwise, by Corollary 3.1, any essential r_{k_i}-redex has an essential residual in t_{i+1}. Thus $(\alpha)_i$ holds for all i, and t is not normalizable. ($\Leftarrow$) For any r-redex u, let the *weight* of u be the length of a maximal essential r-chain. Since the contraction of any essential redex generates essential redexes only with smaller weights, a reduction that in each step contracts an innermost redex among essential ones with maximal weight terminates in a nf.

Lemma 8.6 Let $R = \{r_1, \ldots, r_m\}$ be a PTRS. Then the essentiality indicators of R-rules can be found using the following algorithm:

1st *step:* Mark in the left-hand side of r_1 all variables that do not occur in the right-hand side of r_1.

n^{th} *step* $(n > 1)$: Let r_i be the rule used in the $(n-1)^{th}$ step, and let $r^n : t^n \to s^n$ be r_{i+1} if $i < m$ and r_1 if $i = m$. Mark in s^n the unmarked redex-arguments that correspond to marked variables in left-hand sides of corresponding rules; mark in t^n all unmarked variables that occur in s^n only in marked subterms.

If no occurrences are marked during the last m steps, then the algorithm stops — the unmarked variables in left-hand sides of rules are all the essential arguments. Otherwise proceed by performing the next step.

Proof. Let us show by induction on n that $(\alpha)_n$: the marked occurrences in r^n are inessential. The case $n = 1$ is clear ($r^1 = r_1$). Suppose that $(\alpha)_k$ is valid. By Lemma 8.3, all marked occurrences in s^{k+1} are inessential. Hence, by Corollary 3.1, all marked variables in t^{k+1} are inessential. Thus marked occurrences in R-rules are inessential. Now let $r_i : t' \to s'$, $IE(x,t')$, and let $P : t' \to s' \twoheadrightarrow o'$ be a minimal-length reduction such that x does not have P-descendants. Let us show by induction on $|P|$ that x is marked in t' after termination of the marking procedure. The case $|P| = 1$ is clear — x is an erased argument of t'. Let x' be a descendant of x in s' and let u^* be the minimal redex that contains x' and has a descendant in o'. Then the argument o^* of u^* that contains x' does not have $P\|u^*$-descendants. Hence, by the induction assumption, if u^* is say an r_j-redex, then o^* corresponds to a marked argument of r_j. Thus each occurrence of x in s' is in a marked subterm. Hence, x is marked in t'.

Theorem 8.1 Let t be a term in a PTRS with a finite number of rules. It is decidable whether t has a nf, and the nf can be found using the innermost essential strategy in the minimal number of steps.

Proof. By Lemmas 8.2, 8.5, and 8.6 and Theorem 5.1.

Acknowledgements

The correctness and optimality results were first obtained under the supervision of Sh. Pkhakadze and Kh. Rukhaia for some reductions in the "Notation Theory" of Sh. Pkhakadze [16]. The results were generalized for OTRSs and reported in [8] after I became acquainted with the TRS theory by J. W. Klop's introductory paper [9], pointed to me out by G. Mints. I am grateful to them, as well as to H. Barendregt and J.-J. Lévy, for constant support. I thank H. Barendregt, P.-L. Curien, H. Ganzinger, D. Kesner, J. W. Klop, G. Kucherov, and D. Rösner for organizing my talks at Nijmegen, Paris, Saarbrücken, Orsay, Amsterdam, Nancy, and Ulm. I would like to thank L. Maranget for useful discussions and many helpful comments, and Sh. Aditua, D. Doligez, G. Gonthier, I. Jacobs, B. Pierce, G. Tagviashvili, and K. Urbaitis for their help in the preparation of this paper.

References

1. Berry G., Lévy J.-J. Minimal and optimal computations of recursive programs. JACM 26, 1979, p.148-175.
2. Courcelle B. Recursive Applicative Program Schemes. In: J.van Leeuwen ed. Handbook of Theoretical Computer Science, Chapter 9, vol.B, 1990, p. 459-492.

3. Dershowitz N., Jouannaud J.-P. Rewrite Systems. In: J.van Leeuwen ed. Handbook of Theoretical Computer Science, Chapter 6, vol.B, 1990, p. 243-320.
4. Huet G., Lévy J.-J. Computations in Orthogonal Rewriting Systems. In: Computational Logic, Essays in Honor of Alan Robinson, ed. by J.-L. Lassez and G. Plotkin, MIT Press, 1991. (First appeared in 1979, as IRIA Laboria report number 359).
5. Kennaway J.R. Sequential evaluation strategy for parallel-or and related reduction systems. Annals of Pure and Applied Logic 43, 1989, p.31-56.
6. Kennaway J.R., Sleep M.R. Neededness is hypernormalizing in regular combinatory reduction systems. Preprint, School of Information Systems, University of East Anglia, Norwich, 1989.
7. Kesner D. Free sequentiality in orthogonal order-sorted rewriting systems with constructors, in Proc. 11th Int. Conf. on Automated Deduction, Saratoga Springs, NY, LNAI 607, 1992.
8. Khasidashvili Z. Minimal Normalizing Strategy for Orthogonal and Persistent Term Rewriting Systems (short version). Proceedings of I.Vekua Institute of Applied Mathematics of Tbilisi State University, vol.36, 1990. p.189-199.
9. Klop J.W. Term Rewriting Systems: a tutorial. Bulletin of the EATCS 32, 1987, p. 143-182.
10. Klop J.W. Term Rewriting Systems. In: S.Abramsky, D.Gabby, and T.Maibaum eds. Handbook of Logic in Computer Science, vol. II, Oxford University Press, 1992.
11. Klop J.W., Middeldorp A. Sequentiality in orthogonal term rewriting systems. Report CS-R8932, CWI Amsterdam, 1989.
12. Lévy J.-J. Optimal Reduction in the Lambda-Calculus. In: To H. B. Curry: Essays on Combinatory Logic, Lambda Calculus and Formalism, J. P. Seldin and J. R. Hindley editors, Academic Press, 1980.
13. Maranget L. "La stratégie paresseuse", These de l'Université' de PARIS VII, 1992.
14. O'Donnell M.J. Computing in systems described by equations. Springer LNCS 58, 1977.
15. Oyamaguchi M. Sufficient Sequentiality: A decidability Condition for call-by-need computations in term rewriting systems, Report, Mie University, 1987.
16. Pkhakadze Sh. Some problems of the Notation Theory (in Russian). Proceedings of I.Vekua Institute of Applied Mathematics, Tbilisi 1977.
17. Van Raamsdonk F. A simple proof of confluence for weakly orthogonal combinatory reduction systems. Report CS-R9234, CWI Amsterdam, 1992.
18. Thatte S. A refinement of strong sequentiality for term rewriting with constructors. Information and computation 72, 1987, p.46-65.
19. Toyama Y. Strong sequentiality of left-linear overlapping term rewriting systems. LICS, 1992.

A Graph Reduction Approach to Incremental Term Rewriting

(Preliminary Report)

John Field

IBM T.J. Watson Research Center
P.O. Box 704, Yorktown Heights, NY 10598 USA
jfield@watson.ibm.com

Abstract. Our concern is *incremental term rewriting*: efficient normalization of a sequence of terms that are related to one another by some set of disjoint subterm replacements. Such sequences of *similar* terms arise frequently in practical applications of term rewriting systems. Previous approaches to this problem [9, 10], have applied only to a limited class of reduction systems and rewriting strategies. In this paper, we present a new algorithm, $\text{INC}_{f_\mathcal{R}}$, for carrying out incremental term rewriting in an *arbitrary* left-linear term rewriting system $\mathcal{R}$ possessing a non-parallel normalizing rewriting strategy $f_\mathcal{R}$. This algorithm is based on a novel variant of graph rewriting.

1 Introduction

The logical/computational duality of term rewriting systems (TRSs) makes them attractive for implementing a variety of formal systems. In most such systems, computation is performed by *normalizing* terms—rewriting them until some canonical form is reached. In this paper, we present a general algorithm, $\text{INC}_{f_\mathcal{R}}$, for performing *incremental* reduction in term rewriting systems. Incremental reduction is concerned with normalizing sequences of *similar* terms that are derived from one another by some set of disjoint subterm replacements. Systems that implement term rewriting are often used in settings in which relatively small changes to terms to be reduced are made frequently by the system's user. The advantages of efficient normalization techniques that would obviate reducing each altered term *ab initio* should thus be self-evident.

Consider a TRS that reverses the contents of a list, such that, e.g., Rev([A; B; $\boxed{\text{C}}$; D]) $\longrightarrow^*$ [D; C; B; A]. (We will describe such a rewriting system in more detail in Sect. 2). If we replace the term C (boxed in the initial term above) with E, we get a new list [A; B; E; D]. In this case, we would like to take advantage of the result of having previously reduced [A; B; C; D] to yield a new reversed list, [D; E; B; A], with a minimum of computation. $\text{INC}_{f_\mathcal{R}}$ will produce the final list given the altered initial term without requiring *any* applications of rewriting rules, by *substituting* E for C in the final list.

For a more complicated example, consider the reduction

$$T_0 = \text{Rev}(\text{App}([A; B; C], \boxed{[D; E]})) \longrightarrow^* [E; D; C; B; A]$$

Here, App$(\cdot, \cdot)$ appends two lists. If we replace the boxed list [D; E] in T_0 with, e.g., [G], it is no longer possible to produce the new normal form, [G; C; B; A], with only a simple substitution. INC$_{f\mathcal{R}}$ will instead have the effect of storing the "partially reversed" intermediate term $T_1 = \text{App}(\text{Rev}(\square), [C; B; A])$ and associating it with subsequent replacements at the subterm designated by $\boxed{[D; E]}$ in T_0. The subterm '$\square$' in T_1 corresponds to $\boxed{[D; E]}$ in T_0; by substituting the new term $[G]$ for $\square$ in T_1, we can take advantage of the fact that the sublist $[A; B; C]$ has already been reversed, producing the new list with a minimum of additional reduction steps.

As a final example, consider the following term:

$$\text{Rev}([A; \text{App}(\boxed{\text{Dup}(\text{Hd}(\boxed{[B]}_2), [C; D])}_1, \boxed{[\boxed{E}_4; F]}_3)])$$

Here, we have designated *multiple* points of possible subterm replacement by subscripted boxes. Note that several of these terms are nested within one another. The difficult aspect of this more general problem is to retain the intermediate results of computations that pertain to the context that remains *unchanged* after subterm replacement—and do so with a minimum of computational overhead.

1.1 Outline of Results

Previous work on incremental rewriting by van der Meulen [9, 10] has addressed restricted subsets of TRSs, the *primitive recursive schemes* and *layered primitive recursive schemes*, respectively. The close relation between these systems and *attribute grammars* makes it possible to take advantage of many of the ideas developed for incremental computation in the latter setting [2]. However, van der Meulen's techniques require that an *innermost* reduction strategy be used (at least for a subset of the reduction rules), which is in general incomplete, and is sometimes also inefficient.

By contrast, INC$_{f\mathcal{R}}$ (which is based on a generalization of ideas in [5, 3]) applies to any unconditional left-linear TRS $\mathcal{R}$ having a (non-parallel) normalizing reduction strategy $f_\mathcal{R}$. We will not require that $\mathcal{R}$ be noetherian, nor even that it be confluent. $\mathcal{R}$ may be single- or multi- sorted.

INC$_{f\mathcal{R}}$ relies on four key ideas: First, we systematically transform $\mathcal{R}$ into a new TRS, $\triangle \mathcal{R}$. We also derive a new reduction strategy $f_{\triangle \mathcal{R}}$. These transformations will enable certain key reduction steps to be "checkpointed" for use in subsequent reductions. Second, we use a variant of the well-known technique of *graph reduction*, which will allow the results of appropriate reduction steps to be shared to avoid repeating them when normalizing similar terms. Third, we generalize the algebraic techniques for analysis of rewriting systems developed by Huet and Lévy[6] and extended by Maranget [7, 8] to compare the number of reduction steps performed by INC$_{f\mathcal{R}}$ to corresponding non-incremental reductions. Finally, we achieve a practical implementation of INC$_{f\mathcal{R}}$ by making certain term transformations and adding additional data structures. We will refer to the resulting optimized algorithm as INC$'_{f\mathcal{R}}$.

Although INC$_{f\mathcal{R}}$ requires that certain "administrative" reductions be performed in addition to $\mathcal{R}$-reduction steps, it is possible to show that INC$_{f\mathcal{R}}$ has the property (expressed informally) that no non-administrative reduction step is applied to a term if an equivalent reduction step has been applied to the common part of a term used in a previous reduction, a property we will call *relative optimality*. For *orthogonal* TRSs possessing a *call-by-need* reduction strategy, relative optimality becomes optimality, in the sense that our incremental reductions are of minimum length.

INC$_{f\mathcal{R}}$ requires that those subterms that are replaceable (which we will refer to as *substituends*) be designated *prior* to reduction. The number of administrative operations per non-administrative reduction step required by the optimized algorithm INC$'_{f\mathcal{R}}$ can then be bounded asymptotically by a linear function of the number of substituends in the original term. Although it is possible for *every* subterm to be designated as a substituend, it is often the case in practice that only a subset of terms are replaceable, which would result in reduced administrative complexity. In any event, the worst-case behavior seems very difficult to achieve in all but contrived examples.

Due to space constraints, we will concentrate in the sequel on motivating the concepts we use, formally defining INC$_{f\mathcal{R}}$, and summarizing our correctness and complexity results. Complete details and proofs may be found in [4].

1.2 Definitions

We will require the following definitions:

The set of terms defined over a (single-sorted) signature Σ of function symbols and a denumerable set of variables $\mathcal{V}$ is designated by $Term(\Sigma, \mathcal{V})$. If $\mathcal{V}$ is unspecified, we will assume it to be some fixed set of "standard" variables. For any $F \in \Sigma$, $arity(F)$ denotes the arity of F. The set of variables in a term T is denoted by $vars(T)$.

Given $T \in Term(\Sigma, \mathcal{V})$, an *occurrence* u in T is a *path expression* of the form $[n_1, \ldots, n_m]$, denoting the path from the root of T to one of its subterms. Concatenation of two path expressions is denoted by juxtaposition. The set of occurrences in a term T is denoted by $\mathcal{O}(T)$. For any $u \in \mathcal{O}(T)$, the subterm of T rooted at u is denoted by T/u. The function symbol or variable at the root of T/u is denoted by $T//u$. If T is a term and S is a set of function symbols or variables, $\mathcal{O}_S(T)$ denotes the set $\{u \mid C//u \in S\}$.

Path expression u is a *prefix* of path expression v, notated $u \preceq v$, if there exists w such that $v = uw$; if $w \neq [\ ]$, then $u \prec v$. If $u \preceq v$, then v/u denotes the path w such that $uw = v$. If S is a set of path expressions and u is a path expression, S/u denotes $\{v/u \mid v \in S, u \preceq v\}$; $S \cdot u$ denotes $\{vu \mid v \in S\}$. If S_1 and S_2 are sets of path expressions, $S_1 \cdot S_2 = \{uv \mid u \in S_1, v \in S_2\}$. Path expressions u and v are *disjoint*, notated $u \mid v$, if neither $u \preceq v$ nor $v \preceq u$. A *set* of occurrences S is disjoint if for all $u, v \in S$, $u \mid v$. u is *to the left* of v if there exists w, u', v', i, and j such that $u = w[i]u'$, $v = w[j]v'$, and $i < j$.

Let u be a path expression and T be a term. Then we will refer to pairs of the form $(u \leftarrow T)$ as *fixed terms*. If S is a set of fixed terms, we let $\mathcal{O}(S)$ denote the set $\{u \mid (u \leftarrow T) \in S\}$. We will assume in the sequel that if S is a set of fixed terms, $\mathcal{O}(S)$ is a set of *disjoint* occurrences. If T and S are terms and $u \in \mathcal{O}(T)$, then $T[u \leftarrow S]$ denotes the term T' such that for all $w \in \mathcal{O}(T')$, either $T'//w = T//w$ for $w \prec u$ or $w \mid u$, or there exists $v \in \mathcal{O}(S)$ such that $w = uv$ and $T'//uv = S//v$. If $u \notin \mathcal{O}(T)$, then $T[u \leftarrow S] = T$. If S is a set of fixed terms, then $T[S] = T$ if $S = \emptyset$, otherwise $T[S] = (T[u \leftarrow S])[S - (u \leftarrow S)]$ for any $(u \leftarrow S) \in S$. Since $\mathcal{O}(S)$ is assumed to be disjoint, this definition is unambiguous.

The set of indexed *holes*, $\mathcal{H}$, is a special set of variables of the form $\boxed{i}$, where $i \in \{1, 2, \ldots\}$. Given signature Σ, the set of Σ-*contexts*, denoted by $Cont(\Sigma)$ is the set of terms $Term(\Sigma, \mathcal{V} \cup \mathcal{H})$. For contexts containing a single hole, we will frequently drop the index and simply use '$\Box$'. We will abbreviate $\mathcal{O}_{\boxed{i}}(C)$ by $\mathcal{O}_i(C)$. The set of indices of all indexed holes in a context C, i.e., $\{i \mid \mathcal{O}_i(C) \neq \emptyset\}$, is denoted by $ind(C)$. Our notion of context extends the traditional definition by allowing multiple holes. A context C is *linear* if for all $i \in ind(C)$, $|\mathcal{O}_i(C)| = 1$. C is *simple* if it is linear and $|ind(C)| = 1$. C is *proper* if it is linear, and for all $i \in ind(C)$, $i < j$ implies that for all $u \in \mathcal{O}_i(C)$ and $v \in \mathcal{O}_j(C)$, u is to the left of v. C is a *term context* if $ind(C) = \emptyset$.

A *substitution* σ is specified concretely by postfix expressions of the form $[x_1 := T_1, \ldots, x_n := T_n]$. For $\boxed{i} \in \mathcal{H}$, we abbreviate $[\ldots \boxed{i} := T \ldots]$ by $[\ldots i := T \ldots]$. We abbreviate $[\square := T]$ by $[T]$. If C is not a term context and σ is some substitution such that $\sigma(C)$ is a term context, then we will say that σ *closes* C.

A Σ *term rewriting system* $\mathcal{R}$ is a set of pairs of terms of the form $L \longrightarrow R$ such that $L, R \in Term(\Sigma, \mathcal{V})$ and $vars(L) \subseteq vars(R)$. We denote $\mathcal{R}$'s associated signature by $\Sigma(\mathcal{R})$. If $\alpha = (L \longrightarrow R) \in \mathcal{R}$, then we define $L(\alpha) = L$ and $R(\alpha) = R$. $\mathcal{R}$ is *left-linear* if for all $\alpha \in \mathcal{R}$, and $x \in vars(L(\alpha))$, $|\mathcal{O}_x(L(\alpha))| = 1$. In the sequel, we will assume that every TRS is left-linear.

The one-step $\mathcal{R}$-*contraction relation*, $\longrightarrow_\mathcal{R}$, is defined on elements of $Term(\Sigma, \mathcal{V})$ as follows: $T \longrightarrow_\mathcal{R} T'$ if and only if there exists a pair $A = \langle u, \alpha \rangle$ and a substitution σ such that $\alpha \in \mathcal{R}$, $u \in \mathcal{O}(T)$, $T/u = \sigma(L(\alpha))$, and $T' = T[u \leftarrow \sigma(R(\alpha))]$. We will refer to the pair A as a *contraction*, and the subterm T/u as an α-*redex*. When we want to highlight the role of A, we will write $T \xrightarrow{A}_\mathcal{R} T'$. We will feel free to drop sub- or super- scripts when evident from their equations. Contractions $\langle u, \alpha \rangle$ and $\langle v, \beta \rangle$ are *disjoint* if $u \mid v$; they are *strongly coinitial* if they are disjoint or if there exists $w \neq [\,]$ such that $uw = v$ and $w \notin \mathcal{O}_\Sigma(L(\alpha))$, or such that $vw = u$ and $w \notin \mathcal{O}_\Sigma(L(\beta))$. A left-linear TRS $\mathcal{R}$ is *orthogonal* if for all terms T and contractions A and B of T, A and B are strongly coinitial.

The $\mathcal{R}$-*reduction relation*, $\longrightarrow_\mathcal{R}^*$, is the reflexive, transitive closure of $\longrightarrow_\mathcal{R}$. A term T is an $\mathcal{R}$-*normal form* if there exists no term T' such that $T \longrightarrow_\mathcal{R} T'$. such that $T \longrightarrow^* T'$. A *reduction* ρ is a sequence of contractions $A_1 A_2 \ldots A_n$ such that if ρ is nonempty, there exist terms $T_0, T_1, \ldots, T_n$ where $T_0 \xrightarrow{A_1} T_1 \cdots T_{n-1} \xrightarrow{A_n} T_n$. This reduction is abbreviated by $T_0 \xrightarrow{\rho}^* T_n$. A reduction ρ is a reduction *of* term T if there exists T' such that $T \xrightarrow{\rho}^* T'$, in which case we also will use the applicative notation $(\rho : T) = T'$.

Given TRS $\mathcal{R}$, a partial function $f_\mathcal{R}$ mapping $\Sigma(\mathcal{R})$-terms to $\mathcal{R}$-contractions is a *non-parallel reduction strategy* if for all T such that $f_\mathcal{R}(T) = A$, A is a contraction of T. $f_\mathcal{R}$ is *normalizing* if for all terms T that have a normal form, there is a reduction of the form $T \xrightarrow{f_\mathcal{R}(T)} T_1 \xrightarrow{f_\mathcal{R}(T_1)} T_2 \longrightarrow \cdots \longrightarrow T_{n-1} \xrightarrow{f_\mathcal{R}(T_{n-1})} T_n$ such that T_n is a normal form.

A *parallel contraction* $\mathcal{A}$ is a finite set of mutually disjoint contractions. If $\mathcal{A}$ is a parallel contraction and T is a term, T *contracts in parallel* to T' by $\mathcal{A}$, notated $T \xrightarrow{\mathcal{A}}^{\|} T'$ if either $\mathcal{A} = \emptyset$ and $T = T'$, or there exists a (non-parallel) contraction $A \in \mathcal{A}$ and term T'' such that $T \xrightarrow{A} T'' \xrightarrow{(\mathcal{A} - A)} T'$. Since $\mathcal{A}$ must be disjoint, this definition is unambiguous. The notions of parallel reduction relation, parallel reduction, and parallel reduction strategy are analogous to the non-parallel case.

2 Problem Overview: Single Substituends

To give an overview of the problem of incremental reduction and methods we will use to solve it, we will begin with an instructive (but unrealistically simple) example, in which only single subterm substitutions can be made. In Sect. 4, we will address the general case.

In the sequel, we will use a simple rewriting system on lists, $\mathcal{L}$, whose rules are as follows:

$$\begin{aligned}
\text{Hd}([elt;\ list]) &\longrightarrow elt & \textbf{(Hd)} \\
\text{Tl}([elt;\ list]) &\longrightarrow list & \textbf{(Tl)} \\
\text{App}([elt;\ list_1],\ list_2) &\longrightarrow [elt;\ \text{App}(list_1,\ list_2)] & \textbf{(App1)} \\
\text{App}(\emptyset,\ list) &\longrightarrow list & \textbf{(App2)} \\
\text{Dup}(elt_1,\ [elt_2;\ list]) &\longrightarrow [elt_1;\ \text{Dup}(elt_1,\ list)] & \textbf{(Dup1)} \\
\text{Dup}(elt,\ \emptyset) &\longrightarrow \emptyset & \textbf{(Dup2)} \\
\text{Rev}([elt;\ list]) &\longrightarrow \text{App}(\text{Rev}(list),\ [elt;\ \emptyset]) & \textbf{(Rev1)} \\
\text{Rev}(\emptyset) &\longrightarrow \emptyset & \textbf{(Rev2)} \\
\text{App}(list,\ \emptyset) &\longrightarrow list & \textbf{(App3)} \\
\text{App}(\text{App}(list_1,\ list_2),\ list_3) &\longrightarrow \text{App}(list_1,\ \text{App}(list_2,\ list_3)) & \textbf{(App4)} \\
\text{Rev}(\text{App}(list_1,\ list_2)) &\longrightarrow \text{App}(\text{Rev}(list_2),\ \text{Rev}(list_1)) & \textbf{(RevApp)}
\end{aligned}$$

$[elt;\ list]$ is a list with head elt and tail $list$; $\emptyset$ is the empty list. Lists are constructed over a set of uninterpreted atomic elements A...Z. The intended interpretation of the other function symbols should be evident from their equations. For the sake of brevity, we will adopt the convention that a term of the form [A; B; C] is shorthand for the term [A; [B; [C; $\emptyset$]]]; we will frequently mix the two notations (without ambiguity) when it is appropriate. $\mathcal{L}$ is left-linear, confluent and noetherian; it is not, however, orthogonal.

The $\mathcal{L}$-terms $T_1 = \text{Dup}(\text{Hd}([\text{A}]),\ [\text{B};\ \boxed{[\text{C}]}])$ and $T_2 = \text{Dup}(\text{Hd}([\text{A}]),\ [\text{B};\ \boxed{[\text{D};\text{E}]}])$ are identical except for the subterms enclosed in boxes. We can describe the similarity between T_1 and T_2 more succinctly by decomposing each into two parts: a simple context C and a *substituends* S_1 and S_2 such that $T_1 = C[S_1]$ and $T_2 = C[S_2]$, where $C = \text{Dup}(\text{Hd}([\text{A}]),\ [\text{B};\ \square])$, $S_1 = [\text{C}]$, and $S_2 = [\text{D};\text{E}]$. T_1 and T_2 are related by the common context C; they differ in closing C with terms S_1 and S_2, respectively. We will say that terms related by some common context C are *similar* (modulo C).

Let σ_1 and σ_2 be leftmost-outermost reductions such that $T_1 \xrightarrow{\sigma_1}{}^* T_1' = [\text{A};\text{A}]$ and $T_2 \xrightarrow{\sigma_2}{}^* T_2' = [\text{A};\text{A};\text{A}]$. It is easy to see that many of the contractions in σ_2 effectively repeat contractions already performed in σ_1. Our goal in implementing incremental reduction in TRSs will be to *avoid performing redundant contractions when normalizing similar terms*. $\text{INC}_{f\mathcal{R}}$ will exploit the decomposition of similar terms into context and substituends by *preserving the result of reductions applicable to the common context*.

To see how this would work, consider the leftmost-outermost reduction σ_C of the common context $C = \text{Dup}(\text{Hd}([\text{A}]),\ [\text{B};\ \square])$ such that $C \xrightarrow{\sigma_C}{}^* C' = [\text{A};\text{Dup}(\text{A},\ \square)]$. Note that many of the contractions performed in reductions σ^1 and σ^2 are effectively repeated by σ_C. Finally, let σ_2' be the leftmost-outermost reduction such that $C'[S_2] \xrightarrow{\sigma_2'}{}^* T_2' = [\text{A};\text{A};\text{A}]$. σ_2' has *fewer* contractions than σ_2.

Collecting the observations above, we conclude that we can use the context C to *preserve* the computations of one reduction (say σ_1) by storing the reduced context C'. The fundamental idea behind $\text{INC}_{f\mathcal{R}}$ will be to "project" all the contractions used to normalize a term such as T_1 on a substituend-enclosing context such as C as a *side-effect* of the reduction process, yielding both a normal form T_1' and a reduced context C'. We then use C' to reduce a term containing a new substituend, e.g., S_2, by normalizing $C'[S_2]$ rather than $C[S_2]$. This may cause new contractions to occur in

C' that were not the result of contracting T_1 (C' will not in general be a normal form), yielding C'', which can be used to reduce S_3, etc.

If at each stage we project the contractions performed on the entire term in a *maximal* way on C, we ensure that no contractions are repeated after substituend replacement that would have occurred if the new term were reduced *ab initio*. But how can C' be computed efficiently without requiring undue administrative overhead? What if terms can contain more than one substituend, or substituends can be nested? In the next section, we will outline how to compute C' in the simple case where terms are related by substitution of a single substituend, then generalize these ideas in Sect. 5.

3 Incremental Reduction: Single Substituends

3.1 △-Rules

Given an arbitrary left-linear TRS $\mathcal{R}$, we begin by systematically augmenting it with additional rules and function symbols to yield a new system: $\triangle\mathcal{R}$. The signature of $\triangle\mathcal{R}$ consists of $\Sigma(\mathcal{R})$, plus the following additional function symbols:

Definition 1 (additional symbols of $\Sigma(\triangle\mathcal{R})$). The new function symbols of $\Sigma(\triangle\mathcal{R})$ consist of:

- Annotated *substituend delimiters* of the form $S_u(T^\Delta)$ and $\overline{S}_u(T^\Delta)$, where $T^\Delta \in Term(\Sigma(\triangle\mathcal{R}))$. We will refer to $S_u(\cdot)$ as an *active* delimiter, and $\overline{S}_u(\cdot)$ as a *stable* delimiter.
- An annotated *fork operator* of the form $\triangle(T_L^\Delta, T_R^\Delta)$, where $T_L^\Delta, T_R^\Delta \in Term(\Sigma(\triangle\mathcal{R}))$.
- A *stable* function symbol $\overline{F}$ for every function symbol $F \in \Sigma(\mathcal{R})$ such that $arity(F) > 0$.

We will use the operator $S_u(\cdot)$ to delimit substituends. It has the effect of providing a term representation of the boxes enclosing substituends in previous examples. The overbarred stable function symbols are irreducible copies of their unbarred analogues. The "fork" operator $\triangle(\cdot, \cdot)$ delimits two alternative interpretations (or "forks in the road") of the term being reduced. The left branch of the $\triangle$-node will always refer to a subterm that would have occurred in a corresponding non-incremental reduction; the right branch will refer to a subterm that effectively represents an intermediate state that must be preserved for re-use in subsequent incremental reductions. The annotation u in an annotated symbol will be used to refer to a particular substituend; when the annotations are irrelevant, we will omit them.

The additional rules for $\triangle\mathcal{R}$ are defined schematically as follows:

Definition 2 (additional rules of $\triangle\mathcal{R}$).

$$S_u(T) \longrightarrow \triangle(T, \overline{S}_u(T)) \qquad (\triangle\mathbf{Create})$$
$$F(T_1, \ldots, T_{i-1}, \triangle(T_L, T_R), T_{i+1}, \ldots, T_n) \longrightarrow$$
$$\triangle(F(T_1, \ldots, T_{i-1}, T_L, T_{i+1}, \ldots, T_n), \overline{F}(T_1, \ldots, T_{i-1}, T_R, T_{i+1}, \ldots, T_n)) \quad (\triangle\mathbf{F}_i)$$
(for all function symbols $F \in \Sigma(\mathcal{R})$ such that $arity(F) = n > 0$ and all $0 < i \leq n$)

We will refer to these rules collectively as $\triangle$-*rules*. The definitions above assume that $\mathcal{R}$ is single-sorted; however, the extensions required for the multi-sorted case are sufficiently trivial that we omit them here.

3.2 Application of △-Rules

In transforming $\mathcal{L}$ to yield $\triangle\mathcal{L}$, the stable function symbols have the form $\overline{[}elt;\ list\overline{]}$, $\overline{\text{Hd}}(list)$, etc. Some examples of the newly defined rules of $\triangle\mathcal{L}$ follow:

$$[\triangle(hd_L, hd_R);\ tl] \longrightarrow \triangle([hd_L;\ tl],\ \overline{[}hd_R;\ t\overline{l}\overline{]}) \quad (\triangle\text{ListL})$$
$$[hd;\ \triangle(tl_L, tl_R)] \longrightarrow \triangle([hd;\ tl_L],\ \overline{[}hd;\ tl_R\overline{]}) \quad (\triangle\text{ListR})$$
$$\text{Hd}(\triangle(list_L, list_R)) \longrightarrow \triangle(\text{Hd}(list_L),\ \overline{\text{Hd}}(list_R)) \quad (\triangle\text{Hd})$$
$$\text{Tl}(\triangle(list_L, list_R)) \longrightarrow \triangle(\text{Tl}(list_L),\ \overline{\text{Tl}}(list_R)) \quad (\triangle\text{Tl})$$
$$\vdots$$

We will illustrate the behavior of $\triangle\mathcal{L}$ by example, using it to reduce terms T_1 and T_2 of Sect. 2. For the time being, we can ignore annotations since there is only one substituend. We first require that the substituend in T_1 be delimited using the $S(\cdot)$ operator, yielding the term $T_1^\triangle = \text{Dup}(\text{Hd}([A]), [B; S([C])])$. A leftmost-outermost reduction of $T_1^\triangle$, $\sigma_1^\triangle$, then proceeds as follows:

$$\sigma_1^\triangle:$$
$$T_1^\triangle = \text{Dup}(\text{Hd}([A]),\ [B; S([C])])$$
$$\longrightarrow^* [A;\ \text{Dup}(\text{Hd}([A]),\ \triangle([C],\ \overline{S}([C])))]$$
$$\longrightarrow [A;\ \triangle(\ \text{Dup}(\text{Hd}([A]),\ [C]),\ \overline{\text{Dup}}(\text{Hd}([A]),\ \overline{S}([C]))\ )]$$
$$\longrightarrow^* [A;\ \triangle(A,\ \overline{\text{Dup}}(A,\ \overline{S}([C])))\ ] = T_1^{\triangle'}$$

$T_1^{\triangle'}$ has several interesting properties. First, we note that if we traverse $T_1^{\triangle'}$ from its root and examine only the left argument to the $\triangle$-operator, we can reconstruct the normal form T_1' of T_1. More interestingly, the $\triangle$-operator's right argument looks suspiciously like the reduced context C' we computed in Sect. 2, which we claimed embodied the results of projecting the contractions of reduction σ_1 on C. If we ignore the bars over the function symbols and the stable substituend delimiter $\overline{S}(\cdot)$, this second argument is in fact exactly the term $C'[S_1]$. The "boundary" between the reduced context and the substituend is maintained by the stable delimiter $\overline{S}(\cdot)$. Thus it appears that we are able to "read off" both the normal form of the term T_1 and the reduced context C' from $T_1^{\triangle'}$.

3.3 Interpreting $\Sigma(\triangle\mathcal{L})$-Terms

We formalize the process of "reading off" different interpretations of $\Sigma(\triangle\mathcal{L})$-terms with respect to some annotation set $\mathcal{S}$ using the mappings $interp_\mathcal{S}(\cdot)$ and $interp'_\mathcal{S}(\cdot)$, which are defined inductively as follows:

Definition 3 ($interp_\mathcal{S}(\cdot)$ and $interp'_\mathcal{S}(\cdot)$).

$$\begin{aligned}
interp_\mathcal{S}(\triangle(T_L, T_R)) &= interp_\mathcal{S}(T_R) & u &\in \mathcal{S} \\
interp_\mathcal{S}(\triangle(T_L, T_R)) &= interp_\mathcal{S}(T_L) & u &\notin \mathcal{S} \\
interp_\mathcal{S}(\text{F}(T_1, \ldots, T_n)) &= \text{F}(interp_\mathcal{S}(T_1), \ldots, interp_\mathcal{S}(T_n)) & \text{arity}(\text{F}) &= n > 0 \\
interp_\mathcal{S}(\overline{\text{F}}(T_1, \ldots, T_n)) &= \text{F}(interp_\mathcal{S}(T_1), \ldots, interp_\mathcal{S}(T_n)) & \text{arity}(\text{F}) &= n > 0 \\
interp_\mathcal{S}(\text{F}) &= \text{F} & \text{arity}(\text{F}) &= 0 \\
interp_\mathcal{S}(\overline{\text{F}}) &= \text{F} & \text{arity}(\text{F}) &= 0 \\
interp_\mathcal{S}(S_u(T)) &= interp_\mathcal{S}(T) & * \\
interp_\mathcal{S}(\overline{S}_u(T)) &= interp_\mathcal{S}(T) & * \\[4pt]
interp'_\mathcal{S}(S_u(T)) &= S_u(interp'_\mathcal{S}(T)) & \dagger \\
interp'_\mathcal{S}(\overline{S}_u(T)) &= S_u(interp'_\mathcal{S}(T)) & \dagger
\end{aligned}$$

$interp'_S(\cdot)$ is identical to $interp_S(\cdot)$, except that the clauses denoted by '*' are replaced by clauses denoted by '†'.

For the single-substituend case, the annotation set S will be simply $\{\Box\}$. For any $\Sigma(\triangle\mathcal{L})$-term $T^\triangle$, $interp_{\{\Box\}}(T^\triangle)$ can be read as "interpret $T^\triangle$ with respect to the context defined by the substituend" (which is represented by $\Box$); $interp_\emptyset(T^\triangle)$ can be read as "interpret $T^\triangle$ with respect to *no* substituends" (i.e., with respect to the original term).

From these definitions, we have that $interp_\emptyset(T_1^{\triangle'}) = T_1'$ and $interp_{\{\Box\}}(T_1^{\triangle'}) = C'[S_1]$. We can make a more general observation: Every step of reduction $\sigma_1^\triangle$ maintains the invariants $interp_\emptyset(R_i) = T_1''$ and $interp_{\{\Box\}}(R_i) = C''[S_1'']$, where R_i is an intermediate term of $\sigma_1^\triangle$ and T''', S'', and C'' are such that $T_1 \longrightarrow^* T_1''$, $S_1 \longrightarrow^* S_1''$, $C \longrightarrow^* C''$[1]. It will turn out that this very important invariant holds for *any* rewriting system systematically transformed in the same manner as $\mathcal{L}$. The effect of $\triangle$-rules is thus to lazily replace function symbols that are not parts of redexes in the context with stable copies, while allowing the non-stable redexes that remain in the context to be reduced.

3.4 Incremental Reduction Process

Having computed $C'[S_1]$, we can compute the normal form of $T_2 = C[S_2]$ with the following process:

1. Replace $T_1^{\triangle'}$ by $interp'_{\{\Box\}}(T_1^{\triangle'})$.
2. Replace $S(S_1'')$ in $T_1^{\triangle'}$ by $S(S_2)$ to yield $T_2^\triangle$.
3. Reduce $T_2^\triangle$ to $T_2^{\triangle'}$ such that $interp_\emptyset(T_2^\triangle)'$ is a normal form by reduction $\sigma_2^\triangle$.
4. $interp_\emptyset(T_2^{\triangle'})$ is now equal to T_2'.

This process can be repeated *ad nauseum* for any number of subsequent substituends, and is formalized in its complete generality in Definition 8. It should be clear that $\sigma_2^\triangle$ is closely related to reduction σ_2' of Sect. 2, since we are starting with context C', rather than C. We have already observed that the reduction σ_2' avoids repeating any of the contractions already performed in σ_1. Is this all we need to get efficient incremental reduction? Alas, the answer is no. We are faced with the unpleasant fact that the reduction $\sigma_1^\triangle$ requires *more* contractions than its non-incremental counterpart σ_1, as does $\sigma_2^\triangle$. Not at all the efficient algorithm we were hoping for.

3.5 Graph Reduction

Our problem with excessively lengthy "incremental" reductions is remedied by using the well-known idea of *graph reduction*. Instead of *copying* terms that correspond to repeated instances of variables on the right-hand side of rewrite rules, we *share* them using a graph structure. A single reduction step applied to a shared graph redex can thus replace many separate contractions in its term counterpart.

There are numerous ways to formalize graph reduction; the best for our purposes is Maranget's formulation [7, 8], which identifies a single graph reduction step with the notion of a *complete* parallel contraction. Although space does not permit a detailed

---

[1] In this example, S_1'' is always equal to S_1. In general, some reductions may be performed in S_1 if it is not already a normal form.

review, Maranget's method has the advantage of providing a close connection between graph reduction and the notion of reduction *prefix* defined in [6], which will permit a formal comparison between incremental and non-incremental reductions. Although Maranget treated only orthogonal systems, the results we require can be extended easily to non-orthogonal left-linear systems; details are given in [4]. A more "concrete" formulation of graph reduction which yields equivalent results is given in [1]. Maranget shows that for orthogonal TRSs, complete reductions combined with a *call-by-need* reduction strategy yields reductions that are *optimal* in length. Using this fact, it is be possible define optimal *incremental* reductions for orthogonal TRSs.

Returning to our running example, let us reduce T_1^Δ of Sect. 3.2 using graph reduction. The final terms of such a reduction are depicted in Fig. 1. In this case, all three instances of the redex Hd([A]) are contracted simultaneously, and the result *shared* between $interp_\emptyset(T_1^{\Delta'})$ and $interp_{\{\Box\}}(T_1^{\Delta'})$. Graph reduction allows redexes in the context

$$[\ |\ ,\ \Delta(\ |\ ,\ \overline{Dup}(\ |\ ,\ \overline{S}([C])))\ ] \rightarrow [\ |\ ,\ \Delta(\ |\ ,\ \overline{Dup}(\ |\ ,\ \overline{S}([C])))\ ]$$

Hd([A])   A

Fig. 1. Final steps of graph reduction of T_1^Δ

to be shared with the term being reduced; however, it is critical that we use a reduction strategy for which only function symbols that are in the context and match a redex which "crosses the boundary" between context and substituend are (effectively) copied by application of the Δ-rules. We will formalize this process in Sect. 7.

4 Incremental Reduction with Multiple Substituends

In this section, we consider the problem of incremental reduction in its full generality. When we consider multiple substituends, the utility of our techniques becomes much more evident than in the relatively straightforward single-substituend case. Nonetheless, all the basic ideas of Sect. 2 carry over. Consider the $\Sigma(\mathcal{L})$-term R below:

$$\text{Rev}([A; \text{App}(\boxed{\text{Dup}(\text{Hd}(\boxed{[B]}_2), [C; D])}_1, \boxed{\boxed{E}_4; F]}_3)])$$

R has normal form $R' = [F; E; B; B; A]$. We have enclosed 4 substituends in R in subscripted boxes; note that some of these substituends are nested within one another. From this example, we see that the generalized form of the incremental reduction problem requires that we be able to reduce terms derived from R by replacement of *any* number of disjoint substituends.

4.1 The General Incremental Reduction Problem

Since we can no longer describe each term to be reduced in terms of an appropriate closing of a single context, we use the following notion of *decomposition* to define nested contexts abstractly:

Definition 4 (Decomposition). The set of Σ-*context decompositions*, $Dec(\Sigma)$, is a special subset of $Term((\Sigma \cup \{\langle \cdot, \cdot \rangle, [\cdots]\}), (\mathcal{V} \cup \mathcal{H}))$, with the property that $Cont(\Sigma) \subseteq Dec(\Sigma)$, and for all $D_1, \ldots, D_n \in Dec(\Sigma)$ and $C \in Cont(\Sigma)$ such that $ind(C) \subseteq \{1 \ldots n\}$, $\langle C, [D_1, \ldots, D_n] \rangle \in Dec(\Sigma)$. $\langle \cdot, \cdot \rangle$ has arity 2, and $[\cdots]$ represents a family of symbols of arity ≥ 0. If D is a Σ-decomposition, then $u \in \mathcal{O}(D)$ is a *decomposition occurrence* of D if $u = [\,]$, or if $u = vw$ such that $D/v = \langle C, [D_1, \ldots, D_n] \rangle$, and $w = [\,]$ or $w = [2, i]$ for $i \in \{1 \ldots n\}$. The set of all decomposition occurrences is denoted by $\mathcal{O}_d(D)$.

Given decomposition $D \in Dec(\Sigma)$, the *reification* of D, notated $re(D)$ is the context such that $re(C) = C$ if $C \in Cont(\Sigma)$, otherwise, $re(\langle C, [D_1, \ldots, D_n]\rangle) = C[1 := re(D_1), \ldots, n := re(D_n)]$. If $re(D) = C \in Cont(\Sigma)$, we will say that D is a decomposition *of* C. A decomposition D is *proper* if for all $u \in \mathcal{O}_d(D)$ such that $D/u = \langle C, [D_1, \ldots, D_n]\rangle$, C is proper and $ind(C) = \{1 \ldots n\}$.

Let D be a decomposition and C be a context such that $re(D) = C$. Then for any set $\mathcal{S} \subseteq \mathcal{O}_d(D)$, the set of occurrences in C *induced* by $\mathcal{S}$, denoted $indOccs(D, \mathcal{S})$, is defined inductively by $indOccs(C, \{[\,]\}) = \{[\,]\}$ for $C \in Cont(\Sigma)$, and

$$indOccs(\langle C, [D_1, \ldots, D_n]\rangle, \mathcal{S}) = \mathcal{S}_0 \cup (\mathcal{O}_1(C) \cdot \mathcal{S}_1) \cup \ldots \cup (\mathcal{O}_n(C) \cdot \mathcal{S}_n)$$

otherwise, where $\mathcal{S}_0 = \{[\,]\}$ if $[\,] \in \mathcal{S}$ and $\emptyset$, and for all $i \in \{1 \ldots n\}$, $\mathcal{S}_i = indOccs(D_i, \mathcal{S}/[2, i])$.

Given a decomposition D of C and a set $\mathcal{S} \subseteq \mathcal{O}_d(D)$, it is easy to show that there exists a unique proper decomposition $D_\mathcal{S}$ such that $D_\mathcal{S} = indDec(C, \mathcal{S}')$, where $\mathcal{S}' = indOccs(D, \mathcal{S})$. We denote this decomposition by $indDec(D, \mathcal{S})$.

Decompositions will be used to model in an abstract way the division of a term into "pieces" referred to by multiple substituends. Induced decompositions allow us to single out a particular subset of those pieces that is of interest. Finally, a reified decomposition is the term or context resulting from "re-assembling" the decomposition's pieces. To describe the connection between a decomposition and its term representation using substituend delimiters, we need the following:

Definition 5 ($\triangle Conv(\cdot, \cdot)$). Given a proper decomposition D of a $\Sigma(\mathcal{R})$-term T and path expression $u \in \mathcal{O}_d(D)$, we define the $\Sigma(\triangle \mathcal{R})$-term corresponding to D at u, $\triangle Conv(D, u)$, inductively on the structure of D by

$$\begin{aligned} \triangle Conv(T, u) &= S_u(T) \\ \triangle Conv(\langle C, [D_1, \ldots, D_n]\rangle, u) &= \\ S_u(C[1 := \triangle Conv(D_1, u[2, 1]), &\ldots, n := \triangle Conv(D_n, u[2, n])]) \end{aligned}$$

We will use $\triangle Conv(D)$ to abbreviate $\triangle Conv(D, [\,])$. $\triangle Conv(D)$ has the effect of inserting a substituend delimiter $S_u(\cdot)$ at points in T corresponding to $indOccs(D, \mathcal{O}_d(D))$.

Given the definitions above, we are at last in a position to formalize the general problem of incremental reduction:

Definition 6 (Incremental Reduction Problem). Let $\mathcal{R}$ be a left-linear, single-sorted Σ-TRS. Let $f_\mathcal{R}$ be a normalizing reduction strategy for $\mathcal{R}$. Assume we are given:

- An initial proper decomposition $D_0 \in Dec(\Sigma)$.
- A sequence of *substituend sets* $\mathcal{S}_1, \mathcal{S}_2, \mathcal{S}_3, \ldots$, where each $\mathcal{S}_i$ is a set of fixed contexts of the form $(u \leftarrow D)$, where $D \in Dec(\Sigma)$ is a proper decomposition.

Then for all $i > 0$, we wish to compute normal form terms T_i' such that $T_i \longrightarrow^* T_i'$, $T_i = re(D_i)$, and $D_i = D_{i-1}[\mathcal{S}_i]$.

4.2 Back to the Example

Returning to our example term R above, we demonstrate how it fits into our formal framework. R contains the four substituends $S_1 = \text{Dup}(\text{Hd}([B]), [C; D])$, $S_2 = [B]$, $S_3 = [E; F]$, and $S_4 = E$. The decomposition $D_R = \langle C_\alpha, [\langle C_\beta, [[B]]\rangle, \langle C_\gamma, [E]\rangle]\rangle$, where $C_\alpha = \text{Rev}([A; \text{App}(\boxed{1}, \boxed{2})])$, $C_\beta = \text{Dup}(\text{Hd}(\boxed{1}), [C; D])$, and $C_\gamma = [\boxed{1}; F]$, represents these substituends and their nesting structure. Given D_R and appropriate occurrences therein, we can reconstruct each of the substituends using the identities $S_1 = re(D_R/[2,1])$, $S_2 = re(D_R/[2,1,2,1])$, $S_3 = re(D_R/[2,2])$, and $S_4 = re(D_R/[2,2,2,1])$.

By analogy with the approach taken in Sect. 2, $\text{INC}_{f\mathcal{R}}$ must be able to compute the reduced form of *every* context surrounding any set of substituends. Since the number of such contexts is exponential in the number of substituends, we seem to have reached an impasse. The key to breaching this apparent barrier is to observe that there is an enormous amount of overlap among these contexts. Once again, graph reduction comes to the rescue, by allowing reductions common to contexts to be shared.

The results of reducing the five contexts derived from *induced decompositions* corresponding to various sets of decomposition occurrences in D_R are given below:

$C_{\{[2,1]\}} = \text{Rev}([A; \text{App}(\boxed{1}, [E;F])]) \longrightarrow^* \quad\quad [F;E; \text{App}(\text{Rev}(\boxed{1}), [A])]$

$C_{\{[2,1],[2,1]\}} = \text{Rev}([A; \text{App}(\text{Dup}(\text{Hd}(\boxed{2}), [C;D]), [E;F])]) \longrightarrow^* [F;E; \text{Hd}(\boxed{2}); \text{Hd}(\boxed{2}); A]$

$C_{\{[2,2]\}} = \text{Rev}([A; \text{App}(\text{Dup}(\text{Hd}([B]), [C;D]), \boxed{3})]) \longrightarrow^* \quad \text{App}(\text{Rev}(\boxed{3}), [B;B;A])$

$C_{\{[2,2,2,1]\}} = \text{Rev}([A; \text{App}(\text{Dup}(\text{Hd}([B]), [C;D]), [\boxed{4};F])]) \longrightarrow^* \quad [F;\boxed{4}; B;B;A]$

$C_{\{[2,1],[2,2]\}} = \text{Rev}([A; \text{App}(\boxed{1}, \boxed{3})]) \longrightarrow^* \quad \text{App}(\text{Rev}(\boxed{3}), \text{App}(\text{Rev}(\boxed{1}), [A]))$

The holes in the reduced contexts above are numbered according to the substituends to which they correspond. The first four reductions apply to the contexts enclosing substituends S_1–S_4, respectively. $C_{\{[2,1]\}}$ is derived from the identity $\langle C_{\{[2,1]\}}, [S_1]\rangle = indDec(D_R, \{[2,1]\})$; the others are derived in a similar manner. The last of the contexts above is the context enclosing *both* substituends S_1 and S_3. In general, when performing incremental reduction, we must compute the reduced form of a contexts enclosing any *set* of substituends. These reduced contexts will be computed *simultaneously* as a side effect of reducing R to R', as follows:

We begin by delimiting its substituends with appropriate $S_i(\cdot)$ operators using the function $\triangle Conv(R)$, yielding the term $R^\triangle$:

$$R^\triangle = \triangle Conv(D_R)$$
$$= \text{Rev}([A; \text{App}(S_1(\text{Dup}(\text{Hd}(S_2([B])), [C;D])), S_3([S_4(E); F])])$$

For the sake of clarity, we have taken the liberty of annotating substituend delimiters by numbers corresponding to substituends S_1–S_4, rather than the paths in D_R to which these substituends correspond. That is, annotation 1 corresponds to path $[2,1]$ in D_R, 2 corresponds to path $[2,1,2,1]$, etc. Graph reduction of $R^\triangle$ using $\triangle \mathcal{L}$ yields the graph $R^{\triangle'}$ depicted in Fig. 2 (where shared leaf nodes in the graph are shown unshared for the sake of clarity). The remarkable thing about $R^{\triangle'}$ is that we can "read off" the reduced form C'_S of any context C_S (where S is any subset of the set $\{1,2,3,4\}$) using $interp_S(R^{\triangle'})$. Note in particular that $interp_\emptyset(R^{\triangle'}) = R'$, where R' is the normal form of R. We interpret $C'_\emptyset$ as the reduced context containing *no* substituends, i.e., R'. Thus we see that a single graph gives sufficient information to incrementally reduce the term resulting from replacement of *any* set of substituends. Incremental reduction proceeds as in the single substituend case, by interpreting the term using $interp_S(\cdot)$ and replacing appropriate delimited substituends.

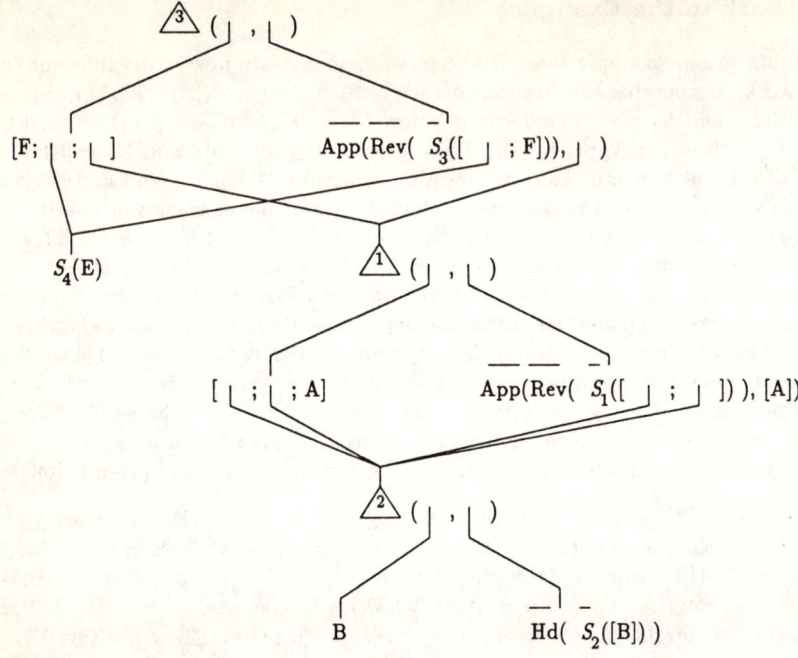

Fig. 2. Graph form of $R^{\Delta'}$

5 Formal Algorithm

We are now in a position to present the general incremental reduction algorithm, $\text{INC}_{f\mathcal{R}}$. Given an instance of the general incremental reduction problem (Definition 6), we define a derived reduction strategy, $f_{\triangle \mathcal{R}}$ and the generalized incremental reduction algorithm, $\text{INC}_{f\mathcal{R}}$, as follows:

Definition 7 (Incremental Strategy).

1. If $f_\mathcal{R}(interp_\emptyset(T))$ is undefined, then $f_{\triangle \mathcal{R}}(T)$ is undefined.
2. If $f_\mathcal{R}(interp_\emptyset(T)) = \langle u, \alpha \rangle$, let u' be the root of the symbol in T which is mapped to u by $interp_\emptyset(T)$. If u' is the root of an α-redex in T, then $f_{\triangle \mathcal{R}}(T) = \langle u', \alpha \rangle$.
3. Otherwise, there is a $\triangle(\cdot, \cdot)$ or $S_u(\cdot)$ operator at some path $u'' = u' \cdot s \cdot i$ that prevents rule r from matching at u. If the function symbol is $\triangle$, then $f_{\triangle \mathcal{R}}(T) = \langle u' \cdot s, \triangle F_i \rangle$ where F is the function symbol at path $u' \cdot s$. Otherwise, $f_{\triangle \mathcal{R}}(T) = \langle u'', \triangle \text{Create} \rangle$.

Definition 8 (Incremental Reduction Algorithm $\text{INC}_{f\mathcal{R}}$).

$T^\Delta := \Delta Conv(D_0);$
FOR $i = 0, 1, 2, \ldots$ DO:
    1)   WHILE $f_{\triangle \mathcal{R}}(T^\Delta)$ is defined DO
          LET $\langle u, \alpha \rangle = f_{\triangle \mathcal{R}}(T^\Delta);$

$$T^\Delta := \langle u, \alpha \rangle : T^\Delta$$
END WHILE

2) $T'_i = interp_\emptyset(T^\Delta)$;
3) $T_{tmp} := interp_{\mathcal{O}(\mathcal{S}_{i+1})}(T^\Delta)$;
4) $\mathcal{S}_{tmp} := \{(w \leftarrow S) \mid (u \leftarrow D) \in \mathcal{S}_{i+1}, w \in (\mathcal{O}_{S_u}(T_{tmp}) \cup \mathcal{O}_{\overline{S}_u}(T_{tmp})),$
   $S = \Delta Conv(D, u) \}$;
5) $T^\Delta := T_{tmp}[\mathcal{S}_{tmp}]$

To extend $\text{INC}_{f\mathcal{R}}$ to graph reduction, we simply extend $f_{\triangle\mathcal{R}}$ to a *complete parallel strategy* $\tilde{f}_{\triangle\mathcal{R}}$ [7, 8]. The graph analogue of $\text{INC}_{f\mathcal{R}}$ can be implemented effectively using the techniques of [1]. A correctness proof for $\text{INC}_{f\mathcal{R}}$ may be found in [4].

6 Administrative Overhead and Optimality Results

In this section, we summarize the our formal results on the complexity of $\text{INC}_{f\mathcal{R}}$; details will be found in [4].

6.1 An Optimized Algorithm: $\text{INC}'_{f\mathcal{R}}$

The administrative overhead of $\text{INC}_{f\mathcal{R}}$ is determined by the number of $\triangle$-rule contractions performed during incremental reduction. Each application of a $\triangle\mathbf{F}_i$ rule propagates a $\triangle$-node and creates a stable copy of a single function symbol. The number of $\triangle\mathbf{F}_i$ rule applications is in turn governed by the reduction strategy $f_{\triangle\mathcal{R}}$, which must propagate $\triangle$-nodes until an $\mathcal{R}$-redex is "uncovered." The key to analyzing and controlling the number of $\triangle$-rule applications is to understand the etiology of *chains* of nested terms headed by adjacent $\triangle$-operators. Formally, such a $\triangle$-*cluster* is a subterm of the form $\triangle(\triangle(\ldots\triangle(T_L^w, T_R^w)\ldots, T_R^v), T_R^u)$. In the worst case, propagating such clusters can cause the number of administrative reductions to be quadratic in the number of non-administrative reductions. Thus, although $\text{INC}_{f\mathcal{R}}$ can itself be implemented without alteration, we will need an alternative representation of $\triangle$-clusters to yield an acceptable bound on administrative overhead.

The basis for an optimized algorithm, $\text{INC}'_{f\mathcal{R}}$, is a data structure which represents $\triangle$-clusters such as the one above by *finite maps* from annotations to terms. Such maps have the form $\{(u \mapsto T_R^u), (v \mapsto T_R^v), (w \mapsto T_R^w)\}$. This representation follows from the observation that multiple instances of the same annotation in a $\triangle$-cluster are redundant, and that adjacent $\triangle$-nodes in a $\triangle$-cluster can be permuted without changing their interpretation with respect to *disjoint* substituends.

We manipulate these maps by transforming each instance of a term of the form $F(\cdots)$ to a term of the form $\langle F(\cdots), map \rangle$. $\langle term, map \rangle$ is a new function symbol which "attaches" the map map to the function symbol at the root of $term$. For example, the term $Rev(App(list_1, list_2))$ becomes $\langle Rev(\langle App(list_1, list_2), map_{App}\rangle), map_{Rev}\rangle$. The map is empty in cases where no $\triangle$-cluster would have been present in the old term structure.

By *uniformly* adding such an extra indirection structure for each function symbol, we can eliminate $\triangle$-rules altogether. Instead, we alter each rule $L \longrightarrow R$ of $\mathcal{R}$ to account for the extra indirection, and to propagate $\triangle$ information represented by maps associated with each function symbol. To properly account for sharing and the manipulation of map structures in an optimized reduction algorithm, it will be convenient to represent transformed equations using *conditional* rules. For example, the rule **Hd** of $\mathcal{L}$ is transformed to:

$$\frac{term_L = \langle \text{Hd}(\langle [\langle elt, map_{elt} \rangle; list'], map_{[]} \rangle), map_{Hd} \rangle}{term_L \longrightarrow \langle elt', (map_{elt} \cup remap(dom(map_{[]}), term_L) \cup map_{Hd}) \rangle} \quad (\textbf{Hd}')$$

The operation '$\cup$' on maps combines two maps; in the case of duplicate domain elements, it gives priority to the pair from the first map. The function $dom(map)$ yields the domain of map (a set of annotations). The function $remap(dom, term)$ creates a map with domain dom, all of whose elements are mapped to term $term$. These operations can all be implemented in linear or sub-linear time (with respect to the size of map domains) using appropriate data structures, e.g., balanced trees. The conditional structure of $\textbf{Hd}'$ is used to bind a variable ($term_L$) to the left-hand side of the rule, which is then saved in the map created on the right-hand side. This operation corresponds to the creation of stable function symbols performed by $\triangle \textbf{F}_i$ rules in $\text{INC}_{f\mathcal{R}}$. An additional benefit of this alternative term structure is the fact that the requisite incremental strategy is now a trivial variant of $f_\mathcal{R}$. Note also that the stable function symbols are no longer required.

It is easy to see that the size of any map is bounded by the number of substituends in the initial term, since the size of any map domain is bounded by the total number of annotations, which are derived from substituends. Since the map operations can be implemented in linear time, and since the number of map operations per simulated $\mathcal{R}$-contraction is bounded by some constant determined by the size of the "largest" rule in $\mathcal{R}$, we have:

Proposition 1. *The worst-case administrative overhead per $\mathcal{R}$-contraction in any simulated reduction carried out by $\text{INC}'_{f\mathcal{R}}$ is a linear function of the number of substituends.*

Due to the uniform structure of $\Sigma(\triangle \mathcal{R})'$-terms, the time required to apply the $\text{INC}'_{f\mathcal{R}}$ variants of $interp_S(\cdot)$ or $interp_S(\cdot)$ (i.e., to derive the "answer" or to start a new reduction in the sequence) is proportional to the size of the resulting term times a log-function of the number of substituends. The latter factor is due to the time required retrieve values from maps (implemented as balanced trees).

6.2 Comparing Reductions; Optimality

In order to make a formal comparison between incremental and non-incremental reductions, we must formalize the notion of "repeated work." As alluded to earlier, we do this using a variant of Huet and Lévy's notion of reduction *prefix* [6], restricted to reductions that are *relative* to one another—roughly, those reductions that contain contractions that are *residuals* of one another, or are residuals of contractions derived from equivalent reductions. A reduction σ is a prefix of reduction τ if τ carries out at least the same contractions as σ, modulo permutation of the order in which contractions are performed.

We then wish to show that two successive reductions simulated by $\text{INC}_{f\mathcal{R}}$ (or $\text{INC}'_{f\mathcal{R}}$) do not *overlap*. For strongly coinitial relative reductions ρ and σ, ρ and σ will be said to overlap if the greatest lower bound of ρ and σ with respect to the relative prefix ordering on strongly coinitial reductions is equivalent to the empty reduction. Put more succinctly, ρ and σ have no reductions in common, which we take as our definition of *relative optimality*. (Greatest lower bounds always exist for *relative* reductions, which is not the case for arbitrary strongly coinitial reductions.) We must also define the *projection* of the reduction of a term on a sub-context of that term, which can be done using an adaptation of of the construction of standard reductions in [6].

Given these definitions, it is possible to show that the projections of the reductions performed on contexts C common to two successive reductions do not overlap when the strategy $f_{\triangle_R}$ of $\text{INC}_{f\mathcal{R}}$ is extended to complete reductions. Thus we have:

Proposition 2. $\text{INC}_{f\mathcal{R}}$ and $\text{INC}'_{f\mathcal{R}}$ *are relatively optimal in general, and optimal in the case of orthogonal systems.*

7 Acknowledgments

I am grateful to Paul Klint, Emma van der Meulen, Mark Wegman, Dan Yellin, and the RTA referees for many helpful comments. I am especially indebted to Frank Tip for his insights and a very careful reading of the paper.

References

1. BARENDREGT, H., VAN EEKELEN, M., GLAUERT, J., KENNAWAY, J., PLASMEIJER, M., AND SLEEP, M. Term graph rewriting. In *Proc. PARLE Conference, Vol. II: Parallel Languages* (Eindhoven, The Netherlands, 1987), Springer-Verlag, pp. 141–158. Lecture Notes in Computer Science 259.
2. DEMERS, A., REPS, T., AND TEITELBAUM, T. Incremental evaluation for attribute grammars with application to syntax-directed editors. *Proc. Eighth ACM Symp. on Principles of Programming Languages* (1981), 105–116.
3. FIELD, J. *Incremental Reduction in the Lambda Calculus and Related Reduction Systems.* PhD thesis, Department of Computer Science, Cornell University, May 1991.
4. FIELD, J. A graph reduction approach to incremental term rewriting. Tech. rep., IBM T.J. Watson Research Center, 1993. (Forthcoming).
5. FIELD, J., AND TEITELBAUM, T. Incremental reduction in the lambda calculus. In *Proc. ACM Conference on LISP and Functional Programming* (Nice, France, June 1990), pp. 307–322.
6. HUET, G., AND LÉVY, J.-J. Computations in orthogonal rewriting systems, I. In *Computational Logic*, J.-L. Lassez and G. Plotkin, Eds. MIT Press, Cambridge, MA, 1991, ch. 11, pp. 395–414.
7. MARANGET, L. Optimal derivations in weak lambda-calculi and in orthogonal term rewriting systems. In *Proc. Eighteenth ACM Symp. on Principles of Programming Languages* (Orlando, FL, January 1991), pp. 255–269.
8. MARANGET, L. *La Stratégie Paresseuse.* PhD thesis, Université de Paris VIII, 1992. (Thèse de Doctorat, in French).
9. MEULEN, E. V. D. Deriving incremental implementations from algebraic specifications. Report CS-R9072, Centrum voor Wiskunde en Informatica (CWI), Amsterdam, 1990. Extended abstract appeared in *AMAST'91: Proceedings of the Second International Conference on Algebraic Methodology and Software Technology*, Workshops in Computing, Springer-Verlag.
10. MEULEN, E. V. D. Fine-grain incremental implementation of algebraic specifications. Report CS-R9159, Centrum voor Wiskunde en Informatica (CWI), Amsterdam, 1991.

Generating tables for bottom-up matching

Ernst Lippe
lippe@serc.nl

Software Engineering Research Centre
Postbus 424
3500 AK Utrecht
The Netherlands

Abstract. Matching forms a bottle-neck in most implementations of rewrite systems. Bottom-up matching is a very fast form of matching. This paper presents a new approach to bottom-up matching that replaces match-sets by their unifiers. In this way it is possible to define a subsumption ordering on the states. A new algorithm is presented that uses the subsumption graph to compute the bottom-up tables. Its time complexity per table entry is $O(rank \times wd)$ where $rank$ is the maximum arity of a function symbol and wd is the maximum number of immediate predecessors in the subsumption graph.

1 Introduction

In implementations of rewriting systems, matching left hand sides of rules with terms is a very important and expensive step. Several different matching algorithms have been described in the literature [HO82, Wal91, PB85, Cha87, CPT90].

Bottom-up matching as introduced in [HO82] is a very fast form of matching. During bottom-up matching every term is visited only once, and no complicated computations are needed because the matching process is completely table-driven. This paper describes a different way of looking at bottom-up matching. Normally, states are described by match-sets. However, it is also possible to represent states by the unifiers of their match-sets. In this way it is much easier to see that it is also possible to compute the subsumption ordering of states. This property is used to derive an efficient algorithm for computation of bottom-up tables.

This paper consists of two parts. First, a description is given of bottom-up matching using unifiers both for the normal approach and for the compressed table approach that was introduced by Chase [Cha87]. In the second part an algorithm for constructing the tables is described.

2 Bottom-up matching

Terms can be seen as labelled trees, with labels from an alphabet $\sum$ of function symbols. Patterns are trees possibly containing variables. In this paper all patterns are linear, each variable occurs only once in a pattern. In effect, this means that the identity of the variables can be ignored. Variables will be written as **X**. The set of all subtrees of a pattern of a given pattern p is called $subterms(p)$. It is useful to include p as well, thus

$p \in subterms(p)$. By abuse of notation (or overloading) the set of subtrees of a set of patterns S is also called $subterms(S) = \{q \mid q \in subterms(t) \land t \in S\}$. A set of patterns P is *subterm-closed* when $subterms(P) \subseteq P$.

The process of finding all matches can be described as: given a set of (linear) patterns $\mathcal{P}$ and a term t, determine which of the patterns in $\mathcal{P}$ match with t. A pattern (a term that may contain variables) p matches a pattern t if there exists a substitution σ that assigns patterns to the variables in p such that $\sigma(p) = t$.

The matching property can be used to define an ordering among patterns. The subsumption ordering $\leq$ is defined as follows. For two patterns A and B, $A \leq B$ holds iff there exists a substitution σ such that $\sigma(A) = B$.[1] The set of patterns with $\leq$ forms a category (or a semi-lattice) with a term consisting of a single variable as initial (or smallest) element.

Central in bottom-up matching is the notion of *states*, that describe the set of patterns that match a given term. A state is constructed in such a way that from the state of a term it can be determined which patterns in $\mathcal{P}$ match this term. Furthermore, the state of a term can be computed from the state of its immediate children and its label. In an implementation states will be coded as numbers, and the state of a term will be computed by a table lookup.

In [HO82] states are described are described by match sets, the maximum subsets of the total set of patterns that match a given pattern. In this paper an alternative formulation is used where each match set is described by the most general unifier of its elements. In section 2.1 it is shown that these notions are equivalent.

Definition 1. The unification closure $un(P)$ of a set P of patterns consists of all patterns in P, plus the single variable pattern **X**, plus all patterns that can be obtained by unification of these elements.

A set of patterns P such that $un(P) = P$ will be called *unification-closed*.

An important property of linear unification, that will be used several times, is the following. Given patterns p_1, p_2, if there exists a pattern t such that $p_1 \leq t \land p_2 \leq t$ then there exists a most general linear unifier u of p_1 and p_2, for which $p_1 \leq u, p_2 \leq u$ and $u \leq t$. In categorical terms, this means that the categorical sum of two terms is equal to their linear unifier.

The first step in the construction process of the tables for bottom-up matching is the computation of the set of states.

Computing states

The set of states $\mathcal{S}$ for a given set of patterns $\mathcal{P}$, can be computed by

$$\mathcal{S} = un(subterms(\mathcal{P}))$$

For readers unfamiliar with bottom-up matching, this definition might seem ad hoc. An intuitive explanation could go as follows. In bottom-up matching, the state of a term t represents the set of patterns that match that term. Thus, if there are two patterns p_1 and $p_2 \in \mathcal{P}$ that both match t, this information must be reflected in the state of the term

---

[1] Notice that this definition is similar to the one in [HO82] but it is the reverse from the ordering that is used by Cai et al [CPT90].

t. This can be achieved by assigning it the state u that is the unifier of p_1 and p_2. This is the reason that the unification closure is used to compute the set of states. Additionally, for bottom-up matching it is not sufficient to perform a unification closure on the set of input patterns $\mathcal{P}$, also the subterms of these patterns must be added. The reason for this is that it must be possible to compute the state of a term by using only the states of its immediate subterms (plus its function symbol). Thus, the state of a subterm must also reflect possible subpatterns that match this subterm. For example, if $\mathcal{P} = \{s(s(\mathbf{X}))\}$ then a separate state is needed for the subpattern $s(\mathbf{X})$.

The size of the set of states depends on the structure of the input patterns. Because linear unification is an associative, commutative, and idempotent operation, the number of states that is computed by the unification closures is at most equal to the number of subsets of $subterms(\mathcal{P})$. It is possible to find pathological examples (see [HO82]) where the number of states is exponential in the number of input patterns. However, such pathological cases appear to be very rare in practical applications of rewrite systems.

Approximation In order to describe the construction of the tables the notion of *approximation* is needed.

Definition 2. For a pattern p and a set of patterns S, the *best approximation* of p with respect to S (notated as $\mathcal{A}_S(p)$) is an $a \in S$ such that $a \leq p$ and there exists no $a' \in S - \{a\}$ such that $a \leq a' \leq p$.

The notion of approximation plays a central role in this description of bottom-up matching. Approximation can be seen as a structured way to reduce information during the bottom-up matching process. When it is known for a term with function symbol c that its i-th subterm t_i matches a pattern p_i, then this term itself must match the pattern $c(p_1, \ldots, p_n)$. However, this pattern may be too specific. As an example, assume that bottom-up matching is used for finding matches with the set of patterns $\{z, s(\mathbf{X})\}$. The state of a given term describes which of these patterns match this term. In bottom-up matching the state of a term is determined by the function symbol of the term plus the states of it immediate subterms. So, when a term with function symbol s is encountered whose only subterm matches with the pattern $s(\mathbf{X})$, then this term will match the pattern $s(s(\mathbf{X}))$. However, this information is too detailed. For the matching process it is sufficient to know that this term matches with the pattern $s(\mathbf{X})$. Approximation replaces a pattern p with a less specific pattern p' such that $p' \leq p$. In effect, the information that the term matched p is thrown away, and only the information that it matched p' is retained. Care must be taken that enough information is retained, therefore the most specific $p' \in S$ is used. For our example term, it would be wrong to use $p' = \mathbf{X}$ because in that case too much information would be lost.

For arbitrary sets of patterns, the best approximation need not be uniquely determined. Fortunately, the following holds for unification-closed sets of patterns.

Theorem 3. *When the set of patterns S is unification-closed then for each pattern p the best approximation of p with respect to S is uniquely determined.*

Proof: First note that there always is at least one element $e \in S$ such that $e \leq p$ because S contains a single-variable pattern. The rest of the proof goes by contradiction.

Assume that there are two distinct patterns e and $e' \in S$ that form the best approximation of p. Then $e \leq p$ and $e' \leq p$ therefore e and e' must have a most general unifier u for which $u \leq p$. Because S is unification closed, $u \in S$. From the properties of unification follows that $e \leq u$ and $e' \leq u$. But this contradicts the assumption that e and e' are distinct best approximations. □

Computing the state of a term
The state of a given term is the best approximation (with respect to the total set of states) of the pattern that is constructed from its label and the states of its subterms.

Example The following example will be used to illustrate this procedure. The example set of patterns consist of:

$$\{plus(X, zero), plus(X, s(X)), fib(s(s(s(X)))), fib(s(s(zero))), fib(s(zero))\}$$

By using the algorithm described above the following set of states is obtained:

0 : X
1 : $plus(X, zero)$
2 : $zero$
3 : $plus(X, s(X))$
4 : $s(X)$
5 : $fib(s(s(s(X))))$
6 : $s(s(s(X)))$
7 : $s(s(X))$
8 : $fib(s(s(zero)))$
9 : $s(s(zero))$
10 : $s(zero)$
11 : $fib(s(zero))$

Now to find the state of a node that has label *plus* and children with state 11 and 6, first construct the corresponding term $plus(fib(s(zero)), s(s(s(X))))$ and then approximate this term with respect to the set of states. In this way its state found to be 3, corresponding with $plus(X, s(X))$. By repeating this procedure the entire table for *plus* can be computed which leads to the following result:

		State of second subterm											
		0	1	2	3	4	5	6	7	8	9	10	11
	0	0	0	1	0	3	0	3	3	0	3	3	0
	1	0	0	1	0	3	0	3	3	0	3	3	0
	2	0	0	1	0	3	0	3	3	0	3	3	0
	3	0	0	1	0	3	0	3	3	0	3	3	0
State	4	0	0	1	0	3	0	3	3	0	3	3	0
of	5	0	0	1	0	3	0	3	3	0	3	3	0
first	6	0	0	1	0	3	0	3	3	0	3	3	0
subterm	7	0	0	1	0	3	0	3	3	0	3	3	0
	8	0	0	1	0	3	0	3	3	0	3	3	0
	9	0	0	1	0	3	0	3	3	0	3	3	0
	10	0	0	1	0	3	0	3	3	0	3	3	0
	11	0	0	1	0	3	0	3	3	0	3	3	0

2.1 Match sets

In [HO82] the states for bottom-up matching are described by match-sets, that are maximum subsets of the subterms of $\mathcal{P}$ that can match a given term t. Similar to [HO82] the set of subterms of $\mathcal{P}$ will be named PF. In this paper match-sets are identified with their most general unifier.

There is a one-to-one relation between these notions. From the definition of match-sets it follows that all elements in a match-set are unifiable, because there must exist a term t that is matched by all patterns in the match-set. As a small technical detail, the unifier of an empty match-set is defined as the pattern consisting of a single variable.

That the match sets are uniquely determined by their unifiers can be seen as follows. Assume that are two different match-sets, m and m' determined by the terms t and t' that have the same unifier u. Because these match-sets are different, one of them must contain a pattern p that is not in the other match-set. Without loss of generalisation it can be assumed that $p \in m$. From the definition of match-sets it follows that p does not match t'. But this leads to a contradiction since from the properties of unification $p \leq u$ and $u \leq t'$. Because the subsumption ordering is transitive $p \leq t'$, in other words p must match t'.

The match-set that corresponds to a given unifier u can be constructed by

$$\{p \in PF \mid p \leq u\}$$

From this definition it can be seen that it is safe to remove elements u from PF that are equal to the unifier of two other elements p and p' in PF. This follows from the fact that for any match set m: $u \in m \Leftrightarrow p \in m \wedge p' \in m$. This property is probably useful for algorithms that are based on match-sets, because in certain cases it can reduce the size of the match-sets.

Match-sets and unifiers represent two different "views" on the concept of a state. Which view is most appropriate depends on the application. For example, The observation that states can be ordered by the subsumption ordering formed the inspiration for the algorithm that is presented in section 4. This observation is not immediately obvious from a description of states in terms in terms of match-sets. Note that Hoffmann and O'Donnell in effect also use unifiers in their discussion of simple pattern forests, where the immediate subsumption graph forms a tree.

3 Reducing the table size

The table that is constructed by the algorithm from the previous section can grow large. When set is the number of states and a is the arity of a label, the size of the table for that label is set^a.

Chase [Cha87] remarked that, when inspecting the constructed tables, it is frequently very obvious that they contain a large amount of repetition, in the form of repeated subtables. In the table of the previous example it can be observed that all rows are equal and that several columns are repeated. Now tables can be made smaller by reducing the amount of repetition.

The basis for this reduction is the observation that for term with a given label the "relevant" set of states for a subterm that occurs at a certain position is normally much smaller than the total set of states. The set of relevant states for the i-th subterm of a term with top-level label c, notated as $\mathcal{S}_{c,i}$, is computed as follows: Select all subpatterns from $\mathcal{P}$ that occur at argument position i for a (sub)pattern with label c, and compute a transitive closure under unification of this set.

Two different types of tables must be computed. The first kind are the tables that describe how each element from $\mathcal{S}$ is mapped unto an element from $\mathcal{S}_{c,i}$. This table will be called $\mu_c^i(p)$ where $p \in \mathcal{S}$. This is done by computing the best approximation with respect to $\mathcal{S}_{c,i}$ for each element in $\mathcal{S}$.

$$\mu_c^i(p) = \mathcal{A}_{\mathcal{S}_{c,i}}(p)$$

Another table is needed to describe how the elements $c(p_1, \ldots))$ where $p_i \in \mathcal{S}_{c,i}$ are mapped onto $\mathcal{S}$. This table will be called $\theta(c(p_1, \ldots))$ where $p_i \in \mathcal{S}_{c,i}$. This definition is slightly different from the ones in [Cha87, CPT90] that define a separate k-dimensional table θ_c for each c, where k is the arity of c. The θ table can be computed by:

$$\theta(p) = \mathcal{A}_\mathcal{S}(p)$$

The correctness of this process follows from the observation that the approximation of a term t that has a top-level label c with respect to $\mathcal{S}$ must be the same as the approximation of a term t' that is obtained from t by replacing its i-th subterm with its approximation with respect to $\mathcal{S}_{c,i}$.

Example

In our example $\mathcal{S}_{plus,0}$ is the following:

$$0 : \mathbf{X}$$

So the table μ_{plus}^0 that maps a state to this substate is trivial:

0	1	2	3	4	5	6	7	8	9	10	11
0	0	0	0	0	0	0	0	0	0	0	0

The second argument of plus is more interesting, $\mathcal{S}_{plus,1}$ is:

$$0 : \mathbf{X}$$
$$1 : zero$$
$$2 : s(\mathbf{X}1)$$

The table μ_{plus}^1 that maps a state to this substate looks as follows:

0	1	2	3	4	5	6	7	8	9	10	11
0	0	1	0	2	0	2	2	0	2	2	0

The table that computes the state for *plus* given the substates of its children is much smaller than the corresponding table for the uncompressed approach. It is as follows:

	0	1	2
0	0	1	3

3.1 Discussion

In the standard bottom-up matching approach the total size of the tables is

$$O(set^{rank} sym)$$

where *set* is the number of elements in the set of states, *rank* is the maximum arity of a label, and *sym* is the number of different labels. Similarly, the size of the tables for the compressed table approach is

$$O(set \times rank \times sym + sset^{rank} sym)$$

where *sset* is the number of elements in the largest substate.

In the worst case, the size of compressed tables is even slightly bigger than in the traditional approach. This is because in the worst case $O(sset) = O(set)$. However, under most practical circumstances, it can be expected that compressed tables are much smaller. Chase [Cha87] gave some empirical evidence that this is indeed the case. There are several reasons why the compressed tables can be expected to be much smaller for most algebraic specifications.

First of all, almost all algebraic specification systems are typed. Thus for a given label, all subterms that can legally occur as subterms at a given argument position must have the same type. The corresponding substate will therefore only contain elements with that type.

Furthermore, functions that have arguments of a given type frequently only match with *constructors* of that type. Constructors are defined as the function symbols that can occur in normal forms. Rules that have arguments that are not constructors are generally avoided because they can easily produce counter-intuitive results, such as the introduction of unwanted "junk" terms or unexpected identifications between terms.

Often, rules for a given function (certainly for functions with many arguments) do not match certain arguments at all. All terms with this function as label will then have a variable at this argument position. In such a case the corresponding size of the substates at this argument position is one.

Therefore, it can be expected that for most algebraic specifications, the compressed table approach will be superior.

4 Computing approximations

It is not immediately obvious that approximations can be computed efficiently. But it turns out that this is possible by using memo-functions.

One of the basic ideas behind the algorithm for computing approximations is to use the subsumption graph. The subsumption graph is determined by the immediate predecessors of each term.

Definition 4. The set of immediate predecessors of a pattern p in a set of patterns S, notated $pred(S, p)$ is defined by:

$$\{t' \in S - \{t\} \mid t' \leq t \land (\neg \exists t'' \in S - \{t, t'\} : t' \leq t'' \land t'' \leq t)\}$$

The following algorithm computes the best approximation of a pattern $t \in U$ with respect to a unification closed set of patterns S, such that $S \subseteq U$. The function $cpred(U, t)$ computes the set of immediate predecessors of t in U. U is added as a special argument to Approx since the immediate predecessors are computed with respect to this set. The function max returns the maximum of its arguments with respect to $\leq$: $max(a, b) = $ if $a \leq b$ then b else a.

$Approx(S, t, U, cpred)$
if $t \in S$ then
    $result := t$;
else
    $result := \mathbf{X}$;
    $forall\ t' \in cpred(U, t)\ do$
        $result := max(result, Approx(S, t', U, cpred))$;
$return(result)$;

Theorem 5. *Under the stated pre-conditions Approx computes the best approximation of t with respect to S.*

Proof: The algorithm always returns a term that matches t and is $\in S$. Now assume that the best approximation of t with respect to S is b. From the definition of approximation $b \leq t$. So there exists a path in the subsumption graph from b to t. Along this path $Approx$ will return v such that $b \leq v$, thus also in t.

It could only return a v that is higher than b, thus $v \not\leq b$, if there was a path in the subsumption graph downwards from t to v. But in that case $v \leq t$ which is not possible, because it was assumed that b was the best approximation. $\square$

Using memo functions A trick that will be used in several of the following algorithms is to use memo-functions. The basic idea of a memo-function is to store the results of all invocations in a table. When a memo-function is invoked again with the same arguments it will return the result that is found in the table without performing further computations. Memo functions require extra storage overhead for the table, and some initialisation overhead because the table must be initialised to a special value that indicates that the corresponding value has not yet been computed.

For the analysis of the time complexity of an algorithm that uses memo-functions first compute the execution time for the algorithm itself under the assumption that the memo-functions can always use table lookup, and then add the time that is needed by the memo-function to really compute all the entries that have been filled in the table.

Thus a memo-function that corresponds with Approx could look as follows:

```
Approxmemo(S, t, U, cpred)
if table(t) ≠ UNCOMPUTED then
    result := table(t);
else if t ∈ S then
    result := t;
else
    result := X;
    forall t' ∈ cpred(U, t) do
        result := max(result, Approxmemo(S, t', U, cpred));
table(t) = result;
return(result);
```

An important parameter for the analysis is wd, the maximum number of immediate predecessors that an element in U can have. This is the maximum number of times that the loop in this algorithm will be executed. As will be shown later, it is possible to use memo-functions for the computation of $cpred$ and max. In a first analysis of $Approxmemo$ it is therefore assumed that they take $O(1)$ time.

Lemma 6. *Assuming that the computation of cpred and max take $O(1)$ time, the computation of Approxmemo(S, t, U, cpred) for all elements $t \in U$ takes $O(|U| \times wd)$ time and $O(|U|)$ space.*

Proof: *Approxmemo* will only do some real computation the first time it is invoked for a specific t. The next time it can do a table lookup that takes $O(1)$ time. On the first invocation it will execute the loop at most wd times. Initialising the *table* will take $O(|U|)$ time. Therefore the total time complexity is $O(|U| \times wd)$.

The storage space needed for *table* is $O(|U|)$. □

4.1 Computing predecessors

For the previous algorithm it is necessary to have a way of computing the immediate predecessors of a term t in a set U. Now if U satisfies certain conditions, this can be done efficiently with a memo-function.

Definition 7. The product-closure of a set of states S, is defined by:

$$pr(S) = \{c(p_1, \ldots) \mid p_i \in S_{c,i}\} \cup \{X\}$$

Definition 8. A set of patterns S is *product-closed* iff $pr(S) = S$.

Lemma 9. *Let S be a product-closed and subterm-closed set of patterns. Let $p = c(t_1, \ldots t_m)$ be a pattern in S such that at least one of $t_1, \ldots t_m$ is not a variable. Now, define the set*

$$Q = \{c(t'_1, \ldots t'_m) \mid \\ \exists! i \in \{1..m\} : (\exists q \in pred(S, t_i) : t'_i = A_{S_{c,i}}(q) \land \forall j \in \{1..m\} - \{i\} : t'_j = t_j)\}$$

Then it holds that:

$$pred(S, p) = Q$$

Proof: First of all, a pattern $p' \in pred(S, p)$ must differ in exactly one subterm from p. Otherwise, if it differs in two subpatterns t'_i and t'_j, a new term p'' can be constructed from p by replacing t_i by t'_i. Because S is product-closed, $p'' \in S$. But since $p' \leq p'' \leq p$, it must be concluded that $p' \notin pred(S, p)$.

Further, $\mathcal{A}_{S_{c,i}}(q) \in S_{c,i}$ thus the constructed term is elements of S. That $\mathcal{A}_{S_{c,i}}(q) \in pred(S_{c,i}, t_i)$ follows from the fact that $S_{c,i} \subseteq S$ and that $S_{c,i}$ is unification-closed.

To show that Q indeed contains all immediate predecessors, assume that there is one $p' \in pred(S, p)$ such that $p' \notin Q$. p' must differ in one subterm t'_i from p. Since $t'_i \notin pred(S, t_i)$ but $t'_i \leq t_i$ there exists a term $t''_i \in S$ different from t_i and t'_i such that $t'_i \leq t''_i \leq t_i$. But then p'' can be constructed from p by replacing t_i with t''_i. Because S is product-closed $p'' \in S$. Now p, p' and p'' only differ in their i-th sub term. Since $t'_i \leq t''_i \leq t_i$, it follows that $p' \leq p'' \leq p$, thus $p' \notin pred(S, p)$. $\square$

This lemma suggests the following memoing algorithm for computing the immediate predecessors of a term p in a product-closed, subterm-closed set of patterns S such that $p \in S$.

```
cpred(S, p)
if predtable(S, p) ≠ UNCOMPUTED then
    result := predtable(S, p);
else if p = X then
    result := ∅;
else if subterms(p) − {X} = ∅ then
    result := {X} ∩ S;
else
    let c(t₁, t₂, ... tₘ) = p;
    result := {c(t'₁, t'₂, ... t'ₘ) | ∃!i ∈ {1..m} :
              (∃ q ∈ cpred(S, tᵢ) : t'ᵢ = A_{S_{c,i}}(q)) ∧ (∀j ∈ {1..m} − {i} : t'ⱼ = tⱼ)};
fi
predtable(S, p) := result;
return(result);
```

Theorem 10. *If a set of patterns S is product-closed and subterm-closed, then for all $p \in S$: $cpred(S, p) = pred(S, p)$.*

Proof: The proof goes by structural induction on p.

When $p = \mathbf{X}$, $cpred(S, \mathbf{X}) = \emptyset = pred(S, \mathbf{X})$.

When $subterms(p) - \{\mathbf{X}\} = \emptyset$, in other words if p only contains variables as arguments, its predecessor is the single-variable pattern if that is a member of S $cpred(S, p) = \{\mathbf{X}\} \cap S = pred(S, p)$.

Now the induction step follows from the previous lemma. $\square$

For the analysis of the algorithm it is convenient to define *rank* as the maximum arity of the symbols in $\sum$.

Theorem 11. *When $\mathcal{A}_{S_{c,i}}(q)$ can be computed in $O(1)$ time, $cpred(S, p)$ can compute immediate predecessors for all elements $p \in S$ in $O(|S| \times rank \times wd))$ time and $O(|S| \times wd)$ space.*

Proof: First observe that cpred will return at most wd elements. Constructing a new term can be done in $O(rank)$ time, and adding a term to the set can be done in $O(1)$ assuming appropriate data-structures. When cpred cannot use a table-lookup, it will take $O(rank \times wd)$ time, since it can be assumed that recursive calls take $O(1)$ time.

The total number of times that it cannot use a table-lookup is $|S|$, therefore its total complexity is $(|S| \times rank \times wd)$. Initialisation of *predtable* can be done in $O(|S|)$ time. Adding this up gives the stated result.

The storage space for $predtable(S)$ is $O(|S| \times wd)$. □

It is possible to compute the immediate predecessors of a pattern t in S by:

$$spred(S,t) = \{Approxmemo(S,p,pr(S),cpred) \mid p \in cpred(pr(S),t)\}$$

As described in section 4.2 $Approxmemo(S,p,pr(S))$ and $cpred(pr(S),t)$ must be computed anyhow for all $p, t \in pr(S)$. Therefore, computing $spred(S,t)$ for all $t \in S$ takes only $O(|S| \times wd)$ extra time.

Computing the unification closure

One step in the computation of the set of states, is the computation of the unifiers of all terms.

By using memo-functions the unifications of all elements in S can be computed in $(O(|S|^2 \times rank)$ time and $O(|S|^2)$ space. Since all patterns are linear, unification is very simple.

$unify(p_1, p_2)$
$\text{if } utable(p_1, p_2) \neq UNCOMPUTED \text{ then}$
      $result := utable(p_1, p_2);$
$\text{else if } p_1 = \mathbf{X} \text{ then}$
      $result := p_2;$
$\text{else if } p_2 = \mathbf{X} \text{ then}$
      $result := p_1;$
$\text{else if } p_1.label \neq p_2.label \text{ then}$
      $result := FAIL;$
else
      $\text{let } c := p_1.label;$
      $\text{for } i := 1 \text{ to } arity(c) \text{ do}$
            $t_i := unify(p_1.child[i], p_2.child[i]);$
      $\text{if } \forall t_i : t_i \neq FAIL \text{ then}$
            $result := c(t_1, \ldots);$
      else
            $result := FAIL;$
$utable(p_1, p_2) = result;$
$return(result);$

It is also possible to use the results of these computations to compute the function *max* that was needed in the definition of *Approx*. This follows from the following property of linear unification: $a \leq b \Leftrightarrow$ the unifier of a and b is equal to b. Thus when all unifications already have been computed *max* can be computed in $O(1)$ time by $max(a, b) = if(unify(a, b) = b)$ then b else a.

4.2 Computing the compressed tables

An algorithm to construct the tables for bottom-up matching can be made by gluing together the algorithms in the previous sections. This paper only describes the algorithm for computing compressed tables. In most cases the corresponding algorithm for computing uncompressed tables by using approximations directly is less efficient than computing the compressed tables first by using approximation, and then computing the uncompressed tables from the compressed ones.

The tables that are needed can be computed by:

$$\theta(p) = Approxmemo(S, p, pr(S), cpred)$$

$$\mu_c^i(p) = Approxmemo(S_{c,i}, p, S, spred)$$

The preconditions for *Approxmemo* are satisfied. Both S and $S_{c,i}$ are unification-closed. Furthermore *cpred* and *spred* return the immediate predecessor in $pr(S)$ and S respectively.

The number of elements in $pr(S)$ is $O(|sset|^{rank} \times sym)$. From the construction of the sets it follows that $S_{c,i} \subseteq S \subseteq pr(S)$. Thus therefore $sset \leq set \leq sym \times sset^{rank}$.
The different steps have the following time complexity.

initialisation + *unify*	$O(sym \times rank^2 \times sset^2)$
$Approxmemo(S_{c,i}, p, S, spred)$	$O(sym \times rank \times set \times wd)$
$Approxmemo(S, p, pr(S), cpred)$	$O(sym \times sset^{rank} \times wd)$
$cpred(pr(S), p)$	$O(sym \times sset^{rank} \times rank \times wd)$
$spred(S, p)$	$O(set \times wd)$

Assuming that $rank \geq 2$ the total time complexity becomes $O(sym \times sset^{rank} \times rank \times wd)$. The complexity for the whole algorithm is determined by the time that is needed for computing *cpred*.

The space needed for the tables has the following complexity:

initialisation + *unify*	$O(sym \times rank \times sset^2)$
$Approxmemo(S_{c,i}, p, S, spred)$	$O(sym \times rank \times set)$
$Approxmemo(S, p, pr(S), cpred)$	$O(sym \times sset^{rank})$
$cpred(pr(S), p)$	$O(sym \times sset^{rank} \times wd)$
$spred(S, p)$	$O(set \times wd)$

The space complexity is also dominated by *cpred*.

Analysis

The algorithm presented has a better time complexity than the the algorithm in [HO82] which was $O(set^{rank} \times sym \times patsize)$.

For simple pattern forests $wd \leq rank$, for elements p of S $|spred(S,p)| \leq 1$ therefore for elements $q \in pr(S) : |cpred(S,p)| \leq rank$. Thus in this case the algorithm probably performs better than Hoffmann and O'Donnell's specialised algorithm for simple pattern forests when $rank < ht$.

The value of wd is very important in the analysis of these algorithms. Unfortunately, it is highly dependent upon the internal structure of the set of patterns. However, in practice wd is low, approximately $O(rank)$. One reason, why it can be expected to be low, is that all computations only find the immediate predecessors in sets that are unification-closed. But, for a unification-closed set of patterns P, $\forall t \in P : \forall p, p' \in pred(P,t) : p \neq p' \Rightarrow unify(p,p') = t$. This follows because $p \leq t$ and $p' \leq t$, therefore the unifier u of p and p' must exist. Furthermore, $u \in P$ since P is unification-closed. Now, because $u \leq t$ and p, p' were immediate predecessors, it follows that $u = t$. Because each pair of elements in $pred(P,t)$ must be pairwise unifiable to t, large sets of predecessors are unlikely to occur.

Cai et al [CPT90] described a fast algorithm for computing the compressed bottom-up tables. In their paper they claim that their algorithm only uses $O(|M(c(t_1,\ldots,t_m))|)$ time for each entry $\theta_c[t_1,\ldots,t_m]$, where $M(p)$ is the match-set for p. However, it would seem that hashing the $t_1,\ldots,t_m$ in step 4 of their algorithm should take $O(m)$ time. If this is indeed true the complexity of their algorithm becomes $O(rank \times |M(p)|)$. For almost all practical applications m is quite low, say < 10, so the difference is quite small.

The time complexity of our algorithm per table entry in θ is $O(rank \times wd)$. It is not immediately possible to compare this value with the results of Cai et al, because wd and the size of the match-sets are highly dependent on the set of patterns. In our (limited) experience, both values tend to be comparable. A good comparison of both algorithms is sadly lacking.

The algorithm in Cai et al is quite sophisticated and uses several rather complicated data-structures. The algorithm presented in this paper is relatively straight-forward, and uses only simple data-structures as arrays and sets. Therefore, it is probably easier to implement, and optimise.

One disadvantage is that the storage requirements are higher than those in [CPT90]. However, in practice this may not be a very serious limitation. A bottom-up matcher normally must be able to store all needed tables in memory, and also needs memory space for storing the terms. Our algorithm requires besides the storage space for the tables at most an extra factor of wd in space. In our experience the average number of immediate predecessors is $O(rank)$, when this is indeed the case, the overhead seems acceptable.

Performance measurements

This section will present some benchmark results for our algorithm. From the foregoing analysis it is clear that the run-time depends in a complicated way on the structure of its inputs. For realistic benchmarks a "representative" sets of input patterns are needed.

The benchmark that is used for this paper has the following structure. For a given value of n the benchmark consisted of the following set of patterns repeated for a_i from $a_1 \ldots a_n$:

$$\{a_i(z, Y), a_i(s(z), Y), a_i(s(s(X)), Y), a_i(z, z), a_i(z, s(X))\}$$

Thus the total number of input patterns is $5n$. This benchmark seems fairly representative for many rewrite systems. The number of patterns with the same top-level function symbol is frequently low (say < 5) and the subterms of the patterns frequently only contain constructor-symbols.

The benchmarks were executed on a DEC 5400, which has a MIPS-RISC based architecture. Its performance has been found to be comparable SUN Sparcstation II. The algorithm has been implemented in C and compiled with the GNU C compiler.

n	Time (s)	$|\mathcal{S}|$	$|\mu|$	$|\theta|$
5	0.16	30	330	66
10	0.24	55	1155	126
100	7.11	505	101505	1206
200	27.32	1005	403005	2406

The size of the μ tables increases as $O(n^2)$, the execution time also increases quadratically. Thus for this set of benchmarks the execution time is linear in the size of the outputs. The implementation that was used for these measurements is not highly optimised. Preliminary experience with an implementation of a modification of this algorithm seems to indicate that performance can be significantly improved.

5 Conclusions and further research

Hoffmann and O'Donnell already noticed in [HO82] that "For the quickest matching time, the bottom-up algorithm, driven by tables, is best". This conclusion still seems valid today. By compiling rewrite rules into a program that uses bottom-up matching it is possible to achieve a performance of 50000-200000 rewrites/s. These results will be described in a forthcoming paper. Using compressed tables is somewhat slower than the traditional approach, due to the additional level of indexing. However, the size of the tables is greatly reduced.

Describing states by unifiers instead of match-sets seems a promising approach. Proofs that uses unifiers are frequently more elegant than their equivalents that use match-sets. Also, the subsumption ordering and the notion of approximation can be manipulated easily. The algorithms in this paper would be more complicated (and probably less efficient) when match-sets were used in their implementation.

It seems that it should be possible to improve the current algorithm. The current algorithm computes predecessors for all $p \in pr(\mathcal{S})$. However, for all $p \in pr(\mathcal{S}) - \mathcal{S}$ it is possible to redefine $pred(pr(\mathcal{S}), p)$ as $\{\mathcal{A}_\mathcal{S}(p)\}$, since all algorithms only effectively use the predecessors that are elements of $\mathcal{S}$. Another observation that may be useful in discovering new algorithms is that it is possible to compute the $\leq$ ordering in $O(set^2 \times rank)$, by using a similar algorithm as for unification.

One of the most interesting aspects of the algorithm in [CPT90] is that it is incremental. In principle, it seems possible to adapt the algorithms in this paper, so they can also be used incrementally.

Several of the idea's in this paper have been inspired by category theory. One of the basic observations was that linear unification could be seen as the categorical sum. Similarly, approximation is a functor to a subcategory. This connection looks like an interesting subject for further research.

Bottom-up matching is a technique that should be seriously considered when implementing systems for rewriting. The compression approach, described by Chase [Cha87] leads to tables with manageable sizes. For the generation of tables, the algorithm described in this paper, and the one in [CPT90], both appear to offer a good performance.

Acknowledgements

I would like to thank Paul Hendriks and Erik Nieuwland for our discussions about rewriting systems. I would also like to thank the referees for their comments.

References

[Cha87] David Chase. An improvement to bottom-up tree pattern matching. In *Proceedings Fourteenth Annual ACM Symposium on Principles of Programming Languages*, pages 168–177, January 1987.

[CPT90] J. Cai, R. Paige, and R. Tarjan. More efficient bottom-up tree pattern matching. In *CAAP '90*, pages 72–86. Springer LNCS 431, 1990.

[HO82] Christoph M. Hoffmann and Michael J. O'Donnel. Pattern matching in trees. *Journal of the ACM*, 29(1):68–95, January 1982.

[PB85] Paul Walton Purdom and Cynthia A. Brown. Fast many-to-one matching algorithms. In *Proceedings Rewriting Techniques and Applications*. Springer LNCS 202, 1985.

[Wal91] Humphrey Robert Walters. *On Equal Terms: Implementing Algebraic Specifications*. PhD thesis, University of Amsterdam, 1991.

On some Algorithmic Problems for Groups and Monoids

Sergei I. Adian[1]

Steklov Mathematical Institute
Vavilov str. 42, 117966 Moscow, RUSSIA
e-mail: si@adian.mian.su

In 1912 Max Dehn [13] formulated three main algorithmic problems for groups presented by defining relations: Word problem, Conjugacy problem and Isomorphism problem. Two years later A.Thue [37] formulated the Word problem for semigroups presented by defining relations (Thue systems).

Let Σ be a finite alphabet and Σ^* be the free monoid generated by Σ, its unit (the empty word) being denoted by 1. For $X, Y \in \Sigma^*$, $X = Y$ means that X and Y are the same word in Σ. A Thue system Π in the alphabet Σ is defined by a set $\{A_i = B_i\}$ of ordered pairs of words of Σ^*. A binary relation $\to$ called *elementary transformation* is defined as follows: for any $X, Y \in \Sigma^*$, $X \to Y$ if and only if there exist $u, v \in \Sigma^*$ such that

$$(X = uA_iv) \ \& \ (Y = uB_iv) \quad \text{or} \quad (X = uB_iv) \ \& \ (Y = uA_iv).$$

The equivalence relation "$X = Y$ in Π" is the reflexive and transitive closure of the elementary transformation relation. This relation is symmetrical, but one can consider so called semi-Thue system with the only elementary transformations of the type $uA_iv \to uB_iv$.

The Word (Conjugacy) Problem for a given Thue system (finitely presented semigroup or group) is a requirement to find an algorithm to recognize, for any two given words in the given alphabet, whether they are equivalent (conjugate) in the system or not. Two elements X and Y of a group (monoid) G are *conjugate* in G if there exist an element T such that $TX = YT$ in G.

The Isomorphism Problem for group (or monoid) presentations is a requirement to find an algorithm to recognize, for any two given finite group (monoid) presentations G_1 and G_2, whether they are isomorphic or not.

In 1947, almost simultaneously, A.Markov [26] and E.Post [32] have constructed finitely presented semigroups (Thue systems) with unsolvable word problem. The undecidability of the word problem (and hence of the conjugacy problem) for groups was obtained for the first time by P.S.Novikov in 1955 [29]. Very soon the unsolvability of the isomorphism problem for arbitrary fixed group has been proved in [2]. Moreover it has been proved that almost all invariant group properties are algorithmically nonrecognisible (*Adian-Rabin theorem*):

Let α be an invariant group property such that there exist a finitely presented group $G_1 \in \alpha$ and a group G_2 that cannot be embedded into a group with the property α. Then the property α isn't algorithmically nonrecognizable (see [1, 2, 3] and [33]).

Similar result for invariant properties of Thue systems has been proved in 1954 by A.A.Markov [26]. For that reason such properties are called Markov properties. A very nice proof of Novikov's theorem has been published in 1959 by W.W.Boone [10]. On the basis of Boone's construction A.A.Fridman [15] proved (under supervision of the author of this talk) that, for any given Turing degree α, there is a finitely presented group with the Word problem of degree α. Similar result for the conjugacy problem has been proved by D.Collins [12]. In 1969 another pupil of the author, V.V.Borisov, proved the existence of a group with unsolvable word problem presented by 12 defining relations [11]. This is the minimal known (April 1993) number of relations for groups with unsolvable word problem. The positive solution of the word problem for one relator groups has been proved by W.Magnus in 1932 [21]. For 2-relator groups the problem is still open. In 1956 G.S.Tseitin [41] constracted a Thue system with unsolvable word problem presented by 7 defining relations. It is the simplest known presentation of a Thue system with unsolvable word problem. Below are the relations of Tseitin's example:

$$ac = ca, \; ad = da, \; bc = cb, \; bd = db, \; eca = ce, \; edb = de, \; cca = ccae.$$

Tseitin proved his result by reducing the word problem for any finitely presented group to the word problem for this particular Thue system. Later on, Yu. Matiyasevich [28] proved the existence of a Thue system presented by 3 defining relations with unsolvable word problem, but one of the relations of his system is very long.

All the results mentioned above has been proved by using some rewriting methods developed in the papers of A.A.Markov, P.S.Novikov, W.W.Boone, S.I.Adian, etc. The authors were studying monoid presentations as Thue systems (finite presentation of a group is a Thue system of a special type). For instance, in the paper [4] the main approch was to study finite sequences of "elementary transformations" in a given Thue system. It is clear that Thue systems were the original prototypes of so called "rewriting systems". The lexicographical ordering of the alphabet can be useful when we study the complexity of simple algorithms. Relationship between word problems for groups and monoids, embedding of monoids into groups, identities in special monoids and some other natural questions for finitely presented monoids (Thue systems) were studied in [4]. Many results in [4] were related to different type of normal forms. I was surprised recently seeing several papers of L.Zhang [42,43,44,45] published in well known mathematical Journals. The large part of these papers looks like a result of rewriting from [4] in the direct meaning of the word. Of course, to refresh the ideas and the technique of [4] may be useful, but it should be done in a more decent way.

More detailed information about the results on algorithmic problems for groups and semigroups obtained at the Steklov Institute and Moscow University one can find in [8]. I would like to mention below only a few of them.

1 The Small Cancellation Property

Suppose we have a finitely presented group $G = <A; D>$, where the set of defining relations is symmetrized, i.e., all relators in D are cyclically reduced and for any $R \in D$ all cyclic shifts of R and their inverses also belong to D. We say that G satisfies the condition $C(k)$ for a rational k, if, for any two relators $R, S \in D$, if $R = BC$ and $S = BH$ then $|B| < k|R|$. We can say that $C(k)$ is the set of finitely presented groups for which the maximal partion of a nontrivial common piece B of two relators is less than k. The algebraic study of small cancellation theory was initiated by V.A.Tartakovskii (see [39, 40]). He solved the Word problem for the groups satisfying the condition $C(1/7)$. Using a geometrization of small cancellation theory in terms of so-called van Kampen diagrams R.Lyndon in [20] solved the word problem for the class $C(1/5)$. A.I.Gol'berg from Moscow University in [16] proved that any finitely presented group have a presentation with the maximal partion of a nontrivial common piece B of any two relators equal to 1/5. So Lindon's result cannot be improved. As Gol'berg's result has a very simple proof we can give it here.

It is easy to check that the following group is trivial.

$$H = <a, b, c, d;\ a^5 = B^6 = C^5 = 1,\ aba^{-1}dc = 1, a^{-1}bcb^{-1}c = 1,\ dad^{-1}cd = 1>$$

As the genarators a^{-1} and b^{-1} conjugate cd and dc they must have the same order, hence we have $a = b = 1$. Then from the first relation of the second line we get $c^2 = 1$ and hence $ca = 1$. Then from the last defining relation we obtain $d = 1$. It is a simple exercise to check that any nontrivial intersection of any two ciclic shifts or conjugates of the relators of the group H has a length 1. So the partion of the maximal nontrivial common piece B of any two relators of H is equal to 1/5.

Now let $G = <A; D>$ be any finitely presented group. We can embed G into a new group G_1 in the following way. We consider the free product G^* of G and sufficiently large number t of copies of H in the alphabet:

$$x_1, x_2, x_3, x_4, x_5, \ldots, x_{5t-4}x_{5t-3}x_{5t-2}x_{5t-1}x_{5t}$$

and rewrite all defining relations from D inserting between any two letters a new letter x_i (each letter x_i we shall use only once!). For example if $a_1 a_2 \ldots a_s \in D$ then the result of insertions of x_i's into it is $a_1 x_1 a_2 x_2 \ldots x_{s-1} a_s x_s$ and it is equal in G^* to the original one because $x_i = 1$ in G^* for any i. Clearly the resulting group is isomorphic to H and the maximal nontrivial common piece B of any two relators of it is equal to 1/5.

A similar result for finitely presented monoids has been proved in 1968 by B.A.Osipova [31], a pupil of the author. We shall formulate also the main result of [31]. Consider a semigroup

$$\Pi = <a, b, \ldots c;\ A_i = B_i>,$$

where all A_i and B_i are nonempty. Let M be the set of all words A_i and B_i. We call the initial (terminal) segment P of a word X from M *regular* if $|P| \geq 1/2|X|$.

We say that Π belongs to the class $K_{1/2}$ if M satisfies the following conditions for any $A, B \in M$:
 a) if $B = RPQ$ and P is a regular initial segment of A, then R is empty,
 b) if $B = RPQ$ and P is a regular terminal segment of A, then Q is empty.
It was proved in [31] that if $\Pi \in K_{1/2}$ and $X = Y$ in Π then

$$|Y| < |X| m^{|X|},$$

where m is the maximum of the lengths of all words from M. So there is an algorithm solving the word problem for all groups of the class $K_{1/2}$.

She also proved that Tseitin's monoid with 7 relations mentioned above is embeddable into a finitely presented monoid that belongs to the class $K^*_{1/2}$ obtained, if we replace in the definition of the class $K_{1/2}$ nonstrict inequality by the strict inequality $|P| > 1/2|X|$. So, her positive result also cannot be improved.

2 Equations and Elementary Theories over Free Groups and Monoids

One of the oldest result on algorithmic problems in algebra was Quine's result on unsolvability of elementary theory of free monoids published in JSL in 1946 (see [33]). Only in 1973 V.G.Durnev [14] proved unsolvability of the positive $\exists\forall\exists^3$-fragment of the elementary theory of free monoids, consisting of negation-free formulas with the quantifier prefix of the form $\exists x \forall y \exists z \exists u \exists v$. In [24] S.S.Marchenkov proved the unsolvability of positive $\forall\exists^4$-theory of free monoids.

To prove positive results in this direction was more complicated. The most important case was the existential fragment, in particular, the problem to recognize for any given system of equations over free monoid, whether there exists a solution or not. A complete positive solution of this problem has been obtained in 1977 by G.S.Makanin, a pupil of the author (see [22]). He introduced a new concept of a generalized equation in order to solve the problem. Using this concept he constructed also an algorithm which, given a coefficient-free eguation over the free monoid, finds the rank of the equation. It has been established in [38] that Makanin's result implies the decidability of the universal theory of a free monoid. Durnev proved the possibility of eliminating the rightmost universal quantifier in the positive elementary theory of free monoid, so positive $\exists\forall$-theory of a free monoid is also decidable.

In 1982 Makanin [23] obtained also a complete solution of the problem of solvability of equations in a free group. This result was based on his method of general equations for free monoids. In [24] Makanin proved also the decidability of the universal and positive theories of a free group. This result is most surprising because, as we mentioned above, the positive theory of the free monoid is undecidable. But the main problem in this area — the Tarsky problem on decidability of the elementary theory of free groups — is still open, as well as another problem of Tarski on elementary equivalence of elementary theories of two free groups of different rank.

Using Makanin's method of generalised equations a pupil of the author A.A.Razborov constructed an algorithm which, given a coefficient-free equation in a free group, finds the rank of the equation. He also obtained a description of the set of solutions for any eguation in a free group by means of arbitrary paths in a certain graph of automorphisms constructed on the base of the given equation [35].

In this short survey I didn't mention many partial results on equations of the other authors, in particular the pioneer results of E.Lentin [18] for elementary equations over free semigroups and of R.Lyndon [19] for equations with one unknown in free groups. The most of other results were covered by the results mentioned above.

3 The Word and Divisibility Problems for 1-Relator Monoids

The word and divisibility problems for one relator semigroups have been studied in [4]. The relation $A = B$ for nonempty A and B is called left side (right side) noncancellable if A and B have distinct initial (distinct internal) letters. In [4] the positive solution of the word problem has been given for 2 substantial classes of relations:

1. $A = 1$ (for arbitrary A),
2. $A = B$ (for nonempty A and B, if $A = B$ is both side noncancellable).

In order to generalize the notion of noncancellable relation to systems of many relations the notion of left side (right side) "cyclefree system of relations" was introduced in [4]. In [5] the author described a simple algorithm $\mathfrak{A}$, which for every semigroup Π presented by left side (right side) cyclefree system of defining relations produces the shortest sequence of elementary transformations of Π transferring a given word X into a word beginning (ending) with a given letter b, provided X was (right side) divisible by b in Π. This algorithm $\mathfrak{A}$ has been used in [6,7,30,36] for a solution of the word and the divisibility problems for some special classes of one relator semi- groups. In particular in [6,7] the word and left side divisibility problems for one-relator semigroups has been reduced to the same problem for semigroups defined by one relation of the sort

$$aAa = bBa.$$

Similar reduction result with a different approch was published in [17].

Let Π be a semigroup presented by a defining relation $A = B$. If $|A| = |B|$, the word problem has a trivial solution, because all words equal in Π to a given word X have the same length $|X|$. If $|A| > |B|$ and the word A is hypersimple the word problem also has a simple solution. In that case for a given word X we can find a reduced form $\overline{X}$ by replacing all occurrences of A into X by B. It is easy to check that, for any 2 given words X and Y $X = Y$ in Π iff $\overline{X}$ coincides with $\overline{Y}$.

I cannot agree with N.V.Book when in the paper [9] (see page 63) he is using this trivial solvable case of the word problem as an argument for a positive conjecture on the real problem of the solvability of the word problem for one-relator monoids. Of course there are more reasonable arguments for such a conjecture.

Much more complicated is the case of a hypersimple B for $|A| > |B|$. For this case the word problem and the left (right) divisibility problem are solvable if B occurs in A (see Theorem 3 in [7]). For the case when B does not occur in A we prove the following partial result.

Theorem 1. *Let $A = aA_1$ and $B = bB_1$, where B doesn't occur in A and no proper initial segment of B is a terminal segment of B or A. Then using the algorithm $\mathfrak{A}$ one can solve the word problem and the left divisibility problem for the semigroup*

$$\Pi =< a, b, ..., c;\ aA_1 = bB_1 > . \qquad (1)$$

In order to prove this theorem we have to remain some definitions and ideas from [4, 5] and prove some lemmata. We recall a definition of the left decompositions $R_l(aX, b)$ of a given word aX relative to b and $R_l(bX, a)$ of a given word bX relative to a (see [5], p.615). These decompositions we define by simultaneous induction on the length of aX (bX). They have the form

$$* H_1 * H_2 * \ldots * H_k * E * Z \qquad (2)$$

where $aX(bX)$ coincides with $H_1H_2...H_kEZ$, the words R_i are called *the components* and E is called *the head* of the decomposition. The components R_i are nonempty and proper initial segments of the left side ($A = aA_1$) or the right side ($B = bB_1$) of the defining relation of Π. We shall call the component H_i in the decomposition (2) the *A-component (B-component)* if it is an initial segment of A (or B). The word Z can be empty and the head E coincides with aA_1 or with bB_1. A decomposition may have no head E. Definition for $R_l(aX, b)$ and for $R_l(bX, a)$ are similar.

Let E be the maximal common initial segment of the words aX and aA_1, i.e. $aX = EZ$ and $aA_1 = ED$ for some Z and D. If $E = aA_1$ then we define $R_l(aX, b) = *E*Z$. Let D be nonempty. In that case we denote E by R_1. If Z is empty then we define $R_l(aX, b) = *R_1*$. If Z is nonempty then either $Z = aY$ for some Y if D begins by b or $Z = bY$ for some Y if D begins by a. As $|Y| < |X|$ by the inductive assumption we have defined a decomposition

$$R_l(aY, b) = *H_2 * H_3 * \ldots * H_k * E * Z_1 \text{ if } Z = aY$$

or

$$R_l(bY, a) = *H_2 * H_3 * \ldots * H_k * E * Z_1 \text{ if } Z = bY.$$

Then we define

$$R_l(aX, b) = *H_1 * H_2 * H_3 * \ldots * H_k * E * Z_1.$$

This is the end of the inductive definition of $R_l(aX, b)$ and $R_l(bX, a)$.

Remark. In a particular case when the semigroup contains only two letters a and b, $R_l(aX, b)$ and $R_l(bX, A)$ always exist. In the case of many generators of Π a decomposition $R_l(aX, B)$ for some pairs (aX, b) may not exist, but there is a simple algorithm $\mathfrak{B}$ to check for any given pair (aX, b) does there exist or not a decomposition $R_l(aX, b)$ and to construct one if it exists (see [5], p. 614).

Algorithm $\mathfrak{A}$ (see [5], p.620):

Let a semigroup Π be presented by a left side cyclefree system of defining relations and a pair (aX, b) be given with distinct letters a and b. Using the algorithm $\mathfrak{B}$ check if there exists a decomposition $R_l(aX, b)$. If $R_l(aX, b)$ doesn't exist or it has no head then aX isn't divisible by b in Π. If $R_l(aX, b)$ has a head, say
$$R_l(aX, b) = *H_1 * H_2 * \ldots * H_k * E * Z,$$
where $E = S$ is a defining relation of Π, replace the head E by S transforming the word $aX = H_1 H_2 \ldots H_k E Z$ into $Y = H_1 H_2 \ldots H_k S Z$. If the initial letter of Y isn't b then repeat the same procedure with the pair (Y, b).

Lemma 2 see [5], p.620. *Let a semigroup Π be presented by a left side cyclefree system of defining relations. If the letters a and b are distinct and aX is left side divisible by b in Π then the algorithm $\mathfrak{A}$ finds the unique sequence of elementary transformations of Π transferring aX into a word bY in the minimal possible number of steps.*

Lemma 3 see [4], p.50. *If*
$$dX = X_0 \to X_1 \to \ldots \to X_t \to \ldots \to X_k = dY \qquad (3)$$
is a shortest sequence of elementary transformations of the semigroup Π where d is a or b, then the first letters of transforming words are not involved in any transformation of this sequence, e.g., in (3) we have a transformation of X into Y.

Remark. Suppose a semigroup Π has a presentation with a cyclefree system of defining relations. If there is an algorithm to recignize for any given word X and letter c if X is left side divisible by c in Π, then one can solve for Π the general left side divisibility and the word problems.

For two given words X, Y this can be proved by induction by the length $|Y|$. Indeed if $Y = cY_1$ then we first check if X is left side divisible by c. If the answer was "no" then X isn't divisible by Y and X isn't equal Y in Π. If the answer was "yes" then we find a word Z such that $X = cZ$. By lemma 1 in such a case X is divisible by Y if and only if Z is divisible by Y_1. Since $|Y_1| < |Y|$ by inductive assumption we have an algorithm to recognize the left side divisibility of Z by Y_1 and hence a left side divisibility of X by Y. If X isn't divisible by Y then X isn't equal to Y in Π. Otherwise we can find an equlity $X = YU$ for some U. Then by lemma 3 we shall have $U = 1$ in Π and the word U must be empty because no elementary transformation of the empty word 1 is possible in Π.

Lemma 4. *Suppose Π is a semigroup presented by (1). Let $A = aA_1$ and $B = bB_1$, where B doesn't occur into A and no proper initial segment of B is an internal segment of B or A. If $X = H_1H_2...H_tAZ$, where A is the head of the left decomposition of X, and $t > 0$, then the algorithm $\mathfrak{A}$ cannot give a positive ansver, that is, for $X = aX_1$ (for $X = bX_1$) X isn't left divisible by b (by a).*

Proof. . Suppose lemma 3 is false. Let X be a contrexample word to the assertion of lemma 3 such that the algorithm $\mathfrak{A}$ gives a positive answer in a minimal possible number r of steps. Let say that X begins b y a. The result of the first step of the algorithm $\mathfrak{A}$ is the word

$$Y = H_1H_2...H_tBZ.$$

H_t may be an A-component or a B-component. As B doesn't occur in A and no initial segment of B is an internal segment of A or B, the word H_tB does not begin by A or B and for maximal nonempty common initial segment R'_t of R_tBZ and A (or B) is longer than R_t and shorter than R_tB. So Y also begins with a and has a decomposition, starting with

$$* H_1 * H_2 * \ldots * H'_t * H_{s+1} * \ldots * H_k *,$$

where H_k is a component that begins with some terminal segment of B in R_tB. Then H_k must be an A-component or an A-head. If $H_k = A$ then Y is also the counterexample to the assertion of Lemma 4 and the algorithm $\mathfrak{A}$ gives the positive answer for Y in $r - 1$ steps. So we have got a contradiction. If H_k is an A-component then after a finite number of steps the algorithm $\mathfrak{A}$ must find a word

$$Y_1 = H_1H_2...H'_tH_{s+1}...H_{k-1}AZ_1,$$

where A is the head of the left decomposition of Y_1 and $k - 1 > 0$ (remember we have an assumption that $\mathfrak{A}$ gives the positive answer in a finite number of steps). Clearly in this case again we get a contradiction with our assumption that X was a counterexample with the minimal possible number of steps of $\mathfrak{A}$. Lemma 4 is proved.

The next lemma is an immediate corrollary of Lemma 4.

Lemma 5. *If $X = H_1H_2...H_tZ$, where $H_1, H_2, \ldots, H_t$ are the first l components of the left decomposition, H_t is an initial segment of A, and $t > 1$, then the algorithm $\mathfrak{A}$ cannot give a positive ansver, i.e., X isn't left divisible by b (or by a). So, if the word X has a decomposition (2) and is divisible by both letters a and b, then all components H_i for $i > 1$ must be B-components and the head E for $t > 0$ must be a B-head.*

Definition. A proper initial segment H of A (of B) is called *almost full* if HA begins with A (with B). It is clear how to check, for a given component, is it almost full or not.

Lemma 6. *Let* (2) *be a left decomposition of the word* X, *where the head* E *is* B.

If all components H_i *in* (2) *are almost full B-components then* X *is divisible by a and by b.*

If there is a component H_i *that is not almost full then* X *isn't divisible by* a *or by* b.

Proof. . The first statement is obvious. Suppose that some component H_i isn't almost full and X is divisible by both letters a and b. Let H_t be the first component in (2) that isn't almost full. By Lemma 5 we can assume that for any $i > 1$ H_i is a B-component. Then the algorithm $\mathfrak{A}$ after a finite number of steps has to get a word

$$Y = H_1 H_2 ... H_t A Z,$$

where $H_t A = H'_t A_1$ and H'_t is the next component after H_{t-1} in the decomposition of the word Y and A_1 is an nonempty proper end of A. As A and B don't occur in A_1, and no initial segment of B can be a terminal segment of A_1 in the decomposition of Y after H'_t must be a component H_k starting by some terminal segment of A_1. By the condition of the theorem 1 such component H_k has to be an A-component. Then by lemma 5 X cannot be divisible by both letters a and b. Again we got a contradiction. Lemma 6 is proved.

Proof. of Theorem 1. As the word B is hypersymple we can assume that $|B| < |A|$. By the remark 3 it is sufficient to have an algorithm to recognize for any word $aX(bX)$, if it is divisible by b (by a) or not. Let a word bX (aX) be given. Using the algorithm $\mathfrak{B}$ we can check the existence of the left decomposition $R_l(bX, a)$ (or $R_l(aX, b)$). If the required decomposition doesn't exist or hasn't got a head, then $bX(aX)$ isn't divisible by a (by b). Let (2) be the required decomposition $R_l(bX, a)$ (or $R_l(aX, b)$). We can assume that it has a head E. If $t = 0$ then clearly $bX(aX)$ is divisible by a (by b). Let $t > 0$. If $E = A$ then by Lemma 4 bX (aX) isn't divisible by a (by b). Let $E = B$. Then by Lemma 6 bX (aX) is divisible by a (by b) if and only if all components H_i in (2) are almost full B-components and this can be checked easily. Theorem 1 is proved.

We illustrate the result of Theorem 1 by some examples. Sure one could mention more such illustrations.

The word and the left divisibility problem are solvable for semigroups

$$1. \Pi = < a, b; aAa = b^r aBa >,$$

where A and B are any words having no subword b^r;

$$2. \Pi = < a, b; bBa = aAa^r >,$$

where B is any word such that bBa is hypersimple and a^r doesn't occur into B.

Next theorem can be proved similarly to Theorem 1.

Theorem 7. *Let*
$$\Pi = < a, b, ..., c; aA_1 = bB_1 >,$$
where $A = aA_1$ and $B = bB1$, B doesn't occur in A and no terminal segment of B is an initial segment of B or A. Then using the algorithm $\mathfrak{A}$ one can solve the word problem and the left divisibility problem for the semigroup Π.

The following theorem is symmetrical to the Theorems 1 and 2.

Theorem 8. *Let $A = A_1 a$ and $B = B1b$, where B doesn't occur into A and no proper terminal (initial) segment of B is an initial (terminal) segment of B or A. Then using the algorithm $\mathfrak{A}$ one can solve the word problem and the right divisibility problem for the semigroup*
$$\Pi = < a, b, \ldots, c; aA_1 = bB_1 >.$$

REFERENCES

1. Adian S.I. (1955). Algorithmic unsolvability of the problem of recognizing certain properties of groups. Dokl. Akad Nauk SSSR **103**, 533-535.
2. Adian S.I. (1957). Unsolvability of certain algorithmic problems of group theory. Trudy Moskov. Mat. Obshch. **6**, 231-298.
3. Adian S.I. (1957). Finitely presented groups and algorithms. Dokl. Akad Nauk SSSR **117**, 9-12.
4. Adian S.I. (1966). Defining relations and algorithmic problems for groups and semigroups. Proc. Steklov Inst. Math. **85**. (English version published by the American Mathematical Society, 1967).
5. Adian S.I. (1976). Word transformations in a semigroup that is given by a system of defining relations. Algebra i Logika **15**, 611-621; English transl. in Algebra and Logic **15** (1976).
6. Adian S.I. and Oganesian G.U. (1978). On the word and divisibility problems in semigroups with a single defining relation. Izv. Akad. Nauk SSSR (Ser. Mat.) **42**, 219-225; English. transl. in Math. USSR Izv. **12** (1978).
7. Adian S.I. and Oganesian G.U. (1987). On the word and divisibility problems in semigroups with one defining relation. Mat. Zametki **41**, 412-421; English transl. in Math. Notes **41** (1987).
8. Adian S.I. and Makanin G.S. (1984). Investigations on algorithmic questions of algebra. Proc. Steklov Inst. Math. **168**, 207-226; English transl. in Proc. Steklov Inst. Math. **3** (1986).
9. Book R.V. (1987) Thue Systems as Rewriting Systems. J. Symbolic Computation **3**, 39-68.
10. Boone W.W. (1959). The word problem. Ann. of Math. (2) **70**, 207-265.
11. Borisov V.V. (1969). Simple examples of groups with unsolvable word problem. Mat. Zametki **6**, 521-532; English transl. in Math. Notes **6** (1969).
12. Collins D.J. (1969) Recursively enumerable degrees and the cojugacy problem. Acta Math. **122**, 115-160.

13. Dehn M. (1912). Über unendliche diskontinuierliche Gruppen. Math. Ann. **71**.

14. Durnev V.G. (1973). Positive theory of a free semigroup. Dokl. Akad Nauk SSSR **211**, 772-774; English transl. in Soviet Math. Dokl. **14** (1973).

15. Fridman A.A. (1962). Turing degrees of the word problem in finitely presented groups. Dokl. Akad Nauk SSSR **147**, 805-808.

16. Gol'berg A.I. (1978). On the impossibility of strenghtening certain results of Greendlinger and Lyndon. Uspekhi Mat. Nauk **33**, 201-202; English transl. in Russian Math. Surveys **33** (1978).

17. Howie J., Pride S. (1986). The word problem for one-relator semigroups. Math. Proc. of the Cambridge Phil. Soc. **99**, 33-44.

18. Lentin A. (1972). Equations dans les monoides libres. Gauthier-Villars and Mouton, Paris.

19. Lyndon R. C. (1960). Equations in free groups. Trans. Amer. Math. Soc. **96**, 445-457.

20. Lyndon R. C. (1966). On Dehn algorithm. Math. Ann. **166**, 208-228.

21. Magnus W. (1932). Das Identitätsproblem für Gruppen mit einer definierenden Relation. Math. Ann. **106**, 295-307.

22. Makanin G.S. (1977). The problem of solvability of equations in a free semigroups. Mat. Sb. **103 (145)**, 147-236; English transl. in Math. USSR Sb. **32** (1977).

23. Makanin G.S. (1982). Eguations in a free group. Izv. Akad. Nauk SSSR (Ser. Mat.) **46**, 1199-1273; English. transl. in Math. USSR Izv. **21** (1983).

24. Makanin G.S. (1984). Decidability of the universal and positive theories in a free groups. Izv. Akad. Nauk SSSR (Ser. Mat.) **48**, 735-749; English. transl. in Math. USSR Izv. **25** (1985).

25. Marchenkov S.S. (1982). Undecidability of the positive AE-theory of a free semigroup. Sibirsk. Mat. Zh. **23**, 196-198.

26. Markov A.A. (1947). On the impossibility of certain algorithms in the theory of associative systems. Dokl. Akad Nauk SSSR **55**, 683-586.

27. Markov A.A. (1954). The theory of Algorithms. Trudy Mat. Inst. Steklov **42**; English transl., Israel Program Sci. Transl., Jerusalem (1961).

28. Matiyasevich Yu.V. (1967). Simple examples of undecidable associative calculi. Dokl. Akad Nauk SSSR **173**, 1264-1266; English transl. in Soviet Math. Dokl. **8** (1967).

29. Novikov P.S. (1955). On the algotithmic unsolvability of the word problem in group theory. Trudy Mat. Inst. Steklov **44**; English transl. in Amer. Math. Soc. Transl. (2) **9** (1958).

30. Oganesian G.U. (1982). On semigroups with one relation and semigroups without cycles. Izv. Akad. Nauk SSSR (Ser. Mat.) **46**, 88-94; English. transl. in Math. USSR Izv. **20** (1983).

31. Osipova V.A. (1968). On the word problem for finitely presented semigroups. Dokl. Akad Nauk SSSR **178**, 1017-1020; English transl. in Soviet Math. Dokl. **9** (1968).

32. Post E.L. (1947). Recursive unsolvability of a problem of Thue. J. Symbolic Logic **12**, 1-11.

33. Quine W.V. (1946). Concatenation as a basis for arithmetic. J. Symbolic Logic **11**, 105-114.

34. Rabin M.O. (1958). Recursive unsolvavility of group theoretic problems. Ann. of Math. (2) **67**, 172-194.

35. Razborov A.A. (1984). On systems of equations in a free groups. Izv. Akad. Nauk SSSR (Ser. Mat.) **48**, 779-832; English. transl. in Math. USSR Izv. **25** (1985).

36. Sarkisian O.A. (1979). Some relations between the word and divisibility problems in groups and semigroups. Izv. Akad. Nauk SSSR (Ser. Mat.) **43**, 909-921; English. transl. in Math. USSR Izv. **15** (1980).

37. Thue A. (1914). Probleme über Verbandenlungen von Zeichenreihen nach gegebenen Regeln. Kristiana Videnskapsselkapets Skr. I: Mat.-Naturvid. Kl. **10**. (Reprinted in his Selected mathematical papers, Universitetsforlaget, Oslo (1977), 493-524).

38. Taimanov A.D. and Khmelevskii Yu.I. (1980). Decidability of the universal theory of a free semigroup. Sibirsk. Mat. Zh. **21**, 228-230.

39. Tartakovskii V.A. (1947). On the word problem for certain types of groups. Dokl. Akad Nauk SSSR **58**, 1909-1910 (Russian).

40. Tartakovskii V.A. (1949). Solution of the word problem for groups with a k-reduced basis for $k > 6$. Izv. Akad. Nauk SSSR (Ser. Mat.) **13**, 483-494; English transl. in Amer. Math. Soc. Transl. (1) **1** (1962).

41. Tseitin G.S. (1956). Associative system with unsolvable word problem. Dokl. Akad Nauk SSSR **107**, 370-371. (The full proof in Trudy Mat. Inst. Steklov **52** (1958), 172-189).

42. Zhang L. (1991). Conjugacy in Special Monoids. J. of Algebra **143**, 487-497.

43. Zhang L. (1992). Applying rewriting methods to special monoids. Math. Proc. Cambridge Phil. Soc. **112**, 495-505.

44. Zhang L. (1992). A short proof of a theorem of Adjan. Proc. of the Amer. Math. Soc. **116**, 1-3.

45. Zhang L. (1992). On the conjugacy problem for one-relator monoids with elements of finite order. International J. of Algebra and Computations. **2**, 209-220.

Combination Techniques and Decision Problems for Disunification

Franz Baader[1] and Klaus U. Schulz[2]

[1] DFKI, Stuhlsatzenhausweg 3, 6600 Saarbrücken 11, Germany, baader@dfki.uni-sb.de
[2] CIS, University Munich, 8000 München 40, Germany, schulz@cis.uni-muenchen.de

Abstract. Former work on combination techniques was concerned with combining unification algorithms for disjoint equational theories $E_1, \ldots, E_n$ in order to obtain a unification algorithm for the union $E_1 \cup \ldots \cup E_n$ of the theories. Here we show that variants of this method may be applied to disunification as well. Solvability of disunification problems in the free algebra of the combined theory $E_1 \cup \ldots \cup E_n$ is shown to be decidable if solvability of disunification problems with linear constant restrictions in the free algebras of the theories E_i is decidable. In order to decide ground solvability (i.e., solvability in the initial algebra) of disunification problems in $E_1 \cup \ldots \cup E_n$ we have to consider a new kind of subproblem for the particular theories E_i, namely solvability (in the free algebra) of disunification problems with linear constant restriction under the additional constraint that values of variables are not E_i-equivalent to variables. The correspondence between ground solvability and this new kind of solvability holds, *(1)* if one theory E_i is the free theory with at least one function symbol and one constant, or *(2)* if the initial algebras of all theories E_i are infinite. Our results can be used to show that the existential fragment of the theory of the (ground) term algebra modulo associativity of a finite number of function symbols is decidable; the same result follows for function symbols which are associative and commutative, or associative, commutative and idempotent.

1 Introduction

In recent years the rôle Robinson's unification—and later unification modulo equational theories—played in theorem proving, term rewriting, and logic programming has more and more been taken on by constraint solving (see e.g., [Bür90, KK89, JL87, Col90]). One advantage of constraint approaches is that it is no longer necessary to compute (a complete set of) solutions; deciding satisfiability of the constraints is usually sufficient. Thus one can, for example, work modulo non-finitary equational theories such as associativity. Another motivation for preferring a constraint approach is that in this setting the expressive power of a formalism can rather naturally be enhanced by considering more general constraints than the equality constraints of unification problems. One of the earliest of these generalizations was Colmerauer's use of equations and negated equations in PROLOG II [Col84]. In the present paper we shall consider solvability of this kind of equational problems (subsequently called disunification problems) modulo equational theories. Possible applications of disunification modulo equational theories in theorem proving and logic programming are sketched in [BB89].

As for unification, the terms in the disunification problems occurring in applications are usually not just built over the signature of the equational theory, but they contain additional free function symbols. More generally, one often wants to solve disunification problems containing function symbols whose properties are defined by different equational theories. For the case of unification, this fact has triggered extensive research on the combination of unification procedures for disjoint equational theories (see, e.g., the introduction of [BS91a] for a brief overview), but until now these approaches have not been generalized to the disunification case. One reason is that until recently the combination methods were restricted to equational theories which are finitary unifying, i.e., they combined algorithms computing finite complete sets of unifiers. In this setting, solvability of disunification problems can be reduced to the unification and the word problem for the equational theory. In fact, to decide solvability of a disunification problem, one simply computes a finite complete set of unifiers for the equations of the problem, and then checks whether one of these unifiers is a solution of the whole disunification problem. This means that for finitary theories it is sufficient to have combination methods for unification. However, if one only has a procedure that decides satisfiability of unification problems, such a reduction of disunification to unification does not seem to be possible. In addition, even if a theory is finitary, the computation of a complete set of unifiers can be of higher complexity than deciding solvability (associativity and commutativity is an example for this phenomenon; see, e.g., [KN92a, KN92b]).

In [BS91a] we have shown how to combine decision procedures for unification, and in the present paper we shall investigate how this method can be generalized to treat solvability of disunification problems. For unification, "solvability" means having a solution in the free algebra (in countably many generators), or equivalently, having a solution in the initial algebra. For disunification, solvability in the initial algebra (called *ground solvability* in the following) implies solvability in the free algebra (simply called *solvability* below), but *not* vice versa. Both types of solvability are considered in the literature (see [Com91, Bür88]), but ground solvability seems to be more interesting for most applications.

For solvability, the adaptation of the combination method to disunification problems is relatively straightforward. The main tool of the method is a decomposition algorithm which transforms every disunification problem Γ in the combination of arbitrary disjoint equational theories $E_1, \ldots, E_n$ into a finite number of tuples $(\Gamma_1, \ldots, \Gamma_n)$, where each Γ_i is an E_i-disunification problem with linear constant restriction.[3] As for unification, Γ is solvable in the combined theory iff for one of these tuples all its components are solvable in the single theories. However, the proof of soundness of the method—which is almost trivial for the

---

[3] This is the obvious adaptation of the notion "unification problem with linear constant restriction," as introduced in [BS91a]; see Section 2 for a definition.

case of unification problems—becomes a lot more involved.

For the ground case, it surprisingly turned out that ground solvability of Γ in the combined theory is not reduced by our method to ground solvability of the components of one of the tuples in the single theories. On the contrary, one has to consider a slightly restricted form of solvability for the E_i-disunification problem with linear constant restriction Γ_i. It should also be noted that for ground solvability to be handled by our method the equational theories have to satisfy an additional condition. This condition holds, however, in various situations which are interesting for applications (see Section 5).

The paper has the following structure. The next section starts with some technical preliminaries. In Section 3 we introduce the decomposition algorithm, show its correctness for the case of solvability, and state some consequences. Section 4 is concerned with ground solvability, and in Section 5 the results are applied to combine disunification algorithms for the free theory and the theories A (associativity), AC (associativity and commutativity), and ACI (associativity, commutativity, and idempotence).

2 Formal preliminaries

For an equational theory E, let $sig(E)$ denote its signature, i.e., the function symbols occurring in the identities of E. We assume that this signature is finite. For disunification it is even more important than for unification to know the signature over which the terms in the formulation of the problem and in the solutions of the problem may be built. For this reason, we shall explicitly talk about (E, Σ)-*disunification* problems, where Σ is a finite superset of $sig(E)$. Such a problem is a finite set of equations and disequations

$$\Gamma = \{s_1 \doteq t_1, \ldots, s_n \doteq t_n\} \cup \{s_{n+1} \not\doteq t_{n+1}, \ldots, s_{n+m} \not\doteq t_{n+m}\},$$

where $s_1, \ldots, t_{n+m}$ are Σ-terms. A *solution* of the (E, Σ)-disunification problem Γ is a Σ-substitution σ such that $s_i \sigma =_E t_i \sigma$ $(i = 1, \ldots, n)$ and $s_{n+j} \sigma \neq_E t_{n+j} \sigma$ $(j = 1, \ldots, m)$. A *ground solution* is a solution that maps all variables occurring in Γ to variable-free Σ-terms. Γ is called *(ground) solvable* iff it has a (ground) solution.

It should be noted that the notion of a disunification problem does not always refer to the same kind of problem in the literature. Our definition coincides with the one of Bürckert [Bür88], who considers existentially quantified equational formulae, but other authors (e.g., Comon [Com91]) allow for arbitrary quantification.

As in the case of unification, one has to distinguish several types of disunification problems. The (E, Σ)-disunification problem is called *elementary*, if $\Sigma = sig(E)$; it is a disunification problem *with constants*, if $\Sigma \setminus sig(E)$ is a set of constants; and it is a *general* disunification problem, if no such restrictions hold.

Solvability of an (E, Σ)-disunification problem obviously means that the equations and disequations can be solved in the E-free Σ-algebra $T(\Sigma, Y)/_{=_E}$ over the countable set of variables Y, whereas ground solvability means that they can be solved in the initial algebra $T(\Sigma, \emptyset)/=_E$. If one has no disequations (i.e., one has a unification problem), then both notions coincide, but this is not the case if disequations are present. For example, let E be the empty theory, and assume that Σ consists of the constant symbol a. The (E, Σ)-disunification problem $\{x \not\doteq a\}$ is solvable, but not ground solvable.

The combination problem for disunification can now formally be defined as follows. Let E_1, E_2 be two equational theories built over the disjoint signatures $\Sigma_1 := sig(E_1)$ and $\Sigma_2 := sig(E_2)$,[4] and let $E = E_1 \cup E_2$ denote their union. We are interested in solving elementary disunification problems for E, i.e., $(E, \Sigma_1 \cup \Sigma_2)$-disunification problems. The terms in such problems are built from variables and symbols of $\Sigma_1 \cup \Sigma_2$. The elements of Σ_1 will be called 1-*symbols* and the elements of Σ_2 2-*symbols*. A term t is called *i-term* iff it is of the form $t = f(t_1, ..., t_n)$ for an i-symbol f ($i = 1, 2$). A subterm s of a 1-term t is called *alien subterm* of t iff it is a 2-term such that every proper superterm of s in t is a 1-term. Alien subterms of 2-terms are defined analogously. An i-term s is *pure* iff it contains only i-symbols and variables. A (dis)equation $s \doteq t$ ($s \not\doteq t$) is pure iff there exists an $i, 1 \leq i \leq 2$, such that s and t are pure i-terms or variables; this (dis)equation is then called an *i*-(dis)equation. Please note that according to this definition equations of the form $x \doteq y$ where x and y are variables are both 1- and 2-equations, and similarly for disequations. In the following, the symbols x, y, z, with or without indices, will always stand for variables.

Solvability of elementary disunification problems in E will be reduced to solvability of disunification problems with constants in the single theories E_1, E_2. But as in the unification case, the solutions of these problems with constants have to satisfy additional restrictions. These restrictions are formalized in the notion of a *disunification problem with linear constant restriction*. For an equational theory F with signature Ω, such a problem consists of two parts:

1. An $(F, \Omega \cup C)$-disunification problem Γ, where C is a finite set of constant symbols not occurring in Ω, and
2. a linear ordering $<$ on $C \cup X$, where X is a finite superset of the set of variables occurring in Γ.

For a given problem of this kind, the sets V_c of *variables which must not use c* are defined as $V_c := \{x \in X; \ x < c\}$, for every $c \in C$. A *solution* of the problem is a substitution σ which assigns terms $x\sigma$ built with variables, symbols from Ω, and constants in C to the variables $x \in X$, solves all equations and disequations of Γ modulo F, and has the additional property that c does not occur in $x\sigma$ for all $c \in C$ and $x \in V_c$. A solution σ is called *restrictive* if for all variables $x \in X$ the value $x\sigma$ is not F-equivalent to a variable. Restrictive solutions will become important if one is interested in ground solvability in the combined theory E.

Disunification problems with linear constant restriction will be denoted in the form $(\Gamma, X, C, <)$, or just as Γ, if no misleading ambiguities are possible.

---

[4] We shall restrict the technical presentation to the combination of two theories. The combination of more than two theories can be treated analogously.

3 Solvability of Disunification Problems

Our first main result says that solvability of disunification problems in the combination of disjoint equational theories can be reduced to solvability of disunification problems with linear constant restriction in the single theories.

Theorem 1. *Let $E_1, \ldots, E_n$ be equational theories over disjoint signatures such that solvability of disunification problems with linear constant restriction is decidable for $E_1, \ldots, E_n$. Then solvability of elementary disunification problems is decidable for the combined theory $E_1 \cup \ldots \cup E_n$.*

This result is analogous to the one for unification given in [BS91a], and it depends on a decomposition algorithm which is very similar to the algorithm presented in that paper. However, the proof of soundness of the method is more complex. As mentioned above, we shall restrict the presentation to the combination of two theories.

The Decomposition Algorithm

The input for this algorithm is an elementary E-disunification problem, i.e., a system

$$\Gamma_0 = \{s_1 \doteq t_1, \ldots, s_n \doteq t_n, s_{n+1} \not\doteq t_{n+1}, \ldots, s_{n+m} \not\doteq t_{n+m}\},$$

where the terms $s_1, \ldots, t_{n+m}$ are built from variables and the function symbols occurring in $\Sigma_1 \cup \Sigma_2$, the signature of $E = E_1 \cup E_2$. The first two steps of the algorithm are deterministic, i.e., they transform the given system into one new system.

Step 1: variable abstraction.
Alien subterms are successively replaced by new variables until all terms occurring in the system are pure. To be more precise, assume that $s \doteq t$ or $t \doteq s$ ($s \not\doteq t$ or $t \not\doteq s$) is an equation (disequation) in the current system, and that s contains the alien subterm s_1. Let x be a variable not occurring in the current system, and let s' be the term obtained from s by replacing s_1 by x. Then the original equation (disequation) is replaced by the two equations $s' \doteq t$ and $x \doteq s_1$ (by the disequation $s' \not\doteq t$ and the equation $x \doteq s_1$). This process has to be iterated until all terms occurring in the system are pure. □

Step 2: split non-variable disequations and non-pure equations.
Each disequation of the form $s \not\doteq t$ (where s or t is not a variable) is replaced by two equations $x \doteq s, y \doteq t$ and a disequation $x \not\doteq y$ where the x, y are always new variables. Each non-pure equation of the form $s \doteq t$ is replaced by two equations $x \doteq s, x \doteq t$ where the x are always new variables. □

It is quite obvious that these two steps do not change solvability of the system. The result is a system which consists of pure equations and of disequations between variables. The third and the fourth step are nondeterministic, i.e., a given system is transformed into finitely many new systems. Here the idea is that the original system is solvable iff at least one of the new systems is solvable.

Step 3: variable identification.
Consider all partitions of the set of all variables occurring in the system such that distinct variables x, y are in distinct classes of the partition if the system contains the disequation $x \neq y$. Each of these partitions yields one of the new systems as follows. The variables in each class of the partition are "identified" with each other by choosing an element of the class as representative, and replacing in the system all occurrences of variables of the class by this representative. In addition, we add a disequation $x \neq y$ for every pair x, y of distinct representatives. □

Step 4: choose ordering and theory indices.
This step does not modify a given system, it just adds some information which will be important in the next step. For a given system, consider all possible strict linear orderings $<$ on the variables of the system, and all mappings *ind* from the set of variables into the set of theory indices $\{1, 2\}$. Each pair $(<, ind)$ yields one of the new systems obtained from the given one. □

For a system obtained by Step 4, let $X_{5,i}$ denote the set of variables of index i ($i = 1, 2$). The last step is again deterministic. It splits each of the systems already obtained into a pair of pure systems.

Step 5: split systems.
A given system Γ is split into two systems $\Gamma = \Gamma_1 \cup \Gamma_2$ such that Γ_1 contains only 1-(dis)equations and Γ_2 only 2-(dis)equations. As an additional restriction, the system Γ_i ($i = 1, 2$) must contain all disequations $x \neq y$ where x or y has index i. This means that disequations between variables of distinct indices are put into both subsystems. The subsystems can now be considered as disunification problems with linear constant restriction $(\Gamma_1, X_{5,1}, X_{5,2}, <)$ and $(\Gamma_2, X_{5,2}, X_{5,1}, <)$ which have to be solved modulo E_1 and E_2, respectively. This means that in the system Γ_i the variables with index i are still treated as variables, but the variables with alien index $j \neq i$ are treated as free constants. □

The output of the algorithm is thus a finite set of pairs (Γ_1, Γ_2) where the first component Γ_1 is an $(E_1, \Sigma_1 \cup X_{5,2})$-disunification problem with linear constant restriction, and the second component Γ_2 is an $(E_2, \Sigma_2 \cup X_{5,1})$-disunification problem with linear constant restriction.

There are three points where this decomposition algorithm is not a totally straightforward adaptation of the one for unification problems. First, we split all non-variable disequations and not only the non-pure ones. This greatly facilitates the proof of correctness of the method, but is not mandatory. Second, we add disequations between all variables that have not been identified with each other in Step 3, and third, disequations involving variables of index i are required to be in Γ_i in Step 5. The latter two points are necessary for the following proposition to hold.

Proposition 2. *The input system Γ_0 is solvable if and only if there exists a pair (Γ_1, Γ_2) in the output set such that Γ_1 and Γ_2 are solvable.*

The proposition shows that the decomposition algorithm can be used to reduce solvability of elementary disunification problems for $E_1 \cup E_2$ to solvability of disunification problems with linear constant restriction for E_1, E_2. Thus Theorem 1 is an immediate consequence of Proposition 2. Before we give a proof of the proposition, let us mention some additional consequences.

Corollary 3. *(1) Let E be an equational theory such that solvability of disunification problems with linear constant restriction is decidable. Then solvability of general E-disunification problems is decidable.*
(2) The result of Theorem 1 can be lifted to general disunification problems, i.e., the assumptions of Theorem 1 are sufficient to get decidability of general disunification problems in the combined theory.
(3) If, for E_1 and E_2, solvability of disunification problems with linear constant restriction can be decided by an NP-algorithm, then solvability of disunification problems in the combined theory is also NP-decidable.

The proof of this corollary is very similar to the one given in [BS91a, BS91b] for the analogous results for unification problems.

In order to prove Proposition 2, some technical background is needed. Without loss of generality, we make the assumption that all equational theories that we consider are consistent. Now let E_1, E_2 be equational theories over disjoint signatures Σ_1, Σ_2. Let $T(\Sigma_1 \cup \Sigma_2, Y)$ be the set of all terms built over the signatures Σ_1, Σ_2 with variables in Y. Applying unfailing completion (see e.g., [DJ90]) to the combined theory $E = E_1 \cup E_2$, but always treating the elements $y \in Y$ as constants, we obtain a possibly infinite ordered-rewriting system R which is confluent and terminating on $T(\Sigma_1 \cup \Sigma_2, Y)$. Thus we eventually obtain, applying R, a unique irreducible normal form $t_{\downarrow R}$ for every term $t \in T(\Sigma_1 \cup \Sigma_2, Y)$. We denote the set of R-irreducible elements of $T(\Sigma_1 \cup \Sigma_2, Y)$ by $T_{\downarrow R}$.

To establish a relationship between impure terms and corresponding pure terms where alien subterms have been replaced by new variables, we consider a bijection $\pi : T_{\downarrow R} \longrightarrow Z$ where Z is a set of variables of appropriate cardinality. This bijection induces mappings π_i of terms in $T(\Sigma_1 \cup \Sigma_2, Y)$ to terms in $T(\Sigma_i, Z)$ as follows. For variables $y \in Y$, $y^{\pi_i} := \pi(y)$ (note that variables are always R-irreducible.) If $t = f(t_1, \ldots, t_n)$ for an i-symbol f, then $t^{\pi_i} := f(t_1^{\pi_i}, \ldots, t_n^{\pi_i})$. Finally, if t is a j-term, $j \neq i$, then $t^{\pi_i} := \pi(t_{\downarrow R})$. The mapping π_i may be regarded as a projection which maps a possibly mixed term to a pure i-term.

A substitution σ is called R-normalized on a finite set of variables X iff $x\sigma \in T_{\downarrow R}$ for all variables $x \in X$. The next lemma was proved—under almost the same assumptions—in [BS91a]: there we additionally assumed that Y and Z are disjoint; but the proof of the lemma does not depend on this property.

Lemma 4. *Let s, t be pure i-terms or variables, and let σ be a substitution which is R-normalized on the variables occurring in s, t. Then*

$$s\sigma =_E t\sigma \quad \text{iff} \quad (s\sigma)^{\pi_i} =_{E_i} (t\sigma)^{\pi_i}.$$

Proof of Proposition 2
Here and in the remainder of this paper, Γ_0 always denotes an input system of the combination algorithm, Γ_j denotes (one of) the system(s) obtained from Γ_0 after Step j of the algorithm ($j = 1, 2, 3, 4$). The two subsystems obtained after Step 5 are denoted by $\Gamma_{5,i}$ ($i = 1, 2$). X_j denotes the set of variables occurring in Γ_j ($j = 0, \ldots, 4$) and $X_{5,i}$ denotes the variables $x \in X_4$ with index i ($i = 1, 2$).

The *proof of completeness* (i.e., of the "only if" part of the proposition) is very similar to the one for the unification case (see [BS91a], proof of Proposition 3.2). There we have shown how a solution σ of the problem Γ_0 can be used to find the correct alternatives in the nondeterministic steps of the decomposition algorithm, and how σ can be split (by projection) into solutions σ_1 and σ_2 of the resulting output systems $\Gamma_{5,1}, \Gamma_{5,2}$. What remains to be shown in the disunification case is that the disequations in the output systems are satisfied as well. But these are just disequations between distinct variables in the systems, and the choice made in the variable identification step is such that the σ-values of all remaining variables are different modulo E. Thus the σ_i solve these disequation, by their definition and by Lemma 4 (see [BS93] for details).

To show *soundness* (i.e., the "if" part of the proposition) we have to demonstrate that Γ_0 is solvable if there exists a pair $(\Gamma_{5,1}, \Gamma_{5,2})$ in the output set such that $\Gamma_{5,1}$ and $\Gamma_{5,2}$ are solvable. In the unification case, this part was almost trivial, but it is a lot more complex here.

Let σ_1 be a solution of $\Gamma_{5,1}$ and σ_2 a solution of $\Gamma_{5,2}$. We may assume that $\sigma_i : X_{5,i} \to T(\Sigma_i \cup X_{5,j}, Y_i)$ ($i, j \in \{1, 2\}, i \neq j$), where Y_1, Y_2 are two disjoint, infinite sets of variables such that X_4 and $Y := Y_1 \cup Y_2$ are disjoint. Let R be a possibly infinite ordered-rewriting system R which is confluent and terminating on $T(\Sigma_1 \cup \Sigma_2, Y)$ (obtained by unfailing completion, as described above).

Using R and the σ_i we shall now define a substitution σ on X_4 which solves Γ_4. It is then trivial to extend σ to a solution of Γ_0. The definition proceeds along the linear order $<$ which was chosen in Step 4 of the algorithm. Assume that $z\sigma \in T_{\downarrow R}$ has been defined for all $z < x$. Without loss of generality we assume that x has index 1. Since σ_1 satisfies the linear constant restriction associated with $<$, we know that all $z \in X_{5,2}$ occurring in $x\sigma_1$ are smaller than x with respect to $<$. For this reason, $x\sigma := (x\sigma_1\sigma)_{\downarrow R}$ is well-defined.

In the corresponding definition for the unification case, the term $x\sigma_1\sigma$ was not R-reduced. This means that the substitution we defined there is not identical to the one defined here, but obviously the two substitution are E-equivalent. For this reason the proof given in [BS91a] to show that σ solves the original unification problem can be taken without change to show that σ solves the equations in Γ_4.

The following two claims, which will be proved by induction on the linear order $<$, establish that σ solves the disequations as well.

(C1) for all $x_1, x_2 \in X_4$ with $x_1 \neq x_2$ we have $x_1\sigma \neq_E x_2\sigma$,
(C2) for each $x_1 \in X_4$: if $\text{ind}(x_1) = i$, then $x_1\sigma \in T_{\downarrow R}$ is an i-term or an element of Y_i.

Without loss of generality, let us consider an element x of index 1. The induction hypothesis that we may use is that Conditions (C1) and (C2) are valid for all $x_1, x_2 < x$. We shall now show that the same is true for all $x_1, x_2 \leq x$. Let $X_{5,2}^{<x} = \{x_1 \in X_{5,2}; x_1 < x\}$. We consider a bijection

$$\pi_x : T_{\downarrow R} \longrightarrow Y \cup X_{5,2}^{<x} \cup Z,$$

where Z is a set of new variables. This bijection has to satisfy the following conditions:

1. $\pi_x(t) \in Y_2 \cup X_{5,2}^{<x} \cup Z$ for every 2-term $t \in T_{\downarrow R}$,
2. $\pi_x(y) \in Y_2 \cup X_{5,2}^{<x} \cup Z$ for every $y \in Y_2$,
3. $\pi_x(y) = y$ for every $y \in Y_1$,
4. $\pi_x(x_2\sigma) = x_2$ for every $x_2 \in X_{5,2}^{<x}$.

It is easy to see that the induction hypothesis guarantees the existence of such a bijection, provided that Z is chosen of appropriate cardinality.

First, let us show that $x\sigma$ is a 1-term in $T_{\downarrow R}$ or an element of Y_1, thus verifying Condition (C2). Obviously $x\sigma \in T_{\downarrow R}$. Let σ_x denote the restriction of σ to the variables $z \leq x$ of X_4. By induction hypothesis, σ_x is R-normalized and we have

$$x\sigma_1\sigma_x = x\sigma_1\sigma =_E (x\sigma_1\sigma)_{\downarrow R} = x\sigma = x\sigma_x.$$

By Lemma 4 we get

$$(x\sigma_1\sigma_x)^{\pi_1} =_{E_1} (x\sigma_x)^{\pi_1},$$

where π_1 is the 1-projection determined by π_x. Now let us show that $(x\sigma_1\sigma_x)^{\pi_1} = x\sigma_1$. If $x\sigma_1 \in Y_1$, this equality obviously holds since σ_x and π_1 do not move these variables, by Condition 3 on π_x. If $x\sigma_1$ is a 1-term, then the "constants" $x_2 \in X_{5,2}$ occurring in this term are in $X_{5,2}^{<x}$. Now σ_x substitutes for these constants R-irreducible 2-terms or elements of Y_2, by induction hypothesis (C2). In both cases, π_1 will reintroduce the old constants again, by Condition 4 on π_x. The variables $y \in Y_1$ occurring in $x\sigma_1$ are not touched, neither by σ_x nor by π_1. Therefore the equality holds again. By our assumption on σ_1 it remains the case where $x\sigma_1 = x_2 \in X_{5,2}$. But this case cannot occur since σ_1 solves the disequation $x \neq x_2 \in \Gamma_{5,1}$, and since $x_2\sigma_1 = x_2$. Combining what we have found so far we get

$$x\sigma_1 =_{E_1} (x\sigma_x)^{\pi_1}.$$

Now suppose that $x\sigma_x$ is a 2-term or an element of Y_2. Then $(x\sigma_x)^{\pi_1} = y$ would be an element of $Y_2 \cup X_{5,2}^{<x} \cup Z$, by Conditions 1 and 2 on π_x, and we have $x\sigma_1 =_{E_1} y$. But $x\sigma_1$ contains only variables from $Y_1 \cup X_{5,2}^{<x}$. Consistency of E_1 implies that $y \notin Z \cup Y_2$. For $y \in X_{5,2}^{<x}$ we get $x\sigma_1 =_{E_1} y = y\sigma_1$. But this is again impossible since σ_1 solves the system $\Gamma_{5,1}$, which contains the disequation $x \neq y$.

By excluding all other cases we have shown that $x\sigma = x\sigma_x$ is a 1-term or a variable in Y_1. Thus (C2) is verified.

Now let us consider Condition (C1). Let $z < x$ and assume that $z\sigma =_E x\sigma$. Since both terms are R-irreducible we have even $z\sigma = x\sigma$. The induction hypothesis and Condition (C2) for x show that z cannot have index 2 since Y_1 and Y_2 are disjoint. Thus x and z both have index 1, and we get

$$z\sigma_1\sigma =_E z\sigma = x\sigma =_E z\sigma_1\sigma.$$

By definition of π_x we have $(z\sigma_1\sigma)^{\pi_1} = z\sigma_1$ and $(x\sigma_1\sigma)^{\pi_1} = x\sigma_1$, as we have seen earlier for x. With Lemma 4 we obtain $z\sigma_1 =_{E_1} x\sigma_1$, which is a contradiction since $x \neq z \in \Gamma_{5,1}$. This concludes the proof of the two claims.

Since all disequations in Γ_4 are disequations between variables, (C1) implies that σ solves these disequations. □

4 Ground Solvability

The preceding section shows that, analogously to the unification case, solvability of disunification problems in the combined theory can be reduced by decomposition to solvability of disunification problems with linear constant restriction in the single theories. An obvious conjecture could be that the same holds for ground solvability, i.e., that *ground solvability* of a disunification problem Γ_0 may be decided by decomposing Γ_0 into a finite set of pairs $(\Gamma_{5,1}, \Gamma_{5,2})$ of E_i-disunification problems with linear constant restriction as described above, and then asking for *ground solvability* of the subproblems. However, this method is only sound, but not complete.

Proposition 5. *Let Γ_0 be an input problem of the decomposition algorithm. Suppose that there exists an output pair $(\Gamma_{5,1}, \Gamma_{5,2})$ such that each $\Gamma_{5,i}$ $(i = 1, 2)$ has a ground solution. Then Γ_0 has a ground solution.*

In fact, assume that the substitution σ is constructed from ground solutions of $\Gamma_{5,1}$ and $\Gamma_{5,2}$ as described in the proof of Proposition 2. It is easy to see that σ is also a ground solution, provided that the simplification ordering used during unfailing completion satisfies the property that at least one ground term is smaller than all variables in Y. (This property can easily be satisfied.)

Conversely, Γ_0 may be ground solvable, even if the decomposition algorithm does not yield a pair of systems which are ground solvable. An example where this situation occurs can be found in [BS93]. The reason for this behaviour is that a ground solution σ of Γ_0 may substitute a variable of index i by an i-term containing alien subterms. When σ is transformed by projection to solutions σ_1, σ_2 of an output pair $(\Gamma_{5,1}, \Gamma_{5,2})$ (see [BS91a], proof of Proposition 3.2), these alien subterms are replaced by variables. In general, for σ_i not all of these variables are elements of $X_{5,j}$, $j \neq i$, i.e., not all of them are considered as constants in $\Gamma_{5,i}$. For this reason, σ_i is not necessarily a ground solution of $\Gamma_{5,i}$.

The next conjecture could thus be that the systems $\Gamma_{5,1}$, $\Gamma_{5,2}$ have to be tested for solvability rather than ground solvability. But a closer look at the solutions σ_1, σ_2 one gets by projection from a ground solution σ of Γ_0 reveals that these solutions satisfy an additional property: since σ substitutes a variable x of index i by an i-term, $x\sigma_i$ is not a variable. In fact, it can easily be shown that σ_i is a restrictive solution of $\Gamma_{5,i}$ (see Section 2 for the definition).

Lemma 6. *Let Γ_0 be an input problem of the decomposition algorithm. If Γ_0 has a ground solution σ, then there exists an output pair $(\Gamma_{5,1}, \Gamma_{5,2})$ of the decomposition algorithm where each subsystem $\Gamma_{5,i}$ has a restrictive solution σ_i.*

A complete proof of this lemma can be found in [BS93]. To get the opposite direction, we need an additional restriction on the equational theories E_1, E_2: the initial algebras have to be infinite.

Lemma 7. *Let E_1, E_2 be equational theories over disjoint signatures Σ_1 and Σ_2 such that $T(\Sigma_i, \emptyset)/=_{E_i}$ is infinite for $i = 1, 2$. Let Γ_0 be a disunification problem in $E_1 \cup E_2$, and suppose that, via decomposition, an output pair $(\Gamma_{5,1}, \Gamma_{5,2})$ is reached such that each system $\Gamma_{5,i}$ has a restrictive solution. Then Γ_0 has a ground solution.*

Proof. Let σ_i be a restrictive solution of $\Gamma_{5,i}$ for $i = 1, 2$. We may assume that

$$\sigma_1 : X_{5,1} \to T(\Sigma_1 \cup X_{5,2}, Y_1)$$
$$\sigma_2 : X_{5,2} \to T(\Sigma_2 \cup X_{5,1}, Y_2)$$

where the sets Y_1 and Y_2 are finite, disjoint and do not contain an element of $X_4 = X_{5,1} \cup X_{5,2}$. Since σ_i is restrictive we know that $x\sigma_i \neq_{E_i} y$ for all $x \in X_{5,i}$ and $y \in Y_i$ ($i = 1, 2$). Let us now consider the following extensions of the systems $\Gamma_{5,i}$:

$$\widehat{\Gamma}_{5,1} := \Gamma_{5,1} \cup \{x \neq y;\ x \in X_{5,1} \cup Y_2, y \in Y_1 \cup Y_2 \cup X_{5,2}, x \neq y\},$$
$$\widehat{\Gamma}_{5,2} := \Gamma_{5,2} \cup \{x \neq y;\ x \in X_{5,2} \cup Y_1, y \in Y_1 \cup Y_2 \cup X_{5,1}, x \neq y\}.$$

The idea is to treat these systems as if they were a new output pair of the decomposition algorithm. For this purpose we choose a linear ordering which extends the linear ordering on X_4 from system Γ_4 and makes all elements $y \in Y_1 \cup Y_2$ smaller than the elements of X_4. We shall treat the elements $y \in Y_i$ as variables with index $j \neq i$. With this indexing and linear order, $(\widehat{\Gamma}_{5,1}, \widehat{\Gamma}_{5,2})$ is in fact an output pair of the algorithm, corresponding to an input system $\widehat{\Gamma}_0$ which is an appropriate extension of Γ_0 by disequations.

In order to show that $\widehat{\Gamma}_0$ (and thus Γ_0) has a ground solution it suffices to prove (by Proposition 5) that each new subsystem $\widehat{\Gamma}_{5,i}$ has a ground solution $\widehat{\sigma}_i$. Without loss of generality, we shall restrict our attention to $\widehat{\Gamma}_{5,1}$. Note that the elements of Y_1 are treated as constants in $\widehat{\Gamma}_{5,1}$. Let Y_2 be the set $\{y_1, \ldots, y_n\}$, and let $t_1, \ldots, t_n \in T(\Sigma_1, \emptyset)$ be pure 1-terms which are ground, and which are not equivalent modulo E_1 to each other and to any term $x\sigma_1$ for $x \in X_{5,1}$. Since $T(\Sigma_1, \emptyset)/=_{E_1}$ is infinite, we can be sure that such terms exist. We define

$$x\widehat{\sigma}_1 := x\sigma_1 \quad \text{for } x \in X_{5,1},$$
$$y_i\widehat{\sigma}_1 := t_i \quad \text{for } i = 1, \ldots, n.$$

It is easy to see that $\widehat{\sigma}_1$ is a ground solution of $\widehat{\Gamma}_{5,1}$. In fact, since the elements of Y_1 are now treated as constants, it is a ground substitution, and it obviously solves the equations and disequations of $\Gamma_{5,1}$. Restrictiveness of σ_1 guarantees that $\widehat{\sigma}_1$ solves the disequations $x \neq y$ for $x \in X_{5,1}$ and $y \in Y_1$; the choice of the t_j guarantees that $\widehat{\sigma}_1$ solves the disequations $x \neq y$ for $x \in X_{5,1} \cup Y_2$ and $x \neq y \in Y_2$. Finally, for $x \in X_{5,2} \cup Y_1$ and $y \in Y_2$, we have $x\widehat{\sigma}_1 = x \neq_{E_1} y\widehat{\sigma}_1 \in T(\Sigma_1, \emptyset)$ since E_1 is consistent. □

Analogously, the two lemmas can be shown for the combination of more than two theories. Thus we obtain the main theorem of this section.

Theorem 8. *Let $E_i, i = 1, \ldots, n$, be equational theories over disjoint signatures Σ_i, and suppose that the initial algebras $T(\Sigma_i, \emptyset)/=_{E_i}$ are infinite. If restrictive solvability of E_i-disunification problems with linear constant restriction is decidable for $i = 1, \ldots, n$, then ground solvability of disunification problems is decidable for $E_1 \cup \ldots \cup E_n$.*

If one of the theories, say E_n, satisfies a stronger restriction, the condition that the initial algebras are infinite can be dropped for the other theories.

Corollary 9. *Let $E_1, \ldots, E_n$ be equational theories over disjoint signatures $\Sigma_1, \ldots, \Sigma_n$. Assume that $T(\Sigma_n, \emptyset)/=_{E_n}$ is infinite, and that every solvable E_n-disunification problem with linear constant restriction has a ground solution. Then ground solvability of disunification problems in $E_1 \cup \ldots \cup E_n$ is decidable if restrictive solvability of E_i-disunification problems with linear constant restriction is decidable for $i = 1, \ldots, n$.*

This can be seen by an inspection of the proof of Lemma 7. Since we know that a solvable system $\Gamma_{5,n}$ has a ground solution, we do not need any alien i-terms ($i \neq n$) to get rid of variables in solutions of $\Gamma_{5,n}$. For the other theories, the assumption that $T(\Sigma_n, \emptyset)/=_{E_n}$ is infinite provides for the required alien terms.

An important case to which this corollary evidently applies is the combination with a free theory. We call an equational theory F the free theory with signature Σ iff $sig(F) = \Sigma$ and $=_F$ is just the syntactic equality of terms. Obviously, considering elementary disunification in the combination of a theory E with free theories corresponds to considering general disunification for E.

Corollary 10. *Let $\Sigma_1, \ldots, \Sigma_n$ be disjoint signatures, $E_1, \ldots, E_{n-1}$ be equational theories over $\Sigma_1, \ldots, \Sigma_{n-1}$, and let E_n be the free theory with signature Σ_n. Assume that Σ_n contains at least one function symbol of arity greater zero and one constant. Then ground solvability of disunification problems in $E_1 \cup \ldots \cup E_n$ is decidable if restrictive solvability of E_i-disunification problems with linear constant restriction is decidable for $i = 1, \ldots, n-1$.*

5 Applications of the Method

The methods developed in the preceding two sections will now be applied to the combination of A, AC, ACI, and free theories. An equational theory is called an A-theory iff its signature consists of a binary function symbol h, and it contains the single axiom $h(h(x,y),z) = h(x,h(y,z))$ (associativity). For AC-theories, one has an additional axiom $h(x,y) = h(y,x)$ (commutativity), and for ACI-theories there is a third axiom $h(x,x) = x$ (idempotence).

Theorem 11. *Solvability of disunification problems is decidable for every theory which is a disjoint combination of finitely many A-, AC-, and ACI-theories and a free theory. To get decidability of ground solvability by our method we have to assume that the free theory contains at least one constant symbol and one function symbol of arity greater than 0.*

By Theorem 1 and Corollary 10 it remains to be shown that solvability and restrictive solvability of disunification problems with linear constant restriction is decidable for these theories. Since this is trivial for the free theory we shall only consider A, AC and ACI in the following.

Firstly, it turns out that restrictiveness of a solution is not a real constraint for these theories. In fact, we can show (see [BS93]) for these theories that a disunification problem with linear constant restriction has a solution iff it has a restrictive solution. In particular, this means that solvability and ground solvability for the combined theory are equivalent.

Secondly, we can show (see [BS93]) that solvability of disunification problems with linear constant restriction for A, AC and ACI can be reduced to ground solvability over a signature which contains $2m+1$ new constants (where m is the number of disequations in the problem).

Thus, eventually one has to consider ground solvability of disunification problems with linear constant restriction for A, AC and ACI. For A one can use a method described by Büchi and Senger [BüS86] to reduce ground solvability of disunification problems with linear constant restriction to ground solvability of unification problems with linear constant restriction. Solvability[5] of unification problems with linear constant restriction for A is treated in [BS91b]. For AC, ground solvability of a disunification problem with linear constant restriction can be reduced to an integer programming problem (which is decidable by an NP-algorithm). Finally, for ACI, ground solvability of a disunification problem with linear constant restriction can be reduced to satisfiability of Boolean formulae (which is NP-decidable as well) (see [BS93] for details).

Since existential equational formulae can be seen as disjunction of disunification problems we have the following immediate consequence of Theorem 11.

Corollary 12. *Let Σ be a signature consisting of $n \geq 1$ binary function symbols $h_1,\ldots,h_n$, and at least one constant and one additional non-constant function*

---

[5] Recall that, for unification, solvability and ground solvability are equivalent.

symbol. Let A_n, AC_n, and ACI_n respectively stand for associativity, associativity and commutativity, and associativity, commutativity and idempotence of the function symbols h_i.

1. The existential theories of the free algebra $T(\Sigma, Y)/=_{A_n}$ and the initial algebra $T(\Sigma, \emptyset)/=_{A_n}$ are decidable.
2. The existential theories of the free algebra $T(\Sigma, Y)/=_{AC_n}$ ($T(\Sigma, Y)/=_{ACI_n}$), and the initial algebra $T(\Sigma, \emptyset)/=_{AC_n}$ ($T(\Sigma, \emptyset)/=_{ACI_n}$) are NP-decidable.

For AC, decidability has already been shown by Comon [Com88]. The result for A seems to be new. There is no real hope to extend these decidability results to equational formulae with more complex quantifier prefix. A recent result by Treinen [Tr92] shows that already the Σ_2 fragment[6] of the theory of the ground term algebra modulo A is undecidable. For AC, Treinen shows that the Σ_3-fragment is undecidable, both for the free algebra and the initial algebra.

6 Conclusion

Since constraint approaches to theorem proving, term rewriting, and logic programming are gaining in importance, constraint solving has become a major research issue in these areas. An important subproblem is the question of how to combine different constraint solving techniques. The present paper can be seen as a contribution to this field, where the constraints are existentially quantified equational formulae that have to be solved in the initial or the free algebra modulo an equational theory. We have seen that the methods developed for the combination of unification algorithms can be applied for disunification as well. For solvability of disunification problems, this was relatively straightforward, even though the proofs become more involved. For ground solvability we surprisingly have to consider a restricted type of solvability (instead of ground solvability) in the single theories.

For the theories A, AC an ACI, solvability and restrictive solvability coincide, which implies that solvability and ground solvability in their combination with a non-trivial free theory are equivalent. However, we have an example of a theory where solvability does not imply restrictive solvability (see [BS93]). An interesting open problem is under what conditions solvability and restrictive solvability coincide, and when solvability and ground solvability refer to the same problem.

Acknowledgement. This work has been supported under Esprit reference EP 6028 and by the German Ministry for Research and Technology (BMFT) under research contract ITW 8903 0.

---

[6] consisting of the closed formulae with quantifier prefix of the form $\exists x \forall y$

References

[BS91a] F. Baader, K.U. Schulz, "Unification in the Union of Disjoint Equational Theories: Combining Decision Procedures," DFKI-Research Report RR-91-33; also in *Proceedings of the 11th International Conference on Automated Deduction, LNCS* 607, 1992.

[BS91b] F. Baader, K.U. Schulz, "General A- and AX-Unification via Optimized Combination Procedures," CIS-Report 92-58, CIS, University Munich; also to appear in the *Proceedings of the Second Workshop on Word Equations and Related Topics IWWERT '91*, Rouen 1991, *LNCS*.

[BS93] F. Baader, K.U. Schulz, "Combination Techniques and Decision Problems for Disunification," DFKI-Research Report RR-93-05, 1993.

[BüS86] J.R. Büchi, S. Senger, "Coding in the Existential Theory of Concatenation," Arch. math. Logik **26**, 1986.

[Bür88] H.J. Bürckert, "Solving Disequations in Equational Theories," *Proceedings of the 9th International Conference on Automated Deduction*, Argonne, LNCS 310, 1988.

[BB89] R. Buntine, H.-J. Bürckert, "On Solving Equations and Disequations," SEKI-Report SR-89-03, University Kaiserslautern, 1989.

[Bür90] H.-J. Bürckert, "A Resolution Principle for Clauses with Constraints," *Proceedings of the 10th International Conference on Automated Deduction*, LNCS 449, 1990.

[Col84] A. Colmerauer, "Equations and Inequations on Finite and Infinite Trees," *Proceedings of the FGCS'84*, pp.85-99.

[Col90] A. Colmerauer, "An Introduction to PROLOG III," *C. ACM* **33**, 1990.

[Com88] H. Comon, "Unification et Disunification. Théorie et Applications," PhD Thesis, Institut National Polytechnique de Grenoble, Grenoble, France, 1988.

[Com91] H. Comon, "Disunification: a Survey," in J.-L. Lassez, G. Plotkin (editors), *Computational Logic*, MIT Press, 1991.

[DJ90] N. Dershowitz, J.P. Jouannaud, "Rewrite Systems," in Volume B of "Handbook of Theoretical Computer Science," North-Holland 1990.

[JK91] J.P. Jouannaud, C. Kirchner, "Solving Equations in Abstract Algebras: A Rule-Based Survey of Unification," in J.-L. Lassez, G. Plotkin (editors), *Computational Logic*, MIT Press, 1991.

[JL87] J. Jaffar, J.L. Lassez, "Constraint Logic Programming," *Proceedings of 14th POPL Conference*, Munich, 1987.

[KN92a] D. Kapur, P. Narendran, "Complexity of Unification Problems with Associative-Commutative Operators," *J. Automated Reasoning* **9**, 1992.

[KN92b] D. Kapur, P. Narendran, "Double Exponential Complexity of Computing Complete Sets of AC-unifiers," *Proceedings of the 7th Annual IEEE Symposium on Logic in Computer Science*, Santa Cruz, California, 1992.

[KK89] C. Kirchner, H. Kirchner, "Constrained Equational Reasoning," *Proceedings of SIGSAM 1989 International Symposium on Symbolic and Algebraic Computation*, ACM Press, 1989.

[SS89] M. Schmidt-Schauß, "Combination of Unification Algorithms," *J. Symbolic Computation* **8**, 1989.

[Tr92] R. Treinen, "A New Method for Undecidability Proofs of First Order Theories," *J. Symbolic Computation* **14**, 1992.

The Negation Elimination from Syntactic Equational Formula is Decidable

Mohamed TAJINE
Université Louis Pasteur
Centre de Recherche en Informatique
7, rue René Descartes
67084 Strasbourg Cedex, France
E-Mail : tajine@dpt-info.u-strasbg.fr

Abstract. In this paper we introduce the property of finitely generated sets in an algebra. This property generalizes several notions of rewriting and logical programming; for example unification and disunification are specific cases of this notion. We use this property to formalize the problem of negation elimination in a syntactic equational formulae (i.e first order formulae whose only predicate is syntactic equality) and we prove that this problem is decidable.

1. Introduction

Solving equations and disequations (i.e particular first-order formulas) in the Herbrand Universe is a fundamental operation in symbolic computation. There are several results for deciding whether an equation or disequation has solutions in Herbrand Universe and if the set of these solutions has an explicit representation : [Rob 65], [PW 78], [MM 82], [LM 87], [Com 88], [Mah 88], [CL 89], [RP 89] and [Taj 92].

In this paper, we generalize these results to any first order formula whose only predicate is syntactic equality (i.e syntactic equational formula). Actually we prove that negation elimination from syntactic equational formula is decidable. This result then allows us to obtain an explicit representation for the set of solutions in Herbrand Universe of syntactic equational formula. We introduce the property of finitely generated sets in an algebra for formalizing this problem. We show the decidability of finitely generated in the Herbrand Universe for a boolean combination[1] of finitely generated sets in Herbrand Universe.

A subset B of an $\mathbb{F}$-algebra $\mathcal{A}$[2] is *finitely generated in* $\mathcal{A}$ if there exist a finite subset G of $T(\mathbb{F}, X)$[3] such that $B = \{ t^{\mathcal{A}} \mid \exists s \in G$ and t is a ground instance of $s \} $ $t^{\mathcal{A}}$ where is the interpretation of the ground term t in the $\mathbb{F}$-algebra $\mathcal{A}$. We note $B = <G>_{\mathcal{A}}$ for short and we say G be a generator of B in $\mathcal{A}$. For example the set $\mathcal{NP} = \{0, 1, 4, 6, 9, ...\}$ of not prime numbers is finitely generated in the $\mathbb{F}$-algebra $\mathcal{A} = T(\mathbb{F})/=_{Ax}$ where $\mathbb{F} = \{+, *, 0, 1\}$, Ax is the set of axioms :

$\{ x+0 = x, x+(y+1) = (x+y)+1, x*0 = 0, x*(y+1) = (x*y)+x \}$ and $=_{Ax}$ is the congruence associate to the set Ax of axioms, actually $\mathcal{NP} = < \{ 0, 1, (x+2)*(y+2) \} >_{\mathcal{A}}$; but the set $\mathcal{P} = \{2, 3, 5, 7, ... \}$ of prime numbers is not finitely generated in $\mathcal{A}$ see [HW 78] for more details. Several notions can be expressed with this notion :

Unification which is the basis of functional programming, consists in solving systems of equations (i.e certain first-order formulas). It can be expressed with the notion of finitely generated :

---

[1] a set of the form $(...(E_1 \cup E_2) \cap (E_3 - E_4) ... \cap E_k)...)$ where $E_1, ..., E_k$ are a finitely generated sets in $T(\mathbb{F})$
[2] the same notion is defined for $\mathcal{A}^k$
[3] $T(\mathbb{F}, X)$ is the set of terms overs the singature $\mathbb{F}$ and a set X of variables; $T(\mathbb{F}) = T(\mathbb{F}, \emptyset)$

$\mathcal{A} \models \exists x_1...x_k \ t = t'$ iff $< \{ (t, t') \}>_{\mathcal{A}} \cap <\{ (y, y) \}>_{\mathcal{A}} \neq \emptyset$ where $\{x_1, ..., x_k\}$ is the set of variables occurring in t, t'. Generally, the set $U_{\mathcal{A}}(t, t') = <\{(t, t')\}>_{\mathcal{A}} \cap <\{(x, x)\}>_{\mathcal{A}}$ is not finitely generated in $\mathcal{A}$ for every $\mathbb{F}$-algebra $\mathcal{A}$, however, it is finitely generated in the Herbrand Universe.
For example, $U_{\mathcal{A}}(x*a, a*x)$ is not finitely generated in the free monoid $\{a, b\}^*$: $\mathcal{A} = \{a, b\}^*$ and $U_{\mathcal{A}}(x*a, a*x) = \{ (a^n, a^n) \mid n \geq 1 \}$; but it is finitely generated in the Herbrand Universe : $\mathcal{A} = T(\{*, a, b\})$ and $U_{\mathcal{A}}(x*a, a*x) = \{ (a*a, a*a) \}$.

There are several algorithms for constructing a finite generator for $U_{\mathcal{A}}(t, t')$ where $\mathcal{A}$ is the Herbrand Universe : Robinson [Rob 65], Martelli and Montanari [MM 76] and [MM 82], Paterson and Wegman [PW 78] and Rezika and Privara [RP 89].

Disunification is solving disequations. Colmerauer [Col 84] did extend equation systems by adding disequations (i.e constrains of the form $t \neq t'$) in the semantics of Prolog II.
$\mathcal{A} \models \exists x_1...x_k \ t \neq t'$ iff $<\{(t, t')\}>_{\mathcal{A}} - <\{(y, y)\}>_{\mathcal{A}} \neq \emptyset$ where $\{x_1, ..., x_k\}$ is the set of variables occurring in t, t' and $t \neq t'$ is not($t = t'$).
Generally, the set $CU_{\mathcal{A}}(t, t') = <\{ (t, t') \}>_{\mathcal{A}} - <\{ (x, x) \}>_{\mathcal{A}}$ is not finitely generated in $\mathcal{A}$ for every $\mathbb{F}$-algebra $\mathcal{A}$ neither in the Herbrand Universe. For example, $CU_{\mathcal{A}}(x*a, a*x)$ is finitely generating in the free monoid $\{a, b\}^*$: $\mathcal{A} = \{a, b\}^*$ and $CU_{\mathcal{A}}(x*a, a*x) = <\{ (xbya, axby) \}>$; and it's also finitely generated in Herbrand Universe : $\mathcal{A}=T(\{*, a, b\})$ and $CU_{\mathcal{A}}(x*a,x*a) =<\{((x*y)*a, a*((x*y), (b*a, a*b)\}>$. But $CU_{\mathcal{A}}(x*y, y*x)$ is not finitely generated neither in $\mathcal{A} = \{a, b\}^*$ nor in $\mathcal{A} = T(\{*, a, b\})$.

There are several algorithms to find a finite generator (if it exists) of $CU_{\mathcal{A}}(t, t')$ in the Herbrand Universe : Lassez and Marriott [LM 87], Comon [Com 88], Comon and Lescanne [CL 89], Tajine [Taj 92].

Several other notions concerning term rewriting systems, namely sufficient completeness : Guttag and Horning [GH 78] or, more generally, inductive reducibility property : Jouannaud and Kounalis [JK 86], can be expressed in terms of this formalism by an appropriate choice of algebra.

In all these cases, the problem is to verify if a specific boolean combination of finitely generated sets in a certain algebra is also finitely generated in the same algebra.

In this paper, we prove the decidability of finitely generated in Herbrand Universe $T(\mathbb{F})$[1] for any boolean combination of any finitely generated sets in $T(\mathbb{F})$.
As a consequence, we prove the decidability of finitely generated for a set of solutions in Herbrand Universe of syntactic equational formulae, and therefore the decidability of negation elimination for syntactic equational formulae.
For example, in the syntactic equational formula A(y) :
$$\forall x \ ((x = y) \vee \exists x_1 (x = s(x_1)) \vee \exists x_2, x_3 \ (x = x_2+x_3))$$
y is the only free variable (i.e $\mathbb{F} = \{ 0, s, + \}$).
0 is a solution of A(y) in $T(\mathbb{F})$ (i.e $T(\mathbb{F}) \models A(0)$), but 0+0 is not a solution of A(y) in $T(\mathbb{F})$ (i.e A(0+0) is false in $T(\mathbb{F})$). Actually 0 is the only solution of A(y) in $T(\mathbb{F})$, thus the set of solutions of A(y) in $T(\mathbb{F})$ is finitely generated in $T(\mathbb{F})$ by $\{ 0 \}$.

Decidability of finitely generated set of solutions of syntactic equational formulae is a generalization of the results of unification and disunification.

In this paper, we consider a finite signature $\mathbb{F}$ with non empty set of constants; the other cases are trivial.

2. Preliminaries

2.1. Some generalities.
We use in this paper the same basic notions of algebraic specifications and term rewriting system theory as N.Dershowitz, J.P.Jouannaud [DJ 90]. $T(\mathbb{F}, X)$ stands for the set of terms (finite labeled trees) over the finite signature $\mathbb{F}$ and an enumerable set of variables X (i.e $T(\mathbb{F}, X)$ is the free $\mathbb{F}$-algebra generated

---

[1] We have the same result in $T(\mathbb{F})^k$ for every k

by X : see Huet and Oppen [H&O 80] for more details). Var(t)⊂X is the set of variables in t∈T($\mathbb{F}$, X). The arity of t, denoted by ar(t) is the cardinal of Var(t). T($\mathbb{F}$) is the set T($\mathbb{F}$, ∅) of ground terms (ground finite labeled trees) over $\mathbb{F}$. For t∈T($\mathbb{F}$, X), Pos(t) denotes the set of positions in t defined as usual as sequences of natural numbers, $t|_u$ is a subterm of t at u, t[u ← s] is the result of replacing $t|_u$ by s in t and VPos(t) = {u∈Pos(t) | $t|_u$∈X}. For two positions u and v, u.v denotes the concatenation of u and v. A term t is said to be linear if there is no x∈Var(t) and u, v∈Pos(t) such that u ≠ v and $t|_u$ = x = $t|_v$. Given u∈Pos(t), len(u) stands for the length of u. For t∈T($\mathbb{F}$, X), maximum-depth, minimum-depth and size of t, denoted by dep(t), mdep(t) and | t |, are respectively defined by dep(t) = 1+ $\max_{u \in Pos(t)}${len(u)}, mdep(t) = 1+min $_{u \in Pos(t)}$ and $t|_u \in X \cup \mathbb{F}_0${len(u)} and | t | = Card(Pos(t)) - Card(VPos(t)) (i.e $\mathbb{F}_0$ is set of constants of $\mathbb{F}$: the elements of arity 0).

A substitution σ on T($\mathbb{F}$, X)[1] is an $\mathbb{F}$-morphism of T($\mathbb{F}$, X) on itself such that the so-called domain of σ, Dom(σ) = {x | x∈X and σ(x) ≠ x } is finite. A substitution σ is said to be linear if ∀x∈Dom(σ) σ(x) is a linear term and ∀x, y∈Dom(σ) x ≠ y ⇒ Var(σ(x))∩Var(σ(y)) = ∅. A substitution σ is said to be ground if ∀x∈Dom(σ) σ(x) is a ground term. Two terms t, t'∈T($\mathbb{F}$, X) are said to be unifiable when there exist a substitution σ such that σ(t) = σ(t'), which is called an unifier of t and t'. It is well know that when t, t'∈T($\mathbb{F}$, X) are unifiable then there exist a unifier σ of t and t' such that for any other unifier τ of t and t', there exists a substitution μ such that ∀x∈X τ(x) = μ(σ(x)); σ is called a most general unifier[2] (i.e σ is unique w.r.t. renaming the variables).

2.2. Others definitions and notations.

The notions and the results proved in this paper are valid for families in T($\mathbb{F}$, X)^k for any k. In order to simplify notations we only study the case k = 1. In the general case *(i.e k ≥ 1)* , the proofs are strictly similar.

Let E⊂T($\mathbb{F}$, X), < E >$_{T(\mathbb{F})}$[3] = { t∈T($\mathbb{F}$) | t = σ(s), for some ground substitution σ and some s in E }.
< E > is called the set generated by E in T($\mathbb{F}$). E⊂T($\mathbb{F}$, X), is said to be a global family when < E > = T($\mathbb{F}$), a free family when ∀t, t'∈E t ≠ t' ⇒ < {t} > ∩ < {t'} > = ∅ and a base when it is both a free and a global family (i.e See Tajine [Taj 92] for more details). For example, for $\mathbb{F}$ = {a, f, g } { f(x), g(x, y) } is a free family of T($\mathbb{F}$), { a, f(x), g(x, x), g(x, f(y), g(x, g(y, z)) } is a global family of T($\mathbb{F}$) and {a, f(x), g(x,y) } is a base of T($\mathbb{F}$).

$\mathcal{P}_f$(T($\mathbb{F}$, X)) be the set of finite subsets of T($\mathbb{F}$, X). Bool(V) is the free Boolean algebra over V where V is an enumerable set .

If H∈Bool($\mathcal{P}_f$(T($\mathbb{F}$, X))), then H∈$\mathcal{P}_f$(T($\mathbb{F}$, X)), H=¬(H'), H=H'∨H" or H = H'∧H" where H', H"∈Bool($\mathcal{P}_f$(T($\mathbb{F}$,X))) and ¬, ∨ and ∧ are the boolean operations.

Definition : (Interpretation)
Let H∈Bool($\mathcal{P}_f$(T($\mathbb{F}$, X)))

- < H > = < G >   if H = G   and   G∈$\mathcal{P}_f$(T($\mathbb{F}$, X))
- < H > = T($\mathbb{F}$) - < H' >   if H = ¬(H')   and   H'∈Bool($\mathcal{P}_f$(T($\mathbb{F}$,X)))
- < H > = < H' > ∪ < H" > if H = H'∨H" and   H', H"∈Bool($\mathcal{P}_f$(T($\mathbb{F}$,X)))
- < H > = < H' > ∩ < H" > if H = H'∧H" and   H', H"∈Bool($\mathcal{P}_f$(T($\mathbb{F}$,X))).

Notation :
Let t∈T($\mathbb{F}$, X), E∈$\mathcal{P}_f$(T($\mathbb{F}$, X)) such that ∀t'∈E var(t)∩var(t') = ∅   then
In(t, E) = { σ(t') | t'∈E, t and t' are unifiables, σ is their mgu. }

Definition :
Let t∈T($\mathbb{F}$, X), B∈$\mathcal{P}_f$(T($\mathbb{F}$, X)) and <B> ⊂ <{t}> then the boolean formula {t}∧¬(B) is called a *elementary clause*.

---

[1] σ can be extend in natural way to k-uples of terms
[2] mgu. for short
[3] in the following, we omet the index T($\mathbb{F}$), because we consider only the Herbrand Universe

Definition :
Let H, H'∈ $Bool(\mathcal{P}_f(T(\mathbb{F},X)))$; H and H' are *equivalent* if $<H> = <H'>$.

In this paper, we will study the following problem :
Let H∈ $Bool(\mathcal{P}_f(T(\mathbb{F},X)))$, is there some G∈ $\mathcal{P}_f(T(\mathbb{F}, X))$ such that H and G are equivalent ?

We will refer to this problem as the finite generator problem.

The finite generator problem generalizes the results of Lassez and Marriott [LM 87] and of syntactic completion [Taj 92].

3. Simplification of Boolean formulas.

From H∈ $Bool(\mathcal{P}_f(T(\mathbb{F},X)))$, we will constructs H'∈ $Bool(\mathcal{P}_f(T(\mathbb{F},X)))$ such that, H and H' are equivalent and H' is a disjunction of elementary clauses.

Property 3.1.1 :
Let $G_1, G_2 \in \mathcal{P}_f(T(\mathbb{F}, X))$ then there is G, G'∈ $\mathcal{P}_f(T(\mathbb{F}, X))$ such that
$<G_1> \cup <G_2> = <G>$ and
$<G_1> \cap <G_2> = <G'>$.

Proof :
We can assume that $\forall t \in G_1, \forall t' \in G_2$ var(t)∩var(t') = ∅, then
$G = G_1 \cup G_2$.
G' = { σ(t) | t∈G_1, ∃t'∈G_2, t and t' are unifiables and σ is their mgu. }.   ■

Corollary 3.1.2 :
Let H∈ $Bool(\mathcal{P}_f(T(\mathbb{F},X)))$ then there is a finite set I such that
$<H> = \bigcup_{i\in I}(<\{t_i\}> - <B_i>)$ where ∀i∈ I, $t_i \in T(\mathbb{F}, X)$, $B_i \in \mathcal{P}_f(T(\mathbb{F}, X))$ and ∀t∈ B_i, ∃σ such that $t = \sigma(t_i)$ (i.e H is equivalent to disjunction of elementary clauses).

Proof :
Let H∈ $Bool(\mathcal{P}_f(T(\mathbb{F},X)))$, then there exists a finite set I such that
$<H> = \bigcup_{i\in I}(<A_i^1> \cap ... \cap <A_i^{ai}> \cap <\neg B_i^1> \cap ... \cap <\neg B_i^{bi}>)$ where
∀i∈ I, ∀j≤ a_i, ∀k ≤ b_i, A_i^j, $B_i^k \in \mathcal{P}_f(T(\mathbb{F}, X))$ because every boolean formula is a disjunction of conjunctions of literals.
Then $<H> = \bigcup_{i\in I}(<A_i> - <B'_i>)$ where
∀i∈ I $<A_i> = <A_i^1> \cap ... \cap <A_i^{ai}>$ and $B'_i = (<B_i^1> \cup ... \cup <B_i^{bi}>) \cap <A_i>$.
(i.e ∀i∈ I, A_i, $B'_i \in \mathcal{P}_f(T(\mathbb{F}, X))$ and $<B'_i> \subset <A_i>$ " property 3.1.1").
∀i∈ I, $<A_i> - <B'_i> = \bigcup_{t\in A_i}(<\{t\}> - <B_{i,t}>)$ where $B_{i,t}$ = In(t, B'_i)
(i.e ∀i∈ I,∀t∈ A_i, $B_{i,t} \in \mathcal{P}_f(T(\mathbb{F}, X))$ and $<B_{i,t}> \subset <\{t\}>$ " property 3.1.1").   ■

In what follows, we set H = { (t_i, B_i) | i∈ I } instead of $\bigvee_{i\in I}(\{t_i\} \wedge \neg(B_i))$.

Remark :
Let t∈ T($\mathbb{F}$, X) with var(t) = {$x_1, ..., x_{ar(t)}$} and let B∈ $\mathcal{P}_f(T(\mathbb{F}, X))$.
We can assume that ∀t'∈ B var(t)∩var(t') = ∅, then
$<\{t\}> \subset <B>$ iff
G = { ($\sigma_{t'}(x_1), ..., \sigma_{t'}(x_{ar(t)})$) | t'∈B, t and t' are unifiable and $\sigma_{t'}$ is their mgu. } is a global family for $T(\mathbb{F})^k$ (i.e $<G> = T(\mathbb{F})^k$). The results of Lassez and Marriott [LM 87] and of syntactic independence [Taj 92] allow whether a finite family is a global family. Actually it is a special case of the finite generator problem.

Corollary 3.1.3 :
Let $H \in \text{Bool}(\mathcal{P}_f(T(\mathbb{F},X)))$ then the problem to know whether $<H>$ is empty is decidable.

Proof :
$H = \{ (t_i, B_i) \mid i \in I \}$, $<H>$ is empty iff $\forall i \in I$, $<\{t_i\}> \subset <B_i>$. ∎

We will now study the process of simplifying an elementary clause c by a set of elementary clauses (*i.e the elimination of the negative part of c wherever it is possible*).

4. Relative simplification.

Before solving the finite generator problem, we propose a procedure allowing to simplify an elementary clause relatively to a set of elementary clauses (*i.e eliminate, if it's possible the negative part of the clause*). In this procedure, we will use the procedure **Compl** which completes for some linear term t^1 the family $\{t\}$ in a base (i.e **Compl** constructs a family C such that $\{t\} \cup C$ is a base).

Procedure Compl (t)

Begin

If $|t| = 0$ then $B^t = \{t\}$ ($t \in X$)
Else (i.e. $t = f(t_1, ..., t_m)$, $ar(f) = m$)

  Begin

  For $i = 1$ to m do
    $B_i = \text{Compl}(t_i) \cup \{t_i\}$;

  $B = \{ f(s_1, ..., s_{j-1}, t_j, s_{j+1}, ..., s_m) \mid (s_1, ..., s_{j-1}, s_{j+1}, ..., s_m) \in B_1 \times ... \times B_{j-1} \times B_{j+1} \times ... \times B_m \}$

  $\cup \{ f(x_1, ..., x_{j-1}, s, x_{j+1}, ..., x_m) \mid s \in (B_j - \{t_j\}) \} \cup \{ g(x_1, ..., x_{ar(g)}) \mid g \in (\mathbb{F} - \{f\}) \}$;

  (where j is such that $\text{Card}(B_j) = \max_{1 \leq k \leq m} \text{Card}(B_k)$ and $\forall k, x_k \in X$)

  End ;

Return$(B - \{t\})$;
End.

There is several other algorithms to find a syntactic complement C : Lassez and Marriott [LM 87], Comon [Com 88], Comon and Lescanne [CL 89], Tajine [Taj 92].

Notation :
Let H, H' be two sets of elementary clauses. Uf(H', H) is the predicate :
$\exists (t', B') \in H'$ s.t. (($\exists t'' \in B'$ s. t. $t'' = \sigma(t')$ and σ is linear over t') or
($\exists (t'', B'') \in H$, $\exists s \in B'$ s. t. t'' and s are unifiables, their mgu. σ is linear over s and $<\{\sigma(s)\}> \not\subset <B''>$)).

Let (t, B) be an elementary clause and H a set of elementary clauses. We give the procedure **Simpl** which simplifies the elementary clause (t, B) relatively to the set of elementary clauses H (*i.e eliminate, if it's possible the negative part B of clause (t, B)*). The procedure is a step when solving the finite generator problem.

---

[1] **Compl** can be extend in natural way to k-uples of terms

Procedure Simpl $((t, B), H)$

Begin

$H' := \{(t, B)\}$;

while $Uf(H', H)$ **do**

Begin

1- **If** ($\exists (t', B') \in H'$, $\exists t'' \in B'$ s. t. $t'' = \sigma(t')$ and σ is linear over t') **then**

    Begin

    $C := Compl(\{(t_1, ..., t_{ar(t')})\})$; (i.e. $t' = K_{t'}(x_1, ..., x_{ar(t')})$ and $\sigma(t') = K_{t'}(t_1, ..., t_{ar(t')})$)

    $H' := (H' - (t', B')) \cup \{(\sigma_c(t'), In(\sigma_c(t'), B')) \mid c \in C$ and $\sigma_c(x_i) = c_i$ where $(c_1, ..., c_{ar(t')}) = c \}$;

    End

2- **Else**

    If

    ($\exists (t', B') \in H'$, $\exists (t'', B'') \in H$, $\exists s \in B'$, t'' and s are unifiables, with mgu. σ linear over s

    and $< \{\sigma(s)\} > \nsubseteq < B'' >$)

    then

      Begin

      $C = Compl(\{(s_1, ..., s_{ar(s)})\})$; ($s = K_s(x_1, ..., x_{ar(s)})$ and $\sigma(s) = K_s(s_1, ..., s_{ar(s)})$)

      $B' := (B' - \{s\}) \cup \{ \sigma_c(s) \mid c \in C$ and $\sigma_c(x_i) = c_i$ where $(c_1, ..., c_{ar(s)}) = c \} \cup$

          $\{ \mu(s') \mid s' \in B''$, s' and $\sigma(s)$ are unifiables with mgu. $\mu \}$;

      End ;

End ;

$Return(H')$;

End.

Remarks :

• The cases 1-, 2- in procedure **Simpl** are the translation of the inclusion exclusion identities :

  - The case 1- is the translation of the law : if $E = A - (A_0 \cup B)$ and $A = A_0 \uplus A_1 \uplus ... \uplus A_k$
then $E = (A_1 - B) \uplus ... \uplus (A_k - B)$ *(i.e $\uplus$ is the disjoint union)*.

  - The case 2- is the translation of the law : if $E = (A - (B \cup C)) \cup ((A' \cup B_0) - C')$ and
$B = B_0 \uplus B_1 \uplus ... \uplus B_k$ then $E = (A - ((B_1 \uplus ... \uplus B_k) \cup C \cup (B_0 \cap C') \cup ((A' \cup B_0) - C')$

• The procedure **Simpl** can be used to verify if $<\{\sigma(s)\}> \nsubseteq <B''>$ (*i.e* $<\{(\sigma(s), In(\sigma(s), B'')\}> \neq \emptyset$).
Actually in this case, the test $Uf(H', H)$ boils down to :

    $\exists (t', B') \in H'$, $\exists t'' \in B'$ such that $t'' = \sigma(t')$ and σ is linear over t' (i.e $H = \emptyset$).

Before proving some properties concerning procedure **Simpl**, we give one example.

Example :

Let $\mathbb{F} = \mathbb{F}_0 \cup \mathbb{F}_2$ with $\mathbb{F}_0 = \{ a \}$ et $\mathbb{F}_2 = \{ f \}$
$t = f(x, f(x, y))$; $B = \{ f(z, f(z, z)) \}$ and $H = \{ (f(f(x_1, y_1), f(f(x_2, y_2), f(x_3, y_3))), \emptyset) \}$

We will simplify (t, B) relatively to H.
Let H'_i be the value of H' at step i
$H'_0 = \{ (f(x, f(x, y)), \{ f(z, f(z, z)) \}) \}$.

Let σ be the substitution defined by $\sigma(x) = z$, $\sigma(y) = z$, then
$f(z, f(z, z)) = \sigma(f(x, f(x, y)))$ and σ is not linear over t.
Let τ be the substitution defined by :
$\tau(z) = f(z_1, z_2)$, $\tau(x_1) = \tau(x_2) = \tau(x_3) = z_1$ and $\tau(y_1) = \tau(y_2) = \tau(y_3) = z_2$
$\tau(f(z, f(z, z))) = \tau(f(f(x_1, y_1), f(f(x_2, y_2), f(x_3, y_3))))$, τ is linear over $f(z, f(z, z))$.
$\{ a \} \cup \{ f(z_1, z_2) \}$ is a base of $T(\mathbb{F})$. $(C = \{ a \})$
$B' = \{ f(a, f(a, a)) \}$
$H'_1 = \{ (f(x, f(x, y)), \{ f(a, f(a, a)) \}) \}$.
Let σ be the substitution defined by $\sigma(x) = \sigma(y) = a$, then $\sigma(f(x, f(x, y))) = f(a, f(a, a))$
$\sigma(f(x, f(x, y))) = K_{f(x, f(x, y))}(a, a)$
$\{ (a, a) \} \cup \{ (a, f(x_1,y_1)), (f(x_2, y_2), a), (f(x_3, y_3), f(x_4, y_4)) \}$ is a base of $T(\mathbb{F})^2$.
$H'_2 = \{ (f(a, f(a, f(x_1,y_1)), \emptyset) ; (f(f(x_2, y_2), f(f(x_2, y_2), a), \emptyset) ;$
$\qquad\qquad\qquad\qquad\qquad (f(f(x_3, y_3), f(f(x_3, y_3), f(x_4, y_4)), \emptyset) \}$

Property 4.1.1 :
the procedure Simpl terminates.

Sketch of proof :

We prove the termination of the procedure Simpl by using the Nœtherian order $\gg \gg$ induced by $>$
over the set $\mathcal{M}$ ($\mathcal{M}(\mathbb{N})$) of finite multi-sets of finite multi-sets of $\mathbb{N}$.

For this, we set for $(t', B') \in H'$ and $s \in B'$, (i.e H' is a set of clauses)

$n_s = 1 + \text{Card}(\{ t'' \mid (t'', B'') \in H, s \text{ and } t'' \text{ unifiable with mgu. } \sigma \text{ and } < \{\sigma(s)\} > \notin < B'' > \})$

Set $\text{MC}((t', B'), H) = \{\{ n_s \mid s \in B' \}\}$ ($\text{MC}((t', B'), H) \in \mathcal{M}(\mathbb{N})$),

and we associate to H' the multi-set $M(H', H) = \{\{\text{MC}((t', B'), H) \mid (t', B') \in H'\}\}$. ($M(H', H) \in \mathcal{M}$ ($\mathcal{M}(\mathbb{N})$)).

If H'_i is the value of H' at step i then we can verify that $\forall i, M(H'_i, H) \gg \gg M(H'_{i+1}, H)$. ∎

Notation :
Let $s \in T(\mathbb{F}, X)$, $\text{Pos}(s, m) = \{ u \mid u \in \text{occ}(s) \text{ and } \text{leng}(u) \leq m \}$

Theorem 4.1 : *(Finite Generator)*
Let (t, B) be an elementary clause and H a set of elementary clauses.
If H' = **Simpl** ((t, B), H) then $< H'> = < \{ (t, B) \} \cup H >$ and
$\exists (t', B') \in H'$ such that $B' \neq \emptyset$ $\Rightarrow$ $\exists G \in \mathcal{P}_f(T(\mathbb{F}, X))$ such that $< G > = < \{ (t, B) \} \cup H >$.

Sketch of proof :
Let H'_i be the value of H' at step i.
1) We can prove, by induction over i that $< H'_i > = < \{ (t, B) \} \cup H >$.
2) Assume that $\exists (t', B') \in H'$ such that $B' \neq \emptyset$ and $\exists G \in \mathcal{P}_f(T(\mathbb{F}, X))$ such that
$< G > = < \{ (t, B) \} \cup H >$.
Let $E = G \cup (\bigcup_{(t'', B'') \in (H \cup \{(t, B)\})} (\{t''\} \cup B''))$, $r = 2*\text{dep}(E)$; and let $s \in B'$ such that :
If ψ is substitution such that $\forall x \in \text{Var}(E), \text{mdep}(\psi(x)) \geq r$ then
$\quad \forall s' \in B', \text{Pos}(\psi(s), r) \subseteq \text{Pos}(\psi(s'), r)$ $\Rightarrow$ $\text{Pos}(\psi(s), r) = \text{Pos}(\psi(s'), r)$

We know that $\exists \tau, s = \tau(t')$ and τ is not linear over t'. Thus, there exists μ, σ such that :
- $\tau = \sigma \mu$
- μ is linear over t' and $\sigma(X) \subset X$

Set $t_1 = \mu(t')$ then $s = \sigma(t_1)$ and σ is not linear over t_1.
Let $l = K(a, ..., a)$ be the well-balanced term built over $\{f, a\}$ where $a \in \mathbb{F}_0$ and $f \in \mathbb{F}_1 \cup ... \cup \mathbb{F}_n$.
(i.e K is a context, $l \in T(\{f, a\})$ and $\text{dep}(l) = \text{mdep}(l)$).
Let σ_1, σ_2 be the substitutions defined by :
- If $\text{Var}(s) = \{ x_1, ..., x_{n1} \}$ then σ_1 is defined by :
$\sigma_1(x_i) = K(u^i_1, ..., u^i_p)$ where p is the number of occurrences of a in l and

$\forall i \leq n_1, \forall j \leq p$, u^i_j is the well-balanced term building over $\{f, a\}$ with $dep(u^i_j) = 3.r.p.(i-1) + 3.r.j$
Set $t'_1 = \sigma_1(s)$ then $t'_1 \in <B'>$, so $t'_1 \notin <\{(t', B')\}>$.
$Var(t_1) = \bigcup_{1 \leq i \leq n_2} \{y^i_1, ..., y^i_{hi}\}$ where $\{y^i_1, ..., y^i_{hi}\} = \sigma^{-1}(x_i)$ because $\sigma(X) \subset X$.
Set $m = dep(t'_1)$ $(m > 3r)$.

• σ_2 is defined by :
$\sigma_2(y^i_j) = K_i(r_{i,j}^1, ..., r_{i,j}^{q_i})$ with $\forall k$, $r_{i,j}^k$ is the well-balanced term building over f and a with $dep(r_{i,j}^k) = (\Sigma_{1 \leq l \leq i-1} h_l.q_l).m + (j-1).q.m + k.m$ where $\sigma_1(x_i) = K_i(a, ..., a)$ and q_i is the number of occurrences of a in $\sigma_1(x_i)$.
Set $t'_2 = \sigma_2(t_1)$.

We can prove that $t'_2 \in (<\{t'\}> - <B'>)$; then $\exists s_1 \in G, \exists \tau_1$ such that $t'_2 = \tau_1(s_1)$ because $<G> = <\{(t, B)\} \cup H>$.
We can prove also that $t'_1 \in <G>$ which is absurd because $t'_1 \notin <\{(t, B)\} \cup H>$. ∎

5. Finite Generator Problem.

We now propose a procedure for solving the finite generator problem :
Let $H \in Bool(\mathcal{P}_f(T(\mathbb{F}, X)))$, is there $G \in \mathcal{P}_f(T(\mathbb{F}, X))$ such that $<H> = <G>$?

Notation :
Let H' be a set of elementary clauses VI(H') is the predicate : $\forall(t', B') \in H'$, $B' = \emptyset$.

We can assume that $H = \{(t_i, B_i), i \in I\}$ (i.e. I is a finite, Corollary 3.1.2)

Procedure GenF (H)

```
Begin
  H' := ∅ ;
  H" := H ;

  For (t, B) ∈ H do

    Begin

      H' := Simpl ( (t, B), (H" - { (t, B) }), F ) ) ;

      H" := (H" - { (t, B) }) ∪ H' ;

      If not(VI(H')) then Break ;

    End ;

  Return(H") ;
End.
```

Example :
Let $\mathbb{F} = \mathbb{F}_0 \cup \mathbb{F}_2$ with $\mathbb{F}_0 = \{a\}$ and $\mathbb{F}_2 = \{f\}$
$H = \{ (f(x_1, f(x_1, y_1)), \{ f(x_2, f(x_2, x_2)) \}) ; (f(x_3, f(y_2, x_3)), \{ f(x_4, f(a, x_4)) \}) \}$
1st step :
$(t, B) = (f(x_1, f(x_1, y_1)), \{ f(x_2, f(x_2, x_2)) \})$
Set (t_i, B_i) the value of (t, B) at step i and $H"_i$ the value of H" at step i.
$H"_0 = \{ (f(x_1, f(x_1, y_1)), \{ f(x_2, f(x_2, x_2)) \}) ; (f(x_3, f(y_2, x_3)), \{ f(x_4, f(a, x_4)) \}) \}$
$(t_0, B_0) = (f(x_1, f(x_1, y_1)), \{ f(x_2, f(x_2, x_2)) \})$
$H' = \{ (f(x_1, f(x_1, y_1)), \{ f(x_2, f(x_2, x_2)) \}) \}$
Let σ be the substitution defined by $\sigma(x_3) = \sigma(y_2) = \sigma(x_2) = z$ then
$\sigma(f(x_1, f(x_1, y_1))) = \sigma(f(x_2, f(x_2, x_2))) = f(z, f(z, z))$.

$\sigma(f(x_2, f(x_2, x_2))) = K_{f(x_2, f(x_2, x_2))}(z)$, $\{z\}$ is a base of $T(\mathbb{F})$ and $f(z, f(z, z))$ et $f(x_4, f(a, x_4))$ are unifiables. Let μ be their mgu. then $\mu(z) = \mu(x_4) = a$.

$H' = \{ (f(x_1, f(x_1, y_1)), \{f(a, f(a, a))\}) \}$

Let σ be the substitution defined by : $\sigma(x_1) = \sigma(y_1) = a$

$f(a, f(a, a)) = \sigma(f(x_1, f(x_1, y_1))) = K_{f(x_1, f(x_1, y_1))}(a, a)$

$\{ (a, a) \} \cup \{ (a, f(x_5, y_5)) ; (f(x_6, y_6), a) ; (f(x_7, y_7), f(x_8, y_8)) \}$ is a base of $T(\mathbb{F})$.

$H' = \{ (f(a, f(a, f(x_5, y_5))), \emptyset) ; (f(f(x_6, y_6), f(f(x_6, y_6), a)), \emptyset) ;$
$\qquad (f(f(x_7, y_7), f(f(x_7, y_7), f(x_8, y_8))), \emptyset) \}$ thus

$H''_1 = \{ (f(a, f(a, f(x_5, y_5))), \emptyset) ; (f(f(x_6, y_6), f(f(x_6, y_6), a)), \emptyset) ;$
$\qquad (f(f(x_7, y_7), f(f(x_7, y_7), f(x_8, y_8))), \emptyset) ; (f(x_3, f(y_2, x_3)), \{f(x_4, f(a, x_4))\}) \}$.

2nd step :

$(t, B) = (f(x_3, f(y_2, x_3)), \{f(x_4, f(a, x_4))\})$.

Let σ be the substitution defined by : $\sigma(x_3) = x_4$ and $\sigma(y_2) = a$ then

$f(x_4, f(a, x_4)) = \sigma(f(x_3, f(y_2, x_3))) = K_{f(x_3, f(y_2, x_3))}(x_4, a)$.

$\{(x_4, a)\} \cup \{(x_4, f(x_9, y_9))\}$ is a base of $T(\mathbb{F})$.

$H''_2 = \{ (f(a, f(a, f(x_5, y_5))), \emptyset) ; (f(f(x_6, y_6), f(f(x_6, y_6), a)), \emptyset) ;$
$\qquad (f(f(x_7, y_7), f(f(x_7, y_7), f(x_8, y_8))), \emptyset) ; (f(x_4, f(f(x_9, y_9), x_4)), \emptyset) \}$.

H''_2 is the final value of H''. Thus :

$<H> = \{ f(a, f(a, f(x_5, y_5))) ; f(f(x_6, y_6), f(f(x_6, y_6), a)) ; f(f(x_7, y_7), f(f(x_7, y_7), f(x_8, y_8))) ;$
$\qquad f(x_4, f(f(x_9, y_9), x_4)) \}$.

Remark :
The procedure **GenF** terminates because the procedure **Simpl** terminates and H is finite set of elementary clauses.

Theorem 5.1 : *(completeness of GenF)*
Let H be a set of elementary clauses. If $H'' = \text{GenF}(H)$ then
$\qquad < H > = < H'' >$ and $\quad VI(H'') \Leftrightarrow \exists G \in \mathcal{P}_f(T(\mathbb{F}, X))$ such that $< H > = < G >$.

Proof :
a) If $VI(H'')$ then $<H''> = <\{t, (t, \emptyset) \in H''\}>$ therefore $G = \{t \mid (t, \emptyset) \in H'\}$ is suitable.
b) If not($VI(H'')$) then by the theorem 4.1 $\nexists G \in \mathcal{P}_f(T(\mathbb{F}, X))$ such that $<G> = <H>$. ∎

6. Syntactic equational formula

A syntactic equational formula is a first order formula whose only predicate is a syntactic equality. Let A be such a formula, Var(A) is the set of free variables of A. A is said to be ground when it has no free variable (i.e Var(A) = $\emptyset$).

If $x_1, ..., x_k$ are the free variables of syntactic equational formula A, then we denote A by $A(x_1, ..., x_k)$.

Definitions :
- $(t_1, ..., t_k) \in T(\mathbb{F})^k$ is a solution of $A(x_1, ..., x_k)$ in $T(\mathbb{F})$ if $T(\mathbb{F}) \models A(t_1, ..., t_k)$ (i.e the ground syntactic equational formula $A(t_1, ..., t_k)$ is true in $T(\mathbb{F})$).
- $[\![A]\!] = \{ (t_1, ..., t_k) \in T(\mathbb{F})^k \mid T(\mathbb{F}) \models A(t_1, ..., t_k) \}$ is the set of solutions of the syntactic equational formula $A(x_1, ..., x_k)$.
- A syntactic equational formula A is said to be valid in $T(\mathbb{F})$ if $[\![A]\!]$ is not the empty set.

Example :
Let $\mathbb{F} = \mathbb{F}_0 \cup \mathbb{F}_1 \cup \mathbb{F}_2$ with $\mathbb{F}_0 = \{0\}$, $\mathbb{F}_1 = \{s\}$ and $\mathbb{F}_2 = \{+\}$.
In the formula " $\forall x (((x = 0) \wedge (x \neq s(y))) \vee \exists z (x = y+z))$ ", y is the only free variable.
The formula " $\forall x ((x = 0) \vee \exists x_1 (x = s(x_1)) \vee \exists x_2, x_3 (x = x_2+x_3))$ " is a ground formula.

Let A(y) be the syntactic equational formula :
$$\forall x\, ((\, (x = y) \vee \exists x_1 (x = s(x_1))\,) \vee \exists x_2, x_3\, (x = x_2+x_3)\,).$$
A(y) is valid in T($\mathbb{F}$), because $\forall x\, ((\,(x = 0) \vee \exists x_1(x = s(x_1))\,) \vee \exists x_2, x_3\, (x = x_2+x_3)\,)$ is true in T($\mathbb{F}$) : T($\mathbb{F}$) $\models$ A(0) (i.e 0 is a solution of A(y) in T($\mathbb{F}$)). Actually, $[\![A(y)]\!] = \{\, 0\, \}$.

Property 6.1.1 :
Let $A(x_1, ..., x_k)$, $A'(x_1, ..., x_k)$ be two syntactic equational formulae then :
- $[\![A(x_1, ..., x_k) \vee A'(x_1, ..., x_k)]\!] = [\![A(x_1, ..., x_k)]\!] \cup [\![A'(x_1, ..., x_k)]\!]$
- $[\![\neg A(x_1, ..., x_k)]\!] = T(\mathbb{F})^k - [\![A(x_1, ..., x_k)]\!]$.

We study in this paragraph the problem of explicit representation of the set $[\![A(x_1, ..., x_k)]\!]$, where $A(x_1, ..., x_k)$ is a syntactic equational formula :
Is there some $G \in \mathcal{P}_f(T(\mathbb{F}, X)^k)$ such that $[\![A(x_1, ..., x_k)]\!] = <G>$? *(i.e the decidability of negation elimination from syntactic equational formula)*

Definition :
A syntactic equational formula b is said to be basic if it is of the form $\exists y\, (x_1 = t_1(y) \wedge ... \wedge x_k = t_k(y))$ where the vectors of variables $x = (x_1, ..., x_k)$ and y have no common variables and $t_1(y), ..., t_k(y)$ are terms whose variables come from y.

We will use the method described in [Mah 88]. He use the following complete axiomatization of the algebra of terms, he describes transformations of syntactic equational formula into equivalent boolean combination of basic formulae.

$\forall x,y\; f(x) = f(y) \Rightarrow x = y.$ *(i.e x, y are a vectors of variables)* (1)

$\forall x,y\; f(x) \neq g(y).$ *(i.e $f, g \in \mathbb{F}$, $f \neq g$ and x, y are a vectors of variables)* (2)

$\forall x\; x \neq t(x).$ *($t \in (T(\mathbb{F}, X) - X)$ and $x \in Var(t)$)* (3)

$\forall x, \exists y\, (\bigvee_{f \in \mathbb{F}} x = f(y)).$ *(y is a vector of variables and $f \in \mathbb{F}$)* (4)

Theorem 6.1 : [Mah 88]
Let A be a syntactic equational formula, then there is a computable boolean combination B of basic formulae such that : (1), (2), (3), (4) $\models$ A $\Leftrightarrow$ B (i.e $[\![A]\!] = [\![B]\!]$).

Remark :
Let $b = (\exists y\, (x_1 = t_1(y) \wedge ... \wedge x_k = t_k(y)))$ be a basic formula then
- $[\![b]\!] = <\{\,(t_1(y), ..., t_k(y))\,\}>$.
- $[\![\neg(b)]\!] = T(\mathbb{F})^k - <\{\,(t_1(y), ..., t_k(y))\,\}> = <\{\,(z_1, ..., z_k)\,\}> - [\![b]\!]$.

Using property 6.1.1 and previous remark, we have :
Corollary :
Let $A(x_1,...,x_k)$ be a syntactic equational formula then there is a computable $H \in Bool(\mathcal{P}_f(T(\mathbb{F},X)^k))$ such that $[\![A(x_1, ..., x_k)]\!] = <H>$.

Theorem 6.2 : *(Decidability of Negation Elimination)*
Let A be a syntactic equational formula, the problem of whether $[\![A]\!]$ is finitely generated is decidable.

Proof :
We have reduced the problem of knowing whether $[\![A]\!]$ is finitely generated in Herbrand Universe to the finite generator problem, which is decidable (See theorems 3.1 and previous corollary). ∎

Corollary 1 :
Let $A(x_1, ..., x_k)$ be a syntactic equational formula. Suppose that :
$\exists G \in \mathcal{P}_f(T(\mathbb{F}, X)^k)$ such that $[\![A]\!] = <G>$ then
(1), (2), (3), (4) $\models$ A $\Leftrightarrow$ $\bigvee_{(t_1(y), ..., t_k(y)) \in G} \exists y\, (x_1 = t_1(y) \wedge ... \wedge x_k = t_k(y))$

Corollary 2 :
The validity of a syntactic equational formula $A(x_1, ..., x_k)$ is decidable.

Proof :
A is not valid $\Leftrightarrow$ $[\![A]\!] = \emptyset = <\emptyset>$. ■

7. Conclusion.

We introduced the property of finitely generated. We showed the decidability of finitely generated in Herbrand Universe for a boolean combination of finitely generated sets and we proved the decidability of negation elimination for syntactic equational formula. In this paper we considered only the Herbrand Universe. It is not difficult to show that the general problem is undecidable. For example the problem of negation elimination is undecidable in free monoid over finite alphabet [Mak 78] and [Mar 82].

References

[Col 84] : A. Colmerauer, " Equations and inequations on finite and infinte trees", in FGCS'84 Proceedings, p 85--99 (1984).

[Com 88] : H. Comon, " Unification and disunification: théorie et applications ", Thèse de doctorat, I.N.P. de Grenoble, France (1988).

[DJ 90] : N.Dershowitz and J.P.Jouannaud, " Rewriting Systems ", In Van Leuven, editor, Handbook of Theoretical Computer Science, Vol. B, p 245--309 (1990).

[GH 78] : J. V. Guttag and J. J. Horning, " The algebraic specification of abstract data types ", Acta infomatica n°10, p 27--52 (1978).

[HO 80] : G. Huet and D. Oppen, " Equations and rewrite rules: a survey ", In R. Book, editor, Formal Language Theory: Perspectives and Open Problems, p 349--405, Academic Press (1980).

[HW 78] : G. H. Hardy and E. M. Wright, " An Introduction to the Theory of Numbers ", Fifth edition, Oxford University Press, p 8 (1978).

[JK 86] : J. P. Jouannaud and E. Kounalis, " Automatic proofs by induction in equational theories without constructors ", In Proceedings of 1st IEEE Symp. Logic in Computer Science, Cambridge, Mass. p 358--366 (1986).

[LM 87] : J. L. Lassez and K. Marriott, " Explicit representation of terms defined by counter examples ", J. Automated Reasoning, 3(3) (1987).

[L&M&M 91] : J.-L. Lassez, M. Maher and K. Marriott, " Elimination of negation in term algebras ", In Proceedings of MFCS, Warsaw (1991).

[Mah 88] : M. J. Maher, " Complete axiomatization of the algebra of finite, rational and infinite trees ", In Proceedings of 3rd IEEE Symposium on Logic in Computer Science, Edinburgh, p 348--357 (1988).

[Mak 78] : G. S. Makanin, " Algorithm decidability of the rank of constant free equations in a free semigroup", Dokl. Akad. Nauk. SSSR 243 (1978).

[Mar 82] : S. S. Marchenko, " Undecidability of the positive $\forall \exists$-theory of a free semigroup ", Sibirskii Matematicheskii Zhurnal, 23 (1), p 437--455 (1982), in Russian.

[M&M 76] : A. Martelli and U. Montanari, " Unification in linear time and space: A structured presentation ", Internal Report B 76-16, Istituto di Elaborazione della Informazione, Pisa, Italy, (1976).

[M&M 82] : A. Martelli and U. Montanari, " An efficient unification algorithm ", ACM Transactions on Progamming Languagages and Systems 4 (2), p 258--282 (1982).

[P&W 78] : M. S. Paterson and M. N. Wegman, " Linear unification ", Jour. Comput. Syst. Sci 16, p 158--167 (1978).

[Rob 65] : J. A. Robinson, " A machine-oriented logic based on the resolution principle ", Journal of the ACM, 12 (1), p 23--41 (1965).

[R&P 89] : P. Ruzicka and I. Privara, " An almost linear Robinson unification algorithm ", Acta Informatica 27, p 61--71 (1989).

[Taj 92] : M. Tajine, " Representation explicite de certains langages de termes : théorie et applications ", Thèse de doctorat, Université Louis Pasteur de Strasbourg, France.

Encompassment Properties and Automata with Constraints

Anne-Cécile CARON   Jean-Luc COQUIDE
Max DAUCHET [*]

Laboratoire d'Informatique Fondamentale de Lille
UFR IEEA. URA 369 CNRS.
Université des Sciences et Technologies de Lille
59655 Villeneuve d'Ascq Cedex. FRANCE.
email: {caronc,coquide,dauchet}@lifl.lifl.fr

Abstract

We introduce a class of tree automata with constraints which gives an algebraic and algorithmic framework in order to extend the theorem of decidability of inductive reducibility. We use automata with equality constraints in order to solve encompassment constraints and we combine such automata in order to solve every first order formulas built up with unary predicates "x *encompasses* t" denoted by $\text{encomp}_t(x)$.

1 Introduction

Our goal is to extend the theorem of decidability of inductive reducibility (Plaisted [19] J.P. Jouannaud and E. Kounalis [12] D. Kapur, P. Narendran and H. Zhang [13])[1] to the class of order-sorted first order formula written with basic predicates "x *encompasses* t" denoted by $\text{encomp}_t(x)$. In a sense, we prove that we can treat sets of terms with equality constraints between subterms like usual sorts (the number of constraints along a path is bounded), and those automata can be seen as solved forms of these constraints (see J.-P. Jouannaud and C. Kirchner [11] and H. Comon [5] about solved forms). Furthermore, automata are classes of algorithms which can be handled for optimization operations on constraints : compilation, partial evaluation. The most technical aspect of our result is emptiness decision, i.e. deciding if the language recognized by an automaton is empty. The idea is to get a pumping lemma generalizing the proof of D. Plaisted. This proof of the "Plaisted's theorem for tree automata" is tedious and does not appear in this paper. See (A.-C. Caron [3]) for more details.

---

[*]This work is supported in part by GDR "Mathématiques et Informatique", by project "Modèles logiques de la programmation" des PRC Informatique and by ESPRIT Basic Research Action 6317 ASMICS 2.

[1] a term is *inductively reducible* if every ground instance of its is reducible. This property is also called *quasi-reducibility* or *ground-reducibility*.

In this introduction, we give

- a formal presentation of our result
- a comparison with other results
- some examples which improve understanding
- some prospects

Let REC be the class of standard bottom-up finite tree automata. A sort is associated with each final state of an automaton of REC. Indeed, a signature of sorts can be translated into an automaton of REC (H. Comon [4]). In the following example, we define the sort Nat (natural numbers), ListInteger (list of integers) and NonEmptyList (non empty list of integers) by the signature :

$$
\begin{aligned}
0 &: \text{Nat} \\
\text{succ} &: \text{Nat} \to \text{Nat} \\
\# &: \to \text{ListInteger} \\
\text{next} &: \text{Nat} \times \text{ListInteger} \to \text{NonEmptyList}
\end{aligned}
$$

NonEmptyList is a subsort of ListInteger.

A term t of sort ListInteger is written :

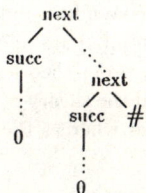

Let $\mathcal{A} = (\Sigma, \mathcal{Q}, \mathcal{Q}_f, \mathcal{R})$ a bottom-up tree automaton where $\Sigma = \{0, \text{succ}(), \#, \text{next}(,)\}$ $\mathcal{Q} = \{q_{\text{Nat}}, q_{\text{LI}}, q_{\text{NEL}}\}$ and $\mathcal{Q}_f = \mathcal{Q}$ and $\mathcal{R}$ is the set of rules :

$0 \to q_{\text{Nat}}(0)$ $\quad \# \to q_{\text{LI}}(\#) \quad q_{\text{NEL}}(x) \to q_{\text{LI}}(x)$
$\text{succ}(q_{\text{Nat}}(x)) \to q_{\text{Nat}}(\text{succ}(x)) \quad \text{next}(q_{\text{Nat}}(x), q_{\text{LI}}(y)) \to q_{\text{NEL}}(\text{next}(x, y))$

Sort NonEmptyList is the language recognized by $\mathcal{A}$ when the final state is q_{NEL}.

We consider unary predicates $\text{encomp}_t(x)$ on T_Σ such that $T_\Sigma \models \text{encomp}_t(s)$ if and only if s encompasses t ($t \in T_\Sigma(\mathcal{X})$). It means that there exists a subterm of s which is an instance of t. We consider the set of first order formulas that we can write with such predicates. These formulas are called *encompassment formulas*.

We define a class of finite bottom-up tree automata EA which have the following properties :

0. for every non deterministic automaton we can construct an equivalent complete and deterministic automaton (one which accepts the same language).
1. the class EA is effectively closed under product and boolean operations.
2. emptiness is decidable.
3. for all t in $T_\Sigma(\mathcal{X})$, atomic formula $\text{encomp}_t(x)$ is definable by an automaton, i.e. we can construct an automaton which recognizes the set of terms of property $\text{encomp}_t(x)$. Moreover, t can be sorted.
4. when t is linear, the automaton which recognizes the set of terms encompassing t is an automaton of REC.

Point 0 is used to express the complement. Points 1, 2 and 3 permit to construct an EA automaton $\mathcal{A}_F$ which specifies the set of terms satisfying an encompassment formula with $F(x)$. For example, if the formula is :

$$\text{encomp}_t(x) \Rightarrow (\text{encomp}_{t_1}(x) \vee \cdots \vee \text{encomp}_{t_p}(x)) \tag{1}$$

we deduce an automaton $\mathcal{A}_F$ from automata specifying $\text{encomp}_t, \text{encomp}_{t_1}, \cdots \text{encomp}_{t_p}$. Using products of sets, we extend the results to formulas $F(x_1, \cdots, x_n)$. These technics are usual (cf M.O. Rabin [20] and W. Thomas [22]). If F contains no free variable, we can decide whether F is true or not. So, automata can be seen as solved forms of formulas.

In short, our automata specify usual sorts and non linear encompassment constraints, in bounded number along every path. If we allow a non bounded number of constraints along a path, we loose the "good" properties 1 and 2 (J. Mongy [18]) and we can describe any recursively enumerable set. Note that in the case of non bounded constraints but only with equalities between sons of a same node, the properties 1 and 2 hold (B. Bogaert and S. Tison [2]). Our result unifies known results : we immediately get decidability of inductive reducibility and related results in the order-sorted case (D. Plaisted [19], J.P. Jouannaud and E. Kounalis [12] D. Kapur, P. Narendran and H. Zhang [13]). In the linear case (when the terms t of the predicates $\text{encomp}_t(x)$ are linear), construction give automata of REC (cf N. Dershowitz [8]). About first order theories and properties on terms, note that the theory of subterm ordering [24] and the theory of tree embedding [23] are undecidable but these two theories concern relations whereas the encompassment theory uses only unary predicates.

The following examples should improve the understanding.

- Let R a rewrite system.

  If $R = \{l_1 \to r_1, \cdots, l_p \to r_p\}$ we can express the inductive reducibility of t by the formula $\forall x F(x)$ where $F(x)$ is the formula (1).
- An equational specification E is said to be sufficiently complete (with respect to a set of constructors C) if and only if for every ground term g in T_Σ there is a ground term c involving only constructors in C such that $g \leftrightarrow^* c$. A set R of rules is sufficiently complete if its associated E is sufficiently complete.

  Theorem ([12] [13]) : a complete constructor-preserving set R of rules is sufficiently complete iff for every f in $\Sigma - C$, $f(x_1, \cdots, x_n)$ is inductively reducible.

  Therefore, if R is a complete constructor-preserving set $\{l_1 \to r_1, \cdots, l_p \to r_p\}$, with $\Sigma - C = \{f_1, \cdots, f_n\}$, R is sufficiently complete iff the following formula is true :

$$\forall x (\text{encomp}_{f_1(x_1, \cdots, x_{i_1})}(x) \vee \cdots \vee \text{encomp}_{f_n(x_1, \cdots, x_{i_n})}(x) \Rightarrow (\text{encomp}_{l_1}(x) \vee \cdots \vee \text{encomp}_{l_p}(x)))$$

Basic linear encompassement constraints can always be suppressed and controled by usual tree automata (see C. Hoffmann and M. O'Donnell [10]). Moreover, those basic encompassment constraints which can be expressed by linear encompassment constraints can be controled by bottom up tree automata. Indeed, given a set of rule $\mathcal{R}$, there exist a bottom-up tree automaton which recognizes the set of (ir)reducible ground terms if and only if $\mathcal{R}$ can be replaced by a set of linear rules without changing the set of (ir)reducible ground terms. See G. Kucherov [14] for more details on relationship between term rewriting systems and recognizable languages.

More generaly, encompassment constraints are boolean combination of basic encompassment constraints and are obtained by automata composition. The functions ω are introduced to control the number of tests of basic constraints along a path. Note the dissymmetry yields by basic constraints (which are sub-terms equalities) and their negations : the automata control a bounded number of test of basic constraints but a non bounded number of tests of their negations. This is necessary to keep the emptiness decision property and can be understood by the fact of transitivity of $=$ and non-transitivity of $\neq$.

We end this introduction with some prospects. We would like to study the notion of minimal automaton, which is linked to the notion of solved constraints. On the other hand, classes of finite automata get a lot of effective properties which could be used for compilation or partial solving of constraints. We wish to generalize our automata to allow a non bounded number of tests between subterms sons of a same node, i.e. to define a class which contains the class defined by B. Bogaert and S. Tison in [2]. We would like also to progress in an axiomatic approach and get the same kind of result for other classes of constraints or in no free algebras. More generally, we are extending our class of automata to merge in a common frame the present work and technics of solving equational formulas (see A. Mal'cev [17], M. J. Maher [16], H. Comon and P. Lescanne [6], H. Comon [4]). Our main prospect is to show the decidability of first order theory of step by step rewriting. In this theory, we consider all the formulas we can write with the first order vocabulary and binary predicates $x \to_{\mathcal{R}} y$ which means "x can be reduced to y using an occurrence of a rule of the system R". In the word case, this result is a consequence of decidability of first order theory of RR-relation (M. Dauchet and S. Tison [7]) or of rational relation with bounded delay (C. Frougny and J. Sakarovitch [9]). In the case of term rewriting systems, it is a generalization of the results we present here.

2 Preliminaries

Let $\mathcal{X}$ be a countably infinite set of *variables*. An *alphabet* is a set Σ of *function symbols*. Associated with every $f \in \Sigma$ is a natural number denoted its *arity*. Function symbols of arity 0 are called *constants*. The set of *terms* $T_\Sigma(\mathcal{X})$ built from an alphabet Σ and a countably infinite set of variables $\mathcal{X}$ with $\Sigma \cap \mathcal{X} = \emptyset$ is the smallest set such that $\mathcal{X} \subset T_\Sigma(\mathcal{X})$ and if $f \in \Sigma$ is an n-ary function symbol and $t_1, \cdots, t_n \in T_\Sigma(\mathcal{X})$ then $f(t_1, \cdots, t_n) \in T_\Sigma(\mathcal{X})$. The set of *ground terms* denoted T_Σ is the set of terms without variables. Identity of terms is denoted by $\equiv$.

A *context* $C[, ...,]$ is a 'term' which contains at least one occurrence of a special constant $\square$. If $C[, ...,]$ is a context with n occurrences of $\square$ and $t_1, \cdots, t_n$ are terms then $C[t_1, \cdots, t_n]$ is the result of replacing from the left to the right the occurrences of $\square$ by $t_1, \cdots, t_n$. A context containing precisely one occurrence of $\square$ is denoted by $C[\ ]$. A term s is a *subterm* of a term t if there exists a context $C[\ ]$ such that $t \equiv C[s]$. A term t is an *instance* of a term s if $t \equiv \sigma(s)$ for a substitution σ. A term t *encompasses* a term s if there exists a subterm of t instance of s.

A *term rewriting system* is a set $\mathcal{R} \subset T_\Sigma(\mathcal{X}) \times T_\Sigma(\mathcal{X})$ of *rewrite rules*. Every rule (l, r) satisfies the following two constraints: the left-hand side l is not a variable and the variables which occur in the right-hand side r also occur in l.

The *rewrite relation* $\rightarrow_\mathcal{R}$ is defined as follows: $s \rightarrow_\mathcal{R} t$ if there exists a rewrite rule $l \rightarrow r$ in $\mathcal{R}$, a substitution σ and a context $C[\ ]$ such that $s \equiv C[l^\sigma]$ and $t \equiv C[r^\sigma]$. The transitive-reflexive closure of $\rightarrow_\mathcal{R}$ is denoted by $\stackrel{*}{\rightarrow}_\mathcal{R}$. If $s \stackrel{*}{\rightarrow}_\mathcal{R} t$ we say that s *reduces to* t.

A *bottom-up tree automaton* is defined by a finite ranked alphabet Σ, a finite set $\mathcal{Q}$ of states, a subset $\mathcal{F}$ of final states and a finite set of transitions.
Transitions are rewrite rules of form :
$f(q_1(x_1), \cdots, q_n(x_n)) \rightarrow q(f(x_1, \cdots, x_n))$ where $f \in \Sigma$, $n \geq 0$ and $(q_1, \cdots, q_n, q) \in \mathcal{Q}^{n+1}$.
The rewrite relation $\vdash_\mathcal{A}$ associated with $\mathcal{A}$ is defined by $t \equiv f(t_1, \cdots, t_n) \stackrel{*}{\vdash}_\mathcal{A} q(t)$ if and only if there exists a rule $f(q_1(x_1), \cdots, q_n(x_n)) \rightarrow q(f(x_1, \cdots, x_n))$ in $\mathcal{R}$ and $\forall i \in [1, n]$, $t_i \stackrel{*}{\vdash}_\mathcal{A} q_i(t_i)$

For every state $q \in \mathcal{Q}$: $L_q(\mathcal{A}) = \{t \in T_\Sigma \mid t \stackrel{*}{\vdash}_\mathcal{A} q(t)\}$
So, the tree language recognized by $\mathcal{A}$ is : $L(\mathcal{A}) = \bigcup_{q \in \mathcal{Q}_f} L_q(\mathcal{A})$
REC denotes the class of bottom-up tree automata.

3 Automata with constraints

3.1 Definition

AC denotes the class of automata with constraints. Basic constraints are unary predicates on T_Σ. We consider boolean composition of basic constraints. For example, $T_\Sigma \models (c \wedge c')(t)$ if and only if $T_\Sigma \models c(t) \wedge c(t')$.

Definition 3.1 *An automaton of AC is a 5-uple* $\mathcal{A} = (\Sigma, \mathcal{Q}, \mathcal{Q}_f, \mathcal{E}, \mathcal{R})$

- Σ *is a finite ranked alphabet.*
- $\mathcal{Q}$ *is a finite set of states.*
- $\mathcal{Q}_f$ *is a finite set of final states* $(\mathcal{Q}_f \subseteq \mathcal{Q})$.
- $\mathcal{E}$ *is a finite set of basic constraints.*
- $\mathcal{R}$ *is a finite set of rules of the following type :*
  $f(q_1(x_1), \cdots, q_n(x_n)) \stackrel{c}{\longrightarrow} q(f(x_1, \cdots, x_n))$ *where c is a boolean combination of elements of $\mathcal{E}$. c is named constraint of the rule.*

REWRITE RELATION ASSOCIATED WITH AUTOMATA WITH CONSTRAINTS :
Consider $t = f(t_1, \cdots, t_n)$.
$t \stackrel{*}{\vdash}_\mathcal{A} q(t)$ if and only if there exists a rule $f(q_1(x_1), \cdots, q_n(x_n)) \stackrel{c}{\longrightarrow} q(f(x_1, \cdots, x_n))$ in $\mathcal{R}$ such that $t \equiv f(t_1, \cdots, t_n)$, $T_\Sigma \models c(t)$ and $\forall i \in [1, n]$, $t_i \stackrel{*}{\vdash}_\mathcal{A} q_i(t_i)$

Definition 3.2 *A tree language $\mathcal{F}$ is AC-recognizable if there exists an automaton $\mathcal{A}$ in AC such that $L(\mathcal{A}) = \mathcal{F}$*

REMARK : The standard bottom-up tree automata are automata with constraints, where $\mathcal{E}$ is empty.

In the following, we consider automata with constraints. $\mathcal{E}_-$ denotes the set $\{\neg P \mid P \in \mathcal{E}\}$.

3.2 Automata transformations

3.2.1 Algorithm suppressing "or"

Lemma 3.3 *Let $\mathcal{A}$ be an automaton with constraints. There exists an equivalent automaton $\mathcal{A}'$ such that its constraints are conjonctions of elements in $\mathcal{E}_- \cup \mathcal{E}$. $\mathcal{A}'$ is called automaton without "or".*

PROOF : The constraints are rewritten in disjonctive normal form and transformed by the rule :

$$\frac{l \xrightarrow{c_1 \vee c_2} r}{l \xrightarrow{c_1} r, \; l \xrightarrow{c_2} r}$$

We get an automaton $\mathcal{A}'$ such that : $\forall q \in Q, L_q(\mathcal{A}) = L_q(\mathcal{A}')$. □

From now, we only consider automata without "or".

Definition 3.4 *An automaton is complete if for every tree t in T_Σ, there exists a state q such that $t \vdash^*_\mathcal{A} q(t)$.*

Definition 3.5 *An automaton $\mathcal{A}$ is deterministic if for every pair of rules with the same left-hand side, constraints are incompatible.(i.e. their conjunction is unsatisfiable)*

Remark : The above definition coincides with the determinism definition for automata of REC.

3.2.2 Completion algorithm

Lemma 3.6 *For every automaton $\mathcal{A} = (\Sigma, \mathcal{Q}, \mathcal{Q}_f, \mathcal{E}, \mathcal{R})$ without "or", there exists a complete automaton $\mathcal{A}' = (\Sigma, \mathcal{Q}', \mathcal{Q}_f, \mathcal{E}, \mathcal{R}')$ without "or" such that for all state q of $\mathcal{Q}$, $L_q(\mathcal{A}) = L_q(\mathcal{A}')$.*

PROOF :

1. For each letter f of Σ_n and $q_1, \cdots, q_n$ in $\mathcal{Q}$,

    (a) If there is no rule with left-hand side $f(q_1(x_1), \cdots, q_n(x_n))$, then we add the rule $f(q_1(x_1), \cdots, q_n(x_n)) \to \text{Dead}(f(x_1, \cdots, x_n))$

(b) In the rules with left-hand side $f(q_1(x_1), \cdots, q_n(x_n))$ are :
$$f(q_1(x_1), \cdots, q_n(x_n)) \xrightarrow{c_1} q_1'(f(x_1, \cdots, x_n)),$$
$$\vdots$$
$$f(q_1(x_1), \cdots, q_n(x_n)) \xrightarrow{c_p} q_p'(f(x_1, \cdots, x_n)),$$
we rewrite the condition $\neg c_1 \wedge \cdots \wedge \neg c_p$ in disjonctive normal form $\gamma_1 \vee \cdots \vee \gamma_k$ where the γ_i are conjunctions of elements of $\mathcal{E} \cup \mathcal{E}_-$ and for every i in [1,k] we add the rule $f(q_1(x_1), \cdots, q_n(x_n)) \xrightarrow{\gamma_i} \mathsf{Dead}(f(x_1, \cdots, x_n))$.

2. For every f in Σ_n, we add all the rules $f(s_1(x_1), \cdots, s_n(x_n)) \to \mathsf{Dead}(f(x_1, \cdots, x_n))$ where at least one s_i is equal to Dead.

$\mathcal{R}'$ is the set of rules obtained by application of the previous transformations on $\mathcal{R}$ and $\mathcal{Q}' = \mathcal{Q} \cup \{\mathsf{Dead}\}$.
Then we have $t \vdash^*_{\mathcal{A}'} \mathsf{Dead}(t) \Leftrightarrow \not\exists q \in \mathcal{Q}\ t \vdash^*_{\mathcal{A}} q(t)$
and $\forall q \in \mathcal{Q}\ (t \vdash^*_{\mathcal{A}} q(t) \Leftrightarrow t \vdash^*_{\mathcal{A}'} q(t))$.
The automaton $\mathcal{A}'$ is without "or" and is complete. □

3.2.3 Determinization algorithm

Lemma 3.7 *If the satisfiability of constraints can be tested, for every complete automaton $\mathcal{A} = (\Sigma, \mathcal{Q}, \mathcal{Q}_f, \mathcal{E}, \mathcal{R})$ without "or", there exists a complete and deterministic automaton $\mathcal{A}' = (\Sigma, \mathcal{Q}', \mathcal{Q}'_f, \mathcal{E}, \mathcal{R}')$ without "or" such that $L(\mathcal{A}) = L(\mathcal{A}')$.*

PROOF : The classical determinization algorithm for automata with constraints does not allow to obtain rules without "or". For example, if we consider two rules with same left-hand side and constraints c and c', the classical algorithm adds rules with constraints $c \wedge c'$, $c \wedge \neg c'$, $\neg c \wedge c'$ and $\neg c \wedge \neg c'$. If c or c' is not a basic constraint, those rules are not without "or". Therefore, in the following construction, we straight considere basic constraints.
As in the determinization algorithm in REC, every state in $\mathcal{Q}'$ is an element of $2^\mathcal{Q}$.

1. initialisation : $\mathcal{Q}' = \emptyset$, $\mathcal{R}' = \emptyset$.

2. iteration : $\forall f \in \Sigma_n$, $\forall k_1, \cdots, k_n \in \mathcal{Q}'$
   We consider all the rules with left-hand sides of the form $f(q_1(x_1), \cdots, q_n(x_n))$ with $\forall i,\ q_i \in k_i$:
   $$f(q_1^1(x_1), \cdots, q_n^1(x_n)) \xrightarrow{c_1} s_1(f(x_1, \cdots, x_n))$$
   $$\vdots$$
   $$f(q_1^p(x_1), \cdots, q_n^p(x_n)) \xrightarrow{c_p} s_p(f(x_1, \cdots, x_n))$$
   Constraints c_i are conjunctions of elements e_i^j of $\mathcal{E}$ and $\mathcal{E}_-$: $\forall i \in [1,p]\ c_i = \bigwedge_{j \in J_i} c_i^j$
   If the rule contains no constraints, we assume that $c_i = e_i^1 = V$.
   In the following, l_i^j denotes e_i^j or its negation. We construct the set Γ of the largest satisfiable conjunctions of constraints l_i^j :

$$\Gamma = \left\{ \begin{array}{l} \gamma = l_{i_1}^{j_1} \wedge \cdots \wedge l_{i_m}^{j_m} \text{ such that } \forall k \in [1,m] \; i_k \in [1,p] \text{ and } j_k \in J_{i_k} \\ \gamma \text{ is satisfiable and } \forall l_i^j \notin \{l_{i_1}^{j_1}, \cdots, l_{i_m}^{j_m}\} \; \gamma \wedge l_i^j \text{ is unsatisfiable} \end{array} \right\}$$

We add in $\mathcal{R}'$ all the rules $f(k_1(x_1), \cdots, k_n(x_n)) \xrightarrow{\gamma} k_\gamma(f(x_1, \cdots, x_n))$ with $\gamma \in \Gamma$ and $k_\gamma = \{s_k \mid \gamma \Rightarrow c_k\}$. We add the k_γ to $\mathcal{Q}'$.

We apply the second point until there is no more new rule.
The set $\mathcal{Q}'_f$ of final states is $\{k \mid \mathcal{Q}_f \cap k \neq \emptyset\}$.
We obtain a deterministic automaton because if γ and γ' are constraints in Γ, $\gamma \wedge \gamma'$ is unsatisfiable.
We prove that for every $t = f(t_1, \cdots, t_n)$, if we suppose there exists one and only one move in $\mathcal{A}'$ such that $t_1 \overset{*}{\vdash}_{\mathcal{A}'} k_1(t_1), \cdots, t_n \vdash_{\mathcal{A}'} k_n(t_n)$, there is one and only one move such that $f(k_1(t_1), \cdots, k_n(t_n)) \vdash_{\mathcal{A}'} k_\gamma(f(t_1, \cdots, t_n))$. So we obtain

$$t \in L_k(\mathcal{A}') \Leftrightarrow t \in (\wedge_{q \in k} L_q(\mathcal{A}))$$

The automaton we obtain contains no rule with "or". □

3.3 Boolean closure

Lemma 3.8 *The class AC is closed under boolean operations : union, intersection, complementation. Moreover, if the satisfiability of constraints is decidable, the closure is effective.*

PROOF :

- It is easy to see that the class AC is closed under complementation :

  For every automaton $\mathcal{A}$ in AC, there exists a complete and deterministic automaton $\mathcal{A}'$ which recognizes the same language. If $\mathcal{A}' = (\Sigma, \mathcal{Q}, \mathcal{Q}_f, \mathcal{E}, \mathcal{R})$ then $\mathcal{B} = (\Sigma, \mathcal{Q}, \mathcal{Q} - \mathcal{Q}_f, \mathcal{E}, \mathcal{R})$ is the complement of $\mathcal{A}'$, (and of $\mathcal{A}$).

- To prove that AC is closed under union and intersection, we prove that the class is closed under product.

We consider $\mathcal{A}$ and $\mathcal{B}$ two automata in AC. $\mathcal{A}'$ and $\mathcal{B}'$ are complete and deterministic automata which recognize respectively L($\mathcal{A}$) and L($\mathcal{B}$).
$\mathcal{A}' = (\Sigma, \mathcal{Q}_A, \mathcal{Q}_A^f, \mathcal{E}_A, \mathcal{R}_A)$ and $\mathcal{B}' = (\Sigma, \mathcal{Q}_B, \mathcal{Q}_B^f, \mathcal{E}_B, \mathcal{R}_B)$
Consider the product $\mathcal{A}' \times \mathcal{B}' = (\Sigma, \mathcal{Q}_A \times \mathcal{Q}_B, F, \mathcal{E}_A \cup \mathcal{E}_B, \mathcal{R})$.

The set $\mathcal{R}$ is constructed as follows :

$\forall f \in \Sigma_n, f((q_1, q'_1)(x_1), \cdots, (q_n, q'_n)(x_n)) \xrightarrow{c \wedge c'} (q, q')(f(x_1, \cdots, x_n)) \in \mathcal{R}$ if and only if $f(q_1(x_1), \cdots, q_n(x_n)) \xrightarrow{c} q(f(x_1, \cdots, x_n)) \in \mathcal{R}_A$ and

$f(q'_1(x_1), \cdots, q'_n(x_n)) \xrightarrow{c'} q'(f(x_1, \cdots, x_n)) \in \mathcal{R}_B$ with $(c \wedge c')$ satisfiable.

We have built a complete and deterministic automaton in AC.

Then $\mathcal{A} \cup \mathcal{B}$ is the automaton $\mathcal{A}' \times \mathcal{B}'$ where $F = (\mathcal{Q}_A^f \times \mathcal{Q}_B) \cup (\mathcal{Q}_A \times \mathcal{Q}_B^f)$ and $\mathcal{A} \cap \mathcal{B}$ is the automaton $\mathcal{A}' \times \mathcal{B}'$ where $F = \mathcal{Q}_A^f \times \mathcal{Q}_B^f$.

Boolean closure is effective when the determinization algorithm can be applied, i.e. when the satisfiability of constraints is decidable. □

Unfortunately, in general, emptiness is not decidable in AC, even for encompassment constraints (see J. Mongy [18]). So, in order to obtain the decidability of emptiness, we must consider the following more restricted class.

4 Encompassment automata

Now we define a subclass of automata with constraints, named encompassment automata, which permit to solve first order formulas built up with predicates $encomp_t$ ($t \in T_\Sigma(\mathcal{X})$). Basic constraints are matching and we wish to bound the number of basic constraints which have been tested and satisfied along the paths. Therefore, we associate with any automaton a function ω from $\mathcal{Q}$ (the set of states) to a finite lattice T.

REMARK : encompassment automata must be "without or". Indeed, in a rule

$$f(q_1(x_1), \cdots, q_n(x_n)) \xrightarrow{c_1 \vee c_2} q(f(x_1, \cdots, x_n))$$

we cannot know whether the rule is applied by satisfaction of c_1 or satisfaction of c_2. Furthermore, for the same reason, the automaton cannot contain rules differing only by constraints. The bounded number of encompassment constraints along every path yields emptiness decidability.

In the following, EA denote the class of encompassment automata.

4.1 Definition and properties of Encompassment Automata

Definition 4.9 *If c is the conjunction of $l_1, \cdots, l_n$ elements in $\mathcal{E} \oplus \mathcal{E}_-$, $\mathcal{E}(c)$ denotes the set of positive properties of c, i.e. $\mathcal{E}(c) = \{l_1 \cdots l_n\} \cap \mathcal{E}$. A rule with constraint c is called positive if $\mathcal{E}(c)$ is not empty.*

We define the class EA, which is a subclass of automata with constraints.

Definition 4.10 *An encompassment automaton $\mathcal{A}$ is a 6-uple $(\Sigma, \mathcal{Q}, \mathcal{Q}_f, \mathcal{E}, \mathcal{R}, \omega)$ such that*

- *$\mathcal{E}$ is a finite set of constraints of form $match_t(x)$ where t is in $T_\Sigma(\mathcal{X})$.*
  $T_\Sigma \models match_t(s)$ iff s matches t. We note $E(\mathcal{E})$ the set $\{t_i | match_{t_i}(x) \in \mathcal{E}\}$.

- *ω is a function from $\mathcal{Q}$ to a finite lattice T such that :*
  for any rule in $\mathcal{R}$ $f(q_1(x_1), \cdots, q_n(x_n)) \xrightarrow{c} q(f(x_1, \cdots, x_n))$,
  $\omega(q)$ is an upper bound of $\{\omega(q_1), \cdots, \omega(q_n)\}$
  $\omega(q)$ is a strict upper bound of $\{\omega(q_1), \cdots, \omega(q_n)\}$ if the rule is positive.

Intuitively, ω is used to bound the number of tests of basic constraints along every path.

In the following, $lub(S)$ is the least upper bound of the set S.

EXAMPLE 1 : $T = [0, \nu + 1]$. We define a function from Q to T which bounds the number of basic constraints on every path.
As we can always suppose that the rules with constant left-hand side contain no constraints,
for any rule $a \rightarrow q(a)$ where a is a constant, $\omega(q)$ is 0.
For any rule $f(q_1(x_1), \cdots, q_n(x_n)) \xrightarrow{c} q(f(x_1, \cdots, x_n))$ where c contains k positive constraints, i.e. $c = \text{match}_{t_1} \wedge \cdots \wedge \text{match}_{t_k} \wedge \neg\text{match}_{t_{k+1}} \wedge \cdots \wedge \neg\text{match}_{t_p}$,
if $lub\{\omega(q_1), \cdots, \omega(q_n)\} + k \leq \nu$ then $\omega(q) = lub\{\omega(q_1), \cdots, \omega(q_n)\} + k$ else $\omega(q) = \nu + 1$.
There is no final state q such that $\omega(q) = \nu + 1$. In other words, $\omega(q) = \nu + 1$ means that q is a dead state.

EXAMPLE 2 : The previous example enables to bound the number of positive constraints along every path. We give now a function ω which globally bounds the number of positive constraints :
for any rule $a \rightarrow q(a)$ where a is a constant, $\omega(q)$ is 0,
for any rule $f(q_1(x_1), \cdots, q_n(x_n)) \xrightarrow{c} q(f(x_1, \cdots, x_n))$ with k positive constraints,
if $\omega(q_1) + \cdots + \omega(q_n) + k \leq \nu$ then $\omega(q) = \omega(q_1) + \cdots + \omega(q_n) + k$ else $\omega(q) = \nu + 1$.
There is no final state q such that $\omega(q) = \nu + 1$.

The constraint encomp_t, where t is a term in $T_\Sigma(\mathcal{X})$ can be broken up into a "recognizable" part (in the meaning of usual finite tree automata) and a "non recognizable" part. For example, if we consider the term $t = b(a(x,y), b(x,y))$, $T_\Sigma \models \text{encomp}_t(s)$ if and only if

- a subterm s' of s is an instance of $b(a(x_1, x_2), b(x_3, x_4))$

- $s'|_{1.1} = s'|_{2.1}$ and $s'|_{1.2} = s'|_{2.2}$

 ($s'|_p$ represents the subterm of s' at position p).

The first point can be expressed by a usual bottom-up tree automaton.
So the rule $f(q_1(x_1), \cdots, q_n(x_n)) \xrightarrow{c} q(f(x_1, \cdots, x_n))$ with $c = \text{match}_{t_1} \wedge \cdots \wedge \text{match}_{t_k} \wedge \neg\text{match}_{t_{k+1}} \wedge \cdots \wedge \neg\text{match}_{t_l}$ can be translated by $f(q'_1(x_1), \cdots, q'_n(x_n)) \xrightarrow{c'} q'(f(x_1, \cdots, x_n))$ where $c' = (e_1, \cdots, e_l)$ is a l-uple of conditions of differences or equalities between positions.

EXAMPLE : Let us consider the constraint $\text{match}_t \wedge \neg\text{match}_{t'}$ with $t \equiv b(a(x,y), b(y,x))$ and $t' \equiv b(a(x', x'), b(y', y'))$.
This constraint will be associated with the rule $b(q_1(x_1), q_2(x_2)) \xrightarrow{(e_1, e_2)} q(b(x_1, x_2))$ where q_1 means an "a" has been encountered at the previous step and q_2 means an "b" has been encountered at the previous step.
$(e_1, e_2) = (1.1 = 2.2 \wedge 1.2 = 2.1, 1.1 \neq 1.2 \vee 2.1 \neq 2.2)$.
It is obvious that an encompassment automaton where E contains only linear terms is equivalent to a usual bottom-up tree automaton.

Lemma 4.11 *For every encompassment automaton* $\mathcal{A} = (\Sigma, \mathcal{Q}, \mathcal{Q}_f, \mathcal{E}, \mathcal{R}, \omega)$ *there exists a complete encompassment automaton* $\mathcal{A}' = (\Sigma, \mathcal{Q}', \mathcal{Q}_f, \mathcal{E}, \mathcal{R}', \omega')$ *such that for all q in $\mathcal{Q}$ $L_q(\mathcal{A}) = L_q(\mathcal{A}')$.*

PROOF : Lemma 3.6 construction is valid, if we add to T a strict upper bound $\omega(\text{Dead})$. Points 1.a and 2 of the algorithm contain no conditional rules, and point 1.b generates a rule such that $\omega(\text{Dead})$ is an upper bound of $\{\omega(q_1), \cdots, \omega(q_n)\}$. □

Lemma 4.12 *For every complete encompassment automaton* $\mathcal{A} = (\Sigma, \mathcal{Q}, \mathcal{Q}_f, \mathcal{E}, \mathcal{R}, \omega)$ *there is a complete and deterministic encompassment automaton* $\mathcal{A}' = (\Sigma, \mathcal{Q}', \mathcal{Q}'_f, \mathcal{E}, \mathcal{R}', \omega')$ *such that $L(\mathcal{A}) = L(\mathcal{A}')$.*

PROOF : We modify the construction of lemma 3.7 :
Using the notations of the proof of lemma 3.7, we introduce the rules

$$f((k_1,i_1)(x_1), \cdots, (k_n,i_n)(x_n)) \xrightarrow{\gamma} (k_\gamma, i_\gamma)(f(x_1, \cdots, x_n))$$

We define ω' by : $\omega'(k_\gamma) = lub\{\omega(s_m) \mid s_m \in k_\gamma\}$ □

Lemma 4.13 *The class of encompassment automata is effectively closed under boolean operation : union, intersection and complementation.*

PROOF : The proof of lemma 3.8 must be modified as follows :

- Let $\mathcal{A}$ be a complete and deterministic automaton. As for automata without constraints, we build the complement of $\mathcal{A}$ changing the set of final states. This does not modify ω.

- EA is closed under product.

    Let $\mathcal{A}$ and $\mathcal{B}$ be two complete and deterministic automata of EA.

    Consider $\mathcal{A} = (\Sigma, \mathcal{Q}_A, \mathcal{Q}_A^f, \mathcal{E}_A, \mathcal{R}_A, \omega_A)$ and $\mathcal{B} = (\Sigma, \mathcal{Q}_B, \mathcal{Q}_B^f, \mathcal{E}_B, \mathcal{R}_B, \omega_B)$

    The product $\mathcal{A} \times \mathcal{B} = (\Sigma, \mathcal{Q}_A \times \mathcal{Q}_B, F, \mathcal{E}_A \cup \mathcal{E}_B, \mathcal{R}, \omega)$ built as in the proof of lemma 3.8 is in EA.

    Indeed, we define ω by : $\forall (q, q') \in \mathcal{Q}_A \times \mathcal{Q}_B, \omega((q, q')) = (\omega_A(q), \omega_B(q'))$. The order is the product of $\preceq_A$ and $\preceq_B$: $(q_1, q_2) \preceq (q'_1, q'_2) \Leftrightarrow (q_1 \preceq_A q'_1 \wedge q_2 \preceq_B q'_2)$.

    So $\mathcal{A} \cup \mathcal{B}$ is the automaton $\mathcal{A}' \times \mathcal{B}'$ where $F = (\mathcal{Q}_A^f \times \mathcal{Q}_B) \cup (\mathcal{Q}_A \times \mathcal{Q}_B^f)$.

    Moreover $\mathcal{A} \cap \mathcal{B}$ is the automaton $\mathcal{A}' \times \mathcal{B}'$ where $F = \mathcal{Q}_A^f \times \mathcal{Q}_B^f$. □

4.2 Emptiness decision

Theorem 4.14 (Plaisted's Theorem for tree automata) *It is decidable whether the language recognized by an encompassment automaton is empty.*

SKETCH OF PROOF : The idea is to generalize the proof of D. Plaisted [19]. We use automata vocabulary and can memorize finite informations in the states. The main extension of the Plaisted's proof is that we must do "parallel pumping", which is sketched in the (simplified) lemma 4.15. A complete (but tedious) proof is in french in [3] and an english detailled proof will be soon available as technical report.

Lemma 4.15 (pumping lemma) *Let $\mathcal{A}$ be an encompassment automaton. There exists B in $\mathbb{N}$ such that if $\mathcal{A}$ recognizes a term t' of depth $d' > B$, it recognizes a term t of depth $d \leq B$, deduced from t' by pumping.*

We define an order relation $\prec_{\mathcal{A}}$ on T_{Σ} named *pumping order*. Intuitively, $t \prec_{\mathcal{A}} t'$ if and only if t is obtained from t' by pumping. When t' is pumped, the equalities checked in t' by $\mathcal{A}$ must be preserved. So we have to associate with each position α in t' the set of all positions $\alpha_1, \cdots, \alpha_p$ linked to α by the equality tests $t'|_\alpha = t'|_{\alpha_1} = \cdots = t'|_{\alpha_p}$ in order to preserve these equalities.

Therefore we define an equivalence relation between positions in t, for all t in T_Σ.

$$\alpha \simeq \beta \Leftrightarrow t|_\alpha = t|_\beta \text{ is an equality checked by } \mathcal{A}.$$

REMARK : $\simeq$ is get by transitivity of the relation $t|_{\alpha.\delta} \sim t|_{\beta.\delta}$ if a positive constraint of a rule checks that $t|_\alpha = t|_\beta$

Note that the cardinality of each equivalence class is bounded. Moreover, if $|\rho|$ denotes the length of position ρ, for all equivalent positions ρ and ρ', $\bigl||\rho| - |\rho'|\bigr|$ is bounded.

Now, we can formally define the pumping order (see figure 1) :

Definition 4.16 $\prec_{\mathcal{A}}$ *is a strict partial order on T_Σ, named pumping order and define by $t \prec_{\mathcal{A}} t'$ if and only if*

- *$t' = u[t'|_{\gamma_1}, \cdots, t'|_{\gamma_p}]$ such that $\{\gamma_1, \cdots, \gamma_p\}$ is an equivalence class for $\simeq$.*

- *there exist $\gamma'_1, \cdots, \gamma'_p$ equivalent positions in t' such that for all i in $[1,p]$, γ_i prefix of γ'_i and $state(t',\gamma_i) = state(t',\gamma'_i)$.*

- *$t = u[t|_{\gamma_1}, \cdots, t|_{\gamma_p}]$ with for all i in $[1,p]$, $t|_{\gamma_i} = t'|_{\gamma'_i}$ and for all position α in u, $state(t,\alpha) = state(t',\alpha)$.*

To prove the pumping lemma, we bound the depth of every minimal term for $\prec_{\mathcal{A}}$. The computation of the bound of a minimal term for this order relation is based on the proof of D. Plaisted [19]. Given a term t minimal for the pumping order, we associate with every path θ in t a tree T_θ. We construct this tree T_θ from nodes of the path θ. We estimate

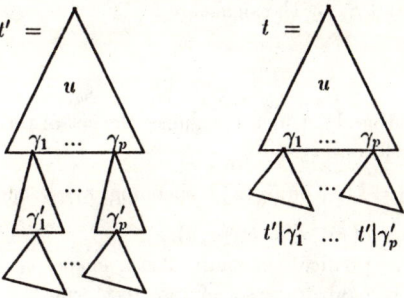

Figure 1: pumping order : $t \prec_{\mathcal{A}} t'$

the size (number of nodes) in T_0. An upper bound on the size of this tree gives an upper bound on the depth of t. The order of magnitude of the bound B is $K^D.\text{Card}(Q).K^{D.K}.H^K$ where K is the maximum number of equalities tested by one rule, D is the cardinality of $E(\mathcal{E})$ and H is the maximum depth of a term in $E(\mathcal{E})$.

This bound B gives a decision algorithm for emptiness. Given an automaton $\mathcal{A}$, $L(\mathcal{A})$ is empty if and only if no term whose depth is lower than B is recognized by $\mathcal{A}$. □

4.3 Encompassment theory

We precise how to build sorted encompassment formulas.

Definition 4.17 *Let Σ be an order-sorted signature.*
- *The set of terms is $T_\Sigma \cup \mathcal{V}$. The elements of $\mathcal{V}$ are called variables.*
- *Encompassment predicates are of the form $\text{encomp}_t(x)$ where t is in $T_\Sigma(\mathcal{X})$.*
- *Sort predicates are of the form $x \in S$ where S is a sort.*
- *Encompassment formulas are generated from terms, sort and encompassment predicates by boolean connectives and quantifiers $\exists, \forall$ (ranging over variables interpreted in T_Σ).*
The sorted first order theory of encompassment is the set of true sentences.

Theorem 4.18 *The sorted first order theory of encompassment is decidable.*

SKETCH OF PROOF : With each predicate we can effectively associate an encompassment automaton which recognizes the set of terms satisfying this predicate (a usual bottom-up automaton is associated with a sort predicate [4]). Moreover, the class of encompassment automata is effectively closed by boolean operations and emptiness is decidable. These properties yield the decidability of the encompassment theory, using old resolution technics (We thank S. Grigorieff who indicates that this result was presented by L. Lowenheim [15] (1915), T. Skolem [21] (1919) and H. Behmann [1] (1922)). □

Appendix

We illustrate the resolution of a formula with an example.
We define two sorts S_1 and S_2 by the signature :
$$\begin{cases} 0: & \to S_1 \\ a: & S_1 \to S_1 \\ b: & S_1 \times S_1 \to S_2,\ S_1 \times S_2 \to S_2,\ S_2 \times S_1 \to S_2,\ S_2 \times S_2 \to S_2 \end{cases}$$
We can build an automaton $\mathcal{A}_{S_2}$ which recognizes the set of terms of sort S_2.
Now, let us consider the formula ϕ :

$$\forall x\ x \in S_2 \Rightarrow ((\text{encomp}_{t_1}(x) \wedge \text{encomp}_{t_2}(x)) \Rightarrow \text{encomp}_{t_3}(x))$$

with $t_1 = b(x,x)$, $t_2 = b(x,a(x))$, $t_3 = b(b(x,y),z)$.
We can associate with the predicate encomp_{t_2} (resp. encomp_{t_1}, encomp_{t_3}) an automaton $\mathcal{A}_2$ (resp $\mathcal{A}_1$, $\mathcal{A}_3$) which recognizes the set of unsorted terms satisfying this predicate. We detail below this automaton : $\mathcal{A}_2 = (\Sigma, \{q_0, q_a, q_b, q_f\}, \{q_f\}, \{\text{match}_{t_2}\}, \mathcal{R}, \omega)$ such that $\omega : \mathcal{Q} \to [0,1]$, $\omega(q_0) = 0$, $\omega(q_a) = 0$, $\omega(q_b) = 0$, $\omega(q_f) = 1$ and $\mathcal{R}$ is the set of rules :

$0 \to q_0(0) \quad a(q_0(x)) \to q_a(a(x)) \quad a(q_a(x)) \to q_a(a(x))$
$a(q_b(x)) \to q_a(a(x)) \quad a(q_f(x)) \to q_f(a(x))$
$b(q_a(x), q_a(y)) \xrightarrow{1=2,1} q_f(b(x,y)) \quad b(q_a(x), q_a(y)) \xrightarrow{1\neq 2,1} q_b(b(x,y))$
$b(q_0(x), q_a(y)) \xrightarrow{1=2,1} q_f(b(x,y)) \quad b(q_0(x), q_a(y)) \xrightarrow{1\neq 2,1} q_b(b(x,y))$
$b(q_b(x), q_a(y)) \xrightarrow{1=2,1} q_f(b(x,y)) \quad b(q_b(x), q_a(y)) \xrightarrow{1\neq 2,1} q_b(b(x,y))$

$b(q_a(x), q_b(y)) \to q_b(b(x,y)) \quad b(q_b(x), q_b(y)) \to q_b(b(x,y)) \quad b(q_0(x), q_b(y)) \to q_b(b(x,y))$
$b(q_0(x), q_0(y)) \to q_b(b(x,y)) \quad b(q_a(x), q_0(y)) \to q_b(b(x,y)) \quad b(q_b(x), q_0(y)) \to q_b(b(x,y))$
$b(q_f(x), q_0(y)) \to q_f(b(x,y)) \quad b(q_f(x), q_a(y)) \to q_f(b(x,y)) \quad b(q_f(x), q_b(y)) \to q_f(b(x,y))$
$b(q_0(x), q_f(y)) \to q_f(b(x,y)) \quad b(q_a(x), q_f(y)) \to q_f(b(x,y)) \quad b(q_b(x), q_f(y)) \to q_f(b(x,y))$
$b(q_f(x), q_f(y)) \to q_f(b(x,y))$

The previous formula ϕ is equivalent to :

$\forall x \; x \notin S_2 \lor \neg(\text{encomp}_{t_1}(x) \land \text{encomp}_{t_2}(x)) \lor \text{encomp}_{t_3}(x)$

We complement ϕ in order to obtain an existential formula :

$$\neg \phi = \exists x \; x \in S_2 \land \text{encomp}_{t_1}(x) \land \text{encomp}_{t_2}(x) \land \neg \text{encomp}_{t_3}(x)$$

The formula ϕ is true if and only if the language recognized by $\mathcal{A}_{S_2} \cap \mathcal{A}_1 \cap \mathcal{A}_2 \cap \bar{\mathcal{A}}_3$ is empty.

Acknowledgments

We acknowledge the referees and H. Comon for useful comments on a preliminary version of this paper.

References

[1] H. Behmann. Beitrage zur algebra der logik, insbesondere zum entscheidungsproblem. *Mathematische Annalen*, 86:163–229, 1922.

[2] B. Bogaert and S. Tison. Equality and disequality constraints on direct subterms in tree automata. In *Lecture Notes in Computer Science Vol 577*, pages 161–171. Symposium on theoretical aspects of computer science, 1992.

[3] A.C. Caron. *Structures et Décision en Réécriture*. PhD thesis, Laboratoire d'Informatique Fondamentale de Lille, Université des Sciences et Technologies de Lille, Villeneuve d'Ascq, France, February 1993.

[4] H. Comon. Equational formulas on order-sorted algebras. In *Proceedings of ICALP'90*, pages 674–688, 1990.

[5] H. Comon. Solving symbolic ordering constraints. *International Journal on Foundations of Computer Science*, 1(4), 1990.

[6] H. Comon and P. Lescanne. Equational problems and disunification. *J. Symbolic Computation*, 7:371–425, 1989.

[7] M. Dauchet and S. Tison. The theory of ground rewrite systems is decidable. Rapport interne I.T. 182, Laboratoire d'Informatique Fondamentale de Lille, Université des Sciences et Technologies de Lille, Villeneuve d'Ascq, France, 1990.

[8] N. Dershowitz. Computing with rewrite systems. *Information and Control*, 65:122–157, 1985.

[9] C. Frougny and J. Sakarovitch. Synchronized rational relations of finite and infinite words. Litp 92.26, Laboratoire d'Informatique Théorique et Programmation, Paris, France, 1992.

[10] C. Hoffmann and M. O'Donnell. Pattern matching in trees. *J.A.C.M.*, 29(1):68–95, January 1982.

[11] J.-P. Jouannaud and C. Kirchner. Solving equations in abstract algebras : A rule-based survey of unification. In J.-L. Lassez and G. Plotkin, editors, *Computational Logic : Essays in Honor of Alan Robinson*. MIT-Press, 1991.

[12] J.P. Jouannaud and E. Kounalis. Automatic proofs by induction in equational theories without constructors. In *Proc. 1st IEEE Symp. Logic in Computer Science*, June 1986.

[13] D. Kapur, P. Narendran, and H. Zhang. On sufficient-completeness and related properties of term rewriting systems. *Acta Informatica*, 24:395–415, 1987.

[14] G. A. Kucherov. On relationship between term rewriting systems and regular tree languages. In *Lecture Notes in Computer Science Vol 488*, pages 299–311. Rewriting Techniques and Applications, April 1991.

[15] L. Lowenheim. Uber moglichkeit im relativkalkul. *Mathematische Annalen*, 76:447–470, 1915. English traduction in "From Frege to Godel". Van Heijenoort editor.

[16] M.J. Maher. Complete axomatizations of the algebras of finite, rational and infinite trees. In *Proc. 3rd IEEE Symposium of Logic in Computer Science*, pages 348–357, July 1988.

[17] A. Mal'cev. On the elementary theories of locally free algebras. Technical report, Soviet Math. Doklady, 1961.

[18] J. Mongy. *Transformation de noyaux reconnaissables d'arbres. Forêts RATEG*. PhD thesis, Laboratoire d'Informatique Fondamentale de Lille, Université des Sciences et Technologies de Lille, Villeneuve d'Ascq, France, 1981.

[19] D.A. Plaisted. Semantic confluence tests and completion method. *Information and Control*, 65:182–215, 1985.

[20] M.O. Rabin. *Handbook of Mathematical Logic*, chapter Decidable theories, pages 595–627. North Holland, 1977.

[21] T. Skolem. Untersuchungen uber die axiome des klassenkalkuls und uber produktations und summationsprobleme, welche gewisse klassen von aussagen betreffen. Skr. Vidensk. Kristiana 3, 1919.

[22] W. Thomas. *Handbook of Theoretical Computer Science*, volume B, chapter Automata on Infinite Objects, pages 134–191. Elsevier, 1990.

[23] R. Treinen. A new method for undecidability proofs of first order theories. Technical Report A 09/90, Universität des Saarlandes, Saarbrücken, May 1990.

[24] K.N. Venkataraman. Decidability of the purely existential fragment of the theory of term algebras. *Journal of the ACM*, 34(2):492–510, 1987.

RECURSIVELY DEFINED TREE TRANSDUCTIONS

Jean-Claude Raoult
IRISA, Campus de Beaulieu
F-35042 RENNES
net: raoult@irisa.fr

Abstract: we give an equational definition for relations over trees, show that they can be described by rational expressions and give sufficient restrictions on the generated relations to ensure the rationality of their domain and range, and their stability under inverse, composition and substitution. We get in this way "rational tree transductions" extending to the case of trees the well known rational transductions over words.

I. INTRODUCTION

Rational word transductions are relations — not just functions — satisfying several pleasant properties:
1) The identity, the inverse and the (associative) composition of rational transductions are again rational transductions.
2) The image of a rational language is again a rational language. Together with (1), this implies that the domain and range of a rational transduction are rational.
3) Several properties are decidable. In particular, given a pair (x, y) and a given a rational transduction R, the membership relation $(x, y) \in R$ is decidable.
4) They accept several equivalent definitions: finite automata with output, grammars of pairs of words, rational subsets of the product of two free monoids, bimorphisms.

All these properties have been known long ago, and are used in several areas of computer science. The definitions by grammars, or by bimorphisms, are adapted to proving further properties. The definition using automata is used in every lexical analyser.

In the case of trees, the situation is far from being so neat (see Raoult [1991]). Surely rational sets of trees are well defined and well-known (see Gecseg & Steinby [1984]). Typical instances of such sets are the parse trees of a context-free grammar, and a few transformations on the parse tree can be realized "rationally". Also, plenty of definitions exist for "rational" tree transformations. To justify the term "rational", most of them use finite mechanical devices: finite automata with output. Engelfriet [1975] sorts some of them into top-down or bottom-up tree transformations which are incomparable in power, and are not preserved by composition. Dauchet et alii [1987] define ground tree transducers by the action of two finite automata, one in the domain and one in the range, and they show their stability under a number of operations, composition and iteration included. But their transformations are too particular to coincide with word transductions, wheN they are restricted to the word case. A nice generalization by Dauchet & Tison [1992] of these ground tree transducers, in which a finite automaton runs on a superposition of two trees, does extend the word transductions to the case of trees. Nevertheless they leave out cases like the following, given in Arnold & Dauchet [1982], which can be obtained with an automaton with output:

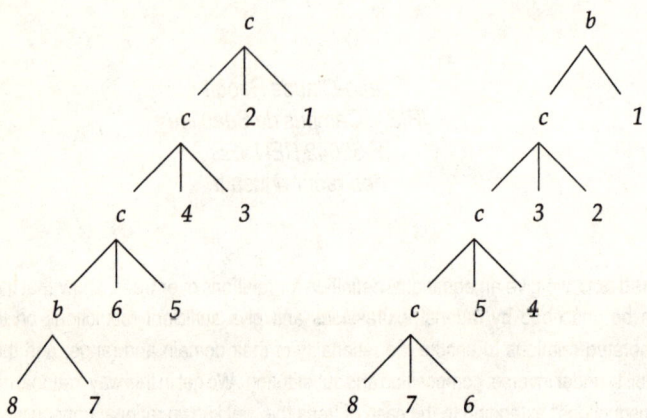

Figure 1. Equal numbers indicate identical subtrees.

Another definition due to Arnold & Dauchet [1982] uses bimorphisms and coincides with non-erasing transductions when restricted to words. For instance, the transformation of figure 1 is definable in this way. Actually, Dauchet defines in his thesis [1977] a very general form of tree transduction using two bimorphisms, which coincides with rational word transductions when restricted to words. We propose here yet another definition, the right one of course, which is akin to Schreiber's syntax connected transductions [1975]. Our definition mimics the definition of word transductions, as in Berstel [1979] for instance. The basic idea is to define tree relations recursively.

Example 1: the relation, say $R(x, y)$, depicted in figure 1 can be defined as being the equality $I(x, y)$ or be true if $y = b(u, v)$, and (x, u, v) satisfy relation L. Now relation $L(x, y, z)$ itself is true if either $x = c(u, v, w)$, $y = c(u', v', w')$ and $w = z$ and $v = w'$ and (u, u', v') satisfy L, or $x = b(u, v)$, and $u = y$ and $v = z$:

$$R(x, y) \Leftarrow x = y \vee y = b(u, v) \wedge L(x, u, v)$$
$$L(x, y, z) \Leftarrow x = c(u, v, w) \wedge y = c(u', v', w') \wedge L(u, u', v') \wedge v = w' \wedge w = z$$
$$\vee x = b(u, v) \wedge u = y \wedge v = z$$

These formulæ can be rewritten in a way looking more algebraic (I denotes the identity):

$$R = I + (x, b(u, v)), Lxuv$$
$$L = (c(u, v, w), c(u', v', w'), z), Luu'v', Ivw', Iwz$$

The transductions that we get are the same as in Dauchet's thesis, but are defined more algebraically and, we hope, more intuitively.

Section II reviews the notation for trees, tuples of trees and graftings and defines the topic of the paper: relations defined by grammars. In section III, we show that these relations can also be described by rational (regular) expressions, and derive a pumping lemma. Nevertheless, unary relations are more general that rational forests, and in section IV, we give a condition for a relation to extend the notion of a rational set of trees. In section V another condition ensures the stability under composition in such a way that the corresponding "rational transductions" extend the rational transductions of words.

II. TREES, TUPLES, RELATIONS AND GRAMMARS

We recall here basic notations. Given a set F of "function symbols", the languages $T(F) = FT^*$ of trees or terms over F and $T^* = \{\varepsilon\} + TT^*$ of lists of trees over F are defined mutually recursively. Note that a list of trees may be empty (ε) but that a tree may not.

If the alphabet is graded by an arity $\rho : F \to \mathbf{N}$, all trees $t = f(t_1, \ldots, t_n)$ are subject to the restriction that the out-degree n of their root equals the arity of the label f: $\rho(f) = n$. Some labels of arity zero are called variables. In this paper, these variables will denote positions, or occurrences, in the list; thus they shall occur only once: the tuple is linear. The tuples of trees will be considered up to variable renaming.

Notation. Suppose that v contains the n variables of a list $x = x_1, \ldots, x_n$ and that w is a n-tuple of trees. We denote by $v \cdot_x w$ the tuple of trees v in which x_1 has been replaced by $w_1, \ldots, x_n$ by w_n and where the variables of w have been indexed by x to avoid conflicts of names with the remaining variables in v.

The result is also $v\sigma$ where $\sigma = [w'_1/x_1, \ldots, w'_n/x_n]$ where $w'_1 \ldots w'_n$ is the result of indexing the variables in w by x. Instead of explicitly indexing the variables of w, we shall assume that the variables of v and w are distinct, unless otherwise stated, even if w is equal to v itself. As a result, if v and w are linear, then $v \cdot_x w$ is linear. This condition is used to prove the following properties.

1) They are commutative: $(u \cdot_x v) \cdot_y w = (u \cdot_y w) \cdot_x v = u \cdot_{xy} (vw)$
2) They are associative: $(u \cdot_x v) \cdot_y w = u \cdot_x (v \cdot_y w)$

It is well-known also that this operation is transitive and yields an ordering called the divisibility ordering or the subsumption ordering (see for instance Dershowitz & Jouannaud [1990] p. 250), denoted by $u \leq v$.

In the sequel, the variables of a tuple will be "bound" by predicate parameters extracted from another graded set P (of predicate symbols or predicate parameters).

Definition 1. *Given a graded set F of function symbols, a set X of nullary variables and a graded set P of predicate symbols, a (parameterized) relation is a linear tuple of trees $v \in T(F, X)^*$ together with a partition of its variables in which each subset of n variables is ordered and labelled by a predicate symbol of arity n in P. In the sequel, parameterized relations will simply be called relations.*

The fact that A labels the ordered set $\{x_1, \ldots, x_n\}$ will be denoted by $Ax_1 \ldots x_n$ and the relation will be denoted by (v, S) where S is the labelled partition, also called the synchronizing set.

Relations can also be viewed as labelled (hyper)graphs: associate a vertex with each subtree different from a variable. With every subtree $t = f(t_1, \ldots, t_n)$ associate a (hyper)arc of arity $n+1$ passing through the vertices associated with $t, t_1, \ldots, t_n$ in this order and labelled by f. Finally, for all formal predicate A having (distinct) arguments $x_1, \ldots, x_n$ define arcs labelled by A and passing through $x_1, \ldots, x_n$ in this order. One may also add a "source" hyperarc passing through all vertices associated to the roots of the tuple. This correspondence is visualized in the figure below, where the graph is drawn and also written as a set of labelled hyperarcs. This last representation is close to a programme with unique assignment, in which the hyperarc $fxyz$ represents the statement $x \leftarrow f(y, z)$.

$$w = (b(b(x, y)z), b(u, b(v, w))), Axu, Ayv, Azw, \text{ or}$$
$$w = \{Rpr, bpqz, bqxy, brus, bsvw, Axu, Ayv, Azw\}$$

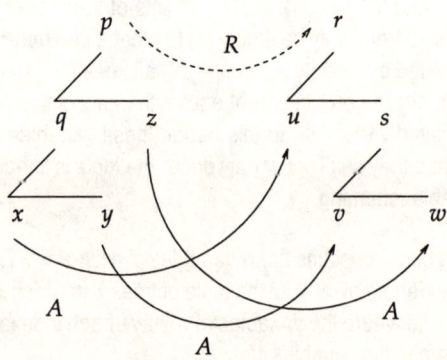

Figure 2. A relation and its associated hypergraph.

The following definition transfers the notion of a (hyper)graph grammar into the framework of relations.

Definition 2. *Given a graded set F of function symbols, a set X of nullary variables, a graded set P of predicate symbols, a production is a pair $A \rightarrow \alpha$ of a predicate symbol A of P and a relation α on n trees, where n is the arity of A; a grammar of relations is a finite set of productions.*

The grammar generates one step of derivation $\alpha \xrightarrow[G]{} \beta$ on relations:

$$\alpha = (v(x_1, \ldots, x_n), Ax_1 \ldots x_n, S)$$
$$\beta = (v(w_1, \ldots, w_n), S \cup S') = (v \cdot_{x_1 \ldots x_n} w_1 \ldots w_n, S \cup S')$$

where $A \rightarrow (w_1 \ldots w_n, S')$ is a production, and we assume here that the variables of $w_1 \ldots w_n$ are distinct (and distinct from the variables of α). It would be different if one would apply the substitution $A \rightarrow (w_1 \ldots w_n, S')$ because (1) only one occurrence of A has been replaced, and (2) there may be several productions of A in the grammar among which a choice can be made freely.

We shall borrow standard terminology and notations from language theory, like non-terminals (those predicate symbols occurring on the left-hand sides of productions), terminals (the others), derivation, derivation trees, generated language, context-free relation, etc. Derivation trees, for instance, can be defined as follows.

Definition 3. *A context-free relation is generated by a grammar, starting from some non-terminal called the axiom. A derivation tree for the grammar G, in which the synchronizing sets of all right-hand sides have been ordered arbitrarily is an ordered tree generated by the grammar*

$$\{A \to p(B_1, ..., B_k); p : A \xrightarrow[G]{} (v, B_1 x_{1,1} \ldots x_{1,n}, \ldots, B_k x_{k,1} \ldots x_{k,m})\}$$

The root of the derivation tree t is denoted by $R(t)$ and the relation yielded by t is denoted by $Y(t)$ and defined recursively:
If $t = B$ then $Y(t) = (x_1 \ldots x_n, B x_1 \ldots x_n)$
If $t = p(t_1, \ldots, t_k)$ then $Y(t) = (v \cdot_{x_{1,1} \ldots x_{1,n}} Y(t_1) \cdots \cdot_{x_{k,1} \ldots x_{k,m}} Y(t_k))$
with the above notations for p.

The synchronizing set is ordered to get an ordered derivation tree. For instance, consider the following grammar, representing a right rotation of AVL-trees.
Example 2:

$$p : A \to (b(b(x, y), z), b(u, b(v, w)), Axu, Ayv, Azw)$$
$$q : A \to (a(x), a(y), Axy)$$
$$r : A \to (e, e)$$

Then $(b(b(a(e), e), e), b(a(e), b(e, e)))$ belongs to $L(G, A)$ and its derivation tree is $p(q(r), r, r)$ (see figure 3 below).

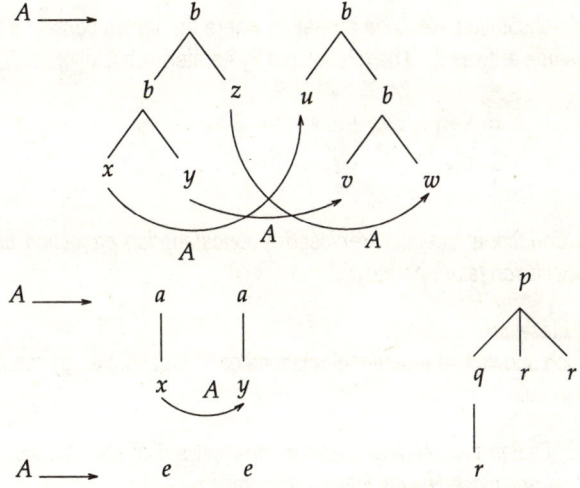

Figure 3. A grammar and a derivation tree for this grammar.

Context-free relations are recursive; it is proved by getting rid of the productions of size zero, as with words.

Proposition 4. *The membership of a given tuple of trees to a given context-free relation is decidable.*

III. EQUATIONAL RELATIONS AND REGULARITY

In free algebras, equations have solutions that can be described by grammars (see Mezei & Wright [1967] for instance, where equationality is called algebraicity); they have also regular characterizations. Regular means, as usual, preserved by union, by some sort of binary operation and by its iteration. In the case of words, this binary operation is concatenation. In the case of free algebras, an analogue of concatenation can be any given binary operator (not even associative: see Steinby [1984]).

In the case of trees, though, the operation that corresponds most closely to concatenation is tree substitution, or grafting. It is also associative, and coincides with concatenation when words are represented by trees. To study substitution, we extend the grafting notation $u \cdot_x v$ defined at the beginning of section II. Consider $(u, Xx, \ldots, Xz, S)$, where the formal predicate X does not occur in S, and $x = x_1 \ldots x_n, \ldots, z = z_1 \ldots z_n$ are $\rho(X)$-tuples of variables. Given a $\rho(X)$-ary relation L, we define $u \cdot_X L$ by

$$u \cdot_X L = \{u \cdot_{x \ldots z} (v \ldots w); v \in L, \ldots, w \in L\}.$$

This definition can be extended additively for all u in some relation L', accounting for the notation $L' \cdot_X L$: suppose that tuples in the relation L' contain a terminal relation symbol X of positive arity and define $L' \cdot_X L$ by

$$L' \cdot_X L = \bigcup_{v \in L'} v \cdot_X L$$

Regarding iterated substitution, let L be a relation where the tuples contain a formal predicate symbol X having same arity as L. The relation got by iterated substitution of L for X is defined by

$$L^{*X} = Xx_1 \ldots x_n + L + \cdots + L \cdot_X \cdots \cdot_X L + \cdots,$$
$$L^{*X} = Xx_1 \ldots x_n + L \cdot_X L^{*X}.$$

Proposition 5. *Equational relations are closed by concatenation, projection, union, substitution and iterated substitution (star operation).*

Proof: reasonably classical.

This proposition shows that equational relations can be described by rational expressions. The converse is also true.

Proposition 6. *Define the set Rat_n as the smallest set of relations containing the finite relations of arity n and closed by the following operations:*
  (i) $L \in Rat_n \wedge M \in Rat_n \Rightarrow L \cup M \in Rat_n$.

(ii) $L \in Rat_n \wedge \rho(X) = m \wedge M \in Rat_m \Rightarrow L \cdot_X M \in Rat_n$.
(iii) $L \in Rat_n \wedge \rho(X) = n \Rightarrow L^{*X} \in Rat_n$.
Then $L \in Rat_n$ if and only if $L = L(G, A)$ for some grammar G and some axiom A of arity n in G.

Proof: the "only if" part has just been proved. The "if" part is easy once one has noticed that the derivation trees for a grammar G make a rational forest, and that the correspondence between a list of trees over R and its derivation tree preserves the rational operations, defined as follows:

$$F + F' = F \cup F' \quad F^{*A} = \{A\} + F \cdot_A F + \cdots + F \cdot_A \cdots \cdot_A F + \cdots$$
$$F \cdot_A F' = \bigcup_{t \in F} t \cdot_A F' = \{A\} + F \cdot_A F^{*A}$$

The yield of a forest is defined as the set of all yields of the trees of the forest. The correspondence is given by the following lemma.

Lemma 7. *Let G be a grammar of relations and F and F' be two sets of derivation trees for G. Then*
$Y(F + F') = Y(F) + Y(F')$, *where the symbol + denotes the union,*
$Y(F \cdot_A F') = Y(F) \cdot_A Y(F')$ *and*
$Y(F^{*A}) = Y(F)^{*A}$.

Proof: the first equality is trivial by definition of $Y(F)$. It is enough to prove the second by induction on a tree t in F, and the third by induction on the iteration.

Consider now an equational relation. The set of all derivation trees is also equational generated by the associated grammar defined in section II. Therefore it is also a rational forest, described by a rational expression (cf. Gecseg & Steinby [1984]). The same expression in which the elementary trees are replaced by their yields describes all the lists of trees having a derivation tree in the forest: the whole language generated by the grammar, QED.

Therefore the equational relations are the elements of $\bigcup_n Rat_n$: they are rational in the sense of proposition 6. This result is a slight refinement of theorem 4.3 of Habel & Kreowski [1987] in the particular case of tree relations.

The correspondence between the derivation trees and their yielded relation gives also an easy pumping lemma: a criterion of non-equationality.

Lemma 8. *Given a grammar, let $M = \sum_A \rho(A)[\rho(A) - 1] \ldots [\rho(A) - k + 1]$. Then every branch $d = (v_1 \rightarrow^+ v_m)$ of a derivation tree where there is a non-terminal in v_m having arguments in k distinct trees can be split, if $|d| \geq M$, into $d = d_1 d_2 d_3$ and $|d_2| < M$ and $|d_1| < M$ or $|d_3| < M$ and d_2 can be iterated: $d_2 = (A \rightarrow^+ t_1 \ldots t_n, A x_1 \ldots x_n, S)$ with $x_{i_1} \in t_{i_1}, \ldots x_{i_k} \in t_{i_k}$.*

The proof is a simple adaptation of the classical proof over trees.

From this lemma, we can deduce for instance that the set of perfect binary trees at all depths is not equational, because X in d_2 has a fixed arity n. Yet it is generated by the following grammar for f:

$$f(x, y) = f(b(x, y), b(x, y)) + x$$

starting from $f(e, e)$. Similarly, the set of trees of the form $a^n b^n(e)$ for $n \geq 0$ not equational, with the same proof as for the non-rationality of the language $\{a^n b^n; n \geq 0\}$. But it is generated by the algebraic grammar $f(x) = a(f(b(x))) + x$ starting from $f(e)$.

IV. DESYNCHRONIZED RELATIONS

There are two reasons why a set of trees fails to be recognizable in the classical sense: unbounded synchronization on different branches, as in the forest $\{f(a^n(e), a^n(e)); n \geq 0\}$, or unbounded synchronization on the same branch, as in $\{a^n b^n(e); n > 0\}$; our grammars cannot generate the second but can generate the first.

Therefore unbounded synchronization on different branches is equational or rational in our sense. To get what is commonly named a recognizable set of trees, we shall forbid unbounded synchronization as defined by the following proposition. In this proposition, the notation $d(x)$ denotes the depth of x, i.e. the length of the path leading from the root to x.

Proposition 9. *For a grammar G, the following properties are equivalent:*
  (i) *For all bound c there exists a derivation $A \to^+ (t_1 \ldots t_n, Bx_1 \ldots x_m, S)$, two indices $r \neq s$ and two variables $x_i \in t_r$ and $x_j \in t_s$ with $d(x_i) \geq c$ and $d(x_j) \geq c$.*
  (ii) *There is a derivation $A \to^k (t_1 \ldots t_n, Ax_1 \ldots x_n, S)$, and two variables x_i and x_j having non-emtpy occurrences in t_i and t_j respectively, where $k < 2 \sum \rho(A)[\rho(A) - 1]$ and $\rho(A)$ is the arity of A.*

The proof is a simple variation on the "pigeon-holes" principle.

Definition 10. *A grammar is desynchronized when it does not satisfy the conditions of proposition 11; it satisfies the negation (D) of assertion (i): for some constant c*

$$A \to^+ (t_1 \ldots t_n, Bx_1 \ldots x_m, S) \land x_i \in t_r \land x_j \in t_s \land r \neq s \Rightarrow d(x_i) < c \lor d(x_j) < c.$$

A relation is "desynchronized" when it is generated by a desynchronized grammar. This property is decidable for the grammars: it suffices to check all derivations of lengths up to $2 \sum \rho(A)[\rho(A) - 1]$.

Corollary 11. *It is decidable whether a grammar is desynchronized or not.*

Beware that dezynchronized relations need not be generated by desynchronized grammars only.

Example 3: the relation $R = (a^*(e), b^*(e))$ is generated by the following desynchronized grammar:

$$R = AB$$
$$A = e + (a(x), Ax)$$
$$B = e + (b(x), Bx)$$

and it is subsequently desynchronized. But it is also generated by the following grammar:

$$R = ((a(x), y), Rxy) + ((x, b(y)), Rxy) + (e, e)$$

which is not desynchronized.

The next proposition uses the following lemma, which reduces the arity of non-terminals by considering the connected components of the right-hand sides. The projection of a sequence s on the indices belonging to an interval I is denoted by $\text{pr}_I(s)$.

Lemma 12. *Let G be a grammar. In a production $p : A \to \alpha$, suppose that the right-hand side is the union of $\text{pr}_I(\alpha)$ and $\text{pr}_J(\alpha)$ (possibly interleaved), where I and J are a partition of $[1, \rho(A)]$. Introduce two non-terminals P_I and P_J with the unique productions $P_I \to \text{pr}_I(\alpha)$ and $P_J \to \text{pr}_J(\alpha)$. Replace p by the production*

$$q : A \to (x_1 \ldots x_n, P_I \, \text{pr}_I(x_1 \ldots x_n), P_J \, \text{pr}_J(x_1 \ldots x_n)).$$

The new grammar is equivalent to G.

Proof: all productions of G are in the new grammar except p, which is equal to

$$A \to (x_1 \ldots x_n, P_I \, \text{pr}_I(x_1 \ldots x_n), P_J \, \text{pr}_J(x_1 \ldots x_n)) \to^2 \alpha$$

so that all terminal relations derived from G are also derived from the new grammar. Conversely, in a derivation of the new grammar yielding a parameterized relation containing no non-terminal from G, each production like q must be followed by a production of P_I and a production of P_J, and both have only one production. The result is that of p, QED.

Proposition 13. *From every desynchronized grammar G we can deduce effectively a grammar H in which all non-terminals have arity one and with each non-terminal A of G a (finite) subset $V(A)$ of N_H^r where r is the arity of A, such that $A \underset{G}{\to}^+ w$ if and only if $\alpha \underset{H}{\to}^+ w$ for some $\alpha \in V(A)$.*

Proof: in two steps. We first build an equivalent grammar in which each non-terminal has all its arguments in a single tree, by induction on the tuple $(n_k, \ldots, n_1)$ in lexicographic order, where n_i is the number of "critical" non-terminals of arity i: those having in some generated relation two arguments belonging to two distinct trees, one of them at depth at least c:

$$C(B) = \exists A \to^+ (t_1 \ldots t_n, B x_1 \ldots x_n, S) \exists ijrs(r \neq s \land x_i \in t_r \land x_j \in t_s \land d(x_i) \geq c.$$

The final step yields the grammar H starting from a grammar in which each non-terminal has all its arguments in a single tree, by repeated use of the lemma.

Proposition 14 can now be rephrased as follows: the desynchronized relations are finite unions of products of recognizable forests, thus showing its analogy with Mezei's theorem.

The following corollary is very cheap.

Corollary 14.
1) *A desynchronized relation which is unary is a rational forest.*
2) *Any projection of a dezynchronized relation is desynchronized.*
3) *The substitution of a desynchronized relation into a desynchronized relation is again desynchronized.*

Proof: The first assertion is a particular case of the definition of recognizable relation. The second assertion is clear, considering proposition 15. To prove the third consider two desynchronized grammars G and H and suppose that the non-terminals of G do not occur in the right-hand sides of H. The grammar $G + H$ is desynchronized. Indeed, suppose a cycle

$$X \to^+ (t_1 \ldots t_n, X x_1 \ldots x_n, S)$$

with two indices $i < j$ with $x_i \in t_i$ and $x_j \in t_j$. Either all non-terminals derived in this cycle are in G; but this is impossible since G is desynchronized. Or they belong to H; this is impossible for the same reason. Or some belong to G and others to H. This is impossible because there is no production of a non-terminal of H containing a non-terminal of G, QED.

V. COMPOSITION OF RELATIONS

Suppose now that two binary relations have desynchronized projections. Their composition need not have the same property, and even need not be context-free, as in the following example.

Example 4: Consider the relations generated by A and X. Here a and b have arity one, e has arity zero, so that these relations are actually relations on words.

$$A = (ax, y, az), A x y z + (e, x, y), B x y$$
$$B = (bx, by), B x y + (e, e)$$

$$X = (bx, y, bz), X x y z + (x, y, e), Y x y$$
$$Y = (ax, ay), Y x y + (e, e)$$

The relations $L(G, X) \subseteq T \times T^2$ and $L(G, A) \subseteq T^2 \times T$ have both projections desynchronized: $L(G, X) = \{(b^m a^n, a^n, b^m); n, m > 0\}$ (the constant e has not been written) and $L(G, A) = \{(a^n, b^m, a^n b^m); m, n > 0\}$. The composition is $\{(b^m a^n, a^n b^m); m, n > 0\}$ which is not equational. What happens here is that two arguments of A (and X) are allowed to occur at unbounded depths while the third argument remains at a fixed depth, remembering in some way a synchronization to come between variables at arbitrarily distant depths. Condition (C) below forbids this situation.

Proposition 15. *For all grammars, conditions (i) and (ii) below are equivalent*
(i) Each projection is desynchronized (D) and in any derivation, two arguments of a non-terminal cannot be arbitrarily deeper than a third one (C):

$$\forall A \xrightarrow{+} (t_1 \ldots t_n, S, B x_1 \ldots x_m) \forall i j p q (p \neq q \land x_i \in t_p \land x_j \in t_q$$
$$\Rightarrow d(x_i) < c \lor d(x_j) < c) \qquad (D)$$

for some constant c and for t_p and t_q in the same projection; and for some other constant b

$$A \xrightarrow{+} (t_1 \ldots t_n, S, B x_1 \ldots x_m) \land p \neq q \neq r \neq p \land x_i \in t_p \land x_j \in t_q \land x_k \in t_r$$
$$\Rightarrow d(x_i) < d(x_k) + b \lor d(x_j) < d(x_k) + b. \qquad (C)$$

(ii) *In each projection, two trees cannot be synchronized past a certain depth: for some constant a, each projection satisfies*

$$A \xrightarrow{+} (t_1 \ldots t_n, S, Bx_1 \ldots x_m) \wedge (d(x_i) \geq a) \Rightarrow \exists p \forall j (x_j \in t_p)). \qquad (T)$$

Proof: $(ii) \Rightarrow (i)$ is clear by taking $b = c = a$. We prove the converse. Suppose that some non-terminal has an argument at depth at least $b + c$ in some tree t_p. Because of the first condition, its arguments in the other trees of the same projection must be at depths less than c. Because of the second condition, they cannot be at depths less than c. Therefore, there can be no argument of this non-terminal in another tree of the same projection. This is the required result with $a = b + c$, QED.

Definition 16. *A context-free relation in $T^p \times T^q$ is called a transduction when it is generated by a grammar satisfying the equivalent conditions of proposition 16.*

The projections of such a grammar generate relations of synchronization degree one, because each projection is equivalent to a grammar in which all non-terminals have dimension one. But the grammar defined in example 1 is not equivalent to a grammar in which all non-terminals are binary. Yet, each projection is desynchronized.

Proposition 17. *Any transduction can be generated by a grammar in which the non-terminals in the right-hand sides have all their arguments in two trees at most, one in each projection.*

Hint of proof: using condition (T) of proposition 16, we consider all possible prefixes of depth at most a of the generated relations in which all variables of a tree are synchronized with all the variables of another tree only if they belong to the two projections of the relation. This is the required condition on the new grammar. Productions are inserted in the grammar when they are compatible with some such prefix. Because of condition (T), we get eventually all the derivations of the initial grammar.

Proposition 18. *Every transduction can be generated by a grammar satisfying the conditions of proposition 17, and in which all non-terminals have only one argument in the left projection (resp. in the right projection).*

Proof: technical! It consists roughly in decomposing each production of G in two steps: in the first step, only the second projection is produced. In the second step the first projection is produced one tree at a time. The resulting grammar satisfies the condition of the proposition.

We are now in a position to prove our main result.

Theorem 19. *The transductions contain the identity and are preserved by converse and composition.*

Proof: The difficult point is the stability under composition. Using proposition 18, we suppose two grammars G and H in which all non-terminals in the right-hand side have at most one argument in the second projection, respectively in the first projection. For all productions $X \longrightarrow (v, S)$ of H, all subterms s of the first tree of v and all non-terminals A of G we introduce a non-terminal

$[Aa_1 \ldots a_n, a_n = s, S|_s]$ where the last argument a_n of A is in the second projection, and where $S|_s$ is the restriction of S to the non-terminals having their first argument in s (and their only argument, recalling the assumption on the grammars):

$$S|_s = \{Xx_1 \ldots x_p \in S; x_1 \in s\}$$

and symmetrically. The productions of $[Aa_1 \ldots a_n, a_n = s, S|_s]$ are the productions of A that are compatible with s at the first argument. The idea is that grammars G and H must produce the same tree, on the second projection for G, on the first projection for H. We generate this tree by letting G and H produce alternately, according to the non-terminal which is "late". See the following examples (6 and 7).

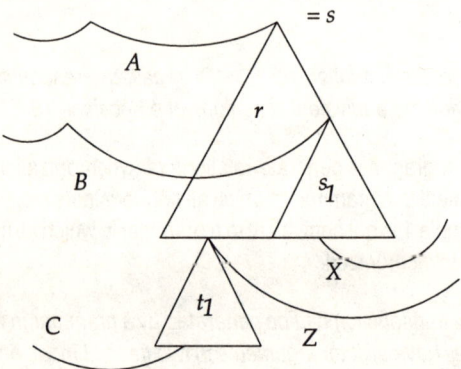

Figure 4. Originally all arguments of X and Z, except the first one, are synchronized with the arguments of A — except the last one. After A has produced $r(x, t_1)$, the arguments of X and Z are desynchronized.

Example 6: Let $I = (e, e) + (ax, ay), Ixy + (bxy, buv), Ixu, Iyv$ and compute first $I \circ I$. Start arbitrarily by letting the left I produce (here $[I \circ I]$ is a shortcut for $[Ixy, y = u, Iuv]$):

$1 : [I \circ I] \longrightarrow (e, y'), [y = e, Iyy']y'$
$2 : [I \circ I] \longrightarrow (bxy, y'), [Ixu, Iyv, x' = buv, Ix'y']xyy'$

These two introduced non-terminal have productions:

$3 : [y = e, Iyy'] \longrightarrow a$
$4 : [Ixu, Iyv, x' = buv, Ix'y'] \longrightarrow (x, y, bu'v'), [I \circ I]xu', [I \circ I]yv'$

Since production 3 is the only production of $[y = e, Iyy']$ we may apply it immediately after production 1, and similarly for production 4 which will follow immediately production 2. We get the grammar of I again, up to the name of variables.

Example 7: consider grammar $G = H$ below

$A = (c(x, y, z), b(b(u, v), w)), Axu, Iyv, Izw + I$
$R = (b(x, y), b(u, v)), Aux, Iyv$

depicted in figure 5. We shall construct the grammar of the relation $A \circ R$ having axiom $S = [Axx', x' = u, Ruu']$. Since R has a unique production, we let it produce first:

$$S = [Axx', x' = u, Ruu'] \longrightarrow (x, b(u', v')), [Axx', x' = b(u, v), Au'u, Ivv']xu'v'$$

Then A is late and we make it produce:

$$[Axx', x' = b(u, v), Au'u, Ivv'] \longrightarrow (b(x, y), u', v'), [Ixx', x' = u, Au'u]xu',$$
$$[Iyy', y' = v, Ivv']yv'$$

coming from $[A, I]$. Knowing $[I \circ I = I]$, the second relation is simply $[Iyv']$. The first one is $(I \circ A^{-1})xu' = A^{-1}xu' = Au'x$ (knowing $I \circ X = X$ for all relation X).

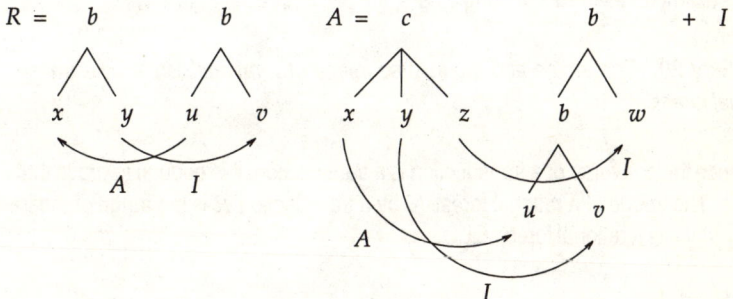

Figure 5. The grammar G.

Coming from the other production of A, we get

$$[Axx', x' = b(u, v), Au'u, Ivv'] \longrightarrow (c(x, y, z), u', v'),$$
$$[Axx', Iyy', b(x', y') = u, Au'u]xu',$$
$$[Izz', z' = v, Ivv']yv'$$

Knowing $I \circ I = I$, the second relation is simply Izv'. The variables in the first relation can be renamed by $[x/u', x'/u, u/x', v/y', u'/x, v'/y]$ and we find, knowing $I \circ I = I$ that this new non-terminal is in fact the same as in the left-hand side of the production. Call this non terminal L. Looking at its arguments, we see that the relation is $Lu'xy$. Thus the composition is complete. We get:

$$S \longrightarrow (x, b(u, v)), Lxuv$$
$$L \longrightarrow (b(x, y), u, v), Aux, Iyv$$
$$L \longrightarrow (c(x, y, z), u, v), Luxy, Izv$$

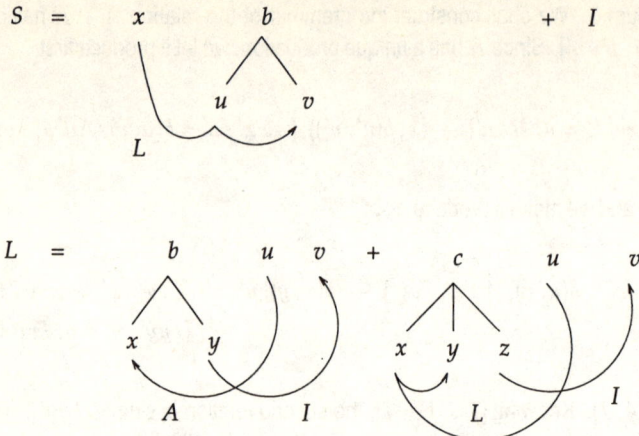

This example generates, among others, the pair of trees in Figure 1.

Corollary 20. *The image and the inverse image of a rational forest by a transduction is a rational forest.*

Proof: since the converse of a transduction is a transduction, it is enough to prove that the image is rational. The image of a rational forest K by a transduction R is the range of the transduction $\text{id}_K \circ R$, which is a rational forest, QED.

VI. CONCLUSION

The family of tree-transformations defined above is general enough to deserve the name of rational transductions:
1) restricted to words, they coincide with rational word transductions;
2) applied to rational forests, they yield rational forests;
3) the family is preserved by composition and inverse.

Note that the image of a single tree by such a rational transduction need not be a finite forest, as in Courcelle [1992], and that they are strictly more general than the ground tree transducers of Dauchet & *al.* [1985]. For instance, our transductions — like word transductions — are not preserved by iteration. An interesting problem is precisely to isolate a good subclass preserved by iteration.

Acknowledgements: many thanks to Anne Grazon who read previous versions of this paper, noted several errors, and had several suggestions.

REFERENCES

A. Arnold & M. Dauchet: Morphismes et bimorphismes d'arbres, *TCS* 20, pp. 33-93 (1982).

M. Dauchet: *Transductions de forêts, bimorphismes de magmoïdes*, Thèse, Université de Lille 1 (1977).

M. Dauchet & S. Tison: Algebraic complexity of Tree Languages. *in Tree automata and languages*, M. Nivat & A. Podelski ed., Elsevier (1992).

M. Dauchet, S. Tison, T. Heuillard, P. Lescanne: Decidability or the Confluence of Ground Term Rewriting Systems, *INRIA* report n. 675 (1987).

N. Dershowitz & J.-P. Jouannaud: Rewriting sytems, *in Handbook of Theoretical Computer Science*, Jan van Leeuwen ed., Elsevier (1990).

J. Engelfriet: Bottom-up and Top-down Tree Transformations — A Comparison, *Math. System Theory* vol. 9 n. 3, pp. 198-231 (1975).

F. Gecseg & M. Steinby: *Tree automata*, Akademiai Kiado, Budapest (1984)

A. Grazon & J.-C. Raoult: Equational sets of tree-vectors, IRISA report n. 563, Rennes (1990).

A. Habel & H.J. Kreowski: Characteristics of graph languages generated by edge replacement, *TCS* 51, pp. 81-115 (1987).

J.-C. Raoult: A survey of tree-transductions, *in Tree automata and languages*, M. Nivat & A. Podelski ed., Elsevier, pp. 311–326 (1992) (and report n. 1410 INRIA-RENNES (1991)).

P. P. Schreiber: Tree-transducers and syntax-connected transductions, *LNCS* 33, pp. 202-208 (1975).

M. Steinby: On certain algebraically defined tree transformations, *in Proc. Coll. Math. Janos Bolyai on Algebra, Combinatorics and Logic in Computer Science*, Györ (Hungary), pp. 745-764 (1983).

AC Complement Problems: Satisfiability and Negation Elimination

Maribel Fernández

LRI, Bât. 490, CNRS / Université de Paris-Sud, 91405 ORSAY Cedex, France.
E-mail: maribel@lri.fr

Abstract. We show that negation elimination is decidable for linear complement problems interpreted in $T(\mathcal{F})/_{=_{AC}}$, where AC is a set of associative and commutative axioms. For this, we present a system of rewrite rules that transforms any linear complement problem into a simple formula, and we give a test for deciding whether a simple formula is satisfiable in $T(\mathcal{F})/_{=_{AC}}$ or not. This test serves as a basis for the development of a negation elimination algorithm.

1 Introduction

Complement problems arise in several domains of Computer Science: in algebraic specifications (the problem of sufficient completeness is nothing but solving a complement problem, and also in the test of ground reducibility complement problems are used [1]), functional programming (e.g. in compilation of pattern matching [12]), logic programming (in particular for the implementation of constructive negation [10]) and constraint programming (to represent solutions of systems of equations and disequations [11]). See [10] for a survey of applications of complement problems.

Actually, as remarked in [1], the computation of term complement is a particular case of solving equational formulae. An *equational formula* is a first order formula constructed over a finite alphabet $\mathcal{F}$ of function symbols and only one relational symbol: equality. The complement problem $t/t_1, \ldots, t_n$, which represents the set of ground instances of t that are not instances of $t_1, \ldots, t_n$, is equivalent to the equational formula $\forall \vec{y} : t \neq t_1 \wedge \ldots \wedge t \neq t_n$ where $\vec{y}$ are the variables appearing in $t_1, \ldots, t_n$. An algorithm for deciding whether an equational formula has solutions in the ground term algebra $T(\mathcal{F})$, which solves in particular complement problems, is given in [1]. See also [11] for a decision algorithm for complement problems in $T(\mathcal{F})$.

Complement problems serve, in general, as a means of representing knowledge, but this way of expressing knowledge has an important drawback: it is not *explicit*. The transformation of a complement problem into an equivalent explicit representation is an interesting problem (see [10]) and it is decidable in $T(\mathcal{F})$: as shown in [11], $t/t_1, \ldots, t_n$ is equivalent to a finite disjunction of terms iff for $1 \leq i \leq n$, the most general unifier of t and t_i is linear in the variables of t.

However, as pointed out in [9] and in [8], in most applications of complement problems the involved function symbols are not free. There exists, in general, some "background knowledge" that provides information about the properties of the operators, i.e. we have to interpret complement problems in the quotient of $T(\mathcal{F})$ by an equational theory E (denoted by $T(\mathcal{F})/_{=_E}$) and not simply in $T(\mathcal{F})$.

Finding a finite explicit representation for a complement problem is the same as *eliminating negation* from the corresponding equational formula. In [3] it is shown that when E is a permutative theory of Mal'cev (a class of equational theories that includes and generalizes commutative theories), negation elimination is decidable for complement problems interpreted in $T(\mathcal{F})/_{=_E}$. Particularly important is the case where E is a set of associative and commutative axioms. AC complement problems, that is, complement problems interpreted in $T(\mathcal{F})/_{=_{AC}}$, have been a subject of active research in the last years (see e.g. [9] and [13], where two different algorithms for solving linear AC complement problems are presented). Here we study negation elimination in AC complement problems.

The necessary and sufficient condition of [11] for negation elimination in $T(\mathcal{F})$ generalizes easily to $T(\mathcal{F})/_{=_E}$ when E is a permutative theory of Mal'cev (by taking E-unifiers instead of syntactical unifiers), but in the case of $T(\mathcal{F})/_{=_{AC}}$ this condition is no longer sufficient. For example, if $+$ is AC, a is a constant and f is a free unary function symbol, we can not eliminate negation from $\forall y : x \neq y + a$. In this paper we show that negation elimination is decidable for *linear* complement problems interpreted in $T(\mathcal{F})/_{=_{AC}}$ (a complement problem $t/t_1, \ldots, t_n$ is linear if $t_1, \ldots, t_n$ do not contain multiple occurrences of the same variable). For non-linear problems the question is still open.

Our approach is inspired by [1], [3] and [2], where the algorithms for solving complement problems in $T(\mathcal{F})$ are expressed by means of *transformation rules*. As a first step towards the definition of an algorithm for negation elimination we give a set of transformation rules over equational formulae, which reduces any complement problem to a *simple formula*. Then, we show how to test whether a simple formula is satisfiable in $T(\mathcal{F})/_{=_{AC}}$ or not, that is, we give an algorithm for solving AC complement problems. Our algorithm is based on "test sets", as the algorithm presented in [9], but it yields some theoretical and practical improvements: on one hand, the use of transformation rules allows simpler proofs of completeness and termination, on the other hand, the complexity of the procedure is smaller. The algorithm for negation elimination is obtained by stepwise refinement of the decision test.

The paper is organized as follows: section 2 contains a concise description of the syntax and semantics of complement problems; section 3 deals with satisfiability and negation elimination in AC complement problems: transformation rules and decision tests; in section 4 we generalize our algorithms to a wider class of equational formulae.

2 AC Complement Problems

2.1 Syntax

We consider a finite set $\mathcal{F}$ of function symbols with fixed arity. $T(\mathcal{F})$ is the infinite set of ground terms over this alphabet. Terms are identified with finite labeled trees as usual. The symbol at position p is denoted $t(p)$, the subterm of t at position p is denoted $t|_p$ and the result of replacing $t|_p$ with u at position p in t is denoted $t[u]_p$. Variables are an infinite distinguished set $\mathcal{X}$ of symbols. The set of terms built on $\mathcal{F}$ and $\mathcal{X}$ is denoted $T(\mathcal{F}, \mathcal{X})$. $Var(t)$ denotes the set of variables appearing in t. A *substitution* σ is a mapping from a finite subset of $\mathcal{X}$, called its *domain* and written $dom(\sigma)$, into $T(\mathcal{F}, \mathcal{X})$. Such a mapping is written $\{x_1 \mapsto t_1, \ldots, x_n \mapsto t_n\}$. Substitutions are extended to $T(\mathcal{F}, \mathcal{X})$ in the usual way. The set of all substitutions is denoted by Σ. By $\sigma|_V$ we denote the restriction of a substitution σ to the set V of variables. Substitutions are used in postfix notation. By $Im(\sigma)$ we denote the set $\{x\sigma \mid x \in dom(\sigma)\}$. We say that $Im(\sigma)$ is *linear* if for all $x \in dom(\sigma)$ $x\sigma$ is linear and $Var(x_i\sigma) \cap Var(x_j\sigma) = \emptyset$ for all $x_i \neq x_j$. A substitution is *idempotent* if $\sigma\sigma = \sigma$. A substitution is *ground* if it assigns ground terms to all the variables in its domain. It is *ground for a set V of variables* if it assigns ground terms to all variables in V. We refer to [5] for missing definitions and notations.

An *equation* is an unordered pair of terms, written $s = t$. A *disequation* is also an unordered pair of terms, written $s \neq t$. An *equational formula* is a first order formula whose atoms are equations or disequations. For simplicity, we will always assume that each variable is bound at most once in a formula and cannot occur both free and bound in the same formula. We use abbreviations such as $\vec{x}, \vec{w}, \vec{y}$ to denote finite sets of variables.

We are interested in two particular fragments of equational formulae: complement problems and unification problems. A *complement problem* is either $\top$, $\bot$ or a formula $\forall \vec{y} : t \neq t_1 \wedge \ldots \wedge t \neq t_n$ where $Var(t) = \vec{x}$ is the set of free variables (or *unknowns*) of the formula and $Var(t_1) \cup \ldots \cup Var(t_n) = \vec{y}$. Without loss of generality we can assume that $Var(t_i) \cap Var(t_j) = \emptyset$ for all $i \neq j$, then a complement problem is *linear* if for $1 \leq i \leq n$, t_i does not contain multiple occurrences of the same variable. A *unification problem* is an existential formula which does not contain any disequation.

2.2 Semantics

Let ϕ be an equational formula and V the set of free variables of ϕ. The set of *solutions* of ϕ is the set of ground substitutions σ such that $V \subseteq dom(\sigma)$ and $T(\mathcal{F}) \models \phi\sigma$. It is denoted by $[\![\phi]\!]$.

We consider interpretations in quotient algebras $T(\mathcal{F})/_{=_{AC}}$ where $=_{AC}$ is the smallest congruence on $T(\mathcal{F}, \mathcal{X})$ such that it is closed by instantiation and contains the set of equations $\{f(x, y) = f(y, x)/f \in \mathcal{F}'\} \cup \{f(x, f(y, z)) = f(f(x, y), z)/f \in \mathcal{F}'\}$ for a subset $\mathcal{F}'$ of $\mathcal{F}$ containing only binary symbols. The symbols in $\mathcal{F}'$ are called AC function symbols and the theory of $=_{AC}$ is called AC equational theory. An AC-*solution* of ϕ is a ground substitution σ for the set of free variables of ϕ, such that $T(F)/_{=_{AC}} \models \phi\sigma$. The set of AC-solutions of ϕ is denoted $[\![\phi]\!]_{AC}$.

In the following the prefix AC will be used to indicate that we are considering the algebra $T(F)/_{=_{AC}}$. Two equational formulae ϕ and ψ are *equivalent* (resp. AC-*equivalent*) if they have the same set of solutions (resp. AC-solutions). This is written $\phi \sim \psi$ iff $[\![\phi]\!] = [\![\psi]\!]$ (resp. $\phi \sim_{AC} \psi$ iff $[\![\phi]\!]_{AC} = [\![\psi]\!]_{AC}$). A *generator* (resp. *AC-generator*) of a set of solutions (resp. *AC*-solutions) of ϕ is a substitution σ whose ground instances are solutions (resp. *AC*-solutions) of ϕ. A set $\mathcal{S}$ of generators is *complete* (for ϕ) if every solution of ϕ is an instance of some generator $\sigma \in \mathcal{S}$. Complete sets of *AC*-generators of unification problems will be called *complete sets of AC-unifiers*[1]. We say that σ is *more general (modulo AC) than* μ *on* V or that μ is an *AC-instance of* σ *on* V, written $\sigma \preceq^V_{AC} \mu$, if there exists a substitution ρ such that $(\sigma\rho)|_V =_{AC} \mu|_V$ (the definition of $=_{AC}$ extends to substitutions as usual). A *complete set of most general AC-unifiers* of a unification problem ϕ (abbreviated $CSU_{AC}(\phi)$) is a *minimal* complete set of *AC*-unifiers of ϕ, that is, for all $\sigma, \mu \in CSU_{AC}(\phi)$, $\sigma \preceq^{Var(\phi)}_{AC} \mu$ implies $\sigma \equiv \mu$. Any unification problem has a finite set of idempotent most general *AC*-unifiers, i.e. *AC*-unification is *decidable and finitary*.

An equational formula ϕ has the E_U (*"equivalent to unification problems"*) *property* (resp. *AC-E_U property*) if there is a unification problem ψ such that $\phi \sim \psi$ (resp. $\phi \sim_{AC} \psi$).

Example 1. The following formulae do have the E_U property and will be used to demonstrate some transformation rules in the following:
- $f(x,y) \neq g(z)$ which is equivalent to T
- $x \neq 0$ which is equivalent to $\exists z : x = s(z)$ if $\mathcal{F} = \{0, s\}$
- $\forall y : x \neq y \lor x \neq f(y)$ which is equivalent to $\forall y : x \neq f(x)$. This formula is in turn equivalent to $x \neq f(x)$ which is equivalent to T
- $\forall y : x \neq y + a$ which is equivalent to $(\exists w_1, w_2, \forall y : x = w_1 + w_2 \land w_1 + w_2 \neq y + a) \lor x = a$ if $\mathcal{F} = \{a, +\}$, and this is equivalent to $x = a$ if $+$ is AC.

We are interested in two semantic properties of *AC*-complement problems: *Satisfiability*, that is, the existence of at least one solution of the problem in the algebra $T(F)/_{=_{AC}}$, and *AC-E_U* (*AC-equivalence to unification problems*), that is, the existence of a unification problem with the same set of solutions in $T(F)/_{=_{AC}}$. Given a complement problem $P \equiv \forall \vec{y} : t \neq t_1 \land \ldots \land t \neq t_n$, if there is a unification problem (a positive formula) ϕ such that $P \sim_{AC} \phi$ we say that *negation can be eliminated from P*.

3 Satisfiability and Negation Elimination

In order to decide if a complement problem is satisfiable in $T(\mathcal{F})/_{=_{AC}}$, or, in general, if a complement problem has the *AC-E_U* property, first we use a set $R_{E_{AC}}$ of rules which transforms any complement problem into an equivalent finite set of simplified formulae, which will be called *simple*, and then we apply a decision test over simple formulae.

Although negation elimination is more general than satisfiability (in that the answer to the latter problem can be deduced from the answer to the former)

---
[1] *AC*-unifiers are usually defined as solutions of unification problems in $T(\mathcal{F}, \mathcal{X})/_{=_{AC}}$. Of course, our *AC*-unifiers are also *AC*-unifiers according to this definition.

we present first a decision test for satisfiability alone, because it illustrates the method and facilitates the presentation and proofs concerning elimination of negation. By refining this test we obtain a test for AC-E_U.

Hereafter we assume that the complement problems we want to decide are linear. However R_{EAC} can also be applied to non-linear problems. The hypothesis of linearity is needed only for the tests.

A *simple formula* is either $\top$, $\bot$ or $\exists \vec{w}, \forall \vec{y} : (\bigwedge_{k \in K} (\bigvee_{l \in L_k} w_{kl} \neq u_{kl})) \wedge x_1 = v_1 \wedge \ldots \wedge x_n = v_n$ where

1. K, L_k are finite sets of indexes,
2. w_{kl} are existentially quantified variables, $Var(v_1) \cup \ldots \cup Var(v_n) = \vec{w}$, and $x_1, \ldots, x_n$ are unknowns,
3. for all k, l, $Var(u_{kl}) \subseteq \vec{y}$; and there exist $k \in K$, $l \in L_k$ such that u_{kl} is not ground,
4. each variable appears at most once in a disjunction.

An *intermediate formula* is either $\top$, $\bot$ or $\exists \vec{w}, \forall \vec{y} : (\bigwedge_{k \in K} (\bigvee_{l \in L_k} t_{kl} \neq u_{kl})) \wedge x_1 = v_1 \wedge \ldots \wedge x_n = v_n$ where

1. K, L_k are finite sets of indexes,
2. for all k, l, $Var(t_{kl}) \cap \vec{y} = \emptyset$; $x_1, \ldots, x_n$ are unknowns and $Var(v_1) \cup \ldots \cup Var(v_n) = \vec{w}$,
3. for all k, l, $Var(u_{kl}) \subseteq \vec{y}$,
4. each variable in $\vec{y}$ appears at most once in a disjunction.

Note that a linear complement problem $\forall \vec{y} : t \neq t_1 \wedge \ldots \wedge t \neq t_n$ with unknowns $x_1, \ldots, x_m$ is equivalent to the intermediate formula $\exists \vec{w}, \forall \vec{y} : (t \neq t_1 \wedge \ldots \wedge t \neq t_n)\{x_1 \mapsto w_1, \ldots, x_m \mapsto w_m\} \wedge x_1 = w_1 \wedge \ldots \wedge x_m = w_m$. In the following *both kinds of formulas will be called complement problems*. It will be clear from the context (when not explicitly written) which is the representation we are considering.

3.1 Transformation Rules

Every intermediate formula (and then every complement problem) which is not a simple formula can be reduced using the set R_{EAC} of transformation rules (see figure 1). It contains some classical rules for eliminating universally quantified variables and reorganizing formulae (see e.g. [4] for more details) and some additional rules which are specific for complement problems modulo AC, such as the explosion and decomposition rules. The empty disjunction is denoted by $\bot$ and the empty conjunction by $\top$.

There are two decomposition rules:

- D_1 is based on the fact that if $t = u$ has an AC-unifier σ which does not instantiate the variables in t then there is no solution for the complement problem $\forall \vec{y} : t \neq u$ where $Var(u) = \vec{y}$, because for all ground substitution ρ for the variables of t, $\sigma\rho$ unifies $t\rho$ and u).

Universal Quantifier Elimination (QE)

$$(QE_1)\ \forall y:\ P \to P$$

If $y \notin Var(P)$.

$$(QE_2)\ \forall \vec{y}:\ P \wedge (y \neq t \vee d) \to \forall \vec{y}:\ P \wedge d\{y \mapsto t\}$$

If d is a disjunction of disequations, $y \in \vec{y}$ and $y \notin Var(t)$.

Reorganization (RO)

$$(RO_1)\ P \wedge \bot \to \bot \qquad (RO_2)\ P \wedge \top \to P$$
$$(RO_3)\ P \vee \top \to \top \qquad (RO_4)\ P \vee \bot \to P$$

Decomposition (D)

$$(D_1)\ \exists \vec{w}, \forall \vec{y}:\ P \wedge (t \neq u \vee d) \to \exists \vec{w}, \forall \vec{y}:\ P \wedge d$$

If $t, u \notin \mathcal{X}$ and there exists $\theta \in CSU_{AC}(t = u \wedge \neg d)$ such that θ does not instantiate existentially quantified variables.

$$(D_2)\ \exists \vec{w} \forall \vec{y}:\ P \wedge (t \neq u \vee d) \to \exists \vec{w} \forall \vec{y} \forall \vec{z}:\ P \wedge \bigwedge_{\theta \in CSU_{AC}(t=u \wedge \neg d)} (\bigvee_{w_j \in dom(\theta)} w_j \neq w_j \theta)$$

If $t, u \notin \mathcal{X}$ and D_1 does not apply. $\vec{z}$ are the variables introduced by $\theta_1, \ldots, \theta_n$.

Explosion (E)

$$(Ex_\forall)\ \exists \vec{w}, \exists w^{(i)}, \forall \vec{y}:\ P \to \bigvee_{f \in \mathcal{F}} \exists \vec{w}, \exists \overrightarrow{w'^{(i+1)}}, \forall \vec{y}:\ P\{w^{(i)} \mapsto f(\overrightarrow{w'^{(i+1)}})\}$$

If the formula is not simple, there is a disequation $w^{(i)} \neq u$ in P such that $Var(u) \cap \vec{y} \neq \emptyset$, $u \notin \mathcal{X}$ and $i \leq K$; no other rule can be applied and $\overrightarrow{w'^{(i+1)}}$ are fresh variables (different for each function symbol).

$$(Ex_{gr})\ \exists \vec{w}, \exists w^{(i)}, \forall \vec{y}:\ P \to \bigvee_{f \in \mathcal{F}} \exists \vec{w}, \exists \overrightarrow{w'^{(i+1)}}, \forall \vec{y}:\ P\{w^{(i)} \mapsto f(\overrightarrow{w'^{(i+1)}})\}$$

If the formula is not simple, there is a disequation $w^{(i)} \neq u$ in P such that u is a ground term, no other rule except $Ex_\forall$ can be applied, $\overrightarrow{w'^{(i+1)}}$ are fresh variables (different for each function symbol), and if $i = K$ there is no disequation $w^{(i)} \neq v$ where $w^{(i)}$ is a variable, v is not a variable and $Var(v) \cap \vec{y} \neq \emptyset$.

Fig. 1. Transformation rules for AC complement problems.

- D_2 decomposes a disequation $t \neq u$ according to the set of AC-unifiers of $t = u$. Unlike for complement problems in $T(\mathcal{F})$, in $T(\mathcal{F})/_{=_{AC}}$ new universally quantified variables may be introduced. To avoid non-termination problems caused by successive applications of explosion and decomposition rules, we add a label 0 to each existentially quantified variable in the starting problem, and we require that the existentially quantified variables introduced in the application of an explosion rule to a variable $x^{(i)}$ have a label $i + 1$. Moreover, we do not allow the application of an explosion rule to a variable labeled by i if $i > K$ for a given K. K can be arbitrarily chosen since it is used only to ensure the termination of the process. If the starting problem is $\forall \vec{y} : t \neq t_1 \wedge \ldots \wedge t \neq t_n$ we will choose $K = max_i\{depth(t_i)\}$, where $depth(t_i)$ is the depth of the tree corresponding to the term t_i. Please note that D_2 applies also when the set of AC-unifiers is empty, and in this case its effect is the elimination of the disjunction containing $t \neq u$.

There are two explosion rules in R_{EAC}. $Ex_\forall$ generalizes the explosion rule of [2]. Besides, now we have to use the domain closure axiom not only for quantifier elimination but also for eliminating disequations whose one member is a ground term. For example, assuming $\mathcal{F} = \{0, s\}$, the formula $x \neq 0$ is equivalent to $\exists w : x = s(w)$. This is expressed by the rule Ex_{gr}. Since complement problems are represented by intermediate formulas where the free variables appear only in equations, we do not need to consider explosions on free variables.

Property 1 (Correctness, termination, irreducible formulae) *Let ϕ be a linear complement problem.*

1. *If $\phi \rightarrow^*_{R_{EAC}} \psi$ then $[\![\phi]\!]_{AC} = [\![\psi]\!]_{AC}$, i.e. R_{EAC} preserves solutions.*
2. *There is no infinite sequence $\phi \rightarrow_{R_{EAC}} \phi_1 \rightarrow_{R_{EAC}} \cdots \phi_n \rightarrow_{R_{EAC}} \cdots$.*
3. *Any irreducible form of ϕ is a finite disjunction of simple formulae.*

As a consequence of property 1:

Theorem 1. *R_{EAC} transforms any linear complement problem into an equivalent finite disjunction of simple formulae.*

3.2 Solving simple formulae

By theorem 1, satisfiability of linear AC complement problems reduces to satisfiability of simple formulae. Let us consider a (non-trivial) simple formula

$$\phi \equiv \exists \vec{w}, \forall \vec{y} : \bigwedge_{k \in K} (\bigvee_{l \in L_k} w_{kl} \neq u_{kl}) \wedge x_1 = v_1 \wedge \ldots \wedge x_n = v_n$$

Then ϕ is satisfiable if and only if $\psi \equiv \exists \vec{w}, \forall \vec{y} : \bigwedge_{k \in K} (\bigvee_{l \in L_k} w_{kl} \neq u_{kl})$ is. The decision test we are going to show is inspired by [9]. We will prove that in order to decide satisfiability of ψ it is sufficient to consider a finite set of substitutions instead of the set of all ground substitutions with domain $\vec{w}$.

Terms involving AC function symbols will be represented in *flattened form* [7]: $flat(t)$, for a term t, is the normal form of (the tree) t for the convergent tree rewriting system

$$f(x_1,\ldots,x_{p-1},f(y_1,\ldots,y_q),x_{p+1},\ldots,x_n) \rightarrow f(x_1,\ldots,x_{p-1},y_1,\ldots,y_q,x_{p+1},\ldots,x_n)$$

for any integers p, q and n such that $1 \leq p \leq n$, and for any AC operator f. In other words, AC-symbols are treated as varyadic symbols in flattened forms.

The notion of flattening extends to substitutions in the natural way: $flat(\sigma)$ assigns $flat(x\sigma)$ to each variable $x \in dom(\sigma)$.

Let $d = max_{k,l}\{depth(flat(u_{kl}))\} + 1$, and let k be the maximal number of arguments of an AC symbol in $flat(u_{kl})$ for u_{kl} in ψ. We can now define the finite sets
$A(\psi) = \{flat(r) \mid r \in T(\mathcal{F}, \mathcal{X}), r$ is linear, $depth(flat(r)) \leq d$, AC-symbols have at most $k+1$ arguments, and variables can occur only at depth $d\}$.
$S_{\vec{w},\vec{y},A(\psi)} = \{\sigma \in \Sigma \mid dom(\sigma) = \vec{w}, Im(\sigma) \subseteq A(\psi)$ and $Im(\sigma)$ is linear, and $Var(Im(\sigma)) \cap \vec{y} = \emptyset\}$.

Lemma 2 below shows that for deciding satisfiability of ψ it is sufficient to consider only the substitutions in $S_{\vec{w},\vec{y},A(\psi)}$.

Lemma 2. *Let ϕ and ψ be as above.* $T(\mathcal{F})/_{=_{AC}} \models \exists \vec{y}: \bigvee_{k \in K} (\bigwedge_{l \in L_k} w_{kl} = u_{kl})$ *iff for all* $\lambda \in S_{\vec{w},\vec{y},A(\psi)}$, $T(\mathcal{F})/_{=_{AC}} \models \exists \vec{y}: \bigvee_{k \in K} (\bigwedge_{l \in L_k} w_{kl}\lambda = u_{kl})$.[2]

Let us call $\mathcal{C}$ the condition of lemma 2: "for all $\lambda \in S_{\vec{w},\vec{y},A(\psi)}, T(\mathcal{F})/_{=_{AC}} \models \exists \vec{y}: \bigvee_{k \in K} (\bigwedge_{l \in L_k} w_{kl}\lambda = u_{kl})$". From lemma 2 we deduce that $\neg \mathcal{C}$, i.e. $\exists \lambda \in S_{\vec{w},\vec{y},A(\psi)}$ such that $T(\mathcal{F})/_{=_{AC}} \not\models \exists \vec{y}: \bigvee_{k \in K} (\bigwedge_{l \in L_k} w_{kl}\lambda = u_{kl})$, is a necessary and sufficient condition for satisfiability of ϕ:

Theorem 3. *Let ϕ be as above. ϕ has solutions in $T(\mathcal{F})/_{=_{AC}}$ iff $\exists \lambda \in S_{\vec{w},\vec{y},A(\psi)}$ such that $T(\mathcal{F})/_{=_{AC}} \not\models \exists \vec{y}: \bigvee_{k \in K} (\bigwedge_{l \in L_k} w_{kl}\lambda = u_{kl})$.*

Theorem 3 suggests an algorithm for solving simple formulae: it is sufficient to test for each $\lambda \in S_{\vec{w},\vec{y},A(\psi)}$ whether $T(\mathcal{F})/_{=_{AC}} \models \exists \vec{y}: \bigvee_{k \in K} (\bigwedge_{l \in L_k} w_{kl}\lambda = u_{kl})$ or, equivalently (since all the variables in $w_{kl}\lambda$ are deeper than those in u_{kl} and the terms u_{kl} are linear), whether there is σ such that $\bigvee_{k \in K} (\bigwedge_{l \in L_k} w_{kl}\lambda =_{AC} u_{kl}\sigma)$. A substitution σ satisfying this condition is usually called an AC-matcher. So, Theorem 3 can be reformulated in other words: a simple formula ϕ is satisfiable iff

---

[2] Linearity is fundamental here. Without linearity we have only a sufficient (but not necessary) condition.

there exists $\lambda \in S_{\vec{w},\vec{y},A(\psi)}$ such that the AC-matching problem $\bigvee_{k \in K} (\bigwedge_{l \in L_k} w_{kl}\lambda =_{AC} u_{kl})$ has no solution. AC-matching is a well-known decidable problem [6].

In fact, this decision test can be applied to the irreducible forms of $R_{AC} = R_{EAC} - \{Ex_{gr}\}$, since Ex_{gr} is only needed to eliminate negation. R_{AC} is correct and terminating. Its irreducible forms differ from those of R_{EAC} in that now all the terms u_{kl} can be ground, i.e. condition 3 in the definition of simple formulae is no longer valid. Now, if for all k, l, u_{kl} is ground, ϕ has solutions since for all terms u, t such that $\neg(u =_{AC} t)$ there is a solution of $u \neq t$ in $T(\mathcal{F})/_{=_{AC}}$. Otherwise we apply the test.

This decision test can be written as a set **T** of transformation rules to be added to R_{AC}, see figure 2.

Test (T)

$(T_0)\ P \equiv \exists \vec{w}: \bigwedge_{k \in K} (\bigvee_{l \in L_k} w_{kl} \neq u_{kl}) \wedge x_1 = v_1 \wedge \ldots \wedge x_n = v_n \rightarrow \top$

If P is irreducible by R_{AC} and all u_{kl} are ground terms.

$(T_1)\ P \equiv \exists \vec{w}, \forall \vec{y}: \bigwedge_{k \in K} (\bigvee_{l \in L_k} w_{kl} \neq u_{kl}) \wedge x_1 = v_1 \wedge \ldots \wedge x_n = v_n \rightarrow \bot$

If P is irreducible by R_{AC}, there exists k, l such that u_{kl} is not ground, and condition $\mathcal{C}$ holds.

$(T_2)\ P \equiv \exists \vec{w}, \forall \vec{y}: \bigwedge_{k \in K} (\bigvee_{l \in L_k} w_{kl} \neq u_{kl}) \wedge x_1 = v_1 \wedge \ldots \wedge x_n = v_n \rightarrow \top$

If P is irreducible by R_{AC}, there exists k, l such that u_{kl} is not ground, and $\neg \mathcal{C}$.

Fig. 2. Decision test for satisfiability of simple formulae

A careful analysis of the algorithm issued from $R_{AC} \cup \mathbf{T}$ shows that the complexity of the transformation process is given by the complexity of AC-unification (which is known to be doubly exponential in general, but is only exponential in the linear case), and that the complexity of **T** is $O(n^{a^d})$, where n is the number of function symbols in the signature and $a = max\{max_{f \in \mathcal{F}}\{arity(f)\}, k\}$. It is then clear that only by decreasing the values of d and k one can improve significantly the efficiency of the algorithm. In this sense, our method improves from [9]: the bounds d and k we use for constructing the test set are smaller than those used in [9], since we do not consider the original problem but the smaller problems obtained after application of R_{AC}[3].

---

[3] For *linear* complement problems the transformation process can not increase the values of k and d.

3.3 Negation Elimination in Simple Formulae

Negation elimination in AC complement problems reduces to negation elimination in finite disjunctions of simple formulae (by theorem 1). Moreover, as a consequence of the following lemma, each disjunct can be treated separatedly, which simplifies even more the problem.

Lemma 4. *Let P be a linear AC-complement problem and $\bigvee_{j \in J} P_j$ be an irreducible form by R_{EAC} of P. Then, P has the AC-E_U property if and only if every P_j has the AC-E_U property.*

Proof. The "if" part is trivial. Let us prove the "only if": Correctness of R_{EAC} implies $[\![P]\!]_{AC} = [\![\bigvee_{j \in J} P_j]\!]_{AC}$, which implies that P has the AC-E_U property if and only if $\bigvee_{j \in J} P_j$ does. Now, the only rules that introduce a disjunction are the explosions, but if $P \to_{\mathbf{E}} \bigvee_{f \in \mathcal{F}} P_f$ then P is AC-E_U if and only if each P_f is (because explosion consists essentially in adding an equation). The thesis follows by a simple induction on the number of applications of $\mathbf{E}$. □

Now we are faced to the problem of negation elimination in *one* simple formula. Let us consider a non-trivial simple formula $\phi \equiv \exists \vec{w}, \forall \vec{y} : \bigwedge_{k \in K} (\bigvee_{l \in L_k} w_{kl} \neq u_{kl}) \wedge x_1 = v_1 \wedge \ldots \wedge x_n = v_n$. The formula $\psi \equiv \forall \vec{y} : \bigwedge_{k \in K} (\bigvee_{l \in L_k} w_{kl} \neq u_{kl})$ will be called the *kernel* of ϕ. Remark that now there are no existentially quantified variables in ψ. The following is a very important property:

Lemma 5. *ϕ has the AC-E_U property if and only if ψ does.*

So, the first step towards obtaining a positive formula equivalent to ϕ is eliminating negation from ψ. If $\exists \vec{w} : \psi \to_{\mathbf{T}} \bot$ then $\phi \sim_{AC} \bot$ and we are finished. But if $\exists \vec{w} : \psi \to_{\mathbf{T}} \top$, i.e. if ϕ has solutions, then it must be rule T_2 which applies (not T_0 because at least one of the u_{kl} is not ground). Then, since the condition of rule T_2 holds, there exists $L = \{\lambda_1, \ldots, \lambda_h\} \subseteq S_{\vec{w},\vec{y},A(\psi)}$ such that for $1 \leq i \leq h$, $T(\mathcal{F})/_{=_{AC}} \not\models \exists \vec{y} : \bigvee_{k \in K} (\bigwedge_{l \in L_k} w_{kl} \lambda_i = u_{kl})$. We will use L to eliminate negation from ψ.

Before presenting the formal description of the negation elimination algorithm, let us illustrate the ideas behind it with an example.

Example 2. Let $\mathcal{F} = \{+, a, b, c, d, e\}$ where $+$ is AC and the other symbols are constants, and let us consider a formula having the AC-E_U property:

$$\psi \equiv \forall y : w \neq a+a+y \wedge w \neq b+b+y \wedge w \neq c+c+y \wedge w \neq d+d+y \wedge w \neq e+e+y$$

In this example $d = 2$ and $k + 1 = 4$.

$A(\psi) = \{\, a, b, c, d, e,$
$a+a, a+b, a+c, a+d, a+e, b+b, b+c, b+d, b+e, \ldots$
$a+a+a, a+a+b, a+a+c, a+a+d, a+a+e, a+b+b, \ldots$
$a+a+a+a, a+a+a+b, \ldots\}$

$S_{w,y,A(\psi)}$ is the set of all the substitutions with domain $\{w\}$ and image in $A(\psi)$.

$L = \{\{w \mapsto P\} \mid P \in \{a,b,c,d,e\}\} \cup$
$\{\{w \mapsto P+Q\} \mid P,Q \in \{a,b,c,d,e\}\} \cup$
$\{\{w \mapsto P+Q+R\} \mid P,Q,R \in \{a,b,c,d,e\},$ and P,Q,R are pairwise distinct constants$\}\cup$
$\{\{w \mapsto P+Q+R+S\} \mid P,Q,R,S \in \{a,b,c,d,e\},$ and P,Q,R,S are pairwise distinct constants$\}$

Since $L \neq \emptyset$, ψ has solutions in $T(\mathcal{F})/_{=_{AC}}$. We are faced now to the problem of deciding whether ψ has the AC-E_U property or not, and in case it has, finding a finite complete set of generators for ψ. L will serve to this purpose. Of course, L is not a complete set of generators of ψ in general. It is clear in this example that the substitutions in L can not generate a solution with more than 4 arguments, as, for instance $\{w \mapsto a+b+c+d+e\}$ (which *is* a solution of ψ).

At first glance one could think that to overcome this incompleteness problem it is sufficient to add new variables as arguments in all the AC symbols with $k+1$ arguments ($k+1$ is the bound used in the construction of $S_{\vec{w},\vec{y},A(\psi)}$). We obtain in this way a set L^* which is certainly complete, but nothing guarantees that the substitutions in L^* will generate only solutions of ψ. In case they are generators the problem is solved, but in the other case we have to consider a bigger test set (with $k' > k$): if in example 2 we take $k' = 4$ and we repeat the construction above, we obtain a set L' of substitutions which is a complete set of generators of ψ.

Formally, let $\psi \equiv \forall \vec{y} : \bigwedge_{k \in K} (\bigvee_{l \in L_k} w_{kl} \neq u_{kl})$ be the kernel of a simple formula ϕ, and assume there exists $L = \{\lambda_1, \ldots, \lambda_h\}$ such that $L \subseteq S_{\vec{w},\vec{y},A(\psi)}$ and for $1 \leq i \leq h$, $T(\mathcal{F})/_{=_{AC}} \not\models \exists \vec{y} : \bigvee_{k \in K} (\bigwedge_{l \in L_k} w_{kl}\lambda_i = u_{kl})$. Since the variables in $Im(\lambda_i)$ are strictly deeper than those in u_{kl} and each variable in u_{kl} appears at most once in a conjunction, for $1 \leq i \leq h$:

(1) $T(\mathcal{F})/_{=_{AC}} \not\models \exists Var(Im(\lambda_i)), \exists \vec{y} : \bigvee_{k \in K} (\bigwedge_{l \in L_k} w_{kl}\lambda_i = u_{kl})$

Let us add a different sub-index to each occurrence of an AC function symbol with $k+1$ arguments in $Im(\lambda_1), \ldots, Im(\lambda_h)$. If $+_{i_1}, \ldots, +_{i_m}$ occur in $Im(\lambda_i)$, we define λ_i^I, $I \subseteq \{i_1, \ldots, i_m\}$, such that it is the same as λ_i but for the subterms whose root is $+_{i_j}$ with $i_j \in I$, where there is one more argument: a new variable x_{i_j}. L^* is the set $\{\lambda_i^I \mid \lambda_i \in L, I \subseteq \{i_1, \ldots, i_m\}\}$.

If there is no $\lambda_i^I \in L^*$ such that

(2) $T(\mathcal{F})/_{=_{AC}} \models \exists Var(Im(\lambda_i^I)), \exists \bar{y}: \bigvee_{k \in K} (\bigwedge_{l \in L_k} w_{kl}\lambda_i^I = u_{kl})$

then $L \cup L^*$ is a complete set of generators for ψ (we will prove below, in theorem 8, its correctness and completeness). Otherwise, we will prove, also in theorem 8 below, that there exists $k' > k$ such that if we define $A'(\psi), S_{\bar{w},\bar{y},A'(\psi)}, L'$ and L'^* in the same way as $A(\psi), S_{\bar{w},\bar{y},A(\psi)}, L$ and L^* but using k' instead of k, then
 - ψ has the $AC\text{-}E_U$ property iff there is no $\lambda'^I_i \in L'^*$ such that (2) holds,
 - if ψ has the $AC\text{-}E_U$ property then $L' \cup L'^*$ is a complete set of generators for ψ.

Let us show now how this k' will be computed. First of all we have to define the notion of "correspondence" between arguments of AC symbols. More precisely, let $+$ be an AC symbol and t, u be flattened terms $t \equiv +(t_1, \ldots, t_n)$ and $u \equiv +(u_1, \ldots, u_m)$, $n \geq m$. An AC-correspondence $\varsigma: [1..n] \to [1..m]$ between t and u is a partial function such that
 - for all $i \in dom(\varsigma)$, t_i and $u_{\varsigma(i)}$ are AC-unifiable and
 - ς is maximal, i.e., there is no $\varsigma': [1..n] \to [1..m]$ such that t_i and $u_{\varsigma'(i)}$ are AC-unifiable and $cardinal(dom(\varsigma')) > cardinal(dom(\varsigma))$.

Property 2 *Given t, u, there exists n such that for all AC-correspondences ς between t and u, $cardinal(dom(\varsigma)) = n$.*

Intuitively, we want k' such that if a substitution $\lambda' \in L'$ has an AC-symbol $+_{ij}$ with $k'+1$ arguments then one of these arguments does not correspond with any of the arguments of $+_{ij}$ in u_{kl}, and as a consequence it can be added an arbitrary number of times in $Im(\lambda')$ and the resulting substitutions are still generators of ψ. (In the example 2, if $\lambda' = \{w \mapsto a+b+c+d+e\}$ then no argument can be repeated, but let $\mathcal{F} = \{a, f, +\}$ where $+$ is AC, f is a unary function symbol and a is a constant, and let us consider the formula $\psi \equiv \forall y: w \neq y+a+a$ which does not have the $AC\text{-}E_U$ property. The substitution $\{w \mapsto f(a)\}$ is a solution of ψ and also $\{w \mapsto f(a)+f(a)\}, \{w \mapsto f(a)+f(a)+f(a)\} \ldots$ are solutions of ψ: $f(a)$ can be added an arbitrary number of times.) We will show that in this case the only finite representation of these solutions of ψ is λ'^*, which has a variable as an argument of $+_{ij}$. Let us develop this idea.

We know that (2) is equivalent to

$$\exists Var(Im(\lambda_i^I)), \exists \bar{y}: \bigvee_{k \in K} (\bigwedge_{l \in L_k} flat(w_{kl}\lambda_i^I) =_P flat(u_{kl}))$$

and (1) is equivalent to

$$\neg(\exists Var(Im(\lambda_i)), \exists \bar{y}: \bigvee_{k \in K} (\bigwedge_{l \in L_k} flat(w_{kl}\lambda_i) =_P flat(u_{kl})))$$

If we have an equality with λ_i^I but not with λ_i it is because some arguments of u_{kl} which do not have a correspondent in $w_{kl}\lambda_i$ are covered by a variable in λ_i^I. That is, there exists $k \in K$ such that for all $l \in L_k$ there are permutations $\overline{flat(w_{kl}\lambda_i)}$,

$\overline{flat(u_{kl})}$ of $\overline{flat(w_{kl}\lambda_i)}$, $\overline{flat(u_{kl})}$ respectively, such that all non variable positions of $\overline{flat(w_{kl}\lambda_i)}$ coincide with those of $\overline{flat(u_{kl})}$ but there are some remaining arguments under some AC symbols $+_{i_j}$ $(i_j \in I)$ without a correspondent. The number n_{ij} of such arguments is computed as follows: if $+_{i_j}$ is an AC symbol with $k+1$ arguments that occurs at position q in $\overline{flat(w\lambda_i)}$ and $\overline{flat(u_{kl})}$ (for all k,l such that $w_{kl} \equiv w$), that is, $\overline{flat(w\lambda_i)}\,|_q \equiv +_{i_j}(t_1\ldots,t_{k+1})$ and $\overline{flat(u_{kl})}\,|_q \equiv +_{i_j}(s_1,\ldots,s_{r_{kl}})$, and for any AC-correspondence ς between $\overline{flat(u_{kl})}\,|_q$ and $\overline{flat(w_{kl}\lambda_i)}\,|_q$, $cardinal(dom(\varsigma)) = c_{kl}$, then $n_{ij} = \sum_{k,l \text{ s.t. } w_{kl}\equiv w} r_{kl} - c_{kl}$.

Let $N_{ij} = n_{ij} - cardinal(\{w_{kl} \mid w_{kl} \equiv w\}) + 1$ and let $N = max_{i,j}\{N_{ij}\}$. We take $k' = k + N$.

Example 3. Let us calculate k' for the problem in the previous example. We have to consider the set of substitutions $\{\{w \mapsto P+Q+R+S\} \mid P,Q,R,S \in \{a,b,c,d,e\}$ and P,Q,R,S are pairwise distinct constants$\} = \{\lambda_1 : \{w \mapsto a+b+c+d\}, \lambda_2 : \{w \mapsto a+b+c+e\}, \lambda_3 : \{w \mapsto a+c+d+e\}, \lambda_4 : \{w \mapsto a+b+d+e\}, \lambda_5 : \{w \mapsto b+c+d+e\}\}$. Each one of these substitutions has one AC symbol with $k+1$ arguments (recall that in this example, $w_k \equiv w$ and $r_k = 3$ for all k). Now, for λ_1, $c_1 = c_2 = c_3 = c_4 = 2$ and $c_5 = 1$ then $n_{11} = 6$ and $N_{11} = 6 - 5 + 1 = 2$. For λ_2, $c_1 = c_2 = c_3 = c_5 = 2$ and $c_4 = 1$ then $n_{21} = 6$ and $N_{21} = 2$. In the same way, $N_{31} = N_{41} = N_{51} = 2$. Then $N = 2$ and $k' = 5$.

Let L' be the set of substitutions computed for ψ by using k' instead of k. By definition, $L' = \{\lambda'_1, \ldots, \lambda'_{h'}\} \subseteq S_{\vec{w},\vec{y},A'(\psi)}$ such that for $1 \leq i \leq h'$,

$$(3) \quad T(\mathcal{F})/_{=_{AC}} \not\models \exists Var(Im(\lambda'_i)), \exists \vec{y} : \bigvee_{k\in K} \left(\bigwedge_{l\in L_k} w_{kl}\lambda'_i = u_{kl} \right)$$

Still we have to prove that $L' \cup L'^*$ is a complete set of generators for ψ when such a set exists. The following properties of L' will be used in the proof.

Lemma 6. *If $\lambda' \in L'$ then $proj_{k+1}(\lambda') \in L$.*

Lemma 7. *If there is $\lambda' \in L'$ such that an AC symbol occurs in $Im(\lambda')$ with $k'+1$ arguments then there is a solution of ψ with $k'+1+i$ arguments for any $i > 0$.*

Let us define by cases the set G of substitutions:
- $G = L \cup L^*$ if there is no $\lambda \in L^*$ such that (2) holds.
- Otherwise, $G = L' \cup L'^*$ if there is no $\lambda' \in L'^*$ such that (2) holds.
- Otherwise, $G = \emptyset$.

For instance, for the problem in example 2 $G = L' \cup L'^*$. Now we are ready to prove the main result of this section.

Theorem 8. *ψ has the AC-E_U property if and only if $G \neq \emptyset$.*
If $G = \{\lambda_1, \ldots, \lambda_n\} \neq \emptyset$ and $\bigcup_i Var(Im(\lambda_i)) = \vec{z}$ then

$$\psi \sim_{AC} \bigvee_{\lambda_i \in G} \exists \vec{z} : \bigwedge_{w_j \in dom(\lambda_i)} w_j = w_j \lambda_i.$$

Proof. First we will prove that if $G \neq \emptyset$ then ψ has the $AC\text{-}E_U$ property, and more precisely, $\psi \sim_{AC} \bigvee_{\lambda_i \in G} \exists \vec{z} : \bigwedge_{w_j \in dom(\lambda_i)} w_j = w_j \lambda_i$, i.e. G is a complete set of AC-generators of ψ when $G \neq \emptyset$.

Correctness: We must show that for all $\lambda \in G$ and for all μ such that $\lambda\mu$ is a ground substitution, $T(\mathcal{F})/_{=_{AC}} \not\models \exists \vec{y} : \bigvee_{k \in K} (\bigwedge_{l \in L_k} w_{kl}\lambda\mu = u_{kl})$. By contradiction: If

$$(*) \quad T(\mathcal{F})/_{=_{AC}} \models \exists \vec{y} : \bigvee_{k \in K} (\bigwedge_{l \in L_k} w_{kl}\lambda\mu = u_{kl})$$

then, $T(\mathcal{F})/_{=_{AC}} \models \exists Var(Im(\lambda)), \exists \vec{y} : \bigvee_{k \in K} (\bigwedge_{l \in L_k} w_{kl}\lambda = u_{kl})$, then, by (1) and (3), $\lambda \notin L$ and $\lambda \notin L'$. Thus, $\lambda \in L^*$, but $(*)$ implies that (2) holds, which contradicts the fact that $G \neq \emptyset$.

Completeness: We must show that for any AC solution μ of ψ there exists $\lambda \in G$ such that μ is an AC-instance of λ. Let μ be a solution of ψ: $T(\mathcal{F})/_{=_{AC}} \not\models \exists \vec{y} : \bigvee_{k \in K} (\bigwedge_{l \in L_k} w_{kl}\mu = u_{kl})$. Let us define $\mu' \equiv proj_{k+1}(\mu)$ in case $G = L \cup L^*$, and $\mu' \equiv proj_{k'+1}(\mu)$ in case $G = L' \cup L'^*$.

1. If $\mu' \equiv \mu$ then there exists $\lambda \in L$ or $\lambda \in L'$ such that $\lambda \preceq_{AC} \mu$, because if $max\{depth(t) \mid t \in Im(\mu')\} \leq d$ then $\mu \in L$ or $\mu \in L'$, and if $max\{depth(t) \mid t \in Im(\mu')\} > d$ then, by definition of $S_{\vec{w},\vec{y},A(\psi)}, S_{\vec{w},\vec{y},A'(\psi)}$, there exists $\lambda \in S_{\vec{w},\vec{y},A(\psi)}$ or $\lambda \in S_{\vec{w},\vec{y},A'(\psi)}$ which coincides with μ till depth $d-1$ and has variables at depth d (i.e., $\lambda \preceq_{AC} \mu$), and $T(\mathcal{F})/_{=_{AC}} \not\models \exists \vec{y} : \bigvee_{k \in K} (\bigwedge_{l \in L_k} w_{kl}\lambda = u_{kl})$ (i.e., $\lambda \in G$) because if there exists θ such that $\bigvee_{k \in K} (\bigwedge_{l \in L_k} w_{kl}\lambda =_{AC} u_{kl}\theta)$, then, since $\mu = \lambda\alpha$, $\bigvee_{k \in K} (\bigwedge_{l \in L_k} w_{kl}\mu = u_{kl}\theta\alpha)$.

2. If $\mu \not\equiv \mu'$, as a consequence of lemma 6, μ' is also a solution, then, using 1, there exists $\lambda \in L$ if $G = L \cup L^*$ or $\lambda \in L'$ if $G = L' \cup L'^*$, such that $\lambda \preceq_{AC} \mu'$. Then, there is $\lambda^* \in L^*$ or $\lambda^* \in L'^*$ such that $\lambda^* \preceq_{AC} \mu$.

Now it remains to prove that if $G = \emptyset$ then ψ does not have the $AC\text{-}E_U$ property. Assume $G = \emptyset$ and ψ has the $AC\text{-}E_U$ property. We will prove that this leads to a contradiction. Since ψ has the $AC\text{-}E_U$ property and AC-unification is finitary, there exists a complete set $\Gamma = \{\gamma_1, \ldots, \gamma_n\}$ of generators of ψ. Since $G = \emptyset$, there is $\lambda^I \in L^*$ and $\lambda'^{I'} \in L'^*$ such that (2) holds. This implies that there exists μ' ground, with $dom(\mu') = Var(Im(\lambda'^{I'}))$, such that

$$(4) \quad \exists \vec{y} : \bigvee_{k \in K} (\bigwedge_{l \in L_k} w_{kl}\lambda'^{I'}\mu' =_{AC} u_{kl})$$

But if $\lambda'^{I'} \in L'^*$ then $\lambda' \in L'$, that is, all the instances of λ' are solutions of ψ, in particular $\lambda'\mu'$. Besides, if $\lambda'^{I'} \in L'^*$, $Im(\lambda')$ has an occurrence of an AC

symbol with $k'+1$ arguments and by lemma 7, there is an argument of an AC symbol $+_j$ that we can add n times for any $n > 0$ yielding always a solution, which we denote by $(\lambda'\mu')^n$.

Since Γ is finite, there exists γ_i such that infinitely many $(\lambda'\mu')^n$ are AC instances of γ_i, that is, there are infinitely many ρ_n such that $\gamma_i\rho_n =_{AC} (\lambda'\mu')^n$.

If $+_j$ occurs in $w_{kl}\lambda'$, $w_{kl}\gamma_i\rho_n =_{AC} w_{kl}(\lambda'\mu')^n$ for infinitely many n.

We assume that all these substitutions are flattened, then $flat(w_{kl}\gamma_i\rho_n) =_P flat(w_{kl}(\lambda'\mu')^n)$ for infinitely many n. Then, they can be made identical by permutations: $\overline{flat(w_{kl}\gamma_i\rho_n)} \equiv \overline{flat(w_{kl}(\lambda'\mu')^n)}$.

Assume $+_j$ occurs at position p in $\overline{flat(w_{kl}(\lambda'\mu')^n)}$. Then either $\exists q \leq p$ such that $\overline{w_{kl}\gamma_i}|_q \equiv x \in \mathcal{X}$ or $\overline{w_{kl}\gamma_i}(p) = +_j$ and $\exists q$ such that $|q|=1$ and $\overline{w_{kl}\gamma_i}(p.q) = x \in \mathcal{X}$, and x occurs only once in $Im(\gamma_i)$ since we are adding arguments only to the $+_j$ occurring at position p. Then, in both cases $\gamma_i \preceq_{AC} \lambda'^{I'}\mu'$. But $\lambda'^{I'}\mu'$ is not a solution of ψ by (4). This contradicts the fact that Γ is a set of generators of ψ. □

As a consequence of theorem 8 *negation elimination is decidable in linear AC complement problems.* We can write the test stated in theorem 8 as a transformation rule to be added to R_{EAC}, see figure 3. The system $R_{EAC} \cup \{T\}$ is obviously terminating. When applied to a linear AC complement problem P, it yields a positive formula equivalent to P whenever such a formula exists, and a finite disjunction of simple formulae equivalent to P in case P does not have the AC-E_U property.

Test (T)

$$(T) \quad P \equiv (\exists \vec{w} \forall \vec{y} : \bigwedge_{k \in K} (\bigvee_{l \in L_k} w_{kl} \neq u_{kl}) \wedge x_1 = v_1 \wedge \ldots \wedge x_n = v_n) \vee d \to$$

$$(\exists \vec{w} : (\bigvee_{\lambda_i \in G} \exists \vec{z} : \bigwedge_{w_j \in dom(\lambda_i)} w_j = w_j\lambda_i) \wedge x_1 = v_1 \wedge \ldots \wedge x_n = v_n) \vee d$$

If P is irreducible by R_{EAC} and $G \neq \emptyset$.

Fig. 3. Negation elimination in simple formulae

4 Generalization

The most interesting generalization of the previous results, the non-linear case, is still open. But our decision algorithms for satisfiability and negation elimination in linear AC complement problems generalize easily to a wider class of formulae in Σ_2: the $\forall$-*linear* formulae. These are formulae $\exists \vec{w}, \forall \vec{y} : t_1 \neq t'_1 \wedge \ldots \wedge t_n \neq t'_n$

where $t'_1, \ldots, t'_n$ do not contain existentially quantified variables and each universally quantified variable occurs only once. R_{EAC} is correct and terminating for ∀-linear formulae. Moreover, the irreducible forms are again simple formulae. Then validity and negation elimination are decidable for ∀-linear formulae in $T(\mathcal{F})/_{=_{AC}}$.

This is another advantage of an approach based on transformation rules: the decomposition of the problem into simplification rules and decision test over irreducible forms allows a very easy generalization of the results when a wider class of formulae has the same irreducible forms.

Acknowledgments: I wish to thank Alexandre Boudet and Hubert Comon for many fruitful discussions about AC-unification algorithms and complement problems. Special thanks to Hubert Comon for reading the first drafts of this article, and for many invaluable comments and suggestions.

References

1. Hubert Comon. Unification et disunification: Théorie et applications. Thèse de Doctorat, Institut National Polytechnique de Grenoble, France, 1988.
2. Hubert Comon. Ground normal forms and inductive proofs. part I: complement problems. Submitted to Journal of Symbolic Computation, 1991.
3. Hubert Comon and Maribel Fernández. Negation elimination in equational formulae. In *Proc. 17th Mathematical Foundations of Computer Science, Praha*, 1992.
4. Hubert Comon and Pierre Lescanne. Equational problems and disunification. *Journal of Symbolic Computation*, 7:371–425, 1989.
5. Nachum Dershowitz and Jean-Pierre Jouannaud. Rewrite systems. In J. van Leeuwen, editor, *Handbook of Theoretical Computer Science*, volume B, pages 243–309. North-Holland, 1990.
6. J.-M. Hullot. Associative commutative patern matching. In *Proc. 6th IJCAI (Vol. I), Tokyo*, pages 406–412, August 1979.
7. Jean-Pierre Jouannaud and Emmanuel Kounalis. Automatic proofs by induction in theories without constructors. *Information and Computation*, 82(1), July 1989.
8. Emmanuel Kounalis. Learning from examples and counterexamples with equational background knowledge. In *Proc. 2nd IEEE Conference on Tools for Artificial Intelligence*, 1990.
9. Emmanuel Kounalis, Denis Lugiez, and L. Pottier. Complement problems modulo associativity and commutativity. In *Proc. 16th Mathematical Foundations of Computer Science, Warsaw*. Springer-Verlag, 1991.
10. Jean-Louis Lassez, M. Maher, and K. Marriott. Elimination of negation in term algebras. In *Proc. 16th Mathematical Foundations of Computer Science, Warsaw*. Springer-Verlag, 1991.
11. Jean-Louis Lassez and K. G. Marriott. Explicit representation of terms defined by counter examples. *Journal of Automated Reasoning*, 3(3):1–17, September 1987.
12. A. Laville. Lazy pattern matching in the ML language. In *Proc. 7th Conf. Found. of Software Technology and Theoretical Computer Science, Pune, INDIA, LNCS 287*, December 1987.
13. D. Lugiez and J.-L. Moysset. Complement problems and tree automata in AC-like theories. In *Proc. Symp. on Theoretical Aspects of Computer Science*, Würzburg, 1993.

A precedence-based total AC-compatible ordering

Albert Rubio and Robert Nieuwenhuis
Technical University of Catalonia
Pau Gargallo 5, 08028 Barcelona, Spain.
E-mail: {rubio,roberto}@lsi.upc.es

April 2, 1993

Abstract

Like Narendran and Rusinowitch [NR91], we define a simplification ordering which is AC-compatible and total on non-AC-equivalent ground terms, without any restrictions on the signature like the number of AC-symbols or free symbols.

An important difference w.r.t. their work is that our ordering is not based on polynomial interpretations, but on a total (arbitrary) precedence on the function symbols, like in LPO or RPO (this solves an open question posed e.g. by Bachmair [Bac91]).

A second difference is that we define an extension to terms with variables, which makes the ordering applicable in practice for complete theorem proving strategies with built-in AC-unification and for orienting non-ground rewrite systems.

Our ordering is defined in a simple way by means of rewrite rules, and can be easily (and efficiently) implemented, since its main component is RPO.

1 Introduction

Automated termination proofs are well-known to be crucial for using rewriting in theorem proving and programming. For the AC-case, cf. e.g. [BP85, BCL87, KSZ90, Bac91]. One of the advantages, wrt. polynomial interpretation-based orderings, of precedence-based orderings is that they allow automatic termination proofs of rewrite systems.

Our main motivation for this work has been the need of a simple ordering for theorem proving strategies with built-in AC-unification. For such strategies, one really needs an extension $\succ_v$ to terms with variables of the total AC-compatible ordering $\succ$ on ground terms such that $s \succ_v t$ implies $s\sigma \succ t\sigma$ for all ground substitutions σ. Furthermore, we believe that a simple ordering like the one defined here can also be a first step towards a decision procedure for the satisfiability of AC-ordering constraints, like done in [Com90, JO91] for LPO and RPO, which has interesting applications in theorem proving [NR92].

Let us first give some intuition about the definition of the ordering $\succ$ on ground terms. In later sections formal definitions of the ordering and of its extension to terms with variables will be given and its properties will be proved in detail.

Let $\succ_1$ be a recursive path ordering with status (RPOS) with a total precedence, and in which AC-symbols have multiset status and where all other symbols have lexicographical left-to-right status. It is well-known that $\succ_1$ is a simplification ordering that is total on non-AC-equivalent terms, but not AC-compatible.

Flattening terms before comparing them under $\succ_1$ solves this problem, but then monotonicity is lost. For example, if $f \succ g$ where $f \in \mathcal{F}_{AC}$, then $f(a,b) \succ_1 g(a,b)$, but for the terms $f(a, f(a,b))$ and $f(a, g(a,b))$, after flattening we get $f(a,a,b) \not\succ_1 f(a, g(a,b))$. Here the problem is caused by the symbol g immediately below the AC-symbol f, with $f \succ g$. These situations are avoided by the ordering defined in this paper, which roughly consists of the following:

Flattening of AC-terms consists of rewriting them to normal form wrt. the set R_F of *flattening* rules of the form:

$$f(\ldots, f(x_1, \ldots, x_n), \ldots) \to f(\ldots, x_1, \ldots, x_n, \ldots)$$

for all AC-function symbols f. We also define a set of *interpretation* rules R_I of the form

$$f(\ldots, g(t_1, \ldots, t_n), \ldots) \to f(\ldots, t, \ldots)$$

for all symbols g and all AC-function symbols f with $f \succ g$, where t is $max_{\succ_1}\{t_1, \ldots, t_n\}$ if $n > 0$ and where t is the smallest constant symbol $\bot$ if $n = 0$ and $g \neq \bot$.

The interpretation $I(t)$ of a term t is defined as the normal form of t wrt. $R_F \cup R_I$, under the strategy (i) leftmost-innermost and (ii) rules of R_F first.

Now given this interpretation, the main ordering $\succ$ described in this paper is defined in a straightforward way (similar to the way it is done by Narendran and Rusinowitch in [NR91]): $s \succ t$ iff:

- $I(s) \succ_1 I(t)$ or

- $I(s) = I(t)$ (up to permutation of arguments of AC-operators; note that then also $f = top(s) = top(t)$) and

    1. $f \in \mathcal{F}_{AC}$ and $\{s_1, \ldots, s_m\} \succ\!\!\!\succ \{t_1, \ldots, t_n\}$ or
    2. $f \notin \mathcal{F}_{AC}$ and $(\exists i : 1..m \ (\forall j < i \ s_j =_{AC} t_j) \ \land \ s_i \succ t_i)$,

where $\succ\!\!\!\succ$ is the extension of $\succ$ to multisets, and where $f(s_1, \ldots, s_m)$ and $f(t_1, \ldots, t_n)$ are the normal forms of s and t wrt. rewriting by R_F only at the topmost position.

Example 1.1 Suppose $a \succ b \succ f \succ g \succ h$ where $f \in \mathcal{F}_{AC}$.

Then for the example above monotonicity is preserved: we have $f(a,b) \succ g(a,b)$ because $I(f(a,b)) = f(a,b) \succ_1 g(a,b) = I(g(a,b))$. Furthermore, $f(a, f(a,b)) \succ f(a, g(a,b))$, since $I(f(a, f(a,b))) = f(a,a,b) \succ_1 f(a,a) = I(f(a, g(a,b)))$.

One may wonder why the interpretation does not keep *all* arguments of g, instead of one maximal one. The answer is that then again monotonicity would be lost. If R_I were:

$$f(\ldots, g(t_1, \ldots, t_n), \ldots) \to f(\ldots, t_1, \ldots, t_n, \ldots)$$

when $f \in \mathcal{F}_{AC}$, $f \succ g$ and $n > 0$, then with the precedence above $g(a,b) \succ h(a,a)$ but $f(b, h(a,a)) \succ f(b, g(a,b))$ because $f(b,a,a) \succ_1 f(b,a,b)$.

The results of this paper are proved in this order:

- $\succ$ is AC-compatible. (lemma 4.2).
- $\succ$ is an ordering relation on $\mathcal{T}(\mathcal{F})$ (it is irreflexive and transitive) (lemma 4.3).
- $\succ$ fulfills the subterm and deletion properties (lemma 4.4).
- $\succ$ fulfills the monotonicity property (lemma 4.7).
- $\succ$ is total on non-AC-equivalent terms (lemma 4.8).

Finally, we define an extension $\succ_v$ to terms with variables and prove that $s \succ_v t$ implies $s\sigma \succ t\sigma$ for all ground substitutions σ (theorem 5.5).

2 Definitions and basic properties

In the following we consider that $\mathcal{F}$ is a finite set of function symbols that is totally ordered by a precedence $\succ_\mathcal{F}$, where $\mathcal{F}_{AC}$ is the subset containing all AC-symbols of $\mathcal{F}$. We assume that $\bot$ is the smallest constant symbol of $\mathcal{F}$.

The arity of a function symbol f, denoted by $\alpha(f)$, is a set of natural numbers that indicates which is the number of arguments that f may take. If $f \in \mathcal{F}_{AC}$ then $\alpha(f)$ contains all natural numbers greater than 1; otherwise, $\alpha(f)$ contains one natural number. $\mathcal{T}(\mathcal{F})$ and $\mathcal{T}(\mathcal{F}, \mathcal{X})$ are defined as usual according to these arities, if $\mathcal{X}$ is a set of variables, whose elements will be denoted by $x, y, z, \ldots$, possibly with subscripts.

We denote by $=_{AC}$ the congruence generated on $\mathcal{T}(\mathcal{F})$ by the associativity and commutativity axioms for the symbols in $\mathcal{F}_{AC}$. Let s, t, s' and t' be arbitrary terms in $\mathcal{T}(\mathcal{F})$, and let u be a non-empty context in $\mathcal{T}(\mathcal{F})$. Then an ordering on $\mathcal{T}(\mathcal{F})$ (a transitive irreflexive relation) $\succ$ fulfills the *subterm property* if $u[t] \succ t$. It fulfills the *deletion property* if $f(\ldots, t, \ldots) \succ f(\ldots\ldots)$ for all f in $\mathcal{F}_{AC}$. Furthermore, it is *monotonic* if $s \succ t$ implies $u[s] \succ u[t]$. A monotonic ordering that fulfills the subterm and deletion properties is called a *simplification ordering* and is *well-founded* ([Der87]): there are no infinite sequences $t_1 \succ t_2 \succ \ldots$. Finally, an ordering $\succ$ is *AC-compatible* if $s' =_{AC} s \succ t =_{AC} t'$ implies $s' \succ t'$.

Let s be a ground term $f(s_1 \ldots s_n)$ with $n \geq 0$. Then the size of s, denoted by $|s|$, is defined as $|s_1| + \ldots + |s_n| + 1$. Furthermore, f is the topmost symbol of s, denoted by $top(s)$.

The extension of the $=_{AC}$ relation to multisets $==_{AC}$ is defined by: $\{s_1, \ldots, s_m\} ==_{AC} \{t_1, \ldots, t_n\}$ iff $m = n$ and if $n > 0$ then there exists some t_j s.t. $s_1 =_{AC} t_j$ and $\{s_1, \ldots, s_m\} \setminus \{s_1\} ==_{AC} \{t_1, \ldots, t_n\} \setminus \{t_j\}$.

Given a relation $\succ$, the AC-multiset extension of $\succ$, denoted by $\succ\!\!\succ$, is defined by: $M = \{s_1, \ldots, s_m\} \succ\!\!\succ \{t_1, \ldots, t_n\} = N$ if

- $M \neq \emptyset$ and $N = \emptyset$ or

- $s_i =_{AC} t_j$ and $M \setminus \{s_i\} \gg N \setminus \{t_j\}$, for some i in $1..m$ and j in $1..n$ or

- $s_i \succ t_{j_1} \wedge \ldots \wedge s_i \succ t_{j_k}$ and
  $(M \setminus \{s_i\} \gg N \setminus \{t_{j_1}, \ldots, t_{j_k}\}$ or $M \setminus \{s_i\} ==_{AC} N \setminus \{t_{j_1}, \ldots, t_{j_k}\})$
  for some i in $1..m$ and $1 \leq j_1 < \ldots < j_k \leq n$ $(k \geq 0)$.

If each symbol f in the set of function symbols $\mathcal{F}$ has a status, denoted by $Stat(f)$, which can be lexicographic (lex) or multiset (mul), then the equivalence relation $=_{ms}$ (equality up to multisets): is defined as follows: $s = f(s_1 \ldots s_m) =_{ms} g(t_1 \ldots t_n) = t$ if and only if $f = g$, $m = n$ and $f \notin \mathcal{F}_{AC}$ and $s_i = t_i$ for all $i : 1..m$ or $f \in \mathcal{F}_{AC}$ and $\{s_1, \ldots, s_m\} =_{ms} \{t_1, \ldots, t_m\}$, where $\{s_1, \ldots, s_m\} =_{ms} \{t_1, \ldots, t_n\}$ iff $m = n$ and if $n > 0$ then there exists some t_j s.t. $s_1 =_{ms} t_j$ and $\{s_1, \ldots, s_m\} \setminus \{s_1\} =_{ms} \{t_1, \ldots, t_n\} \setminus \{t_j\}$.

The *recursive path ordering with status* is defined as an extension of an ordering $\succ_\mathcal{F}$ on the set of function symbols $\mathcal{F}$:

$$s = f(s_1 \ldots s_m) \succ_{rpos} g(t_1 \ldots t_n) = t \quad \text{if}$$

1. $f \succ_\mathcal{F} g$ and $\forall j : 1..n \; s \succ_{rpos} t_j$,

2. $g \succ_\mathcal{F} f$ and $\exists i : 1..m \; s_i \succ_{rpos} t$ or $s_i =_{ms} t$,

3. $f = g$, $Stat(f) = lex$ and

    (a) $\exists i : 1..m \; s_i \succ_{rpos} t$ or $s_i =_{ms} t$, or else

    (b) $\exists i : 1..m \; \forall j < i \; s_j =_{ms} t_j$ and $s_i \succ_{rpos} t_i$, and $\forall k : 1..m \; s \succ_{rpos} t_k$

4. $f = g$, $Stat(f) = mul$ and $\{s_1, \ldots, s_m\} \gg_{rpos} \{t_1, \ldots, t_n\}$

where $\gg_{rpos}$ is the multiset extension of $\succ_{rpos}$.

We define the ordering $\succ_1$ on ground terms as the recursive path ordering with status, where a symbol f has status multiset if $f \in \mathcal{F}_{AC}$ and lexicographic status left-to-right if $f \notin \mathcal{F}_{AC}$. By $s \succeq_1 t$, we mean $s \succ_1 t$ or $s =_{ms} t$. By $max_{\succ_1}(S)$ we denote a maximal element wrt. $\succ_1$ of the set of terms S.

Lemma 2.1 *The ordering $\succ_1$ is a simplification ordering on $T(\mathcal{F}, \mathcal{X})$, where $s \succ_1 t$ or $t \succ_1 s$ for all terms s and t in $T(\mathcal{F})$ with $s \neq_{ms} t$. Furthermore, $s' =_{ms} s \succ_1 t =_{ms} t'$ implies $s' \succ_1 t'$, for all terms s, s', t and t' in $T(\mathcal{F}, \mathcal{X})$.*

3 The interpretation

Definition 3.1 The rewrite system R_F contains the following rules:

$$f(x_1 \ldots x_m, f(y_1 \ldots y_r), z_1 \ldots z_n) \to f(x_1 \ldots x_m, y_1 \ldots y_r, z_1 \ldots z_n)$$

for all $f \in \mathcal{F}_{AC}$ and $m + n \geq 1$ and $r \geq 2$.

Definition 3.2 The rewrite system R_I consists of the rules:

$$f(x_1 \ldots x_m, g(t_1 \ldots t_r), y_1 \ldots y_n) \rightarrow f(x_1 \ldots x_m, t_j, y_1 \ldots y_n)$$

for all $f \in \mathcal{F}_{AC}$, non-constant symbols g s.t. $f \succ_\mathcal{F} g$, $m + n \geq 1$ and $r \in \alpha(g)$ and for all $t_1, \ldots, t_r, \in \mathcal{T}(\mathcal{F})$ s.t. $t_j = max_{\succ_1}\{t_1, \ldots, t_r\}$, and the rules:

$$f(x_1 \ldots x_m, c, y_1 \ldots y_n) \rightarrow f(x_1 \ldots x_m, \bot, y_1 \ldots y_n)$$

for all $f \in \mathcal{F}_{AC}$, constant symbols c s.t. $f \succ_\mathcal{F} c$ and $c \neq \bot$, and $m + n \geq 1$.

If $f(s_1, \ldots, s_m)$ is the normal form of a ground term s wrt. rewriting by R_F only at the topmost position, then the *top-flattening of s*, denoted by $tf(s)$, is $\langle s_1, \ldots s_m \rangle$.

Definition 3.3 Let s be a ground term in $\mathcal{T}(\mathcal{F})$. An interpretation of s is a normal form of s wrt. $R_F \cup R_I$ under the strategy (i) leftmost-innermost and (ii) the rules of R_F are applied first.

Lemma 3.4 For each ground term s, the interpretation of s always exists, is in $\mathcal{T}(\mathcal{F})$ and is unique up to $=_{ms}$ for all terms equivalent under $=_{ms}$, i.e. if $s =_{ms} t$, s' is an interpretation of s, and t' is an interpretation of t then $s' =_{ms} t'$. In the following, we denote the interpretation of s by $I(s)$.

Definition 3.5 Given a term s and a function symbol f, the interpretation of s wrt. to f, denoted by $I_f(s)$, is defined as follows:

- if $f \notin \mathcal{F}_{AC}$ then $I_f(s) = \langle I(s) \rangle$ and
- if $f \in \mathcal{F}_{AC}$ then $I_f(s) = \langle s_1, \ldots, s_m \rangle$, if $I(f(\bot, s)) = f(\bot, s_1, \ldots, s_m)$.

Lemma 3.6 Let s and t be ground terms. The interpretation I satisfies the following properties:

1. $I(I(s)) = I(s)$
2. $top(I(s)) = top(s)$
3. If $I(s) =_{ms} I(t)$ then $top(s) = top(t)$.
4. If $s = f(s_1 \ldots s_m)$ and $I(s) = f(s'_1 \ldots s'_n)$ then $n \geq m$.
5. $I(f(\ldots s \ldots)) =_{ms} I(f(\ldots I(s) \ldots))$.
6. If $I(s) =_{ms} I(t)$ then $I(f(\ldots s \ldots)) =_{ms} I(f(\ldots t \ldots))$.
7. If $s = f(s_1 \ldots s_m)$ and $f \notin \mathcal{F}_{AC}$ then $I(s) = f(I(s_1) \ldots I(s_m))$.
8. $s = I(s)$ or $s \succ_1 I(s)$.
9. $I(f(s_1, \ldots, s_m)) = f(I_f(s_1), \ldots, I_f(s_m))$.

Lemma 3.7 Let s be a ground term s.t. $top(s) = f \in \mathcal{F}_{AC}$ and $tf(s) = \langle s_1, \ldots, s_m \rangle$. Then $I(s) =_{ms} I(f(s_1 \ldots s_m))$.

Proof By induction on $|s|$. Suppose $s = f(s'_1 \ldots s'_n)$. We distinguish two cases:

1. $top(s'_i) \neq f$ for all i in $1..n$. Then $s'_i = s_i$ for all i in $1..n$ and therefore $I(s) = I(f(s'_1 \ldots s'_n)) = I(f(s_1 \ldots s_m))$.

2. $top(s'_i) = f$ for some i in $1..n$. Suppose $s'_i = f(s''_1 \ldots s''_r)$ and therefore $I(s'_i) = f(I_f(s''_1) \ldots I_f(s''_r))$. Now $I(s) = I(f(s'_1, \ldots, s'_i, \ldots s'_n)) =_{ms}$
(by prop. 5 of the previous lemma)
$I(f(s'_1, \ldots, I(s'_i), \ldots s'_n)) = I(f(s'_1, \ldots, f(I_f(s''_1) \ldots I_f(s''_r)), \ldots s'_n)) = I(f(s'_1, \ldots, I_f(s''_1) \ldots I_f(s''_r), \ldots s'_n)) = I(f(s'_1, \ldots, s''_1 \ldots s''_r, \ldots s'_n))$.
Now since $tf(s) = tf(f(s'_1, \ldots, s''_1 \ldots s''_r, \ldots s'_n))$ by induction hypothesis $I(s) =_{ms} I(f(s'_1, \ldots, s''_1 \ldots s''_r, \ldots s'_n)) =_{ms} I(f(s_1 \ldots s_m))$. □

Lemma 3.8 Let s and t be two ground terms. If $s =_{AC} t$ then $I(s) =_{ms} I(t)$

Proof By induction on $|s| + |t|$. Assume $tf(s) = \langle s_1, \ldots, s_m \rangle$ and $tf(t) = \langle t_1, \ldots, t_n \rangle$. Since $s =_{AC} t$ we have $top(s) = top(t) = f$ and $m = n$. If $f \notin \mathcal{F}_{AC}$ then $s_i =_{AC} t_i$ for all i in $1..m$ and by inductive hypothesis $I(s_i) =_{ms} I(t_i)$. Since if $f \notin \mathcal{F}_{AC}$ then $I(s) = f(I(s_1), \ldots, I(s_m))$ and $I(t) = f(I(t_1), \ldots, I(t_m))$ it follows that $I(s) =_{ms} I(t)$.

If $f \in \mathcal{F}_{AC}$ then for all s_i in $\{s_1, \ldots, s_m\}$ there is a t_j in $\{t_1, \ldots, t_m\}$ s.t. $s_i =_{AC} t_j$. Then by induction hypothesis $I(s_i) =_{ms} I(t_j)$, which implies $f(I(s_1) \ldots I(s_m)) =_{ms} f(I(t_1) \ldots I(t_m))$. Then by lemma 3.4 $I(f(I(s_1) \ldots I(s_m))) =_{ms} I(f(I(t_1) \ldots I(t_n)))$. But by the previous lemma and by prop. 5 of lemma 3.6 $I(s) =_{ms} I(f(s_1 \ldots s_m)) =_{ms} I(f(I(s_1) \ldots I(s_m)))$, and similarly $I(t) =_{ms} I(f(I(t_1) \ldots I(t_n)))$, which altogether implies $I(s) =_{ms} I(t)$. □

4 The ordering

Definition 4.1 Let s and t be terms in $\mathcal{T}(\mathcal{F})$, s.t. $tf(s) = \langle s_1, \ldots, s_m \rangle$ and $tf(t) = \langle t_1, \ldots, t_n \rangle$. Then $s \succ t$ if and only if:

- $I(s) \succ_1 I(t)$ or

- $I(s) =_{ms} I(t)$ and
  1. $top(s) \in \mathcal{F}_{AC}$ and $\{s_1, \ldots, s_m\} \succ\!\!\succ \{t_1, \ldots, t_n\}$ or
  2. $top(s) \notin \mathcal{F}_{AC}$ and $\exists i : 1..m \ \forall j < i \ s_j =_{AC} t_j \wedge s_i \succ t_i$,

Lemma 4.2 The relation $\succ$ is AC-compatible.

Proof Let s, s', t and t' be terms in $\mathcal{T}(\mathcal{F})$ with $s' =_{AC} s \succ t =_{AC} t'$. We prove by induction on $|s| + |t|$ that $s' \succ t'$. There are two possibilities:

1. $I(s) \succ_1 I(t)$. Then by lemma 3.8 we have $I(s') =_{ms} I(s)$ and $I(t) =_{ms} I(t')$ and therefore $I(s') \succ_1 I(t')$.

2. $I(s) =_{ms} I(t)$. Now we suppose $tf(s) = \langle s_1, \ldots, s_m \rangle$, $tf(s') = \langle s'_1, \ldots, s'_m \rangle$, $tf(t) = \langle t_1, \ldots, t_n \rangle$ and $tf(t') = \langle t'_1, \ldots, t'_n \rangle$. We distinguish two cases:

(a) $top(s) \in \mathcal{F}_{AC}$. By induction hypothesis for all pairs (s_i, t_j) s.t. $s_i \succ t_j$ there is a pair $(s'_{i'}, t'_{j'})$ s.t. $s_i =_{AC} s'_{i'}$, $t_j =_{AC} t'_{j'}$ and $s'_{i'} \succ t'_{j'}$, and for all pairs (s_i, t_j) s.t. $s_i =_{AC} t_j$ there is a pair $(s'_{i'}, t'_{j'})$ s.t. $s'_{i'} =_{AC} t'_{j'}$. Therefore, since $\{s_1, \ldots, s_m\} \succcurlyeq \{t_1, \ldots, t_n\}$, we have $\{s'_1, \ldots, s'_m\} \succcurlyeq \{t'_1, \ldots, t'_n\}$ and $s' \succ t'$.

(b) $top(s) \notin \mathcal{F}_{AC}$. Then $s_1 =_{AC} t_1 \wedge \ldots \wedge s_{i-1} =_{AC} t_{i-1} \wedge s_i \succ t_i$ for some i in $1..m$. Moreover $s'_j =_{AC} s_j =_{AC} t_j =_{AC} t'_j$ for all j in $1..i-1$. Now by induction hypothesis $s'_i =_{AC} s_i \succ t_i =_{AC} t'_i$ implies $s'_i \succ t'_i$. Therefore $s' \succ t'$. □

Lemma 4.3 The relation $\succ$ is an ordering relation on $\mathcal{T}(\mathcal{F})$ i.e. it is irreflexive and transitive.

Proof Transitivity: For all $s, t, u \in \mathcal{T}(\mathcal{F})$ we prove by induction on $|s| + |t| + |u|$ that $s \succ t$ and $t \succ u$ implies $s \succ u$. We distinguish four cases depending on the interpretations of s, t and u:

1. $I(s) \succ_1 I(t)$ and $I(t) \succ_1 I(u)$. Then by transitivity of $\succ_1$ we have $I(s) \succ_1 I(u)$ and therefore $s \succ u$.

2. $I(s) \succ_1 I(t) =_{ms} I(u)$. Then $I(s) \succ_1 I(u)$ and $s \succ u$.

3. $I(s) =_{ms} I(t) \succ_1 I(u)$. Then $I(s) \succ_1 I(u)$ and $s \succ u$.

4. $I(s) =_{ms} I(t) =_{ms} I(u)$. Then $top(s) = top(t) = top(u)$. Suppose $tf(s) = \langle s_1, \ldots, s_m \rangle$, $tf(t) = \langle t_1, \ldots, t_n \rangle$ and $tf(u) = \langle u_1, \ldots, u_r \rangle$. If $top(s) \in \mathcal{F}_{AC}$ then we have $\{s_1, \ldots, s_m\} \succcurlyeq \{t_1, \ldots, t_n\}$ and $\{t_1, \ldots, t_n\} \succcurlyeq \{u_1, \ldots, u_r\}$. By induction hypothesis, $s_i \succ t_j$ and $t_j \succ u_k$ implies $s_i \succ u_k$, and by AC-compatibility $s_i =_{AC} t_j$ and $t_j \succ u_k$ implies $s_i \succ u_k$ and $s_i \succ t_j$ and $t_j =_{AC} u_k$ implies $s_i \succ u_k$ for all i in $1..m$, j in $1..n$ and k in $1..r$. Now, from the definition of $\succcurlyeq$ it follows that $\{s_1, \ldots, s_m\} \succcurlyeq \{u_1, \ldots, u_r\}$. If $top(s) \notin \mathcal{F}_{AC}$ then:
$s_1 =_{AC} t_1 \wedge \ldots \wedge s_{p-1} =_{AC} t_{p-1} \wedge s_p \succ t_p$, for some p in $1..m$, and
$t_1 =_{AC} u_1 \wedge \ldots \wedge t_{q-1} =_{AC} u_{q-1} \wedge t_q \succ u_q$, for some q in $1..m$.
But then by induction hypothesis and AC-compatibility, for $k = min(p, q)$ we have $s_1 =_{AC} u_1 \wedge \ldots \wedge s_{k-1} =_{AC} u_{k-1} \wedge t_k \succ u_k$.

Irreflexivity: We prove $s \not\succ s$ by induction on $|s|$. $I(s)$ is unique up to $=_{ms}$, so $I(s) \not\succ_1 I(s)$, since $t =_{ms} t'$ implies that $t \not\succ_1 t'$. If $tf(s)$ is $\langle s_1, \ldots, s_n \rangle$, then by inductive hypothesis, the relation $\succ$ is irreflexive for all s_i with i in $1..n$. If $top(s) \in \mathcal{F}_{AC}$ then, since $\succ$ is irreflexive, transitive and AC-compatible for all s_i, from the definition of $\succcurlyeq$ it follows that $\{s_1, \ldots, s_n\} \not\succcurlyeq \{s_1, \ldots, s_n\}$. On the other hand if $top(s) \notin \mathcal{F}_{AC}$ then $s \not\succ s$, since $s_i \not\succ s_i$ for all i in $1..n$. □

Lemma 4.4 The ordering $\succ$ fulfills the subterm and deletion properties.

Proof Subterm property: We obtain $f(\ldots s \ldots) \succ s$ for all $s \in \mathcal{T}(\mathcal{F})$ by proving directly $I(f(\ldots s \ldots)) \succ_1 I(s)$ by induction on $|s|$. Let $f(\ldots s \ldots)$ be a term $f(u_1 \ldots u_p, s, u'_1 \ldots u'_q)$, with $u_1 \ldots u_p, u'_1 \ldots u'_q$ in $\mathcal{T}(\mathcal{F})$ and $p + q + 1 \in \alpha(f)$.

By lemma 3.6 (prop. 5) $I(f(u_1 \ldots u_p, s, u'_1 \ldots u'_q)) = I(f(u_1 \ldots u_p, I(s), u'_1 \ldots u'_q))$. If we suppose $I_f(u_1) \ldots I_f(u_p) = v_1 \ldots v_{p'}$ and $I_f(u'_1) \ldots I_f(u'_q) = v'_1 \ldots v'_{q'}$ then $I(f(\ldots s \ldots)) = I(f(v_1 \ldots v_{p'}, I(s), v'_1 \ldots v'_{q'}))$.

Assume $I(s) = g(s_1 \ldots s_m)$ (and therefore $I(s_i) = s_i$ for all i in $1..m$). Then we prove that $I(f(v_1 \ldots v_{p'}, g(s_1 \ldots s_m), v'_1 \ldots v'_{q'})) \succ_1 g(s_1 \ldots s_m)$. We consider the following four cases:

1. $f \notin \mathcal{F}_{AC}$ or $g \succ_\mathcal{F} f$. Then $I(f(v_1 \ldots v_{p'}, g(s_1 \ldots s_m), v'_1 \ldots v'_{q'})) = f(v_1 \ldots v_{p'}, g(s_1 \ldots s_m), v'_1 \ldots v'_{q'})$, which (by the subterm property of $\succ_1$) is greater wrt. $\succ_1$ than $g(s_1 \ldots s_m)$.

2. $f \in \mathcal{F}_{AC}$ and $f = g$. Then we know that $I(f(v_1 \ldots v_{p'}, f(s_1 \ldots s_m), v'_1 \ldots v'_{q'})) = f(v_1 \ldots v_{p'}, s_1 \ldots s_m, v'_1 \ldots v'_{q'})$, which is greater wrt. $\succ_1$ than $f(s_1 \ldots s_m)$ since $p' + q' > 0$.

3. $f \in \mathcal{F}_{AC}$, $f \succ_\mathcal{F} g$ and $m > 0$. Then $I(f(v_1 \ldots v_{p'}, g(s_1 \ldots s_m), v'_1 \ldots v'_{q'})) = I(f(v_1 \ldots v_{p'}, s_j, v'_1 \ldots v'_{q'}))$, where $s_j = max_{\succ_1}\{s_1 \ldots s_m\}$. By induction hypothesis we have $I(f(v_1 \ldots v_{p'}, s_j, v'_1 \ldots v'_{q'})) \succ_1 I(s_j) = s_j$. Moreover, since $s_j = max_{\succ_1}\{s_1, \ldots, s_m\}$, it holds that $I(f(v_1 \ldots v_{p'}, s_j, v'_1 \ldots v'_{q'})) \succ_1 s_i$ for all i in $1..m$. Therefore, by definition of $\succ_1$ and since $top(I(f(v_1 \ldots v_{p'}, s_j, v'_1 \ldots v'_{q'})))$ is f, it follows that $I(f(v_1 \ldots v_{p'}, s_j, v'_1 \ldots v'_{q'})) \succ_1 g(s_1 \ldots s_m)$.

4. $f \in \mathcal{F}_{AC}$, $f \succ_\mathcal{F} g$ and $m = 0$. Then we know that $I(f(v_1 \ldots v_{p'}, g, v'_1 \ldots v'_{q'})) = f(v_1 \ldots v_{p'}, \bot, v'_1 \ldots v'_{q'})$, which is greater wrt. $\succ_1$ than g.

Deletion property: We obtain $f(\ldots s \ldots) \succ f(\ldots \ldots)$ for all $f \in \mathcal{F}_{AC}$ by proving directly $I(f(\ldots s \ldots)) \succ_1 I(f(\ldots \ldots))$. Let $f(\ldots s \ldots)$ be a term $f(u_1 \ldots u_p, s, u'_1 \ldots u'_q)$, with $u_1 \ldots u_p, u'_1 \ldots u'_q$ in $\mathcal{T}(\mathcal{F})$ and $p + q + 1 \in \alpha(f)$.

If we suppose $I_f(u_1) \ldots I_f(u_p) = v_1 \ldots v_{p'}$ and $I_f(u'_1) \ldots I_f(u'_q) = v'_1 \ldots v'_{q'}$ then $I(f(\ldots s \ldots)) = f(v_1 \ldots v_{p'}, I_f(s), v'_1 \ldots v'_{q'})$ and $I(f(\ldots \ldots)) = f(v_1 \ldots v_{p'}, v'_1 \ldots v'_{q'})$. But since $I_f(s)$ is non-empty and $f \in \mathcal{F}_{AC}$, we have $f(v_1 \ldots v_{p'}, I_f(s), v'_1 \ldots v'_{q'}) \succ_1 f(v_1 \ldots v_{p'}, v'_1 \ldots v'_{q'})$. □

The following two lemmas will be used to prove the monotonicity of $\succ$.

Lemma 4.5 Let s and t be ground terms where $top(s) = f$ and $f \in \mathcal{F}_{AC}$. If $I(s) \succ_1 I(t)$ then $I(f(\ldots s \ldots)) \succ_1 I(f(\ldots t \ldots))$.

Proof The proof is by induction on $|s| + |t|$. Let $I(s)$ be $f(s_1 \ldots s_m)$ with $m \geq 2$ and let $I(t)$ be $g(t_1 \ldots t_n)$. We will show that
$I(s') = I(f(u_1 \ldots u_p, s, u'_1 \ldots u'_q)) \succ_1 I(f(u_1 \ldots u_p, t, u'_1 \ldots u'_q)) = I(t')$
for any $u_1 \ldots u_p, u'_1 \ldots u'_q$ in $\mathcal{T}(\mathcal{F})$ s.t. $p + q + 1 > 1$. If we suppose $I_f(u_1) \ldots I_f(u_p) = v_1 \ldots v_{p'}$ and $I_f(u'_1) \ldots I_f(u'_q) = v'_1 \ldots v'_{q'}$ then we have

$$I(s') = I(f(u_1 \ldots u_p, I(s), u'_1 \ldots u'_q)) = f(v_1 \ldots v_{p'}, s_1 \ldots s_m, v'_1 \ldots v'_{q'})$$
$$I(t') = I(f(u_1 \ldots u_p, I(t), u'_1 \ldots u'_q)) = I(f(v_1 \ldots v_{p'}, g(t_1 \ldots t_n), v'_1 \ldots v'_{q'}))$$

We consider three cases:

1. $g \succ_\mathcal{F} f$. Then $s_i \succeq_1 g(t_1 \ldots t_n)$ for some i in $1..m$, by definition of $\succ_1$, and $I(t') = f(v_1 \ldots v_{p'}, g(t_1 \ldots t_n), v'_1 \ldots v'_{q'})$. Therefore, since $m \geq 2$, it follows that $I(s') = f(v_1 \ldots v_{p'}, s_1 \ldots s_m, v'_1 \ldots v'_{q'}) \succ_1 f(v_1 \ldots v_{p'}, g(t_1 \ldots t_n), v'_1 \ldots v'_{q'}) = I(t')$.

2. $g = f$. Then we have $\{s_1, \ldots, s_m\} \gg_1 \{t_1, \ldots, t_n\}$ and
$I(s') = f(v_1 \ldots v_{p'}, s_1 \ldots s_m, v'_1 \ldots v'_{q'}) \succ_1 f(v_1 \ldots v_{p'}, t_1 \ldots t_n, v'_1 \ldots v'_{q'}) = I(t')$.

3. $f \succ_{\mathcal{F}} g$. Then we distinguish two more cases:

   (a) g is a constant symbol. In this case we have $I(t') = f(v_1 \ldots v_{p'}, \bot, v'_1 \ldots v'_{q'})$. Since $m \geq 2$ and $\bot$ is the smallest term, it follows (by definition of $\succ_1$) that $I(s') = f(v_1 \ldots v_{p'}, s_1 \ldots s_m, v'_1 \ldots v'_{q'}) \succ_1 f(v_1 \ldots v_{p'}, \bot, v'_1 \ldots v'_{q'}) = I(t')$.

   (b) g is not a constant symbol. Then by definition of $\succ_1$ $I(s) \succ_1 t_i = I(t_i)$ for all i in $1..n$ and therefore by induction hypothesis
   $f(v_1 \ldots v_{p'}, s_1 \ldots s_m, v'_1 \ldots v'_{q'}) = I(f(v_1 \ldots v_{p'}, I(s), v'_1 \ldots v'_{q'}) \succ_1$
   $I(f(v_1 \ldots v_{p'}, I(t_i), v'_1 \ldots v'_{q'})) = I(f(v_1 \ldots v_{p'}, t_i, v'_1 \ldots v'_{q'}))$ for all i in $1..n$.
   Since $I(t') = I(f(v_1 \ldots v_{p'}, t_j, v'_1 \ldots v'_{q'}))$ for some j in $1..n$, it follows that $I(s') \succ_1 I(t')$. □

Lemma 4.6 Let s and t be ground terms. If $I(s) \succ_1 I(t)$ then $I(f(\ldots s \ldots)) \succeq_1 I(f(\ldots t \ldots))$.

Proof The proof is by induction on $|s| + |t|$. Let $I(s)$ be $g(s_1 \ldots s_m)$ and let $I(t)$ be $h(t_1 \ldots t_n)$. We will show that
$$I(s') = I(f(u_1 \ldots u_p, s, u'_1 \ldots u'_q)) \succeq_1 I(f(u_1 \ldots u_p, t, u'_1 \ldots u'_q)) = I(t')$$
for any $u_1 \ldots u_p, u'_1 \ldots u'_q$ in $\mathcal{T}(\mathcal{F})$ s.t. $p + q + 1 \in \alpha(f)$. If we suppose $I_f(u_1) \ldots I_f(u_p) = v_1 \ldots v_{p'}$ and $I_f(u'_1) \ldots I_f(u'_q) = v'_1 \ldots v'_{q'}$ then we have
$I(s') = I(f(u_1 \ldots u_p, I(s), u'_1 \ldots u'_q)) = I(f(v_1 \ldots v_{p'}, g(s_1 \ldots s_m), v'_1 \ldots v'_{q'}))$
$I(t') = I(f(u_1 \ldots u_p, I(t), u'_1 \ldots u'_q)) = I(f(v_1 \ldots v_{p'}, h(t_1 \ldots t_n), v'_1 \ldots v'_{q'}))$

If $f \notin \mathcal{F}_{AC}$ then
$I(s') = f(v_1 \ldots v_{p'}, g(s_1 \ldots s_m), v'_1 \ldots v'_{q'})$ and $I(t') = f(v_1 \ldots v_{p'}, h(t_1 \ldots t_n), v'_1 \ldots v'_{q'})$.
Therefore $I(s') \succ_1 I(t')$ by monotonicity of $\succ_1$, since $g(s_1 \ldots s_m) \succ_1 h(t_1 \ldots t_n)$.

If $f \in \mathcal{F}_{AC}$, then the case $f = g$ holds by the previous lemma. If $g \succ_{\mathcal{F}} f$, then by monotonicity of $\succ_1$ and by lemma 3.6(8), $I(s') = f(v_1 \ldots v_{p'}, g(s_1 \ldots s_m), v'_1 \ldots v'_{q'}) \succ_1 f(v_1 \ldots v_{p'}, h(t_1 \ldots t_n), v'_1 \ldots v'_{q'}) \succeq_1 I(t')$. If $f \succ_{\mathcal{F}} g$, there are three more cases depending on how g and h are related:

1. $g = h$. Then $\{s_1, \ldots, s_m\} \gg_1 \{t_1, \ldots, t_n\}$. So if $s_j = max_{\succ_1} \{s_1, \ldots, s_m\}$ and $t_k = max_{\succ_1} \{s_1, \ldots, s_m\}$ then $s_j \succeq_1 t_k$ must be true. Then:

   (a) $I(s_j) = s_j \succ_1 t_k = I(t_k)$. In this case by induction hypothesis we have that
   $I(s') = I(f(v_1 \ldots v_{p'}, s_j, v'_1 \ldots v'_{q'}) = I(f(v_1 \ldots v_{p'}, I(s_j), v'_1 \ldots v'_{q'}) \succeq_1$
   $\succeq_1 I(f(v_1 \ldots v_{p'}, I(t_k), v'_1 \ldots v'_{q'})) = I(f(v_1 \ldots v_{p'}, t_k, v'_1 \ldots v'_{q'})) = I(t')$.

   (b) $s_j =_{ms} t_k$. Then it follows that $I(s') = I(f(v_1 \ldots v_{p'}, s_j, v'_1 \ldots v'_{q'}) =_{ms}$
   $=_{ms} I(f(v_1 \ldots v_{p'}, t_k, v'_1 \ldots v'_{q'})) = I(t')$.

2. $h \succ_{\mathcal{F}} g$. Then $s_i \succeq_1 h(t_1 \ldots t_n)$ for some i in $1..m$ and if $s_j = max_{\succ_1} \{s_1, \ldots, s_m\}$ then obviously $s_j \succeq_1 h(t_1 \ldots t_n)$. We treat separately the cases $\succ_1$ and $=_{ms}$:

   (a) $I(s_j) = s_j \succ_1 h(t_1 \ldots t_n) = I(t)$. Then by induction hypothesis we have
   $I(s') = I(f(v_1 \ldots v_{p'}, s_j, v'_1 \ldots v'_{q'}) = I(f(v_1 \ldots v_{p'}, I(s_j), v'_1 \ldots v'_{q'}) \succeq_1$
   $\succeq_1 I(f(v_1 \ldots v_{p'}, I(t), v'_1 \ldots v'_{q'})) = I(t')$.

(b) $I(s_j) = s_j =_{ms} h(t_1 \ldots t_n) = I(t)$. Then, by lemma 3.6 (prop. 6), $I(s') = I(f(v_1 \ldots v_{p'}, s_j, v'_1 \ldots v'_{q'})) =_{ms} I(f(v_1 \ldots v_{p'}, t, v'_1 \ldots v'_{q'})) = I(t')$.

3. $g \succ_{\mathcal{F}} h$. Then we distinguish two more cases:

   (a) h is a constant symbol. In this case $I(t') = f(v_1 \ldots v_{p'}, \perp, v'_1 \ldots v'_{q'})$, and, since $\perp$ is the smallest term we can replace $g(s_1 \ldots s_m)$ for, it follows that $I(s') = I(f(v_1 \ldots v_{p'}, g(s_1 \ldots s_m), v'_1 \ldots v'_{q'})) \succeq_1 f(v_1 \ldots v_{p'}, \perp, v'_1 \ldots v'_{q'}) = I(t')$.

   (b) h is not a constant symbol. Then we have $I(s) \succ_1 t_i = I(t_i)$ for all i in $1..n$ and therefore by induction hypothesis $I(f(v_1 \ldots v_{p'}, I(s), v'_1 \ldots v'_{q'})) \succeq_1$ $\succeq_1 I(f(v_1 \ldots v_{p'}, I(t_i), v'_1 \ldots v'_{q'})) = I(f(v_1 \ldots v_{p'}, t_i, v'_1 \ldots v'_{q'}))$ for all i in $1..n$. Since $I(t') = I(f(v_1 \ldots v_{p'}, t_j, v'_1 \ldots v'_{q'}))$ for some j in $1..n$, it follows that $I(s') \succ_1 I(t')$. □

Lemma 4.7 The ordering $\succ$ is monotonic, i.e. $s \succ t$ implies $f(\ldots s \ldots) \succ f(\ldots t \ldots)$ for all $s, t \in T(\mathcal{F})$.

Proof We will show that $s' = f(u_1 \ldots u_p, s, u'_1 \ldots u'_q) \succ f(u_1 \ldots u_p, t, u'_1 \ldots u'_q) = t'$ for all $u_1 \ldots u_p, u'_1 \ldots u'_q$ in $T(\mathcal{F})$ s.t. $p + q + 1 \in \alpha(f)$.

If $s \succ t$ then $I(s) \succ_1 I(t)$ or $I(s) =_{ms} I(t)$. By lemmas 4.6 and 3.6 (prop. 6) we have $I(s') \succ_1 I(t')$ or $I(s') =_{ms} I(t')$. If $I(s') \succ_1 I(t')$ then obviously $s' \succ t'$ holds. Then from now on in this proof we assume $I(s') =_{ms} I(t')$. We distinguish two cases:

1. $f \notin \mathcal{F}_{AC}$. Then $tf(s') = \langle u_1 \ldots u_p, s, u'_1 \ldots u'_q \rangle$ and $tf(t') = \langle u_1 \ldots u_p, t, u'_1 \ldots u'_q \rangle$, and since $u_i =_{AC} u_i$ for all i in $1..p$ and $s \succ t$ it follows that $s' \succ t'$.

2. $f \in \mathcal{F}_{AC}$. Assume $tf(s) = \langle s_1 \ldots s_m \rangle$, $tf(t) = \langle t_1 \ldots t_n \rangle$, $top(s) = g$ and $top(t) = h$. Suppose $tf(f(u_1 \ldots u_p)) = \langle v_1 \ldots v_{p'} \rangle$ and $tf(f(u'_1 \ldots u'_q)) = \langle v'_1, \ldots, v'_{q'} \rangle$. Now if:

   (a) $I(s) =_{ms} I(t)$. Then $top(s) = top(t)$ and $\{s_1, \ldots, s_m\} \succ\!\!\succ \{t_1, \ldots, t_n\}$. Now, if $f \neq top(s)$ then
   $tf(s') = \{v_1, \ldots, v_{p'}, s, v'_1, \ldots, v'_{q'}\} \succ\!\!\succ \{v_1, \ldots, v_{p'}, t, v'_1, \ldots, v'_{q'}\} = tf(t')$
   If $f = top(s)$ then $s' \succ t'$ since
   $\{v_1, \ldots, v_{p'}, s_1, \ldots, s_m, v'_1, \ldots, v'_{q'}\} \succ\!\!\succ \{v_1, \ldots, v_{p'}, t_1, \ldots, t_n, v'_1, \ldots, v'_{q'}\}$.

   (b) $I(s) \succ_1 I(t)$. Then $top(s) \neq f$ since otherwise by lemma 4.5 $I(s') \succ_1 I(t')$. Now if $f = top(t)$ then, since t_i is a proper subterm of t for all i in $1..n$, by the subterm property (lemma 4.4) we have $t \succ t_i$, and, since $s \succ t$, by transitivity (lemma 4.3) $s \succ t_i$ for all i in $1..n$. Now it follows that $tf(s') = \{v_1, \ldots, v_{p'}, s, v'_1, \ldots, v'_{q'}\} \succ\!\!\succ \{v_1, \ldots, v_{p'}, t_1, \ldots, t_n, v'_1, \ldots, v'_{q'}\} = tf(t')$.
   Otherwise if $f \neq top(t)$ then
   $tf(s') = \{v_1, \ldots, v_{p'}, s, v'_1, \ldots, v'_{q'}\} \succ\!\!\succ \{v_1, \ldots, v_{p'}, t, v'_1, \ldots, v'_{q'}\} = tf(t')$.
   □

Lemma 4.8 The ordering $\succ$ is total on non-AC-equivalent terms, i.e. given two ground terms s and t, either $s =_{AC} t$ or s and t are related by $\succ$ ($s \succ t$ or $t \succ s$).

Proof The proof is done by induction on $|s| + |t|$. Suppose $s \neq_{AC} t$.

1. If $I(s) \neq_{ms} I(t)$ then $s \succ t$ or $t \succ s$ holds since $\succ_1$ is total on terms non-equivalent under $=_{ms}$.

2. If $I(s) =_{ms} I(t)$ then $top(s) = top(t)$. Suppose $\mathit{tf}(s) = \langle s_1, \ldots, s_m \rangle$ and $\mathit{tf}(t) = \langle t_1, \ldots, t_n \rangle$. There are two possibilities:

   (a) $top(s) \in \mathcal{F}_{AC}$. Then by induction hypothesis all terms s_i and t_j are comparable for i in $1..m$ and j in $1..n$, i.e. $u =_{AC} v$ or $u \succ v$ or $v \succ u$ for all $u, v \in \{s_1, \ldots, s_m, t_1, \ldots, t_n\}$. It cannot be the case that $\{s_1, \ldots, s_m\} ==_{AC} \{t_1, \ldots, t_n\}$ because $s \neq_{AC} t$, and therefore $m + n \neq 0$. We now prove that such sets $\{s_1, \ldots, s_m\}$ and $\{t_1, \ldots, t_n\}$ are comparable wrt. $\succcurlyeq$ by induction on $m + n$. If $m = 0$ or $n = 0$ then this is the case. Otherwise, since $\succ$ is total on $\{s_1, \ldots, s_m\}$ and on $\{t_1, \ldots, t_n\}$, there are terms u and v that are maximal wrt. $\succ$ in $\{s_1, \ldots, s_m\}$ and in $\{t_1, \ldots, t_n\}$ respectively. If $u =_{AC} v$ then by induction hypothesis $\{s_1, \ldots, s_m\} \setminus \{u\}$ is comparable (wrt. $\succcurlyeq$) with $\{t_1, \ldots, t_n\} \setminus \{v\}$ (since they are not equivalent wrt. $==_{AC}$) and therefore $\{s_1, \ldots, s_m\}$ and $\{t_1, \ldots, t_n\}$ are comparable wrt. $\succcurlyeq$. If $u \neq_{AC} v$ then suppose wlog. that $u \succ v$. By transitivity and AC-compatibility $u \succ t_j$ for all j in $1..n$ and therefore $\{s_1, \ldots, s_m\} \succcurlyeq \{t_1, \ldots, t_n\}$.

   (b) $top(s) \notin \mathcal{F}_{AC}$. Then if i is the minimum in $1..m$ s.t. $s_i \neq_{AC} t_i$, by induction hypothesis $s_i \succ t_i$ or $t_i \succ s_i$ and therefore $s \succ t$ or $t \succ s$. □

Theorem 4.9 The ordering $\succ$ is a simplification ordering on $\mathcal{T}(\mathcal{F})$ that is total on non-AC-equivalent terms in $\mathcal{T}(\mathcal{F})$.

5 The extension to terms with variables

Here we consider the same rules R_F and R_I as before, except that now the first type of rules of R_I is adapted to terms with variables, i.e.:

$$f(x_1 \ldots x_m, g(t_1 \ldots t_r), y_1 \ldots y_n) \to f(x_1 \ldots x_m, t_j, y_1 \ldots y_n)$$

for all $f \in \mathcal{F}_{AC}$, non-constant symbols g s.t. $f \succ_\mathcal{F} g$, $m + n \geq 1$ and $r \in \alpha(g)$ and for all $t_1, \ldots, t_r \in \mathcal{T}(\mathcal{F}, \mathcal{X})$ s.t. $t_j = max_{\succ_1}\{t_1, \ldots, t_r\}$.

The only difference is that here the terms $t_1, \ldots, t_r$ belong to $\mathcal{T}(\mathcal{F}, \mathcal{X})$, and therefore there may by more than one maximal element in $\{t_1, \ldots, t_r\}$ wrt. $\succ_1$, which implies that the interpretation in this case is a *set* of normal forms:

Definition 5.1 Let s be a term in $\mathcal{T}(\mathcal{F}, \mathcal{X})$. We define the variable interpretation of s, denoted by $I_v(s)$, as the set of normal forms of s wrt. $R_F \cup R_I$ under the strategy (i) leftmost-innermost and (ii) rules of R_F first.

Definition 5.2 Let s and t be terms in $\mathcal{T}(\mathcal{F}, \mathcal{X})$, s.t. $I_v(s) = \{s_1, \ldots, s_m\}$ and $I_v(t) = \{t_1, \ldots, t_n\}$, and $\mathit{tf}(s) = \langle s'_1, \ldots, s'_p \rangle$ and $\mathit{tf}(t) = \langle t'_1, \ldots, t'_q \rangle$. Then $s \succ_v t$ if and only if:

1. $\forall j : 1..n \; \exists i : 1..m \; s.t. \; s_i \succ_1 t_j$ or

2. $\forall j : 1..n \; \exists i : 1..m \; s.t. \; s_i \succeq_1 t_j$ and

   (a) $top(s) \in \mathcal{F}_{AC}$ and $\{s'_1, \ldots, s'_p\} \succ\!\!\succ_v \{t'_1, \ldots, t'_q\}$ or

   (b) $top(s) \notin \mathcal{F}_{AC}$ and $\exists i : 1..p \; \forall j < i \; s_j =_{AC} t_j \wedge s_i \succ_v t_i$.

In theorem 5.5 we will prove that for all s and t in $T(\mathcal{F}, \mathcal{X})$ and for all ground substitutions σ, if $s \succ_v t$ then $s\sigma \succ t\sigma$. But, before we are able to do that, we need the following lemmas:

Lemma 5.3 Let s and t be terms in $T(\mathcal{F}, \mathcal{X})$ with $s' \in I_v(s)$ and $t' \in I_v(t)$. Then $s' \succ_1 t'$ implies $I(s'\sigma) \succ_1 I(t'\sigma)$ for all ground substitutions σ.

Proof By induction on $|s'| + |t'|$.

If t' is some variable x, then, since $s' \succ_1 t'$, x is a proper subterm of s' and $t'\sigma$ is a proper subterm of $s'\sigma$. Then by lemma 4.4 $I(s'\sigma) \succ_1 I(t'\sigma)$.

If t' is not a variable, then it is of the form $g(t_1 \ldots t_n)$ with $n \geq 0$. Since $s' \succ_1 t'$, the term s' cannot be a variable, so let s be of the form $f(s_1 \ldots s_m)$ with $m \geq 0$. Then we have the following cases:

1. $g \succ_\mathcal{F} f$. Then there is an s_i s.t. $s_i \succeq_1 t'$. If $s_i =_{ms} t'$ then, by lemma 4.4, $I(s'\sigma) = I(f(s_1\sigma, \ldots, s_m\sigma)) \succ_1 I(s_i\sigma) =_{ms} I(t'\sigma)$. Similarly, if $s_i \succ_1 t'$ then by induction hypothesis $I(s_i\sigma) \succ_1 I(t'\sigma)$ and also $I(s'\sigma) \succ_1 I(t'\sigma)$.

2. $f \succ_\mathcal{F} g$. Then $s' \succ_1 t_i$ for all i in $1..n$ and by induction hypothesis $I(s'\sigma) \succ_1 I(t_i\sigma)$ for all i in $1..n$. Now $top(I(s'\sigma)) = f$ and thus $I(s'\sigma) \succ_1 g(I(t_1\sigma), \ldots, I(t_n\sigma)) \succeq_1 I(g(I(t_1\sigma), \ldots, I(t_n\sigma))) = I(t'\sigma)$.

3. $f = g$.

   If $f \notin \mathcal{F}_{AC}$ then $n = m$ and $s_1 =_{ms} t_1 \wedge \ldots \wedge s_{i-1} =_{ms} t_{i-1} \wedge s_i \succ_1 t_i$ for some i in $1..m$. Now we have
   $I(s'\sigma) = I(f(s_1, \ldots, s_m)) = f(I(s_1), \ldots, I(s_m))$ and
   $I(t'\sigma) = I(f(t_1, \ldots, t_n)) = f(I(t_1), \ldots, I(t_n))$, so we need to prove that
   $f(I(s_1), \ldots, I(s_m)) \succ_1 f(I(t_1), \ldots, I(t_n))$, which means
   $I(s_1\sigma) =_{ms} I(t_1\sigma) \wedge \ldots \wedge I(s_{i-1}\sigma) =_{ms} I(t_{i-1}\sigma) \wedge I(s_i\sigma) \succ_1 I(t_i\sigma)$, which holds by the induction hypothesis.

   If $f \in \mathcal{F}_{AC}$ then $\{s_1 \ldots s_m\} \succ\!\!\succ_1 \{t_1 \ldots t_n\}$. Now we have
   $I(s'\sigma) = I(f(s_1\sigma, \ldots, s_m\sigma)) = f(I_f(s_1\sigma), \ldots, I_f(s_m\sigma))$ and
   $I(t'\sigma) = I(f(t_1\sigma, \ldots, t_n\sigma)) = f(I_f(t_1\sigma), \ldots, I_f(t_n\sigma))$, so we need to prove that
   $f(I_f(s_1\sigma), \ldots, I_f(s_m\sigma)) \succ_1 f(I_f(t_1\sigma), \ldots, I_f(t_n\sigma))$, which means
   $\{I_f(s_1\sigma), \ldots, I_f(s_m\sigma)\} \succ\!\!\succ_1 \{I_f(t_1\sigma), \ldots, I_f(t_n\sigma)\}$. We proceed by induction on m.

   If no s_i in $1..m$ is a variable, then $I(s_i\sigma) = I_f(s_i\sigma)$ for all i, since s' is in normal form wrt. the interpretation I_v. But then, since by the induction hypothesis $s_i \succ_1 t_j$ implies $I(s_i\sigma) \succ_1 I(t_j\sigma)$, also $\{s_1 \ldots s_m\} \succ\!\!\succ_1 \{t_1 \ldots t_n\}$ implies

$\{I(s_1\sigma)\ldots I(s_m\sigma)\}\gg_1\{I(t_1\sigma)\ldots I(t_n\sigma)\}$ and therefore
$\{I_f(s_1\sigma),\ldots,I_f(s_m\sigma)\}\gg_1\{I_f(t_1\sigma),\ldots,I_f(t_n\sigma)\}$.

If some s_i (wlog. s_1) is a variable x and $x = t_j$ for some t_j (wlog. t_1) then $\{s_2\ldots s_m\}\gg_1\{t_2\ldots t_n\}$ and by induction hypothesis
$\{I_f(s_1\sigma),\ldots,I_f(s_m\sigma)\}\gg_1\{I_f(t_1\sigma),\ldots,I_f(t_n\sigma)\}$.

If some s_i (wlog. s_1) is a variable x and $x \neq t_j \ \forall\, t_j$, then $\{s_2\ldots s_m\}\succeq_1\{t_1\ldots t_n\}$.
If $\{s_2\ldots s_m\} =_{ms} \{t_1\ldots t_n\}$, then $\{I_f(s_1\sigma)\ldots I_f(s_m\sigma)\}\gg_1\{I_f(t_1\sigma)\ldots I_f(t_n\sigma)\}$.
If $\{s_2\ldots s_m\}\gg_1\{t_1\ldots t_n\}$, then $\{I_f(s_1\sigma)\ldots I_f(s_m\sigma)\}\gg_1\{I_f(t_1\sigma)\ldots I_f(t_n\sigma)\}$ by induction hypothesis. □

Lemma 5.4 Let s be a term in $T(\mathcal{F},\mathcal{X})$ with $I_v(s) = \{t_1,\ldots,t_n\}$. Then, for all ground substitutions σ, $I(s\sigma) =_{ms} max_{\succ_1}\{I(t_1\sigma),\ldots,I(t_n\sigma)\}$.

We omit here the proof of the previous lemma. It essentially holds because the variable interpretations keep at any point all the terms that could be maximal for some σ.

Theorem 5.5 For all s and t in $T(\mathcal{F},\mathcal{X})$ and for all ground substitutions σ, if $s \succ_v t$ then $s\sigma \succ t\sigma$.

Proof By induction on $|s| + |t|$. Suppose $I_v(s) = \{s_1,\ldots,s_m\}$ and $I_v(t) = \{t_1,\ldots,t_n\}$, and $tf(s) = \langle s'_1,\ldots,s'_p\rangle$ and $tf(t) = \langle t'_1,\ldots,t'_q\rangle$.

If $s \succ_v t$ by case 1. of the definition of $\succ_v$, then $\forall j : 1..n \ \exists i : 1..m \ s.t. \ s_i \succ_1 t_j$. Now by lemma 5.3 $s_i \succ_1 t_j$ implies $I(s_i\sigma) \succ_1 I(t_j\sigma)$ for all i, j and σ. But then, if for some i and j we have $I(s_i\sigma) = max_{\succ_1}\{I(s_1\sigma),\ldots,I(s_m\sigma)\}$ and $I(t_j\sigma) = max_{\succ_1}\{I(t_1\sigma),\ldots,I(t_n\sigma)\}$, it holds that $I(s_i\sigma) \succ_1 I(t_j\sigma)$, which implies $I(s\sigma) \succ_1 I(t\sigma)$, and $s\sigma \succ t\sigma$, since by lemma 5.4 $I(s\sigma) = I(s_i\sigma)$ and $I(t\sigma) = I(t_j\sigma)$.

If $s \succ_v t$ by case 2. of the definition of $\succ_v$, then $\forall j : 1..n \ \exists i : 1..m \ s.t. \ s_i \succeq_1 t_j$. But then, again, if for some i and j we have $I(s_i\sigma) = max_{\succ_1}\{I(s_1\sigma),\ldots,I(s_m\sigma)\}$ and $I(t_j\sigma) = max_{\succ_1}\{I(t_1\sigma),\ldots,I(t_n\sigma)\}$, it holds that $I(s_i\sigma) \succeq_1 I(t_j\sigma)$, which implies $I(s\sigma) \succeq_1 I(t\sigma)$, since by lemma 5.4 $I(s\sigma) = I(s_i\sigma)$ and $I(t\sigma) = I(t_j\sigma)$. Now if $I(s\sigma) \succ_1 I(t\sigma)$, then $s\sigma \succ t\sigma$ directly. Otherwise, $I(s\sigma) =_{ms} I(t\sigma)$. Now if $top(s) \in \mathcal{F}_{AC}$ then, since $\{s'_1,\ldots,s'_p\}\gg_v\{t'_1,\ldots,t'_q\}$, by induction hypothesis in the usual way we get $\{s'_1\sigma,\ldots,s'_p\sigma\}\gg\{t'_1\sigma,\ldots,t'_q\sigma\}$, and therefore $s\sigma \succ t\sigma$.

If $top(s) \notin \mathcal{F}_{AC}$ then $s'_1 =_{AC} t'_1 \wedge \ldots \wedge s'_{i-1} =_{AC} t'_{i-1} \wedge s'_i \succ_v t'_i$ implies by induction hypothesis $s'_1\sigma =_{AC} t'_1\sigma \wedge \ldots \wedge s'_{i-1}\sigma =_{AC} t'_{i-1}\sigma \wedge s'_i\sigma \succ t'_i\sigma$, which implies $s\sigma \succ t\sigma$. □

Example 5.6 Suppose $*, + \in \mathcal{F}_{AC}$ and $* \succ_\mathcal{F} +$. Then $x * y + x * z \succ_v x * (y + z)$, since $I_v(x * y + x * z) = \{x * y, x * z\}$, $I_v(x * (y + z)) = \{x * y, x * z\}$ and $x * y + x * z \succ_1 x * y$ and $x * y + x * z \succ_1 x * z$.

Example 5.7 *Rings.* With $+ \in \mathcal{F}_{AC}$ and $* \succ_\mathcal{F} I \succ_\mathcal{F} + \succ_\mathcal{F} 0$, we have the following canonical term rewrite system:

$$
\begin{aligned}
0 + x &\rightarrow x \\
I(x) + x &\rightarrow 0 \\
I(0) &\rightarrow 0 \\
I(I(x)) &\rightarrow x \\
I(x + y) &\rightarrow I(x) + I(y) \\
x * (y + z) &\rightarrow (x * y) + (x * z) \\
(x + y) * z &\rightarrow (x * z) + (y * z) \\
0 * x &\rightarrow 0 \\
x * 0 &\rightarrow 0 \\
x * I(y) &\rightarrow I(x * y) \\
I(x) * y &\rightarrow I(x * y)
\end{aligned}
$$

Note that we can not define $*$ as an AC-symbol, since, as shown in the previous example, the distributivity axiom is then oriented reversely.

Example 5.8 *Milners's nondeterministic machines.* With $+ \in \mathcal{F}_{AC}$ and $L \succ_{\mathcal{F}} T \succ_{\mathcal{F}} + \succ_{\mathcal{F}} 0$ we have the following canonical term rewrite system:

$$
\begin{aligned}
0 + x &\rightarrow x \\
x + x &\rightarrow x \\
T(T(x)) &\rightarrow T(x) \\
L(T(x)) &\rightarrow L(x) \\
T(x) + x &\rightarrow T(x) \\
T(x + y) + x &\rightarrow T(x + y) \\
T(T(y) + x) &\rightarrow T(x + y) + T(y) \\
L(T(y) + x) &\rightarrow L(x + y) + L(y)
\end{aligned}
$$

6 Conclusions and further work

As shown in example 5.6, in many cases the orientations obtained are different from the usual ones. For instance, two AC-compatible orderings that can orient the distributivity axiom in the "appropriate" way are the *associative path ordering* [BP85] and the one given in [KSZ90]. The problem with these orderings is that they are not total on ground terms. In the first one (APO) two AC-symbols cannot be compared in the precedence unless they are related by a distributivity law. The second one, on the other hand, is not total because of its way of comparing terms with the same AC-top-symbol, since sometimes the subterms can be used only once in the comparison. For instance, if $a \succ_{\mathcal{F}} b \succ_{\mathcal{F}} c$ then the terms $f(a,c)$ and $f(b,b)$ are incomparable. Again, this is not the case in our ordering, since we compare multisets in the usual way (as RPO, for instance) which preserves the totality property.

However, note that in unfailing strategies, where the totality property is essential, it is less crucial in which direction the rules are oriented. Furthermore, as mentioned in the introduction, we believe that the ordering defined here could lead to a procedure for deciding the satisfiability of AC-ordering constraints, which, as shown for the case modulo the empty theory, has interesting applications in automated theorem proving.

Finally we want to point out that the strategy chosen in the definition 3.3 is the only one we have found that allows to get a unique interpretation for any ground term and to fulfil the properties of lemma 3.6, and that other similar or at first sight better strategies, may not satisfy the uniqueness or some other property. For instance, if the interpretation were defined by means of the rewrite relation $\to_{R_I} \to^!_{R_F}$ on flattened terms without any strategy, i.e. rewriting by R_I at arbitrary positions, and keeping the terms flattened after each step, then some terms have no unique interpretation: with $f \succ g \succ h \succ i \succ a \succ b$ and $h, i \in \mathcal{F}_{AC}$, $h(b, i(f(a,b), g(f(b,b))))$ would have two possible interpretations: $h(b, g(f(b,b)))$ and $h(b, f(b,b))$.

References

[Bac91] Leo Bachmair. Associative-commutative reduction orderings. Technical Report MPI-I-91-209, Max-Planck-Institut für Informatik, Saarbrücken, November 1991. To appear in IPL.

[BCL87] Ahlem Ben-Cherifa and Pierre Lescanne. Termination of rewriting systems by polynomial interpretations and its implementation. *Science of Computer Programming*, 9:137–160, 1987.

[BP85] L. Bachmair and D. A. Plaisted. Termination orderings for associative-commutative rewriting systems. *Journal of Symbolic Computation*, 1:329–349, 1985.

[Com90] Hubert Comon. Solving symbolic ordering constraints. *International Journal of Foundations of Computer Science*, 1(4):387–411, 1990.

[Der87] Nachum Dershowitz. Termination of rewriting. *Journal of Symbolic Computation*, 3:69–116, 1987.

[JO91] J-P. Jouannaud and M. Okada. Satisfiability of systems of ordinal notations with the subterm property is decidable. In *Automata, Languages and Programming, 18th International Colloquium*, LNCS 510, Madrid, Spain, July 16–20 1991. Springer-Verlag.

[KSZ90] D. Kapur, G. Sivakumar, and H. Zhang. A new method for proving termination of ac-rewrite systems. In *Conf. Found. of Software Technology and Theoretical Computer Science*, LNCS 472, pages 134–148, New Delhi, India, December 1990. Springer-Verlag.

[NR91] Paliath Narendran and Michael Rusinowitch. Any ground associative commutative theory has a finite canonical system. In *Fourth int. conf. on Rewriting Techniques and Applications*, LNCS 488, pages 423–434, Como, Italy, April 1991. Springer-Verlag.

[NR92] Robert Nieuwenhuis and Albert Rubio. Theorem proving with ordering constrained clauses. In Deepak Kapur, editor, *11th International Conference on Automated Deduction*, LNAI 607, pages 477–491, Saratoga Springs, New York, USA, June 15–18, 1992. Springer-Verlag.

Extension of the associative path ordering to a chain of associative commutative symbols

Catherine DELOR and Laurence PUEL

Laboratoire de Recherche en Informatique
Bat.490, Université de Paris Sud
91405 ORSAY cedex, France.
E-mail puel@lri.lri.fr

Abstract. In this paper, we give a generalization of the *associative path ordering*. This ordering has been introduced by Bachmair and Plaisted [5] and is a restricted variant of the *recursive path ordering* which can be used for proving the termination of associative-commutative term rewriting systems. This ordering requires strong conditions on the precedence on the alphabet. In this article, we treat the case of a precedence which contains a chain of AC symbols. We also introduce some unary symbols comparable with AC symbols.

Introduction

A classical means of computing with equations is to use rewriting systems. The main prerequisite of the methods based on this principle is the termination of the reduction defined by the rewriting system. When we consider a classical term algebra, we dispose of many techniques for proving termination. A very powerful tool is for example the *recursive path ordering* which extends a precedence on the set of symbols. But when we have to deal with an equational theory, it is much more difficult to find a suitable method for proving termination.

In this paper, we study an ordering for proving termination in an associative-commutative theory. This ordering is a restricted variant of the *rpo*, and is also based on a precedence on the symbols. In section 2, we recall the definition of the *associative path ordering* introduced by Bachmair and Plaisted [5]. Then in sections 3 to 5, we extend this ordering by softening the conditions imposed on the precedence on the symbols. The main result is expressed in section 3, where we present an ordering which can be used when there exists in the alphabet an arbitrarily long chain of AC symbols. We give also another extension in which we allow unary symbols between the associative-commutative ones. Then we give some examples where these two extensions are mixed.

1 Notations and tools

1.1 Basic definitions

In this document we work on a set of terms. The set of the symbols on which the terms are built is called F. This set of symbols contains some associative and commutative symbols which belong to the subset F_{AC} of F. The set of the variables is X. The set of the terms is denoted by $T(F,X)$. For a term $t \in T(F,X)$, $t(\Lambda)$ is the head symbol of t, $\bar{t}$ is the flattened form of t.

A rewrite rule is a couple of terms $(l,r) \in T(F,X)^2$ such that the set of the variables of r is included in the set of the variables of l. We also use the notation $l \to r$ for such a rule. A *rewriting system* is a finite set of rewrite rules. We associate to a rewriting system R a reduction relation on the terms, $\to_R$, defined by $t \to_R t'$ if there exist a node u of the term t, a rule $l \to r$ and a substitution σ such that $\sigma(l) = t/u$ and $t' = t[u \leftarrow \sigma(r)]$. A rewriting system *terminates* (or is *nœtherian*) if there exists no infinite sequence of reductions by this relation. A rewriting system is *confluent* if, for every term t such that $t \to_R^* t_1$ and $t \to_R^* t_2$, there exists a term t' such that $t_1 \to_R^* t' \leftarrow_R^* t_2$. When a system R is confluent and terminating, we can associate to a term a unique normal form for R. It will be denoted by $t \downarrow_R$. Two rules R_1 and R_2 commute if $t_1 \to_{R_1} t_2 \to_{R_2} t_3$ implies that there exists a term t_4 such that $t_1 \to_{R_2} t_4 \to_{R_1} t_3$

1.2 Termination of rewriting systems

Polynomial interpretations Each symbol of the alphabet is interpreted as a polynomial on a commutative ring A on which a well-founded ordering $<_A$ is defined. Let A be a commutative ring. Let n be the biggest arity of a symbol in F. Let us define Φ, a morphism from $T(F,X)$ into $A[X_1,...,X_n]$, such that for all $f \in F$, $\Phi(f) \in A[X_1,...,X_{ar(f)}]$ and $\Phi(X) \subset A$. Then we define the ordering by $t_1 <_{pol} t_2$ iff $\Phi(t_1) <_A \Phi(t_2)$. We restrict ourselves to the polynomials the coefficients of which are in **N**. M.Rusinovitch and P.Narendran [8] introduced a new ordering for proving AC-termination using a polynomial interpretation where the commutative ring A is $\mathbf{Z}[X_1,\cdots,X_n]$. This ordering is a reduction ordering which is total on the classes generated by the associativity and commutativity axioms.

It is possible to use polynomial interpretations to prove AC-termination [6].

The recursive path ordering Let $>$ be a precedence ordering on F.

Definition 1. The *recursive path ordering* $>_{rpo}$ corresponding to $>$ is defined by:
$$s = f(s_1,\cdots,s_n) >_{rpo} t = g(t_1,\cdots,t_m)$$
if and only if one of the following conditions holds:

1. $f = g$ and $\{s_1,\cdots,s_n\} \gg_{rpo} \{t_1,\cdots,t_m\}$
2. $f > g$ and $s >_{rpo} t_i$ for all i, $1 \le i \le m$
3. $f \not> g$ and $s_i \ge_{rpo} t$, for some i, $1 \le i \le n$

where $\gg_{rpo}$ is the extension of $>_{rpo}$ to multisets.

2 The associative path ordering

2.1 Associative-commutative theory

An associative-commutative theory is a particular case of an equational theory. The set F of operator symbols contains some symbols which satisfy the following equations:
$$f(x,y) = f(y,x)$$
$$f(x,f(y,z)) = f(f(x,y),z)$$
that is to say $F_{AC} \neq \emptyset$.

In order to compare terms in such a theory, we have to use an ordering which is independent of the representative of the equivalence class of the equational relation. A way to do this is to compare unique representatives of the equivalence class instead of the terms themselves.

In the case of an AC theory, this representative is the flattened form of the terms: we consider that the elements of F_{AC} have a variable arity and we normalize the terms by the system Fl:
$$f(f(x,y),z) \rightarrow f(x,y,z)$$
$$f(x,f(y,z)) \rightarrow f(x,y,z)$$
for every $f \in F_{AC}$.

This "flattened" form of the term denoted by $\bar{t}$, is unique up to the order of the sons of the AC symbols. We say that two flattened terms are *permutation equivalent* if they are equal except for the order of the sons of their AC symbols.

Unfortunately, the recursive path ordering cannot be used to compare the flattened forms of terms, because the ordering that we obtain this way is not monotonic.

Example 1. Let us consider the alphabet $\{f,g\}$, where f and g are AC, and the precedence $f > g$. Then $f(x,y) > g(x,y)$. But if these two terms are embedded in the context $f(\_,z)$, after normalization we have to compare $f(x,y,z)$ and $f(g(x,y),z)$ which are ordered in the wrong way.

2.2 The associative path ordering

Bachmair and Plaisted [5] have introduced a new ordering which adapts the *rpo* when there exist some AC symbols. It is based on the idea of normalizing the terms by a convergent set of rules before flattening them and comparing them with the *rpo*. This ordering is called *apo* for "Associative Path Ordering".

Let us give a formal definition of this ordering. Let us suppose that the associative-commutative part of our system, F_{AC}, is not empty. For the moment, the arity of the symbols of F_{AC} is supposed to be 2. Let $>$ be a precedence ordering on F.

Definition 2 Associative Path Condition. $>$ satisfies the associative path condition if for each AC symbol f either

- f is minimal in F or
- there exists $g \in F_{AC}$ such that f is minimal in $F\backslash\{g\}$.

For this set F and this precedence $>$ we define the set of rewrite rules $\mathcal{D}$ consisting of the rules
$$f(g(x,y),z) \rightarrow g(f(x,z),f(y,z))$$
for all AC symbols f and g such that $g < f$. These rules express the distributivity of non minimal AC symbols on the others.

<u>Property</u>([5]) This system $\mathcal{D}$ terminates and is confluent.

We can define the application A from $T(F,X)$ to $T(F,X)$ which associates to a term its normal form for $\mathcal{D}$, flattened by Fl. ($A(t) = \overline{t \downarrow_\mathcal{D}}$.)

Definition 3. A partial ordering is admissible for a transform A if it is monotonic, compatible with AC, and well founded on every set $[t] = \{s | A(s) \sim A(t)\}$ where $\sim$ is the permutation equivalence.

Definition 4. Let $>$ be a precedence on F which satisfies the associative path condition and $\triangleright$ a partial ordering admissible for the transform A associated to $>$. The corresponding *apo* is defined by:
$$s >_{apo} t \text{ iff } A(s) >_{rpo} A(t) \text{ or } A(s) \sim A(t) \text{ and } s \triangleright t$$

Proposition 5. *[5] The apo is a simplification ordering.*

The associative path condition imposed on the precedence is very restrictive. In practice, we usually have to use more general orderings on the alphabet. Here, we soften the conditions on the precedence, and see the properties of the corresponding ordering. We use the same principle as in the *apo*: we transform the terms by a terminating rewriting system with respect to a strategy and then compare the normal forms.

First, we allow more than two comparable AC symbols and then we introduce unary symbols. At last we mix the two cases.

3 Apo with more than two comparable AC symbols

There are two subcases: either there exist n AC symbols h_k, $0 \leq k \leq n$ such that $h_k < h_0$ for all $k > 0$ or there exist n AC symbols h_k, $0 \leq k \leq n$ such that $h_k < h_{k+1}$ for all $k < n$

3.1 n AC symbols h_k, $0 \leq k \leq n$, and $h_k < h_0$ for all $k > 0$

We treat the particular case of three AC symbols f, g, h, such that $f < h$ and $g < h$. The method can be extended straightaway to the case of n symbols. The first idea is to extend the normalization system with other distributivity rules. More precisely, let $\mathcal{D}$ be the following system of distributivity rules, based on the same remark as in [5].

$$h(f(x,y),z) \to f(h(x,z),h(y,z))$$
$$h(g(x,y),z) \to g(h(x,z),h(y,z))$$

- The termination is proved by a polynomial interpretation [5].
- This system is not confluent: let us consider the term $h(f(x,y),g(z,t))$.

$$h(f(x,y),g(z,t)) \to f(h(x,g(z,t)),h(y,g(z,t)))$$
$$\to^* f(g(h(x,z),h(x,t)),g(h(y,z),h(y,t)))$$
$$h(f(x,y),g(z,t)) \to g(h(f(x,y),z),h(f(x,y),t))$$
$$\to^* g(f(h(x,z),h(y,z)),f(h(x,t),h(y,t)))$$

Nevertheless, it is possible to find a strategy which leads to a unique normal form. Let us define a priority on the distributivity rules. The strategy consists in normalizing by the first(with respect to the priority) rule (because one rule alone is confluent and terminating) and then by the other and repeating this operation in the same order until no rule can be applied. This strategy terminates because the system is noetherian.

Let us denote $t \downarrow$ the normal form that we obtain using such a given convergent strategy. As in the previous section we define the ordering by:

Definition 6. $t_1 < t_2$ iff $\overline{t_1 \downarrow} <_{rpo} \overline{t_2 \downarrow}$

Proposition 7. *This ordering is a simplification ordering.*

Proof. We do not give the proof but just point out the differences between this case and the proof made in [5]. As in the proof by Bachmair and Plaisted, we consider the different contexts in which the terms are embedded and the normalizations that are introduced by these contexts. The difference is that we cannot restrict ourselves to irreducible contexts because the strategy may impose to make the normalization with the term that is embedded first. Let us give an example of a context that has to be studied here and not in the Bachmair and Plaisted's proof.

Example 2. Let us suppose that we impose to normalize in priority by the distributivity of h on f. Let us suppose that we study the subterm property. Let us consider the term $t = f(x,y)$.

- If it is embedded in $h(\_,u)$ where $u \neq g(\alpha,\beta)$, it is normalized into $f(h(x,u),h(y,u))$ and $f(x,y) <_{rpo} f(h(x,u),h(y,u))$.

- The context $h(\_, g(\alpha, \beta))$ can be normalized into $g(h(\_, \alpha), h(\_, \beta))$. We still have $f(x, y) <_{rpo} g(f(h(x, \alpha), h(y, \alpha)), f(h(x, \beta), h(y, \beta)))$.
- If t is embedded in the context before the normalization, we have:

$$h(f(x,y), g(\alpha, \beta)) \to f(h(x, g(\alpha, \beta)), h(y, g(\alpha, \beta)))$$
$$\to f(g(h(x, \alpha), h(x, \beta)), g(h(y, \alpha), h(y, \beta)))$$

We still have $f(x, y) <_{rpo} f(g(h(x, \alpha), h(x, \beta)), g(h(y, \alpha), h(y, \beta)))$, but it is not a consequence of the two first inequalities. This particular reducible context had to be tested. Only a finite number of contexts have to be tested: the contexts of height 1 in which two different rules apply at the root. In every case, we prove the required inequalities. It is the same for the monotonicity.

3.2 n AC symbols l_k ($1 \le k \le n$) such that $l_k < l_{k+1}$ ($k < n$)

If we have three symbols such that $f < g < h$, the normalization system is

$$h(f(x, y), z) \to f(h(x, z), h(y, z))$$
$$h(g(x, y), z) \to g(h(x, z), h(y, z))$$
$$g(f(x, y), z) \to f(g(x, z), g(y, z))$$

This system is not confluent, and we can show that no polynomial interpretation can prove its termination.

Anyway, as we saw in section 3.1, the system containing the first two rules terminates and we can use a strategy leading to a unique normal form. We are going to normalize by the system containing the first two rules and then by the last rule, which is also confluent and nœtherian. An application of this last rule on a term which is normalized by the first system cannot create an application of this first system. Therefore this strategy terminates and is confluent.

In the normalization system containing the two rules dealing with h, we have to choose which rule we use first. There is an ambiguity when we have a term of the form $h(g(x, y), f(u, v))$. The two possible derivations are

$$h(g(x, y), f(u, v)) \to g(h(x, f(u, v)), h(y, f(u, v)))$$
$$\to g(f(h(x, u), h(x, v)), f(h(y, u), h(y, v)))$$
$$\to^* f(g(h(x, u), h(y, u)), g(h(x, u), h(y, v)),$$
$$g(h(x, v), h(y, u)), g(h(x, v), h(y, v)))$$

and
$$h(g(x, y), f(u, v)) \to f(h(g(x, y), u), h(g(x, y), v))$$
$$\to f(g(h(x, u), h(y, u)), g(h(x, v), h(y, v)))$$

On this example, we notice that it is better to use first the distributivity of h on f, that is to say on the smallest AC symbol, in order to restrict the size of the normalized terms and to need less rules to reach the normal form. This is general: to normalize at once by the distributivity on the smallest AC symbol makes this symbol go up first, and then no other rule, created by this normalization can make it go back down.

This method applies when we have an arbitrarily long chain of AC symbols:

The method Let us consider a precedence ordering $<$ on the set of symbols F such that for all f in F_{AC}, there exist a chain l in F_{AC} of length n_l, and an index i_0, such that $f = l_{i_0}$ and, for all i, $1 \le i < n_l$, $l_i < l_{i+1}$ and l_{i+1} is minimal in $F \setminus \{l_1, \cdots l_i\}$.

We define an extension of the *apo* using the same method as in [5]:

$$t_1 <_{eapo} t_2 \Leftrightarrow \overline{t_1 \downarrow_{\mathcal{D}}} <_{rpo} \overline{t_1 \downarrow_{\mathcal{D}}}$$

The following work is composed of two parts. First, we define the normalization system $\mathcal{D}$ and study its properties. Then with this system, we define the extension of the associative path ordering and prove that it is a simplification ordering.

The normalization system Let us define for a chain l and two indices $1 \le j < k \le n_l$, the rule

$$R_{k,j}(l) = l_k(l_j(x, y), z) \to l_j(l_k(x, z), l_k(y, z))$$

We use the notation, for $2 \le k \le n_l$, $\mathcal{D}_k(l) = \{R_{k,j} | 1 \le j < k\}$ and $\mathcal{D}(l) = \bigcup_{2 \le k \le n_l} \mathcal{D}_k(l)$. The set $\mathcal{D}$ is defined by $\mathcal{D} = \bigcup \mathcal{D}(l)$ for all the chains of AC symbols in F.

Each set of rules $\mathcal{D}(l)$ only uses the symbols of l. Therefore, for two different chains l and l', the rules of the sets $\mathcal{D}(l)$ and $\mathcal{D}(l')$ commute. We can normalize separately by one system and then by the other. Therefore, we can easily derive the case of several chains from the case where there is only one chain.

In the following, we consider only one chain of AC symbols, and we simplify our notation by suppressing the references to the name l of the chain.

Lemma 8. *For all indices k and j, the rule $R_{k,j}$ terminates and is confluent.*

Lemma 9. *For every index k such that $2 \le k \le n$, $\mathcal{D}_k$ terminates.*

Proof. The proof is made using the following polynomial interpretation:

- $\Phi(l_k) = x * y$
- $\Phi(l_j) = x + y + 2$ for $1 \le j < k$

Unfortunately, $\mathcal{D}_k$ is not confluent. But there exists a strategy which leads to a unique normal form for $\mathcal{D}_k$. normalization by the rule $R_{k,1}$, normalization by the rule $R_{k,2}, \ldots$, normalization by the rule $R_{k,k-1}$, then we repeat this loop until the problem is irreducible for $\mathcal{D}_k$.

This terminates because $\mathcal{D}_k$ is nœtherian and leads to a unique normal form for a term, because each step leads to a unique normal form. Extending our notation $t \downarrow_{\mathcal{D}_k}$ denotes this normal form. As we work with AC symbols, we often have to deal with flattened forms of the terms for one of the symbols in F_{AC}. We use the notation $\overline{t}^i$ for the flattened form of the term t for the symbol l_i ($1 \le i \le n$).

Let us give some properties of the normalization by the system $\mathcal{D}_k$:

Lemma 10. *If $k < i$, $\overline{t \downarrow_{\mathcal{D}_k}}^i = \overline{t}^i \downarrow_{\mathcal{D}_k}$.*

Here is a lemma which expresses the commutation of the rules of $\mathcal{D}_k$, in spite of the strategy:

Lemma 11. *Let u and v be two terms, f be a symbol of the alphabet and k be an integer such that $2 \leq k \leq n$, $f(u,v) \downarrow_{\mathcal{D}_k} = f(u \downarrow_{\mathcal{D}_k}, v \downarrow_{\mathcal{D}_k}) \downarrow_{\mathcal{D}_k}$*

In order to obtain an irreducible form for the whole system $\mathcal{D}$, we need to define a strategy for which any derivation is finite and the normal form is unique.

Lemma 12. *Let i be an index such that $1 \leq i \leq n$. Let t be a term. If for all k such that $i < k \leq n$, t is irreducible for $\mathcal{D}_k$ and $t \to^*_{\mathcal{D}_i} t'$, then t' is also irreducible for $\mathcal{D}_k$, ($i < k \leq n$).*

This lemma ensures that for every term $t \in T(F, X)$, $t \downarrow_{\mathcal{D}_n} \downarrow_{\mathcal{D}_{n-1}} \cdots \downarrow_{\mathcal{D}_2}$ is irreducible for every rule $R_{k,i}$, $1 \leq i < k \leq n$. This gives a strategy for normalizing by the whole set of rules. The normal form will be denoted by $t \downarrow$. ($t \downarrow = t \downarrow_{\mathcal{D}_n} \downarrow_{\mathcal{D}_{n-1}} \cdots \downarrow_{\mathcal{D}_2}$.)

Now that the normalization by the system $\mathcal{D}$ is defined, half of the problem is solved. Before defining precisely the ordering, let us give two commutation properties of the normalization system. These lemmata prove that sometimes we can normalize the terms as if there was no strategy at all: first, we normalize the subterms and then we use the normalization rules at the root.

Lemma 13. *For all $1 < i < j \leq n$, if $t \to^+_{\mathcal{D}_j} t'_1 \to^+_{\mathcal{D}_i} t''$, and if one of the rules of $\mathcal{D}_i$ was applicable before the application of $\mathcal{D}_j$ then $t \to^+_{\mathcal{D}_i} t'_2 \to^*_{\mathcal{D}_j} t'_3 \to^*_{\mathcal{D}_i} t''$.*

Lemma 14. *For all i, $1 \leq i \leq n$ and for all terms u and v, $l_i(u,v) \downarrow = l_i(u \downarrow, v \downarrow) \downarrow$.*

Corollary 15. *Let n be the number of AC symbols. Then $l_n(u,v) \downarrow = l_n(u \downarrow, v \downarrow) \downarrow_{\mathcal{D}_n}$.*

The ordering We can now define the extended associative path ordering.

Definition 16. $t_1 <_{eapo} t_2$ iff $\overline{t_1 \downarrow} <_{rpo} \overline{t_2 \downarrow}$ where $\overline{t}$ is the flattened form of the term for all the AC symbols $(l_i)_{1 \leq i \leq n}$.

Definition 17. $(<_p)_{1 \leq p \leq n}$ is the sequence of orderings defined by induction on p by:

- $t_1 <_1 t_2$ iff $\overline{t_1}^1 <_{rpo} \overline{t_2}^1$.
- If $p > 1$, $t_1 <_p t_2$ iff $\overline{t_1 \downarrow_{\mathcal{D}_p}}^p <_{p-1} \overline{t_2 \downarrow_{\mathcal{D}_p}}^p$.

It is easy to prove that each relation $<_p$ is a partial ordering. An immediate consequence of the definition of these orderings is the following: if t_1, t'_1, t_2, t'_2 are terms such that $t_1 \to^*_{\mathcal{D}'} t'_1$ and $t_2 \to^*_{\mathcal{D}'} t'_2$, according to the previous strategy, where $\mathcal{D}' = \bigcup_{1 < i \leq k} \mathcal{D}_i$,

$$t_1 <_k t_2 \Leftrightarrow t'_1 <_k t'_2$$

Lemma 18. *For a chain of AC symbols of length n, $<_{eapo}$ is the ordering $<_n$.*

In order to prove that the *eapo* is a simplification ordering, we need some properties of the sequence of orderings $(<_p)_{1 \leq p \leq n}$.

Lemma 19. *Let u and v be two terms. Let f and g be two symbols such that $f < g$, f and g AC. If $<_{k-1}$ is a simplification ordering and $<_k$ satisfies the subterm property, $f(u,v) <_k g(u,v)$.*

Lemma 20. *Let f and g be two symbols. If $<_k$ satisfies the subterm property, $f(u,v) <_k g(u',v')$, $f(u,v)$ is irreducible for $\mathcal{D}$, and $f \not\leq g$, then $f(u,v) <_k u'$ or $f(u,v) <_k v'$.*

In this lemma, the irreducibility of $f(u,v)$ is essential. If we consider h such that $h < f$ and $h < g$, f and h AC symbols, $f(h(x,y),a)$ is normalized into $h(f(x,a),f(y,a))$. This normal form is smaller than $g(f(x,a),f(y,a))$, but not than $f(x,a)$ or $f(y,a)$.

Definition 21. Let P_k be the following property: if $f < g$ and $g(u',v')$ is irreducible for all $\mathcal{D}_p$, $2 \leq p \leq k$, then

$$f(u,v) <_k g(u',v') \Leftrightarrow u <_k g(u',v') \text{ and } v <_k g(u',v')$$

By the definition of the *rpo*, P_1 is true.

Lemma 22. *If $<_{k-1}$ is a simplification ordering which satisfies the property P_{k-1} and if $<_k$ satisfies the subterm property, then $<_k$ satisfies the property P_k.*

A consequence of the previous lemma and remark is the following:

Lemma 23. *If for all $p < k$, $<_p$ is a simplification ordering and $<_k$ satisfies the subterm property, then $<_k$ satisfies the property P_k.*

Corollary 24. *Let f and g be two symbols such that $f < g$. If $<_p$ is a simplification ordering for all $p < k$ and $<_k$ satisfies the subterm property and $t \downarrow_{\mathcal{D}_k} \ldots \downarrow_{\mathcal{D}_2} (\Lambda) = g$, $f(u,v) <_k t$ iff $u <_k t$ and $v <_k t$.*

If we do not suppose in P_k that $g(u',v')$ is irreducible, $<_k$ does not satisfy P_k.

We come back to the main result which gives its interest to the eapo.

Proposition 25. *The "extended associative path ordering" is a simplification ordering.*

Proof. We prove by induction on n that, when we have a chain of AC symbols of length n, $<_n$ is a simplification ordering.

If $n = 2$, this ordering is the *apo* and we know that it is a simplification ordering.

Otherwise, let us suppose that the result is proved for all $p < n$. Let $l_1 < \ldots < l_n$ be a chain of n AC symbols.

<u>Subterm property</u> We only have to consider what happens when the term is embedded in a context $f(\_, t')$ for every symbol f in the alphabet. In order to prove that $t <_n f(t, t')$, we prove that $\overline{t \downarrow_{\mathcal{D}_n}}^n <_{n-1} \overline{f(t, t') \downarrow_{\mathcal{D}_n}}^n$, considering in the last inequality that only the $n-1$ smaller symbols l_i ($1 \leq i \leq n$) are AC.

<u>Monotonicity</u> We suppose that $u <_n v$ and we want to show that for every context C, $C(u) <_n C(v)$. We can restrict ourselves to contexts of height 1. Therefore, we suppose that the context is of the form $f(\_, t')$ where f is a symbol of the alphabet. If $f \neq l_n$, then for every term t, $C(t) \downarrow_{\mathcal{D}_n} = C \downarrow_{\mathcal{D}_n} (t \downarrow_{\mathcal{D}_n})$ and we conclude thanks to the monotonicity of $<_{n-1}$.

If $C = l_n(\_, t')$, then we make an induction on the sum of the lengths of the longest derivations in $\mathcal{D}_n$ that we can do on the two terms, with respect to the strategy. As $l_n(u, t') \downarrow = l_n(u \downarrow, t' \downarrow) \downarrow_{\mathcal{D}_n}$ and $l_n(v, t') \downarrow = l_n(v \downarrow, t' \downarrow) \downarrow_{\mathcal{D}_n}$ (lemme 15), we restrict ourselves to the case where the three subterms u, v, t' are irreducible for $\mathcal{D}$. We use the notation $u = l_i(u_1, u_2)$, $v = l_j(v_1, v_2)$ and $t' = l_k(t'_1, t'_2)$ when the head symbol of these subterms are AC. We give below the cases where previous lemmata are useful.

- If $l_n(v, t')$ is irreducible, and $l_n(u, t')$ is reducible, $u = l_i(u_1, u_2)$, $i < n$ and $l_n(u, t') \to l_i(l_n(u_1, t'), l_n(u_2, t'))$. We know that $u_1 <_n l_i(u_1, u_2) <_n v$ and then by induction hypothesis $l_n(u_1, t') <_n l_n(v, t')$, and $l_n(u_2, t') <_n l_n(v, t')$. As $l_i < l_n$ and $l_n(v, t')$ is irreducible, $l_i(l_n(u_1, t'), l_n(u_2, t')) <_n l_n(v, t')$ (lemma 23).
- If $l_n(u, t')$ is irreducible and $l_n(v, t')$ is reducible at the root, then $v = l_j(v_1, v_2)$ and $j < n$, $l_n(v, t') \to l_j(l_n(v_1, t'), l_n(v_2, t'))$. As $l_n(u, t')$ is irreducible, $u(\Lambda) \not< l_n$, so $u(\Lambda) \not\leq l_j$. $u <_n l_j(v_1, v_2)$ and $l_j(v_1, v_2)$ is irreducible, so thanks to lemma 20, $u <_n v_1$ or $u <_n v_2$. Thanks to the induction hypothesis, $l_n(u, t') <_n l_n(v_1, t')$ or $l_n(u, t') <_n l_n(v_2, t')$ and we conclude thanks to the subterm property.
- If both terms are reducible and $i < k < j$, $l_n(u, t') \to l_i(l_n(u_1, t'), l_n(u_2, t'))$ and $l_n(v, t') \to l_k(l_n(v, t'_1), l_n(v, t'_2))$. By induction hypothesis, $l_n(u_1, t') <_n l_n(v, t')$ and $l_n(u_2, t') <_n l_n(v, t')$. As v and t' are irreducible and because of the strategy, $l_n(v, t') \downarrow = l_k(l_n(v, t'_1) \downarrow, l_n(v, t'_2) \downarrow)$. $l_i < l_k$. Therefore, $l_i(l_n(u_1, t'), l_n(u_2, t')) <_n l_n(v.t')$ (corollary of lemma 23).

We proved that this ordering is a simplification ordering. Let us study what happens when we apply substitutions to the terms that we want to compare.

<u>Stability by an application of a substitution</u> the example below shows that the ordering defined is not stable by application of a substitution.

Example 3. If we consider three AC symbols, l_0, l_1, l_2, the two terms $l_2(l_1(x, y), z)$ and $l_1(l_2(x, z), l_2(y, z))$ are interpreted in the same way. But if we substitute $l_0(u, v)$ to z, the two terms are normalized respectively into

$$l_0(l_1(l_2(x, u), l_2(y, u)), l_1(l_2(x, v), l_2(y, v)))$$

and

$$l_0(l_1(l_2(x, u), l_2(y, u)), l_1(l_2(x, u), l_2(y, v)), l_1(l_2(x, v), l_2(y, u)), l_1(l_2(x, v), l_2(y, v)))$$

and the second term contains the first one.

But as in the case of the *apo*, it is sufficient to test a finite number of substitutions to ensure the decreasingness for all substitutions. In fact, these substitutions are the ones which can introduce a new application of a normalization rule.

As $l_i(u,v) \downarrow = l_i(u\downarrow, v\downarrow)\downarrow$ for every AC symbol l_i and every couple of terms (u,v) (lemma 14), we can restrict ourselves to the substitutions σ such that $\sigma(x)$ is irreducible for $\mathcal{D}$, for all x.

If a variable x occurs in a subterm $l_k(x,t')$ of a term t, this occurrence is critical because there are substitutions on x which can lead to some normalizations in this subterm. These are the substitutions σ_i, for $1 \leq i < k$, defined by $\sigma_i(x) = l_i(u,v)$, where u and v are new variables. We say in this case that x has a critical occurrence in t for l_k and we use the notation $\Sigma_k(x) = \{\sigma_i(x) = l_i(u,v) | 1 \leq i < k\}$.

For a term t and a variable x, $K(x,t)$ is the set of the indices k such that x has a critical occurrence in t for l_k. Then we define $\Sigma(x,t) = \bigcup_{k \in K(x,t)} \Sigma_k(x)$, and $\Gamma(t) = \bigcup_{x \in X} \Sigma(x,t)$.

We can now give a description of the substitutions that we have to test when we want to compare two terms.

Proposition 26. *Let t and t' be two terms. If $t <_{eapo} t'$ and for every substitution $\sigma \in \Gamma(t) \cup \Gamma(t')$, $\sigma(t) <_{eapo} \sigma(t')$, then for every substitution τ, $\tau(t) <_{eapo} \tau(t')$.*

Proof. The proof is the same as in [5], adding an induction on the number of comparable symbols.

4 Some unary symbols smaller than AC operators

Let us consider the case where the signature contains some unary symbols comparable with AC symbols. In this section, we restrict ourselves to the case where there exist only two comparable AC symbols, but we allow a chain of unary symbols extending the case of one unary symbol [1]. We will see in a following section how to mix the two extensions.

4.1 Some unary symbols between two AC symbols

We suppose that there exist a chain of unary symbols $(u_i)_{1 \leq i \leq n}$ such that $g < u_1 < ... < u_n < f$, where f and g are AC symbols.

The method We define a normalization system $\mathcal{D}$ which contains the rules expressing the distributivity of the non minimal AC symbols on the minimal ones. Some special rules dealing with the unary symbols are also necessary. This system is confluent and terminating.

Then we define the ordering *mapo* ("modified associative path ordering") by

$$t_1 <_{mapo} t_2 \text{ iff } \overline{t_1}\downarrow_\mathcal{D} <_{rpo} \overline{t_2}\downarrow_\mathcal{D}$$

where $\overline{t}$ is the flattened form of the term t. We prove that this ordering is a simplification ordering.

First, we define a new condition which plays the same part as the associative path condition, but is a little more general.

Definition 27. A precedence satisfies the condition *mapo* if for each symbol $f \in F_{AC}$ either

- f is minimal in F or
- there exist a symbol $g \in F_{AC}$ and a chain of unary symbols $U = \{u_i | 1 \leq i \leq n\}$ such that f is minimal in $F\backslash(U \cup \{g\})$ and $g < u_1 < ... < u_n < f$.

In the following, we only consider precedence orderings which satisfy this condition.

The normalization system The system $\mathcal{D}$ is composed of:

$$f(g(x,y),z) \rightarrow g(f(x,z), f(y,z)) \qquad (1)$$
$$f(u_i(x),y) \rightarrow u_i(f(x,y)) \qquad (2)$$
$$u_i(g(x,y)) \rightarrow g(u_i(x), u_i(y)) \qquad (3)$$
$$u_k(u_j(x)) \rightarrow u_j(u_k(x)) \qquad (4)$$

for all symbols f, g AC, u_i, u_k, u_j unary, such that $g < u_i < f, g < u_j < u_k < f$. The rules (1) and (2) are needed to ensure the monotonicity of the ordering. The other ones are compulsory for the normalization system to be confluent.

Proposition 28. *The system $\mathcal{D}/\mathcal{AC}$ terminates and is confluent.*

Proof. <u>Termination</u> We prove the termination of this system thanks to the AC-compatible polynomial ordering based on the interpretation defined by:

- $\Phi(f(P,Q)) = \Phi(P) * \Phi(Q)$, if f is a non minimal AC operator.
- $\Phi(g(P,Q)) = \Phi(P) + \Phi(Q) + 2$, if g is a minimal AC symbol.
- $\Phi(u_i(P)) = (i+1) * \Phi(P) + (i+1)$, if u_i is a unary symbol smaller than an AC operator.

This interpretation is compatible with the AC theory because the polynomials $axy+b(x+y)+c$ which interpret the AC symbols satisfy the condition $ac+b = b^2$ (see section 1.2). This interpretation decreases when one of the rules dealing with AC operators or when a rule which permutes two unary symbols is applied.

<u>Confluence</u> Now, to show the AC confluence of the system, we only have to show that it is locally AC-confluent and locally AC-coherent. We compute all the critical pairs in the system and between the system and the equations of the theory and we can see that they are confluent. This computation that is not detailed here is confirmed by the use of an AC version of the Knuth-Bendix algorithm.

The ordering The ordering is defined as in the former section. We call it *mapo* for modified associative path ordering.

Definition 29 Modified associative path ordering.

$$t_1 <_{mapo} t_2 \Leftrightarrow \overline{t_1 \downarrow_{\mathcal{D}}} <_{rpo} \overline{t_2 \downarrow_{\mathcal{D}}}$$

Proposition 30. *Every mapo is a simplification ordering.*

Proof. Irreflexivity and transitivity come from the corresponding properties of the *rpo*.
Subterm property: we have to show that for every term $t \in T(F, X)$ and every symbol $h \in F$, $t <_{apo} h(\cdots, t, \cdots) = s$. As the normalization system is confluent, we can suppose without any loss of generality that t and the context, $h(\cdots, \_, \cdots)$, in which it is embedded are irreducible for $\mathcal{D}$. The proof is by induction on the number n of the symbols of t which are smaller than h in the precedence.
Monotonicity: We have to prove that for all terms s and t and every symbol f, $s >_{mapo} t$ implies $f(\cdots, s, \cdots) >_{mapo} f(\cdots, t, \cdots)$. Without loss of generality, we can restrict ourselves to the case where the two terms s and t are irreducible for $\mathcal{D}$. We have to discuss all the possible cases in which a rule can be applied in one term or another.

If $u = f(..., s, ...)$ and $v = f(..., t, ...)$ are irreducible for $\mathcal{D}$, we can conclude just as in [4] thanks to the *mapo* condition. Otherwise, as for the subterm property, we are going to prove the result by induction on the number of symbols smaller than f occurring in s and t.

We showed that the ordering that we defined was a simplification ordering. Now we study the stability properties of this ordering when a substitution is applied to the terms to be compared.

The *rpo* is obviously stable by application of a substitution. But this property is not immediate for the *mapo* because $\overline{\sigma(t) \downarrow} \neq \sigma(\overline{(t) \downarrow})$.

Lemma 31. *For every $s \in T(F, X)$, $\overline{s} <_{rpo} s$, $s \downarrow_{\mathcal{D}} <_{rpo} s$.*

Lemma 32. *Let x be a variable, t be a term such that $f(t, x)$ is irreducible. Then for every t', there exists a term τ such that $f(t, \tau)$ is a subterm of $\overline{f(t, t')} \downarrow$.*

Proposition 33. *Let s and t be two terms such that $s <_{mapo} t$. Then, for every substitution σ, $\sigma(s) <_{mapo} \sigma(t)$.*

Proof. As $\mathcal{D}$ is confluent, for every term u, $\sigma(u) \downarrow = \sigma \downarrow (u \downarrow) \downarrow$. Without loss of generality, we can suppose that s and t are irreducible as well as $\sigma(x)$ for every x. The proof by induction on the sum of the size of s and t has been carefully done using previous corollary and lemma.

Of course, the *mapo* also applies when there is only one AC symbol f, and some unary symbols are smaller than f for the precedence. It cannot be used when there are unary symbols smaller than two AC symbols, because in this case, the normalization system is not confluent.

5 Other extensions

5.1 A unary symbol smaller than two comparable AC symbols

Let the precedence be $u < g < f$, where f and g are AC and u is unary.

We define a *mapo* in this particular case by

$$t_1 <_{mapo} t_2 \text{ iff } \overline{t_1 \downarrow} <_{rpo} \overline{t_2 \downarrow}$$

where $t \downarrow$ is the normal form of the term t for the system:

$$g(u(x), y) \rightarrow u(g(x, y))$$
$$f(u(x), y) \rightarrow u(f(x, y))$$
$$f(g(x, y), z) \rightarrow g(f(x, z), f(y, z))$$
$$u(u(x)) \rightarrow u(x)$$

It is confluent and we prove its termination with the same polynomial interpretation as in the previous case.

Let us suppose that u is minimal for the precedence. Then we can prove the following properties for the deduced ordering:

Proposition 34. *If t is a subterm of t', then $t \leq_{mapo} t'$. If $t_1 \leq_{mapo} t_2$ and C is a context, then $C(t_1) \leq_{mapo} C(t_2)$.*

This method does not apply when there exist n unary symbols $u_1 ... < u_n < g < f$, where f and g are AC, with $n \geq 2$.

5.2 Some unary symbols in a chain of AC symbols

Let us mix the two orderings that we have defined: the *eapo* and the *mapo*.

Let us consider a precedence of the form $u_0 < l_1 < ... < l_{n-1} < u_1 < ... < u_p < l_n$, where no other symbol is smaller than l_n.

Let us consider the systems

- $\mathcal{D}_{n,i}$, for $1 \leq i \leq n-1$, which contains the rules

$$l_n(l_i(x, y), z) \rightarrow l_i(l_n(x, z), l_n(y, z))$$
$$l_n(u_j(x), y) \rightarrow u_j(l_n(x, y)) \quad \text{for } 0 \leq j \leq p,$$
$$u_j(l_i(x, y)) \rightarrow l_i(u_j(x), u_j(y)) \quad \text{for } 1 \leq j \leq p,$$
$$l_i(u_0(x), y) \rightarrow u_0(l_i(x, y))$$
$$u_0(u_0(x)) \rightarrow u_0(x)$$
$$u_i(u_k(x)) \rightarrow u_j(u_k(x)) \quad \text{for } 1 \leq j < k \leq p,$$

- $\mathcal{D}'$ the normalization system of the *eapo* for the $n-1$ symbols $l_1 < \cdots < l_{n-1}$.

$\mathcal{D}'$ and the $\mathcal{D}_{n,i}$ terminate and are confluent. Let $\mathcal{D}_n = \cup_{1 \leq i \leq n-1} \mathcal{D}_{n,i}$. We define, for a term t, $t \downarrow_{\mathcal{D}_n} = (t \downarrow_{\mathcal{D}_{n,1}} \cdots \downarrow_{\mathcal{D}_{n,n-1}}) \downarrow_{\mathcal{D}_n}$.

We define the *meapo* by:

$$t_1 <_{meapo} t_2 \text{ iff } \overline{t_1 \downarrow_{\mathcal{D}_n} \downarrow_{\mathcal{D}'}} <_{rpo} \overline{t_2 \downarrow_{\mathcal{D}_n} \downarrow_{\mathcal{D}'}}$$

After the normalization by $\mathcal{D}_n$, every unary symbol is either under l_n, or above all a sequence of the l_i, $1 \leq i \leq n$. The permutation of the l_i by $\mathcal{D}'$ does not create applications of the rules dealing with the unary symbols. $t \downarrow_{\mathcal{D}_n} \downarrow_{\mathcal{D}'}$ is irreducible for all the rules of $\mathcal{D}_n$ and $\mathcal{D}'$.

Proposition 35. *The meapo is a simplification ordering.*

6 Applications

First example In [5], Bachmair and Plaisted explain that they cannot prove by the *apo* the termination of the standard confluent equational term rewriting system for associative commutative rings. We can prove this termination with the *mapo*. Let us consider the system:

$$\begin{array}{ll} x + 0 \to x & I(x+y) \to I(x) + I(y) \\ x + I(x) \to 0 & x * (y+z) \to (x*y) + (x*z) \\ I(0) \to 0 & x * 0 \to 0 \\ I(I(x)) \to x & x * I(y) \to I(x*y) \end{array}$$

The *mapo* corresponding to the precedence $0 < + < I < *$ shows the termination. The rules equivalent with respect to the *mapo* are the distributivity rules the termination of which is already proved.

Second example Let us consider the following example of rewrite system:

$$comp(comp(x,y),z) \to comp(x, comp(y,z))$$
$$comp(lift(s), comp(lift(t), u)) \to comp(lift(comp(s,t)), u)$$

Let us prove its termination, using a *mapo*.

- First, supposing that *comp* is associative and commutative, we prove that the second rule terminates modulo AC. Let us consider the precedence $lift < comp$ and the normalization rule $comp(x, lift(y)) \to lift(comp(x,y))$. If we use the corresponding *mapo* to compare the two members of our rewriting system, we have to compare thanks to the *rpo* the two terms $lift(lift(comp(x,y,z)))$ and $lift(comp(x,y,z))$. Obviously, the first one is larger because the second one is embedded in it.
- Then, as the first rule expressing the associativity of *comp* terminates, we use the following lemma to conclude.

Lemma 36. *Let R_1 and R_2, two rewrite systems such that R_1 terminates and R_2/R_1 terminates. Then $R = R_1 \cup R_2$ terminates.*

Here, R_1 consists of the first rule, R_2 of the second. R_1 and R_2/R_1 terminate. Then the whole system terminates.

7 Conclusion

In this article we gave some extensions of the associative path ordering. These extensions are obtained either by softening the control imposed on the precedence to use such an ordering, or by using an ordering different from the recursive path ordering as a basis of the comparison.

We think that the conditions that we obtained on the precedence are very general and that it would be difficult to soften them anymore. But the use of other orderings can lead to some new results.

References

1. Leo Bachmair. *Proof Methods for Equational Theories*. PhD thesis, University of Illinois at Urbana-Champaign, 1987.
2. Leo Bachmair. Associative-commutative reduction orderings. *Info Proc. Letters*, 1992.
3. Leo Bachmair and Nachum Dershowitz. Commutation, transformation, and termination. In Jorg H. Siekmann, editor, *Proc. 8th Int. Conf. on Automated Deduction, Oxford, England, LNCS 230*, pages 5–20, July 1986.
4. Leo Bachmair and Nachum Dershowitz. Completion for rewriting modulo a congruence. In Pierre Lescanne, editor, *Proceedings of the Second International Conference on Rewriting Techniques and Applications*, pages 192–203, Bordeaux, France, May 1987. Vol. 256 of *Lecture Notes in Computer Science*, Springer, Berlin.
5. Leo Bachmair and David A. Plaisted. Termination orderings for associative-commutative rewriting systems. *Journal of Symbolic Computation*, 1(4):329–349, December 1985.
6. Ahlem Ben Cherifa and Pierre Lescanne. Termination of rewriting systems by polynomial interpretations and its implementation. Research Report 677, INRIA, June 1987.
7. Hubert Comon and Catherine Delor. Equational formulas with membership constraints. Technical report, Laboratoire de Recherche en informatique, March 1991. To appear in Information and Computation.
8. Paliath Narendran and Michaël Rusinowitch. Any ground associative-commutative theory has a finite canonical system. In Ronald V. Book, editor, *Proc. 4th Rewriting Techniques and Applications, Como, LNCS 488*. Springer-Verlag, April 1991.
9. Joachim Steinbach. AC-termination of rewrite systems: a modified Knuth-Bendix ordering. In H. Kirchner and W. Wechler, editors, *Proc. 2nd Int. Conf. on Algebraic and Logic Programming, LNCS 463*, pages 372–386, October 1990.
10. Joachim Steinbach. Improving associative path orderings. In *Proc. 10th Int. Conf. on Automated Deduction, Kaiserslautern, LNCS 449*. Springer-Verlag, July 1990.

Polynomial Time Termination and Constraint Satisfaction Tests *

David A. Plaisted

Department of Computer Science
University of North Carolina at Chapel Hill
Chapel Hill, NC 27599-3175
e-mail: plaisted@cs.unc.edu

Abstract. We show that the termination of ground term-rewriting systems is decidable in polynomial time. This result is extended to ground rational term-rewriting systems. We apply this result to show that the problem of determining whether there exists a simplification ordering over a possibly extended signature, satisfying a set of stict inequalities between terms, is decidable in polynomial time. As a simple consequence, it is decidable in polynomial time whether there exists a simplification ordering which shows that a ground term rewriting system terminates.

1 Introduction

It was shown in [HL78] that the question whether a ground term rewriting system terminates is decidable, but the exact complexity of this test has been unknown until now. Motivated by a constraint satisfaction problem, we recently began to look at this problem. Here we show how to decide termination of ground systems in polynomial time. The algorithm is rule-based, exhaustively generating all relations $r \rightarrow^* s$ where r is a subterm of the right hand side of a rule and s is a subterm of the left hand side of a rule. We show that this can be done in polynomial time, and that this information can be used to decide termination in polynomial time. A related problem is polynomial time solvable, and the extension to rational terms is also solvable in polynomial time.

We apply this result to a constraint satisfaction problem. A number of constraint satisfaction tests have been presented recently, including [KKR90, Com90, JO91, NR92]. Most of these seem to require double exponential time. These are concerned with deciding if there are values for variables satisfying a collection of inequalities involving the lexicographic path ordering or the recursive path ordering. Our result differs, in that we are interested in whether there is any simplification ordering and any instantiation of the variables, satisfying a collection of inequalities. We also allow a possibly extended signature; this is important to insure that our satisfiability results still hold if new function symbols are later introduced. With these criteria, we obtain a polynomial time

---

* This research was partially supported by the National Science Foundation under grant CCR-9108904

decision procedure. In order to obtain this result, we first show that a particular termination problem is decidable in polynomial time. This result is an extension of the polynomial time decision procedure for ground term-rewriting systems mentioned above. The extension is to allow embedding as part of the rewrite relation.

This result is surprising to us. We initially thought that this problem had a non-primitive-recursive complexity. Also, we implemented a decision procedure for this problem in one version of our prover [AP92]. However, this implementation was sometimes very slow. Now we could do a much faster implementation. It's not clear yet whether the constraint satisfaction problem as discussed here is useful for a theorem prover. For example, it does not appear possible to implement the popular lexicographic path orderings [KL80] directly using this method. Typical theorem provers use a fixed ordering, also, while our constraint satisfaction test considers many orderings at the same time. This means that much longer rewrite sequences can be generated, since a rewrite sequence can be allowed if there is a simplification ordering permitting it. In addition, when using our constraint test, it is only possible to allow rewrite rules $s \rightarrow t$ in which $s > t$ in *every* simplification ordering, that is, in which t is properly embedded in s. The cost for not specifying a particular ordering is an increase in the strength of the test for ordering equations into rewrite rules. However, this decision procedure provides a useful starting point in the search for efficient constraint satisfaction tests that are applicable to theorem proving or term rewriting. Also, it would seem that this procedure could be used for a theorem prover by requiring that there exist a single simplification ordering consistent with all the choices made so far in the run of the theorem prover, or even in any given rewrite sequence.

We begin the presentation of the decision procedure for termination of ground term-rewriting systems with a sketch of the method. Suppose R is a ground term rewriting system and we can generate all true statements of the form $r \rightarrow^* s$ where r is a subterm of the right hand side of a rule in R and s is a subterm of the left hand side of a rule in R. From this, we can generate all sequences

$r_1 \rightarrow^* s_1[\alpha_1 \leftarrow r_2]$
$r_2 \rightarrow^* s_2[\alpha_2 \leftarrow r_3]$
...

where $r_i \rightarrow s_i$ are rules in R and α_i are positions. These represent ways in which a subterm of the left-hand-side of a rule in R can be rewritten. From this information, we can decide termination; it turns out that R fails to terminate iff there is some such sequence where r_k is r_1 for $k > 1$. So we have three theorems.

2 Using relations on subterms

Theorem 1. *Suppose R is a ground term-rewriting system. In polynomial time we can exhaustively generate all relations $r \rightarrow^* s$ where r is a subterm of the right hand side of a rule in R and s is a subterm of the left hand side of a rule in R.*

Proof. Given below.

Theorem 2. *Given the relations in Theorem 1, we can in polynomial time generate all (true) statements $s|_\alpha \to^* t$ such that there exists a rule $r \to s$ in R and t is the left-hand side of a rule in R.*

Proof. The term $s|_\alpha$ is a subterm of a right-hand side of a rule in R and t is a left-hand side of a rule in R.

Definition 3. We say $\alpha \leq \beta$ if α is a prefix of β, that is, $t|_\beta$ is a subterm of $t|_\alpha$.

Definition 4. Let $t_1 \, t_2 \, \ldots \, t_n \, \ldots$ be a rewrite sequence. Suppose α_i is the position of the rewrite applied to t_i. We say i is *forwards maximal* in this sequence if for all $k > i$, $\alpha_i \leq \alpha_k$.

Definition 5. We say a rewrite sequence $t_1 \, \ldots \, t_m$ is *constricting* if it has the following property: Let α_i be the position at which the rewrite is applied in t_i. Then, for all i, j, if i and j are forwards maximal in this sequence and $i < j$ then $\alpha_i \leq \alpha_j$.

Lemma 6. *Suppose R is non-terminating. Then R has an infinite constricting rewrite sequence. (This applies to general R, not just to ground R, incidentally.)*

Proof. Suppose R is non-terminating. We construct a constricting rewrite sequence $t_1 \, t_2 \, t_3 \, \ldots$ as follows. Let us say $N(t)$ if there is an infinite R-rewrite sequence starting from t. Let t_1 be a minimal term such that $N(t_1)$. Let t_2 be some term such that $t_1 \to_R t_2$ and $N(t_2)$. Note that this rewrite may be at the top level, or at a subterm of t_1. Now, t_2 must have at least one subterm t_2' such that $N(t_2')$. Possibly t_2' is t_2. Let u_2 be the leftmost innermost subterm of t_2 such that $N(u_2)$. Let t_3 be obtained from t_2 by applying a rewrite rule to some subterm of u_2 to obtain a term v_2 such that $N(v_2)$. Thus t_2 is $t_2[u_2]$ and a rewrite is applied to a subterm of u_2 to obtain t_3, which is $t_2[v_2]$. Continue this process, at each stage rewriting the leftmost innermost term u such that $N(u)$, to some term v such that $N(v)$. This is done by applying a rewrite rule to u or to one of its subterms.

We claim that the rewrite sequence so constructed is constricting. Suppose not. Let α_i be the position at which the rewrite is applied to t_i. Then there must exist i and j such that i and j are forward maximal and not $(\alpha_i \leq \alpha_j)$. However, we can't have $\alpha_j < \alpha_i$ since i is forward maximal. Therefore, α_i and α_j are incomparable. Since $t_i|_{\alpha_i}$ is the redex, $t_i|_{\alpha_i}$ must be a subterm of some leftmost innermost subterm u_i of t_i such that $N(u_i)$. Suppose u_i is at the β_i position of t_i. Then $\beta_i \leq \alpha_i$. Now, no proper subterm v of u_i satisfies $N(v)$. Thus eventually all rewrite sequences from proper subterms of u_i will stop, and a rewrite must be applied to a subterm at the u_i position of t_i, that is, at the β_i position. But α_i is forwards maximal, so $\beta_i = \alpha_i$. Thus in fact, $N(t_i|_{\alpha_i})$, and $t_i|_{\alpha_i}$ is the leftmost innermost subterm of t_i such that $N(t_i|_{\alpha_i})$. This means that all further rewrites will be applied at positions greater than or equal to α_i. This implies that $\alpha_i \leq \alpha_j$.

Let's say $S(r,t)$ if r,t are as in theorem 2, that is, there exists a rule $r \to s$ in R and t is the left-hand side of a rule in R and there exists a position α such that $s|_\alpha \to^* t$.

Theorem 7. *The system fails to terminate iff there is a sequence of left-hand sides $r_1 r_2 ... r_k$ such that $S(r_i, r_{i+1})$ for all i and r_k is r_1. Furthermore, the existence of such a sequence can be tested in polynomial time.*

Proof. If such a sequence exists then the system is easily seen to fail to terminate. For the other direction, assume that the system R fails to terminate. By the lemma, we can assume that there is an infinite constricting rewrite sequence t_1 $t_2 ... t_n ...$. Let α_i be the position at which the rewrite is applied to t_i. Let i_1 i_2 ... be the forward maximal elements. Consider the rewrites $t_{i_1} \to t_{i_1+1}$, $t_{i_2} \to t_{i_2+1}, ...$. These replace the α_{i_1}, α_{i_2}, ... subterms, respectively. Let u_{i_1}, u_{i_2} be the left hand sides of the rules used, so that $t_{i_1}|_{\alpha_{i_1}}$ is u_{i_1} et cetera. We claim that $S(u_{i_1}, u_{i_2})$, $S(u_{i_2}, u_{i_3})$, et cetera, satisfying the theorem. This follows because $t_{i_1}|_{\alpha_{i_1}}$ is the left-hand side of a rule in R, and i_1 is forward maximal, so that all succeeding rewrites occur at positions greater than or equal to α_{i_1}. Thus $t_{i_2}|_{\alpha_{i_1}}$ is $t_{i_1}|_{\alpha_{i_1}}$ with some subterm replaced by u_{i_2}, as required by the definition of S. Since the u_{i_j} are left hand sides of rules, and there are only finitely many of them, there must be a repetition among $u_{i_1}, u_{i_2}, u_{i_3}, ...$, so we get a sequence as in the theorem.

For the polynomial time part, given theorem 7 we can generate the assertions $S(r, t)$. Then we can check whether there is a sequence $r_1 r_2 ... r_n$ as in the theorem using a topological sort algorithm.

3 Finding relations on subterms

Now the main work is to prove theorem 1. For this we use a collection of inference rules for deriving such statements, and show that they are complete.

Let T be the set of terms t such that t is a subterm of a left or right hand side of a rule in R. We write $r \mathrel{\TP}^* s$ if $r \to^* s$ and r is in T and s is in T. We give a collection of inference rules for $\mathrel{\TP}^*$:

1. If $r \to s$ is a rule in R then $r \mathrel{\TP}^* s$.
2. If $r \mathrel{\TP}^* s$ and $s \mathrel{\TP}^* t$ then $r \mathrel{\TP}^* t$.
3. If $r_1 \mathrel{\TP}^* s_1$ and ... and $r_n \mathrel{\TP}^* s_n$ then $f(r_1 ... r_n) \mathrel{\TP}^* f(s_1 ... s_n)$.

Theorem 8. *If $r \to^* s$ and r is in T and s is in T then $r \mathrel{\TP}^* s$ can be derived using these inference rules (in polynomial time).*

Proof. By induction. We order derivation sequences $r \to^* s$ by the number of rewrites done. If the number of rewrites are the same, we order them by the size of r. Now, suppose $r \to^* s$ and r, s are in T. Suppose that some rewrite rule $u \to v$ is used at the top level in the derivation. Then $r \to^* u$ and $v \to^* s$ and u, v are in T. The lengths of the derivations $r \to^* u$ and $v \to^* s$ are smaller than that of $r \to^* s$ so by induction we can assume that $r \mathrel{\TP}^* s$ and $s \mathrel{\TP}^* t$

are derivable using the above three rules. Note that $u \twoheadrightarrow^* v$ is derivable using rule 1. Then $r \twoheadrightarrow^* v$ is derivable using rule 2, and $r \twoheadrightarrow^* s$ is derivable using rule 2. Suppose now that no rewrite rule is used at the top level in the derivation $r \to^* s$. Let r be $f(r_1...r_n)$, then s is $f(s_1...s_n)$ where $r_i \to^* s_i$ for all i. Note that all r_i and s_i are also in T. The lengths of the derivations $r_i \to^* s_i$ are no greater than that of $r \to^* s$, and the sizes of r_i are smaller. So these derivations $r_i \to^* s_i$ are smaller in the ordering on derivations we are using. So we can assume by induction that $r_i \twoheadrightarrow^* s_i$. Then, using rule 3, we can derive $r \twoheadrightarrow^* s$.

We now show that this derivation can be found in polynomial time. Suppose N is $|T|$, that is, the number of terms in the set T. Note that N is bounded by the size of R, written as a character string, since T is the set of all subterms of rules in R. In fact, N is sometimes much smaller than the size of R, since a given subterm can occur many times in R. There is one rule of type 1 for each rule of R. There are $O(N^3)$ rules of type 2, and $O(N^2)$ rules of type 3, in which r, s, and t are all instantiated to particular elements of T. A simple iteration can therefore derive all statements of the form $r \twoheadrightarrow^* s$ in polynomial time. However, we show that it can be done in $O(N^3)$ time.

There are a small number of rules of type 1. Each one introduces a statement of type $r \twoheadrightarrow^* s$. The total work for rules of type 1 is thus $O(N)$.

We now consider the rules of type 2. In general, whenever we add a statement $r \twoheadrightarrow^* s$, we process it. We look for t such that $s \twoheadrightarrow^* t$ has been added already, and we look for u such that $u \twoheadrightarrow^* r$ has been added already. We then add all statements $r \twoheadrightarrow^* t$ and $u \twoheadrightarrow^* s$. The time to do all of this is $O(N)$ since there are at most $O(N)$ such terms t and u and we can keep a list of all such t and u with s and r, respectively. Thus it takes time $O(N)$ to process each new statement $r \twoheadrightarrow^* s$ this way. Since there will be at most $O(N^2)$ statements $r \twoheadrightarrow^* s$ processed, the total work for rules of type 2 is $O(N^3)$.

We also process the statement $r \twoheadrightarrow^* s$ by looking for instances of rule 3 to which it can contribute. Thus, if r is r_i and s is s_i, we are making progress towards the rule "If $r_1 \twoheadrightarrow^* s_1$ and ... and $r_n \twoheadrightarrow^* s_n$ then $f(r_1...r_n) \twoheadrightarrow^* f(s_1...s_n)$." To reduce the work, we have to do the bookkeeping here carefully. With each rule of type 3, we record which $r_i \twoheadrightarrow^* s_i$ it is currently looking for. Initially, it is looking for $r_1 \twoheadrightarrow^* s_1$. When this is found, it is looking for $r_2 \twoheadrightarrow^* s_2$, and so on. When all hypotheses have been found, the conclusion is recorded. Whenever a statement $r \twoheadrightarrow^* s$ is recorded, we see if it is being looked for by some rule or rules of type 3. If so, we go through all such rules of type 3 and indicate that the next hypothesis is being looked for. Now, a given rule $r \twoheadrightarrow^* s$ may be looked for by $O(N^2)$ rules of type 3. However, the total work to process rules of type 3 is $O(nN^2)$, since there are $O(N^2)$ such rules and each one looks for at most n hypotheses, where n is the arity of f. Note that n is less than N so the work for rules of type 3 is $O(N^3)$. A more careful analysis shows that the actual work for rules of type 3 is $O(N^2)$. Let N_f be the number of term occurrences in r that have f as the top-level function symbol. If f has arity n then there are at most N_f/n term occurrences that appear as first arguments to f, at most N_f/n term occurrences that appear as second arguments to f, and so on, so the number of conclusions $r \twoheadrightarrow^* s$ looked for by rules of type 3 with f at the top level is bounded by $(N_f/n) * (N_f/n) + ... + (N_f/n) * (N_f/n)$ or $n * (N_f/n) * (N_f/n)$, which is $(1/n) * N_f^2$. By noting that the sum over N_f for various f is bounded by N, and noting that $x_1^2 + ... + x_n^2 \leq (x_1 + ... + x_n)^2$ for $x_i \geq 0$, we obtain that the sum of $(1/n) * N_f^2$ over f, is bounded by N^2. In fact, it is bounded by $(1/n) * N^2$. Therefore the total work for rules of type 3 is $O(N^2)$.

Therefore, the total work to process all inference rules and derive all possible conclusions is $O(N^3)$, as stated in the theorem. It wouldn't be surprising if this exponent could be reduced using asymptotically fast matrix multiplication algorithms.

4 An application

The same techniques yield a polynomial time decision procedure for the following problem:

Given a possibly nonterminating ground term-rewriting system R and two ground terms s and t, to decide if $s \to_R^* t$.

Theorem 9. *This problem is decidable in polynomial time.*

Proof. We can use the decision procedure for $\twoheadrightarrow^*$ to solve this problem in $O(N^3)$ time, by adding all subterms of s and t to the set T.

In fact, we note that the time bound does not depend on the size of R but rather on the number of distinct subterms of terms in R. This is because it is the number of distinct subterms that determines the number of applications of rules in the inference system of theorem 1. If we represent these terms as directed acyclic graphs then the time to process them depends on the number of distinct subterms and not on the size of the subterms.

5 Rational terms

We now show that this result can be extended to rational terms, that is, term rewriting systems in which the left and right hand sides of terms are rational terms.

Definition 10. A *rational term* is a possibly infinite term with only finitely many distinct subterms. We view a rational term as an infinite term, although it can also be represented as a finite graph.

Definition 11. A *rational term-rewriting system* is a term-rewriting system where the left and right-hand sides of the rules may be rational terms.

Lemma 12. *Suppose R is a non-terminating rational term-rewriting sequence. Then R has an infinite constricting rewrite sequence.*

Proof. Suppose R is non-terminating. Then it has an infinite rewrite sequence $t_1\ t_2\ t_3\ ...\ $. Suppose this sequence has infinitely many rewrites at the top level. Then it is constricting. Otherwise, there is some i such that all rewrites past t_i occur at proper subterms. Suppose t_i is $f(s_1...s_n)$. Then the rewrites past t_i occur within one of the subterms s_i, so the rewrite sequence is really made up of n separate rewrite sequences on the n top-level subterms. At least one of these subterms, say s_j, has an infinite rewrite sequence. We eliminate all rewrites on the terms s_i for $i \neq j$. We then apply the same construction recursively to the rewrite sequence beginning with s_j. In the limit this results in an infinite constricting rewrite sequence. Note that this same construction could be used for the case of finite terms.

Definition 13. Suppose $t_1\ t_2\ t_3\ ...$ is a finite (rational) term rewriting sequence. Then the *depth* of this sequence is the maximal depth of a redex that is rewritten in any t_i.

Theorem 14. *Suppose R is a ground rational term-rewriting sequence. Then we can test whether R is terminating in polynomial time.*

Proof. R fails to terminate iff R has an infinite constricting rewrite sequence, by the lemma. By using inference rules as above, adapted to rational ground terms, we can test for the existence of such a sequence in polynomial time. To show completeness of the rules, we use induction on the length (number of rewrite steps) of the rewrite sequence, and for sequences of the same length, we use induction on the depth of the rewrite sequence.

Note that this proof actually hides some interesting behavior. In a rational rewrite sequence, we have infinite terms, and it is possible that a rewrite can occur very deep within a term. However, the inference rules only capture rewrite sequences that are relatively shallow relative to the length of the rewrite sequence. This shows that if there is a very deep rewrite sequence on rational terms then there is a shallow rewrite sequence that is in some sense similar.

6 Constraint Satisfaction

We return to the application of this termination result to constraint satisfaction, starting with some examples. Consider the constraint

$$f(f(x)) > f(g(f(x))).$$

The question is whether there exists a ground term t and a simplification ordering $>$ such that $f(f(t)) > f(g(f(t)))$. The answer is no, since $f(f(x))$ is embedded in $f(g(f(x)))$ for any x. Now consider the constraint

$$(f(f(x)) > k(x)) \land (k(x) > f(g(f(x)))).$$

Likewise, this is unsatisfiable, since the variable x is global and must be replaced by the same ground term t in both inequalities. Finally, consider the constraint

$$(f(f(x)) > k(x)) \wedge (k(y) > f(g(f(y)))).$$

This constraint is satisfiable, since we can replace x by the term a and y by the term b and there is a simplification ordering satisfying the constraint

$$(f(f(a)) > k(a)) \wedge (k(b) > f(g(f(b)))).$$

Also, we consider an extended signature. So the constraint

$$(f(a) > x) \wedge (x > a)$$

is satisfiable. This is because we can replace x by a new symbol b and it is possible to find a simplification ordering satisfying

$$(f(a) > b) \wedge (b > a).$$

The reason for considering an extended signature is so that our satisfiability results will not be changed if new symbols are added. This is often the case in theorem proving, where Skolem functions may be introduced, or if a system is embedded in a larger system having more symbols. In addition, the decision to allow an extended signature significantly simplifies the decision procedure.

In order to handle these constraints, we first consider ground constraints and their associated term-rewriting systems. For the constraint

$$(f(f(a)) > k(a)) \wedge (k(b) > f(g(f(b)))),$$

the associated term-rewriting system R is

$$f(f(a)) \to k(a)$$

$$k(b) \to f(g(f(b))).$$

The system S corresponding to the embedding relation is

$$f(x) \to x$$

$$g(x) \to x$$

$$k(x) \to x.$$

It turns out that the constraint is satisfiable iff the rewrite system $R \cup S$ is terminating. Above we showed how to test termination of ground systems like R in polynomial time. By extending the methods used there, we now show that termination of systems like $R \cup S$ can also be decided in polynomial time. This yields a polynomial time decision procedure for the associated constraint satisfaction problem.

7 The embedding relation

Suppose R is a ground term-rewriting system. Let S be the rules $f(x_1...x_n) \to x_i$ for f appearing in R. We say u *is embedded in* t if $t \to_S^* u$. Define the rewrite relation $\to_{e,R}$ (embedding rewriting) by $r \to_{e,R} s$ if r is $r[t]$ and u is embedded in t and $u \to v$ is a rule in R and s is $r[v]$.

Note that if $r_1 \to_{e,R} r_2$ then $r_1 \to_{R \cup S} r_2$. We now show that if $r_1 \to_{R \cup S}^* r_2$ then there is a term r_3 such that $r_1 \to_{e,R}^* r_3$ and $r_3 \to_S^* r_2$. This is done by showing that if $r \to_S^* t$ and $t \to_R u$ then there is a term v such that $r \to_{e,R} v$ and $v \to_S^* u$.

Theorem 15. *If $r \to_S^* t$ and $t \to_R u$ then there is a term v such that $r \to_{e,R} v$ and $v \to_S^* u$.*

Proof. By induction on the length of the $r \to_S^* t$ rewrite sequence, and for sequences of the same length, by induction on the size of r. Suppose the length is zero. Then r and t are identical, and so $r \to_R u$ and we can pick v to be u. Suppose the theorem is true when the rewrite sequence $r \to_S^* t$ has n or fewer rewrites. We prove it when this sequence has $n+1$ rewrites. In that case, $r \to_S r_1$ and $r_1 \to_S^* t$. By induction, there is a term v such that $r_1 \to_{e,R} v$ and $v \to_S^* u$. Suppose the rewrite $r_1 \to_{e,R} v$ occurs at the top level, that is, there is a rule $t_1 \to v$ in R and $r_1 \to_S^* t$. Then $r \to_S^* t$ also, so $r \to_{e,R} v$. Suppose the rewrite $r_1 \to_{e,R} v$ does not occur at the top level. Now, consider whether the rewrite $r \to_S r_1$ occurs at the top level. If not, both rewrites occur at proper subterms, so we can obtain the theorem by induction on the size of r. If the rewrite $r \to_S r_1$ occurs at the top level, then r is $f(s_1,...s_k)$ and one of the top-level subterms, say s_i, is r_1. In this case note that $f(...s_i...) \to_{e,R} f(...v...) \to_S^* f(...u...) \to_S u$. Thus $r \to_{e,R} f(...v...) \to_S^* u$. This completes the proof.

Corollary 16. *If $r_1 \to_{R \cup S}^* r_2$ then there is a term r_3 such that $r_1 \to_{e,R}^* r_3$ and $r_3 \to_S^* r_2$.*

Proof. By repeated application of the theorem.

Corollary 17. *The system $R \cup S$ is nonterminating iff the rewrite relation $\to_{e,R}$ is nonterminating.*

Proof. If $\to_{e,R}$ is nonterminating then so is $R \cup S$ since every $\to_{e,R}$ rewrite is also an $R \cup S$ rewrite. Suppose $R \cup S$ is nonterminating. Then there is an infinite $R \cup S$ rewrite sequence. By repeated applications of the theorem, we can change the $\to_R$ rewrites into $\to_{e,R}$ rewrites and repeatedly move the $\to_S$ rewrites to the end. Continuing this process yields in the limit an infinite $\to_{e,R}$ rewrite sequence.

8 Using the relations on subterms

We now show how to test for termination of $R \cup S$ in polynomial time using the relation $\to_{e,R}$. The proof is similar to that for the termination of ground term rewriting systems.

Theorem 18. *In polynomial time we can exhaustively generate all relations $r \to_{e,R}^* s$ where r is a subterm of the right hand side of a rule in R and s is a subterm of the left hand side of a rule in R.*

Proof. Given below.

Theorem 19. *Given the relations in Theorem 18, we can in polynomial time generate all statements $s|_\alpha \to_{e,R}^* t$ where α is a position in s such that there exists a rule $r \to s$ in R and t is the left-hand side of a rule in R.*

Proof. The term $s|_\alpha$ is a subterm of a right-hand side of a rule in R and t is a left-hand side of a rule in R.

Let's say $S(r,t)$ if r,t are as in the above theorem, that is, there exists a rule $r \to s$ in R and t is the left-hand side of a rule in R and there exists a position α such that $s|_\alpha \to_{e,R}^* t$.

Theorem 20. *The system $R \cup S$ fails to terminate iff there is a sequence of left-hand sides $r_1\ r_2\ \ldots\ r_k$ such that $S(r_i, r_{i+1})$ for all i and r_k is r_1. Furthermore, the existence of such a sequence can be tested in polynomial time*

Proof. Similar to that for ground term rewriting systems, given in theorem 7.

9 Generating the relations in polynomial time

We now prove theorem 18.

Proof. We need to generate all rewrite relations $s \to_{e,R}^* t$ where s is a subterm of the right hand side of a rule and t is a subterm of the left hand side of a rule. Let T be the set of subterms of left or right hand sides of rules in R, as before. Write $s \mathrel{\overset{*}{\looparrowright}_S} t$ if s is in T and t is in T and $s \to_S^* t$. We give a set of inference rules for the relation $\overset{*}{\looparrowright}_S$, and show that they are complete and that all consequences can be found in polynomial time. Then we give a set of inference rules for the relation $\to_{e,R}^*$ on T. The rules for $\overset{*}{\looparrowright}_S$ are as follows:
1. $s \mathrel{\overset{*}{\looparrowright}_S} s$ for all s in T.
2. If $s \mathrel{\overset{*}{\looparrowright}_S} t$ and $f(...s...)$ is in T then $f(...s...) \mathrel{\overset{*}{\looparrowright}_S} t$.
3. If $s \mathrel{\overset{*}{\looparrowright}_S} t$ and $t \mathrel{\overset{*}{\looparrowright}_S} u$ then $s \mathrel{\overset{*}{\looparrowright}_S} u$.
4. If $s_i \mathrel{\overset{*}{\looparrowright}_S} t_i$ for all i and $f(s_1...s_k)$ and $f(t_1...t_k)$ are in T then $f(s_1...s_k) \mathrel{\overset{*}{\looparrowright}_S} f(t_1...t_k)$.

We show completeness as follows: If $s \to_S^* t$ then we can show that there is an outermost rewrite sequence, that is, all the rewrites at the top-level occur first, if there are any, then the subterms are recursively rewritten. The top level rewrites are handled by rule 2. The rewrites to proper subterms are handled by rule 4. The transitivity is handled by rule 3. The proof then can be obtained by a simple induction, using rule 1 as the base case. The proof for the polynomial time bound is about the same as for the ground termination proof, given in theorem 7, and the $O(N^3)$ bound holds here as well, where N is the number of elements in T (and therefore N is bounded by the length of R, written as a character string).

Let us write $r \twoheadrightarrow_{e,R}^* s$ to denote that $r \to_{e,R}^* s$ and r, s are in T. Now, after deriving all statements $r \twoheadrightarrow_S^* s$, we then derive all statements $r \twoheadrightarrow_{e,R}^* s$ using the following inference rules:
1. $r \twoheadrightarrow_S^* s$ and $s \to_R t$ imply $r \twoheadrightarrow_{e,R}^* t$
2. If $r \twoheadrightarrow_{e,R}^* s$ and $s \twoheadrightarrow_{e,R}^* t$ then $r \twoheadrightarrow_{e,R}^* t$
3. If $r_1 \twoheadrightarrow_{e,R}^* s_1$ and ... and $r_n \twoheadrightarrow_{e,R}^* s_n$ then $f(r_1...r_n) \twoheadrightarrow_{e,R}^* f(s_1...s_n)$.

We claim that all (true) statements of the form $r \twoheadrightarrow_{e,R}^* s$ can be derived from these rules in $O(N^3)$ time, as before. The proof is similar to the proof for ground term-rewriting systems, except that rule 1 processes $O(N)$ rules $s \to t$, and each one may require $O(N)$ work, since there may be $O(N)$ statements of the form $r \twoheadrightarrow_S^* s$ for a fixed term s. Thus rule 1 may require in all $O(N^2)$ work. The analysis for rules 2 and 3 is as in the proof for ground term-rewriting systems. The proof of completeness is also similar to that for ground term-rewriting systems.

Now, consider the following two problems:

Given a possibly nonterminating ground term-rewriting system R and two ground terms s and t, to decide if $s \to_{e,R}^* t$.

Given a possibly nonterminating ground term-rewriting system R and two ground terms s and t, to decide if $s \to_{R \cup S}^* t$.

Theorem 21. *These problems are decidable in polynomial time.*

Proof. We can use the above inference rules to solve the first problem in $O(N^3)$ time, by adding all subterms of s and t to the set T. For the second problem, we use corollary 1 of theorem 15; we need to find a term u such that $s \to_{e,R}^* u$ and $u \to_S^* t$. For this we also need to add subterms of s and t to T and need to add some more inference rules, namely the following:
5. If $r_1 \twoheadrightarrow_{e,R}^* u$ and $u \to_S^* r_2$ then $r_1 \to_{R \cup S}^* r_2$.
6. If $s_i \to_{R \cup S}^* t_i$ for all i and $f(s_1...s_k)$ and $f(t_1...t_k)$ are in T then $f(s_1...s_k) \to_{R \cup S}^* f(t_1...t_k)$.

The proof of completeness and polynomial time is similar to the one for $\twoheadrightarrow_S^*$ given above.

10 A constraint satisfaction problem

We now show how this result can be applied to obtain a polynomial time decision procedure for a certain kind of constraint.

Suppose we have a set of constraints
$s_1 > t_1$
$s_2 > t_2$
$s_3 > t_3$
...

where the s_i and t_i are terms. Note that the terms may just be variables. We want to know whether there exists a simplification ordering and an instantiation of the variables by ground terms, such that all of these inequalities are satisfied. The variables are global, so that a given variable has the same meaning in all inequalities. We allow the variables to be instantiated over a possibly extended signature, so that the satisfiability of the constraint will not be affected by the introduction of new symbols. So we have the following definition.

Definition 22. A conjunction $s_1 > t_1 \land s_2 > t_2 \land ... \land s_n > t_n$ of inequalities is *satisfiable* if there exists a ground substitution Θ (over a possibly extended signature) and a simplification ordering $>$ such that for all i, $s_i\Theta > t_i\Theta$.

Recall that a simplification ordering is transitive, irreflexive, and has the subterm and replacement property; the subterm property states that $f(r_1...r_n) > r_i$ for all terms r_i and the replacement property states that if $r > s$ then $f(...r...) > f(...s...)$. We do not require the ordering to be total. This is because the axiom for a total ordering is not a Horn clause and seems to make the decision procedure harder.

In order to develop a decision procedure, we need to consider a more general notion of satisfiability, namely, satisfiability of a set of first-order formulas. Let A be a first-order formula; we say A is satisfiable if there is a structure I which is a model of A, that is, $I \models A$. Let $C(>)$ be a conjunction of inequalities $\land_i(s_i > t_i)$ as above. Suppose $\overline{x}$ are the variables in C. Let $A(>)$ be the axioms for a simplification ordering, namely, for the subterm property, the replacement property, and the axioms for a partial ordering. Suppose $\overline{y}$ are the variables in A. Let $F(>)$ be the formula $(\exists \overline{x})C(>) \land (\forall \overline{y})A(>)$.

Theorem 23. $F(>)$ *is first-order satisfiable iff there is a simplification ordering $>$ over a possibly extended signature and a ground substitution Θ over that signature such that $C(>)\Theta$.*

Proof. If such a simplification ordering exists, then from it we can construct a structure I such that I satisfies F, by interpreting $>$ to be that simplification ordering, and noting that this will satisfy A since $>$ is a simplification ordering and it will satisfy $(\exists \overline{x})C(>)$ since $>$ satisfies $C(>)\Theta$. If $F(>)$ is first-order satisfiable then there is a first-order structure I satisfying F. From I we can construct a simplification ordering as specified. For this part it may be necessary to use an extended signature, since we may need Skolem constants to refer to the interpretations of the existentially quantified variables in C. We note that these Skolem constants may need to be explicitly mentioned in the ground substitution Θ.

Now, $F(>)$ can be Skolemized and converted to a set $F'(>)$ of first-order clauses, containing $C(>)$, with variables replaced by new Skolem constants, together with $A(>)$, with variables implicitly universally quantified. This will be a set of Horn clauses. (Note that we do not include the axiom for totality $x > y \lor x = y \lor x < y$ which is not a Horn clause.) Let $C'(>)$ be $C(>)$ with variables replaced by new Skolem constants. Then $F'(>)$ is $C'(>) \land A(>)$. Now, $F(>)$ is satisfiable iff $F'(>)$ is satisfiable, since Skolemization is satisfiability-preserving. Thus the original set $C(>)$ of constraints is satisfiable (relative to simplification orderings) iff $F'(>)$ is satisfiable. The only negative clause is irreflexivity, which states that not$(y < y)$. It turns out by the properties of Horn clauses that $F'(>)$ is unsatisfiable iff some instance $t < t$ is a logical consequence of $F'(>) - \{\neg(y < y)\}$.

Let Inf be the set of inference rules obtained as follows: For each Horn clause $(L_1 \wedge ... \wedge L_n \supset L)$ in $F'(>)$ (where L_i and L are positive literals) we have the inference rule
$$\frac{L_1 \quad ... \quad L_n}{L}$$
which permits L to be derived if $L_1, ..., L_n$ are already known, and also we have all instances of this rule obtained by replacing the variables by arbitrary terms in a consistent way.

Theorem 24. *$F(>)$ is (first-order) unsatisfiable iff there exists a term r such that $r < r$ is derivable from $F'(>) - \{\neg(y < y)\}$ using these rules.*

Proof. $F(>)$ is unsatisfiable iff $F'(>)$ is, by properties of Skolemization. Using properties of Horn clauses, we can show that $F'(>)$ is unsatisfiable iff such a term r exists. The relevant fact is that only one instance of a negative clause needs to be used in any proof of unsatisfiability, and that the proofs can be obtained entirely by forward reasoning, as embodied by the rules Inf. This follows because sets of Horn clauses have a unique minimal model, containing all their logical consequences, and this minimal model can be obtained by iterating the logical consequence operation given above.

11 Relating the constraint problem to term rewriting

Now, we can further relate this derivation to term-rewriting systems, as follows: Given a conjunction C of inequalities $\wedge_i s_i > t_i$, we consider the associated term rewriting system $R(C)$ which has the rules

$\hat{s_1} \to \hat{t_1}$
$\hat{s_2} \to \hat{t_2}$
$\hat{s_3} \to \hat{t_3}$
...

where $\hat{s_i}$ and $\hat{t_i}$ are s_i and t_i with variables replaced by distinct new constants, together with rules of the form $f(x_1...x_n) \to x_i$ for all f appearing in s_i and t_i.

Theorem 25. *The formula $F(>)$ is unsatisfiable iff there is a term r such that $r \to^* r$ using the set $R(C)$ of rewrite rules.*

Proof. $F(>)$ is unsatisfiable iff for some r, $r > r$ is derivable using the above inference rules obtained from the Horn clauses. From such a derivation, one can construct a rewrite sequence $r \to^* r$. For the other direction, if $r \to^* r$ then one can show that $r > r$ is derivable using the inference rules given above, so $F(>)$ is unsatisfiable.

We now show that it is sufficient to test termination of $R(C)$.

Theorem 26. *The formula $F(>)$ is satisfiable iff the set $R(C)$ of rewrite rules terminates.*

Proof. We show that $R(C)$ does not terminate iff there is a term r such that $r \to^* r$. If such a term r exists, clearly $R(C)$ is nonterminating. Suppose R is nonterminating. Then by Kruskal's theorem [Kru60], we know that there are terms r and s such that $r \to^* s$ and such that r is embedded in s. But, if r is embedded in s, then $s \to^* r$ since $R(C)$ contains the rules $f(x_1...x_n) \to x_i$ for all f and all x_i. Thus $r \to^* r$. Combining this result with the previous theorem, we obtain the desired conclusion.

Corollary 27. *Satisfiability of $F(>)$, and hence of the set $C(>)$ of constraints, can be tested in polynomial time.*

Proof. By theorem 18, we can test in polynomial time whether a ground term rewriting system together with rules of the form $f(x_1...x_n) \to x_i$ terminates.

This result is interesting, because known methods for solving constraint satisfaction problems for particular orderings such as the recursive path ordering and the lexicographic path ordering take double exponential time.

12 Extensions

We now give some extensions of this result to formulas involving negation and equality.

Recall that a conjunction $s_1 > t_1 \land s_2 > t_2 \land ... \land s_n > t_n$ of inequalities is *satisfiable* if there exists a ground substitution Θ (over a possibly extended signature) and a simplification ordering $>$ such that for all i, $s_i\Theta > t_i\Theta$. Also recall that $A(>)$ are the axioms for a simplification ordering. Let us consider more general formulas involving $>$ and $=$ and negation. Suppose $C(>,=)$ is a formula built up from terms, the binary relation symbols $>$ and $=$, and Boolean connectives. We say $C(>,=)$ is satisfiable if there is a simplification ordering $>$ and a ground substitution Θ such that $C\Theta$ is true when $=$ is interpreted in the usual way, that is, two terms are equal if they are syntactically identical. Suppose $B(>,=)$ are the axioms for equality and simplification orderings. The equality axioms are reflexivity, transitivity, symmetry, and the axioms stating that the replacement of equals by equals is equal. Then we can also define satisfiability by saying C is satisfiable iff there is a ground instance $C\Theta$ of C such that $(\forall \overline{y})B(>,=) \land C\Theta$ is satisfiable (in first-order logic), where $\overline{y}$ are the variables in B. We want to determine the complexity of deciding satisfiability of such formulas. Let $B'(>,=)$ be the axioms other than $(\forall x \neg(x > x))$. So the problem is equivalent to showing that $(\forall \overline{y})B'(>,=) \land C \land (\forall x \neg(x > x))$ is satisfiable in first-order logic. Let C' be $C \land (\forall x \neg(x > x))$. Then we want to determine if $(\forall \overline{y})B'(>,=) \land C'$ is satisfiable.

Note that C' can be converted to formula $C_1 \lor ... \lor C_n$ in disjunctive normal form, and $(\forall \overline{y})B'(>,=) \land C'$ is satisfiable iff some formula $(\forall \overline{y})B'(>,=) \land C_i$

is. Also, since $B'(>,=)$ is a Horn set without negative clauses, a conjunction $(\forall \bar{y})B'(>,=) \wedge C_i \wedge e_1 \wedge e_2$ is satisfiable iff $(\forall \bar{y})B'(>,=) \wedge C_i \wedge e_1$ is satisfiable and $(\forall \bar{y})B'(>,=) \wedge C_i \wedge e_2$ is satisfiable, where e_1 and e_2 are of the form $\neg(s > t)$ or $\neg(s = t)$. In this way we can reduce the problem to that of deciding satisfiability of a conjunct containing at most one negative formula.

We now discuss determining satisfiability of a single such disjunct. Equalities can be eliminated from C_1 in standard ways; see for example [JK91]. The idea is to use rules that replace $f(r_1...r_n) = f(s_1...s_n)$ by $r_1 = s_1, ..., r_n = s_n$ and a few other rules. Especially significant is the replacement rule, which replaces variable x by term t everywhere if $x = t$ has been derived and t does not contain x. These rules may make the terms bigger, but since the running time only depends on the number of distinct subterms, it will still be polynomial. Each application of the replacement rule eliminates all subterms containing the variable x and adds corresponding subterms containing t, so the total number of subterms does not change. If a conjunction contains only one formula of the form $\neg(s = t)$ and no equalities, then this conjunction is satisfiable iff s and t are not syntactically identical. (This follows because we can replace the variables by new Skolem constants.) If the conjunction contains only the negative formula $(\forall x \neg(x > x))$ then satisfiability can be tested using the corollary to theorem 26. If the conjunction contains only one formula of the form $\neg(s > t)$ and no inequalities, where s and t are ground terms, then we first have to determine if the conjunction without $\neg(s > t)$ is satisfiable, interpreting $>$ as a simplification ordering; this is the same as testing termination, which can be tested using the corollary to theorem 26. If this is unsatisfiable, then the conjunction is unsatisfiable by a simplification ordering, since the termination test fails. Otherwise, we can test satisfiability with $\neg(s > t)$ included in the conjunction, using the methods of theorem 21. Also, all of this can be done in polynomial time for a single conjunct. The reason is that the elimination of equalities may introduce big terms but the number of distinct subterms is polynomial, and the time bound depends only on the number of distinct subterms, if terms are represented as directed acyclic graphs. The directed acyclic graph structure represents each distinct subterm t by a data structure t', and with each occurrence of t there is a pointer to t'. Therefore we obtain the following result.

Theorem 28. *The satisfiability of a conjunction of formulas of the form $s = t$, $s > t$, $\neg(s = t)$, $\neg(s > t)$, or $\neg(s \geq t)$ can be tested in polynomial time.*

Proof. Given above, except for the $\neg(s \geq t)$ case. For this, we note that $\neg(s \geq t)$ reduces to $\neg(s > t) \wedge \neg(s = t)$, which are covered already. We note that we are not assuming that the ordering is total. Thus we can have $\neg(s > t) \wedge \neg(t > s)$ without having $s = t$.

We also note that we can get a slightly different result by considering quasi-simplification orderings. A *quasi-simplification ordering* is a transitive and reflexive relation that satisfies the replacement property and the subterm property.

Theorem 29. *The satisfiability of a conjunction of formulas of the form $s_i > t_i$ and one formula of the form $\neg(s > t)$, interpreting $>$ as a quasi-simplification ordering, can be tested in polynomial time. (The question is whether there exists a quasi-simplification ordering and a ground instance of the formula satisfying the quasi-simplification ordering).*

Proof. We can use the polynomial time method for the second problem of theorem 21. We note that the rewrite relation $\to^*_{R \cup S}$ is the minimal quasi-simplification ordering satisfying the inequalities $s_i > t_i$ for rules $s_i \to t_i$ in R, if such a quasi-simplification ordering exists.

Theorem 30. *The satisfiability of an arbitrary formula involving the predicates $=, >$, and Boolean connectives is NP-complete.*

Proof. We can in nondeterministic polynomial time extract a conjunct of the conjunctive normal form and test it for satisfiability. This shows that the problem is in NP. NP-hardness is easy to show by reducing from satisfiability, translating a predicate P_i into the inequality $x_i > false$.

References

[AP92] G. Alexander and D. Plaisted. Proving equality theorems with hyper-linking. In *Proceedings of the 11th International Conference on Automated Deduction*, pages 706–710, 1992. system abstract.

[Com90] H. Comon. Solving inequations in term algebras. In *Proceedings of 5th IEEE Symposium on Logic in Computer Science*, pages 62–69, 1990.

[HL78] G. Huet and D. Lankford. On the uniform halting problem for term rewriting systems. Technical Report Rapport Laboria 283, IRIA, Le Chesnay, France, 1978.

[JK91] Jean-Pierre Jouannaud and Claude Kirchner. Solving equations in abstract algebras: A rule-based survey of unification. In J.-L. Lassez and G. Plotkin, editors, *Computational Logic: Essays in Honor of Alan Robinson*. MIT Press, Cambridge, MA, 1991. To appear.

[JO91] Jean-Pierre Jouannaud and Mitsuhiro Okada. Satisfiability of systems of ordinal notations with the subterm property is decidable. In *Proceedings of the Eighteenth EATCS Colloquium on Automata, Languages and Programming*, Madrid, Spain, July 1991. Vol. 510 in *Lecture Notes in Computer Science*, Springer, Berlin.

[KKR90] C. Kirchner, H. Kirchner, and M. Rusinowitch. Deduction with symbolic constraints. *Revue Francaise d'Intelligence Artificielle*, 4(3):9 – 52, 1990.

[KL80] S. Kamin and J.-J. Levy. Two generalizations of the recursive path ordering. Unpublished, February 1980.

[Kru60] J.B. Kruskal. Well-quasi-ordering, the tree theorem, and vazsonyi's conjecture. *Transactions of the American Mathematical Society*, 95:210–225, 1960.

[NR92] R. Nieuwenhuis and A. Rubio. Theorem proving with ordering constrained clauses. In *Proceedings of the 11th International Conference on Automated Deduction*, pages 477–491, July 1992.

Linear interpretations by counting patterns

Ursula Martin

Department of Mathematical and Computational Sciences, University of St Andrews,
St Andrews, Fife KY16 6SX, Scotland Email:um@cs.st-and.ac.uk

Abstract. We introduce a new family of well-founded monotonic orderings on terms, constructed bu counting certain patterns in terms called zig-zags. These extend the familiar Knuth Bendix orderings, providing in general continuum many distinct new orderings with a given choice of Knuth-Bendix weight.

1 Introduction

The purpose of this paper is to introduce a new family of well-founded monotonic orderings on terms, constructed by counting certain patterns in terms, called zig-zags. Counting the patterns generates a sequence of vectors, and the orderings are induced from orderings on the vectors. Thus we call these orderings linear interpretations.

Intuitively the idea is this. Knuth and Bendix in [8] (see also [5]) introduced the idea of ordering terms by assigning a weight to each function symbol and then to a term by adding up the weight of the function symbols it contains. So for example if f has weight 2 and a weight 3 then $f(a, f(a, a))$ has weight 13. If two terms have the same weight their order is determined by an operator precedence. The Knuth Bendix ordering is monotonic and well-founded. In effect the term s is represented by a vector $v_1(s) = (\#(a, s), \#(f, s))$ which records the multiplicity of each function symbol, and terms are ordered by a lifting a well-founded monotonic ordering $>_1$ on vectors. In our example this vector ordering is given by $(u, v) >_1 (u', v')$ if and only if $3u + 2v > 3u' + 2v'$.

Our orderings are constructed by counting occurrences of certain patterns within terms, thus representing a term by a sequence of vectors $v_1(s), v_2(s), \ldots,$ then ordering this sequence lexicographically by combining monotonic orderings $>_i$ on each component. If $>_1$ is well-founded these new orderings will be well-founded. As we shall see there may be continuum many choices for $>_i$, and hence continuum many ways of extending a given $>_1$ in this way to get a new well-founded monotonic ordering on terms. This means that we have continuum many ways of extending a given choice of weights, whereas the Knuth-Bendix ordering only has finitely many.

The main purpose of this paper is to formalise all this, and to show that we obtain in this way a large family of new well-founded monotonic orderings, together with an effective algorithm for determining whether or not a given rewrite system can be proved terminating with one of these orderings.

The rest of the paper is organised as follows. We conclude this section with some examples. Section two contains definitions. In section three we define the patterns we shall use, called zig-zags, and prove the lemma which is at the heart of the technique. In section four we define the orderings, prove that they are monotonic and well-founded and indicate some further properties. Section five contains some more examples.

A full survey of the theory of term orderings is given in [4] or [14], where descriptions of known simplification orderings such as the Knuth Bendix ordering [8, 5], recursive path ordering [3], recursive decomposition ordering [7] and polynomial orderings [1] may be found. The use of interpretations on known mathematical structures for termination proofs has been pioneered in in [10, 15]; the most widely used techniques are the polynomial orderings [1, 4, 9].

Example 1

The terms $s = f(a, f(a,a))$ and $t = f(f(a,a), a)$ contain the same function symbols with the same multiplicities, so $v_1(s) = v_1(t)$. To compare them we will work with the "patterns"

$$f \xrightarrow{1} f, f \xrightarrow{2} f, f \xrightarrow{1} a, f \xrightarrow{2} a$$

where $f \xrightarrow{1} f$ denotes an f occurring in a term with a further f occurring somewhere in its first argument, $f \xrightarrow{2} a$ denotes an f occurring in a term with an a occurring somewhere in its second argument and similarly for $f \xrightarrow{2} f$ and $f \xrightarrow{1} a$. Now s contains no patterns $f \xrightarrow{1} f$, one pattern $f \xrightarrow{2} f$, two $f \xrightarrow{1} a$ and three $f \xrightarrow{2} a$. We represent s by the vector $v_2(s) = (0, 1, 2, 3)$. Similarly for t we get the vector $v_2(t) = (1, 0, 3, 2)$. Now let $>_1, >_2$ be any monotonic orderings on vectors of length 2 and 4 respectively, and suppose that $>_1$ is well-founded. Then the lexicographic combination $\succ$ of $>_1, >_2$ induces a well-founded monotonic ordering $>$ on terms by $s > t$ if and only if $(v_1(s), v_2(s)) \succ (v_1(t), v_2(t))$. So for example if $>_2$ is the lexicographic order from the right we get $s > t$, and if $>_2$ is the lexicographic order from the left we get $t > s$.

In general we will count patterns of length two as here, and order the resulting vectors. If these vectors are the same we use similar patterns of length three, like $f \xrightarrow{1} f \xrightarrow{2} f$, and so on for larger and larger patterns.

Example 2

For another example consider terms in two unary operators f, g and a variable x. Let $v_1(s) = (\#(f, s), \#(g, s))$. We compute $v_2(s)$ by counting occurrences of the patterns $f \xrightarrow{1} g$, and $v_3(s)$ by counting $f \xrightarrow{1} g \xrightarrow{1} f$ and $g \xrightarrow{1} f \xrightarrow{1} g$. Now let $>_1, >_2, >_3$ be any monotonic orderings on vectors of length 2, 1 and 2 respectively, with $>_1$ well-founded, and $>$ the ordering induced by their lexicographic combination as before. Then $>$ is well-founded and monotonic.

For example the table below shows $>$ on terms containing f twice and g three times, where $>_2$ is the lexicographic ordering, and we show the results of four alternatives for $>_3$. These are lexicographic from the left ($>_{3(L)}$) and from the right($>_{3(R)}$), the ordering $(u, v) >_{3(3,2)} (u', v')$ if and only if $3u + 2v > 3u' + 2v'$ and the ordering $(u, v) >_{3(-3,-2)} (u', v')$ if and only if $3u + 2v < 3u' + 2v'$. For comparison we show the results of the recursive path orderings and Knuth Bendix orderings with $f > g$; on these terms they reduce to the lexicographic ordering from the left and right respectively. Observe that all six orderings are different.

Term s	$v_2(s)$	$v_3(s)$	$>_{3(L)}$	$>_{3(R)}$	$>_{3(3,2)}$	$>_{3(-3,-2)}$	RPO	KBO
$f(f(g(g(g(x)))))$ (6)	(0,0)	10	10	10	10	10	10
$f(g(f(g(g(x)))))$ (5)	(1,2)	9	9	9	9	9	9
$f(g(g(f(g(x)))))$ (4)	(2,2)	8	7	8	7	7	8
$g(f(f(g(g(x)))))$ (4)	(0,4)	7	8	7	8	8	6
$f(g(g(g(f(x)))))$ (3)	(3,0)	6	5	5	6	4	7
$g(f(g(f(g(x)))))$ (3)	(1,4)	5	6	6	5	6	5
$g(f(g(g(f(x)))))$ (2)	(2,2)	4	3	4	3	3	4
$g(g(f(f(g(x)))))$ (2)	(0,4)	3	4	3	4	5	3
$g(g(f(g(f(x)))))$ (1)	(1,2)	2	2	2	2	2	2
$g(g(g(f(f(x)))))$ (0)	(0,0)	1	1	1	1	1	1

The notation in the orderings columns indicates $10 > 9 > 8 \cdots$.

Example 3

As in the previous example we consider terms in two unary operators f, g and a variable x. We compute v_1, v_2 and v_3 as before, and let $>_1, >_2, >_3$ be any monotonic orderings on vectors of length 2, 1 and 2 respectively with $>_1$ well-founded, and $>$ the ordering induced by their lexicographic combination, which is well-founded and monotonic. For any positive real number λ we define $>_{3(\lambda)}$ by $(u, v) >_{3(\lambda)} (u', v')$ if and only if $u + \lambda v > u' + \lambda v'$. Let N, M be positive integers. We have

Term s	$v_1(s)$	$v_2(s)$	$v_3(s)$
$f^N g^{2M} f^N(x)$	$(2N, 2M)$	$2MN$	$(2MN^2, 0)$
$g^M f^{2N} g^M(x)$	$(2N, 2M)$	$2MN$	$(0, 2M^2 N)$

and hence

$$f^N g^{2M} f^N(x) >_{3(\lambda)} g^M f^{2N} g^M(x) \text{ if } N/M > \lambda$$

and

$$g^M f^{2N} g^M(x) >_{3(\lambda)} f^N g^{2M} f^N(x) \text{ if } \lambda > N/M.$$

We shall show in Lemma 4 that each choice of λ gives a different ordering, and hence even for fixed $>_1, >_2$ there are continuum many such orderings $>$.

2 Background

An ordering on a set S is a transitive irreflexive relation. If $\{S_i, >_i \mid i \in \{1 \ldots n\}\}$ is a sequence of ordered sets then the lexicographic combination $\succ$ on the direct product $S_1 \times S_2 \times \ldots S_n$ is defined by $s = (s_1, s_2, \ldots) \succ t = (t_1, t_2, \ldots)$ if and only if there is an i with $s_i >_i t_i$, and $s_j = t_j$ for all $j < i$. Then $\succ$ is an ordering. If each $>_i$ is total then $\succ$ is total. If each $>_i$ is well-founded then $\succ$ is well-founded. The lexicographic combination of an infinite sequence is defined similarly; however $\succ$ need not be well-founded even if each component is, as $(s_1, 0, \ldots) \succ (0, s_2, \ldots) \succ \ldots$. We shall need to consider the well-foundedness of the restriction of $\succ$ to certain subsets.

Lemma 1. *Let $S = S_1 \times S_2$, and let $>_1$ and $>_2$ any orderings on S_1 and S_2 respectively. Let $P \subseteq S$ satisfy*

$$\forall x \in S_1 . (|P \cap \{(x,y) | y \in S_2\}| < \infty)$$
and
the restriction of $>_1$ to $\{x | \exists y . (x, y) \in P\}$ is well founded.

Then the restriction to P of the lexicographic combination $\succ$ of $>_1$ and $>_2$ is well-founded.

If $\succ$ is an ordering on T and $f : S \longrightarrow T$ is a function then the relation $>$ defined on S by $s > t$ if and only if $f(s) \succ f(t)$ is an ordering on S, called the ordering induced by f. If the restriction of $\succ$ to the image of f is well-founded then so is $>$.

We work throughout with a finite set of function symbols $\mathcal{G}$, and consider the set $\mathcal{T}(\mathcal{G})$ of all terms constructed from symbols in $\mathcal{G}$. The arity of a function symbol f is fixed, denoted by $\alpha(f)$. Function symbols with arity 0 are called constants.

An ordering $>$ on terms is called monotonic if it satisfies

$$s > t \Rightarrow f(s_1, \ldots, s, \ldots, s_n) > f(s_1, \ldots, t, \ldots, s_n),$$

and a simplification ordering if further

$$s = f(s_1, \ldots, s_i, \ldots, s_n) > s_i$$

for all terms s, t and $i = 1, \ldots, n = \alpha(f)$.

It can be shown [6] that any simplification ordering $>$ on $\mathcal{T}(\mathcal{G})$ has the finite basis property, that is if $u_1, u_2, \ldots$ is an infinite sequence of terms then there exist i, j with $i < j$ and $u_i < u_j$ or $u_i = u_j$. It follows in particular that any simplification ordering is well-founded. Any total well-founded monotonic ordering on $\mathcal{T}(\mathcal{G})$ is a simplification ordering.

3 Definitions of zig-zags

To describe the orderings we first need to formalise the notion of a zig-zag. The definitions are best understood by keeping in mind examples like those of the introduction.

A zig-zag $Z = (z, \eta)$ of length n on a set of function symbols $\mathcal{F}$ consists of a non-empty sequence $z = [f_1, \ldots, f_n]$ of elements of $\mathcal{F}$ and a total function $\eta : [1, \ldots, n-1] \to \mathbf{N}$ such that $1 \leq \eta(i) \leq \alpha(i)$ for each $1 \leq i \leq n-1$. For convenience we represent Z as a labeled directed graph

$$f_1 \stackrel{\eta(1)}{\to} f_2 \stackrel{\eta(2)}{\to} \cdots f_{n-1} \stackrel{\eta(n-1)}{\to} f_n.$$

We denote by $Z_\mathcal{F}(n)$ or $Z(n)$ the set of all zig-zags of length n on $\mathcal{F}$.
For example if $\mathcal{F} = \{f, a\}$ where $\alpha(f) = 2, \alpha(a) = 0$, then

$$Z_\mathcal{F}(2) = \{f \xrightarrow{1} f, f \xrightarrow{2} f, f \xrightarrow{1} a, f \xrightarrow{2} a\}.$$

For any alphabet $\mathcal{F}$ we have

$$Z_\mathcal{F}(1) = \{f | f \in \mathcal{F}\}$$

as each η has domain the empty set. Observe that it follows from the definition that if a constant appears in a zig-zag of length n it must be the last element, f_n.

We shall need to consider the ring of formal sums over $\mathbf{R}$, the real numbers[1], of elements of $Z_\mathcal{F}(n)$, denoted by $\mathbf{R}[Z_\mathcal{F}(n)]$. This may be regarded as the vector space over $\mathbf{R}$ with basis the finite set $Z_\mathcal{F}(n)$, so that $\mathbf{R}[Z_\mathcal{F}(n)] = \{\sum r_p p | r_p \in \mathbf{R}, p \in Z_\mathcal{F}(n)\}$.

We may join two zig-zags together by a "bridge" if we are given a suitable label for the bridge, so that informally $f \xrightarrow{2} f$ and $f \xrightarrow{1} a$ may be joined by a bridge labelled 1 to get

$$(f \xrightarrow{2} f) \circ_1 (f \xrightarrow{1} a) = f \xrightarrow{2} f \xrightarrow{1} f \xrightarrow{1} a.$$

where $\circ_1$ is a labelled concatenation operator defined formally below. We extend our definition to define $q1 \circ_j q2$ for any zig-zag $q1$ and any element $q2$ of $\mathbf{R}[Z_\mathcal{F}(n)]$. (Notice for example that) is not a zig-zag.)

Definition 2. i Let $Z = ([f_1, \ldots, f_n], \eta)$, $Z' = ([g_1, \ldots, g_m], \eta')$ be zig-zags, and suppose that $1 \leq \alpha(f_n)$, and let $1 \leq p \leq \alpha(f_n)$. We define a new zig-zag

$$Z \circ_p Z' = ([f_1, \ldots, f_n, g_1, \ldots, g_m], \theta),$$

where

$$\theta : f_i \to \eta(f_i), 1 \leq i \leq (n-1)$$
$$\theta : f_n \to p$$
$$\theta : g_i \to \eta'(g_i), 1 \leq i \leq (m-1)$$

If $\alpha(f_n) = 0$ then let $Z \circ_p Z' = 0$.

ii If $q1 \in Z_\mathcal{F}(n)$, j is at most the arity of the last element of $q1$ and $q2 = \sum r_p p \in \mathbf{R}[Z_\mathcal{F}(n)]$, with $r_p \in \mathbf{R}, p \in Z_\mathcal{F}(m)$ then we define

$$q1 \circ_j q2 = \left(\sum r_p q1 \circ_j p\right) \in \mathbf{R}[Z_\mathcal{F}(m+n)],$$

where the sum is well-defined since $Z_\mathcal{F}(m)$ is finite. (If $q2 = 0$ then $q1 \circ_j q2 = 0$).

---

[1] All we have done would work of course if we restricted to natural numbers here. However Example 3 shows us that, even in the case of monadic terms, distinct positive reals give us distinct orderings, and so we get more orderings by considering all reals and not just natural numbers.

3.1 The number of zig-zags in a term

We now formalise the notion of counting the number of zig-zags on a set of function symbols $\mathcal{F}$ in a term of $\mathcal{T}(\mathcal{G})$, where $\mathcal{F}$ is a subset of $\mathcal{G}$. Informally the zig-zag $f_1 \xrightarrow{i} f_2 \xrightarrow{j} \cdots \xrightarrow{k} f_n$ occurs in a term s if s has a subterm $f_1(s_1, \ldots, s_i, \ldots, s_m)$ where $f_2 \xrightarrow{j} \cdots \xrightarrow{k} f_n$ occurs in s_i. Thus $f \xrightarrow{1} a$ occurs three times in $f(f(a,a),a)$. We denote the number of occurrences of the zig-zag p in the term s by $\#(p,s)$, and record the total number of occurrences of each zig-zag in s by the map

$$\zeta_n : s \to \sum \#(p,s)p$$

where the sum is over all $p \in Z_{\mathcal{F}}(n)$. Thus

$$\zeta_n : \mathcal{T}(\mathcal{G}) \to \mathbf{R}[Z_{\mathcal{F}}(n)].$$

Thus for example if $\mathcal{F} = \mathcal{G} = \{f, a\}$ then

$$\zeta_2(f(f(a,a),a)) = 1.(f \xrightarrow{1} f) + 0.(f \xrightarrow{2} f) + 3.(f \xrightarrow{1} a) + 2.(f \xrightarrow{2} a),$$

and if $\mathcal{F} = \{f\}$, $\mathcal{G} = \{f, a\}$, then

$$\zeta_2(f(f(a,a),a)) = 1.(f \xrightarrow{1} f).$$

This is still not a formal definition of $\#(p,s)$ or ζ_n. This is given as follows.

We give a recursive definition of the formal sum $\zeta_n(s)$, where

$$\zeta_n : \mathcal{T}(\mathcal{G}) \to \mathbf{R}[Z_{\mathcal{F}}(n),]$$

for $\mathcal{F} \subseteq \mathcal{G}$. If $h \in \mathcal{G}$ then define δ_h to be 1 if $h \in \mathcal{F}$ and 0 otherwise. Then $\#(p,s)$ is defined as the coefficient of p in $\zeta_n(s) \in \mathbf{R}[Z_{\mathcal{F}}(n)]$, and p is said to occur in s if $\#(p,s) \geq 1$.

Definition 3. Let $s \in \mathcal{T}(\mathcal{G})$.
For $s = a$ a constant let

$$\zeta_1(s) = \delta_a a.$$

For $s = f(s_1, s_2, \ldots, s_n)$ let

$$\zeta_1(s) = \delta_f f + \sum_{i=1}^{i=n} \zeta_1(s_i)$$

For $m \geq 2$ and $s = a$ a constant let

$$\zeta_m(s) = 0.$$

For $m \geq 2$ and $s = f(s_1, s_2, \ldots, s_n)$ let

$$\zeta_m(s) = (\sum_{i=1}^{i=n} \delta_f(f \circ_i \zeta_{m-1}(s_i))) + (\sum_{i=1}^{i=n} \zeta_m(s_i))$$

Notice that $\zeta_m(s) = 0$ when m is greater than the depth of s. We may combine the maps ζ_i to obtain a map

$$\zeta : T(\mathcal{G}) \to \Pi_{n=1}^{n=\infty} \mathbf{R}[Z_\mathcal{F}(n)],$$

given by

$$\zeta(s) = (\zeta_1(s), \zeta_2(s), \ldots);$$

the entries of $\zeta(s)$ become 0 after some point.

3.2 A property of ζ_m

The property of the map ζ_m which makes our ordering work is this. Intuitively we want to compare two terms u, v by comparing $\zeta_r(u) = \zeta_r(v)$ lexicographically for each r. We need the subterm property, which asserts that if u is obtained from v by replacing the subterm s by the subterm t, and $s > t$, then $u > v$. Thus we need to calculate the relationship between $\zeta_r(u), \zeta_r(v), \zeta_r(s), \zeta_r(t)$ for each r; the next Lemma does this.

We need to show that if $\zeta_r(s) = \zeta_r(t)$ then $\zeta_r(u) = \zeta_r(v)$, and if $\zeta_r(s) > \zeta_r(t)$ then $\zeta_r(u) > \zeta_r(v)$. This follows from the calculations in the next Lemma.

Lemma 4. *Let $s, t \in T(\mathcal{G})$, and $k \geq 1$, and suppose that $\zeta_r(s) = \zeta_r(t)$ for all $r < k$. Let $n = \alpha(f) \geq 1$. Then for any terms*

$$u = f(s_1, \ldots, \underset{\underset{j}{\uparrow}}{s}, \ldots, s_n) \quad v = f(s_1, \ldots, \underset{\underset{j}{\uparrow}}{t}, \ldots, s_n)$$

and for all $r \leq k$ we have

$$\zeta_r(u) - \zeta_r(v) = \zeta_r(s) - \zeta_r(t)$$

Proof. If $r = 1$ then

$$\zeta_1(u) - \zeta_1(v) = (\delta_f f + \sum_{i=1}^{i=n} \zeta_1(s_i)) - (\delta_f f - \zeta_1(s) + \zeta_1(t) + \sum_{i=1}^{i=n} \zeta_1(s_i))$$

$$= \zeta_1(s) - \zeta_1(t).$$

If $2 \leq r \leq k$ then

$$\zeta_r(u) = (\sum_{i=1}^{i=n} \delta_f f \circ_i \zeta_{r-1}(s_i)) + (\sum_{i=1}^{i=n} \zeta_r(s_i))$$

$$= \zeta_r(v) - \delta_f f \circ_j \zeta_{r-1}(t) + \delta_f f \circ_j \zeta_{r-1}(s) + \zeta_r(s) - \zeta_r(t).$$

Now by assumption $\zeta_{r-1}(s) = \zeta_{r-1}(t)$, so $\delta_f f \circ_j \zeta_{r-1}(s) = \delta_f f \circ_j \zeta_{r-1}(t)$ and hence $\zeta_r(u) - \zeta_r(v) = \zeta_r(s) - \zeta_r(t)$ as required.

3.3 Orderings on $\mathbf{R}[Z_{\mathcal{F}}(n)]$

Our orderings on $\mathcal{T}(\mathcal{G})$ will be induced from orderings on the real vector space $\mathbf{R}[Z_{\mathcal{F}}(n)]$, whose dimension is just the size of $Z_{\mathcal{F}}(n)$. The elements of $Z_{\mathcal{F}}(n)$ form a basis for $\mathbf{R}[Z_{\mathcal{F}}(n)]$ over $\mathbf{R}$, which we will call the standard basis.

We recall from [11] and [12] the information that we need.

Definition 5. An ordering on $\mathbf{R}^n$ is called monotonic if $u > v$ implies $u + w > v + w$ for all $u, v, w \in \mathbf{R}^n$, and a division ordering if further each coordinate vector $E_x > 0$.

We have

Lemma 6. *If $>$ on $\mathbf{R}^n$ is monotonic then its restriction to $\mathbf{N}^n$ is well-founded if and only if each $E_x \not< 0$. If $>$ on $\mathbf{R}^n$ is a monotonic division ordering then the restriction of $>$ to $\mathbf{N}^n$ has the finite basis property.*

In what follows we identify the coordinate vectors with the standard basis vectors of $\mathbf{R}[Z_{\mathcal{F}}(n)]$.

Any total ordering $>$ on $Z_{\mathcal{F}}(n)$ may be extended lexicographically to a total ordering on $\mathbf{R}[Z_{\mathcal{F}}(n)]$ by $u = \Sigma_p r_p p > v = \Sigma_p s_p p$ if and only if for some $p \in Z_{\mathcal{F}}(n)$ we have $r_p > s_p$ and $r_q = s_q$ for all $q < p$. This is a total division ordering.

To obtain further orderings on $\mathbf{R}^n$ we proceed as follows (see [11]). Let A be any matrix with n rows, where vectors are regarded as row vectors. The ordering $u > v$ if and only if uA is greater than vA in the lexicographic ordering from the left is a monotonic ordering. If further each coordinate vector is greater than 0 it is a division ordering. All total monotonic and division orderings arise in this way, and each orthogonal matrix A gives rise to a distinct such ordering. The $n!$ orthogonal matrices obtained by permuting the rows of the identity matrix give $n!$ lexicographic orderings on $\mathbf{R}^n$, which are just the lexicographic extensions of the $n!$ possible permutations of the standard basis vectors.

Vectors are sometimes "ordered by weight", that is a fixed vector w of weights is given and the ordering is defined by $u > v$ if and only if $u.w > v.w$, where . is the usual dot product of vectors. Ordering by weight corresponds to a matrix whose first column is the corresponding vector of weights.

If D is a finite subset of $\mathbf{R}^n \times \mathbf{R}^n$ it is decidable whether or not there is a monotonic or division ordering $>$ such that $u > v$ for all $(u, v) \in D$. The procedure is essentially that of solving linear inequalities; more details are given in [5].

4 Orderings on $\mathcal{T}(\mathcal{G})$

We may now define orderings on $\mathcal{T}(\mathcal{G})$ as follows.

Definition 7. Let $\mathcal{F} \subseteq \mathcal{G}$, and for each $m \geq 1$ let $>_m$ be a monotonic ordering on $\mathbf{R}[Z_\mathcal{F}(m)]$. Let $\succ$ be the lexicographic combination of $(>_1, >_2, \ldots)$ on $\Pi_{n=1}^{n=\infty} \mathbf{R}[Z_\mathcal{F}(n)]$. Let $>$ be the ordering on $\mathcal{T}(\mathcal{G})$ induced from $\succ$ by

$$\zeta : \mathcal{T}(\mathcal{G}) \to \Pi_{n=1}^{n=\infty} \mathbf{R}[Z_\mathcal{F}(n)].$$

Thus $u > v$ if and only if there is an $i \geq 1$ such that $\zeta_i(u) >_i \zeta_i(v)$ and $\zeta_j(u) = \zeta_j(v)$ for all $j < i$.

We call $>$ a linear interpretation of $\mathcal{T}(\mathcal{G})$ over $\mathcal{F}$.

Notice that if we take $>_j$ to be the empty ordering, which is well-founded, for $j > N$ then the ordering $>$ is just that obtained by taking the lexicographic combination of $(>_1, >_2, \ldots, >_N)$ on $\Pi_{n=1}^{n=N} \mathbf{R}[Z_\mathcal{F}(n)]$, as in the examples.

We have

Theorem 8. *1. The relation $>$ is monotonic ordering.*
2. If each coordinate vector u in $\mathbf{R}[Z_\mathcal{F}(1)]$ satisfies $u \not<_1 0$ then $>$ is well-founded.
3. If $\mathcal{F} = \mathcal{G}$ and $>_1$ is a division ordering then $>$ is a simplification ordering.

Proof. 1. We show that if $s > t$ then

$$u = f(s_1, \ldots, \underset{\underset{k}{\uparrow}}{s}, \ldots, s_n) > u = f(s_1, \ldots, \underset{\underset{k}{\uparrow}}{t}, \ldots, s_n).$$

Suppose that $s > t$, so that there is an $i \geq 1$ such that $\zeta_j(s) = \zeta_j(t)$ for all $j < i$ and $\zeta_i(s) >_i \zeta_i(t)$. Then by the previous lemma $\zeta_j(u) - \zeta_j(v) = \zeta_j(s) - \zeta_j(t)$ for all $j \leq i$. Thus $\zeta_j(u) - \zeta_j(v) = 0$ for all $j < i$, and $\zeta_i(u) = \zeta_i(s) + (\zeta_i(v) - \zeta_i(t)) >_i \zeta_i(t) + (\zeta_i(v) - \zeta_i(t)) = \zeta_i(v)$ since $>_i$ is monotonic. Thus $u > u$.

2. It suffices to show that the restriction of $\succ$ to the image of ζ is well-founded. Let P be the image of ζ. Now if $x_1 \in \mathbf{R}[Z_\mathcal{F}(1)]$ and $\zeta(x) = (x_1, x_2, \ldots) \in P$ then there are only finitely many choices for $(x_2, x_3, \ldots) \in \Pi_{n=2}^{n=\infty} \mathbf{R}[Z_\mathcal{F}(n)]$, since x_1 determines the function symbols from $\mathcal{F}$ that occur in x and their multiplicities, and hence upper bounds on the sizes and numbers of zig-zags which can occur in x. It follows from Lemma 1 that it suffices to show that the restriction of $>_1$ to the set of possible x_1 is well-founded. But this set is a subset of $\mathbf{N}[Z_\mathcal{F}(1)]$, and since each coordinate vector u in $\mathbf{R}[Z_\mathcal{F}(1)]$ satisfies $u \not<_1 0$ it follows from Lemma 6 that the restriction of $>_1$ to $\mathbf{N}[Z_\mathcal{F}(1)]$ is well-founded.

3. We need to show that $f(s_1, \ldots, s_i, \ldots, s_n) > s_i$ for all i. Now

$$\begin{aligned}\zeta_1(f(s_1, \ldots, s_i, \ldots, s_n)) &= \delta_f f + \Sigma_{k=1}^{k=n} \zeta_1(s_k) \\ &= \zeta_1(s_i) + (\delta_f f + \Sigma_{k \neq i} \zeta_1(s_k)) \\ &\geq_1 \zeta_1(s_i) + \delta_f f \\ &>_1 \zeta_1(s_i)\end{aligned}$$

as $\delta_f f = f >_1 0$ and $>_1$ is a monotonic division ordering.

4.1 Some further properties of the ordering

We record some further properties of the ordering.

We observe that there are continuum many distinct orderings of this kind, even when $\mathcal{F} = \mathcal{G}$, as there are continuum many choices for $>_1$, each of which gives rise a distinct ordering $>$ when combined with the empty ordering for each $>_i, i \geq 1$.

In fact there are many choices for each $>_i$.

Lemma 9. *Let $>$ be as defined as in Example 3. Then each choice of λ gives a distinct value for $>_{3(\lambda)}$, and hence for $>$.*

Proof. Let λ_1, λ_2 be distinct positive reals, with $\lambda_1 > \lambda_2$. Then there exist positive integers M, N and a rational number N/M with $\lambda_1 > N/M > \lambda_2$. It follows that

$$f^N g^{2M} f^N(x) >_{3(\lambda_2)} g^M f^{2N} g^M(x) \text{ as } N/M > \lambda_2$$

and

$$g^M f^{2N} g^M(x) >_{3(\lambda_1)} f^N g^{2M} f^N(x) \text{ as } \lambda_1 > N/M.$$

Hence $>_{3(\lambda_1)}$ and $>_{3(\lambda_2)}$ are different, and give rise to distinct orderings $>$. It follows that there are continuum many distinct orderings $>$ which arise in this way.

Observe that this Lemma shows that there are "more" orderings of this kind than there are Knuth Bendix orderings, as after choosing the weights, that is choosing $>_1$, the Knuth Bendix ordering is totally determined by choosing a precedence on the operator symbols, which, if there are n operator symbols, can be done in $n!$ ways.

Similarly there are "more" orderings of this kind than there are recursive path orderings, since there are only finitely many recursive path orderings on a given signature.

It is straightforward to extend our proofs to show that that if $\mathcal{F} = \mathcal{G}$ and consists of unary function symbols and a single variable x then our orderings are preserved under variable substitutions. The situation in general is less clear.

We note that if each $>_i$ is a total ordering then $>$ is a total ordering on $\mathcal{T}(\mathcal{G})/\approx$, the set of $\approx$-equivalence classes of $\mathcal{T}(\mathcal{G})$ under the relation $s \approx t$ if and only if $\zeta(s) = \zeta(t)$. This follows as for any terms s, t with $s \not\approx t$ there is an i with $\zeta_i(s) \neq \zeta_i(t)$ and $\zeta_j(s) = \zeta_j(t)$ for $j < i$. Thus since $>_i$ is total, $\zeta_i(s)$ and $\zeta_i(t)$ are comparable under $>_i$, so that s and t must be comparable. Thus $>$ is total on $\mathcal{T}(\mathcal{G})/\approx$.

It follows that if $\mathcal{F} = \mathcal{G}$ and each $>_i$ is total then $>$ is a total ordering on $\mathcal{T}(\mathcal{G})$. It follows from Lemma 1 that the order type is then just that of $>_1$, which is at most $\omega^{|\mathcal{F}|}$.

Lemma 10. *Let $R = \{l_i \to r_i\}$ be a finite set of pairs of elements of $\mathcal{T}(\mathcal{G})$. Then it is decidable whether or not there is a subset $\mathcal{F}$ of $\mathcal{G}$ and a simplification/well-founded linear interpretation $>$ with respect to $\mathcal{F}$ with $l_i > r_i$ for each i.*

Proof. We consider the simplification case; the well-founded case is similar. Let $\mathcal{F}$ be a subset of $\mathcal{G}$. For each i let $\phi(i)$ be such that $\zeta_j(l_i) = \zeta_j(r_i)$ for $j < \phi(i)$, and $\zeta_{\phi(i)}(l_i) \neq \zeta_{\phi(i)}(r_i)$. Now let $P_k = \{(l_i, r_i) | \phi(i) = k\}$. For each $k > 1$ it is decidable whether or not there is a monotonic ordering $>_k$ on $\mathbf{R}[Z_{\mathcal{F}}(k)]$ with $\zeta_k(l_i) >_k \zeta_k(r_i)$ for all $(l_i, r_i) \in P_k$, and it is decidable whether or not there is a division ordering $>_1$ on $\mathbf{R}[Z_{\mathcal{F}}(1)]$ with $\zeta_1(l_i) >_k \zeta_1(r_i)$ for all $(l_i, r_i) \in P_1$. Now R is finite, so $max\{\phi(i)\}$ exists. If for each $max\{\phi(i)\} \geq k \geq 1$ the appropriate ordering $>_k$ exists then we can use these $>_k$ to define an ordering $>$ with $l_i > r_i$ for each i; otherwise no such ordering exists. We may repeat this process for each subset of $\mathcal{G}$, since $\mathcal{G}$ is finite.

5 Some examples

We calculate some examples. In each case $\mathcal{F} = \mathcal{G} = \{f, a, b, c\}$ so $\mathbf{R}[Z_{\mathcal{F}}(1)]$ has dimension 4, $\mathbf{R}[Z_{\mathcal{F}}(2)]$ has dimension 8 with basis elements $\{f \xrightarrow{2} f, f \xrightarrow{1} f, f \xrightarrow{2} a, f \xrightarrow{1} a, f \xrightarrow{2} b, f \xrightarrow{1} b, f \xrightarrow{2} c, f \xrightarrow{1} c\}$ and $\mathbf{R}[Z_{\mathcal{F}}(3)]$ has dimension 16. We let $>$ be any linear interpretation over $\mathcal{G}$.

Example 4

Let $s = f(f(a, b), c), t = f(a, f(b, c))$. Then

$$\zeta_1(s) = \zeta_1(t) = 2f + a + b + c$$
$$\zeta_2(s) = 2f \xrightarrow{1} f + 2f \xrightarrow{1} a + f \xrightarrow{1} b + f \xrightarrow{2} c + f \xrightarrow{2} b$$
$$\zeta_2(t) = f \xrightarrow{2} f + f \xrightarrow{1} a + f \xrightarrow{2} b + 2f \xrightarrow{2} c + f \xrightarrow{1} b$$
$$\zeta_2(s) - \zeta_2(t) = 2f \xrightarrow{1} f - f \xrightarrow{2} f + f \xrightarrow{1} a - f \xrightarrow{1} c$$

Thus $s > t$ with any choice of monotonic ordering for $>_1$, and any choice of monotonic ordering for $>_2$ which has $\zeta_2(s) >_2 \zeta_2(t)$, for example a lexicographic ordering with $f \xrightarrow{1} f >_2 f \xrightarrow{2} f$.

Example 5

We consider the twelve terms obtained by permuting a, b, c in the terms s, t of the previous example. We calculate ζ_2 in each case, and represent the result as a vector on the basis vectors taken in the order $\{f \xrightarrow{2} f, f \xrightarrow{1} f, f \xrightarrow{2} a, f \xrightarrow{1} a, f \xrightarrow{2} b, f \xrightarrow{1} b, f \xrightarrow{2} c, f \xrightarrow{1} c\}$. Permuting these basis vectors gives 8! possible lexicographic orderings on vectors as candidates for $>_2$ and we show the effect of four of these.

Term s	Vector $\zeta_2(s)$	$>^1$	$>^2$	$>^3$	$>^4$
$f(f(a,b),c)$	$(0,1,0,2,1,1,1,0)$	1	2	9	11
$f(f(a,c),b)$	$(0,1,0,2,1,0,1,1)$	7	1	6	12
$f(f(b,c),a)$	$(0,1,1,0,0,2,1,1)$	10	3	4	1
$f(f(c,b),a)$	$(0,1,1,0,1,1,0,2)$	12	4	8	2
$f(f(c,a),b)$	$(0,1,1,1,1,0,0,2)$	11	6	5	10
$f(f(b,a),c)$	$(0,1,1,1,0,2,1,0)$	2	5	3	9
$f(a,f(b,c))$	$(1,0,0,1,1,1,2,0)$	4	7	10	5
$f(a,f(c,b))$	$(1,0,0,1,2,0,1,1)$	8	8	12	6
$f(b,f(c,a))$	$(1,0,2,0,0,1,1,1)$	9	11	1	4
$f(c,f(b,a))$	$(1,0,2,0,1,1,0,1)$	6	12	7	3
$f(c,f(a,b))$	$(1,0,1,1,2,0,0,1)$	5	10	11	7
$f(b,f(a,c))$	$(1,0,1,1,0,1,2,0)$	3	9	2	8

Here the orderings $>^i$ are the lexicographic orderings on vectors generated by the following orderings on basis vectors

$>^1: f \xrightarrow{1} c >^1 f \xrightarrow{2} c >^1 f \xrightarrow{1} b >^1 f \xrightarrow{2} b >^1 f \xrightarrow{1} a >^1 f \xrightarrow{2} a >^1 f \xrightarrow{1} f >^1 f \xrightarrow{2} f$

$>^2: f \xrightarrow{2} f >^2 f \xrightarrow{1} f >^2 f \xrightarrow{2} a >^2 f \xrightarrow{1} a >^2 f \xrightarrow{2} b >^2 f \xrightarrow{1} b >^2 f \xrightarrow{2} c >^2 f \xrightarrow{1} c$

$>^3: f \xrightarrow{2} b >^3 f \xrightarrow{1} b >^3 f \xrightarrow{2} c >^3 f \xrightarrow{1} c >^3 f \xrightarrow{2} f >^3 f \xrightarrow{1} f >^3 f \xrightarrow{2} a >^3 f \xrightarrow{1} a$

$>^4: f \xrightarrow{1} a >^4 f \xrightarrow{2} a >^4 f \xrightarrow{1} f >^4 f \xrightarrow{2} f >^4 f \xrightarrow{1} c >^4 f \xrightarrow{2} c >^4 f \xrightarrow{1} b >^4 f \xrightarrow{2} b$

The notation in the orderings columns indicates $12 >^i 11 >^i 10 \cdots$.

Example 6

We consider the terms

$$s = f(f(a, f(a,a)), f(a,a)), t = f(f(a,a), f(f(a,a),a)).$$

We have $\zeta_1(s) = \zeta_1(t), \zeta_2(s) = \zeta_2(t)$ and

$\zeta_3(s) = 1.(f \xrightarrow{1} f \xrightarrow{2} f) + 0.(f \xrightarrow{2} f \xrightarrow{1} f) +$ (terms in $f \xrightarrow{?} f \xrightarrow{?} a, f \xrightarrow{?} f \xrightarrow{?} b, f \xrightarrow{?} f \xrightarrow{?} c$)

$\zeta_3(t) = 0.(f \xrightarrow{1} f \xrightarrow{2} f) + 1.(f \xrightarrow{2} f \xrightarrow{1} f) +$ (terms in $f \xrightarrow{?} f \xrightarrow{?} a, f \xrightarrow{?} f \xrightarrow{?} b, f \xrightarrow{?} f \xrightarrow{?} c$)

Thus $s > t$ if we take for $>_3$ any monotonic ordering in which $\zeta_3(s) >_3 \zeta_3(t)$; for example a lexicographic ordering on vectors generated by any ordering on the basis vectors in which $f \xrightarrow{1} f \xrightarrow{2} f$ is greater than all other basis vectors, similarly we can make $s < t$ by making $\zeta_3(s) <_3 \zeta_3(t)$.

Acknowledgements

Much of the work was done while on sabbatical leave at MIT and the University of Edinburgh, and the author thanks John Guttag, of MIT, and Alan Bundy and Rod Burstall of Edinburgh, for their generous hospitality.

The author also thanks the conference referees for several perceptive and helpful comments.

References

1. A. Ben Cherifa and P. Lescanne, Termination of rewriting systems by polynomial interpretations and its implementation, Science of Computer Programming 9 (1987) 137-160
2. M. Dauchet, Simulation of Turing machines by a left linear rewrite rule, in Proc 3rd International Conference on Term Rewriting Systems, Springer Lecture Notes in Computer Science 355 (1989) 109-120
3. N. Dershowitz, Orderings for term-rewriting systems, Theoretical Computer Science 17 (1982) 279-301
4. N. Dershowitz, Termination of Rewriting, Journal of Symbolic Computation, 3 (1987) 69-116
5. A. J. J. Dick, J. R. Kalmus and U. Martin, Automating the Knuth Bendix ordering, Acta Informatica 28 (1990) 95-119
6. G. Higman, Ordering by divisibility in abstract algebras, Proceedings of the London Mathematical Society 2 (1952) 326-336
7. J-P. Jouannaud, P. Lescanne and F. Reinig, Recursive decomposition ordering, in Formal description of programming concepts 2, Elsevier 1982, ed· D Bjorner, pp 331-348
8. D. Knuth and P. Bendix, Simple Word Problems in Universal Algebras, *in* Computational Problems in Abstract Algebra, Pergamon Press 1970, ed J. Leech.
9. D. S. Lankford, On proving term rewriting systems are Noetherian, Tech report MTP-3, Louisiana Technical University, Ruston 1979
10. P. Lescanne, Termination of rewrite systems by elementary interpretations, *in* H Kirchner and G Levi, editors, Proc 3rd International Conference on Algebraic and Logic Programming, Springer Lecture Notes in Computer Science 463 (1992) 21-36
11. U. Martin, A geometrical approach to multiset orderings, Theoretical computer Science 67 (1989) 37-54
12. U. Martin, On the diversity of orderings on strings, RHBNC Technical Report 1991
13. P. Narendran and M. Rusinowitch, Any ground associative-commutative theory has a finite canonical system, Proc 4th International Conference on Term Rewriting Systems, Springer Lecture Notes in Computer Science 488, 423-434
14. J. Steinbach, Extensions and comparison of simplification ordering, in Proc 3rd International Conference on Term Rewriting Systems, Springer Lecture Notes in Computer Science 355 (1989) 434-448
15. H. Zantema, Termination of term rewriting by interpretation, Utrecht University technical report RUU-CS-92-14, April 1992

SOME UNDECIDABLE TERMINATION PROBLEMS FOR SEMI-THUE SYSTEMS

Géraud Sénizergues

LaBri
Université de Bordeaux I
351, Cours de la Libération 33405 Talence, France ** ***

Abstract. We show that the uniform termination problem is undecidable for length-preserving semi-Thue systems having 10 rules. We then give an explicit uniformly-terminating semi-Thue system T having 9 rules which is "universal with respect to termination problems" in some sense. It follows that there exists a fixed rule (u_0, v_0) such that $T \cup \{(u_0, v_0)\}$ has 10 rules and undecidable termination problem.

** mailing adress:LaBri and UER Math-info, Université Bordeaux1
351 Cours de la libération -33405- Talence Cedex.
email:ges@geocub.greco-prog.fr
fax: 56-84-66-69

*** This work has been supported by the ESPRIT Basic Research Working Group "COMPUGRAPH II"

System Descriptions

Saturation of first-order (constrained) clauses with the *Saturate* system

Pilar Nivela and Robert Nieuwenhuis[*]

1 Background

During the last years, several *saturation*-based theorem proving techniques for full first-order (*ordering and/or equality constrained*) clauses with equality have emerged.

Bachmair and Ganzinger ([BG91]) apply a *model construction* technique for proving the refutational completeness of *strict superposition* (paramodulation restricted to maximal terms of maximal equations of clauses). Their powerful *abstract redundancy notions* allow *redundant inferences* to be ignored and *redundant clauses* to be deleted without loosing completeness.

This makes it also possible to compute finite *saturated* sets of axioms. Using saturated sets, more efficient proof strategies become (refutationally) complete, like the set-of-support strategy, which is normally incomplete for ordered inference systems and also for equality clauses. Saturated sets fulfilling some syntactic properties even provide decision procedures, like rewrite proofs do in the equational case. Moreover, if a finite saturated set (not containing the empty clause) is obtained for a given theory, then its *consistency* has been proved (normally one can only prove *inconsistencies*).

Strict *basic* superposition, where no inferences have to be computed on subterms generated in previous inferences (like in basic *narrowing*), is proved complete in [NR92a] by means of *equality constraints* and independently also in [BGLS92]. In [NR92b] it is shown that by means of *ordering constrained clauses* one can restrict the search space even further without loosing completeness, by keeping the ordering restrictions of the inferences in constraints which are *inherited*.

As far as we know, no effort had been spent up to now on solving the problems that appear when putting this kind of techniques into practice: designing feasible redundancy provers that fit into the theoretical concepts (a key aspect of the system), choosing adequate inference systems for each case, designing fair control strategies and heuristics depending on whether the purpose is to find a saturated set or a refutation proof, how (and when) to do constraint solving in practice, etc.

---

[*] We would like to thank Harald Ganzinger for inviting us to the Max-Planck-Institut für Informatik, Saarbrücken, Germany, and for his encouragement to carry out this work during our half-year stay there. Permanent address of both authors: Technical University of Catalonia, Pau Gargallo 5, 08028 Barcelona, Spain. E-mail: {nivela,roberto}@lsi.upc.es.

These problems are addressed here in the context of their implementation in our *Saturate* system, written in Quintus-Prolog, a modular toolkit for experimenting with such methods. The system includes several inference calculi (including the ones for equality and ordering constrained clauses), as well as facilities for recovering proofs of axioms, for computing timing statistics, for choosing in a user-friendly way the different settings of the system (control strategies, redundancy provers, inference systems, amount of user interaction...), etc. All this is built by combining modules like constraint solvers or redundancy provers, starting from the abstract data types for elements like clauses, constraints or terms.

2 A short overview of the system

The system has been designed with the aim of being flexible both in its *use* and in its *extension/modification*. We first describe the main facilities for a flexible *use* of the system:

1. Settings of the system

The user can choose *any* subset of all parts of the system according to the problem at hand. These settings can be interactively modified at any moment, i.e. it is not necessary to fix them initially, although the user may define different own default settings.

Inference rules currently available are: *strict superposition left, strict superposition right, equality factoring, equality resolution, merging paramodulation, ordered factoring*. For instance, a minimal complete set of inference rules can be chosen according to the kind of axioms to be treated.

Constraint inheritance strategies can be added in order to restrict the search space: no constraint inheritance (then completeness is preserved under more powerful redundancy criteria), ordering constraint inheritance, equality constraint inheritance (basic strategies), or the combination of both. Clauses with unsatisfiable constraints are tautologies and can be eliminated. Also when no constraints are inherited, checking the ordering restrictions of each inference amounts to deciding the satisfiability of an ordering constraint (this is decidable for LPO, cf. [Com90] and also [NR92b] under extended signatures). The constraint solving module is therefore an important part of the system, which is also frequently used in redundancy proofs.

Marking strategies. In [BG91] their completeness is proved: one can mark an arbitrary negative equation in clauses and compute no inferences with the non-marked equations of such marked clauses (cf. also [NN91]).

Selection strategies: it is possible to choose between various (fair) selection strategies for determining at every stage of the saturation process which clause is used to compute inferences with.

Redundancy proofs: the amount of effort dedicated to redundancy proofs can be set depending on the purpose of the process (refutation or completion). If the purpose is to obtain a saturated system for a given theory, then probably more redundancy provers should be activated. Cf. the following section for more details on redundancy provers.

Output information: this allows to observe particular aspects of a saturation process without getting lost in a huge amount of information. For instance, if we are interested in the behaviour of a particular redundancy prover, a particular inference rule, or the number of consequences generated by a particular clause, then the active output messages can be set in such a way that only this information is provided.

Timing points: they exist for the important tasks of the system, and allow to determine e.g. the time spent on a particular inference rule, a redundancy prover, or on constraint solving, by just activating the corresponding subset of timing points. Timing statistics are displayed when a saturated system is obtained or when the user requires this explicitly.

2. Saturation of a set of axioms

A saturation process of the current data base of clauses can be initiated according to the previous options. It is possible to read in user axioms and also already saturated sets (generated and saved from previous runs of the system). Saturated sets will be treated by the system accordingly, i.e. no unnecessary inferences will be computed. The saturation process finishes when a saturated set of clauses is found (e.g. when the empty clause is generated). If the empty clause is not in the final set, then the *consistency* of the theory has been proved and the saturated set can be saved for further efficient theorem proving in this theory. Also the state of a saturation process can be saved between executions.

3. Proof recovering

The system displays a complete proof for a given clause number, starting from the initial axioms and indicating the intermediate steps.

4. Help facilities

There are also on-line help facilities about all the possibilities of the system.

The main facilities for a flexible *extension/modification* of the system are due to the fact that the system has been developed following the traditional principles of module abstraction.

Let us consider for instance the integration of a new inference rule into the system. First, the rule has to be expressed in terms of the operations of the already existing abstract data types, for clauses or equations, like get_max_eq_in_succedent_clause(C, E), or is_greater_eq(E1, E2). After this, the rule, eventual messages, timing points, proof recovering information, etc. only have to be declared in one table. This updates the menus and all aspects of the control. Modifying the internal representation of a data type needs of course only local changes.

3 Redundancy provers and examples

One of the limits of the current theorem proving systems is caused by the retention of too many clauses. Many theorem proving systems apply heuristics to decide which clauses are generated first, which ones are kept, and which ones are disregarded. However, in our opinion it is also worthwhile to investigate the practical applicability of the existing restrictive ordered inference systems and the powerful *abstract redundancy notions* for inferences and clauses in which completeness is not sacrificed *a priori*.

A clause is redundant (in a set of clauses S) if all its ground instances can be (semantically) deduced from *smaller* ground instances of clauses in S. Similarly, an *inference* is redundant if, for all its ground instances, its conclusion follows from instances smaller than the maximal premise. It is of course not possible to directly implement these abstract redundancy notions. Instead one should look for practical, sufficiently general methods that fit into these abstract notions. We have been able to do so and obtain some full first-order saturated systems that had not been obtained before by means of practical methods.

Classical redundancy proofs. Among the more classical methods, the system of course includes elimination of redundant literals, tautology detection, subsumption, condensement and demodulation. For clauses with equality, tautology checking amounts to deciding whether the succedent is deducible from the equations in the antecedent. We do this by rewriting the skolemized succedent with the (ground) completed skolemized antecedent. This method is also applied in the elimination of redundant literals. If for a clause C there exists a substitution σ such that $C\sigma \subseteq C$ then it is said that C is *condensed* into $C\sigma$. For example $p(x,y), p(a,b) \to q(a), q(x)$ is logically equivalent to (and can be replaced by) the smaller clause $p(a,b) \to q(a)$.

Case analysis on variables is applied in [MN90] for proving the ground confluence of ordered rewrite systems. For example, adding to the axioms for associativity and commutativity of $+$ the consequence $x + (y + z) = y + (x + z)$ yields a ground confluent ordered rewrite system. This can be shown by analyzing all possible orderings (with $=$ and $\succ$) between $x\sigma$, $y\sigma$ and $z\sigma$: $x\sigma = y\sigma = z\sigma$, $x\sigma = y\sigma \succ z\sigma$, $x\sigma \succ y\sigma = z\sigma$, etc. This method is extended in the *Saturate* system for proving redundancy properties on all ground instances of general clauses. However, this method has two main drawbacks: on one hand, it is expensive. The number of $f(n)$ of possible orderings for n variables is given by $f(3) = 13$, $f(4) = 75$, $f(5) = 541$, $f(6) = 4683, \ldots$ On the other hand, this case analysis is not always powerful enough: Suppose we have the equation $(x + y) + z = z + (x + y)$. Then the critical pair $(z + (x + y)) + u = u + ((x + y) + z)$ can be eliminated only by considering a case analysis on $x\sigma + y\sigma \succ z\sigma$, $x\sigma + y\sigma = z\sigma$, and $z\sigma \succ x\sigma + y\sigma$.

Constrained rewriting on ordering constrained equations $t = t'$ [C], where $t = t'$ is an equation and C is an LPO-ordering constraint. If all instances of $t = t'$ for which C is satisfied are of the form $s = s$ then $t = t'$ [C] is a tautology (this is decidable).

Let $u = v$ and e [C] be an equation and a constrained equation whit $e|_p = u\sigma$. Then e [C] rewrites by our notion of *constrained rewriting* into $e[v\sigma]_p$ [$C \wedge u\sigma \succ v\sigma$] and into the *complementary* equation e [$C \wedge u\sigma \not\succ v\sigma$]. If we apply such constrained rewrite steps whenever $C \wedge u\sigma \succ v\sigma$ is satisfiable in the sense of [NR92b], then this rewriting process terminates and provides a redundancy method that can handle examples like the previous one, as the search is more directed towards the different cases that have to be considered.

Clausal rewriting applying clauses of a set S, denoted $\to_S$, is defined on sets of clauses as follows: if C is a tautology or C is subsumed by a clause in S then $N \cup \{C\} \to_S N$. If $t = t' \vee l_1 \vee \ldots \vee l_n$ is a clause in S, and $C|_p = t\sigma$, then $N \cup \{C\} \to_S N \cup \{\, C[t'\sigma]_p \vee l_1\sigma \vee \ldots \vee l_n\sigma,\ C \vee \neg l_1\sigma,\ \ldots,\ C \vee \neg l_n\sigma\,\}$.

In a similar way, not only equality literals $t = t'$, but also positive and even negative atoms can be applied in clausal rewrite steps. In this generalization of demodulation, if $\{C\} \to_S^* \emptyset$ then $S \models C$. It is checked that the instances of the clauses of S used are smaller than C.

By this redundancy method and some restrictions of it, the *Saturate* system automatically obtains, among many other examples, a saturated set for total orderings:

$$\to p(x,x) \qquad \to p(x,y) \vee p(y,x)$$
$$p(x,y) \wedge p(y,z) \to p(x,z) \qquad p(x,y) \wedge p(y,x) \to x = y$$

References

[BG91] Leo Bachmair and Harald Ganzinger. Rewrite-based equational theorem proving with selection and simplification. Technical Report MPI-I-91-208, Max-Planck-Institut für Informatik, Saarbrücken, August 1991. To appear in Journal of Logic and Computation.

[BGLS92] Leo Bachmair, Harald Ganzinger, Christopher Lynch, and Wayne Snyder. Basic paramodulation and superposition. In Deepak Kapur, editor, *11th International Conference on Automated Deduction*, LNAI 607, pages 462–476, Saratoga Springs, New York, USA, June 15–18, 1992. Springer-Verlag.

[Com90] Hubert Comon. Solving symbolic ordering constraints. *International Journal of Foundations of Computer Science*, 1(4):387–411, 1990.

[MN90] Ursula Martin and Tobias Nipkow. Ordered rewriting and confluence. In Mark E. Stickel, editor, *10th International Conference on Automated Deduction*, LNAI 449, pages 366–380, Kaiserslautern, FRG, July 24–27, 1990. Springer-Verlag.

[NN91] Robert Nieuwenhuis and Pilar Nivela. Efficient deduction in equality horn logic by horn-completion. *Information Processing Letters*, 39(1):1–6, July 1991.

[NR92a] Robert Nieuwenhuis and Albert Rubio. Basic superposition is complete. In B. Krieg-Brückner, editor, *European Symposium on Programming*, LNCS 582, pages 371–390, Rennes, France, February 26–28, 1992. Springer-Verlag.

[NR92b] Robert Nieuwenhuis and Albert Rubio. Theorem proving with ordering constrained clauses. In Deepak Kapur, editor, *11th International Conference on Automated Deduction*, LNAI 607, pages 477–491, Saratoga Springs, New York, USA, June 15–18, 1992. Springer-Verlag.

MERILL: An Equational Reasoning System in Standard ML

Brian Matthews,
Department of Computing Science, University of Glasgow, Glasgow, G12 8QQ, U.K.[1]
brian@dcs.glasgow.ac.uk.

1 Introduction

MERILL [2] is a general purpose order-sorted equational reasoning system. Written in Standard ML, it has been developed by the author at the S.E.R.C. Rutherford Appleton Laboratory (RAL) and the Dept of Computing Science at Glasgow University[3]. The development of MERILL was inspired and influenced by the ERIL (Equational Reasoning: an Interactive Laboratory) system developed by Jeremy Dick at RAL and Imperial College, London [1, 2]. This system used order-sorted reasoning in a practical equational reasoning system. However, ERIL was slow, and lacked significant features. MERILL is a entirely new implementation which retains the major features of ERIL whilst being significantly faster and having new facilities. An important extension is the integration of order-sorted reasoning with associative-commutative (AC) operators. The system incorporates more recent work on this, for example [4]. Thus it is comparable with ELIOS-OBJ [3] rather than more traditional rewriting systems. Standard ML was chosen due to its ease of use and of modifiability through its module system, together with the emergence of efficient implementations.

Part of the philosophy of MERILL is that the user is in control. Thus the system has few built in assumptions and does little reasoning in the background. This means that the user has to define the object syntax themselves and control the reasoning strategy. The user is thus given a great deal of freedom. Significant features of the MERILL system include: terms defined over an order-sorted signature; user defined mixfixed syntax for terms; associative-commutative operators; equality sets for data separation; a variety of term orderings; several completion algorithms; a menu based interface.

2 The MERILL System

The central database of the system is divided into three major components: the signature, the equality sets and the environment. These mutually dependent data sets combine to provide the raw material for the computational tools, and provide the storage for their results. Computational tools include rewriting,

---

[1] On leave from SERC, Rutherford Appleton Laboratory, Didcot, OXON, OX11 0QX, U.K.
[2] Available via anonymous ftp from the University of Glasgow ftp.dcs.glasgow.ac.uk (130.209.240.50)
[3] Sponsored by the SERC/DTI IEATP project "Verification Techniques for LOTOS Specifications" between RAL, Glasgow University and Royal Holloway and Bedford New College.

unification and completion. We shall explore the features of MERILL by means of the specification of arithmetic over the natural numbers, given in figure 1.

Sorts:	$Bool,\ nat,\ zero,\ posnat$	
Subsorts:	$zero\ <\ nat,\quad posnat\ <\ nat$	
Operators:	$ff :\to Bool$	$tt :\to Bool$
	$0 :\to zero$	$succ : nat \to posnat$
	$\_ > \_ : nat\ nat \to Bool$	$pred : posnat \to nat$
	$\_ + \_ : nat\ nat \to nat$	$\_ * \_ : nat\ nat \to nat$
	$\_ + \_ : zero\ zero \to zero$	$\_ * \_ : posnat\ posnat \to posnat$
	$\_ + \_ : nat\ posnat \to posnat$	$\_ * \_ : zero\ nat \to zero$
	$\_ + \_ : posnat\ nat \to posnat$	$\_ * \_ : nat\ zero \to zero$
Variables	$n, n_1, n_2 : nat,\ p : posnat$	
Equations:	$pred(succ(n)) = n$	$n_1 + n_2 = n_2 + n_1$
	$n + 0 = n$	$(n + n_1) + n_2 = n + (n_1 + n_2)$
	$n + (succ(n_1)) = succ(n + n_1)$	$n_1 * n_2 = n_2 * n_1$
	$n * 0 = 0$	$(n * n_1) * n_2 = n * (n_1 * n_2)$
	$(succ(n)) * n_1 = (n * n_1) + n_1$	$n * (n_1 + n_2) = (n * n_1) + (n * n_2)$
	$0 > n = ff$	$p > 0 = tt$
	$succ(n) > succ(n_1) = n > n_1$	

Figure 1: Example Order-Sorted Specification

This example shows the distinctive features of order-sorted equational logic as implemented in MERILL : a hierarchy of sorts; partial functions such as pred made total on subsorts; multiple ranks for function overloading; the use of sorted variables in equations to define the behaviour of functions restricted to certain input domains. Further, the addition and multiplication functions are declared to be AC and require special handling.

2.1 The Signature

The signature is a database which defines the object language. It comprises of four components: sorts; a sort ordering over the declared sorts; operators with their ranks; and sorted variable forms. The user would move to the sort menu, add the sorts, then move to the sort-ordering menu to declare the ordering, and then move on to the operator menu. Figure 2 gives a screen from the MERILL system showing operators being added, a typical example of the style of interface to the system. The top of the screen shows the operators already added to the system, together with their ranks. In the middle of the screen is a menu, from which the user selects an item, in this case "a" for adding new operators. Each rank is added individually; here we declare the ">" operator, and two ranks of the multiplication operator. The addition function is annotated "(ASSOC COMM)" to mark that this operator has been declared AC.

Classes of variable names are similarly declared with their sort. Using these declarations, the system builds a mixfix parser used by the rest of the system to parse terms. In addition to declaring operators, the system allows the user to declare that certain operators are commutative or associative-commutative. The

```
---------------------------------OPERATORS---------------------------------
1         0 :   -> zero
2         succ( _ ) :   nat -> posnat
3         pred( _ ) :   posnat -> nat
4         _ + _ :   nat nat -> nat (ASSOC COMM)
                :   zero zero -> zero (ASSOC COMM)
                :   nat posnat -> posnat (ASSOC COMM)
                :   posnat nat -> posnat (ASSOC COMM)
---------------------(h - help, Control-C - Interrupt)---------------------
                                                Operator Options
                                                a    Add Operators
                                                d    Delete Operators
                                                e    Equational Theory
                                                >>   a
Enter Operators:
>> _ > _ :   nat nat -> Bool
>> _ * _ :   nat nat -> nat
>> _ * _ :   posnat posnat -> posnat
```

Figure 2: Adding operators to MERILL

system generates a representation of this equational theory for use in completion.

2.2 Equality Sets

The user declares equalities which are associated together in equality sets. This allows separation of data into various sets which the user can handle independently. Equalities can be of four varieties:

Equations	=	Axioms not used for rewriting.
Rewrite Rules	=>	Rules used for rewriting.
Conjectures	=?=	Declared true when rewritten to identity.
Conditionals	$e_1, \ldots, e_n ==> e$	Used for rewriting, but not completion.

2.3 Environment

The environment allows the selection of term orderings and completion strategies which are independent of the nature of the signature used. Currently, the Recursive Path ordering, and Knuth-Bendix ordering are available, as well as an Associative Knuth-Bendix ordering suitable for use with AC operators. To use these orderings the user must set up appropriate precedences and weights on operators, also within the environment.

2.4 Tools in MERILL

A series of tools are available which use this database of signature, equality sets and environment. There are tests on the signature available for regularity, monotonicity and inhabitedness, although these properties are not enforced.

Tools available for using on terms and equalities include explicit unification of terms; rewriting of terms and equations; generating critical pairs, and three order-sorted completion algorithms. The first is ordinary completion without AC-operators; the second uses AC-operators, but restricted to left-linear rules; the third is full AC-completion which generates associative extensions to rules. The second method is faster than the third, but can only be used in restricted cases.   In the example we wish to complete the given set of rules using the full

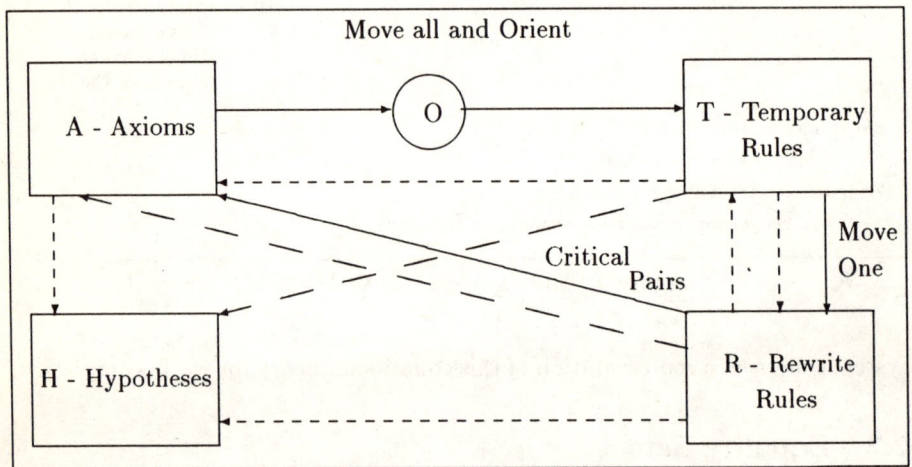

Figure 3: Running AC-Completion.

AC-completion algorithm. The organisation of the algorithm is given in figure 3. Boxes represent equality sets, solid arrows movement of equalities during completion, and dashed arrows rewriting of equalities in one set by another set. Equality sets are selected from the database and completion proceeds by interaction between them: equations and critical pairs are placed in A, oriented into rules by the ordering O and placed into T for rewriting. From there, rules are selected for moving into R where critical pairs are generated. Hypotheses to be considered during completion can be placed in H. Termination occurs when the sets A and T are empty; R will then be confluent.

Figure 4 gives the results of running this strategy upon the example from a MERILL session. No new non-trivial pairs were created. However, rule 10 is new, and marked with an asterisk. This is an extended rule, added to ensure confluence of AC-rewriting, generated by superposing rules on the associative laws of the equational theory, defined by the operators.

3 Conclusions

MERILL has achieved it initial goals. Its completion algorithm for example is approximately 14 times faster than its predecessor ERIL and also combines AC operators with order-sorted rewriting. However, MERILL is also designed to

```
------------------------------Rewrite Rules------------------------------
10 Equalities
  1    pred(succ(n)) => n
  2    n + 0 => n
  3    n * 0 => 0
  4    0 > n => ff
  5    p > 0 => tt
  6    n + (succ(n1)) => succ( n + n1 )
  7    (succ(n)) > (succ(n1)) => n > n1
  8    (succ(n))* n1 => ( n * n1 )+ n1
  9    n *( n1 + n2 ) => ( n * n1 )+( n * n2 )
 10    ((succ(n))* n1 )* n1 => ( n1 * n1 )+( n1 *( n * n1 )) (*)
----------------------(h - help, Control-C - Interrupt)----------------------
                                              Equality Set Options
                                          a     Add Equations
                                          d     Delete Equations
                                          o     Orient Equations
                                         >>
```

Figure 4: Complete Set of Rewrite Rules for Arithmetic on Naturals.

be a vehicle for experimenting with new techniques in equational reasoning. A major extension is dynamic order-sorted rewriting, a new presentation of order-sorted logic which overcomes some of the short comings of order-sorted reasoning [5]. The flexibility of Standard ML is demonstrated by the ease which this new method can implemented. Other extensions planned include a tactic language, new proof techniques, and further orderings.

References

[1] A J J Dick. ERIL. Equational reasoning: an interactive laboratory. In B. Buchberger, editor, *Proceedings of the EUROCAL conference*. Springer-Verlag, 1985.

[2] A J J Dick. *Order-Sorted Equational Reasoning and Rewrite Systems*. PhD thesis, Imperial College, University of London, 1987.

[3] I Gnaedig. ELIOS-OBJ. Theorem proving in a specification language. In B. Kreig-Bruckner, editor, *Proc. of ESOP'92*, volume 582 of *Lecture Notes in Computer Science*, pages 182–199. Springer-Verlag, 1992.

[4] I Gnaedig, C Kirchner, and H Kirchner. Equational completion in order-sorted algebras. *Theoretical Computer Science*, 72:169–202, 1990.

[5] Brian Matthews and Phil Watson. Dynamic order-sorted rewriting. University of Glasgow, in Preparation., 1993.

Reduce the Redex → ReDuX

Reinhard Bündgen

Wilhelm-Schickard-Institut, Universität Tübingen
D-7400 Tübingen, Fed. Rep. of Germany
phone: x7071/295459 — fax: x7071/295958
e-mail: ⟨buendgen@informatik.uni-tuebingen.de⟩

The ReDuX[1]-system is a work-bench for programming and experimenting with term rewriting systems. It is focused towards the implementation of completion procedures with special emphasis on inductive completion. From the programmer's point of view ReDuX provides a large library of data types and algorithms (over 450) which allows for high level programming. The experimentalist also finds a collection of ready-to-run programs (see Table 1, and [WB91]). ReDuX has been developed as an extension of the TC- and IC-systems [Küc82a, Bün87] and has been used as a research tool over the last years. For the last two years it has also been employed as a tutorial system for courses on term rewriting systems at the University of Tübingen.

program	program description	main authors
TO	test environment for term orderings	Küchlin, Schwärzler, Bündgen, Wendel
TC0	Knuth-Bendix completion procedure	Küchlin, Bündgen
TCS	Knuth-Bendix completion procedure applying subconnectedness criterion	Küchlin
TCT	Knuth-Bendix completion procedure applying transformation criterion	Bündgen, Küchlin
TCTS	Knuth-Bendix completion procedure applying transformation and subconnectedness criteria	Küchlin, Bündgen
TST	top set tree and ground normal form grammar computation	Bündgen, Eckhardt
ITST	interleaved top set computation based on top set trees	Bündgen
PTST	pruned top set tree and ground normal form grammar computation	Eckhardt
LAB	inductive completion laboratory including ground normal form analysis, positional ground reducibility test and an inductive completion procedure according to [Küc89]	Bündgen
AC	rewrite laboratory for term rewriting systems with associative-commutative operators: AC-matching, AC-unification, AC-normalization, AC-critical pair computation and AC-completion	Bündgen

Table 1. Preassembled ReDuX programs

---

[1] ReDuX is not really an acronym. It is a mixture of the words 'reduce' and 'redex' and is the result of a brain storming of the author and Wolfgang Küchlin — it just sounds good.

ReDuX Features

The standard ReDuX algorithms are designed for many sorted algebraic specifications. In addition there are extensions which allow to include operators which are implicitly known to be associative and commutative.

The core of the system is a collection of algorithms providing important data structures (term, equation, rewrite rule, substitution, ...) and the basic operations on these data structures (input/output, term-equality, match, unification, copy algorithms, reduction, ...).

The Knuth-Bendix completion procedures [KB70] of the ReDuX system use a strategy which gives least priority to the orientation of rules. Even though this strategy may be inferior in some cases [Les89], it is a good candidate to investigate critical pair criteria. Completion procedures using two critical pair criteria (alone or together) are implemented: the subconnectedness criterion [Küc86] and the transformation criterion [Bün91d].

So far two term orderings have been implemented: the Knuth-Bendix ordering and a lexicographic path ordering with lexicographical status [KNS85]. Term orderings can be combined lexicographically. Interpretation functions from ReDuX terms to SAC-2 polynomials allow easy implementation of polynomial orderings.

The major part of the ReDuX system concerns inductive completion procedures. There is an automatic ground normal form analysis [BE92] and a positional ground reducibility test [BK89] for term rewriting systems which describe a regular tree language of irreducible ground terms. Based on this positional ground reducibility test, the inductive completion procedure of Küchlin [Küc89] has been implemented. If a left-hand side of a new lemma, proposed by this completion procedure, has more than one inductively complete position set the choice is done interactively. The completion procedure also suggests to use an inessential critical pair if the rule derived by this pair reduces an essential critical pair. Otherwise only essential critical pairs are considered. The ground normal form analysis is run as a preprocessing step of the inductive completion procedure. If it determines that the set of irreducible ground terms is freely generated by a set of constructors the inductive lemmas with top constructors may be decomposed as proposed in [HH80].

There is a Peterson-Stickel completion procedure [PS81] for algebraic specifications which may include associative and commutative operators. This program includes a 'fertilization' procedure to selectively compute critical pairs between a new equation and a rule. The fertilization procedure may be controlled interactively thus providing a powerful means for the experimentalist to control the completion process. Fertilization has been proven very helpful for the completion of finitely presented algebraic structures which allow for symmetrization (see [Bün91d]).

ReDuX comes with a set of Unix-tools to support documentation (like kwic-index generators for all algorithms and pattern search in all sources), automatic testing and installation.

Design goals

The ReDuX system is written in the ALDES language [LC92] and uses the low level libraries of the SAC-2 computer algebra system [Col80] to take advantage of both a procedural programming style and a built-in list and symbol processing system. All ReDuX procedures are compatible with the ALDES/SAC-2 system. Therefore

```
TYPE LISTZ.
CONSTS    0-INT.              nil-LIST
VARS      A,B,C-INT.          L,M,N-LIST
OPS
    s(INT)-INT.    p(INT)-INT.   +(INT,INT)-INT    FIX:INFIX.
    @(LIST,LIST)-LIST    FIX: INFIX.              # append
    [(INT,LIST)-LIST     FIX: ROUNDFIX ROUND: ].  # cons
AXIOMS
    1) (nil @ L) == L
    2) ([ A,L ] @ M) == [ A,(L @ M) ]
    3) s(p(A)) == A
    4) p(s(A)) == A
    5) (0 + A) == A
    6) (s(A) + B) == s((A + B))
    7) (p(A) + B) == p((A + B))
END
...

GNF grammar for data type LISTZ
<LISTZ>  ::= <A>    | <L>
<L>      ::= [ <A>,<L> ]    | nil
<s(A)>   ::= s(<s(A)>)    | s(0)
<p(A)>   ::= p(<p(A)>)    | p(0)
<A>      ::= s(<s(A)>)    | s(0)   | p(<p(A)>)   | p(0)   | 0
```

Fig. 1. The ground normal form grammar computed for LIST by ITST

ReDuX and SAC-2 build a platform to combine concepts from term rewriting and computer algebra.

ReDuX is an *open system*. That means *all* sources are available and all ReDuX algorithms include SAC-2-style specifications and documentation. A general description of the structure of the ReDuX system, together with an introduction to programming with ReDuX is given in [Bün91b].

Portability is another important consideration. ALDES is a very portable language which is compiled to highly portable subsets of C or Fortran. In addition there is an extremely small interface to the operating system (only 3 procedures). At the University of Tübingen there are ReDuX installations on Sun3, Sun4, and IBM RS6000. An installation on a Linux system has also been reported.

Experimental systems like ReDuX must of course provide the possibility to *measure the performance* of algorithms w.r.t. both the time and the space consumed. In addition there are trace mechanisms to study the behavior of an algorithm.

Other goals are of course to *avoid hidden bounds* and to provide *extensibility* to new problems (e.g. development of parallel or distributed term rewriting systems).

Experiments and Applications

ReDuX has been used for many applications, only a selection of which is mentioned here. Among others we studied the impact (and interdependence) of the subconnectedness and the transformation critical pair criteria and there rôle in symmetrization procedures [Küc86, Bün91d]). Experimental results with different reduction strategies have been reported in [Küc82b]. Implementations of different methods to analyze ground normal forms, and compute (positional) ground reducibility tests were presented in [Bün87, BK89, BE92, Bün92]. Inductive properties of finitely presented groups have been investigated for groups represented by canonical TRS [Bün90]. The AC-system has been used to discover a canonical term rewriting system describing multivariate polynomials in distributive normal form and to simulate Buchberger's algorithm [Bün91d, Bün91c, Bün91a].

Table 2 shows some experiments which were run on a Sun ELC under Sun OS 4.1.3. All times are given in ms.

no.	input specification	experiment	program	time
1	lr-system [KB70]	completion	TCT	408
2	free group [KB70]	completion	TCT	204
3	$\langle a, b; a^4 = b^2, a^{-1} = bab^{-1}\rangle$	completion	TCT	6698
4	free group $\langle a, b, c, d; \rangle$	ground normal form analysis	PTST	1174
5	result of 3	ground normal form analysis	PTST	68
6	R2 [KNZ86]	ground normal form analysis	PTST	51
7	distr. lattice [PS81]	AC-completion	AC	43554
8	comm. ring with 1 [PS81]	AC-completion	AC	15980
9	RX [Bün91a]	AC-confluence test	AC	32606

Table 2. Some benchmarks

Availability

ReDuX is available from the author and may be used for non-commercial purposes. It comes with all sources, an ALDES compiler and documentation for programmers and users.

Acknowledgements

The author is grateful to Wolfgang Küchlin for introducing him to programming term rewriting systems and providing his TC-system, to all those mentioned in Table 1 for contributing to the system and to Jochen Walter for help with the documentation and numerous installation tools.

References

[BE92] Reinhard Bündgen and Hasko Eckhardt. A fast algorithm for ground normal form analysis. In H. Kirchner and G. Levi, editors, *Algebraic and Logic Programming*, pages 291 – 305, 1992.

[BK89] Reinhard Bündgen and Wolfgang Küchlin. Computing ground reducibility and inductively complete positions. In Nachum Dershowitz, editor, *Rewriting Techniques and Applications*, pages 59–75. Springer-Verlag, 1989.

[Bün87] Reinhard Bündgen. Design, implementation, and application of an extended ground-reducibility test. Master's thesis, Computer and Information Sciences, University of Delaware, Newark, DE 19716, 1987.

[Bün90] Reinhard Bündgen. Applying term rewriting methods to finite groups. In H. Kirchner and W. Wechler, editors, *Algebraic and Logic Programming*. Springer-Verlag, 1990.

[Bün91a] Reinhard Bündgen. Completion of integral polynomials by AC-term completion. In Stephen M. Watt, editor, *International Symposium on Symbolic and Algebraic Computation*, pages 70 – 78, 1991.

[Bün91b] Reinhard Bündgen. The ReDuX system documentation. Technical Report 91-5, Wilhelm-Schickard-Institut, Universität Tübingen, D-7400 Tübingen, 1991.

[Bün91c] Reinhard Bündgen. Simulating Buchberger's algorithm by Knuth-Bendix completion. In Ronald V. Book, editor, *Rewriting Techniques and Applications*, pages 386–397. Springer-Verlag, 1991.

[Bün91d] Reinhard Bündgen. *Term Completion Versus Algebraic Completion*. PhD thesis, Universität Tübingen, D-7400 Tübingen, Germany, May 1991.

[Bün92] Reinhard Bündgen. Test sets for AC-ground reducibility. Unpublished manuscript, Universität, Tübingen, 1992.

[Col80] G. E. Collins. ALDES and SAC-2 now available. *SIGSAM Bull.*, 12(2):19, 1980.

[HH80] Gérard Huet and Jean-Marie Hullot. Proofs by induction in equational theories with constructors. In *Proc. 21st FoCS*, pages 96–107, Los Angeles, CA, 1980.

[KB70] Donald E. Knuth and Peter B. Bendix. Simple word problems in universal algebra. In J. Leech, editor, *Computational Problems in Abstract Algebra*. Pergamon Press, 1970.

[KNS85] D. Kapur, P. Narendran, and G. Sivakumar. A path ordering for proving termination of term rewriting systems. In *Mathematical Foundation of Software Developement*, pages 173–187. Springer-Verlag, 1985.

[KNZ86] Deepak Kapur, Paliath Narendran, and Hantao Zhang. Proof by induction using test sets. In J. Siekmann, editor, *8th International Conference on Automated Deduction*, pages 99–117. Springer-Verlag, 1986.

[Küc82a] Wolfgang Küchlin. An implementation and investigation of the Knuth-Bendix completion algorithm. Master's thesis, Informatik I, Universität Karlsruhe, D-7500 Karlsruhe, W-Germany, 1982.

[Küc82b] Wolfgang Küchlin. Some reduction strategies for algebraic term rewriting. *ACM SIGSAM Bull.*, 16(4):13–23, November 1982.

[Küc86] Wolfgang Küchlin. A generalized Knuth-Bendix algorithm. Technical Report 86-01, Mathematics, Swiss Federal Institute of Technology (ETH), CH-8092 Zürich, Switzerland, January 1986.

[Küc89] Wolfgang Küchlin. Inductive completion by ground proof transformation. In H. Aït-Kaci and M. Nivat, editors, *Resolution of Equations in Algebraic Structures*, volume 2 of *Rewriting Techniques*, chapter 7. Academic Press, 1989.

[LC92] Rüdiger G. K. Loos and George E. Collins. Revised report on the algorithm description language ALDES. Technical Report 92-14, Wilhelm-Schickard-Institut für Informatik, Tübingen, 1992.

[Les89] Pierre Lescanne. Completion procedures as transition rules + control. In M. Diaz and F. Orejas, editors, *TOPSOFT '89*, pages 21–41. Springer-Verlag, 1989.

[PS81] G. Peterson and M. Stickel. Complete sets of reductions for some equational theories. *Journal of the ACM*, 28:223–264, 1981.

[WB91] Jochen Walter and Reinhard Bündgen. The ReDuX user guide. Technical Report 91-9, Wilhelm-Schickard-Institut, Universität Tübingen, D-7400 Tübingen, 1991.

AGG — An Implementation of Algebraic Graph Rewriting *

Michael Löwe and Martin Beyer

Computer Science Department, Technical University of Berlin,
Franklinstr. 28/29, Sekr. FR 6-1 / FR 5-6, D 1000 Berlin 10, Germany,
{loewe,beyer}@cs.tu-berlin.de

Abstract. The AGG-system (Algebraic Graph Grammar System) is a prototype implementation of the algebraic approach to graph transformation [Ehr79]. It has been programmed in EIFFEL and runs on SUN workstations under X Window 11.5. It consists of a flexible graphical editor and a derivation component. The editor allows the graphical manipulation of rules, redices and derivation results. The derivation component performs direct transformation steps for user-selected rules and redices.

Introduction

The methods and techniques of graph rewriting have been advantageously applied to the efficient implementation of functional and declarative languages [Ken90, BvEvLP87, PJ87]. The algebraic approach to this issue is well known under the notion of Jungle Rewriting introduced by Plump et.al. [HKP91]. Recently, even some original algorithms in this field have been formulated as graph transformation systems, for example Lamping's algorithm for the optimal reduction of λ-expressions [Lam91].

Although graph transformation is quite successful in this area, there have been little efforts in the rewriting community to offer *graph rewriting itself* as a rule based specification and programming paradigm and to design supporting systems. This is mainly due to the fact that graph rewriting looses a lot of its intuitiveness if graphs and productions are coded in string languages. Thus, a suitable system for graph rewriting needs to be graphical and has to handle not only the static layout problems for graphical representations which have been addressed for example in the EDGE-system [NP91] but also dynamic layout and arrangement problems which are created by rule applications. Therefore the few existing systems are restricted either to context-free rewriting, e.g. Graph^{Ed}[Him91, ER91], or to the graphical representation and manipulation of the rules and do not offer graphical support for rule application and inspection of the resulting graphs, e.g. PAGGED [Göt87] and IPSEN [Nag87, Sch91].

By contrast the implementation project for algebraic graph transformation which has been started in 1989 at the Technical University of Berlin aims at an implementation of general algebraic graph rewriting with graphical representation and editing

---

* This work has been partially supported by the German Ministry of Research and Technology (BMFT), project "KORSO (Korrekte Software)" and by the ESPRIT BRWG 3299 "COMPUGRAPH (Computing by Graph Transformation)".

support on all levels, i.e. rules, redices, derivation steps and derivation sequences with resulting structures. The AGG–system described in this paper is the first stable version in this implementation project. It provides the graphical editor of the final system and a derivation component which performs direct derivation with rules at user-selected redices.

Informal Introduction to Algebraic Graph Transformation

The AGG–system implements the algebraic approach to single pushout graph rewriting [LE90], a further development of the double pushout approach [Ehr79]. This section introduces the main ideas of this rewriting approach using, for space limitations, a simple but typical example, i.e. a graph model for a queue structure.[2] The head of the queue is marked by a $\triangle$-vertex and the end by a $\square$-vertex. Fig. 1 (a) depicts the structure representing the empty queue.

A rewrite rule consists of three components, i.e. the left-hand side, the right-hand-side and a partial mapping from the left-hand side to the right-hand-side. Fig. 1 (b) and (c) shows two rules which model insertion and removal of items in the queue structure. Left-hand and right-hand sides of rules are just graphs specifying the pre- and postcondition of rule application. The mapping of a rule, indicated by dotted edges in fig. 1, specifies the items which are not changed by rule application.

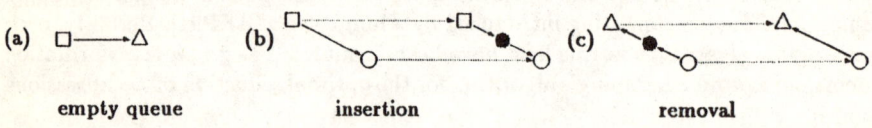

(a) empty queue (b) insertion (c) removal

Fig. 1. Queue structure

For a rule to be applicable, the pattern of the left-hand side has to be found in a mother graph which results in a graph morphism from a left-hand side into the mother graph. For example, fig. 2 (a) depicts a situation in which the queue insertion rule can be applied. Note that the pattern matching realizes a simple variable concept. It allows to match $\bigcirc$-nodes with nodes of any shape and color while all other nodes must be mapped to nodes with the same shape or color. With this concept the insertion rule can be uniformly applied to empty and non-empty queues.

The operational effect of the rule application is first the deletion of all items in the mother graph corresponding to objects in the left-hand side which are not in the domain of the rule mapping and second the addition of all items in the rule's right-hand side which are not in the image of the rule mapping. Fig. 2 (b) shows the result of applying the insertion rule at the redex of fig. 2 (a). Note that the

---

[2] More complex examples can be found in [Löw93].

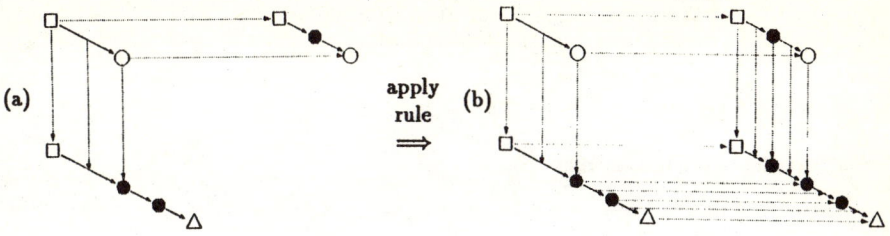

Fig. 2. A transformation step (graph level)

application induces two additional morphisms, one which embeds the rule's right-hand side in the resulting structure and another partial one which specifies the items of the mother graph which are preserved by the rule application.

The theoretical elegance of this concept for rewriting lies in the fact that rewriting situations, like the one in fig. 2 (b), can be characterized as pushouts in the category of partial graph morphisms. The resulting theory concerning issues like confluence, parallel and synchronized rewriting is comprehensively described in [Löw93].

The Editor

The editor of the AGG-system provides comprehensive functionality for the input and modification of graphs, for example: (1) Each graph can be edited in any number of windows offering different views. The view in a window can be influenced by a set of parameters including node size, zooming factor, grid functionality and a specification of node and edge types visible in the view. (2) Nodes and edges are easily drawn by simple mouse clicks. Each object can be moved by mouse dragging and, due to the graph-orientedness of the system, incident objects are adjusted appropiately. (3) Selection of objects is possible (among others) by specifying a rectangular area of the graph as well as by graph-oriented operations. (4) Copying, resizing, moving, mirroring of any subgraph. (5) "Save" and "Load" functionality.

Since AGG has been designed to support the editing and simulation of graph rewriting, it realizes a bunch of useful features that go beyond the standard functionality of graphical editing systems. Due to space limitations we focus on the two most important concepts of this kind, i.e. the concept of "higher-order edges" (edges between edges) and the AGG–abstraction mechanism. The latter provides the user interface for the derivation component and is explained in the next section. The former allows to draw morphisms $f : A \to B$ between graphs A and B as bundles of edges from the objects of A to the objects of B, compare fig. 2 (a) and (b). The editor functionality mentioned above has been orthogonally adjusted to this more general graph concept.

And, furthermore, AGG offers special operations supporting this concept which can be advantageously used to input and edit rewrite rules, for example the *copy connect* function. Fig. 3 (a) – (d) illustrates how the rules shown in fig. 1 can be drawn with AGG.

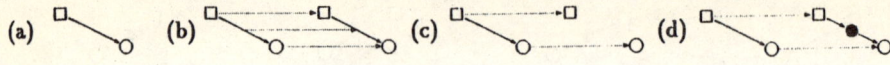

Fig. 3. Drawing a production rule

First the left-hand side is drawn (a). Second, an application of *copy connect* provides the *identity mapping* (b). In the third step, all items which shall be deleted by rule application can be removed from the right-hand side (c) resulting in a partial morphism from left to right (i.e. deletion is propagated to incident edges). What a rule application shall add, can now be specified by just drawing these items in the right–hand side (d). With this pragmatics of rule input, a rule can be considered as a "macro" storing the set of user interactions of the third and fourth step.

Direct Derivations

The main mechanism that enables the user to trigger direct derivations is the abstraction mechanism of AGG. It allows to abstract graphs to nodes and bundles of edges in the same direction between graphs to single edges. AGG presents the different abstraction levels of a graph in different windows. The edit functions like move, copy, delete, etc. are orthogonally generalized to the abstraction concept. Hence, the movement of an abstract object triggers a movement of all "refinements" by the same distance. Fig. 4 shows an abstraction of the two graphs of fig. 2. The three vertices on the left correspond to the sides of the rule and to the mother graph. The two edges represent the two morphisms between these graphs.

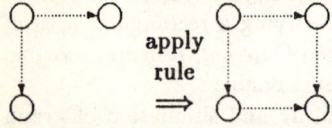

Fig.4. A transformation step (diagram level)

With the abstraction concept, the user is able to decompose large graphs into smaller units, i.e. subgraphs and morphisms, and to represent these units as a diagram in the sense of category theory on the abstract level. These diagrams are used as the interface to the derivation component. Therefore the application of the insertion rule at the redex of fig. 2 can be triggered by invoking the *apply rule*-operation (via menu or keyboard) and clicking on the edge representing the redex in any window displaying the diagram level.[3] Fig. 5 depicts a screen where this operation has been invoked in the window at the bottom right. Note that both abstraction levels are displayed in two windows each offering different views. The result is the graph in fig. 2 (b) on the concrete level and the corresponding completion of the diagram in fig. 4 on the abstract level. The new node is the abstraction of the resulting graph and the two new edges represent the pushout morphisms.

---

[3] For the time being, the only way to trigger rule application is by user interaction.

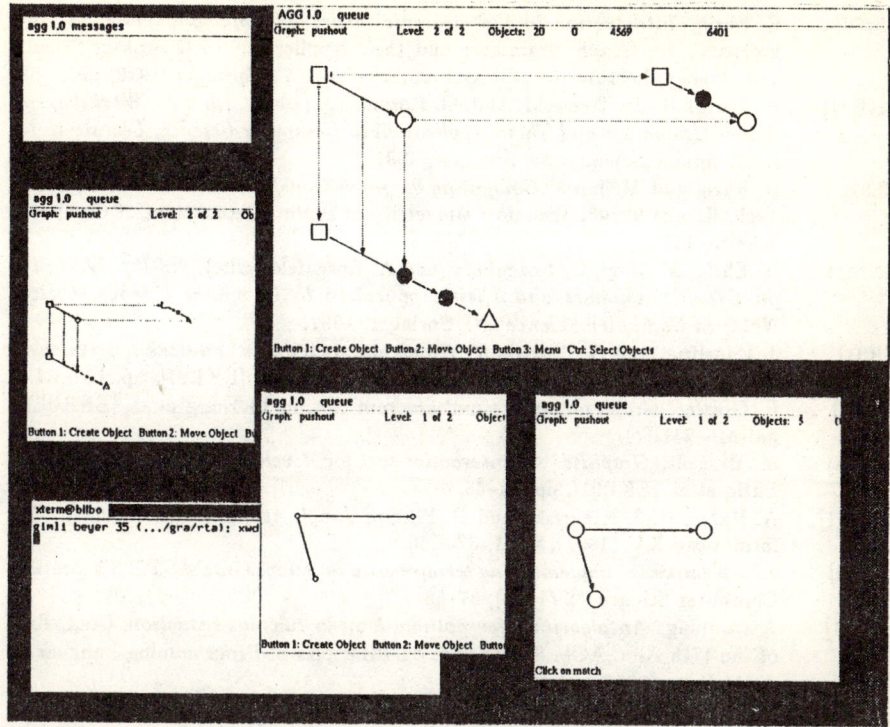

Fig. 5. Using the AGG-system

Outlook

The AGG-system was designed and implemented by Martin Beyer [Bey93] as part of the COMPUGRAPH project [EL92]. It has been used in courses on graph grammars at Berlin. Later versions of AGG are intended to be used for case studies on "graph transformation for specification and programming", which is one of three subjects within the current COMPUGRAPH II project. The next versions of the system will include textual labels as well as automatic finding of redices and in-place transformation, so that transformation sequences can be triggered and executed. The resulting challenges concerning the layout of graphs will be addressed in later stages of the project.

References

[Bey93]   M. Beyer, *AGG, An Algebraic Graph Grammar System, User Manual*, Technical University of Berlin, Department of Computer Science, 1993.

[BvEvLP87] T. Brus, M.C.J.D. van Eekelen, M. van Leer, and M.J. Plasmeijer, *Clean – a language of functional graph rewriting*, Proc. 3rd Int. Conf. on Functional Programming Languages and Computer Architecture (FPCA '87) Portland, Oregon, USA, Springer LNCS 274, 1987, pp. 364 – 384.

[Ehr79]   H. Ehrig, *Introduction to the algebraic theory of graph grammars*, 1st Int. Workshop on Graph Grammars and their Application to Computer Science and Biology, Lecture Notes in Computer Science 73, Springer, 1979, pp. 1–69.

[EKR91]   H. Ehrig, H.-J. Kreowski, and G. Rozenberg (eds.), *4th Int. Workshop on Graph Grammars and Their Application to Computer Science*, Lecture Notes in Computer Science 532, Springer, 1991.

[EL92]   H. Ehrig and M. Löwe, *Computing by graph transformation — Final Report*, Tech. Report 92/08, Technical University of Berlin, Department of Computer Science, 1992.

[ENRR87]   H. Ehrig, M. Nagl, G. Rozenberg, and A. Rosenfeld (eds.), *3rd Int. Workshop on Graph Grammars and Their Application to Computer Science*, Lecture Notes in Computer Science 291, Springer, 1987.

[ER91]   J. Engelfriet and G. Rozenberg, *Graph grammars based on node rewriting: An introduction to NLC graph grammars*, In Ehrig et al. [EKR91], pp. 12 – 23.

[Göt87]   H. Göttler, *Graph grammars and diagram editing*, In Ehrig et al. [ENRR87], pp. 216–231.

[Him91]   M. Himsolt, *GraphEd: An interactive tool for developing graph grammars*, In Ehrig et al. [EKR91], pp. 61–65.

[HKP91]   A. Habel, H.-J. Kreowski, and D. Plump, *Jungle evaluation*, Fundamenta Informaticae XV (1991), no. 1, 37 – 60.

[Ken90]   J.R. Kennaway, *Implementing term rewrite languages in DACTL*, Theoretical Computer Science 72 (1990), 37–58.

[Lam91]   J. Lamping, *An algorithm for optimal lambda calculus reduction*, Conf. Rec. of the 17th Ann. ACM Symposium on Principles of Programming Languages, ACM Press, 1991, pp. 16 – 30.

[LE90]   M. Löwe and H. Ehrig, *Algebraic approach to graph transformation based on single pushout derivations*, Graph-Theoretic Concepts in Computer Science, WG '90 (R.H. Möhring, ed.), Springer LNCS 484, 1990, pp. 338–353.

[Löw93]   M. Löwe, *Algebraic approach to single-pushout graph transformation*, Theoretical Computer Science 109 (1993), 181–224.

[Nag87]   M. Nagl, *A software development environment based on graph technology*, In Ehrig et al. [ENRR87], pp. 458–478.

[NP91]   F. Newbery Paulisch, *The Design of an Extendible Graph Editor*, Ph.D. thesis, University of Karlsruhe, Department of Informatics, March 1991.

[PJ87]   S.L. Peyton Jones, *The Implementation of Functional Programming Languages*, Prentice Hall, 1987.

[Sch91]   A. Schürr, *Operationales Spezifizieren mit programmierten Graphersetzungssystemen*, Deutscher Universitätsverlag GmbH, Wiesbaden, 1991.

Smaran[1] : A Congruence-Closure Based System for Equational Computations[2]

Rakesh M. Verma

Department of Computer Science, University of Houston, Houston, TX 77004

1 Introduction

In this paper we describe *Smaran*, an efficient system for normalizing terms with respect to given rewrite systems (TRS's), based on the congruence closure approach. Several variants of straight reduction (SR), which differ on the choice of the redex to be replaced, have been proposed for normalization ([9], [5], [10], etc). A major problem with these methods is the danger of repeated computations, because they do not store the history of their previous computations. There are several ways of introducing history in SR to overcome this problem. For example, one can build a table of terms and their normal forms, but, unless they are carefully implemented, such methods may be inefficient. Some mixed success with such a method is reported in [3].

In 1980, by extending the congruence closure algorithm CCA ([2], [7], [8]) Chew[1] proposed an elegant normalization algorithm (hereafter called CCNA) for an important subclass of left-linear systems. In 1989, this was extended by [11], [14] to an important subclass of non left-linear systems, viz., nonoverlapping and noetherian systems. Recently, we have shown that this algorithm can also be used for several classes of priority rewrite systems, and for all confluent and noetherian systems in which root overlaps are consistent [12]. Besides its wide applicability, CCNA has other important advantages also (see [1] or [12]).

Because of its many advantages and its wide applicability, we have been studying the theory and practice of CCNA at the University of Houston for the last several years. On the practical front, our goals are: to explore whether an efficient implementation of CCNA is possible and to get a concrete idea of the extent to which history helps in faster normalization. With these goals in mind, we present, *Smaran*, an efficient implementation of CCNA in this paper. Two interesting features of *Smaran* are: 1. In contrast with other implementations such as Hoffmann and O'Donnell's equational interpreter [4], *ML*, etc., we do not impose restrictions such as left-linearity or left sequentiality to gain efficiency. 2. A "lazy" version of our program is easily obtained by changing the definition of the normal form.

Smaran is written in C and assumes as input a confluent system for which leftmost-outermost strategy is normalizing. Some optimizations of our implementation are discussed in section *2*. Our results are encouraging and show that the method is of practical significance. Some results are presented in section *3* but to save space the TRS's are given in [13]. We were keen on finding out how the program performs when it cannot reuse history. Hence we have included programs for list reversal, tree insertion, and prime number generation in our experiments,

---
[1] From Sanskrit, meaning to remember
[2] Partially supported by NSF grant CCR-9010366

where history cannot be reused, except when the same term is repeated. The basic concept in CCNA is that of an *unreduced signature*. There is at most one in each class and it has been shown, for various systems ([1], [14], [12]), that it is sufficient to examine only these signatures to find normal forms. For a detailed description of CCNA see [13].

2 A Brief Description of *Smaran*

Central to the implementation of the congruence closure approach are (1) matching, (2) identification of key matches, (3) the search of signatures, and (4) the manipulation of signatures and classes. The following operations on classes and signatures are needed: Union(set1, set2), Find(signature), Membership(signature, set) Set_unred_sig(signature, class), Get_unred_sig(class), Delete(signature), Insert(signature, set).

In *Smaran*, a signature is a record containing the first character of the root label (used primarily for signatures representing integers), some flags, and a list of integer parameters. For efficiency, all strings appearing in the input TRS are encoded by numbers and the encoding of the root label is stored in the signature. This eliminates a lot of string operations replacing them with integer operations. Each class is given a number. For identification of signatures that must be changed when a class is unioned with another class, we associate a dependency list with each class. All signatures that must be modified in the event that this class changes are put on the class's dependency list. Every time a union occurs and a class disappears, signatures on the class's dependency list are examined and modified. Modification of a signature may result in further union operations being executed.

Optimizing Matching & Identification of Key Matches. The normalization process requires extensive matching. Speeding up normalization involves optimizing each matching operation as well as rapid identification of key matches that lead to the normal form faster.

Optimization of the matching operation is even more crucial in *Smaran* because it allows nonlinearity in the rules. Consequently *Smaran*'s matching procedure is more complicated because of consistency requirements, i.e., different occurrences of the same variable are instantiated to the same class. The substitution for a variable has to be accessed for creating the right hand side instance of a successful match, and for consistency checking. Since the number of variables can be quite large, this access should be very efficient. *Smaran* achieves this using the integer encodings of variables. The code of a variable is computed only once for each variable, and is stored along with the variable so that access to the code takes only unit time. Accessing the value of the variable is now done in constant time by direct indexing using the integer encoding of the variable.

One way to implement normalization would be to compare the unreduced signature of every class with each rule until a match is found. The problem with this strategy is that as the number of classes and signatures increase, the time taken for normalization increases dramatically. To solve this problem, *Smaran* tries to find a match first with the *unreduced signature* of the class containing the term to be normalized. If no matches are found against this signature, then *Smaran* recursively tries matching the subterms of this signature against the rules. This is in effect equivalent to integrating the leftmost outermost strategy with congruence closure, and gives much better results. (We have also designed a parallel-outermost implementation for nonterminating left-linear and confluent TRS.) For example, all other things being equal, if every rule is matched against every class the time taken to reverse a list of size 110 is 4 times more than with our strategy. As expected this difference in timing widens further as the number of classes grow. Another optimization in *Smaran* is to match rules against the most recently created class after a successful match. This cuts down the timings by a factor of two.

Implicit Evaluation and Deletion of Trivial Signatures. To speed up the normalization process, arithmetic and relational operations on integers are done automatically in *Smaran*. This means that the user does not have to specify any rules for these functions. Our experience has shown that implementing these operations using rules is grossly inefficient. An important optimization in *Smaran* is that signatures with arithmetic or relational root operators whose operands are directly available are evaluated and only their results are stored, the signatures themselves are discarded. This cuts down further the number of signatures and speeds up normalization. However, for arithmetic and relational signatures whose operands are not available *Smaran* applies a brute-force top-down evaluation procedure. A bottom-up strategy will give better results and will be incorporated in future versions.

Union and Find. There are several data structures for representing sets, e.g., lists, trees, arrays, etc. We chose to implement sets by a collection of arrays both for efficiency and simplicity. This aspect of our implementation is mostly standard (for details see [13]).

Space. Due to the potentially large number of signatures and classes, we decided to reuse integers representing signatures and class names as they were freed by delete and union operations respectively. Memory management is primitive and expensive in *Smaran*, space is requested of the needed size whenever it is needed.

3 Experimental Results

Smaran is implemented in *C* under *Unix* on a diskless Sun Sparc 1 workstation. Brief results of some experiments are given in Table 1 (all timings for BST insertion are averages for random order insertion). The sample TRS's are omitted to save space (see [13]). Timings for the sieve program are given when $sieve(from(2, 200), n)$ is called, where n is the desired number of primes. Construction of the list $from(2, 200)$ takes 0.75s and *is* included in the timings. Timings for Tree isomorphism are averages and do not include nonisomorphic trees as input, since this case runs much faster than the isomorphic case. Also, the trees are constructed with different node values so that there is only minimal sharing. The only history that is stored for each program is that acquired while calculating the normal form for a single term. When the calculation for the next term starts all previous history is discarded.

How Much Does History Help? Our experiments validate that history can be beneficial in two ways. The first benefit is illustrated by the Fibonacci Program, where history acquired during the calculation of the normal form of a single term is reused. The other advantage is when the normal form of a term is already present. For the list reversal case if the result is already present, the retrieval time is 0.04s, for the tree insertion program it is 0.03s, and for the prime number generator it is 0.02s (these timings are for the largest sized terms in each case). These timings are significantly smaller than the time taken for recomputing their normal forms. As expected, we found a tradeoff between the time saved due to storage of history and the increase in the time for matching and other operations due to an increase in the number of classes and signatures. However, *Smaran* does not get swamped by the overhead. The reasons are the reliance of the method on a distinguished signature in each class, and *Smaran*'s efficient normalization strategy.

program	size	time (s)	number of classes
Fibonacci	60	1.58	124
Tree Isomorphism	20 each	0.24	122
	40 each	0.82	202
List Reversal	150	0.25	457
BST Insertion	40	0.82	282
	60	1.84	565
Sieve for primes	first 30	5.46	773
	first 50	6.33	773

Table 1: Timing results for five programs (Timings include 0.75s for list construction in the Sieve program.)

Where is all the Time Spent? Table 2 shows the runtime profiles of *Smaran* for prime-number generation and search tree insertion. These are based on information provided by the *prof* profiling utility. Several operations consume significant percentages of time. For systems involving more symbolic computing, the three most expensive groups of operations in *Smaran* are hash table routines, dependency list manipulation, and dynamic memory management. Direct comparisons of *Smaran* with other systems involving normalization is difficult because most of these do not have built-in arithmetic, source languages are different, etc. However, on a Sun 3/60, we compared *Smaran* with SbReve (sb3c and sb3d) for TRS's without arithmetic operations (list reversal and tree isomorphism without descendant calculations) and found that *Smaran* was faster for lists with more than 20 elements and trees of size 15. Both were close for smaller sizes.

Prog.	Hash	Dep. list	Mem. manag.	Match	Sig. manip.	Classes
Primes	17%	15%	11.6%	9%	8%	773
Insert	15%	15%	11.7%	10.5%	11%	865

Table 2: Runtime profiles for normalization of sieve(from(2,200),50) and worst-case insertion of 40 nodes.

4 Future Plans

Any efficient implementation should incorporate a fast deletion strategy, so that excessive amounts of history can be pruned. We have already made a beginning in this direction by discarding signatures corresponding to arithmetic and relational expressions that can be directly evaluated. Future versions of *Smaran* will include matching algorithms using indexing and better integer manipulation and memory allocation schemes. We feel that *Smaran* will be beneficial in the Knuth-Bendix completion procedure [6]. The reason is that in this procedure a lot of medium-sized terms are created and normalized. Also, terms are often repeated during completion. Therefore, we plan to optimize *Smaran* further and then integrate it with the completion procedure.

Acknowledgements. The author is grateful to Jieh Hsiang for helpful discussions on SbReve and for providing versions of SbReve.

References

[1] P. Chew. An improved algorithm for computing with equations. In *Proc. of the IEEE Symp. on Foundations of Computer Science*, volume 21, pages 108–117, 1980.

[2] P.J. Downey, R. Sethi, and R.E. Tarjan. Variations on the common subexpression problem. *JACM*, 27(4):758–771, 1980.

[3] J. Goguen, J. Meseguer, and D. Plaisted. *Programming with Parametrized Abstract Objects in OBJ*, pages 163–193. North-Holland, 1983.

[4] C.M. Hoffmann and M.J. O'Donnell. Programming with equations. *ACM Transactions on Programming Languages and Systems*, 4:83–112, 1982.

[5] G. Huet and J.J. Levy. Call by need computations in non-ambiguous linear term rewriting systems. In *Rapport Laboria 283*. IRIA, 1979.

[6] D.E. Knuth and P. Bendix. Simple word problems in universal algebra. In J. Leech, editor, *Computational Problems in Abstract Algebra*, pages 263–297. Oxford, Pergammon Press, 1970.

[7] D. Kozen. Complexity of finitely presented algebras. In *Proc. Ninth ACM Symposium on Theory of Computing*, pages 164–177, 1977.

[8] G. Nelson and D.C. Oppen. Fast decision algorithms based on congruence closure. *JACM*, 27:356–364, 1980.

[9] M.J. O'Donnell. *Computing in Systems Described by Equations*, volume 58 of *Lecture Notes in Computer Science*. Springer-Verlag, 1977.

[10] R.C. Sekar and I.V. Ramakrishnan. Equational logic programming: Beyond strong sequentiality. In *Proc. of the IEEE Conf. on Logic in Computer Science*, 1990.

[11] Rakesh M. Verma. *Equations, Nonoblivious Normalization, and Term Matching Problems*. PhD thesis, State University of New York at Stony Brook, 1989.

[12] Rakesh M. Verma. A theory of using history for equational systems with applications. In *Proc. of IEEE FOCS*, pages 348–357, 1991.

[13] Rakesh M. Verma. Smaran: A congruence closure based system for equational computations. Technical Report UH-CS-92-25, University of Houston, 1992.

[14] R.M. Verma and I. V. Ramakrishnan. Nonblivious normalization algorithms for nonlinear systems. *Proc. ICALP*, vol. 443, pages 370–385. Springer-Verlag, 1990.

LAMBDALG: Higher Order Algebraic Specification Language

Yexuan Gui and Mitsuhiro Okada
Department of Computer Science, Concordia University
Montreal, Quebec H3G 1M8, CANADA

1 Introduction

Typed functional languages like ML or Haskel do not allow algebraic definitions of abstract data types and operators although they may employ a very rich machinery for defining polymorphic recursive functions of a higher type. On the other hand, equational languages like OBJ allow arbitrary (first order) algebraic definitions, but they do not have the full-power of parametricity given by ML polymorphism nor functional definitions of higher types. Under these circumstances it is very attractive to combine these two different kinds of languages to host both features. Then the unified language would allow easy definitions of quite complex objects in a simple declarative style. LAMBDALG is a specification language which hosts both these features. The computation model (operational semantics) of LAMBDALG is based on the combined system of polymorphic typed lambda calculus and the first order and restricted higher order term rewriting in [Jouannaud-Okada 91]. As a high-level specification language, LAMBDALG integrates the OBJ3 style module based algebraic specification language [Goguen-Winkler 88] and the ML style polymorphic typed functional language [Harper et al. 86]. Algebraic definition of abstract data types and context–sensitive definitions of operators can be directly executed, in contrast to the traditional ML–style languages or their extensions (eg. [Mitchell et al. 90]). In particular, LAMBDALG offers the followings;
1. The usual higher order rewrite definition of recursive functionals (of higher types) is extended to a non-recursive scheme of functionals defined on algebraic data types. Mixed terms composed of lambda terms, first order algebraic terms and higher order functional constants can be used.
2. The order-sorted structure of the algebraic base types is extended to the higher type structure, which accommodates higher order rewriting with sub–typing, based on the underlying sub–typing mechanism of [Cardelli-Mitchell 89].
3. The first order polymorphic type ranges over the algebraic base types (the basic data types), while the second order polymorphic type ranges over higher types. A first order and a higher order sub-typing are defined in terms of a bounded first order and a bounded second order polymorphism, respectively.

2 Examples

The followings are some simple examples of programming with LAMBDALG. In the examples, a variable with one quote (like 'a) is a first order polymorphic type variable (ranging over the algebraic base types), and a variable with two quotes (like "a) is a second order polymorphic type variable (ranging over all the types).

The first example is a non-recursive *mapcar* based on associative *append*.

```
module LIST is
        types nelist < list .   ***> non-empty list is a subtype of list.
        var x    : 'a .          ***> x is of base type.
        vars l, m, n : list .
        op nil : list .
        op cons  : ('a , list) -> nelist .
        op append : (list, list) -> list [assoc] .
                                 ***> append is an associative operator.
        eq append(l, nil) = l .
        eq append (nil, l) = l .
        eq append(cons(x, l), m) = cons(x, append(l, m)) .
endm
```

(Note the attribute [assoc] in the declaration of *append* has the same effect as defining "eq append(l, append(m, n)) = append(l, append(m, n))" .)

```
module MAPCAR is
        using LIST .  ***> MAPCAR module imports LIST module.
        types 'a, 'b .  ***> first-order type variables
        var X : 'a -> 'b .
        var x : 'a .  ***> x is of base type.
        vars l, m : list .
        eq mapcar(X, nil) = nil .
        eq mapcar(X, cons(x, l)) = cons(X(x), mapcar(X, l)) .
        eq mapcar(X, append(l, m)) = append(mapcar(X, l), mapcar(X, m))
endm
```

To check whether the equations terminate, we enter the following at top level

```
LA> terminating_check(MAPCAR)
ok.
```

The termination and the ground Church–Rosser of higher type functionals are tested by the syntactic criteria given in [Jouannaud-Okada 91]. To execute the program, we enter the following expression at top level.

```
LA> reduce in MAPCAR : mapcar(lambda(x)(x + 1), cons(9, (cons 3 nil)))
value: [10, 4]
Type: list
```

If the user does not specify any specific option of reduction strategies, the default is the outer–most. The β–reduction always has higher priority than term rewriting.

The following is an example of second order sub–typing defined by bounded polymorphism.

```
module SORTING is
    using LIST .
    typevar 'a < real . ***> 'a ranges over the subtypes of real.
    var   e, e' : 'a .
    var   l : list .
    var   LT    : ('a , 'a) -> bool .
    op sorting  : ('a , 'a -> bool) * list -> list .
    op sorted   : (('a , 'a -> bool), list) -> bool .
    cq sorting(LT, l) = l if sorted l .
    cq sorting(LT, [e | e' | l]) =
        sorting(LT, [e' | sorting(LT, [e | l])]) if LT(e',e) == true .
    cq sorting(LT, [e | e' | l]) =
        sorting(LT, [e | sorting(LT, [e' | l])]) if LT(e',e) =/= true .
    eq sorted(LT, nil) = true .
    eq sorted(LT, e) = true .
    cq sorted(LT, [e | e' | l]) =
                sorted(LT, [e' | l]) if LT(e, e') == true .
endm
```

The built-in types *integer* and *real* have the subtype relation *integer* < *real*, and the operator *less_than* is of type $real * real \rightarrow bool$. After entering

```
LA> reduce in SORTING: sorting(less_than, [9, 17, 2])
```

The interpreter first checks the type correctness by the type inference rules, then, if there is no type error, starts the execution. In our example, type variable 'a gets type *integer* which is a subtype of type *real*. And the type of operator *less_than* (which is $real * real \rightarrow bool$) is also a subtype of the type $integer * integer \rightarrow bool$. So all the arguments of sorting are of correct types.

3 Parametrization Power

LAMBDALG supports algebraic specifications which the usual functional languages can not support. It also directly supports both sub-sort polymorphism and parametric polymorphism. In sub-sort polymorphism, operation overloading is consistent under sub-sort restriction [Goguen-Winkler 88], and in parametric polymorphism, an overloading operator's type can be determined automatically from the context. By contrast, usual algebraic equational specification languages, like OBJ, do not directly support parametric polymorphism. LAMBDALG also supports modularization. LAMBDALG's modules encapsulate relating code, thus make it more reusable.

Although some algebraic specification languages such as OBJ allow parameterized module which can be "reused" for a variety of applications by choosing different parameter values, they can not have the full power of parametric polymorphism; in parameterized modular programming, all the formal type parameters of a module must be instantiated. Each operator in an instantiated module is only of a particular single type. This makes the language less expressive, and sometimes results in complicated code. For example, consider the following LAMBDALG expression

```
reduce in MAPCAR : mapcar(not.FNS, mapcar(is_even.FNS, [2, 7, 6]))
```

The definitions of *is_even* and *not* are contained in module FNS. There are two occurrences of mapcar. They are of different types. The type of the inner occurrence of *mapcar* is (*integer* → *bool*) * *list* → *list*, and the type of the outer occurrence of *mapcar* is (*bool* → *bool*) * *list* → *list*.

In order to write a similar expression in a parameterized modular programming language, we have to make two instantiations of the same module. In OBJ, for example, we should write the following code (cf [Goguen–Winkler 88], p. 35):

```
make MAP1 is MAP[(is-even_).FNS] endm
make MAP2 is MAP[(not_).FNS] endm
reduce in MAP2 :   map(map.MAP1(2 7 6))
```

We have to inform the system the inner occurrence of map comes from module MAP1 and the outer occurrence of map comes from module MAP2. Hence LAMBDALG specification is simpler. This is more clear when one considers the following LAMBDALG specification:

```
mapcar(lambda(l)mapcar(twice.FNS, l), [[1, 2, 3], [2, 1, 8, 4]])
```

There is even no obvious way to express it in OBJ or other usual first-order parameterized modular programming languages.

4 Implementation Techniques

The current implementation is an interpreter written in ANSI–C under Unix (SunOs) environment. The language parser detects all syntactic and lexical errors. Many semantic errors (mostly typing errors) are also detected at program entry time. Several novel techniques are employed for efficient term rewriting.

The basic units manipulated by the LAMBDALG language are modules. A module is semantically an environment defined by the declarations inside it. In LAMBDALG, terms are represented internally as directed acyclic graphs (DAG). An interior node represents an operator and its children its operands. Common sub–terms are represented by the same node. It is known that DAG–rewriting ensures the correctness of the theory of computational models for the combined languages in [Jouannaud-Okada 91], hence this computation model is considered as the operational semantics of LAMBDALG. The language interpreter performs type inference and type checking while building DAGs.

Equations are used as pattern–directed rewrite rules. Termination test by using several path orderings as well as the test for higher-order rewrite rules in [Jouannaud–Okada 91] have been built into the interpreter. It issues a warning message whenever a rule fails the test.

The language's inference engine performs two different kinds of reductions: algebraic reductions (term rewriting) and β–reductions. Performing a reduction consists of a local transformation of the graph representing the term, so the process of reduction successfully modifies the graph until it reaches its normal form, the result of the computation. In order to achieve efficient term rewriting, the interpreter uses the top operator of the term being rewritten as a key to retrieve appropriate rules. The language's rule-base is organized in a 2-3 tree data structure for efficient rule retrieval. The key of a rule in the 2-3 tree is the top operator on its left-hand side. There are a few built-in operators (such as arithmetic operator +). The interpreter carries out the evaluation using C functions when it is possible.

5 Further Development

The framework of LAMBDA is now being extended to offer the following:
1. Strong type inference rules are available. A type declaration of an algebraic constructor produces a type inference rule.
2. The "proof-as-program" principle and the "formula-as-specification" principle of the traditional polymorphic type theory are used in a modified form, based on recursive type inference rules, instead of impredicative type inference rules. A rewrite program can be automatically extracted from a higher-order equational logic proof of an equational specification, based on [Okada 93].
3. Type inferences can be used for proving an algebraic property. This is used for checking the ground Church-Rosser of the additional algebraic rule.
4. The record calculus of higher types, which can capture more features of the object-oriented type theory based on the OBJ3-style first-order theory.

6 References

[Breazu-Tannen 91]. Val Breazu-Tannen and Jean Gallier. Polymorphic rewriting conserves algebraic strong normalization. *Theoretical Computer Science*, 1991. (The former version in ICALP'89)
[Cardelli-Mitchell 89]. L. Cardelli and J.C. Mitchell. Operations on records. In *Math. Foundations of Prog. Lang. Semantics*. 1989.
[Dershowitz-Okada 91]. N. Dershowitz and M. Okada. A rationale for conditional equational programming. *Theoretical Computer Science*, 1991.
[Futatsugi et al. 85]. K. Futatsugi, J. Goguen, Jean-Pierre Jouannaud, and J. Meseguer. Principles of OBJ2. *ACM POPL'85*
[Goguen et al. 85]. J. Goguen, Jean-Pierre Jouannaud, and J. Meseguer. Operational semantics for first-order algebra. *ICALP'85, Springe LNCS 194*
[Goguen-Winkler 88]. J. A. Goguen, T. Winkler. Introducing OBJ3. *Technical Report SRI-CSL-88-9*, Computer Science Laboratory, SRI International, 1988.
[Gui-Okada 93]. Y. Gui and M. Okada. System Description of LAMBDALG, LPAR'93. (Logic Programming and Automated Reasoning, 1993).
[Harper et al. 86]. R. Harper, D. MacQueen, and R. Milner. Standard ML, *LFCS Report Series, ECS-LFCS-86-2*, 1986.
[Jouannaud-Okada 91]. J. Jouannaud, M. Okada. A computation model for executable higher-order algebraic specification languages. *Proc. 6th IEEE LICS*. 1991.
[Mitchell et al. 90]. J. Mitchell, S. Meldal, and N. Madhav. An extension of standard ML modules with subtyping and inheritance. *ACM POPL'90*
[Okada 89]. M. Okada. Strong normalizability for the combined system of the typed lambda calculus and an arbitrary convergent term rewriting system. *ACM ISSAC 89*
[Okada 93]. M. Okada. Type-theoretic term rewriting theory. *Technical report*. Logic & Formal Methods Lab, Concordia University, 1993.

Open problems

More Problems in Rewriting*

Nachum Dershowitz[1], Jean-Pierre Jouannaud[2], and Jan Willem Klop[3]

[1] Department of Computer Science, University of Illinois, 1304 West Springfield Avenue, Urbana, IL 61801, U.S.A, nachum@cs.uiuc.edu
[2] Laboratoire de Recherche en Informatique, Bat. 490, Université de Paris Sud, 91405 Orsay, France, jouannau@lri.lri.fr
[3] CWI, Kruislaan 413, 1098 SJ Amsterdam, The Netherlands
Department of Mathematics and Computer Science, Free University, de Boelelaan 1081, 1081 HV Amsterdam, The Netherlands, jwk@cwi.nl

1 Introduction

Two years ago, in the proceedings of the previous conference, we presented a list of open problems in the theory of rewriting [Dershowitz et al., 1991a]. This time, we report on progress made during the intervening time, and then list some new problems. (A few additional questions on the subject appear in the back of [Diekert, 1990].) We also mention a couple of long-standing open problems which have recently been answered. The last section contains a partisan list of interesting areas for future research. A new, comprehensive survey of the field is [Klop, 1992].

Please send any contributions by electronic or ordinary mail to any of us. We hope to continue periodically publicizing new problems and solutions to old ones. We thank all the individuals who contributed questions, updates and solutions.

2 Old Problems

Five of the forty-four problems listed in [Dershowitz et al., 1991a] have been solved and some progress has been made on ten more. For convenience, we repeat the problems (in small type) about which we are able to report progress.

Problem 1. An important theme that is largely unexplored is definability (or implementability, or interpretability) of rewrite systems in rewrite systems. Which rewrite systems can be directly defined in lambda calculus? Here "directly defined" means that one has to find lambda terms representing the rewrite system operators, such that a rewrite step in the rewrite system translates to a reduction in lambda calculus. For example, Combinatory Logic is directly lambda definable. On the other hand, not every orthogonal rewrite system can be directly defined in lambda calculus. Are there universal rewrite systems, with respect to direct definability? (For alternative notions of definability, see [O'Donnell, 1985].)

---

* The first author was supported in part by the National Science Foundation under Grants CCR-90-07195 and CCR-90-24271 and by a Meyerhoff Visiting Professorship at the Weizmann Institute of Science; the second author was partially supported by the ESPRIT working groups COMPASS and CCL; the third author's work was partially supported by ESPRIT BRA project 6454: Confer.

Some progress has been made in [Berarducci and Böhm, 1992].

Problem 7 (H. Comon, M. Dauchet).   Is it possible to decide whether the set of ground normal forms with respect to a given (finite) term-rewriting system is a regular tree language? See [Gilleron, 1991; Kucherov, 1991].

This has been answered in the affirmative [Vágvölgyi and Gilleron, 1992; Kucherov and Tajine, 1993; Hofbauer and Huber, 1993].

Problem 20 (Y. Métivier [1985]).   What is the best bound on the length of a derivation for a one-rule length-preserving string-rewriting (semi-Thue) system? Is it $O(n^2)$ (n is the size of the initial term) as conjectured in [Métivier, 1985], or $O(n^k)$ (k is the size of the rule) as proved there.

Rumor has it that the conjecture has been shown true.

Problem 21 (M. Dauchet).   Is termination of one linear (left and right) rule decidable? Left linearity alone is not enough for decidability [Dauchet, 1989].

A less ambitious, long-standing open problem (mentioned in [Dershowitz and Jouannaud, 1990]) is decidability for *one* (length-increasing) monadic (string, semi-Thue) rule. Termination is undecidable for non-length-increasing monadic systems of rules [Caron, 1991]. For one monadic rule, confluence is decidable [Kurth, 1990; Wrathall, 1990]. What about confluence of one non-monadic rule?

Problem 24.   The existential fragment of the first-order theory of the "recursive path ordering" (with multiset and lexicographic "status") is decidable when the precedence on function symbols is total [Comon, 1990; Jouannaud and Okada, 1991b], but is undecidable for arbitrary formulas. Is the existential fragment decidable for partial precedences?

The Σ_4 ($\exists^*\forall^*\exists^*\forall^*$) fragment is undecidable, in general [Treinen, 1992]. The positive existential fragment for the empty precedence (that is, for homeomorphic tree embedding) is decidable [Boudet and Comon, 1993]. One might also ask whether the first-order theory of *total* recursive path orderings is decidable. Related results include the following: The existential fragment of the subterm ordering is decidable, but its Σ_3 ($\exists^*\forall^*\exists^*$) fragment is not [Venkataraman, 1987]. The first-order theory of encompassment (the instance-of-subterm relation) is claimed decidable [Caron et al., 1993]. Once we're at it, we might as well ask what the complexity of the satisfiability test for the existential fragment is—in the total case.

Problem 25 (R. Treinen [1990]).   Is the theory of multisets (AC) completely axiomatizable? In other words, is it decidable whether a first-order formula containing only equality as predicate symbol is valid in the algebra $T(\mathcal{F})/AC(F)$? It is known that the Σ_3 fragment is undecidable when there are at least one unary function symbol (besides the AC one) and one constant; the Σ_1 fragment is decidable; the full theory is decidable even when there are no other symbols (besides constants) [Treinen, 1990].

Whether the Σ_2 ($\exists^*\forall^*$) fragment is decidable remains open; see [Treinen, 1992]. A positive answer was given for the important special case of "complement problems" in [Kounalis et al., 1991]. One might also consider the case where one is given terms $t_1, \ldots, t_n$ and a term t containing associative-commutative symbols and free symbols, and are to decide whether all ground instances of t are ground instances of some t_i. Special cases of the latter question have been studied in [Kounalis and Lugiez, 1991; Kounalis et al., 1991; Fernández, 1993; Lugiez and Moysset, 1993].

Problem 27 (P. Lescanne). In [Lescanne, 1990] an extension of term embedding, called "well-rewrite orderings", was introduced, leading to an extension of the concept of simplification ordering. How can those ideas best be extended to form the basis for some new kind of "recursive path ordering"?

Progress in this direction has been reported in [Weiermann, 1992].

Problem 28 (P. Lescanne). Polynomial and exponential interpretations have been used to prove termination. For the former there are some reasonable methods [Ben Cherifa and Lescanne, 1987; Lankford, 1979] that can help determine if a particular interpretation decreases with each application of a rule. Are there other implementable methods suitable for exponential interpretations?

Some work on this problem has been reported in [Lescanne, 1992].

Problem 29. Any rewrite relation commutes with the strict-subterm relation; hence, the union of the latter with an arbitrary terminating rewrite relation is terminating, and also "fully invariant" (closed under instantiation). Which is the finest (maximal) relation with these properties? (It is not subterm.) Is "encompassment" ("containment", the combination of subterm and subsumption) the finest relation which preserves termination (without full invariance)?

The finest relation we know of which could answer the first question is the variant of subterm that allows multiple occurrences of variables to be renamed apart.

Problem 33. Completion modulo associativity and commutativity (AC) [Peterson and Stickel, 1981] is probably the most important case of "extended completion"; the general case of finite congruence classes is treated in [Jouannaud and Kirchner, 1986]. Adding an axiom (Z) for an identity element, however, gives rise to infinite classes. This case was viewed as conditional completion in [Baird et al., 1989], and solved completely in [Jouannaud and Marché, 1990]. The techniques, however, do not carry over to completion with idempotence (I) added; how to handle ACZI-completion effectively is open.

C. Marché [1993] has used rewriting techniques to show decidability of the word problem for any theory comprised of a set of ground equations, associativity and commutativity laws for arbitrarily many operators, plus identity and idempotency laws for any number of those operators.

Problem 34. Ordered rewriting computes a given convergent set of rewrite rules for an equational theory E and an ordering $>$ whenever such a set R exists for $>$, provided $>$ can be made total on ground terms. Unfortunately, this is not always possible, even if $>$ is derivability ($\rightarrow_R^+$) in R. Is there a set of inference rules that will always succeed in computing R whenever R exists for $>$?

A proposal appears in [Devie, 1991]; more work is called for.

Problem 38 (J. Siekmann). Is satisfiability of equations in the theory of distributivity (unification modulo a distributivity axiom) decidable?

The question should read "modulo one right- and one left-distributivity axiom". (With just one of these, the problem had already been solved in [Tiden and Arnborg, 1987].) A partial positive solution is given in [Contejean, 1993], based on a striking result on the structure of certain proofs modulo distributivity. Although many more cases are described in [Contejean, 1992; Contejean, 1993], the general case remains open.

Problem 39. Rules are given in [Jouannaud and Kirchner, 1991] for computing dag-solved forms of unification problems in equational theories. The *Merge* rule $x \approx s, x \approx t \Rightarrow x \approx s, s \approx t$ given there assumes that s is not a variable and its size is less than or equal to that of t. Can this condition be improved by replacing it with the condition that the rule *Check** does not apply? (In other words, is *Check** complete for finding cycles when *Merge* is modified as above?)

The problem has been solved by H. Comon [1993] using an extended *Check* rule (requiring a congruence closure step). The original question—for whatever it may be worth—stands.

Problem 42 (H. Comon). Given a first-order formula with equality as the only predicate symbol, can negation be effectively eliminated from an arbitrary formula ϕ when ϕ is equivalent to a positive formula? Equivalently, if ϕ has a finite complete set of unifiers, can they be computed? Special cases were solved in [Comon, 1988; Lassez and Marriott, 1987].

A positive solution is given in [Tajine, 1993].

Problem 43. Design a framework for combining constraint solving algorithms.

Some particular cases have been attacked: In [Baader and Schulz, 1992] it was shown how decision procedures for solvability of unification problems can be combined. In [Baader and Schulz, 1993] a similar technique is applied to (unquantified) systems of equations and disequations. In [Ringeissen, 1992] the combination of unification algorithms is extended to the case where alphabets share constants. In related work [Boudet, 1992], unification is performed in the combination of an equational theory and membership constraints.

3 New Problems

Problems 45–50 appeared (with minor variations) in our technical report [Dershowitz et al., 1991b]. In the meantime, one (no. 48) has been answered.

Problem 45 (M. Venturini-Zilli). Some reduction graphs in λ-calculus [Venturini-Zilli, 1984] are isomorphic to ordinals. For example, the reduction graph of $(\lambda x.y)((\lambda z.zzz)(\lambda z.zzz))$ is isomorphic to $\omega + 1$. Which ordinals appear in this way as reduction graphs? It is known that all ordinals less than ϵ_0 can be so represented.

Problem 46 (D. Kapur). Ground reducibility of extended rewrite systems, modulo congruences like associativity and commutativity (AC), is undecidable [Kapur *et al.*, 1987]. For left-linear AC systems, on the other hand, it is decidable [Jouannaud and Kounalis, 1989]. What can be said more generally about restrictions on extended rewriting that give decidability?

This problem is related to number 25.

Problem 47. For reductions of transfinite length, a version of the Parallel Moves Lemma can be proved if one consider only "strongly converging" infinite reductions in the sense of [Kennaway *et al.*, 1991]. However, if one wants to consider converging reductions, as in [Dershowitz *et al.*, 1991c], then it is not difficult to construct a counterexample, not to the infinite Parallel Moves Lemma itself, but to the method of proof (cf. [Kennaway *et al.*, 1990]). An infinite Parallel Moves Lemma might involve a different notion of "descendant".

Problem 48 (H.-C. Kong). Consider the following relation on strings over an infinite set $\mathcal{X}$ of variables: $x_1 x_2 \cdots x_m \hookrightarrow y_1 y_2 \cdots y_n$ if there exists a renaming $\rho : \mathcal{X} \to \mathcal{X}$ such that $x_i \rho = y_{j_i}$ for $1 \leq j_1 < j_2 < \cdots < j_m \leq n$. Is this "embedding" relation $\hookrightarrow$ a well-quasi-ordering (that is, must every infinite sequence of strings contain two strings, such that the first embeds in the second)?

The answer is "yes". (Map each variable to the position of its leftmost occurrence and use the fact that strings of natural numbers are well-quasi-ordered by the embedding extension of $\leq$ to strings.)

Problem 49 (M. Hermann). Suppose ordinary completion (as in [Dershowitz and Jouannaud, 1990], for example) is non-terminating for some initial set of equations E, completion strategy, and reduction ordering. Must there be a finite depth N for E such that for any $n > N$ restricting the generation of critical pairs to overlaps at positions that are no deeper than n in the overlapped left-hand side (but otherwise not changing the strategy) also produces a non-terminating completion sequence?

Problem 50. Combinations of typed λ-calculi with term-rewriting systems have been studied extensively in the past few years [Barbanera, 1990; Breazu-Tannen and Gallier, 1989; Dershowitz and Okada, 1990; Dougherty, 1991]. The strongest termination result allows first-order rules as well as higher-order rules defined by a generalization of primitive recursion. Suppose all rules for functional constant F follow the schema:

$$F(\bar{l}[\bar{X}], \bar{Y}) \to v[F(\bar{r}_1[\bar{X}], \bar{Y}), ..., F(\bar{r}_m[\bar{X}], \bar{Y}), \bar{Y}]$$

where the (not necessarily disjoint) variables in $\bar{X}$ and $\bar{Y}$ are of arbitrary order, each of $\bar{l}, \bar{r}_1, ..., \bar{r}_m$ is in $\mathcal{T}(\mathcal{F}, \{\bar{X}\})$, $v[\bar{z}, \bar{Y}]$ is in $\mathcal{T}(\mathcal{F}, \{\bar{Y}, \bar{z}\})$, for new variables $\bar{z}$ of appropriate types, and $\bar{r}_1, \ldots, \bar{r}_m$ are each less than $\bar{l}$ in the multiset extension of the strict subterm ordering. If $\mathcal{T}(\mathcal{F}, \mathcal{X})$ is the term-algebra which includes only *previously* defined functional constants—forbidding the use of mutually recursive functional constants—termination is ensured [Jouannaud and Okada, 1991a]. Does termination also hold when there are mutually recursive definitions? Does this also hold when the subterm assumption is unfulfilled? (In [Jouannaud and Okada, 1991a] an alternative schema is proposed, with the subterm assumption weakened at the price of having only first-order variables in $\bar{X}$.) Questions of confluence of combinations of typed λ-calculi and higher-order systems also merit investigation.

These results have been extended to combinations with more expressive type systems [Barbanera and Fernandez, 1993a; Barbanera and Fernandez, 1993b].

Problem 51 (H. Comon, M. Dauchet). Is the first order theory of one-step rewriting ($\to_R$) decidable? Decidability would imply the new result on the decidability of the first-order theory of encompassment (that is, being an instance of a subterm), based on pumping properties [Caron *et al.*, 1993]. (It is well known that the theory of $\to_R^*$ is in general undecidable.)

Problem 52 (R. Statman). It has been remarked by C. Böhm [Barendregt, 1984] that Y is a fixed point combinator if and only if $Y \leftrightarrow^* (SI)Y$ (Y and SIY are convertible). Also, if Y is a fixed point combinator, then so is $Y(SI)$. Is there is a fixed point combinator Y for which $Y \leftrightarrow^* Y(SI)$?

Problem 53 (R. Statman). A term M in Combinatory Logic or λ-calculus is *recurrent* if $N \to^* M$ whenever $N \leftrightarrow^* M$ (this notion is due to M. Venturini-Zilli.) Let's call M *hyper-recurrent* if N is recurrent for all $N \leftrightarrow^* M$. (Equivalently, M is hyper-recurrent if $P \to^* Q \to^* P$ whenever $P \leftrightarrow^* Q \leftrightarrow^* M$.) Are there any hyper-recurrent combinators? (The problem comes up immediately when the Ershov-Visser theory [Visser, 1980] for $\leftrightarrow^*$ is applied to $\to^*$. It is known that hyper-recurrent combinators don't exist for Combinatory Logic [Statman, 1991].)

Problem 54 (R. Statman). Recall that M is a *universal generator* if each combinator P has a superterm Q such that $M \to^* Q$. Call M a *uniform universal generator* if there exists a context $C[\cdot]$ such that, for each combinator P, we have $M \to^* C[P]$. Is there a uniform universal generator? (For Combinatory Logic, if we restrict the context $C[\cdot]$ to be of the form $(N\cdot)$, no such term exists [Statman, 1992].)

Problem 55 (R. Statman). It has been proved that (in λ-calculus or Combinatory Logic) every recursively enumerable set of ground terms that is closed under conversion has the form $\{M|PM \leftrightarrow^* Q\}$ for some P and Q. Which sets have the form $\{M|Q \to^* PM\}$?

Problem 56 (V. van Oostrom). An abstract reduction system is "decreasing Church-Rosser", if there exists a labelling of the reduction relation by a well-founded set of labels, such that all local divergences can be completed to form a "decreasing diagram" (see [Oostrom, 1992] for precise definitions). Does the Church-Rosser property imply decreasing Church-Rosser? That is, is it always possible to localize the Church-Rosser property? This is known to be the case for (weakly) normalizing and finite systems.

Problem 57 (F. Baader [1990]). Does there exist a semigroup theory (without constants in the equations) for which there is a reduced canonical term-rewriting system (with the right-hand side and subwords of the left in normal form) which is not length decreasing?

Problem 58 (M. Oyamaguchi). Is any "strongly" non-overlapping right-linear term-rewriting system confluent? ("Strong" in the sense that left-hand sides are non-overlapping even when the occurrences of variables have been renamed apart [Chew, 1981].) On the one hand, strongly non-overlapping systems need not be confluent [Huet, 1980]; on the other hand, strongly non-overlapping right-ground systems are [Oyamaguchi and Ohta, 1993].

Problem 59 (M. Kurihara, M. Krishna Rao). One of the earliest results established on modularity of combinations of term-rewriting systems is the confluence of the union of two confluent systems which share no symbols [Toyama, 1987]; if symbols are shared modularity is not preserved by union [Kurihara and Ohuchi, 1992]. Some sufficient conditions for modularity of confluence of constructor-sharing systems that are terminating have been found [Kurihara and Ohuchi, 1992; Middeldorp and Toyama, 1991]. Are there interesting sufficient conditions that are independent of termination?

Problem 60 (H. Zantema). Let R be a many-sorted term-rewriting system and R' the one-sorted system consisting of the same rules, but in which all operation symbols are considered to be of the same sort. Any rewrite in R is also a rewrite in R'. The converse does not hold, since terms and rewrite steps in R' are allowed that are not well-typed in R. In [Zantema, 1993] it was shown that termination of R is in general not equivalent to termination of R', but it is if R does not contain both collapsing and duplicating rules. Are termination of R and of R' equivalent in the case where all variables occurring in R are of the same sort? If this statement holds, it would follow that simulating operation symbols of arity n greater than 2 by $n-1$ binary symbols in a straightforward way does not affect termination behavior.

Problem 61 (T. Nipkow, M. Takahashi). For higher-order rewrite formats as given by combinatory reduction systems [Klop, 1980] and higher-order rewrite systems [Nipkow, 1991; Takahashi, 1993], confluence has been proved in the restricted case of orthogonal systems. Can confluence be extended to such systems when they are weakly orthogonal (all critical pairs are trivial)? When critical pairs arise only at the root, confluence is known to hold.

Problem 62 (V. van Oostrom). Let R and S be two left-linear, confluent combinatory reduction systems with the same alphabet. Suppose the rules of R do not overlap the rules of S. Is $R \cup S$ confluent? This is true for the restricted case when R is a term-rewriting system (an easy generalization of a result by F. Müller [1992]), or if neither system has critical pairs. (The restriction to the same alphabet is essential, since confluence is in general not preserved under the addition of function symbols, not even for left-linear systems.)

Problem 63 (M. Oyamaguchi). Is confluence of right-ground term-rewriting systems decidable? Compare [Oyamaguchi, 1987; Dauchet et al., 1990; Dauchet and Tison, 1990; Oyamaguchi and Ohta, 1993].

Problem 64. Is confluence of ordered rewriting (using the intersection of one step replacement of equals and a reduction ordering that is total on ground terms) decidable when the (existential fragment of the) ordering is? This question was raised in [Nieuwenhuis, 1993], where some results were given for the lexicographic path ordering.

Problem 65 (D. Cohen, P. Watson [1991]). An interesting system for doing arithmetic by rewriting was presented in [Cohen and Watson, 1991]. Unfortunately, its termination has not been proved.

Problem 66 (F. Baader, K. Schulz [1992]). Is there an equational theory for which unification with constants is decidable, but general unification (where free function symbols of arbitrary arity may occur) is undecidable? From the results in [Baader and Schulz, 1992] it follows that this question can be reformulated as follows: Is there an equational theory for which unification with constants is decidable, but unification with linear constant restrictions is undecidable? Another way of formulating the question is: Consider *positive* first-order formulæ containing equality as the only predicate symbol, and function symbols from a given alphabet $\mathcal{F}$. Is there an equational theory E with alphabet $\mathcal{F}$ such that whether $E \models \phi$ is decidable for closed formulae ϕ with quantifier prefix $\forall^* \exists^*$, but undecidable for arbitrary quantifier prefixes.

Problem 67 (F. Baader, K. Schulz [1992]). It was shown in [Baader and Schulz, 1992] that being able to solve unification problems with linear constant restrictions is a necessary and sufficient condition for the possibility of combining unification algorithms. Other approaches [Schmidt-Schauß, 1989; Boudet, 1990] require solvability of constant elimination problems, which was shown to be equivalent to presupposing solvability of unification problems with arbitrary constant restrictions [Baader and Schulz, 1992]. Is there an equational theory for which solvability of unification problems with linear constant restrictions is decidable, but solvability of unification problems with arbitrary constant restrictions is undecidable? Is there an equational theory for which unification problems with linear constant restrictions always have a finite complete set of solutions, but unification problems with arbitrary constant restrictions sometimes don't?

Problem 68 (H. Comon). Consider the existential fragment of the theory defined by a binary predicate symbol $\subseteq$, a finite set of function symbols $f_1, \ldots, f_n$, the function symbols $\cap, \cup, \neg$, and the projection symbols $f_{i,j}^{-1}$ for $j \leq arity(f_i)$. Variables are interpreted as subsets of the Herbrand Universe. With the obvious interpretation of these symbols, is satisfiability of such formulæ decidable? Special cases have been solved in [Heintze and Jaffar, 1990; Aiken and Wimmers, 1992; Bachmair et al., 1993; Gilleron et al., 1993].

Problem 69 (C. Kirchner, J. Zhang). What is the syntactic type (maximum number of top-level steps needed in an equational proof [Boudet and Contejean, 1992]) of the distributivity axiom? What is the syntactic type of "three-way" commutativity:

$$f(x,y,z) = f(x,z,y) = f(y,x,z) = f(y,z,x) = f(z,x,y) = f(z,y,x)$$
$$f(f(x,y,z),u,x) = f(x,y,f(z,u,x))$$

What are the unification type, decidability, and syntactic type of "mid-commutativity": $(x+y)+(u+v) = (x+u)+(y+v)$?

Problem 70 (J.-C. Raoult). There exist finite automata for words, trees, and dags. No really good comparable notion is available for graphs. (Perhaps there is one akin to the ideas in [Litovski et al., to appear] on label rewriting.)

Problem 71 (J.-C. Raoult). There are good algorithms for pattern-matching for words and trees, but not yet for graphs.

Problem 72 (J.-C. Raoult). Graph rewritings, like term or word rewritings, are usually finitely branching. There are relations that are not finitely branching, yet satisfy good properties: rational transductions of words, tree-transductions. A good definition of graph transduction, that extends rational word transductions is still lacking.

Problem 73 (J.-C. Raoult). Termination is, as we know, undecidable. Yet, there are several sufficient conditions ensuring termination for word and term rewritings. Most are suitable extensions of Higman's or Kruskal's embeddings [Kruskal, 1960]. Robertson and Seymour [Robertson and Seymour, 1982] have achieved a similar theorem for undirected graphs. However, no embedding theorem has yet been proved for directed graphs, and (consequently?) powerful termination orderings remain to be designed.

Problem 74 (D. Plump). Graph rewriting systems that implement term rewriting systems (see, for example, [Barendregt et al., 1987; Hoffmann and Plump, 1991]) are terminating whenever term rewriting is. The converse, however, does not hold [Plump, 1991]. How can termination orderings for term rewriting be adapted to cover those cases in which graph rewriting is terminating although term rewriting is not?

Problem 75 (D. Plump). In contrast to term rewriting, confluence of general (hyper-)graph rewriting—in the "Berlin approach"—is undecidable, even for terminating systems [Plump, 1993]. What sufficient conditions make confluence decidable?

4 New Solutions

Two old problems (omitted from our previous list) which have recently been solved are the following:

Problem 76. Cycle unification [Bibel et al., 1992] is undecidable [Devienne, 1993; Hanschke and Würtz, 1993]. This was a long standing open problem, related to the non-termination of simple logic programs.

Problem 77. J. Jezek, J. B. Nation, and R. Freese [Freese, 1993] have shown that there is no finite, normal form, associative-commutative term-rewriting system for lattices. This is somewhat surprising because every lattice term is equivalent under lattice theory to a shortest term which is unique up to associativity and commutativity (known as "Whitman canonical form").

5 Research Areas

Current research topics in rewriting include the following ten:

Typed Rewriting Under reasonable assumptions, virtually everything in ordinary (untyped) rewriting extends to the multisorted case. Adding subsorts supports inheritance and allows functions to be completely defined without having to introduce error elements for when they are applied outside their intended domains. But deduction in such "order-sorted" algebras presents some difficulties. The most popular approach is to insist that the sort of the right-hand side is always contained in that of the left; see [Dick and Watson, 1991]. A general approach requires a subcase of second-order unification [Comon, 1992]. A subject of vigorous investigation is that of typed λ-calculi [Bezem and Groote, 1993]. Though the relevance of this subject resides largely in the fields of automated deduction and of proof theory, a considerable segment pertains to term rewriting. For example, much attention has been devoted to termination proofs of typed λ-calculi.

Higher-order rewriting Beginning with [Breazu-Tannen and Gallier, 1989], researchers have been looking at ways of combining terminating confluent calculi with first-order ("algebraic") rewriting in such a way as to preserve their convergence, thereby endowing rewriting with higher-order capabilities. Recent contributions are [Jouannaud and Okada, 1991a; Barbanera and Fernandez, 1993a; Barbanera and Fernandez, 1993b]. Of a more general nature, proposals have been made for quite general rewriting formats that include rewriting with bound variables as in typed λ-calculi, yielding pleasant mixtures of pattern matching and variable binding. The suggestions in [Klop, 1980; Nipkow, 1991; Takahashi, 1993] are quite close, which is encouraging, as it may hint at a canonical framework for higher-order rewriting.

AC termination Recent work on proving termination of associative-commutative rewriting (the most prevalent extension of term rewriting) includes [Kapur *et al.*, 1990; Rubio and Nieuwenhuis, 1993; Delor and Puel, 1993]. It would be nice to somehow combine these results in an ordering that could orient distributivity the right way and be total when the precedence is. The ordering in [Kapur *et al.*, 1990] was incorporated in the RRL system, but most of this work has yet to filter down into widespread implemented tests that can be used within those rewrite-based theorem provers which support associativity and commutativity.

Hierarchical systems From the point of view of software engineering, it is important that properties of rewrite programs, like termination and confluence, be modular. That is, we would like to be able to combine two terminating systems, or two convergent systems, and to have the same properties hold for the combined system. This is not true in general, not even when one system makes no reference to the function symbols and constants used in the other. Finding useful cases when systems may safely be combined is a current area of study; see, for example, [Toyama, 1987; Toyama *et al.*, 1989; Middeldorp, 1990; Middeldorp and Toyama, 1991; Kurihara and Ohuchi, 1992; Dershowitz, 1993].

Logic programming Rewriting techniques have found applications in logic programming and constraint-based programming (besides their obvious application to functional programming). Semantic unification using rewrite-rules has been proposed by a number of people ([Reddy, 1986; Dershowitz and Plaisted, 1988], among others) as an ideal basis for a synthesis of functional and logic programming; the SLOG language [Fribourg, 1985] is a case in point. Refinements of universal unification for when a rewrite system is available have been found (see [Jouannaud and Kirchner, 1991]). Combining constraints with deduction, whether equational [Kirchner and Kirchner, 1989] or full first order [Kirchner et al., 1990], is another potential growth area.

Theorem proving and symbolic computation Since the pioneering work of Lankford [1975], research on the application of ideas from rewriting to more traditional refutational theorem provers for first-order predicate calculus has proceeded in bits and spurts. Recent work has shown that using orderings on terms and formulæ helps restrict deduction and increase the amount of simplification and redundancy elimination that can be incorporated without forfeiting completeness. For a survey, see [Hsiang et al., 1992]. These successes ought to be extended to higher-order calculi, which have been enjoying success in their own right. Ad-hoc rewriting has always been present in symbolic computation systems (e.g. Reduce, Macsyma); Gröbner-basis techniques are an integral part of some modern systems. The time appears ripe—indeed some projects have been initiated—to pursue significant applications of rewriting and typed calculi (supporting inheritance) in computer algebra and proof checking.

Complexity issues There is a dearth of results on the complexity of problems in rewriting and unification. (This, despite the problems posed in our lists.) One of the handful of exceptions (this one on AC-unification) is [Kapur and Narendran, 1992]. There is room for a lot more work on this side of theory.

Rewriting, automata and symbolic constraints Rewriting ground terms has much to do with formal language theory. In particular, bottom-up tree automata can be represented naturally by rewrite systems. The language of ground terms in normal form for a given system appears to be a key to many problems. Automata are also useful for solving symbolic constraints, following up on an idea pioneered by Büchi and Rabin. By encoding the set of solutions of an atomic constraint by some kind of automaton (closed under the usual Boolean operations), it is possible to solve arbitrary quantifier-free constraints. This technique has been widely used extensively in the past few years [Dauchet et al., 1990; Dauchet and Tison, 1990; Gilleron, 1991; Kucherov, 1991; Kucherov and Tajine, 1993; Gilleron et al., 1993; Caron et al., 1993].

Concurrency Confluent systems, in general, and orthogonal ones, in particular, are natural candidates for parallel processing, since rewrites at different positions are more or less independent of each other. Work is being undertaken on language and implementation issues raised by this possibility; see, for example, [Goguen et al., 1987; Meseguer, 1992; Berry and Boudol, 1992]. Much work is being done on combinations of λ-calculus and process calculi. A well-known example is the π-calculus, which extends Milner's CCS, as well as λ-calculus; see [Milner et al., 1992].

Graph rewriting The notion of rewriting (as it appeared already in Thue's [1914] work) can profitably be applied to structures other than finite terms. Graph rewriting is one such (graphs allow one to represent structure-sharing); another is infinite terms (see [Dershowitz *et al.*, 1991c; Inverardi and Nesi, 1991; Kennaway *et al.*, 1991]). Graph rewriting is often called "term-graph rewriting" to distinguish it from the more general approach of graph grammars. At present, (term) graph rewriting is only beginning to enjoy the attention of researchers in term rewriting. The lack of popularity thus far may be due to the intrinsic difficulty of finding workable formalisms for graph rewriting, avoiding on the one hand overly abstract category-theoretic formulations, and on the other hand overly implementation-oriented formulations with pointers, redirections, and the like.

References

[Aiken and Wimmers, 1992] A. Aiken and E. Wimmers. Solving systems of set constraints. In *Proceedings of the Seventh Symposium on Logic in Computer Science*, pages 329–340, Santa Cruz, CA, June 1992. IEEE.

[Baader, 1990] Franz Baader. Rewrite systems for varieties of semigroups. In M. Stickel, editor, *Proceedings of the Tenth International Conference on Automated Deduction (Kaiserslautern, West Germany)*, volume 449 of *Lecture Notes in Computer Science*, pages 381–395, Berlin, July 1990. Springer-Verlag.

[Baader and Schulz, 1992] Franz Baader and Klaus Schulz. Unification in the union of disjoint equational theories: Combining decision procedures. In D. Kapur, editor, *Proceedings of the Eleventh International Conference on Automated Deduction (Saratoga Springs, NY)*, volume 607 of *Lecture Notes in Artificial Intelligence*, Berlin, June 1992. Springer-Verlag.

[Baader and Schulz, 1993] Franz Baader and Klaus Schulz. Combination techniques and decision problems for disunification. In C. Kirchner, editor, *Proceedings of the Fifth International Conference on Rewriting Techniques and Applications (Montreal, Canada)*, Lecture Notes in Computer Science, Berlin, 1993. Springer-Verlag.

[Bachmair *et al.*, 1993] Leo Bachmair, Harald Ganzinger, and Uwe Waldmann. Set constraints are the monadic class. In *Proceedings of the Symposium on Logic in Computer Science (Montreal, Canada)*. IEEE, 1993.

[Baird *et al.*, 1989] Timothy Baird, Gerald Peterson, and Ralph Wilkerson. Complete sets of reductions modulo Associativity, Commutativity and Identity. In N. Dershowitz, editor, *Proceedings of the Third International Conference on Rewriting Techniques and Applications (Chapel Hill, NC)*, volume 355 of *Lecture Notes in Computer Science*, pages 29–44, Berlin, April 1989. Springer-Verlag.

[Barendregt, 1984] Henk P. Barendregt. *The Lambda Calculus, its Syntax and Semantics*. North-Holland, Amsterdam, second edition, 1984.

[Barbanera, 1990] F. Barbanera. Combining term rewriting and type assignment systems. *International J. of Foundations of Computer Science*, 1:165–184, 1990.

[Barbanera and Fernandez, 1993a] F. Barbanera and M. Fernandez. Modularity of termination and confluence in combinations of rewrite systems with λ_ω. In *Proceedings of the 20th International Colloquium on Automata, Languages, and Programming*, Lund, Sweden, 1993.

[Barbanera and Fernandez, 1993b] F. Barbanera and M. Fernandez. Combining first and higher order rewrite systems with type assignment systems. In *Proceedings of the International Conference on Typed Lambda Calculi and Applications*, Utrecht, Holland, 1993.

[Barendregt et al., 1987] H. P. Barendregt, M. C. J. D. van Eekelen, J. R. W. Glauert, J. R. Kennaway, M. J. Plasmeijer, and M. R. Sleep. Term graph rewriting. In *Proceedings of the European Workshop on Parallel Architectures and Languages*, volume 259 of *Lecture Notes in Computer Science*, pages 141–158, Berlin, 1987. Springer-Verlag.

[Ben Cherifa and Lescanne, 1987] Ahlem Ben Cherifa and Pierre Lescanne. Termination of rewriting systems by polynomial interpretations and its implementation. *Science of Computer Programming*, 9(2):137–159, October 1987.

[Berarducci and Böhm, 1992] Alessandro Berarducci and Corrado Böhm. A self-interpreter of lambda calculus having a normal form. Rapporto tecnico 16, Dip. di Matematica Pura ed Applicata, Universita di L'Aquila, October 1992.

[Berry and Boudol, 1992] G. Berry and G. Boudol. The chemical abstract machine. *Theoretical Computer Science*, 96:217–248, 1992.

[Bezem and Groote, 1993] M. Bezem and J. F. Groote, editors. *Proceedings of the International Conference on Typed Lambda Calculi and Applications (Utrecht, The Netherlands)*, volume 664 of *Lecture Notes in Computer Science*, Berlin, 1993. Springer-Verlag.

[Bibel et al., 1992] W. Bibel, S. Hölldobler, and J. Würtz. Cycle unification. In D. Kapur, editor, *Proceedings of the Eleventh International Conference on Automated Deduction (Saratoga Springs, NY)*, volume 607 of *Lecture Notes in Artificial Intelligence*, pages 94–108, Berlin, June 1992. Springer-Verlag.

[Boudet, 1990] Alexandre Boudet. Unification in combination of equational theories: An efficient algorithm. In *Proceedings of the Tenth International Conference on Automated Deduction (Kaiserslautern, Germany)*, volume 449 of *Lecture Notes in Computer Science*. Springer-Verlag, 1990.

[Boudet, 1992] Alexandre Boudet. Unification in order-sorted algebras with overloading. In D. Kapur, editor, *Proceedings of the Eleventh International Conference on Automated Deduction (Saratoga Springs, NY)*, volume 607 of *Lecture Notes in Artificial Intelligence*, Berlin, June 1992. Springer-Verlag.

[Boudet and Comon, 1993] Alexandre Boudet and Hubert Comon. About the theory of tree embedding. In J.-P. Jouannaud, editor, *Proceedings of the Colloquium on Trees in Algebra and Programming (Orsay, France)*, Lecture Notes in Computer Science, Berlin, April 1993. Springer-Verlag.

[Boudet and Contejean, 1992] Alexandre Boudet and E. Contejean. On n-syntactic equational theories. In H. Kirchner and G. Levi, editors, *Proceedings of the Third International Conference on Algebraic and Logic Programming (Pisa, Italy)*, volume 632 of *Lecture Notes in Computer Science*, pages 446–457, Berlin, September 1992. Springer-Verlag.

[Breazu-Tannen and Gallier, 1989] Val Breazu-Tannen and Jean Gallier. Polymorphic rewriting conserves algebraic strong normalization. In *Proceedings of the Sixteenth International Colloquium on Automata, Languages and Programming (Stresa, Italy)*, volume 372 of *Lecture Notes in Computer Science*, pages 137–150, Berlin, July 1989. European Association of Theoretical Computer Science, Springer-Verlag.

[Caron, 1991] A.-C. Caron. Linear bounded automata and rewrite systems: Influence of initial configurations on decision properties. In *Proceedings of the International Joint Conference on Theory and Practice of Software Development, volume 1: Colloquium on Trees in Algebra and Programming (Brighton, U.K.)*, volume 493 of *Lecture Notes in Computer Science*, pages 74–89, Berlin, April 1991. Springer-Verlag.

[Caron et al., 1993] A.-C. Caron, J.-L. Coquidé, and M. Dauchet. Encompassment properties and automata with constraints. In C. Kirchner, editor, *Proceedings of the Fifth International Conference on Rewriting Techniques and Applications (Montreal, Canada)*, Lecture Notes in Computer Science, Berlin, 1993. Springer-Verlag.

[Chew, 1981] Paul Chew. Unique normal forms in term rewriting systems with repeated variables. In *Proceedings of the Thirteenth Annual Symposium on Theory of Computing*, pages 7–18. ACM, 1981.

[Cohen and Watson, 1991] D. Cohen and P. Watson. An efficient representation of arithmetic for term rewriting. In R. Book, editor, *Proceedings of the Fourth International Conference on Rewriting Techniques and Applications (Como, Italy)*, volume 488 of *Lecture Notes in Computer Science*, pages 240–251, Berlin, April 1991. Springer-Verlag.

[Comon, 1988] Hubert Comon. *Unification et Disunification: Théorie et Applications*. PhD thesis, l'Institut National Polytechnique de Grenoble, 1988.

[Comon, 1990] Hubert Comon. Solving inequations in term algebras (Preliminary version). In *Proceedings of the Fifth Annual Symposium on Logic in Computer Science*, pages 62–69, Philadelphia, PA, June 1990. IEEE.

[Comon, 1992] Hubert Comon. Completion of Rewrite Systems with Membership Constraints. In *Proceedings of the 19th International Conference on Automata, Languages and Programming*, Vienna, Austria, 1992.

[Comon, 1993] H. Comon, personal communication, 1993.

[Contejean, 1992] Evelyne Contejean. Eléments pour la Décidabilité de l'Unification Modulo la Distributivité. PhD thesis, Université de Paris-Sud, Orsay, France, 1992.

[Contejean, 1993] Evelyne Contejean. A partial solution for D-unification based on a reduction to AC1-unification. In *Proceedings of the Twentieth International Colloquium on Automata, Languages and Programming (Lund, Sweden, July 1993)*, Lecture Notes in Computer Science, Berlin. Springer-Verlag.

[Contejean, 1993] Evelyne Contejean, personal communication, 1993.

[Dauchet, 1989] M. Dauchet. Simulation of Turing machines by a left-linear rewrite rule. In N. Dershowitz, editor, *Proceedings of the Third International Conference on Rewriting Techniques and Applications (Chapel Hill, NC)*, volume 355 of *Lecture Notes in Computer Science*, pages 109–120, Berlin, April 1989. Springer-Verlag.

[Dauchet et al., 1990] Max Dauchet, Thierry Heuillard, Pierre Lescanne, and Sophie Tison. Decidability of the confluence of finite ground term rewriting systems and of other related term rewriting systems. *Information and Computation*, 88(2):187–201, October 1990.

[Dauchet and Tison, 1990] M. Dauchet and S. Tison. The theory of ground rewrite systems is decidable. In *Proceedings of the Fifth Symposium on Logic in Computer Science*, pages 242–248, Philadelphia, PA, June 1990.

[Delor and Puel, 1993] C. Delor and L. Puel. Extension of the associative path ordering to a chain of associative commutative symbols. In C. Kirchner, editor, *Proceedings of the Fifth International Conference on Rewriting Techniques and Applications (Montreal, Canada)*, Lecture Notes in Computer Science, Berlin, 1993. Springer-Verlag.

[Dershowitz, 1993] Nachum Dershowitz. Hierarchical termination. Technical report 93-?, Leibnitz Center for Research in Computer Science, Hebrew University, Jerusalem, Israel, 1993.

[Dershowitz and Jouannaud, 1990] Nachum Dershowitz and Jean-Pierre Jouannaud. Rewrite systems. In J. van Leeuwen, editor, *Handbook of Theoretical Computer Science*, volume B: Formal Methods and Semantics, chapter 6, pages 243–320. North-Holland, Amsterdam, 1990.

[Dershowitz et al., 1991a] Nachum Dershowitz, Jean-Pierre Jouannaud, and Jan Willem Klop. Open problems in rewriting. In R. Book, editor, *Proceedings of the Fourth International Conference on Rewriting Techniques and Applications (Como, Italy)*, volume 488 of *Lecture Notes in Computer Science*, pages 445–456, Berlin, April 1991. Springer-Verlag.

[Dershowitz et al., 1991b] Nachum Dershowitz, Jean-Pierre Jouannaud, and Jan Willem Klop. Open problems in rewriting. Technical report CS-R9114, Computer Science, CWI, Amsterdam, December 1991.

[Dershowitz et al., 1991c] Nachum Dershowitz, Stéphane Kaplan, and David A. Plaisted. Rewrite, rewrite, rewrite, rewrite, rewrite,.... *Theoretical Computer Science*, 83(1):71–96, 1991.

[Dershowitz and Okada, 1990] Nachum Dershowitz and Mitsuhiro Okada. A rationale for conditional equational programming. *Theoretical Computer Science*, 75:111–138, 1990.

[Dershowitz and Plaisted, 1988] Nachum Dershowitz and David A. Plaisted. Equational programming. In J. E. Hayes, D. Michie, and J. Richards, editors, *Machine Intelligence 11: The logic and acquisition of knowledge*, chapter 2, pages 21–56. Oxford Press, Oxford, 1988.

[Devie, 1991] Hervé Devie. Une Approche Algébrique de la Réécriture et son Application à la dérivation de Procédures de Complétion. PhD thesis, Université de Paris-Sud, Orsay, France, 1991.

[Devienne, 1993] Phillipe Devienne, personal communication, 1993.

[Dick and Watson, 1991] A. J. J. Dick and P. Watson. Order-sorted term rewriting. *Computing J.*, 34(1):16–19, February 1991.

[Diekert, 1990] Volker Diekert, editor. *Proceedings of the Workshop of the ERBA-Working-Group No. 3166 on Algebraic and Syntactic Methods in Computer Science (ASMICS)*, Munich, Germany, January 1990. Technische Universität München.

[Dougherty, 1991] Daniel Dougherty. Adding algebraic rewriting to the untyped lambda calculus (extended abstract). In Ron Book, editor, *Proceedings of the Fourth International Conference on Rewriting Techniques and Applications (Como, Italy)*, volume 488 of *Lecture Notes in Computer Science*, pages 37–48, Berlin, April 1991. Springer-Verlag.

[Fernández, 1993] Maribel Fernández. AC-complement problems: Validity and negation elimination. In C. Kirchner, editor, *Proceedings of the Fifth International Conference on Rewriting Techniques and Applications (Montreal, Canada)*, Lecture Notes in Computer Science, Berlin, 1993. Springer-Verlag.

[Freese, 1993] R. Freese, personal communication, 1993.

[Fribourg, 1985] Laurent Fribourg. SLOG: A logic programming language interpreter based on clausal superposition and rewriting. In *Proceedings of the Symposium on Logic Programming*, pages 172–184, Boston, MA, July 1985. IEEE.

[Gilleron, 1991] R. Gilleron. Decision problems for term rewriting systems and recognizable tree languages. In *Proceedings of the Eighth Symposium on Theoretical Aspects of Computer Science*, February 1991.

[Gilleron et al., 1993] Rémy Gilleron, Sophie Tison, and Marc Tommasi. Solving systems of set constraints using tree automata. In *Proceedings of the Symposium on Theoretical Aspects of Computer Science (Würzburg, Germany)*, Lecture Notes in Computer Science, Berlin, 1993. Springer-Verlag.

[Goguen et al., 1987] J. A. Goguen, C. Kirchner, and J. Meseguer. Concurrent term rewriting as a model of computation. In R. Keller and J. Fasel, editors, *Proceedings of Graph Reduction Workshop (Santa Fe, NM)*, volume 279 of *Lecture Notes in Computer Science*, pages 53–93. Springer-Verlag, 1987.

[Hanschke and Würtz, 1993] Philipp Hanschke and Jörg Würtz. Satisfiability of the smallest binary program. *Information Processing Letters*, 45(5):237–241, April 1993.

[Heintze and Jaffar, 1990] Nevin Heintze and Joxan Jaffar. A decision procedure for a class of set constraints. In *Proceedings of the Fifth Symposium on Logic in Computer Science (Philadelphia, PA)*, pages 42–51. IEEE, June 1990.

[Hofbauer and Huber, 1993] D. Hofbauer and M. Huber. Computing linearizations using test sets. In M. Rusinowitch, editor, *Proceedings of the Third International Workshop on Conditional Rewriting Systems (Pont-a-Mousson, France, July 1992)*, volume 656 of *Lecture Notes in Computer Science*, pages 287–301, Berlin, January 1993. Springer-Verlag.

[Hoffmann and Plump, 1991] B. Hoffmann and D. Plump. Implementing term rewriting by jungle evaluation. *RAIRO Theoretical Informatics and Applications*, 25(5):445–472, 1991.

[Hsiang et al., 1992] Jieh Hsiang, Helene Kirchner, Pierre Lescanne, and Michael Rusinowitch. The term rewriting approach to automated theorem proving. *J. Logic Programming*, 14(1&2):71–99, October 1992.

[Huet, 1980] Gérard Huet. Confluent reductions: Abstract properties and applications to term rewriting systems. *J. of the Association for Computing Machinery*, 27(4):797–821, October 1980.

[Huet and Oppen, 1980] Gérard Huet and Derek C. Oppen. Equations and rewrite rules: A survey. In R. Book, editor, *Formal Language Theory: Perspectives and Open Problems*, pages 349–405, New York, 1980. Academic Press.

[Inverardi and Nesi, 1991] P. Inverardi and M. Nesi. Infinite normal forms for non linear term rewriting systems. Technical Report B4-41, Istituto di Elaborazione, Pisa, Italy, October 1991.

[Jouannaud and Kirchner, 1991] Jean-Pierre Jouannaud and Claude Kirchner. Solving equations in abstract algebras: A rule-based survey of unification. In J.-L. Lassez and G. Plotkin, editors, *Computational Logic: Essays in Honor of Alan Robinson*. MIT Press, Cambridge, MA, 1991.

[Jouannaud and Kirchner, 1986] Jean-Pierre Jouannaud and Hélène Kirchner. Completion of a set of rules modulo a set of equations. *SIAM J. on Computing*, 15:1155–1194, November 1986.

[Jouannaud and Kounalis, 1989] Jean-Pierre Jouannaud and Emmanuel Kounalis. Automatic proofs by induction in equational theories without constructors. *Information and Computation*, 81(1):1–33, 1989.

[Jouannaud and Marché, 1990] Jean-Pierre Jouannaud and Claude Marché. Completion modulo associativity, commutativity and identity. In Alfonso Miola, editor, *Proceedings of the International Symposium on the Design and Implementation of Symbolic Computation Systems (Capri, Italy)*, volume 429 of *Lecture Notes in Computer Science*, pages 111–120, Berlin, April 1990. Springer-Verlag.

[Jouannaud and Okada, 1991a] Jean-Pierre Jouannaud and Mitsuhiro Okada. Executable higher-order algebraic specification languages. In *Proceedings of the Sixth Symposium on Logic in Computer Science*, pages 350–361, Amsterdam, The Netherlands, 1991. IEEE.

[Jouannaud and Okada, 1991b] Jean-Pierre Jouannaud and Mitsuhiro Okada. Satisfiability of systems of ordinal notations with the subterm property is decidable. In J. Leach Albert, B. Monien, and M. Rodríguez Artalejo, editors, *Proceedings of the Eighteenth EATCS Colloquium on Automata, Languages and Programming (Madrid, Spain)*, volume 510 of *Lecture Notes in Computer Science*, pages 455–468, Berlin, July 1991. Springer-Verlag.

[Kapur and Narendran, 1992] Deepak Kapur and Paliath Narendran. Double-exponential complexity of computing a complete set of AC-unifiers (Preliminary report). In *Proceedings of the Seventh Symposium on Logic in Computer Science*, pages 11–21, Santa Cruz, CA, June 1992. IEEE.

[Kapur et al., 1987] Deepak Kapur, Paliath Narendran, and Hantao Zhang. On sufficient completeness and related properties of term rewriting systems. *Acta Informatica*, 24(4):395–415, August 1987.

[Kapur et al., 1990] Deepak Kapur, G. Sivakumar, and Hantao Zhang. A new method for proving termination of AC-rewrite systems. In *Proceedings of the Tenth International Conference of Foundations of Software Technology and Theoretical Computer Science*, volume 472 of *Lecture Notes in Computer Science*, pages 133–148, Berlin, 1990. Springer-Verlag.

[Kennaway et al., 1990] J. R. Kennaway, J. W. Klop, M. R. Sleep, and F. J. de Vries. Transfinite reductions in orthogonal term rewriting systems. Technical Report CS-R9041, CWI, Amsterdam, 1990.

[Kennaway et al., 1991] J. R. Kennaway, J. W. Klop, M. R. Sleep, and F. J. de Vries. Transfinite reductions in orthogonal term rewriting systems (Extended abstract). In Ron Book, editor, *Proceedings of the Fourth International Conference on Rewriting Techniques and Applications (Como, Italy)*, Lecture Notes in Computer Science, pages 1–12, Berlin, April 1991. Springer-Verlag.

[Kirchner and Kirchner, 1989] C. Kirchner and H. Kirchner. Constrained equational reasoning. In *Proceedings of the ACM-SIGSAM 1989 International Symposium on Symbolic and Algebraic Computation, Portland (Oregon)*, pages 382–389. ACM Press, July 1989. Report CRIN 89-R-220.

[Kirchner et al., 1990] C. Kirchner, H. Kirchner, and M. Rusinowitch. Deduction with symbolic constraints. *RAIRO Theoretical Informatics and Applications*, 4(3):9–52, 1990. Special issue on Automatic Deduction.

[Klop, 1980] Jan Willem Klop. *Combinatory Reduction Systems*, volume 127 of *Mathematical Centre Tracts*. Mathematisch Centrum, Amsterdam, 1980.

[Klop, 1992] Jan Willem Klop. Term rewriting systems. In S. Abramsky, D. M. Gabbay, and T. S. E. Maibaum, editors, *Handbook of Logic in Computer Science*, volume 2, chapter 1, pages 1–117. Oxford University Press, Oxford, 1992.

[Kounalis and Lugiez, 1991] Emmanuel Kounalis and Denis Lugiez. Compilation of pattern matching with associative commutative functions. In *Proceedings of the International Joint Conference on Theory and Practice of Software Development, volume 1: Colloquium on Trees in Algebra and Programming (Brighton, U.K.)*, Lecture Notes in Computer Science, pages 57–73, Berlin, April 1991. Springer-Verlag.

[Kounalis et al., 1991] Emmanuel Kounalis, Denis Lugiez, and L. Pottier. A solution of the complement problem in associative-commutative theories. In A. Tarlecki, editor, *Proceedings of the Sixteenth International Symposium on Mathematical Foundations of Computer Science (Kazimierz Dolny, Poland)*, volume 520 of *Lecture Notes in Computer Science*, pages 287–297, Berlin, September 1991. Springer-Verlag.

[Kruskal, 1960] Joseph B. Kruskal. Well-quasi-ordering, the Tree Theorem, and Vazsonyi's conjecture. *Transactions of the American Mathematical Society*, 95:210–225, May 1960.

[Kucherov, 1991] G. Kucherov. On relationship between term rewriting systems and regular tree languages. In Ron Book, editor, *Proceedings of the Fourth International Conference on Rewriting Techniques and Applications (Como, Italy)*, volume 488 of *Lecture Notes in Computer Science*, pages 299–311, Berlin, April 1991. Springer-Verlag.

[Kucherov and Tajine, 1993] G. Kucherov and M. Tajine. Decidability of regularity and related properties of ground normal form languages. In M. Rusinowitch, editor, *Proceedings of the Third International Workshop on Conditional Rewriting Systems (Pont-a-Mousson, France, July 1992)*, volume 656 of *Lecture Notes in Computer Science*, pages 272–286, Berlin, January 1993. Springer-Verlag. To appear in *Information and Computation*.

[Kurihara and Ohuchi, 1992] M. Kurihara and A. Ohuchi. Modularity of simple termination of term rewriting systems with shared constructors. *Theoretical Computer Science*, 103:273–282, 1992.

[Kurth, 1990] W. Kurth. *Termination und Konfluenz von Semi-Thue-Systems mit nur einer Regel*. PhD thesis, Technische Universität Clausthal, Clausthal, Germany, 1990.

[Lankford, 1975] Dallas S. Lankford. Canonical inference. Memo ATP-32, Automatic Theorem Proving Project, University of Texas, Austin, TX, December 1975.

[Lankford, 1979] Dallas S. Lankford. On proving term rewriting systems are Noetherian. Memo MTP-3, Mathematics Department, Louisiana Tech. University, Ruston, LA, October 1979.

[Lassez and Marriott, 1987] J.-L. Lassez and K. G. Marriott. Explicit representation of terms defined by counter examples. *J. Automated Reasoning*, 3(3):1–17, September 1987.

[Lescanne, 1990] P. Lescanne. Well rewrite orderings. In J. Mitchell, editor, *Proceedings of the Fifth Symposium on Logic in Computer Science*, pages 239–256, Philadelphia, PA, 1990.

[Lescanne, 1992] Pierre Lescanne. Termination of rewrite systems by elementary interpretations. In H. Kirchner and G. Levi, editors, *Proceedings of the Third International Conference on Algebraic and Logic Programming (Pisa, Italy)*, volume 632 of *Lecture Notes in Computer Science*, pages 21–36. Springer-Verlag, September 1992.

[Litovski et al., to appear] Igor Litovski, Yves Métivier, and Eric Sopena. Definitions and comparisons of local computations on graphs. *Mathematical Systems Theory*, to appear. Available as internal report 91-43 of LaBRI, University of Bordeaux 1.

[Lugiez and Moysset, 1993] D. Lugiez and J.-L. Moysset. Complement problems and tree automata in AC-like theories. In *Proceedings of the Symposium on Theoretical Aspects of Computer Science (Würzburg, Germany)*, Lecture Notes in Computer Science, Berlin, 1993. Springer-Verlag.

[Marche, 1993] Claude Marché, personal communication, 1993.

[Meseguer, 1992] José Meseguer. Conditional rewriting logic as a unified model of concurrency. *Theoretical Computer Science*, 96:73–155, 1992.

[Métivier, 1985] Yves Métivier. Calcul de longueurs de chaînes de réécriture dans le monoïde libre. *Theoretical Computer Science*, 35(1):71–87, January 1985.

[Middeldorp, 1990] Aart Middeldorp. *Modular Properties of Term Rewriting Systems*. PhD thesis, Vrije Universiteit, Amsterdam, The Netherlands, 1990.

[Middeldorp and Toyama, 1991] Aart Middeldorp and Yoshihito Toyama. Completeness of combinations of constructor systems. In R. Book, editor, *Proceedings of the Fourth International Conference on Rewriting Techniques and Applications (Como, Italy)*, volume 488 of *Lecture Notes in Computer Science*, pages 174–187, Berlin, April 1991. Springer-Verlag.

[Milner et al., 1992] R. Milner, J. Parrow, and D. Walker. A calculus of mobile processes, I and II. *Information and Computation*, 100:1–77, 1992.

[Müller, 1992] Fritz Müller. Confluence of the lambda calculus with left-linear algebraic rewriting. *Information Processing Letters*, 41:293–299, April 1992.

[Nipkow, 1991] Tobias Nipkow. Higher-order critical pairs. In *Proceedings of the Sixth Symposium on Logic in Computer Science*, pages 342–349, Amsterdam, The Netherlands, 1991. IEEE.

[Nipkow, 1991] Tobias Nipkow. Higher-order critical pairs. In *Proceedings of the Sixth Symposium on Logic in Computer Science*, pages 342–349, Amsterdam, The Netherlands, 1991. IEEE.

[Nieuwenhuis, 1993] Robert Nieuwenhuis. A new ordering constraint solving method and its applications. In C. Kirchner, editor, *Proceedings of the Fifth International Conference on Rewriting Techniques and Applications (Montreal, Canada)*, Lecture Notes in Computer Science, Berlin, 1993. Springer-Verlag.

[O'Donnell, 1985] Michael J. O'Donnell. *Equational Logic as a Programming Language*. MIT Press, Cambridge, MA, 1985.

[Oostrom, 1992] V. van Oostrom. Confluence by decreasing diagrams. IR 298, Vrije Universiteit, Amsterdam, The Netherlands, August 1992. To appear in *Theoretical Computer Science*.

[Oyamaguchi, 1987] M. Oyamaguchi. The Church-Rosser property for ground term rewriting systems is decidable. *Theoretical Computer Science*, 49(1):43–79, 1987.

[Oyamaguchi and Ohta, 1993] M. Oyamaguchi and Y. Ohta. On the confluent property of right-ground term rewriting systems. *Trans. IEICE*, J76-D-I:39–45, 1993.

[Peterson and Stickel, 1981] Gerald E. Peterson and Mark E. Stickel. Complete sets of reductions for some equational theories. *J. of the Association for Computing Machinery*, 28(2):233–264, April 1981.

[Plump, 1991] D. Plump. Implementing term rewriting by graph reduction: Termination of combined systems. In S. Kaplan and M. Okada, editors, *Proceedings of the Second International Workshop on Conditional and Typed Rewriting Systems (Montreal, Canada, June 1990)*, volume 516 of *Lecture Notes in Computer Science*, pages 307–317, Berlin, 1991. Springer-Verlag.

[Plump, 1993] D. Plump. Hypergraph rewriting: Critical pairs and undecidability of confluence. In M. R. Sleep, M. J. Plasmeijer, and M. C. van Eekelen, editors, *Term Graph Rewriting: Theory and Practice*, chapter 15. Wiley, 1993. To appear.

[Reddy, 1986] Uday S. Reddy. On the relationship between logic and functional languages. In D. DeGroot and G. Lindstrom, editors, *Logic Programming: Functions, Relations, and Equations*, pages 3–36. Prentice-Hall, Englewood Cliffs, NJ, 1986.

[Ringeissen, 1992] Christophe Ringeissen. Unification in a combination of equational theories with shared constants and its application to primal algebras. In A. Voronkov, editor, *Proceedings of the Conference on Logic Programming and Automated Reasoning (St. Petersburg, Russia)*, volume 624 of *Lecture Notes in Artificial Intelligence*, Berlin, July 1992. Springer-Verlag.

[Robertson and Seymour, 1982] Neil Robertson and P. D. Seymour. Graph minors IV. Tree-width and well-quasi-ordering. Submitted 1982; revised January 1986.

[Rubio and Nieuwenhuis, 1993] A. Rubio and R. Nieuwenhuis. A precedence-based total AC-compatible ordering. In C. Kirchner, editor, *Proceedings of the Fifth International Conference on Rewriting Techniques and Applications (Montreal, Canada)*, Lecture Notes in Computer Science, Berlin, 1993. Springer-Verlag.

[Schmidt-Schauß, 1989] M. Schmidt-Schauß. Unification in a combination of arbitrary disjoint equational theories. *J. Symbolic Computation*, 8(1&2):51–99, 1989.

[Statman, 1991] R. Statman. There is no hyperrecurrent S,K combinator. Research Report 91-133, Department of Mathematics, Carnegie Mellon University, Pittsburgh, PA, 1991.

[Statman, 1992] R. Statman. A short note on a problem of Ray Smullyan. Rapport, Institut National de Rechereche en Informatique et en Automatique, Le Chesnay, France, 1992.

[Tajine, 1993] M. Tajine. Negation elimination for syntactic equational formula. In C. Kirchner, editor, *Proceedings of the Fifth International Conference on Rewriting Techniques and Applications (Montreal, Canada)*, Lecture Notes in Computer Science, Berlin, 1993. Springer-Verlag.

[Takahashi, 1993] M. Takahashi. λ-calculi with conditional rules. In M. Bezem and J. F. Groote, editors, *Proceedings of the International Conference on Typed Lambda Calculi and Applications (Utrecht, The Netherlands)*, volume 664 of *Lecture Notes in Computer Science*, pages 406–417, Berlin, 1993. Springer-Verlag.

[Thue, 1914] A. Thue. Probleme über veranderungen von zeichenreihen nach gegeben regeln. *Skr. Vid. Kristianaia I. Mat. Naturv. Klasse*, 10/34, 1914.

[Tiden and Arnborg, 1987] Erik Tiden and Stefan Arnborg. Unification problems with one-sided distributivity. *J. of Symbolic Computation*, 3:183–202, 1987.

[Toyama, 1987] Yoshihito Toyama. On the Church-Rosser property for the direct sum of term rewriting systems. *J. of the Association for Computing Machinery*, 34(1):128–143,

January 1987.

[Toyama et al., 1989] Yoshihito Toyama, Jan Willem Klop, and Hendrik Pieter Barendregt. Termination for the direct sum of left-linear term rewriting systems. In Nachum Dershowitz, editor, *Proceedings of the Third International Conference on Rewriting Techniques and Applications (Chapel Hill, NC)*, volume 355 of *Lecture Notes in Computer Science*, pages 477–491, Berlin, April 1989. Springer-Verlag.

[Treinen, 1990] Ralf Treinen. A new method for undecidability proofs of first order theories. In K. V. Nori and C. E. Veni Madhavan, editors, *Proceedings of the Tenth Conference on Foundations of Software Technology and Theoretical Computer Science*, volume 472 of *Lecture Notes in Computer Science*, pages 48–62. Springer-Verlag, 1990.

[Treinen, 1992] Ralf Treinen. A new method for undecidability proofs of first order theories. *J. Symbolic Computation*, 14(5):437–457, November 1992.

[Vágvölgyi and Gilleron, 1992] S. Vágvölgyi and R. Gilleron. For a rewrite system it is decidable whether the set of irreducible, ground terms is decidable. *Bulletin of the European Association for Theoretical Computer Science*, 48:197–209, October 1992.

[Venkataraman, 1987] K. N. Venkataraman. Decidability of the purely existential fragment of the theory of term algebras. *J. of the Association for Computing Machinery*, 34(2):492–510, 1987.

[Venturini-Zilli, 1984] M. Venturini-Zilli. Reduction graphs in the Lambda Calculus. *Theoretical Computer Science*, 29:251–275, 1984.

[Visser, 1980] A. Visser. Numerations, lambda calculus, and arithmetic. In Hindley and Seldin, editors, *Essays on Combinatory Logic, Lambda-Calculus, and Formalism*, pages 259–284. Academic Press, 1980.

[Weiermann, 1992] Andreas Weiermann. Well-rewrite orderings and the induced recursive path orderings. Unpublished note, Institut für Mathematische Logik und Grundlagenforschung, 1992.

[Wrathall, 1990] C. Wrathall. Confluence of one-rule Thue systems. In *Proceedings of the First International Workshop on Word Equations and Related Topics (Tubingen)*, volume 572 of *Lecture Notes in Computer Science*, pages 237–246, Berlin, 1990. Springer-Verlag.

[Zantema, 1993] H. Zantema. Type removal in term rewriting. In M. Rusinowitch, editor, *Proceedings of the Third International Workshop on Conditional Rewriting Systems (Pont-a-Mousson, France, July 1992)*, volume 656 of *Lecture Notes in Computer Science*, pages 148–154, Berlin, January 1993. Springer-Verlag.

Authors Index

Fernàndez, M., 358

Adian, S.I., 289
Agusti, J., 17
Ariola, Z.M., 183
Asperti, A., 152
Avenhaus, J., 62

Baader, F., 301
Bachmair, L., 1
Backofen, R., 121
Beyer, M., 451
Bündgen, R., 446

Caron, A.-C., 328
Chakrabarti, S., 77
Coquide, J.-L., 328

Dauchet, M., 328
Delor, C., 389
Denzinger, J., 62
Dershowitz, N., 198, 468
Dougherty, D.J., 137

Ferreira, M.C.F., 213
Field, J., 259

Gallier, J., 136
Gramlich, B., 228
Gui, Y., 462

Hoot, C., 198

Jouannaud, J.-P., 468

Khasidashvili, Z., 243
Klop, J.W., 468

Laneve, C., 152
Levy, J., 17
Lippe, E., 274
Lynch, C., 2
Löwe, M., 451

Martin, U., 421

Matthews, B., 441
Middledorp, A., 228
Moser, M., 92

Niehren, J., 106
Nieuwenhuis, R., 374, 436
Nivela, P., 436

Okada, Y., 462

Plaisted, D., 405
Podelski, A., 106
Puel, L., 389

Raoult, J.-C., 343
Rubio, A., 374

Schulz, K.U., 301
Senizergue, G., 434
Snyder, W., 2

Tajine, M., 316
Treinen, R., 106

Van Raamsdonk, F., 168
Verma, R., 457

Werner, A., 47

Yelick, K., 77

Zantema, H., 213
Zhang, H., 32

Springer-Verlag and the Environment

We at Springer-Verlag firmly believe that an international science publisher has a special obligation to the environment, and our corporate policies consistently reflect this conviction.

We also expect our business partners – paper mills, printers, packaging manufacturers, etc. – to commit themselves to using environmentally friendly materials and production processes.

The paper in this book is made from low- or no-chlorine pulp and is acid free, in conformance with international standards for paper permanency.

Lecture Notes in Computer Science

For information about Vols. 1–610
please contact your bookseller or Springer-Verlag

Vol. 611: M. P. Papazoglou, J. Zeleznikow (Eds.), The Next Generation of Information Systems: From Data to Knowledge. VIII, 310 pages. 1992. (Subseries LNAI).

Vol. 612: M. Tokoro, O. Nierstrasz, P. Wegner (Eds.), Object-Based Concurrent Computing. Proceedings, 1991. X, 265 pages. 1992.

Vol. 613: J. P. Myers, Jr., M. J. O'Donnell (Eds.), Constructivity in Computer Science. Proceedings, 1991. X, 247 pages. 1992.

Vol. 614: R. G. Herrtwich (Ed.), Network and Operating System Support for Digital Audio and Video. Proceedings, 1991. XII, 403 pages. 1992.

Vol. 615: O. Lehrmann Madsen (Ed.), ECOOP '92. European Conference on Object Oriented Programming. Proceedings. X, 426 pages. 1992.

Vol. 616: K. Jensen (Ed.), Application and Theory of Petri Nets 1992. Proceedings, 1992. VIII, 398 pages. 1992.

Vol. 617: V. Mařík, O. Štěpánková, R. Trappl (Eds.), Advanced Topics in Artificial Intelligence. Proceedings, 1992. IX, 484 pages. 1992. (Subseries LNAI).

Vol. 618: P. M. D. Gray, R. J. Lucas (Eds.), Advanced Database Systems. Proceedings, 1992. X, 260 pages. 1992.

Vol. 619: D. Pearce, H. Wansing (Eds.), Nonclassical Logics and Information Proceedings. Proceedings, 1990. VII, 171 pages. 1992. (Subseries LNAI).

Vol. 620: A. Nerode, M. Taitslin (Eds.), Logical Foundations of Computer Science – Tver '92. Proceedings. IX, 514 pages. 1992.

Vol. 621: O. Nurmi, E. Ukkonen (Eds.), Algorithm Theory – SWAT '92. Proceedings. VIII, 434 pages. 1992.

Vol. 622: F. Schmalhofer, G. Strube, Th. Wetter (Eds.), Contemporary Knowledge Engineering and Cognition. Proceedings, 1991. XII, 258 pages. 1992. (Subseries LNAI).

Vol. 623: W. Kuich (Ed.), Automata, Languages and Programming. Proceedings, 1992. XII, 721 pages. 1992.

Vol. 624: A. Voronkov (Ed.), Logic Programming and Automated Reasoning. Proceedings, 1992. XIV, 509 pages. 1992. (Subseries LNAI).

Vol. 625: W. Vogler, Modular Construction and Partial Order Semantics of Petri Nets. IX, 252 pages. 1992.

Vol. 626: E. Börger, G. Jäger, H. Kleine Büning, M. M. Richter (Eds.), Computer Science Logic. Proceedings, 1991. VIII, 428 pages. 1992.

Vol. 628: G. Vosselman, Relational Matching. IX, 190 pages. 1992.

Vol. 629: I. M. Havel, V. Koubek (Eds.), Mathematical Foundations of Computer Science 1992. Proceedings. IX, 521 pages. 1992.

Vol. 630: W. R. Cleaveland (Ed.), CONCUR '92. Proceedings. X, 580 pages. 1992.

Vol. 631: M. Bruynooghe, M. Wirsing (Eds.), Programming Language Implementation and Logic Programming. Proceedings, 1992. XI, 492 pages. 1992.

Vol. 632: H. Kirchner, G. Levi (Eds.), Algebraic and Logic Programming. Proceedings, 1992. IX, 457 pages. 1992.

Vol. 633: D. Pearce, G. Wagner (Eds.), Logics in AI. Proceedings. VIII, 410 pages. 1992. (Subseries LNAI).

Vol. 634: L. Bougé, M. Cosnard, Y. Robert, D. Trystram (Eds.), Parallel Processing: CONPAR 92 – VAPP V. Proceedings. XVII, 853 pages. 1992.

Vol. 635: J. C. Derniame (Ed.), Software Process Technology. Proceedings, 1992. VIII, 253 pages. 1992.

Vol. 636: G. Comyn, N. E. Fuchs, M. J. Ratcliffe (Eds.), Logic Programming in Action. Proceedings, 1992. X, 324 pages. 1992. (Subseries LNAI).

Vol. 637: Y. Bekkers, J. Cohen (Eds.), Memory Management. Proceedings, 1992. XI, 525 pages. 1992.

Vol. 639: A. U. Frank, I. Campari, U. Formentini (Eds.), Theories and Methods of Spatio-Temporal Reasoning in Geographic Space. Proceedings, 1992. XI, 431 pages. 1992.

Vol. 640: C. Sledge (Ed.), Software Engineering Education. Proceedings, 1992. X, 451 pages. 1992.

Vol. 641: U. Kastens, P. Pfahler (Eds.), Compiler Construction. Proceedings, 1992. VIII, 320 pages. 1992.

Vol. 642: K. P. Jantke (Ed.), Analogical and Inductive Inference. Proceedings, 1992. VIII, 319 pages. 1992. (Subseries LNAI).

Vol. 643: A. Habel, Hyperedge Replacement: Grammars and Languages. X, 214 pages. 1992.

Vol. 644: A. Apostolico, M. Crochemore, Z. Galil, U. Manber (Eds.), Combinatorial Pattern Matching. Proceedings, 1992. X, 287 pages. 1992.

Vol. 645: G. Pernul, A M. Tjoa (Eds.), Entity-Relationship Approach – ER '92. Proceedings, 1992. XI, 439 pages, 1992.

Vol. 646: J. Biskup, R. Hull (Eds.), Database Theory – ICDT '92. Proceedings, 1992. IX, 449 pages. 1992.

Vol. 647: A. Segall, S. Zaks (Eds.), Distributed Algorithms. X, 380 pages. 1992.

Vol. 648: Y. Deswarte, G. Eizenberg, J.-J. Quisquater (Eds.), Computer Security – ESORICS 92. Proceedings. XI, 451 pages. 1992.

Vol. 649: A. Pettorossi (Ed.), Meta-Programming in Logic. Proceedings, 1992. XII, 535 pages. 1992.

Vol. 650: T. Ibaraki, Y. Inagaki, K. Iwama, T. Nishizeki, M. Yamashita (Eds.), Algorithms and Computation. Proceedings, 1992. XI, 510 pages. 1992.

Vol. 651: R. Koymans, Specifying Message Passing and Time-Critical Systems with Temporal Logic. IX, 164 pages. 1992.

Vol. 652: R. Shyamasundar (Ed.), Foundations of Software Technology and Theoretical Computer Science. Proceedings, 1992. XIII, 405 pages. 1992.

Vol. 653: A. Bensoussan, J.-P. Verjus (Eds.), Future Tendencies in Computer Science, Control and Applied Mathematics. Proceedings, 1992. XV, 371 pages. 1992.

Vol. 654: A. Nakamura, M. Nivat, A. Saoudi, P. S. P. Wang, K. Inoue (Eds.), Prallel Image Analysis. Proceedings, 1992. VIII, 312 pages. 1992.

Vol. 655: M. Bidoit, C. Choppy (Eds.), Recent Trends in Data Type Specification. X, 344 pages. 1993.

Vol. 656: M. Rusinowitch, J. L. Rémy (Eds.), Conditional Term Rewriting Systems. Proceedings, 1992. XI, 501 pages. 1993.

Vol. 657: E. W. Mayr (Ed.), Graph-Theoretic Concepts in Computer Science. Proceedings, 1992. VIII, 350 pages. 1993.

Vol. 658: R. A. Rueppel (Ed.), Advances in Cryptology – EUROCRYPT '92. Proceedings, 1992. X, 493 pages. 1993.

Vol. 659: G. Brewka, K. P. Jantke, P. H. Schmitt (Eds.), Nonmonotonic and Inductive Logic. Proceedings, 1991. VIII, 332 pages. 1993. (Subseries LNAI).

Vol. 660: E. Lamma, P. Mello (Eds.), Extensions of Logic Programming. Proceedings, 1992. VIII, 417 pages. 1993. (Subseries LNAI).

Vol. 661: S. J. Hanson, W. Remmele, R. L. Rivest (Eds.), Machine Learning: From Theory to Applications. VIII, 271 pages. 1993.

Vol. 662: M. Nitzberg, D. Mumford, T. Shiota, Filtering, Segmentation and Depth. VIII, 143 pages. 1993.

Vol. 663: G. v. Bochmann, D. K. Probst (Eds.), Computer Aided Verification. Proceedings, 1992. IX, 422 pages. 1993.

Vol. 664: M. Bezem, J. F. Groote (Eds.), Typed Lambda Calculi and Applications. Proceedings, 1993. VIII, 433 pages. 1993.

Vol. 665: P. Enjalbert, A. Finkel, K. W. Wagner (Eds.), STACS 93. Proceedings, 1993. XIV, 724 pages. 1993.

Vol. 666: J. W. de Bakker, W.-P. de Roever, G. Rozenberg (Eds.), Semantics: Foundations and Applications. Proceedings, 1992. VIII, 659 pages. 1993.

Vol. 667: P. B. Brazdil (Ed.), Machine Learning: ECML – 93. Proceedings, 1993. XII, 471 pages. 1993. (Subseries LNAI).

Vol. 668: M.-C. Gaudel, J.-P. Jouannaud (Eds.), TAPSOFT '93: Theory and Practice of Software Development. Proceedings, 1993. XII, 762 pages. 1993.

Vol. 669: R. S. Bird, C. C. Morgan, J. C. P. Woodcock (Eds.), Mathematics of Program Construction. Proceedings, 1992. VIII, 378 pages. 1993.

Vol. 670: J. C. P. Woodcock, P. G. Larsen (Eds.), FME '93: Industrial-Strength Formal Methods. Proceedings, 1993. XI, 689 pages. 1993.

Vol. 671: H. J. Ohlbach (Ed.), GWAI-92: Advances in Artificial Intelligence. Proceedings, 1992. XI, 397 pages. 1993. (Subseries LNAI).

Vol. 672: A. Barak, S. Guday, R. G. Wheeler, The MOSIX Distributed Operating System. X, 221 pages. 1993.

Vol. 673: G. Cohen, T. Mora, O. Moreno (Eds.), Applied Algebra, Algebraic Algorithms and Error-Correcting Codes. Proceedings, 1993. X, 355 pages 1993.

Vol. 674: G. Rozenberg (Ed.), Advances in Petri Nets 1993. VII, 457 pages. 1993.

Vol. 675: A. Mulkers, Live Data Structures in Logic Programs. VIII, 220 pages. 1993.

Vol. 676: Th. H. Reiss, Recognizing Planar Objects Using Invariant Image Features. X, 180 pages. 1993.

Vol. 677: H. Abdulrab, J.-P. Pécuchet (Eds.), Word Equations and Related Topics. Proceedings, 1991. VII, 214 pages. 1993.

Vol. 678: F. Meyer auf der Heide, B. Monien, A. L. Rosenberg (Eds.), Parallel Architectures and Their Efficient Use. Proceedings, 1992. XII, 227 pages. 1993.

Vol. 683: G.J. Milne, L. Pierre (Eds.), Correct Hardware Design and Verification Methods. Proceedings, 1993. VIII, 270 Pages. 1993.

Vol. 684: A. Apostolico, M. Crochemore, Z. Galil, U. Manber (Eds.), Combinatorial Pattern Matching. Proceedings, 1993. VIII, 265 pages. 1993.

Vol. 685: C. Rolland, F. Bodart, C. Cauvet (Eds.), Advanced Information Systems Engineering. Proceedings, 1993. XI, 650 pages. 1993.

Vol. 686: J. Mira, J. Cabestany, A. Prieto (Eds.), New Trends in Neural Computation. Proceedings, 1993. XVII, 746 pages. 1993.

Vol. 687: H. H. Barrett, A. F. Gmitro (Eds.), Information Processing in Medical Imaging. Proceedings, 1993. XVI, 567 pages. 1993.

Vol. 688: M. Gauthier (Ed.), Ada - Europe '93. Proceedings, 1993. VIII, 353 pages. 1993.

Vol. 689: J. Komorowski, Z. W. Ras (Eds.), Methodologies for Intelligent Systems. Proceedings, 1993. XI, 653 pages. 1993. (Subseries LNAI).

Vol. 690: C. Kirchner (Ed.), Rewriting Techniques and Applications. Proceedings, 1993. XI, 488 pages. 1993.

Vol. 691: M. Ajmone Marsan (Ed.), Application and Theory of Petri Nets 1993. Proceedings, 1993. IX, 591 pages. 1993.

Vol. 692: D. Abel, B.C. Ooi (Eds.), Advances in Spatial Databases. Proceedings, 1993. XIII, 529 pages. 1993.

Vol. 694: A. Bode, M. Reeve, G. Wolf (Eds.), PARLE '93. Parallel Architectures and Languages Europe. Proceedings, 1993. XVII, 770 pages. 1993.